MOLECULAR ASPECTS
OF
MEMBRANE TRANSPORT

PROGRESS IN CLINICAL AND BIOLOGICAL RESEARCH

MOLECULAR ASPECTS
OF
MEMBRANE TRANSPORT

Proceedings of the ICN—UCLA Symposium
held at Keystone, Colorado
March 13—18, 1977

Editors
DALE OXENDER
University of Michigan, Ann Arbor

C. FRED FOX
University of California, Los Angeles

Alan R. Liss, Inc. • New York

©1978 Alan R. Liss, Inc.

Address all inquiries to the publisher:
Alan R. Liss, Inc.
150 Fifth Avenue
New York, New York 10011

Library of Congress Cataloging in Publication Data

Main entry under title:

Molecular aspects of membrane transport.

(Progress in clinical and biological research; v. 22)
Includes bibliographical references and indexes.
1. Biological transport – Congresses. 2. Membrane (Biology) – Congresses.
3. Molecular biology – Congresses. I. Oxender, Dale. II. Fox, C. Fred. III. ICN Pharmaceuticals, inc. IV. California. University. University at Los Angeles.
V. Series.
QH509.   574.8'75   78-541   ISBN 0-8451-0022-X

Printed in the United States of America

NOTE

Pages 1–544 are reprinted from Journal of Supramolecular Structure, Volumes 6 and 7, 1977. The page numbers in the Table of Contents, Author Index, and Subject Index of this volume correspond to the page numbers at the foot of these pages.

Contents

x   Contents

The paper, "Purification and Characteristics of Hydrophobic Membrane Protein(s) Required for DCCD
Sensitivity of ATPase in Mycobacterium phlei," by Natalie S. Cohen, Soon-Ho Lee, and Arnold F. Brodie,
which was presented at this symposium does not appear in this volume. It will be published in Journal
of Supramolecular Structure, Volume 8, Number 1, 1978.

Preface

The ICN—UCLA conference on Molecular Aspects of Membrane Transport follows previous meetings in this series held on topics related to membrane structure and function. The conference was organized to bring together basic research scientists in the field of membrane transport who are utilizing both procaryotic and eucaryotic systems, along with scientists who are studying diseases arising from defects in transport. Focusing on molecular approaches to the area, the conference considered a variety of topical issues such as energy coupling mechanisms and mechanochemical processes which may play roles in barrier transit, transport of macromolecular species in secretion, reconstitution of transport systems from highly purified and well characterized preparations of isolated components, and the limits of knowledge in technical areas which provide the tools exploited in these explorative studies.

Three formal modes of presentation were utilized at this conference: eight plenary sessions, approximately 100 poster sessions, and four workshops, which contributed to an active participatory exchange. This volume includes papers arising from each of these modalities. Many of the papers submitted to these proceedings were also considered for publication in the Journal of Supramolecular Structure, either as standard articles describing original research, or as reveiws. The reviews were solicited by invitation to the plenary session lecturers. Those articles which were referred and accepted for publication in the Journal are identified in this volume as being reprinted from the Journal.

The travel and subsistence expenses of the invited speakers were defrayed in part by contract number 263-77C-0251CC, awarded by the National Institute of Allergy and Infectious Diseases and the National Institute of Arthritis, Metabolism, and Digestive Diseases. Endowments from Hoffman-La Roche, Inc., Nutley, New Jersey; Eli Lilly Research Laboratories, Indianapolis, Indiana; and Sandoz, Inc., East Hanover, New Jersey; also contributed greatly to the overall support of the meeting. We would like to acknowledge the continued support which ICN Pharmaceuticals, Inc., provides to this conference series in general. Finally, the valuable assistance of the members of the ICN—UCLA Symposia staff, Fran Stusser, Robert Williams, and Marie Deyl, is gratefully acknowledged.

Dale Oxender
C. Fred Fox

Acknowledgment

The editors would like to acknowledge the assistance of the following persons who helped review the manuscripts included in these proceedings.

K. Altendorf
G. F. Ames
D. Aminoff
J. J. Anderson
G. Ashwell
R. D. Berlin
P. D. Boyer
D. Branton
M. S. Brown
G. Cecchini
H. N. Christensen
R. K. Crane
D. Cunningham
D. S. Eisenberg
W. Epstein
T. D. Gelehrter
F. M. Harold
L. A. Heppel
R. W. Hogg
L. E. Hokin

U. Hopfer
H. R. Kaback
R. G. Langdon
J. H. Law
A. Martinosi
E. F. Neufeld
A. B. Pardee
T. R. Riggs
B. Rosen
T. Roth
G. A. Scarborough
P. Siekovitz
R. D. Simoni
T. Steck
L. Stryer
R. Sweet
E. W. Taylor
M. Weber
T. H. Wilson
C. G. Winter
J. Yu

Journal of Supramolecular Structure 6:61−75 (1977)
Molecular Aspects of Membrane Transport 1−15

Differential Phosphorylation of Band 3 and Glycophorin in Intact and Extracted Erythrocyte Membranes

M. Marlene Hosey and Mariano Tao

Department of Biological Chemistry, University of Illinois at the Medical Center, Chicago, Illinois 60612

This report presents an analysis of the phosphorylation of human and rabbit erythrocyte membrane proteins which migrate in $NaDodSO_4$-polyacrylamide gels in the area of the Coomassie Blue-stained proteins generally known as band 3. The phosphorylation of these proteins is of interest as band 3 has been implicated in transport processes. This study shows that there are at least three distinct phosphoproteins associated with the band 3 region of human erythrocyte membranes. These are band 2.9, the major band 3, and PAS-1. The phosphorylation of these proteins is differentially catalyzed by solubilized membrane and cytoplasmic cyclic AMP-dependent and -independent erythrocyte protein kinases. Band 2.9 is present and phosphorylated in unfractionated human and rabbit erythrocyte ghosts but not in NaI- or dimethylmaleic anhydride (DMMA)-extracted membranes. These latter membrane preparations are enriched in band 3 and in sialoglycoproteins. The NaI-extracted ghosts contain residual protein kinase activity which can catalyze the autophosphorylation of band 3 whereas the DMMA-extracted ghosts are usually devoid of any kinase activity. However, both NaI- and DMMA-extracted ghosts, as well as Triton X-100 extracts of the DMMA-extracted ghosts, can be phosphorylated by various erythrocyte protein kinases. The kinases which preferentially phosphorylate the major band 3 protein are inactive towards PAS-1 while the kinases active towards PAS-1 are less active towards band 3. The band 3 protein in the DMMA-extracted ghosts can be cross-linked with the Cu^{2+}-o-phenanthroline complex. The cross-linking of band 3 does not affect its capacity to serve as a phosphoryl acceptor nor does phosphorylation affect the capacity of band 3 to form cross-links. In addition to band 2.9, the major band 3 and PAS-1, another minor protein component appears to be present in the band 3 region in human erythrocyte membranes. This protein is specifically phosphorylated by the cyclic AMP-dependent protein kinases isolated from the cytoplasm of rabbit erythrocytes. The rabbit erythrocyte membranes lack PAS-1 and the cyclic AMP-dependent protein kinase substrate.

Key words: erythrocyte membranes, protein phosphorylation, band 2.9, band 3, glycophorin
(PAS-1 and PAS-2)

Previous studies from this (1−4) and other laboratories (5−8) have established that human and rabbit erythrocyte membranes contain multiple protein substrates for membrane-bound and soluble cyclic AMP-dependent and -independent erythrocyte pro-

Received March 14, 1977; accepted March 16, 1977

tein kinases. While erythrocyte membrane phosphorylation occurs in isolated ghosts (1–3, 5, 6) as well as in intact red cells (4, 9, 10), the significance of these reactions remains undocumented. This is largely due to the fact that the identity and functions of many of the membrane proteins which can undergo phosphorylation are unknown.

Of considerable interest in the study of membrane phosphorylation is the reaction involving band 3 (nomenclature of Steck, Ref. 11) as the band 3 protein of human erythrocyte membranes has been implicated in various transport processes (12–18). Band 3 is a lightly glycosylated protein that accounts for approximately 25% of the total Coomassie Blue-stained peptides of the erythrocyte membrane resolved by $NaDodSO_4$-polyacrylamide gel electrophoresis. However, the diffuse region in the gel which is generally designated as band 3 contains one or more substrates for erythrocyte protein kinases (1, 3). That the area stained with Coomassie Blue as band 3 contains more than one peptide species has been suggested for both human (19) and rabbit (20) erythrocytes. In this study we report on an analysis of the phosphorylation of partially purified preparations of human and rabbit band 3 area proteins by erythrocyte protein kinases derived from the membrane and the cytoplasmic fractions. Our results show that particulate and soluble preparations of band 3-enriched membranes of human erythrocytes contain at least 2 distinct peptides which can be differentially phosphorylated by various erythrocyte protein kinases. One of these phosphopeptides is the major band 3 protein while the other appears to be the major erythrocyte sialoglycoprotein, glycophorin (PAS-1). Comparisons are made of the phosphorylation of intact ghosts and the band 3-enriched ghosts.

MATERIALS AND METHODS

Hemoglobin-free ghosts of human and rabbit erythrocytes were prepared from fresh or outdated blood according to the method of Dodge et al. (21) as previously described (22). The cyclic AMP-independent erythrocyte membrane protein kinases were extracted from human and rabbit red cell ghosts with 0.5 M NaCl and partially purified by $(NH_4)_2 SO_4$ fractionation and gel filtration (23). These kinases are referred to as HMK, the human erythrocyte membrane kinase, and MK-I and MK-II, the rabbit erythrocyte membrane kinases. The enzymes have been characterized in detail with regards to their activity towards exogenous (23) and membrane substrates (24). In the presence of 0.4 M salt MK-I has a molecular weight (MW) of ∼ 100,000 and can catalyze the phosphorylation of casein and membrane proteins using ATP or GTP. In contrast, MK-II and HMK, which appear to be similar in nature, are ∼ 30,000 daltons in the presence of salt and can use only ATP to phosphorylate casein and some membrane proteins. Under certain conditions, MK-II and HMK can also use GTP as a phosphoryl donor (23, 24). The cyclic AMP-dependent protein kinases from rabbit erythrocyte lysates, which use only ATP as a phosphoryl donor, were purified according to Tao and Hackett (25). The soluble casein kinases, which can use either ATP or GTP as phosphoryl donor and are independent of cyclic nucleotides, were obtained according to Kumar and Tao (26).

"Purification" of band 3 was achieved by extracting whole ghosts with either 1.0 M NaI (12) or with 2,3-dimethylmaleic anhydride (DMMA) using 1.5–2.0 mg DMMA/mg protein (27). In some instances the latter procedure was repeated a second time. Triton-extraction of the DMMA-extracted ghosts was performed with 0.125% Triton X-100 in 5 mM Tris-HCl, pH 7.5, according to Zala and Kahlenberg (13). The DMMA-, NaI-, or

Triton-extracted ghosts were washed 2–3 times with either 10 mM Tris-HCl, pH 7.5, or with deionized H_2O. The reaction mixture contained (in a final volume of 50 μl): 0.1 M glycine-NaOH, pH 8.5; 10 mM $MgCl_2$; 0.2 mM [γ-^{32}P] ATP or [γ-^{32}P] GTP (150–400 cpm/pmol); ± 10μM cyclic AMP; ± kinases; and 10–30 μg of membrane proteins. Phosphorylation was initiated by the addition of ATP (or GTP), buffer, and Mg^{2+}, and terminated with 20 volumes of a KCl stopping solution (150 mM KCl; 10 mM Tris-HCl, pH 7.5;1 mM EDTA) followed by centrifugation (3). The phosphorylated ghosts were applied to 4% polyacrylamide slab gels (unless otherwise specified) containing 0.2% $NaDodSO_4$ (27, 28), electrophoresed at 75 mA/slab, and stained with Coomassie Brilliant Blue (CBB). The destained gels were dried and radioautograms prepared and scanned using a Zeineh Soft-Laser densitometer (1, 3). The glycoprotein content of membrane extracts was determined in gels stained with periodic acid-Schiff's reagent (PAS) (27, 28).

Cross-linking of DMMA-extracted human erythrocyte membranes was performed before and after phosphorylation. The proteins to be cross-linked before phosphorylation were mixed with an equal volume of 280 μM $CuSO_4$/140 μM o-phenanthroline (CuP) and incubated for 20 min at $0°C$. The reaction was terminated by 20 volumes of the KCl stopping solution followed by centrifugation at 27,000 $\times$ g for 20 min. The pellet was washed once with 20 mM Tris-HCl, pH 7.5. After centrifugation the cross-linked membranes were resuspended and phosphorylated as described above. However, in these studies the phosphorylation reactions (control, CuP treated before and after) were all terminated with 20 volumes of 20 mM Tris-HCl, pH 7.5, and centrifuged. For studies of cross-linking after phosphorylation, the pellets were resuspended to ~20 μl and 20 μl of 280 μM $CuSO_4$/140 μM o-phenanthroline was added. After a 20 min incubation at $0°C$ the sample was dissolved in the $NaDodSO_4$ diluent. In these experiments the samples were all dissolved in the $NaDodSO_4$ diluent minus dithiothreitol and applied to $NaDodSO_4$-polyacrylamide gels containing 3.2% acrylamide.

The [γ-^{32}P] ATP and [γ-^{32}P] GTP were obtained from Amhersham/Searle. Electrophoresis supplies were from Bio-Rad. Radioautograms were prepared from Kodak NoScreen Medical X-Ray film. All other reagents were obtained from Sigma Chemical Co. or other commercial sources.

RESULTS

Initial experiments designed to analyze the phosphorylation of band 3-enriched membranes were performed using ghosts which had been extracted with either NaI (1.0 M) or with DMMA (1.5–2.0 mg DMMA/mg protein). As determined by $NaDodSO_4$-polyacrylamide gel electrophoresis, both particulate preparations were enriched with respect to band 3. However, the ghosts extracted with NaI contained a greater number of "other" protein bands than the DMMA pellets (data not shown). Both preparations were analyzed in order to determine if they contained endogenous kinase activity or substrates for the solubilized cyclic AMP-independent human erythrocyte membrane kinase (HMK). Table I shows that autophosphorylation of band 3 occurs in the NaI-extracted but not in the DMMA-extracted ghosts. The autophosphorylation reaction was observed in the presence of ATP but not GTP and was slightly enhanced by cyclic AMP. The results suggest that the NaI-extracted ghosts may contain residual activity of the human erythrocyte membrane cyclic AMP-dependent and -independent kinases. It is evident from Table I that both preparations served equally well as substrates for the solubilized HMK in the presence of ATP. Usage of GTP was poor or nonexistent.

TABLE I. Phosphorylation of Band 3 Protein(s) in NaI- and Dimethylmaleic Anhydride (DMMA)-Extracted Human Erythrocyte Ghosts

Experiment	ATP	ATP + Cyclic AMP	GTP
		(arbitrary densitometric units)	
NaI-Ghosts			
Control	30	44	0.6
+ HMK	170	148	22
DMMA-Ghosts			
Control	0	0	0
+ HMK	150	168	0.8

NaI- or DMMA-extracted ghosts were phosphorylated as described under Methods and electrophoresed in a 4% slab gel. The radioautogram was traced on a desitometer and the areas under the peaks corresponding to band 3 were integrated. The values are expressed as arbitrary densitometric units.

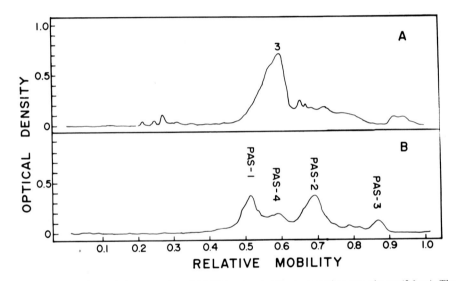

Fig. 1. The proteins and glycoproteins of DMMA-extracted human erythrocyte ghosts (26 μg). The electrophoresis was carried out in NaDodSO$_4$ (0.2%)-polyacrylamide (4%) gels; the gels were stained for protein with Coomassie Blue (A) and for sialoglycoprotein with periodic acid-Schiff's reagent (B).

Since the DMMA-ghosts were considerably more enriched in band 3 and appeared to contain suitable substrate(s) for phosphorylation, further analysis of band 3 phosphorylation was performed using these ghosts. Figure 1 shows the electrophoretic profile of a typical DMMA ghost preparation on a 4% polyacrylamide slab gel. In this preparation, the band 3 protein accounts for greater than 60% of the total CBB-stained protein (Fig. 1A) vs less than 30% in unfractionated human erythrocyte ghosts (data not shown). The DMMA-extracted ghosts appear to retain all the major sialoglycoproteins. The electrophoretic mobilities of the PAS-stained sialoglycoproteins in the 4% gel (Fig. 1B) are slower than in the more commonly used 5 or 5.6% gels (for example, see Refs. 10, 11, 28). This is consistent with the observation that the mobilities of glycoproteins in NaDodSO$_4$

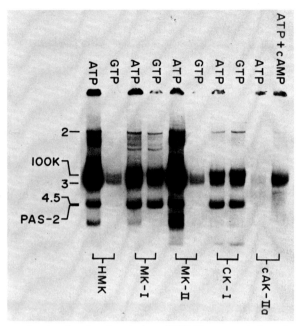

Fig. 2. Radioautogram showing the phosphorylation of DMMA-extracted ghosts by erythrocyte protein kinases. The membranes (13 μg) were phosphorylated in the presence of: HMK (4.8 μg); MK-I (36 μg); MK-II (13 μg); CK-I (0.75 μg); or cAK-IIa (15 μg). The conditions for phosphorylation and electrophoresis were as described in Methods. [γ-^{32}P] ATP:244 cpm/pmol; [γ-^{32}P] GTP:226 cpm/pmol. The radioautogram was exposed for 4 days.

gels are dependent on the percentage of acrylamide cross-linking (29, 30). Analysis of the DMMA pellet of the rabbit erythrocyte membranes gave a CBB-staining profile similar to that of Fig. 1A, although the band 3 peak was more symmetrical. As expected, no PAS-staining component was detected in the rabbit erythrocyte membranes. These membranes do not appear to contain major sialoglycoproteins (1, 20, 31).

Previous studies of membrane phosphorylation by membrane-bound (1) and soluble (3) kinases indicated that more than one peptide migrating in the band 3 area of human and rabbit erythrocyte membranes could serve as substrates for phosphorylation. These proteins were tentatively identified as bands 2.9 and 3, or collectively as area 3 (1, 3). The evidence for the presence of the phosphoproteins in the area of band 3 was based on the study of the effects of pH and/or kinases on their phosphorylation (1, 3). In view of this, it was necessary to determine whether the band 3-enriched ghosts contained multiple protein substrates for phosphorylation. Figure 2 shows a radioautogram depicting the phosphorylation of DMMA-extracted human erythrocyte ghosts by the solubilized membrane kinases HMK, MK-I, and MK-II, by the erythrocyte cytoplasmic casein kinase-I (CK-I), and by cyclic AMP- dependent protein kinase IIa (cAK-IIa). In order to facilitate comparison of the activities of these enzymes toward membrane substrates, the specific activities of the kinases used in these studies were calibrated on the basis of their ability to phosphorylate casein or histone. The amounts of HMK, MK-I, MK-II, and CK-I used would catalyze equivalent amounts of ^{32}P incorporation into casein, while the amount of cAK-IIa used would catalyze twice the amount of ^{32}P incorporation into histone. From Fig. 2 several points are evident. The solubilized membrane kinases HMK and MK-II

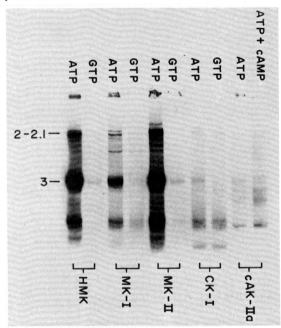

Fig. 3. Radioautogram depicting the phosphorylation of DMMA-extracted rabbit erythrocyte ghosts (6 μg) by erythrocyte protein kinases. Conditions were as described in Fig. 2.

catalyze the phosphorylation of band 3 equally well in the presence of ATP (columns 1 and 5), while neither enzyme uses GTP effectively (columns 2 and 6). However, it is not clear from Fig. 2 whether the phosphorylation in this area is entirely due to band 3 or whether the protein designated as 100K, which migrates in the proximity of band 3, is also phosphorylated by these enzymes. Moreover, MK-I appears to behave differently in that it catalyzes the phosphorylation of 100K and band 3 to the same extent with ATP. However, with GTP, the 100K protein is the preferred substrate. As this enzyme preparation is not homogenous, it is possible that the phosphorylation pattern differences seen with ATP and GTP may reflect the activity of multiple kinases. The cytoplasmic CK-I catalyzes the phosphorylation of the 100K protein and, to a lesser extent, band 3 in the presence of either ATP or GTP. The cyclic AMP-dependent protein kinase IIa (cAK-IIa) catalyzes the phosphorylation of a protein with a mobility similar to band 3 in a reaction that is dependent on cyclic AMP.

Figure 3 shows a similar experiment using DMMA-extracted rabbit erythrocyte membranes. The patterns of phosphorylation obtained with HMK and MK-II are similar to those observed in the DMMA-extracted human erythrocyte ghosts. This appears to be the extent of similarity between the two systems. In the DMMA-extracted rabbit erythrocyte membranes, no comparable phosphorylation of a 100K protein was detected in the presence of MK-I. Furthermore, the phosphorylation of the band 3 area by either CK-I or cAK-IIa is much less prominent in the DMMA-extracted rabbit erythrocyte membranes.

Although the phosphoprotein with a mobility corresponding to 100,000 daltons is not demonstrated in the DMMA-extracted rabbit erythrocyte membranes, we have previously demonstrated that intact rabbit, as well as human, erythrocyte ghosts contain a protein behaving similarly to that designated as 100K in Fig. 2 (1, 3). This protein was tentatively identified as band 2.9 (1, 3). Since the DMMA-extracted human and rabbit

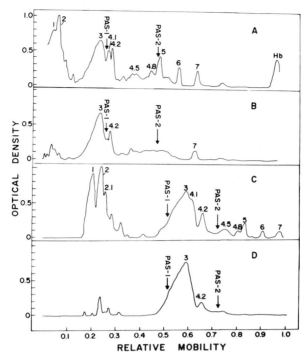

Fig. 4. Mobilities of human erythrocyte membrane proteins and glycoproteins in NaDodSO₄ gels containing 7% or 4% polyacrylamide. Duplicate samples of whole ghosts (A and C, 23 μg) or DMMA-extracted ghosts (B and D, 14 μg) were applied to 7% (A and B) or 4% (C and D) gels which were stained with Coomassie Blue or periodic acid-Schiff's reagent.

erythrocyte membranes exhibit a major difference in the phosphorylation in the area of 100K, it becomes necessary to determine whether the human erythrocyte membranes contain, in addition to band 2.9, a distinct protein component with the same mobility, but which is not easily extractable by DMMA. It is of interest to note that PAS-1, which is present in the human but not in the rabbit erythrocyte membranes, also migrates in the area corresponding to 100K. Furthermore, the sialoglycoproteins remain bound to the membranes after extraction with DMMA. In view of these characteristics, the possibility that the 100K phosphoprotein may represent PAS-1 seems likely.

In order to explore the possible relationship of band 2.9 and PAS-1 to the 100K phosphoprotein, the electrophoretic mobilities of these components were examined in gels containing different amounts of acrylamide. The rationale behind this approach is based on the finding that the membrane PAS components behave anomalously during NaDodSO₄-polyacrylamide gel electrophoresis when compared to standard proteins (29, 30). This anomalous behavior is due to a decreased binding of NaDodSO₄ by the glycoproteins as compared to nonglycosylated, or lightly glycosylated, proteins. As shown by Segrest and Jackson (30), the glycoproteins exhibit a decreased mobility and hence a higher apparent MW in NaDodSO₄ gels of low percentage of acrylamide cross-linking. However, in gels of higher acrylamide cross-linking, the electrophoretic mobilities of the glycoproteins reflect more closely their true molecular weight. Figures 4 and 5 compare the CBB- and PAS-staining profiles of human and rabbit erythrocyte membrane proteins in NaDodSO₄ gels containing 7% and 4% acrylamide. Figure 4, A and B, shows the electro-

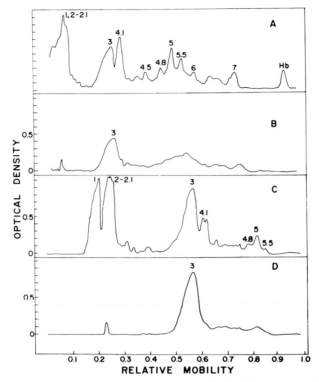

Fig. 5. Mobilities of rabbit erythrocyte membrane proteins in NaDodSO₄ gels containing 7% or 4% polyacrylamide. Whole ghosts (A and C, 18 μg) or DMMA-extracted ghosts (B and D, 15 μg) were electrophoresed on 7% (A and B) and 4% (C and D) gels which were stained with Coomassie Blue or periodic acid-Schiff's reagent. No PAS-staining glycopeptides were evident.

phoretic pattern of human erythrocyte membrane proteins and sialoglycoproteins in 7% gels. PAS-1 migrates faster than band 3 but slower than band 4.1, while PAS-2 migrates between 4.8 and 5.0. However, in the 4% gels (Fig. 4, C and D), PAS-1 and PAS-2 exhibit a higher apparent molecular weight and migrate slower than band 3 and band 4.5, respectively. Figure 5 illustrates the mobilities of rabbit erythrocyte membrane proteins in the 7% (Fig. 5, A and B) and 4% (Fig. 5, C and D) gels. No PAS-stained bands are evident.

Figure 6 compares the radioautograms showing the phosphorylation profiles of intact and DMMA-extracted ghosts of human erythrocytes in 4% (top) and 7% (bottom) gels. Several observations can be made. In column 3, which illustrates the phosphorylation of intact ghosts with CK-I and GTP, note that the mobility of the protein designated as 2.9, relative to the other phosphopeptides, is the same in the 2 different gel concentrations; that is, band 2.9 in intact erythrocytes migrates with an apparent MW of approximately 100,000 daltons regardless of the acrylamide concentration. The results suggest that band 2.9 is either lightly glycosylated or not a glycoprotein and is distinct from PAS-1 (see below). On the other hand, the electrophoretic profiles of the phosphoproteins of DMMA-extracted ghosts phosphorylated in the presence of CK-I and ATP (column 4) behave anomalously in NaDodSO₄ gels. Of particular interest is the electrophoretic behavior of the phosphopeptide which migrates in the region of 100K in the 4% gel (column 4, top). This phosphopeptide exhibits a different mobility on 7% gel (column 4, bottom). The

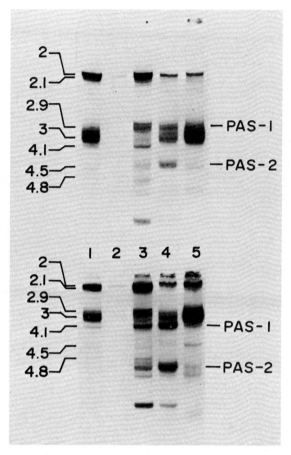

Fig. 6. Differential phosphorylation of bands 2.9, 3, PAS-1, and PAS-2 of human erythrocyte membranes. Phosphorylated proteins were applied to either a 4% (top) or 7% (bottom) gel. (1) whole ghosts + ATP; (2) whole ghosts + GTP; (3) whole ghosts + GTP + CK-I; (4) DMMA-extracted ghosts + ATP + CK-I; (5) DMMA-extracted ghosts + ATP + HMK. Whole ghosts, 23 μg; DMMA-extracted ghosts, 14 μg; CK-I, 1.9 μg; HMK, 4.9 μg; ATP, 350 cpm/pmol; GTP, 390 cpm/ pmol. The radioautograms were exposed for 1 day.

results show that the 100K phosphopeptide of the DMMA-extracted human erythrocyte membranes has the same characteristic anomalous electrophoretic property as the sialoglycoprotein PAS-1 and may be identified with PAS-1 and PAS-2. The two sialoglycoproteins are interconvertible (32, 33). Thus, it appears that PAS-1 and PAS-2 are phosphorylated in the DMMA-extracted ghosts. On the other hand, band 2.9 appears to represent a separate phosphopeptide which is not present in the DMMA-extracted ghosts. It is somewhat more difficult to analyze the phosphorylation of PAS-1 and PAS-2 by CK-I in intact ghosts based on changes in the apparent MW of the PAS-staining components in 4% and 7% gels. The intact ghosts contain many more phosphopeptides than the DMMA-extracted membranes. However, a direct comparison of the phosphopeptides of intact and DMMA-extracted membranes resolved in the 4% and 7% gels (column 3 vs 4) indicates that PAS-1 and PAS-2 in the intact ghosts may be also phosphorylated by CK-I.

The study with HMK reveals a somewhat different phosphorylation pattern; HMK

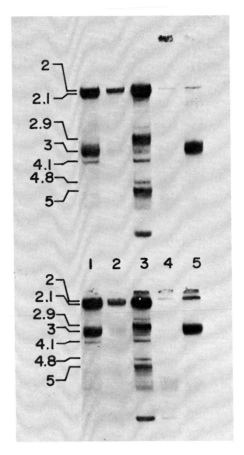

Fig. 7. Phosphorylation of intact (18 μg) and DMMA-extracted (10 μg) rabbit erythrocyte ghosts. Conditions were as described in Fig. 6.

catalyzes the phosphorylation of band 3 and several other minor components in the DMMA-extracted ghosts (column 5). Based on the analysis of the electrophoretic mobilities of the phosphopeptides in 4% and 7% gels (column 5, top vs bottom), it is concluded that the sialoglycoproteins, PAS-1 and PAS-2, in the DMMA-extracted ghosts are not phosphorylated by the solubilized human erythrocyte membrane cyclic AMP-independent protein kinase.

Figure 7 is a similar experiment comparing the phosphorylation patterns of intact and DMMA-extracted rabbit erythrocyte ghosts. As seen in column 3 the unfractionated rabbit ghosts contain a band 2.9 protein that is phosphorylated in the presence of CK-I and GTP. This phosphopeptide is similar to that of the human band 2.9 in that it exhibits no anomalous behavior in NaDodSO$_4$ gels and migrates as a 100,000 dalton protein. In addition, band 2.9 in the rabbit, as in the human, is not retained in the DMMA-extracted ghosts as evidenced by the lack of such a phosphoprotein in these membranes (column 4). No phosphoproteins corresponding to PAS-1 and PAS-2, or to other glycopeptides, are evident in the rabbit erythrocyte membranes; the apparent MWs of the phosphopeptides in these membranes do not vary with the gel concentrations.

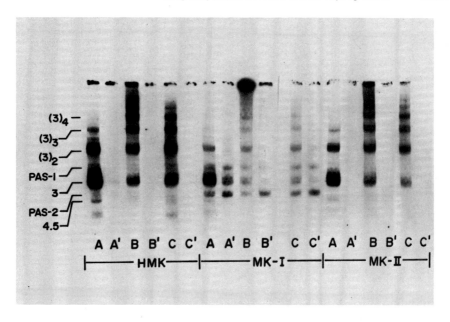

Fig. 8. Phosphorylation of DMMA-extracted human erythrocyte ghosts by HMK, MK-I, and MK-II before and after cross-linking of the membrane proteins with o-phenanthroline-Cu^{2+} complex (CuP). (A) control + ATP; (B) proteins were cross-linked with CuP, then phosphorylated with ATP; (C) proteins were phosphorylated with ATP then cross-linked with CuP. The experimental conditions of the A', B', and C' gels correspond respectively to A, B, and C except with GTP as the phosphoryl donor. DMMA-extracted ghosts, 20 μg; HMK, 7.4 μg; MK-I, 36 μg; MK-II, 13 μg; ATP, 200 cpm/pmol; GTP, 208 cpm/pmol. The radioautogram was exposed for 4 days.

   The phosphorylation of the proteins of DMMA-extracted membranes was further examined after extraction with 0.125% Triton X-100 in 5 mM Tris-HCl, pH 7.5 (13). According to Kahlenberg (13), extraction of DMMA-treated human erythrocyte membranes with Triton X-100 yields a membrane preparation containing band 3 and minor components and a supernatant fraction containing both band 3 and the PAS-staining glycoproteins. The residual band 3 associated with the Triton pellets of both the human and rabbit erythrocyte membranes can be phosphorylated by HMK, MK-I, and MK-II in the presence of ATP (data not shown). Similarly, the proteins in the Triton supernatants also can serve as phosphoryl acceptors. The phosphorylation patterns of both the rabbit and human Triton supernatants were identical to those illustrated in Fig. 2.

   Recently, Shapiro and Marchesi (10) have examined the labeling of the band 3 proteins with ^{32}P in the intact human erythrocytes. They showed that not all the radioactivity in the band 3 region could be attributed to either band 3 protein or PAS-1 and suggested that the (Na^+, K^+)-ATPase, or other unidentified membrane components, might account for some of the radioactivity. In our present study, it is unlikely that the phosphorylated intermediate of the (Na^+, K^+)-ATPase contributes any labeling to the band 3 region since the acyl phosphate intermediate (34) would not be expected to remain intact during the staining-destaining conditions (1, 5). That the band 3 protein is the major phosphopeptide found in the band 3 area is further substantiated by our cross-linking studies. Figure 8 illustrates the phosphorylation of DMMA-extracted ghosts by

HMK, MK-I, or MK-II before and after treatment with the Cu^{2+}-o-phenanthroline complex (CuP) which induces cross-linking of membrane proteins (11, 35). The A gels represent the phosphorylation pattern of band 3 and other components with ATP in the absence of cross-linking. As the samples applied to these gels were not treated with dithiothreitol the formation of the band 3 dimer is evident. Phosphorylation of DMMA-extracted ghosts by HMK, MK-I, and MK-II before (B gels) and after (C gels) cross-linking of proteins with CuP, leads to the formation of trimer, tetramer, and perhaps larger aggregates of the band 3 phosphopeptides. These cross-linked products were further analyzed by 2-dimensional gel electrophoresis (data not shown) in which the cross-links were cleaved by dithiothreitol in the second dimension (35). No significant difference exists between preparations treated with CuP before phosphorylation (B columns) and those treated with CuP after phosphorylation (C columns). The results indicate that phosphorylation of band 3 in the DMMA-extracted ghosts by the solubilized kinases is not inhibited by cross-linking of the band 3 proteins into higher MW aggregates. We also have utilized the 2-dimensional gel electrophoresis procedure to study the phosphorylation of band 3 cross-linked in unfractionated ghosts and found similarly that the major band 3 protein is phosphorylated. However, in contrast to results obtained using the DMMA-extracted ghosts and the isolated kinase, autophosphorylation of band 3 in unfractionated ghosts is inhibited somewhat by the cross-linking of the proteins [(36), and unpublished observations]. The experiments with MK-I show that the phosphorylated peptides corresponding to PAS-1 and PAS-2 do not appear to be cross-linked, an observation consistent with previously reported data (11, 12, 14). The A', B', C' gels show the phosphorylation of DMMA-extracted ghosts in the presence of GTP. We see that HMK and MK-II fail to utilize GTP as the phosphoryl donor. However, MK-I can utilize GTP to phosphorylate band 3, PAS-1 and PAS-2. From the data it is evident that the phosphorylation and the mobility of the PAS components are not affected by treatment with CuP.

DISCUSSION

The results presented in this paper establish the substrate specificity of erhthrocyte protein kinases towards membrane proteins migrating in the CBB-stained area known as band 3. Unfractionated human and rabbit erythrocyte ghosts contain at least 2 proteins, designated as bands 2.9 and 3, which migrate in this area. In addition, the band 3 area of human erythrocyte membranes also contains the sialoglycoprotein PAS-1. Bands 2.9, 3, and PAS-1 are differentially phosphorylated by erythrocyte protein kinases.

Band 2.9, which migrates with the diffusely stained, trailing "tail" of band 3, is phosphorylated in unfractionated human and rabbit erythrocyte ghosts by the cytoplasmic casein kinases, CK-I (Figs. 6 and 7) and CK-II (3). A phosphoprotein with a mobility similar to 2.9 is also detected in the autophosphorylation of rabbit and human erythrocyte membranes (1), and in the phosphorylation of heat-inactivated rabbit erythrocyte membranes by HMK, MK-I, and MK-II (24). Whether this overlap in substrate specificity between MK-I and MK-II or HMK is real or represents an artefact due to the inhomogeneity of the MK-I preparation is not known. Band 2.9 does not remain associated with band 3 after DMMA extraction. Further characterization of band 2.9 with regards to its identity and function as a phosphopeptide awaits isolation of the protein.

The band 3 protein in DMMA-extracted ghosts of human and rabbit erythrocytes is phosphorylated by HMK and MK-II in the presence of ATP but not GTP. That GTP is not used in the phosphorylation of band 3 in DMMA-extracted ghosts is in contrast to the results obtained for the autophosphorylation of band 3 in unfractionated ghosts [(1, 3) and Figs. 6, 7] and for the phosphorylation of band 3 in heat-inactivated ghosts by HMK and MK-II (24). Under these conditions GTP is used for band 3 phosphorylation, although to a lesser extent than ATP. Band 3 also appears to be a substrate for MK-I and CK-I in the presence of either ATP or GTP; however, these enzymes are much less active towards band 3 in comparison to MK-II or HMK (Fig. 2).

A protein migrating with band 3 in human erythrocyte ghosts is phosphorylated in the presence of cyclic AMP-dependent protein kinase I (data not shown) or IIa (Fig. 2). This reaction requires cyclic AMP and ATP. The fact that this reaction does not occur in DMMA-extracted ghosts of rabbit erythrocytes and that the mobility of the cyclic AMP-dependent substrate appears to differ slightly from the substrate of HMK and MK-II leads us to question whether the cyclic AMP-dependent protein kinase substrate is identical to that of HMK and MK-II. The cyclic AMP-dependent phosphopeptide appears to migrate slightly slower than the leading edge of the HMK or MK-II substrate and does not coincide with PAS-1. These results suggest that another protein may be present in the band 3 of DMMA-extracted membranes which is specifically phosphorylated by cAK-I or cAK-IIa.

The PAS-stained glycopeptides, PAS-1 and PAS-2, are phosphorylated in DMMA-extracted human erythrocyte ghosts by MK-I and CK-I, but not by HMK, MK-II, cAK-IIa (Fig. 2), or cAK-I (data not shown). The phosphorylation of PAS-1 and PAS-2 in the DMMA-extracted ghosts catalyzed by either MK-I or CK-I occurs to the same extent with ATP or GTP. The enzymes which phosphorylate glycophorin are independent of cyclic nucleotides, but can be inhibited by the red cell metabolite 2,3-diphosphoglyceric acid (23, 26). The enzymes MK-I and CK-I are derived from rabbit erythrocyte membranes and cytoplasm, respectively. Salt extracts of human erythrocyte membranes contain very little, if any, kinase activity corresponding to MK-I (23). Whether or not the soluble casein kinases CK-I and CK-II also are present in human erythrocyte lysates has not been investigated. It would seem paradoxical that the human erythrocyte would contain the substrates but not the kinases whereas the rabbit erythrocyte, the kinases but not the glycoprotein substrates. However, that glycophorin appears to be phosphorylated in intact erythrocytes (10) would tend to argue for the presence of kinases similar to MK-I and/or CK-I or CK-II in human erythrocytes.

The physiological role of the phosphorylation of the band 3 area peptides has not been determined but is of great interest in view of the membrane functions attributed to these proteins. Although the identity and function of band 2.9 is unknown, the properties of band 3 and glycophorin, both of which span the membrane asymmetrically, have been extensively studied. Band 3 has been implicated in the transport of anions (14–17) and water (18) across the red cell membrane. That anion transport may be occurring through band 3 has been suggested on the basis of the binding of inhibitors of anion transport to band 3 (14–17). However, Lepke et al. (17) show that binding of the inhibitors also occurs to some degree at sites other than band 3, one of which may correspond to the phosphopeptide we term 2.9. Further evidence suggesting that band 3 contains the anion transport channel comes from reconstitution of the transport by incorporating Triton extracts containing band 3 into lecithin vesicles (16). The possibility that phosphorylation may provide a positive or negative regulation of anion transport has not been determined.

Band 3 is a sulfhydryl-containing protein which is thought to exist in the membrane as a dimer (11, 35). It is of interest that the cross-linking of the band 3 protein by the oxidation of thiol groups with CuP does not inhibit the phosphorylation of band 3 by the solubilized membrane protein kinases (Fig. 8). Since the phosphorylation of band 3 is not affected by cross-linking of the proteins and vice versa, the data suggest that the cross-linking sites may be remote from the phosphorylation site(s). It has been proposed that the band 3 subunits aggregate to form the aqueous channel for anion transport (15). It seems attractive to postulate that phosphorylation-dephosphorylation of band 3 may play a role in the aggregation-disaggregation of this protein.

Band 3 contains cytoplasmic binding sites for other membrane proteins, including bands 4.2 and 6 (11, 37). Phosphorylation of band 3 has been reported to occur at 2 sites: one within the 10,000 dalton region of the NH_2 terminal (cytoplasmic) and the second site, in an undefined portion of the polypeptide (38). It would be of interest to determine if the binding of the membrane polypeptides to band 3 is dependent on the phosphorylation state of the band 3 protein.

Although band 3 has been implicated previously in glucose transport (12, 13, 39), recent studies in a more defined system indicate that it is not band 3 but a protein migrating in the area designated as 4.5 that may be responsible for the sugar transport (40, 41). It is of interest that a 4.5 area protein in DMMA-extracted human erythrocyte ghosts also is phosphorylated by HMK and MK-II (Fig. 6). The phosphorylation of the protein labelled 4.5 in the DMMA-extracted human erythrocyte ghosts is not catalyzed by the cyclic AMP-dependent protein kinases (Fig. 2). On the other hand, a protein in the intact erythrocyte with similar mobility is phosphorylated in rabbit and human erythrocyte membranes by cyclic AMP-dependent enzymes. That more than one protein may be present in the 4.5 area in intact ghosts is evident in the 7% gel shown in Fig. 4A. The results suggest that the proteins phosphorylated in the 4.5 area in the DMMA-extracted vs the intact erythrocyte membranes may be physically distinct. However, the possibility that there is only one protein substrate for phosphorylation which is modified during the DMMA extraction should not be ruled out.

It has been suggested that glycophorin (PAS-1 and PAS-2), which is phosphorylated in the DMMA-extracted ghosts (Fig. 6) as well as in intact erythrocytes (10), contains receptors or other surface recognition sites of the erythrocyte (42). That PAS-1 and PAS-2 represent the dimer and monomer, respectively, of glycophorin has been adequately demonstrated (32, 33). Under certain conditions used to solubilize membranes for electrophoresis PAS-1 is converted to PAS-2. The results obtained in this paper do not indicate whether the state of aggregation affects the phosphorylation of glycophorin. Although the conformation of glycophorin in the lipid bilayer is not known, it has been suggested that the glycoproteins exist as aggregates in the membrane in order to facilitate receptor function (33). Recently, Shapiro and Marchesi have demonstrated that the site of phosphorylation of glycophorin is located on the cytoplasmic COOH terminal end. It remains to be determined if phosphorylation plays a role in the formation of aggregates of the protein.

ACKNOWLEDGMENTS

This work was supported by the American Cancer Society grant #BC-65C. M. M. Hosey is a National Research Service Act postdoctoral fellow. M. Tao is an Established Investigator of the American Heart Association.

REFERENCES

1. Hosey MM, Tao M: Biochemistry 15:1561, 1976.
2. Hosey MM, Tao M: Nature 263:424, 1976.

3. Hosey MM, Tao M: J Biol Chem 252:102, 1977.
4. Plut DA, Hosey MM, Tao M: Fed Proc 36:320, 1977.
5. Avruch J, Fairbanks G: Biochemistry 13:5507, 1974.
6. Fairbanks G, Avruch J: Biochemistry 13:5514, 1974.
7. Guthrow CE, Allen JE, Rasmussen H: J Biol Chem 247:8145, 1972.
8. Rubin C, Erlichman J, Rosen OM: J Biol Chem 247:6135, 1972.
9. Palmer FBSC, Verpoorte JA: Can J Biochem 49:337, 1971.
10. Shapiro DL, Marchesi VT: J Biol Chem 252:508, 1977.
11. Steck TL: J Mol Biol 66:295, 1972.
12. Kahlenberg A: J Biol Chem 251:1582, 1976.
13. Zala CA, Kahlenberg A: Biochem Biophys Res Commun 72:866, 1976.
14. Cabantchik ZI, Rothstein A: J Memb Biol 15:207, 1974.
15. Rothstein A, Cabantchik ZI, Knauf P: Fed Proc 35:3, 1976.
16. Rothstein A, Cabantchik ZI, Balshin M, Juliano R: Biochem Biophys Res Commun 64:144, 1975.
17. Lepke S, Fasold H, Pring M, Passow H: J Membr Biol 29:147, 1976.
18. Brown PA, Feinstein MB, Sha'afi RI: Nature 254:523, 1975.
19. Steck TL: J Cell Biol 62:1, 1974.
20. Vimr ER, Carter JR; Biochem Biophys Res Commun 73:779, 1976.
21. Dodge JT, Mitchell C, Hanahan DJ: Arch Biochem Biophys 100:119, 1963.
22. Hosey MM, Tao M: Biochem Biophys Res Commun 64:1263, 1975.
23. Hosey MM, Tao M: Biochim Biophys Acta (In press).
24. Hosey MM: Fed Proc 36:641, 1977.
25. Tao M, Hackett P: J Biol Chem 248:5324, 1973.
26. Kumar R, Tao M: Biochim Biophys Acta 410:87, 1975.
27. Steck TL, Yu J: J Supramol Struct 1:220, 1973.
28. Fairbanks G, Steck TL, Wallach DFH: Biochemistry 10:2606, 1971.
29. Segrest JP, Jackson RL, Andrews EP, Marchesi VT: Biochem Biophys Res Commun 44:390, 1971.
30. Segrest JP, Jackson RL: Methods Enzymol 28:54, 1972.
31. Hamaguchi H, Cleve H: Biochem Biophys Res Commun 47:459, 1972.
32. Marton LSG, Garvin JE: Biochem Biophys Res Commun 52:1457, 1973.
33. Furthmayr H, Marchesi VT: Biochemistry 15:1137, 1976.
34. Avruch J, Fairbanks G: Proc Natl Acad Sci USA 69:1216, 1972.
35. Wang K, Richards FM: J Biol Chem 249:8005, 1974.
36. Plut DA, Hosey MM, Tao M: J Supramol Struct (Suppl 1)) 6:662, 1977.
37. Marchesi VT, Furthmayr H, Tomito M: Ann Rev Biochem 45:667, 1976.
38. Drickamer LK: J Biol Chem 251:5115, 1976.
39. Kasahara M, Hinkle PC: Proc Natl Acad Sci USA 73:396, 1976.
40. Hinkle PC, Kasahara M: J Supramol Struct (Suppl 1) 6:614, 1977.
41. Kahlenberg A, Zala CA: J Supramol Struct (Suppl 1) 6:696, 1977.
42. Marchesi VT, Tillack TW, Jackson RL, Segrest JP, Scott RE: Proc Natl Acad Sci USA 69:1445, 1972.

Journal of Supramolecular Structure 6:77—84 (1977)
Molecular Aspects of Membrane Transport 17—24

Isolation of the Alanine Carrier From the Membranes of a Thermophilic Bacterium and Its Reconstitution Into Vesicles Capable of Transport

Hajime Hirata, Nobuhito Sone, Masasuke Yoshida, and Yasuo Kagawa

Department of Biochemistry, Jichi Medical School, Tochigi-ken, Japan 329-04

A carrier protein mediating alanine transport was purified from the membranes of the thermophilic bacterium PS3, by ion exchange chromatography in the presence of both Triton X-100 and urea.

The alanine carrier was recovered in the nonadsorbed fraction from either DEAE- or CM-cellulose columns, suggesting that its isoelectric point was in the neutral pH region.

The final preparation contained virtually no electron transfer components, ATPase, or NADH dehydrogenase. Polyacrylamide gel electrophoresis in the presence of sodium dodecyl sulfate revealed that the final preparation consisted of two major protein components with molecular weights of 36,000 and 9,400.

Active transport of alanine after incorporation of the alanine carrier into reconstituted proteoliposomes was driven not only by an artificial membrane potential generated by potassium ion diffusion via valinomycin but also by mitochondrial cytochrome oxidase incorporated into the same liposomes and supplemented with both cytochrome c and ascorbic acid.

The membrane-integrated portion (TF_0) of the ATPase complex uncoupled alanine transport by conducting protons across the membrane.

Key words: thermophilic bacterium, transporting proteoliposome, proteoliposome reconstitution, alanine carrier

Isolated bacterial membranes have been widely used for studies on various energy coupling mechanisms including oxidative phosphorylation and active transport (1, 2). Since they are devoid of the many other complicated biochemical reactions occurring in the cytoplasm, such preparations are suitable for studying membrane-bound reactions. However, a much more simplified system is obviously required for detailed studies on the molecular mechanisms of these energy transformation processes. Recently components necessary for oxidative phosphorylation have been isolated and reconstituted into proteoliposomes capable of energy transformation (3—6). On the other hand, carrier proteins mediating movement of substrates across the permeability barrier have not been solubilized, except in a few cases, including the glucose carrier of erythrocytes (7), that from small intestine (8) or the proline carrier from Bacillus subtilis (9).

Received March 14, 1977; accepted March 18, 1977

In our earlier reports (10), we described the solubilization and partial purification of an alanine carrier from the membranes of the thermophilic bacterium PS3 and its reconstitution into proteoliposomes capable of active transport of alanine dependent on a membrane potential induced by K^+ diffusion mediated by valinomycin. However, those preparations were contaminated by minute quantities of cytochromes which might have introduced unnecessary complications of interpretation. The present report describes more satisfactory purification procedures and a further characterization of the active transport of alanine in reconstituted proteoliposomes.

MATERIALS AND METHODS

Materials

The thermophilic bacterium PS3 (kindly donated by Dr. T. Oshima) and preparation of membranes were described previously (11). The preparation of PS3 phospholipids was also described previously (3). Phospholipid fractionation was carried out by silicic acid column chromatography using Unisil (100–200 mesh, Clarkson Chemical Company, Inc., Pennsylvania). We used L-[U-^{14}C]-alanine purchased from Daiichi Radiochemicals. Cholic acid (Sigma Chemical Company) was recrystallized as described by Kagawa (12). Cytochrome oxidase of beef heart mitochondria was prepared according to the method of Yonetani (13). Horse heart cytochrome c was purchased from Sigma Chemical Company and the other compounds used were commercial preparations.

Phospholipid-Detergent Mixture

The phospholipid-detergent mixture was prepared as described elsewhere (14).

Reconstitution of Proteoliposomes

Reconstitution of proteoliposomes and K^+-loading were carried out by either the dialysis method or the dilution method as described elsewhere (14).

Proteoliposomes containing both the alanine carrier and cytochrome oxidase were prepared by the method of Hinkle et al. (15), except that the carrier and deoxycholate were added to the phospholipid, oxidase, and cholate as described in the Figure legends.

Assay of Alanine Carrier Activity

The alanine carrier activity was assayed at 40°C as described elsewhere (10, 14).

Other Analytical Methods

Gel electrophoresis was carried out on 7.5% polyacrylamide gel in the presence of 0.05% sodium dodecyl sulfate under the conditions described by Weber and Osborn (16). Before application to the gels, proteins were incubated at 95°C for 5 min in a solution containing 1% sodium dodecyl sulfate, 5% 2-mercaptoethanol, and 25 mM potassium phosphate buffer, pH 7.0.

Other analytical methods were described previously (3).

Purification of Alanine Carrier

The extraction of membranes by cholate-deoxycholate (Step 1) and the first DEAE-cellulose column chromatography in the presence of 0.25% Triton X-100 (Step 2) have been described in detail elsewhere (14). After removal of Triton X-100 from the DEAE eluate by ammonium sulfate fractionation in the presence of 1% Na-cholate (10, 14), the resulting precipitates were collected by centrifugation (20,000 × g, 10 min), washed twice with 50 mM Tris-SO$_4$ (pH 8) and resuspended in the same buffer. This preparation was named DE-1.

Step 3: DEAE-cellulose column chromatography in the presence of Triton and urea. The DE-1 preparation was diluted fivefold with distilled water to which Triton X-100 and solid urea were added to final concentrations of 0.5% and 6 M, respectively. The solution was centrifuged for 10 min at 140,000 × g, and the supernatant liquid was applied to a DEAE-cellulose column equilibrated with 10 mM Tris-SO$_4$ (pH 8) containing 0.25% Triton X-100 and 4 M urea. The nonadsorbed fractions were collected, combined, and immediately dialyzed overnight at room temperature against 25 mM Tris-SO$_4$ (pH 8) containing 0.25 mM EDTA. The dialysate was subjected to ammonium sulfate fractionation in the presence of 1% Na-cholate. The resulting precipitates were collected by centrifugation (20,000 × g, 10 min), washed twice with 50 mM Tris-SO$_4$ (pH 8), and resuspended in the same buffer. This preparation was named UDE-1.

Step 4: CM-cellulose column chromatography in the presence of Triton and urea. The UDE-1 preparation was diluted 2.5-fold with distilled water and treated with Triton X-100 and urea as described for Step 3. After centrifugation the supernatant liquid was applied to a CM-cellulose column equilibrated with 20 mM Tris-Cl (pH 7.4) containing 0.25% Triton X-100 and 4 M urea. The nonadsorbed fractions were collected and treated as described for Step 3. The final preparation was named UDE-CM-1.

RESULTS AND DISCUSSION

Purification of Alanine Carrier

In our earlier reports (10) we described the solubilization of the alanine carrier from membranes by the use of a cholate-deoxycholate mixture and its partial purification by DEAE-cellulose column chromatography and gel filtration in the presence of Triton X-100. However, the final preparation contained small amounts of cytochromes and numerous other protein contaminants detectable by gel electrophoresis in the presence of sodium dodecyl sulfate. Furthermore, the prolonged exposure of the alanine carrier to a high concentration of Triton X-100 (more than 1%), which happened at the gel filtration step, resulted in considerable inactivation. After various efforts to remove the contaminants, we found that ion exchange cellulose column chromatography in the presence of both Triton X-100 and urea was most effective. Results of a typical purification of the alanine carrier are summarized in Table I. In this particular case, CDE-P preparations (cholate-deoxycholate extracts, see Ref. 10 for details) from several batches were combined and used as the starting material. The specific activity of alanine transport by the original membranes was around 1 nmole per min per mg of protein. Since the alanine transport activity of the reconstituted vesicles was dependent on the species of phospholipids used to form them (10), meaningful comparison of the specific activities with those of the original membrane was not possible. However, the specific activity of the final preparation was more than 10-fold that of the original membranes.

In Table II the amounts of the various contaminants in the final preparation are shown. The preparation contained virtually no electron transfer components and no ATPase. During the first DEAE-cellulose column chromatography, most of the ATPase and NADH dehydrogenase activities were removed. However, a considerable amount of

TABLE I. Summary of Purification of Alanine Carrier

| Preparation[a] | Protein mg | Alanine carrier activity[b] | | |
		Specific U/mg	Total U	Yield %
CDE-P	326	4.29	1,399	100
DE-1	35.1	7.51	263.6	18.8
UDE-1	14.7	14.35	210.9	15.1
UDE-CM-1	3.0	18.00	53.8	3.8

[a]Abbreviations are as in text.
[b]One unit of alanine carrier activity is defined as the amount transporting 1 nmole of alanine under the standard assay condition (10, 14).

TABLE II. Contaminants in UDE-CM-1 Preparation

Component	Membranes	CDE-P[a]	UDE-CM-1[a]
		units/mg. protein	
ATPase	1.19	3.78	< 0.01
NADH dehydrogenase	3.24	0.10	< 0.01
NADH oxidase	1.12	0.10	0.00
		nmoles/mg protein	
Cytochrome b	4.0	0.89	0.00
Cytochrome c + c_1	2.6	0.93	0.00
Cytochrome a	0.6	0.24	0.00
Flavin	3.1	0.21	0.00
		mg/mg protein	
Phospholipids	0.11	0.21	< 0.01

[a]Abbreviations are as in text.

cytochromes b and c + c_1 and approximately one-third of the phospholipids remained in the DE-1 preparation (data not shown). Removal of the phospholipids resulted in aggregation of the proteins, which were hard to dissolve in the Triton solution. Thus, for solubilization of these aggregates, more than 6 M urea was required in addition to Triton. However, since the exposure of proteins to high concentrations of urea inactivated the activity (data not shown), a rather rapid manipulation was necessary during the urea treatment. At the second DEAE-cellulose column chromatography in the presence of both Triton and urea, the activity appeared in the nonadsorbed fractions, which were completely free of phospholipids and electron transfer components except for a minute quantity of cytochrome b (measureable only by its absorbance in the Soret region). The final preparation after CM-cellulose column chromatography, however, was completely devoid of cytochrome. The fact that the activity adsorbed on neither DEAE- nor CM-cellulose columns in the presence of Triton and urea indicated that the alanine carrier had a neutral isoelectric point.

Gel Electrophoresis

Figure 1A shows the electrophoretic pattern of the final preparation on polyacrylamide gel containing 0.05% sodium dodecyl sulfate, pH 7.0. There were 2 major

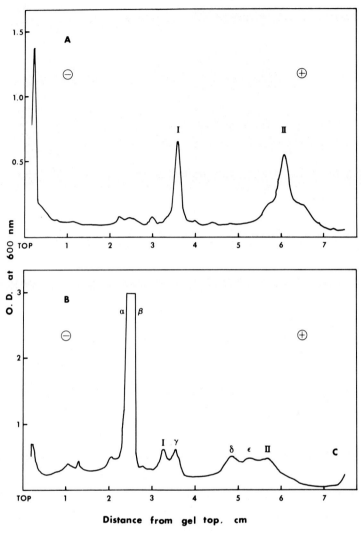

Fig. 1. Electrophoresis of the final preparation (UDE-CM-1) in sodium dodecyl sulfate-polyacrylamide gels. The gels were loaded with A) UDE-CM-1, 20 μg or B) UDE-CM-1, 20 μg plus TF$_1$ 30 μg. The staining was carried out using Coomassie Brilliant Blue. α, β, γ, δ, and ϵ correspond to subunits of TF$_1$ (19).

bands (I and II) and a small number of minor bands. The latter seem to be due to contaminants since their relative intensities varied in different preparations. In Fig. 1B the electrophoretic pattern of the final preparation mixed with purified TF$_1$, so as to provide subunits as molecular weight markers, is shown. The α, β, γ, δ, and ϵ subunits of TF$_1$ have molecular weights of 56,000, 53,000, 32,000, 15,500 and 11,000, respectively. From these results, the molecular weights of bands I and II were estimated to be 36,000 and 9,400, respectively. Although we have no evidence at the moment whether these 2 major bands are the essential components (or subunits) of the alanine carrier protein, these 2 bands appear to relate to its activity since they can be observed in every preparation so far tested. In our earlier reports, we described the apparent molecular weight of the alanine carrier to be 150,000, as determined by gel filtration in the presence of 0.5%

Triton X-100 (10). Possible explanations for the discrepancy in the two values for the molecular weight are: a) the protein may be in its aggregated form when only Triton X-100 is present, b) the protein may be a large oligomer, or c) bound surfactant may increase the apparent molecular weight of the protein when estimated by gel filtration.

Alanine Transport in Proteoliposomes With Integral Cytochrome Oxidase

As described previously, alanine transport by the reconstituted proteoliposomes is dependent on a membrane potential generated by K^+ diffusion mediated by valinomycin (10). However, this uptake is transient and dependent on the magnitude of the K^+ concentration gradient across the membrane. Thus, the extent of loading of K^+ into the proteoliposomes was the limiting factor in the alanine transport. For better analysis of

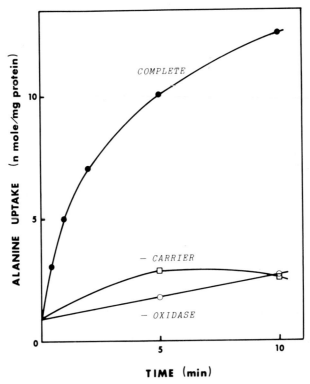

TIME (min)

Fig. 2. Uptake of alanine by proteoliposomes containing both the alanine carrier and cytochrome oxidase. To 0.2 ml of PS3 phosphatidylethanolamine-detergent mixture (50 mg of phosphatidylethanolamine, 20 mg Na-cholate, and 10 mg deoxycholate per ml) was added 20 μg of protein of UDE-CM-1 and 0.4 mg protein of purified cytochrome oxidase and the mixture dialyzed against 10 mM Tricine-NaOH (pH 8) containing 150 mM KCl and 5 mM $MgSO_4$ for 20 hr at 4°C. The reconstituted proteoliposomes were collected by centrifugation (140,000 × g, 60 min), washed once with 50 mM Tricine-NaOH (pH 8) containing 5 mM $MgSO_4$, and resuspended in the same buffer. The assay mixture (0.1 ml) contained proteoliposomes, 1.7 μg of UDE-CM-1 proteins; cytochrome c, 0.375 mg; Tris-maleate, pH 7.0, 0.188 M; $MgSO_4$, 0.019 M; Na-ascorbate, 0.025 M; and ^{14}C-alanine, 0.021 mM, 1.25 μCi/ml. The reaction was started by the addition of ^{14}C-alanine and ascorbate with vigorous shaking at 37°C, and at an appropriate time it was terminated by dilution with 2 ml of 0.05 M LiCl. Then the mixture was filtered and washed with 2 ml of 0.05 M LiCl. ●) proteoliposomes containing both the alanine carrier and cytochrome oxidase; ○) proteoliposomes without cytochrome oxidase; □) proteoliposomes without the alanine carrier.

the molecular mechanisms of active alanine transport, a continuous supply of electro-chemical energy might be desirable.

Mitochondrial cytochrome oxidase has been known to generate a proton-motive force when reconstituted into proteoliposomes and supplemented with cytochrome c and ascorbic acid (15, 17). The direction of generation of the proton-motive force is dependent on the side to which these electron donors are added; an interior negative potential is obtained when cytochrome c and ascorbic acid are added to the external medium.

Figure 2 shows the alanine uptake driven by ascorbic acid oxidation by the recon-stituted proteoliposomes containing both the alanine carrier and the purified cytochrome oxidase of beef heart mitochondria. On addition of ascorbic acid, an accumulation of alanine by the reconstituted proteoliposomes was observed. Control experiments with proteoliposomes lacking either the carrier or cytochrome oxidase showed no uptake of alanine. The presence of KCN or carbonylcyanide p-trifluoromethoxyphenylhydrazone completely abolished the uptake (data not shown). The results indicated that energization of the alanine transport can be accomplished by an electrical generator of a quite different nature, animal mitochondrial cytochrome oxidase.

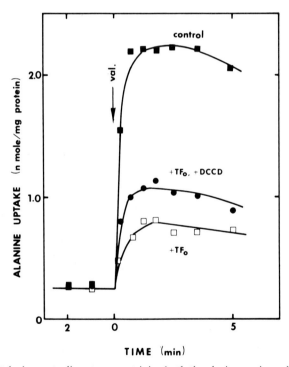

Fig. 3. Alanine uptake by proteoliposomes containing both the alanine carrier and TF_0. Proteolipo-somes were reconstituted by the dialysis method and loaded with 0.5 M potassium phosphate buffer (pH 8), as described alsewhere (14). In the case of preparations of proteoliposomes containing TF_0, 50 μg of TF_0 proteins and 40 μg of UDE-1 proteins were reconstituted with 10 mg of PS3 phospho-lipids. Assay methods and conditions were described previously (10, 14). ■) proteoliposomes con-taining only the alanine carrier; □ and ●) proteoliposomes containing both the alanine carrier and TF_0. Dicyclohexylcarbodiimide (0.1 mM) was added as indicated. An arrow indicates the addition of valinomycin (2 μg per ml).

TF$_0$ Integrated Proteoliposomes

The membrane-integrated portion (TF$_0$) of the proton-translocating ATPase complex of the thermophilic bacterium PS3 has been shown to be a specific proton conductor (18). As shown on Fig. 3, proteoliposomes containing both the alanine carrier and TF$_0$ show reduced alanine accumulation driven by a membrane potential, which is partially restored by the addition of dicyclohexylcarbodiimide. These results are consistent with the fact that the proton-conducting activity of TF$_0$ is known to be inhibited by dicyclohexylcarbodiimide (18). Thus TF$_0$ uncouples the active transport of alanine by conducting protons across the membranes.

ACKNOWLEDGMENTS

The authors are grateful to Dr. Tairo Oshima of Mitsubishi Life Science Laboratory, Machida City, Tokyo, for kindly providing the strain used. They also thank the members of Kyowa Hakko Co., Ltd. for large-scale cultures of the bacterium. The technical assistance of Ms. Toshiko Kanbe and Ms. Keiko Ikeba is acknowledged.

REFERENCES

1. Harold FM: In Sanadi DR, Packer L (eds): "Current Topics in Bioenergetics," New York and London: Academic Press, Vol 6, p 83, 1976.
2. Kaback HR: Science 186:882, 1974.
3. Sone N, Yoshida M, Hirata H, Kagawa Y: J Biol Chem 250:7917, 1975.
4. Sone N, Yoshida M, Hirata H, Kagawa Y: J Biol Chem (In press).
5. Ryrie IJ, Blackmore PF: Arch Biochem Biophys 176:127, 1976.
6. Serrano R, Kanner BI, Racker E: J Biol Chem 251:2453, 1976.
7. Kasahara M, Hinkle PC: Proc Natl Acad Sci USA 73:396, 1976.
8. Crane RK, Malathi P, Preiser H: Biochem Biophys Res Commun 71:1010, 1976.
9. Kusaka I, Hayakawa K, Kanai K, Fukui S: Eur J Biochem 71:451, 1976.
10. Hirata H, Sone N, Yoshida M, Kagawa Y: Biochem Biophys Res Commun 69:665, 1976.
11. Yoshida M, Sone N, Hirata H, Kagawa Y: J Biol Chem 250:7910, 1975.
12. Kagawa Y, Racker E: J Biol Chem 246:5477, 1971.
13. Yonetani T: In Estabrook RW, Pullman ME (eds): "Methods in Enzymology," New York and London: Academic Press, Vol 10, p 332, 1967.
14. Hirata H: In Fleischer S, Packer L (eds): "Methods in Enzymology," New York and London: Academic Press. (In press).
15. Hinkle PC, Kim JJ, Racker E: J Biol Chem 247:1338, 1972.
16. Weber K, Osborn M: J Biol Chem 244:4406, 1969.
17. Drachev LA, Jasaites AA, Kaulen AD, Kondrashin AA, Chu LV, Semenov AY, Severina II, Skulachev VP: J Biol Chem 251:7072, 1976.
18. Okamoto H, Sone N, Hirata H, Yoshida M, Kagawa Y: J Biol Chem (In press).
19. Yoshida M, Sone N, Hirata H, Kagawa Y: J Biol Chem (In press).

Journal of Supramolecular Structure 6:85—94 (1977)
Molecular Aspects of Membrane Transport 25—34

Mutations Affecting the Binding, Internalization, and Lysosomal Hydrolysis of Low Density Lipoprotein in Cultured Human Fibroblasts, Lymphocytes, and Aortic Smooth Muscle Cells

Michael S. Brown, Richard G. W. Anderson, and Joseph L. Goldstein

Departments of Internal Medicine and Cell Biology, University of Texas Health Science Center at Dallas, Dallas, Texas 75235

Studies comparing the metabolism of low density lipoprotein (LDL) in normal cells and in cells cultured from patients with homozygous familial hypercholesterolemia have disclosed the existence of a receptor for plasma LDL. This receptor has been identified on the surface of human fibroblasts, lymphocytes, and aortic smooth muscle cells. An extension of these studies to cell strains derived from patients with other single gene defects in cholesterol metabolism has provided additional insight into the normal mechanisms by which cells regulate their cholesterol content and how alterations in these genetic control mechanisms may predispose to atherosclerosis in man.

Key words: lipoproteins, receptors, smooth muscle cells, cholesterol, atherosclerosis, familial hypercholesterolemia

When cultured human fibroblasts are deprived of cholesterol, they synthesize a specific cell surface receptor that binds low density lipoprotein (LDL), the major cholesterol-carrying lipoprotein of human plasma (1—3). Binding of LDL to the receptor is the first step in a pathway — the LDL pathway — by which cells take up the lipoprotein and utilize its cholesterol. By regulating the number of cell surface LDL receptors, cells are able to control the rate of entry of cholesterol, thereby assuring themselves an adequate supply of the sterol while at the same time preventing its overaccumulation. Since plasma LDL is derived ultimately from lipoproteins synthesized in the liver or intestine, the LDL pathway constitutes a mechanism in vivo by which cholesterol can be delivered to peripheral tissues from the liver (the main site of cholesterol synthesis) or the intestine (the site of cholesterol absorption from the diet).

SEQUENTIAL STEPS IN THE LDL PATHWAY AS DELINEATED IN FIBROBLASTS

Once LDL has bound to the LDL receptor in normal fibroblasts the lipoprotein is internalized by adsorptive endocytosis and delivered to lysosomes (4—6) where the protein and cholesteryl ester components of the lipoprotein are hydrolyzed (7, 8). The resulting free cholesterol is then available to be used by the cell for membrane synthesis. When

Abbreviations: FH — familial hypercholesterolemia; HDL — high density lipoprotein; HMG CoA reductase — 3-hydroxy-3-methylglutaryl coenzyme A reductase; LDL — low density lipoprotein

Received March 14, 1977; accepted March 18, 1977

sufficient cholesterol has accumulated to satisfy this requirement, 3 regulatory events occur: 1) cholesterol synthesis is suppressed through a reduction in the activity of the rate-controlling enzyme, 3-hydroxy-3-methylglutaryl coenzyme A reductase (HMG CoA reductase) (9); 2) excess lipoprotein-derived free cholesterol is reesterified for storage as cholesteryl esters through an activation of an acylCoA:cholesterol acyltransferase (10, 11); and 3) synthesis of the LDL receptor itself is diminished, thereby preventing further entry of LDL-cholesterol into the cell (12). The sequential steps in the LDL pathway are illustrated diagrammatically in Fig. 1.

MUTATIONS AFFECTING THE LDL PATHWAY IN FIBROBLASTS

At each step in the delineation of the LDL pathway in cultured fibroblasts, interpretation of the data has been clarified by the analysis of mutant fibroblasts derived from patients with genetic defects involving specific steps in the pathway. To date, 6 such mutations have been identified, 5 affecting the LDL pathway at the cellular level and one affecting the secretion of plasma LDL itself.

The mutation that has proved to have the greatest explanatory potential is the one found in patients with the homozygous form of receptor-negative familial hypercholesterolemia (FH), an autosomal dominant disorder (mutation No. 2 in Fig. 1). Fibroblasts from these FH homozygotes lack functional LDL receptors as determined by assays that are sufficiently sensitive to detect about 2% of the normal number (1, 13). As a result, these cells fail to bind and take up the lipoprotein with high affinity and, therefore, fail to hydrolyze either its protein or cholesteryl ester components. Because they are unable to utilize LDL-cholesterol, these homozygote cells must satisfy their cholesterol require-

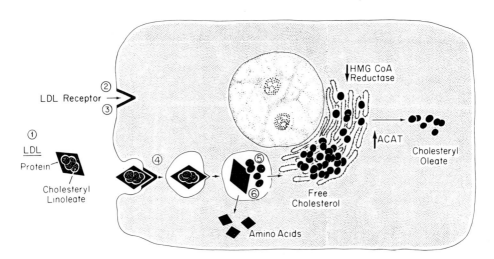

Fig. 1. Sequential steps in the LDL pathway in cultured human fibroblasts. The numbers indicate the sites at which mutations have been identified: 1) abetalipoproteinemia; 2) familial hypercholesterolemia, receptor-negative; 3) familial hypercholesterolemia, receptor-defective; 4) familial hypercholesterolemia, internalization defect; 5) Wolman disease; and 6) cholesteryl ester storage disease. HMG CoA reductase denotes 3-hydroxy-3-methylglutaryl coenzyme A reductase, and ACAT denotes fatty acyl-coenzyme A:cholesterol acyltransferase. (Modified from M.S. Brown and J.L. Goldstein, Science 191: 150–154, 1976. Copyright 1976 by the American Association for the Advancement of Science. Reprinted with permission.)

ment by synthesizing large amounts of cholesterol de novo even when high levels of LDL are present in the culture medium. Moreover, in these mutant cells LDL does not stimulate the formation of cholesteryl esters.

In addition to the cells in which LDL receptor activity is not detectable (receptor-negative), a second class of mutant fibroblasts has been observed in which the maximal number of functional LDL receptors is reduced to about 5–20% of normal (mutation No. 3 in Fig. 1). Patients with this mutation have been designated as having the receptor-defective type of homozygous FH (13). Both the receptor-defective and the receptor-negative mutations have been observed in fibroblasts obtained from subjects with the clinical phenotype of homozygous FH. Such patients manifest extremely high levels of plasma LDL (their LDL-cholesterol levels are about six- to tenfold above normal), accumulation of free and esterified cholesterol in interstitial spaces and within phagocytic cells of the skin and tendons, and severe atherosclerosis with myocardial infarction occurring as early as 18 months of age.

The heterozygous parents of both the receptor-negative and receptor-defective FH homozygotes are more mildly affected than the homozygotes. These heterozygotes manifest plasma LDL-cholesterol levels that are two- to fourfold above the normal level and they usually develop clinical signs of atherosclerosis between the ages of 30 and 60. Fibroblasts from heterozygotes with the receptor-negative mutation have been shown to synthesize about one-half the normal number of LDL receptors and thus their cells take up and degrade LDL at one-half the normal rate (14). The consequences that this 50% deficiency in LDL receptors creates for the regulation of cholesterol metabolism in this genetically dominant syndrome are discussed elsewhere (15).

A third type of mutation in LDL uptake has been described in a patient who manifests a clinical syndrome indistinguishable from that of homozygotes with the receptor-negative and receptor-defective mutations (mutation No. 4 in Fig. 1). Fibroblasts from this patient are unique in that they are able to bind normal amounts of LDL at the receptor site but are unable to internalize the receptor-bound lipoprotein (16). As a result, in these cells, just as in the cells from receptor-negative FH homozygotes, high affinity degradation of LDL does not occur, and the lipoprotein does not suppress cholesterol synthesis nor does it stimulate cholesteryl ester formation. The existence of this internalization mutation indicates that the adsorptive endocytosis of LDL by cells requires at least 2 functionally distinct active sites, each of which can be altered by mutation, namely, a site that is required to bind LDL and a site that participates in the internalization of the lipoprotein after it is bound to the receptor site. Our preliminary genetic studies suggest that these 2 sites may reside on the same peptide chain.

To date, 22 strains of fibroblasts derived from patients in 11 countries who manifest the clinical phenotype of classic homozygous FH have been studied in detail in our laboratory. Of these 22 cell strains, 12 were found to be receptor-negative, 9 were receptor-defective, and 1 exhibited the internalization defect (Table I).

In addition to the 3 known mutations that produce the syndrome of familial hypercholesterolemia, 2 other mutations have been shown to affect the LDL pathway in fibroblasts (1–3). These mutations (mutations No. 5 and No. 6 in Fig. 1) occur in patients with the Wolman disease and cholesteryl ester storage disease, two autosomal recessive disorders in which cholesteryl esters accumulate abnormally in lysosomes of cells throughout the body. Studies by Patrick and Lake disclosed that the primary defect in the Wolman disease involves the absence of a lysosomal acid lipase that normally hydrolyzes both cholesteryl esters and triglycerides (17). In the related, but clinically less severe, syndrome

TABLE I. Biochemical Analysis of Fibroblast Strains Derived From 22 Subjects With the Clinical Syndrome of Homozygous FH

Biochemical Phenotype	Number of Subjects
Receptor-negative	12
Receptor-defective	9
Internalization defect	1

of cholesteryl ester storage disease the activity of this same lysosomal acid lipase is reduced to about 1—5% of normal (18). When incubated with LDL, fibroblasts from patients with both of these disorders bind and take up the lipoprotein normally and degrade its protein component at a normal rate (19, 20). However, the defect in lysosomal acid lipase activity prevents normal hydrolysis of the cholesteryl ester component of LDL. As a result, the cholesteryl esters of LDL accumulate within lysosomes and the lipoprotein fails acutely to suppress HMG CoA reductase or to activate the acyl-CoA:cholesterol acyltransferase (19, 20).

Clinically, in the Wolman disease the absence of the lysosomal acid lipase produces a massive accumulation of cholesteryl esters and triglycerides in nearly all body tissues and death ensues within the first year of life (21). In cholesteryl ester storage disease, the residual acid lipase activity is sufficient to allow survival to young adulthood. It is of importance that in the few of these young patients whose tissues have been studied at autopsy cholesteryl ester accumulation was particularly marked in the arterial wall and advanced atherosclerosis was noted (18, 21).

Another mutation that has proved useful in working out the LDL pathway is the one that produces the autosomal recessive disease abetalipoproteinemia (mutation No. 1 in Fig. 1). This genetic defect causes a block in either the synthesis or secretion of apoprotein B so that the plasma of these patients is devoid of LDL (22). Although their plasma contains cholesterol bound to other lipoproteins (mainly, high density lipoprotein [HDL]), HDL-cholesterol is not taken up efficiently by normal fibroblasts in tissue culture (7) and hence it fails to suppress HMG CoA reductase activity or stimulate cholesteryl ester formation in normal cells (9, 10). These observations support the conclusion that only those human cholesterol-carrying lipoproteins that contain apoprotein B — namely, LDL and VLDL — are able to bind to the LDL receptor and deliver cholesterol to fibroblasts.

EXPRESSION OF THE LDL PATHWAY IN LYMPHOCYTES

All of the steps of the LDL pathway in human fibroblasts also have been identified in long-term human lymphoid cells maintained in suspension culture (23, 24) and in human lymphocytes freshly isolated from the bloodstream (25, 26). In particular, studies by Kayden et al. (23) and Ho et al. (24—26) have shown that the processes of cholesterol synthesis and esterification in lymphoid cells and in circulating lymphocytes are regulated through the LDL receptor. These investigators have also shown that lymphoid cells and lymphocytes from patients with the receptor-negative form of homozygous FH are markedly deficient in cell surface LDL receptor activity. Therefore, these cells exhibit the same constellation of secondary defects (i.e., defective LDL uptake and degradation, over-production of cholesterol, and failure to induce cholesteryl ester formation) as do the receptor-negative FH homozygote fibroblasts.

EVIDENCE FOR AN LDL RECEPTOR IN AORTIC SMOOTH MUSCLE CELLS AND ITS ABSENCE IN CELLS FROM A PATIENT WITH HOMOZYGOUS FH

Figure 2 shows the electron microscopic appearance of human aortic smooth muscle cells maintained in monolayer culture. As described for monkey smooth muscle cells (27), these human cells show abundant myofilaments and peripheral dense bodies and are there-fore morphologically distinct from cultured human fibroblasts. Moreover, the human aortic smooth muscle cells, like the monkey smooth muscle cells, grow in a distinctive pattern characterized as "hills and valleys."

Previous studies have shown that cultured human aortic smooth muscle cells take up and degrade ^{125}I-LDL by a process that is saturable with respect to LDL concentration (28, 29). The data in Fig. 3 confirm that this uptake process is dependent on a high af-finity cell surface receptor that is specific for LDL. In this experiment normal aortic smooth muscle cells were subjected to prior growth in the absence of lipoproteins. The cells were then incubated at $37°C$ with 20 μg protein/ml of ^{125}I-LDL either alone or in the presence of increasing concentrations of unlabeled LDL (density, 1.019–1.063 g/ml) or HDL (density, 1.085–1.215 g/ml). After 2 hr, the cell monlayers were washed extensively and the amount of ^{125}I-LDL bound to the receptor was determined by releasing it with heparin, a sulfated glycosaminoglycan that removes ^{125}I-LDL from its surface binding site (4). Whereas unlabeled LDL at a concentration of about 20 μg protein/ml produced a 50% competitive reduction of the binding of ^{125}I-LDL, unlabeled HDL did not com-pete significantly at concentrations as high as 900 μg protein/ml (Fig. 3A). Unlabeled LDL, but not unlabeled HDL, competitively inhibited the proteolytic degradation of ^{125}I-LDL (Fig. 3B). These data indicate that, as in human fibroblasts and lymphocytes, the proteolytic degradation of ^{125}I-LDL by aortic smooth muscle cells is dependent on high affinity cell surface binding.

Evidence that the LDL receptor on cultured human aortic smooth muscle cells is genetically the same as the LDL receptor on fibroblasts and lymphocytes is provided by study of aortic smooth muscle cells obtained from a patient with homozygous FH. Table II shows that the binding, uptake, and degradation of ^{125}I-LDL is markedly deficient in these mutant cells ($<$ 2% of normal values).

Consistent with their deficiency in LDL binding, internalization, and degradation was the finding that the cells from the FH homozygote were markedly deficient in their ability to respond to LDL by suppressing the activity of HMG CoA reductase (Fig. 4) or by enhancing the rate of cholesteryl ester formation (Table III). That the latter defect in the mutant cells was not due to a primary abnormality in the cholesterol esterification system itself was evident from the finding that the FH homozygote cells were able to develop a normal rate of cholesteryl [^{14}C]oleate formation when stimulated with 25-hydroxycholesterol (Table III).

In fibroblasts the binding and uptake of LDL leads to an increase in the cellular content of esterified cholesterol (30). A similar phenomenon occurs in smooth muscle cells. Thus, in the experiment shown in Fig. 5, normal aortic smooth muscle cells were grown for 9 days in the absence of lipoproteins so as to achieve nearly complete depletion of their content of esterified cholesterol. When LDL was then added to

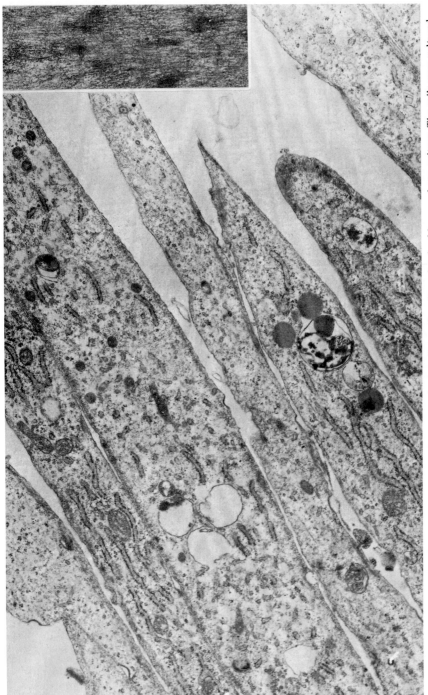

Fig. 2. Electron microscopic appearance of normal human aortic smooth muscle cells maintained in monolayer culture. The cells were cultured by the method of Ross (32) from explants of the thoracic aorta of a normal 6-month-old human fetus (28, 31). A cell monolayer was prepared as described in the legend to Table II. On day 19 of cell growth, the cells were fixed in situ for 30 min with 3% glutaraldehyde in 0.1 M sodium phosphate at pH 7.3. The cells were then scraped from the dish and centrifuged (5 min, 4°C, 12,000 rpm). The resulting cell pellet was postfixed with 2% OsO_4 in 0.1 M sodium phosphate at pH 7.3, dehydrated, embedded in araldite, and stained with uranyl acetate and lead citrate. Note the abundant myofilaments (inset) and peripheral dense bodies. Magnifications, × 14,850; inset, × 37,980.

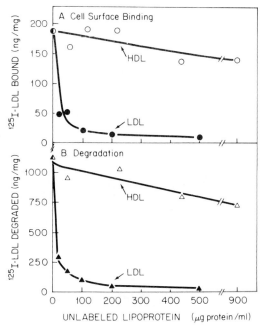

Fig. 3. Comparison of the ability of unlabeled LDL and HDL to compete with ^{125}I-LDL for cell surface binding (A) and degradation (B) in monolayers of normal human aortic smooth muscle cells. On day 0, 1 × 10^5 cells from stock flasks were seeded into each 60-mm petri dish containing 3 ml of growth medium with 10% fetal calf serum (16). On days 2, 4, 7, and 10, the cells received fresh growth medium containing 10% calf lipoprotein-deficient serum. On day 11, the medium was replaced with 2 ml of fresh medium containing 5% human lipoprotein-deficient serum, 20 μg protein/ml of ^{125}I-LDL (224 cpm/ng), and the indicated amount of either unlabeled LDL (density, 1.019–1.063 g/ml) (●, ▲) or unlabeled HDL (density, 1.085–1.215) (○, △). After incubation for 2 hr at 37°C, the medium was removed and its total content of ^{125}I-labeled, trichloroacetic acid-soluble material was measured (7). The cell monolayers were then washed by a standard procedure and the total amount of heparin-releasable ^{125}I-LDL was determined (4). Each value represents the average of duplicate incubations.

these cells, the cellular content of esterified cholesterol increased (Fig. 5B). The presence of HDL at concentrations up to 400 μg protein/ml did not affect the basal level of either free or esterified cholesterol nor did it influence the increment in esterified cholesterol achieved with LDL at either 10 or 50 μg protein/ml.

Of particular relevance to atherosclerosis are the studies on the regulation of the LDL pathway in cultured human aortic smooth muscle cells (3, 31). These studies have demonstrated that aortic smooth muscle cells, like fibroblasts and lymphocytes, regulate the number of LDL receptors so that the cells take up and degrade only enough LDL to supply cellular needs for cholesterol. Because of this regulation, the addition of large amounts of native LDL to smooth muscle cells in culture does not induce an overaccumulation of cholesteryl esters of the type that is observed in the atherosclerotic lesion in vivo (31). These data thus emphasize the critical role of the regulation of the LDL receptor in protecting aortic smooth muscle cells against such an overaccumulation of cholesteryl esters.

TABLE II. Comparison of the Binding, Internalization, and Degradation of ^{125}I-LDL in Monolayers of Aortic Smooth Muscle Cells From a Normal Subject and a Patient With Homozygous FH

Genotype of aortic smooth muscle cells	Heparin-releasable ^{125}I-LDL	Heparin-resistant ^{125}I-LDL	Rate of degradation ^{125}I-LDL
	ng/mg	ng/mg	ng·4hr^{-1}·mg^{-1}
Normal	125 (114)	504 (473)	1941 (1895)
FH homozygote	4.3 (2.3)	17 (7.5)	37 (21)

The homozygous FH smooth muscle cell strain was derived from an 8-year-old female (M.C.) whose clinical features have been described previously (33). Segments of M.C.'s thoracic aorta were obtained at the time that M.C. underwent aortic valve replacement in October, 1976. Smooth muscle cells were cultured as described in the legend to Fig. 2. On day 0, 1.5×10^5 cells from stock flasks were seeded into each 60-mm Petri dish containing 3 ml of growth medium with 10% fetal calf serum. On day 3, each cell monolayer received 3 ml of fresh medium with 10% fetal calf serum. On day 5, the medium was replaced with 3 ml of fresh medium containing 5% human lipoprotein-deficient serum. On day 7, each monolayer received 2 ml of medium containing 5% human lipoprotein-deficient serum and 10 μg protein/ml of ^{125}I-LDL (233 cpm/ng) in the absence and presence of 340 μg protein/ml of unlabeled LDL. After incubation for 4 hr at 37°C, the total amounts of heparin-releasable ^{125}I-LDL (4), heparin-resistant ^{125}I-LDL (4), and ^{125}I-LDL degradation (7) were determined. Each value represents the average of triplicate incubations. The numbers in parentheses represent the amount of binding, internalization, or degradation that was due to the specific, high affinity process. These values were determined from the difference between values obtained in the absence and presence of the excess unlabeled LDL (7, 15).

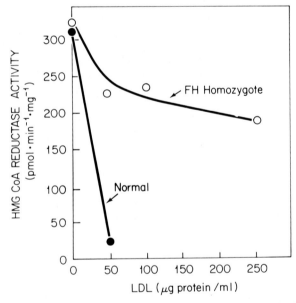

Fig. 4. Comparison of LDL-mediated suppression of HMG CoA reductase activity in monolayers of aortic smooth muscle cells from a normal subject and a patient with homozygous FH. Monolayers were prepared as described in the legend to Table II. On day 7, after incubation for 48 hr in the absence of lipoproteins, each monolayer received 2 ml of medium containing 5% human lipoprotein-deficient serum and the indicated concentration of LDL. After incubation for 6 hr at 37°C, the cells in each dish were harvested for measurement of HMG CoA reductase activity (9). Each value represents the average of duplicate incubations.

TABLE III. Comparison of LDL-Mediated Stimulation of Cholesteryl Ester Formation in Monolayers of Aortic Smooth Muscle Cells From a Normal Subject and a Patient With Homozygous FH

Genotype of aortic smooth muscle cells	Addition to medium	[^{14}C] Oleate incorporated into cholesteryl [^{14}C] oleate
		pmol·hr^{-1}·mg^{-1}
Normal	None	17
	LDL, 50 μg protein/ml	867
FH homozygote	None	1.5
	LDL, 200 μg protein/ml	1.0
	25-Hydroxycholesterol, 5 μg/ml	1,620

Monolayers of smooth muscle cells were prepared as described in the legend to Table II. On day 7, after incubation for 48 hr in the absence of lipoproteins, each monolayer received 2 ml of medium containing 5% human lipoprotein-deficient serum and the indicated addition. After incubation for 5 hr at 37°C, the cells in each dish were pulse-labeled for 2 hr at 37°C with 0.1 mM [^{14}C] oleate-albumin (8,700 cpm/nmol), after which each monolayer was harvested for determination of the cellular content of cholesteryl [^{14}C] oleate (10). Each value represents the average of duplicate incubations.

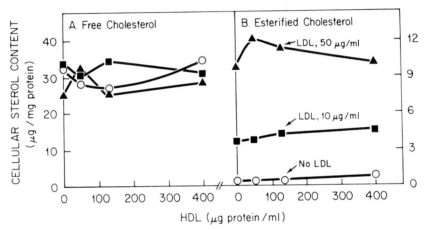

Fig. 5. Inability of HDL to prevent the LDL-mediated accumulation of esterified cholesterol in normal human aortic smooth muscle cells. On day 0, 1.5 × 10^5 cells were seeded into each 60-mm petri dish containing 3 ml of growth medium with 10% fetal calf serum (16). On days 2, 4, and 7, the cells received 3 ml of growth medium containing 10% calf lipoprotein-deficient serum. On day 9, the medium was replaced with 2 ml of fresh medium containing 5% human lipoprotein-deficient serum, the indicated concentration of LDL (density, 1.019–1.063 g/ml) and the indicated concentration of HDL (density, 1.085–1.215 g/ml). After incubation at 37°C for 48 hr, the cell monolayers were washed, harvested, and pooled (2 dishes per sample), and their content of free (A) and esterified (B) cholesterol was determined (30). Each value represents the average of duplicate samples.

ACKNOWLEDGMENTS

These studies were supported by research grants from the National Institutes of Health (GM 19258, HL 16024, and GM 21698). M.S.B. is an Established Investigator of the American Heart Association. J.L.G. is the Recipient of a U.S. Public Health Service Research Career Development Award (GM 70,277).

REFERENCES

1. Brown MS, Goldstein JL: Science 191:150, 1976.
2. Goldstein JL, Brown MS: Curr Top Cell Regul 11:147, 1976.
3. Goldstein JL, Brown MS: Annu Rev Biochem 46:897, 1977.
4. Goldstein JL, Basu SK, Brunschede GY, Brown MS: Cell 7:85, 1976.
5. Anderson RGW, Goldstein JL, Brown MS: Proc Natl Acad Sci USA 73:2434, 1976.
6. Anderson RGW, Grown MS, Goldstein JL: Cell 10:351, 1977.
7. Goldstein JL, Brown MS: J Biol Chem 249:5153, 1974.
8. Brown MS, Dana SE, Goldstein JL: Proc Natl Acad Sci USA 72:2925, 1975.
9. Brown MS, Dana SE, Goldstein JL: J Biol Chem 249:789, 1974.
10. Goldstein JL, Dana SE, Brown MS: Proc Natl Acad Sci USA 71:4288, 1974.
11. Brown MS, Dana SE, Goldstein JL: J Biol Chem 250:4025, 1975.
12. Brown MS, Goldstein JL: Cell 6:307, 1975.
13. Brown MS, Goldstein JL: N Engl J Med 294:1386, 1976.
14. Brown MS, Goldstein JL: Science 185:61, 1974.
15. Goldstein JL, Sobhani MK, Faust JR, Brown MS: Cell 9:195, 1976.
16. Brown MS, Goldstein JL: Cell 9:663, 1976.
17. Patrick AD, Lake BD: Nature 222:1067, 1969.
18. Beaudet AL, Ferry GD, Nichols BL Jr, Rosenberg HS: J Pediatr (In press).
19. Goldstein JL, Dana SE, Faust JR, Beaudet AL, Brown MS: J Biol Chem 250:8487, 1975.
20. Brown MS, Sobhani MK, Brunschede GY, Goldstein JL: J Biol Chem 251:3277, 1976.
21. Sloan HR, Fredrickson DS: In Stanbury JB, Wyngaarden JB, Fredrickson DS (eds): "The Metabolic Basis of Inherited Disease." New York: McGraw-Hill, 1972, pp 493–530.
22. Fredrickson DS, Gotto AM, Levy RI: In Stanbury JB, Wyngaarden JB, Fredrickson DS (eds): "The Metabolic Basis of Inherited Disease." New York: McGraw-Hill, 1972, pp 493–530.
23. Kayden HJ, Hatam L, Beratis NG: Biochemistry 15:521, 1976.
24. Ho YK, Brown MS, Kayden HJ, Goldstein JL: J Exp Med 144:444, 1976.
25. Ho YK, Brown MS, Bilheimer DW, Goldstein JL: J Clin Invest 58:1465, 1976.
26. Ho YK, Faust JR, Bilheimer DW, Brown MS, Goldstein JL: J Exp Med (In press).
27. Ross R, Glomset JA: Science 180:1332, 1973.
28. Goldstein JL, Brown MS: Arch Pathol 99:181, 1975.
29. Bierman EL, Albers JJ: Biochim Biophys Acta 388:198, 1975.
30. Brown MS, Faust JR, Goldstein JL: J Clin Invest 55:783, 1975.
31. Goldstein JL, Anderson RGW, Buja M, Basu SK, Brown MS: J Clin Invest (In press).
32. Ross R: J Cell Biol 50:172, 1971.
33. Bilheimer DW, Goldstein JL, Grundy SM, Brown MS: J Clin Invest 56:1420, 1975.

Journal of Supramolecular Structure 6:95–101 (1977)
Molecular Aspects of Membrane Transport 35–41

The Transport of Lysosomal Enzymes

Elizabeth F. Neufeld, Gloria N. Sando*, A. Julian Garvin, and
Leonard H. Rome

*National Institute of Arthritis, Metabolism, and Digestive Diseases, National Institutes of
Health, Bethesda, Maryland 20014*

This paper reviews the experimental evidence for the proposal that hydrolytic enzymes
are introduced into lysosomes of cultured fibroblasts only after secretion and receptor-
mediated recapture.

Key words: lysosomes, lysosomal enzymes, pinocytosis, secretion, α-L-iduronidase

How are hydrolytic enzymes transferred from the site of their synthesis to lysosomes? In the usual concept of lysosome formation, the hydrolases proceed from rough to
smooth endoplasmic reticulum, then either to the Golgi apparatus or to a specialized
region of the smooth endoplasmic reticulum known as GERL. There they are concentrated
into small vesicles (primary lysosomes) which detach from the Golgi or from GERL.
Eventually these vesicles fuse with phagocytic, pinocytic, or autophagic vacuoles which
contain macromolecules to be hydrolyzed but not the necessary enzymes. The organelles
that result from the fusion — i.e., the secondary lysosomes — contain enzymes, substrates,
and the appropriately acid environment for hydrolysis to take place (for reviews, see Refs.
1–4). The best evidence for the transport of the hydrolytic enzymes by way of primary
lysosomes has been obtained in polymorphonuclear leukocytes. Particles which are rich in
acid hydrolases and which correspond to primary lysosomes — the azurophilic granules —
have been isolated by centrifugation and have been seen to fuse with phagocytic vacuoles
after ingestion of bacteria (2). However, the evidence is weaker for nonphagocytic mammalian cells, since primary lysosomes have not been isolated and their role in enzyme
transport has been deduced from static morphological studies by electron microscopy.

THE SECRETION-RECAPTURE HYPOTHESIS

An alternative hypothesis would have hydrolytic enzymes secreted to the cell exterior (perhaps through secretory vesicles), recaptured by a receptor-mediated endocytosis, and only then packaged into lysosomes (Fig. 1). Prior to secretion, the enzymes would
be equipped with a structural feature (a "recognition marker") to insure binding to the

*Postdoctoral Fellow of the Arthritis Foundation
Received March 14, 1977; accepted March 18, 1977

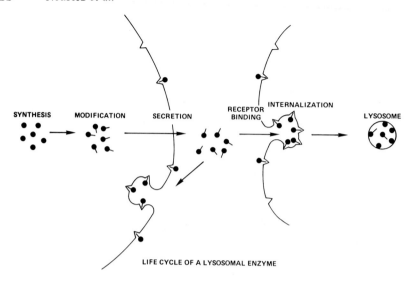

SYNTHESIS   MODIFICATION   SECRETION   RECEPTOR BINDING   INTERNALIZATION   LYSOSOME

LIFE CYCLE OF A LYSOSOMAL ENZYME

Fig. 1. The secretion-recapture hypothesis for introducing hydrolytic enzymes into lysosomes.

receptors on the cell surface. In this hypothesis, the pinocytic vacuoles could be considered either as primary lysosomes (if no substrate were taken in at the same time) or as secondary lysosomes (if substrate were taken in with the enzymes, or if the internalized membrane were itself the substrate to be hydrolyzed).

This proposal provides a unifying theory for the following observations made on cultured human skin fibroblasts: a) some hydrolytic enzymes, when introduced into the culture medium, are taken into the fibroblasts by an efficient and selective mechanism which depends on a recognition marker on the enzyme and a receptor on the fibroblasts; and b) in fibroblasts from certain human mutants, hydrolytic enzymes appear to lack a recognition marker and are inappropriately located in the extracellular fluid rather than within the lysosomes (5).

Selectivity and efficiency of uptake of some hydrolytic enzymes were first noted in the case of α-L-iduronidase (6) and later for a number of other enzymes which participate in the hydrolysis of mucopolysaccharides, glycoproteins, and glycolipids (Table I). Each of the hydrolases listed also exists in low uptake form, the distinction between "high" and "low" uptake being in part one of degree. Since fibroblasts can probably take in almost any macromolecular substrate introduced into the culture medium (e.g., dextran, ^{125}I-albumin, or horseradish peroxidase) to a slight extent, uptake is considered "low" if it represents internalization of about 1% of the amount present in the medium (over a period of 1–2 days) and "high" if it substantially exceeds that value. Uptakes of 25% or greater have been reported.

More important than the extent of uptake is its saturability, which must be expected of a receptor-mediated mechanism. Saturability of uptake has been shown for urinary α-N-acetylglucosaminidase (14), testicular β-galactosidase (17), platelet β-glucuronidase (19), and urinary α-L-iduronidase (Fig. 2). An apparent K_m of 10^{-9}M has been calculated for the uptake of iduronidase (20).

It is apparent from Table I that there is no correlation between the source of a hydrolytic enzyme or the reaction catalyzed, and its occurrence in high uptake form. The

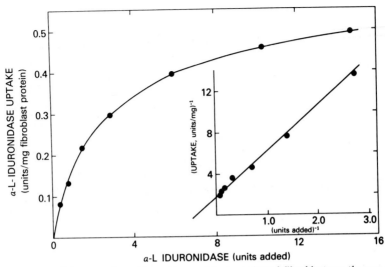

Fig. 2. Dependence of the rate of α-L-iduronidase uptake by cultured fibroblasts on the concentration of enzyme in the medium. Hurler fibroblasts were incubated with increasing levels of iduronidase activity in serum-free medium, at 35°C, for 4 hr. Internalized enzyme activity was assayed in homogenates of the trypsinized cells after 3 cycles of freezing and thawing (21). The assumptions implicit in treating pinocytosis by Michaelis-Menten kinetics are discussed in Ref. 20.

TABLE I. Hydrolytic Enzymes Known to Exist in High or Low Uptake Form

Enzyme	Sources of high uptake form[a]	Sources of low uptake form[a]
α-L-Iduronidase	Urine (6, 7)	Urine (7), kidney (8)
β-Glucuronidase	Platelets (9), spleen (10), liver (11), placenta (11), fibroblasts (11)	Same tissues as high uptake form (10, 11)
β-Hexosaminidase	Fibroblasts (12)	Placenta (12), liver (13)
α-N-Acetylglucosaminidase	Urine (14)	Placenta (15)
Arylsulfatase A	Urine (16)	
β-Galactosidase	Bovine testes (17)	
Iduronate sulfatase	Urine (18)	Serum (18)

[a]All sources are of human origin unless otherwise indicated.

high uptake form of β-hexosaminidase has been converted to the low uptake form (i.e., the recognition marker was destroyed) without affecting the catalytic activity of the enzyme, by treatment with dilute $NaIO_4$ (12). The recognition marker of β-galactosidase was destroyed by digestion with a partially purified preparation of mannosidase from Aspergillus niger (17). These experiments led to the suggestion that a carbohydrate residue is the recognition marker for uptake into fibroblasts, perhaps mannose or some sequence of sugars containing mannose.

Modification of the recognition marker of the enzyme is but one of the approaches to discover its structure; another is the use of inhibitors to compete for the receptor site. Inhibition experiments have ruled out a terminal β-galactose and β-N-acetylglucosamine residue as the recognition marker for α-L-iduronidase (20), thereby differentiating the recognition of hydrolases by fibroblasts from recognition of circulatory glycoproteins by hepatocytes (22–24). D-Mannose, L-fucose, and some mannosides inhibit the uptake of several enzymes (17, 19, 20). A potent glycoprotein inhibitor of iduronidase uptake, of $K_i \sim 10^{-8}$ M, has been purified from normal human urine; it is thought to be a mixture of hydrolases and denatured hydrolases that use the same receptor. An exciting and unexpected development has been the finding that mannose-6-phosphate and some phosphomannans inhibit the uptake of β-glucuronidase, and that the enzyme itself can be converted to the low uptake form by treatment with alkaline phosphatase (19). The uptake of α-L-iduronidase is likewise competitively inhibited by mannose-6-phosphate and diminished by pretreatment with phosphatase (20). Sly and co-workers (19) have proposed that the recognition marker is a phosphorylated carbohydrate residue, probably phosphomannose, on the high uptake form of hydrolases.

Although most of the uptake systems studied have involved enzymes and fibroblasts of human origin, there appears to be considerable cross-species interaction; bovine and rat fibroblasts recognize the marker on human β-glucuronidase (25) whereas human fibroblasts recognize bovine β-galactosidase (17).

Lysosomal enzymes injected intravenously into rats are rapidly cleared from the circulation, primarily into the liver and spleen (26–28). The clearance of circulating lysosomal enzymes, which has been studied most thoroughly with β-glucuronidase, is mediated by a recognition system different from that of fibroblasts, since enzymes which are of the low uptake form with respect to fibroblasts can be rapidly taken out of the plasma. Periodate reduces the rate of clearance, and from competition experiments it is thought that the recognition is through N-acetylglucosamine residues (29, 30) or through mannose (30a).

GENETIC DISEASES ASSOCIATED WITH INAPPROPRIATE LOCALIZATION OF HYDROLYTIC ENZYMES

As first noted by Wiesmann and colleagues (31, 32), fibroblasts of patients with mucolipidosis II (I-cell disease) are deficient in several lysosomal glycosidases and sulfatases. These intracellular deficiencies are accompanied by an excess of the same enzymes in the patient's body fluids. These observations have been extended to patients with mucolipidosis III (pseudo-Hurler polydystrophy), a clinically milder condition (for reviews of the clinical and biochemical findings in the two disorders see Refs. 5 and 33).

Fibroblasts from such patients are not "leaky" and appear to have normal receptors for hydrolases, for when presented with high uptake α-L-iduronidase, they internalize it with the same velocity and kinetic constant as do other fibroblasts and retain it with the same 9-day half-life (20, 34). On the other hand, the enzymes secreted by fibroblasts of patients with mucolipidoses II or III are of the low uptake form (34). As seen from Fig. 1, the secretion-recapture hypothesis provides a simple explanation for these diverse findings: a mutation in the synthesis of the marker would cause secretion of low uptake enzymes, which would fail to be recaptured and, if stable, would accumulate outside the cell. The many other abnormalities of these mutant fibroblasts [e.g., presence of unusual isozymes (35), excessive sialic acid content of some hydrolases (36), and increased fragility of the fibroblast membranes to freezing and detergent (37)] , may be viewed as additional effects of the primary enzyme defect, or as secondary effects of the many enzyme deficiencies, particularly of the recently discovered deficiency of sialidase (38).

Determination of lysosomal deficiencies, whether by direct measurement of enzyme activity or by observation of storage vacuoles by electron microscopy, shows that the mutation of mucolipidoses II and III is manifested primarily in cells of connective tissue (as well as in certain kidney cells and in Schwann cells) but is not shown by leucocytes, hepatocytes, and neurons. Even in cultured fibroblasts, acid phosphatase and β-glucosidase are not depressed. Thus the effects of the mutation appear limited to some cells and to some hydrolytic enzymes. The secretion-recapture hypothesis must likewise be limited until it is known whether these variations are caused by the existence of more than one recognition system or more than one mechanism for transporting hydrolytic enzymes into lysosomes.

PHYSIOLOGICAL IMPLICATIONS

Although a mechanism involving secreted enzyme as a transport form has not been proposed for any other group of intracellular enzymes, it has an analogy in the secretion and subsequent endocytosis of thyroglobulin by epithelial cells of the thyroid gland (39).

From Fig. 1, it is clear that enzymes may be synthesized in one cell and packaged into the lysosomes of its neighbors. Intercellular exchange of lysosomal enzymes has been invoked to explain the cross-correction of defective mucopolysaccharide catabolism by fibroblasts cultured from patients with genetically distinct mucopolysaccharidoses (5). Intercellular transfer of β-hexosaminidase from normal to deficient cells has been demonstrated by direct assay of the enzyme in single fibroblasts, before and after cocultivation (40); however, no transfer of β-galactosidase or of α-glucosidase was demonstrated in similar experiments, and the reason for the apparent difference is not clear. Transfer of β-glucuronidase has been observed to occur in vivo between cells of many tissues of tetraparental mice (41, 42).

In an attempt to test the secretion-recapture hypothesis, we have grown normal human skin fibroblasts in the presence of goat antibody to human α-L-iduronidase (8). This treatment resulted in a drop of up to one-half of the intracellular iduronidase activity, and the effect was completely reversed when the antibody was withdrawn (Fig. 3). Three other lysosomal enzymes, β-galactosidase, β-glucuronidase, and arylsulfatase A were unaffected. The antibody does not act by inhibiting the catalytic activity of α-L-iduronidase. These data may be interpreted as the result of competition between the antibody and the fibroblast receptors for the extracellular iduronidase. In view of the high affinity of the fibroblasts for the enzyme (see above), it is not surprising that the antibody was least effective at high cell density.

Tulkens et al. (43) previously showed that fibroblasts cultured in the presence of antibodies to liver lysosomal enzymes took on the appearance of cells from mucolipidosis II patients. Although the authors attributed this effect to an inhibition of lysosomal enzymes by endocytosed antibodies, they noted that the quantity of ingested antibody seemed insufficient to explain the observed reduction in hydrolytic activity. A plausible explanation for their experiments, as for ours, is an inhibition by the antibodies of the packaging, rather than of the activity, of lysosomal enzymes.

The pathway for hydrolytic enzymes shown in Fig. 1 suggests several ways in which the level of intracellular and extracellular enzyme could be influenced: by the rates of secretion and internalization, as well as by the rates of synthesis and degradation of the

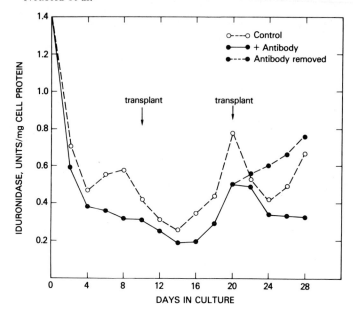

Fig. 3. Effect of anti-iduronidase in the culture medium on the level of intracellular α-L-iduronidase. Normal human skin fibroblasts were grown in 100 mm plastic petri plates, either in modified Eagle's Minimal Essential Medium with 10% fetal calf serum and antibiotics (o- - - - - - - -o) or in the same medium to which goat anti-iduronidase (8) had been added to a concentration of 0.7 mg antibody per ml medium (●——●). Cells were transplanted to low density on days 0, 10, and 20, and medium was changed every other day. On day 20, a set of plates which had been grown in the presence of antibody was transferred to medium without antibody (●---●). Intracellular α-L-iduronidase was measured as previously described (21). Other experimental details will be described elsewhere.

enzyme or receptor. There are physiological and pathological conditions in which the activity of extracellular hydrolases is markedly increased: for instance, in bone exposed to parathyroid hormone (44), in cartilage exposed to Vitamin A (45), and in synovium of patients with rheumatoid arthritis (45). In each case, the increase in extracellular enzyme has been attributed to increased exocytosis from lysosomes. These conditions should be reexamined in the light of the secretion-recapture pathway.

REFERENCES

1. DeDuve C, Wattiaux R: Annu Rev Physiol 28:435. 1966.
2. Cohn ZA, Fedorko ME: In Dingle JT, Fell HB (eds): "Lysosomes in Biology and Pathology." Amsterdam: North-Holland Publishing Co., Vol I, 1969, pp 44–63.
3. Novikoff AB: In Hers HG, van Hoof F (eds): "Lysosomes and Storage Diseases." New York: Academic Press, 1973, pp 2–37.
4. Novikoff AB: Proc Natl Acad Sci USA 73:2781, 1976.
5. Neufeld EF, Lim TW, Shapiro LJ: Ann Rev Biochem 44:357, 1975.
6. Bach G, Friedman R, Weissmann B, Neufeld EF: Proc Natl Acad Sci USA 69:2048, 1972.
7. Shapiro LJ, Hall CW, Leder IG, Neufeld EF: Arch Biochem Biophys 172:156, 1976.
8. Rome L, Garvin AJ, Neufeld EF: Fed Proc 36:749, 1977.
9. Brot FE, Glaser JH, Roozen KJ, Sly WS, Stahl PD: Biochem Biophys Res Commun 57:1, 1974.
10. Nicol DM, Lagunoff D, Pritzl P: Biochem Biophys Res Commun 59:941, 1974.
11. Glaser JH, Roozen KH, Brot FE, Sly WS: Arch Biochem Biophys 166:536, 1975.
12. Hickman S, Shapiro LJ, Neufeld EF: Biochem Biophys Res Commun 57:55, 1974.

13. Schneck L, Amsterdam D, Brodes SE, Rosenthal AL, Volk BW: Pediat 52:221, 1973.
14. von Figura K, Kresse H: J Clin Invest 53:85, 1974.
15. O'Brien JS, Miller AL, Loverde AW, Veath ML: Science 181:753, 1973.
16. Wiesmann UN, DiDonato S, Herschkowitz NN, Biochem Biophys Res Commun 66:1338, 1975.
17. Hieber V, Distler J, Myerowitz R, Schmickel RD, Jourdian GW: Biochem Biophys Res Commun 73:710, 1976.
18. Liebaers I, Neufeld EF: unpublished data.
19. Kaplan A, Achord DT, Sly WS: Proc Natl Acad Sci USA (In press).
20. Sando GN, Neufeld EF: Manuscript in preparation.
21. Hall CW, Neufeld EF: Arch Biochem Biophys 158:817, 1973.
22. Ashwell G: Morell AG, Adv Enzymol 41:99, 1974.
23. Lunney J, Ashwell G: Proc Natl Acad Sci USA 73:341, 1976.
24. Stockert RJ, Morell AG, Scheinberg IH: Biochem Biophys Res Commun 68:988, 1976.
25. Frankel HA, Glaser JH, Sly WS: Pediatr Res (In press).
26. Schlesinger P, Rodman JS, Frey M, Lang S, Stahl P: Arch Biochem Biophys 177:606, 1976.
27. Stahl P, Rodman JS, Schlesinger P: Arch Biochem Biophys 177:594, 1976.
28. Achord D, Brot F, Gonzalez-Noriego A, Sly W, Stahl P: Pediatr Res (In press).
29. Stahl P, Six H, Rodman JS, Schlesinger P, Tulsiani DRP, Touster O: Proc Natl Acad Sci USA 73:4045, 1976.
30. Stahl P, Schlesinger PH, Rodman JS, Doebber T: Nature 264:86, 1976.
30a. Achord DT, Brot FE, Bell CE, Sly WS: Fed Proceed 36:653, 1977.
31. Wiesmann UN, Lighbody J, Vassella F, Herschkowitz NN: N Engl J Med 284:109, 1971.
32. Wiesmann U, Vassella F, Herschkowitz N: N Engl J Med 285:1090, 1971.
33. McKusick VA, Neufeld EF, Kelly TE: In Stanbury JB, Wyngaarden JB, Fredrickson DS (eds): "Biochemical Basis of Inherited Disease." 4th Ed. New York: McGraw-Hill (In press).
34. Hickman S, Neufeld EF: Biochem Biophys Res Commun 49:992, 1972.
35. Lie KK, Thomas GH, Taylor HA, Sensenbrenner JA: Clin Chim Acta 45:243, 1973.
36. Vladutiu GD, Rattazzi MC: Biochem Biophys Res Commun 67:956, 1975.
37. Sly WS, Lagwinska E, Schlesinger S: Proc Natl Acad Sci USA 73:2443, 1976.
38. Thomas GH, Tiller GE Jr, Reynolds LW, Miller CS, Bace JW: Biochem Biophys Res Commun 71:188, 1976.
39. Wollman SH: In Dingle JT, Fell HB (eds): "Lysosomes in Biology and Pathology." Amsterdam: North Holland Publishing Co., Vol I, pp 483–508, 1969.
40. Reuser A, Halley D, De Wit E, Hoogeveen A, van der Kamp M, Mulder M, Galjaard H: Biochem Biophys Res Commun 69:311, 1976.
41. Feder N: Nature 263:67, 1976.
42. Herrup K, Mullen RJ, Feder N: Fed Proc 35:1371, 1976.
43. Tulkens P, Trouet A, van Hoof F: Nature 228:1282, 1970.
44. Vaes, G: In Dingle JT, Fell HB (eds): "Lysosomes in Biology and Pathology." Amsterdam: North Holland Publishing Co, Vol I, pp 217–253, 1969.
45. Dingle JT: ibid, vol II, pp 420–436, 1969.
46. Weissmann G: Arthr Rheum 9:834, 1966.

Journal of Supramolecular Structure 6:103−124 (1977)
Molecular Aspects of Membrane Transport 43−64

Neutral Amino Acid Transport in Surface Membrane Vesicles Isolated from Mouse Fibroblasts: Intrinsic and Extrinsic Models of Regulation

Julia E. Lever

Department of Cell Regulation, Imperial Cancer Research Fund Laboratories, Lincoln's Inn Fields, London WC2A 3PX, England

Membrane transport carrier function, its regulation and coupling to metabolism, can be selectively investigated dissociated from metabolism and in the presence of a defined electrochemical ion gradient driving force, using the single internal compartment system provided by vesiculated surface membranes. Vesicles isolated from nontransformed and Simian virus 40-transformed mouse fibroblast cultures catalyzed carrier-mediated transport of several neutral amino acids into an osmotically-sensitive intravesicular space without detectable metabolic conversion of substrate.

When a Na^+ gradient, external Na^+ > internal Na^+, was artifically imposed across vesicle membranes, accumulation of several neutral amino acids achieved apparent intravesicular concentrations 6- to 9-fold above their external concentrations. Na^+-stimulated alanine transport activity accompanied plasma membrane material during subcellular fractionation procedures. Competitive interactions among several neutral amino acids for Na^+-stimulated transport into vesicles and inactivation studies indicated that at least 3 separate transport systems with specificity properties previously defined for neutral amino acid transport in Ehrlich ascites cells were functional in vesicles from mouse fibroblasts: the A system, the L system and a glycine transport system. The pH profiles and apparent K_m values for alanine and 2-aminoisobutyric acid transport into vesicles were those expected of components of the corresponding cellular uptake system.

Several observations indicated that both a Na^+ chemical concentration gradient and an electrical membrane potential contribute to the total driving force for active amino acid transport via the A system and the glycine system. Both the initial rate and quasi-steady-state of accumulation were stimulated as a function of increasing concentrations of Na^+ applied as a gradient (external > internal) across the membrane. This stimulation was independent of endogenous Na^+, K^+-ATPase activity in vesicles and was diminished by monensin or by preincubation of vesicles with Na^+. The apparent K_m for transport of alanine and 2-aminoisobutyric acid was decreased as a function of Na^+ concentration. Similarly, in the presence of a standard initial Na^+ gradient, quasi-steady-state alanine accumulation in vesicles increased as a function of increasing magnitudes of interior-negative membrane potential imposed across the membrane by means of K^+ diffusion potentials (internal > external) in

Abbreviations used: iso-Abu − 2-aminoisobutyric acid; $TPMP^+$ − triphenylmethylphosphphonium ion; SV 40 − Simian virus 40; FCCP − carbonyl cyanide p-trifluoromethoxyphenylhydrazone.

Julia E. Lever is now at the Salk Institute for Biological Studies, P.O. Box 1809, La Jolla, CA 92112.

Received March 14, 1977; accepted March 22, 1977

the presence of valinomycin; the magnitude of this electrical component was estimated by the apparent distributions of the freely permeant lipophilic cation triphenylmethylphosphonium ion. Alanine transport stimulation by charge asymmetry required Na^+ and was blocked by the further addition of either nigericin or external K^+. As a corollary, Na^+-stimulated alanine transport was associated with an apparent depolarization, detectable as an increased labeled thiocyanate accumulation. Permeant anions stimulated Na^+-coupled active transport of these amino acids but did not affect Na^+-independent transport. Translocation of K^+, H^+, or anions did not appear to be directly involved in this transport mechanism. These characteristics support an electrogenic mechanism in which amino acid translocation is coupled to an electrochemical Na^+ gradient by formation of a positively charged complex, stoichiometry unspecified, of Na^+, amino acid, and membrane component.

Functional changes expressed in isolated membranes were observed to accompany a change in cellular proliferative state or viral transformation. Vesicles from Simian virus 40-transformed cells exhibited an increased V_{max} of Na^+-stimulated 2-aminoisobutyric acid transport, as well as an increased capacity for steady-state accumulation of amino acids in response to a standard Na^+ gradient, relative to vesicles from nontransformed cells. Density-inhibition of nontransformed cells was associated with a marked decrease in these parameters assayed in vesicles. Several possibilities for regulatory interactions involving gradient-coupled transport systems are discussed.

Key words: transport mechanisms, amino acids, mouse fibroblasts, plasma membrane vesicles, regulation, SV40 transformation

The noncovalent coupling of active transport systems for certain nutrients to metabolic energy stored in the form of electrochemical ion circuits across the membrane has emerged as a useful concept to visualize an integral feature of the functional specialization of a diversity of biological membranes (1–9). Thus the alkali-ion gradient hypothesis that Na^+-dependent active transport systems for certain organic solutes in intestinal and renal brush border membranes, erythrocytes, and tumor cells (1–4, 6) are driven by electrochemical Na^+ gradients generated by active Na^+ extrusion by the Na^+, K^+-ATPase shares this fundamental biological principle with Mitchell's subsequent proposal that certain respiration-coupled active transport systems in bacterial cytoplasmic membranes are driven by electrochemical proton gradients generated across the membrane by active proton extrusion (5, 7, 8).

Although much of the information which led to these concepts was obtained using intact cells, an important step was provided by the use of isolated membrane vesicles as a system for transport studies (10). This approach, extended to isolated membranes from animal cells (11–16), obviates ambiguities due to such factors as unknown compartmentalization of ions and substrates, reliance on inhibitors, analogs and metabolic depletion, interaction of ionophore probes with intracellular membranes, and possible coupled or passive movement of other ions which complicate the interpretation of cellular uptake data. Such cell-free transport studies, one example of which will be discussed here, are precursors to more detailed future descriptions of these processes at the molecular level, which would require the use of purified membrane macromolecules.

When surface membranes of fibroblasts grown in culture are disrupted using a nitrogen cavitation procedure (17), unilaminar sacs are formed of average diameter 0.1–0.2 μm (18) which can be identified by characteristic enzymatic and other markers and separated from fragments of intracellular membranes, organelles, and soluble contaminants by centrifugation methods (15, 17). These vesiculated plasma membranes are selectively permeable and retain functional transport carriers for ions and nutrients (15, 16, 18, 19), properties which permit their use to investigate membrane transport mechanisms at the level of flux of molecules between 2 bulk aqueous phases.

Outlined here are some characteristics of the coupling between electrochemical Na^+ gradients, membrane potential, and accumulation of neutral amino acids via functional components of cellular uptake systems, isolated in membrane vesicles prepared from the surface membranes of mouse fibroblasts. As well as providing an experimental confirmation of general predictions of the alkali-ion gradient hypothesis, these observations reveal additional details of the translocation process.

At one level, this interaction between carrier, ion, substrate, and charge asymmetry forms an intrinsic, fixed mode of regulation of solute translocation, dynamically mediated by the structure of the transport component itself. Superimposed on this are extrinsic modes of regulation mediated by genetic expression and cellular and external agents which can also influence these membrane functions. It was noted that alterations in cellular uptake of neutral amino acids, either by cell density-inhibition in nontransformed fibroblasts or accompanying transformation by Simian virus 40, were expressed in the transport properties of their isolated membrane vesicles (16, 19). Possible intrinsic and extrinsic controls of Na^+-dependent amino acid transport are discussed. Some of the experiments discussed here have been published previously (16, 19, 20).

MATERIALS AND METHODS

Cell Culture

Swiss and Balb/c 3T3 mouse fibroblasts transformed by Simian virus 40 (SV 40), Balb/c tertiary mouse embryo cells, and Swiss 3T3 fibroblasts were propagated in Dulbecco's modified Eagle's medium with 10% serum as described previously (16). Cell lines were routinely monitored for absence of mycoplasma contamination.

Preparation of Membrane Vesicles

Cells $(1-9 \times 10^9)$ were washed twice with phosphate-buffered saline at room temperature, harvested by scraping with a rubber blade, and washed again. The cell pellet was resuspended in 0.25 M sucrose, 0.2 mM $MgCl_2$, 0.01 M Tris-HCl, pH 7.5, centrifuged at $4,000 \times g$ for 20 min and the cell pellet was resuspended in 100 ml of this solution with $CaCl_2$ substituted for $MgCl_2$. Cells were disrupted by nitrogen cavitation, and previously described centrifugation methods were used to isolate either purified mixed vesicles (16, 20) or purified plasma membrane vesicles (20, 21).

Purified mixed vesicles contained 39–68% of the total 5′-nucleotidase activity, $20 \pm 2\%$ of the total NADH oxidase activity, $1.5 \pm 0.5\%$ of the total succinate-cytochrome c reductase activity, and 11% of the total cellular protein. Purified plasma membranes from SV 3T3 fractionated after dextran-110 discontinuous gradients (20) contained 20% of the total 5′-nucleotidase activity, 2.6% of cellular NADH oxidase activity, 0.10% of total succinate-cytochrome c reductase activity, and 2% of cellular protein.

Aliquots of 1 ml of vesicles were stored suspended in 0.25 M sucrose, 0.01 M Tris-HCl, pH 7.5, in liquid nitrogen.

Enzyme Assays

Subcellular fractions of vesicle preparations were assayed as described previously

(20) for the following marker enzymes: $5'$ nucleotidase ($5'$-ribonucleotide phosphohydrolase; EC 3.1.3.5), a marker for plasma membrane (17, 22), NADH oxidase (NADH: oxidoreductase, EC 1.6.99.3), a marker for endoplasmic reticulum and outer mitochondrial membrane (23), and succinate-cytochrome c reductase (EC 1.3.99.1), a marker for the inner mitochondrial membrane (24). Na^+, K^+, Mg^{2+}-dependent ATPase activity (ATP phosphohydrolase, EC 3.6.1.3), a plasma membrane marker (22), was estimated as the increase in initial rate of ATP hydrolysis relative to mixtures lacking KCl, using the method described previously (20). Protein was determined by the method of Lowry et al. (25).

Membrane Vesicle Transport Assays

Incubations were carried out at $21°C$ in $100~\mu l$ volumes containing $40-250~\mu g$ membrane vesicle protein, 0.125 M sucrose, 10 mM Tris-phosphate, pH 7.4, 5 mM $MgCl_2$ and other additions as indicated. Uptake was terminated by dilution with 5 ml of 0.8 M NaCl, 0.01 M Tris-HCl, pH 7.5 (wash buffer) at $2°C$, immediate filtration through a 0.2 μm or 0.45 μm pore size nitrocellulose filter (Schleicher & Schuell, 2.5 cm diameter), and filtration of an additional 5 ml of wash buffer, as described previously (16, 20). Nonspecific adsorption to vesicles was determined at zero-time by adding labeled substrate after dilution of vesicles. Radioactivity of dried filters was measured by scintillation counting in toluene-Liquifluor (New England Nuclear Corp.). The apparent initial rates of transport were proportional to the amount of vesicle protein up to 250 μg and not affected by addition of nonspecific protein. Minimal leakage of accumulated substrate was observed after 50-fold dilution of vesicles in wash buffer at $2°C$, but efflux was appreciable after dilution in wash buffer at $37°C$.

For assay of triphenylmethylphosphonium ion ($TPMP^+$) uptake, 2.5 cm diameter, 0.5 μm pore size cellulose acetate filters (Millipore type EH) were used because of high blank values due to binding of this substrate to nitrocellulose.

Binding of $TPMP^+$ to membranes was estimated after osmotic lysis of vesicles as less than 8 pmol per mg. Efflux of $TPMP^+$ during collection of vesicles within 20 s was minimal; $t\frac{1}{2}$ of efflux was 2.5 min under these conditions. Accumulated $TPMP^+$ was sensitive to the osmolarity of the suspension and proportional to the amount of vesicle protein.

Analysis of Accumulated Radioactivity

Radioactive amino acids accumulated in vesicles after collection by filtration were eluted with H_2O, concentrated by lyophilization and analyzed by thin-layer chromatography as described previously (20).

Intravesicular Volume

The internal volume of vesicles was measured by accumulation of 0.5 mM 3-0-methyl-^{3}H-glucose using the filtration assay. The measured efflux of 3-0-methylglucose in wash buffer indicated less than 20% loss of internal solute during collection of vesicles. An average value of $1 \pm 0.2~\mu l/mg$ protein of purified mixed vesicles was used in calculations of internal solute concentrations. Estimated internal volumes were proportional to the amount of vesicle protein.

Materials

^{3}H-TPMP$^+$ bromide*, 114 Ci/mol and 440 Ci/mol (26), and unlabeled TPMP$^+$ bromide were generous gifts from Dr. H. R. Kaback, Roche Institute of Molecular Biology,

Nutley, New Jersey. Potassium ^{14}C-thiocyanate (11.8 fmol/cpm), L-2,3-^{3}H-alanine (39 fmol/cpm), 2-amino-1-^{14}C-isobutyric acid (iso-Abu) (91 fmol/cpm or 61 fmol/cpm), L-(methyl-^{3}H) methionine (30 fmol/cpm), L-4,5-^{3}H-leucine (33 fmol/cpm), 2-^{3}H-glycine (30 fmol/cpm) and L-(G-^{3}H) glutamine (39 fmol/cpm) were purchased from the Radiochemical Centre, Amersham, England.

Monensin and nigericin were donated by Dr. G. L. Smith, Lilly, U.K. and carbonyl-cyanide p-trifluoromethoxyl phenylhydrazone was from Dr. P. Heytler, du Pont de Nemours, Delaware. Valinomycin, ouabain, and oligomycin were purchased from Sigma.

RESULTS

Criteria for Carrier-Mediated Amino Acid Transport in Vesicles

Several lines of evidence indicated that specific plasma membrane carriers in vesicles mediate amino acid influx and efflux. When Na^+ was preequilibrated across the membrane to abolish active accumulation, countertransport (27), a standard criterion for carrier-mediated exchange diffusion, could be demonstrated both by increased uptake of labeled alanine when vesicles were preloaded with unlabeled alanine (Table I) and by increased efflux of labeled alanine accumulated in vesicles when unlabeled alanine was added to the suspension medium (20). Stimulation was specific for pairs of substrates of this carrier (Table I). Accumulated amino acid was retained in a selectively permeable compartment enclosed by vesicle membranes with negligible binding to membranes, as shown by its sensitivity to variation in the osmotic pressure of the external suspension medium (16). Both influx and efflux were temperature-dependent (20) such that the rate of solute translocation in both directions was increased as the temperature was increased over the range 2–37°C. Transport of alanine, leucine, glutamine, or methionine was not associated with metabolic conversion of substrate detectable by thin-layer chromatography of labeled material extracted after accumulation in vesicles.

TABLE I. Countertransport Stimulation of Amino Acid Transport in Vesicles

Amino acid		Intravesicular labeled amino acid			
Preloaded[2]	External	30 sec	1 min	5 min	15 min
		nmol/mg			
L-alanine, 10 mM	^{3}H-ala, 1.7 mM	9.5	14	15	13.5
–	^{3}H-ala, 1.7 mM	3.1	3.7	6.3	7.1
iso-Abu, 9 mM	^{14}C-iso-Abu, 1.1 mM	2.0	2.8	5.8	6.3
–	^{14}C-iso-Abu, 1.1 mM	2.4	3.1	3.2	4.9
iso-Abu, 9 mM	^{3}H-ala, 0.2 mM, plus iso-Abu, 0.9 mM	0.50	0.51	0.27	0.42
–	^{3}H-ala, 0.2 mM, plus iso-Abu, 0.9 mM	0.08	0.27	0.25	0.59
glycine, 10 mM	^{3}H-ala, 0.2 mM, plus glycine, 1 mM	0.33	0.68	0.81	0.41
–	^{3}H-ala, 0.2 mM, plus glycine, 1 mM	0.26	0.41	0.61	0.50
Leucine, 10 mM	^{3}H-ala, 0.2 mM, plus leu, 1 mM	0.47	0.38	0.62	0.57
–	^{3}H-ala, 0.2 mM, plus leu, 1 mM	0.08	0.45	0.44	0.55
Methionine, 10 mM	^{3}H-ala, 0.2 mM, plus met, 1 mM	0.43	0.45	0.57	0.72
–	^{3}H-ala, 0.2 mM, plus met, 1 mM	0.47	0.51	0.20	0.30

[2]Vesicles were incubated 15 min with or without the indicated unlabeled amino acid and 50 mM NaCl and then aliquots (44–52 μg) were diluted 10-fold to a 100 μl volume containing 50 mM NaCl and the indicated external labeled and unlabeled amino acids.

The Na[+]-stimulated transport activity was associated with plasma membrane fragments. Plasma membrane fractions recovered from dextran-110 discontinuous gradients represented 20% of the total plasma membranes, purified 21-fold as estimated by the marker enzyme 5′-nucleotidase, and were associated with 20% of the total Na[+]-stimulated iso-Abu transport activity, 9.2-fold increased in specific activity (20). Only 1.4% of the total transport activity was associated with endoplasmic reticulum, 5.5-fold reduced in specific activity with respect to the homogenate and 50-fold reduced with respect to plasma membranes. Other fractions, including nuclear and mitochondrial, showed no detectable Na[+]-stimulated transport activity. Similar results were obtained for the subcellular distribution of Na[+]-stimulated alanine activity.

Competitive interactions among neutral amino acids for transport into vesicles, summarized in Table II, resembled those previously described for uptake in Ehrlich

TABLE II. Specificity of Na[+] Gradient-Stimulated Amino Acid Uptake in Vesicles

Unlabeled amino acid 5 mM	Initial rate of uptake of radioactive amino acid[2]			
	^{3}H-ala	^{3}H-met	^{3}H-leu	^{14}C-iso-Abu
	% control[c]			
No addition	100	100	100	100
L-alanine	14	37	90	18
D-alanine	79	75	85	71
L-methionine	15	26	74	27
D-methionine	83	22	28	78
L-leucine	66	42	42	68
D-leucine	83	38	30	86
iso-Abu	31	78	71	31
D-glutamine	87	71	79	82
N-methyl-DL-alanine	21	68	62	20
Glycine	80	90	82	78
L-phenylalanine	87	51	37	70
L-glutamine	27	36	59	38
L-isoleucine	92	48	40	89
L-histidine	58	28	53	48
L-serine	18	32	62	31
L-valine	63	31	38	73
L-proline	29	34	50	26
L-threonine	56	45	47	75
L-tryptophan	69	34	8	66
L-cysteine	22	23	33	25
L-sarcosine	44	36	81	32
L-homoserine	25	28	46	22
1-aminocyclopentane-1-carboxylic acid	20	31	32	30
minus Na[+][b]	15	14	50	20

[a]The initial rate of uptake was measured 30 sec after the addition of 50 mM NaSCN and either 0.2 mM L-[methyl-^{3}H] methionine (30 fmol/cpm), 0.2 mM L-[2,3-^{3}H] alanine (43 fmol/cpm), 0.2 mM L-[4,5-^{3}H] leucine (33 fmol/cpm) or 0.17 mM [1-[1]C] iso-Abu (91 fmol/cpm).

[b]50 mM KSCN was substituted for 50 mM NaSCN.

[c]100% represented 3 nmol/min·mg for alanine uptake, 1.7 nmol/min·mg for methionine uptake, 1 nmol/min·mg for leucine uptake and 1 nmol/min·mg for iso-Abu uptake. Results were averaged from duplicates with a range of ± 15%.

ascites cells (28, 29) and indicated that the A and L systems and a glycine system were functional in vesicles. N-methyl-DL-alanine was as effective as L-alanine in inhibiting L-^{3}H-alanine transport, which suggests that the ASC system (29) does not contribute appreciably to alanine transport in vesicles at this pH. Furthermore, pH profiles for the initial rate of transport of iso-Abu and alanine into vesicles (20) resembled those of cellular uptake in Ehrlich ascites tumor cells (28), with optimal uptake at pH 7.4. The similarity between K_m values obtained for Na^+-dependent iso-Abu uptake in intact cells (30) and those measured in vesicles further confirmed the identity of the transport carrier activity assayed in vesicles as a component of the corresponding cellular uptake system (Table III). The maximal velocity (V_{max}) values for iso-Abu transport into vesicles, after correction for recovery of membrane protein, resembled V_{max} values reported for iso-Abu uptake in these cell lines (30), an indication that vesicles retain intact a large proportion of functional transport carriers which operate in living cells.

TABLE III. Kinetic Parameters of Sodium Gradient-Stimulated Amino Acid Transport in Vesicles*

Substrate	Na^+ concentration	Amino acid transport	
		K_m	V_{max}
	(mM)	(mM)	(nmol/min per mg)
Balb/c SV3T3 membranes:			
iso-Abu	0	5	2.4
	50	1.33	2.9
Swiss SV3T3 membranes:			
Alanine	0	10 ± 2	20
	1	2.4	12
	10	1.7	15
	50	1.2	15
Balb/c tertiary mouse embryo fibroblasts (subconfluent):			
iso-Abu	50	1.1 ± 0.4	1 ± 0.2

*Vesicles were incubated 30 sec with various concentrations of ^{3}H-alanine or iso-^{14}C-Abu at the indicated Na^+ concentrations. For alanine uptake, SCN^- was used as counterion, at a constant concentration of 50 mM maintained by addition of KSCN. For iso-Abu uptake, Cl^- was the counterion, maintained at 50 mM by the addition of choline Cl. Results were analyzed by inverse-reciprocal plots. The transport activity of vesicles from confluent mouse embryo fibroblasts was too low to permit accurate determination of these parameters.

Concentrative Amino Acid Transport in Vesicles is Coupled to an Electrochemical Na^+ Gradient

The alkali-ion gradient hypothesis, reviewed recently by Crane (9) and summarized in Fig. 1, leads to the expectation that amino acid accumulation is dictated directly as a function of the vectorial sum of the direction and magnitude of the electrochemical Na^+ gradient and membrane potential; in intact cells, this driving force is maintained by the electrogenic Na^+ pump but in vesicles these gradients can be artificially manipulated and tend to dissipate with time.

When a Na^+ gradient (external > internal) was artificially created across the membrane at zero-time by adding Na^+ together with substrate to the external suspension medium, both the initial rate and steady state of alanine and glycine uptake were marked-

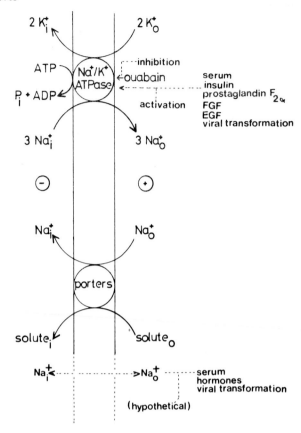

Fig. 1. Characteristics of Na⁺-dependent transport into vesicles predicted by the "sodium gradient" hypothesis: some possible targets for regulation.

ly increased, as shown in Fig. 2, A and B. No appreciable stimulation of accumulation was observed after similar imposition of gradients of chloride salts of K^+, choline$^+$, Li^+, $Tris^+$, Mg^{2+}, Rb^+, or Ca^{2+} across the membrane (20); Na^+ gradients similarly enhanced active transport of iso-Abu, glutamine, and methionine (16, 20); maximal accumulation driven by external 50 mM NaCl achieved apparent internal concentrations 6- to 7-fold above external concentrations of these amino acids. By contrast leucine uptake showed much less stimulation by a Na^+ gradient (Fig. 2C). System L may partially mediate Na^+-independent transport of alanine, as recently described for iso-Abu transport (31).

Conditions expected to dissipate the Na^+ gradient, such as the addition of monensin, an ionophore which catalyzes an electroneutral Na^+/H^+ exchange across the membrane (32), or incubation of vesicles with Na^+ for 15 min before adding amino acid (16), decreased both the initial rate and steady state of accumulation of these amino acids, as shown for alanine and glycine in Fig. 2.

The nature of the major counteranion increased the initial rate and maximal extent of Na^+-stimulated amino acid accumulation in the order $NO_3^- > SCN^- > Cl^- > SO_4^{2-}$ (20), which reflects their relative permeability to biological membranes (33). This stimulation by SCN^- is illustrated in Fig. 2.

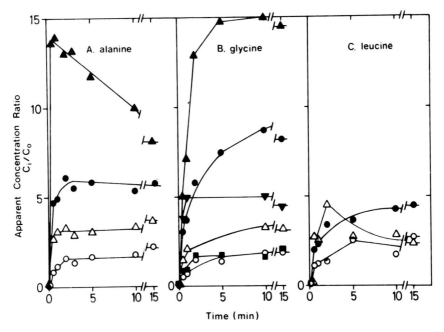

Fig. 2. Na$^+$ gradient-dependence of A) L-[2,3-^{3}H] alanine, B) [2-^{3}H] glycine, and C) L-[4,5-^{3}H] leucine transport into vesicles. Na$^+$, choline, or K$^+$ salts and the indicated 0.2 mM labeled amino acid were added at zero-time: ▲) 50 mM NaSCN; ●) 50 mM NaCl; ○) 50 mM choline chloride; △) 50 mM NaCl plus 14 μM monensin in 1% Me$_2$SO; ▼) 50 mM NaSCN plus 14 μM monensin in 1% Me$_2$SO; ■) 50 mM KSCN. Control experiments (not shown) indicated that 1% Me$_2$SO had no effect on uptake.

Although vesicles retained functional Na$^+$, K$^+$-ATPase activity, the Na$^+$ gradient-stimulated amino acid transport could be dissociated from this activity (20). Addition of ouabain for 20 min to vesicle suspensions completely blocked K$^+$-dependent ATP hydrolysis in the presence of concentrations of ATP, KCl, MgCl$_2$ and NaCl required for optimal Na$^+$, K$^+$-ATPase activity yet did not inhibit Na$^+$-stimulated alanine transport in similarly treated aliquots of the same vesicle preparation. Addition of 5 mM KCl and 2 mM ATP did not stimulate amino acid transport.

Sensitivity of amino acid transport to variation in external Na$^+$ was most pronounced at low amino acid concentrations, as shown previously (20). Interpretation of these data by Lineweaver-Burk analyses, summarized in Table III, revealed that substitution of Na$^+$ for choline$^+$ decreased the apparent K$_m$ for alanine or iso-Abu transport into vesicles with minimal effects on V$_{max}$ when a counteranion was present at a constant concentration. The apparent K$_m$ was decreased as the Na$^+$ concentration was increased.

At 0.2 mM alanine, both the initial rate and quasi-steady state of alanine transport increased as a function of increasing NaSCN gradients when the SCN$^-$ concentration was maintained constant by the addition of KSCN, as illustrated in Figs. 3 and 4. When the Na$^+$ gradient was short-circuited by the addition of monensin, quasi-steady-state accumulation was greatly diminished and was rendered independent of variation in Na$^+$ concentration. By contrast, the increased initial rate of alanine uptake as a function of Na$^+$, a measure of carrier activity plus driving force, persisted although diminished in the presence of monensin and showed an apparent saturation above 50 mM Na$^+$. Therefore, in the absence

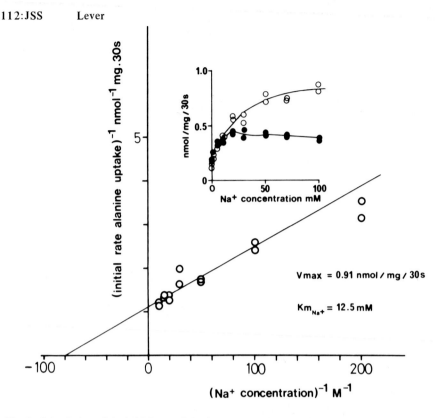

Fig. 3. Stimulation of the initial rate of alanine transport into vesicles as a function of external Na⁺ concentration. At zero-time, the indicated concentrations of NaSCN and 0.2 mM L-[2,3-^{3}H] alanine were added to aliquots of 120 μg vesicles (Balb SV3T3) treated with either 1% Me$_2$SO (○) or 14 μm monensin in 1% Me$_2$SO (●). The total SCN⁻ concentration was maintained constant at 100 mM by addition of KSCN so that the Na⁺/K⁺ ratio was varied. Uptake was determined after 30 sec.

of its chemical concentration difference across the membrane, Na⁺ influences the affinity of alanine for the membrane transport component. A predictable uncertainty in the quantitative interpretation of these results is the unmeasured dissipative Na⁺ flow resulting from the expected coupled and uncoupled leaks during the course of amino acid uptake; whereas the external Na⁺ concentration is constant, the internal concentration, and thus the magnitude of the Na⁺ gradient, varies. However, several observations pertinent to this point indicate that the relationship between quasi-steady-state solute accumulation in vesicles and the external Na⁺ concentration can roughly approximate a thermodynamic equilibrium between the amino acid accumulation ratio and the driving force.

Dissipation of driving force should allow only transient accumulation, with efflux of accumulated amino acid resembling an overshoot. At true steady state, this would lead to collapse of gradients of Na⁺ and amino acid across the membrane. No overshoot of amino acid accumulation was observed within the 15 min time interval shown, either with purified mixed vesicles or purified plasma membrane vesicles, in the presence of a NaCl gradient. When monensin was added to vesicles at 15 min after amino acid transport was initiated to accelerate dissipative Na⁺ flow, amino acid efflux was observed (Fig. 5). This indicated that a residual Na⁺ gradient persisted at 15 min and that the apparent amino

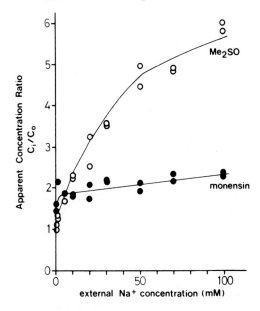

Fig. 4. Dependence of quasi-steady-state levels of alanine accumulation in vesicles on the external Na^+ concentration imposed as a Na^+ gradient. Symbols and conditions as in Fig. 4, but uptake was measured after 10 min.

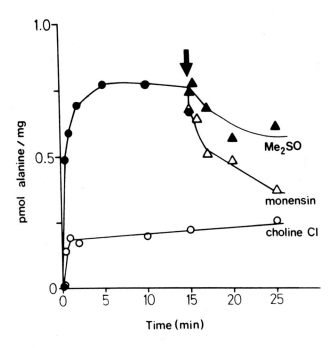

Fig. 5. Efflux of accumulated iso-Abu induced by the addition of monensin. Aliquots of 120 μg vesicles were incubated with 0.2 mM ^{3}H-alanine and either 50 mM NaCl (●) or 50 mM choline Cl (○). Arrow indicates addition at 15 min of either 2% Me_2SO (▲) or 30 μm monensin in 2% Me_2SO (△).

acid accumulation in vesicles was in equilibrium with the Na^+ gradient. By contrast, amino acid accumulation stimulated either by a transient K^+ diffusion potential facilitated by valinomycin or by a gradient of the Na^+ salt of a highly permeant anion such as SCN^- did show a gradual overshoot pattern. Surprisingly, Lineweaver-Burk inverse-reciprocal plots of the initial rates of Na^+-stimulated amino acid uptake in vesicles as a function of amino acid concentration did not show the curvature expected from increased coupled Na^+ influx at increased amino acid concentrations. Furthermore, the logarithm of the observed quasi-steady-state accumulation of alanine was a linear function of the logarithm of the external Na^+ concentration at a constant external amino acid concentration (20). Presumably, Na^+ influx under certain experimental conditions is not fast enough in these membranes to contribute a measurable source in variability, but can be accelerated when charge compensation across the membrane is greatly facilitated by experimental manipulation. Other membrane systems, such as intestinal brush border membranes, which exhibit rapid overshoot of Na^+ stimulated glucose and alanine uptake (12) may contain endogenous systems to facilitate Na^+ influx, such as the proposed Na^+/H^+ antiport system (37).

Membrane Potential and Amino Acid Transport

Amino acid translocation into vesicles was stimulated by an interior-negative membrane potential. This response specifically required the presence of Na^+. Figure 6 shows that when the internal lumen of vesicles was loaded with K^+ in the presence of valinomycin, a K^+-specific ionophore (32), and then vesicles were diluted to achieve a 10-fold lower external K^+ concentration at the time of addition of amino acid and external Na^+, a

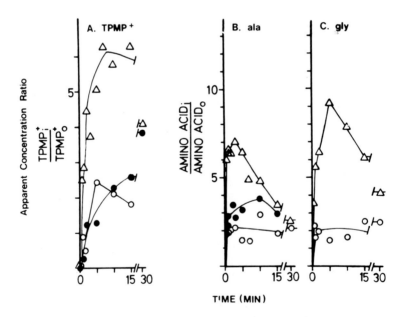

Fig. 6. Stimulation of Na^+-dependent amino acid and lipophilic cation accumulation by imposition of an interior-negative membrane potential. A) 0.1 mM ^{3}H-triphenylmethylphosphonium bromide; B) 0.2 mM ^{3}H-alanine; C) 0.2 mM ^{3}H-glycine. Aliquots of 10 μl of vesicles (6.3 mg per ml) were preincubated 15 min with: •) 2% ethanol and 3% MeSO; △) 90 μM valinomycin, 2% ethanol, 3% Me$_2$SO, and 50 mM KCl; or ○) 90 μM valinomycin; 40 μm nigericin, 2% ethanol, 50 mM KCl, and 3% Me$_2$SO. Then, vesicles were diluted 10-fold into 100 μl volumes containing 50 mM NaCl and the indicated labeled substrate, for the times shown.

striking transient enhancement of alanine and glycine accumulation was observed. Such a K^+ diffision potential (internal $>$ external) and the large selective increase in K^+ permeability mediated by translocation of the positively-charged K^+-valinomycin complex (32) would be expected to create a transient internal-negative membrane potential in vesicles which were otherwise relatively impermeable to ions, of a magnitude dictated mainly by the K^+ potential, (34, 35).

This assumption was verified in these preparations using the apparent distributions of the lipophilic cation $TPMP^+$ as shown in Fig. 6A. Theoretically, from the distribution of a cation which freely distributes across the membrane according to electrical effects, membrane potential, $\Delta\psi$, can be calculated according to the Nernst equation:

$$\Delta\psi \frac{-RT}{nF} \ln \frac{TPMP^+_i}{TPMP^+_0} = -58.8 \log \frac{TPMP^+_i}{TPMP^+_0}$$

From the 6-fold apparent accumulation of lipophilic cation, a membrane potential of -45.8 mV could be calculated.

Several observations indicated that this lipophilic cation distributed according to electrical effects rather than binding to the membrane or specific transport (Lever JE, in preparation). The accumulation of lipophilic cation showed a positive correlation with the K^+ dilution factor under these conditions (Lever JE, in preparation). Furthermore, the combination of nigericin and valinomycin has been shown to collapse membrane potential in similar situations (36) and would be expected to abolish $TPMP^+$ accumulation. When nigericin plus valinomycin were added to K^+-loaded vesicles before dilution, $TPMP^+$ did not accumulate beyond an apparent internal concentration equal to its external concentration. Similarly, amino acid accumulation was dissipated. The nonelectrogenic ionophore nigericin added to K^+-loaded vesicles did not induce $TPMP^+$ accumulation above its internal concentration nor did it stimulate amino acid transport. Thus K^+ influences this distribution electrically, by its electrogenic translocation only, but is not chemically coupled either to lipophilic cation or amino acid accumulation. Finally, when K^+-loaded vesicles were diluted with valinomycin in the presence of an external K^+ concentration equal to its internal concentration, no lipophilic cation accumulation was observed. Retention of $TPMP^+$ in membranes due to binding was negligible as estimated after osmotic lysis.

Significantly, the response of amino acid translocation to stimulation by an interior-negative membrane potential specifically required Na^+. Since Na^+ was not required for generation of an interior-negative membrane potential by these manipulations, as shown by the lack of Na^+-dependence of $TPMP^+$ accumulation (Table IV), the Na^+ requirement must be at the level of amino acid translocation.

Amino acid accumulation increased as a function of the magnitude of membrane potential generated under these conditions, estimated by $TPMP^+$ accumulation, as shown in Fig. 7. Since in this experiment, assay of amino acid accumulation and generation of membrane potential were independent of the metabolic functions of the cell, amino acid transport stimulated by an electrochemical Na^+ gradient involves a direct and obligatory coupling of the translocation step to membrane potential at the plasma membrane.

Although this positive correlation demonstrated that the energy from the electrical potential difference across the membrane contributes to the driving force for amino acid transport, amino acid accumulation driven by both a Na^+ gradient and an imposed electrical difference always exceeded the energy provided by the total electrical component alone. Thus, Table IV shows that in the presence of an initial driving force

TABLE IV. Sodium Requirement for Electrogenic Amino Acid Translocation*

Additions		Apparent concentration gradient, C_i/C_o	
		TPMP+	Alanine
Valinomycin	NaCl	4.7	8.5
Valinomycin	Choline Cl	6.4	2.1
Valinomycin	NaCl + KCl	1.3	4.8
Valinomycin	Choline Cl + KCl	1.4	1.6
Valinomycin + monensin	NaCl	5.9	4.6
Valinomycin + monensin	Choline Cl	–	2.6
Valinomycin + nigericin	NaCl	–	2.6
Valinomycin + nigericin	Choline Cl	–	1.6

*Vesicles (7.6 mg/ml) were preincubated 15 min with 50 mM KCl and the indicated ionophores in 1% Me_2SO: valinomycin, 90 μM; nigericin, 27 μM; and sodium monensin, 14 μM. Then, 38 μg aliquots were diluted 10-fold with respect to KCl and ionophore concentrations and incubated with 20 μM ^{3}H-triphenyl-methylphosphonium and 50 mM of the indicated chloride salt; 76 μg aliquots were similarly diluted and incubated with 0.2 mM ^{3}H-alanine and 50 mM chloride salts. Maximum accumulation was estimated at 2–5 min after dilution. C_i/C_o refers to the ratio of the observed intravesicular solute concentration to its external concentration.

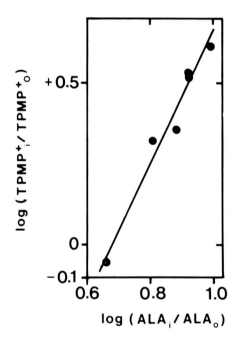

Fig. 7. Correlation of levels of quasi-steady-state alanine accumulation stimulated by a Na+ gradient with magnitudes of interior-negative membrane potential estimated by ^{3}H-TPMP accumulation in vesicles. Vesicles were incubated with valinomycin and 50 mM KCl as in Fig. 7, and then 10 μl aliquots were diluted to various final volumes in the range 20–200 μl which contained various final external KCl concentrations in the range 50–2.5 mM, 50 mM NaCl, and either 0.1 mM ^{3}H-triphenylmethyl-phosphonium bromide or 0.2 mM L-^{3}H-alanine. Maximal accumulation was estimated by the filtration assay.

provided by 50 mM external Na^+ plus a 10-fold K^+ dilution potential a maximal apparent alanine accumulation of 8.5-fold was achieved representing −54.6 mV, whereas a membrane potential of −39.5 mV was estimated from the apparent 4.7-fold triphenyl-methylphosphonium ion accumulation assuming a coupling ratio of 1.

When monesin was added to short-circuit the Na^+ gradient without affecting membrane potential, alanine accumulation decreased from 8.5-fold to 4.6-fold, a value compatible with the electrical driving force estimated by the apparent 5.9-fold $TPMP^+$ accumulation under these conditions. This clearly demonstrates that amino acid accumulation is driven by the sum of contributions from an electrical potential and a chemical difference in Na^+ concentration across the membrane. These parameters can vary independently but Na^+ is required for the response to charge asymmetry. Under these conditions the electrical potential does not arise from the Na^+ gradient itself; in whole cells this potential may arise from electrogenic gradients of other ions. This would be in contrast with proton-coupled transport systems, in which proton gradients contribute to membrane potential (7, 38).

Inactivation of Carrier Activity

Alanine transport activity stimulated by a Na^+ gradient was specifically inactivated by treatment of vesicles with p-chloromercuribenzenesulfonate, as shown in Fig. 8. The Na^+-independent alanine transport and Na^+-stimulated glycine transport into vesicles were relatively insensitive to this reagent. Both the initial rate and quasi-steady state of

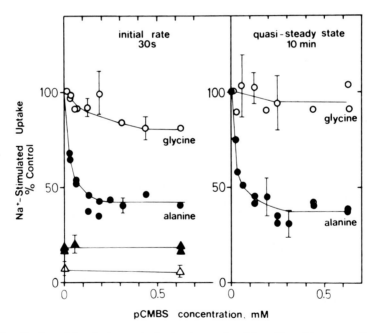

Fig. 8. Inactivation of the alanine carrier by treatment of vesicles with p-chloromercuribenzenesulfonate. Aliquots of 56 μg vesicles were treated 15 min with the indicated concentrations of p-chloromercuribenzenesulfonate, then diluted 2-fold into incubation mixtures containing 0.2 mM ^{3}H alanine (●, △) or 0.2 mM ^{3}H-glycine (○, ▲) and either 50 mM NaCl (○, ●) or 50 mM choline chloride (▲, △). Uptake was measured after 30 sec and 10 min. Initial rates of 100% represented 2.1 nmol/mg/30 sec for glycine and 820 pmol/mg/30 sec for alanine. Quasi-steady-state accumulation of 100% represented 2.5 nmol/mg for glycine and 1.2 nmol/mg for alanine.

alanine transport showed parallel inactivation as a function of inhibitor concentration.

The relative insensitivity of quasi-steady-state glycine accumulation driven by a Na^+ gradient to this sulfhydryl reagent indicates that the ability of the membrane to maintain a Na^+ gradient is not affected by this reagent. These results suggest that the presence of Na^+ induces a selective conformational change in the alanine carrier to expose sulfhydryl groups. Inactivation of Na^+-stimulated alanine transport was also observed in the presence of monensin (not shown).

Physiological Regulation of Amino Acid Transport

The activity of certain nutrient transport systems fluctuates in a complex array with changes in cellular physiological state in many instances, such as the stimulation of proliferation of resting nontransformed mouse fibroblasts by serum or mitogenic hormones (39–41), after viral transformation (41, 42) and the triggering of the blast transformation of lymphocytes (43, 44). The V_{max} of iso-Abu transport activity of nontransformed mouse fibroblast cultures diminishes as cultures become quiescent and increases accompanying viral transformation (42).

Several possibilities for cellular regulation of neutral amino acid transport are suggested by the demonstrated characteristics of energy coupling of this process. Notably, the activity of the Na^+, K^+-ATPase transport system, which directly utilizes cellular ATP to maintain cellular K^+ and Na^+ gradients and generate an interior-negative membrane potential as demonstrated in certain cell systems (45, 46), appears to be a primary target for regulation as summarized in Fig. 1. The "Na^+ pump" of nontransformed mouse fibroblast cultures is rapidly stimulated by serum (47), physiological concentrations of insulin, and prostaglandin $F_{2\alpha}$ (48) which initiate cell proliferation (49), or by low, mitogenic concentrations of fibroblastic growth factor (FGF) or epidermal growth factor (EGF) (Lever JE, and Jimenez de Asua L, in preparation), by a mechanism which does not require cyclic nucleotide fluctuation or protein synthesis (48). Also, increases in Na^+, K^+-ATPase transport activity accompany viral transformation in certain cases (50). On the basis of this indication that cytoplasmic membrane potential and Na^+ gradients are regulated by certain hormones, possibly by their direct interaction with the membrane, it can be postulated that rapid and transient regulation of those transport systems which are coupled to Na^+ or electrical gradients can be accomplished either by this means or by other unidentified electrogenic membrane systems.

Amino acid transport activity in vesicles can be evaluated in the presence of a standard Na^+ gradient, independently of Na^+, K^+-ATPase activity. This evaluation should permit detection of regulatory changes in the amino acid carrier itself or in other membrane constituents which may affect uptake. Furthermore, the contribution of proteins, lipids or hormones removed during membrane isolation which are necessary to maintain increased carrier activity might be detected by reconstitution to membrane vesicles.

Transport activities of vesicles isolated from nontransformed and SV 40-transformed mouse fibroblasts were compared (16), as shown in Fig. 9 and Tables III and V. Both an increased maximal velocity of Na^+-dependent iso-Abu transport and increased quasi-steady-state accumulation were noted in vesicles from SV 40-transformed cells compared with those from proliferating nontransformed cells (Fig. 9). These parameters were decreased in vesicles from confluent nontransformed cells. The initial rates of adenosine uptake, relatively invariant with growth or transformed state (51), were compared in addition to marker enzymes, to indicate variation in transport-specific activity due to unsealed vesicles or different distributions of transport-competent vesicles during purification from

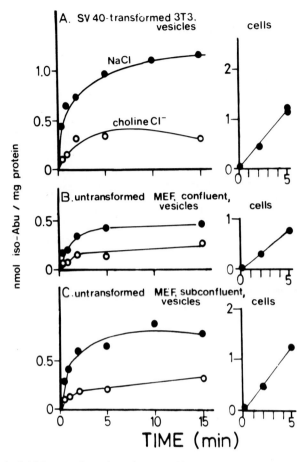

Fig. 9. Changes in the initial rate and quasi-steady state of iso-Abu accumulation in isolated membrane vesicles and uptake by intact cells which accompany cell-density growth inhibition or SV 40 transformation of mouse embryo fibroblasts. Vesicles were incubated for the indicated times with 0.17 mM ^{14}C-iso-Abu and either 50 mM NaCl (●) or 50 mM choline chloride (○). Uptake of 0.06 mM ^{3}H-iso-Abu in Dulbecco's modified Eagle's medium minus amino acids and minus serum by cells attached to 30 mm petri dishes at 37°C was determined according to a procedure described previously (40). Total cellular protein of each dish was measured (25). M.E.F.) Balb/c tertiary mouse embryo fibroblasts.

each cell type. Furthermore, rates of uridine uptake in vesicles, a cell density-dependent transport function (51) expressed in isolated membranes (52), were also compared. These transport specific activities, summarized in Table V, indicate that Na$^+$-independent iso-Abu transport activities showed much less variation than Na$^+$-dependent activities.

The changes in iso-Abu transport activity of cell populations expressed in vesicles as alterations in V_{max} could represent either changes in the number or mobility of A system carrier molecules and/or the net driving force acting on the system. Such changes would alter cellular uptake of these amino acids even at high, saturating amino acid concentrations. The changes which also affect quasi-steady-state accumulation should be largely independent of the number of carriers, but represent the net driving force acting on the carriers. Since a standard Na$^+$ gradient driving force was artificially applied to the membrane, the latter changes probably represent changes in membrane Na$^+$ permeability.

TABLE V. Specific Activities of Transport Systems in Membrane Vesicles*

Source of membranes	No. of preparations	Transport specific activity[a]			
		iso-Abu			
		Na+-dependent	Na+-independent	Adenosine	Uridine
		pmol/min per mg protein[b]			
Balb/c M.E.F.[c]					
(confluent)	5	440	560	1.7	0.10
(subconfluent)	3	1,300	810	2.6	0.20
Balb/c SV3T3					
(confluent)	3	960	870	2.9	0.19
(subconfluent)	2	2,300	930	2.0	0.19
Swiss 3T3 K	2				
(confluent)	2	190	210	1.3	0.04
(subconfluent)	2	440	600	1.7	0.26
Swiss SV3T3					
(confluent)	2	660	530	2.2	0.32
(subconfluent)	2	480	687	1.9	0.13

*Recovery of protein and marker enzymes for mixed plasma membrane preparations from each source was described previously (16, 19).

[a]Incubations were 30 sec and 1 min with 50 mM NaCl and either 1 mM iso-^{14}C-Abu, (0.061 pmol/cmp), 0.5 μM ^{3}H-adenosine (9×10^{-5} pmol/cpm), or 0.2 μM ^{3}H-uridine (3.9×10^{-5} pmol/cpm), using 70–170 μg vesicle protein per 100 μl of incubation mixture. Initial rates of uptake for each substrate were assayed at concentrations below their K_m values to minimize entry by simple diffusion. The Na+-independent rate of iso-Abu uptake was determined using 50 mM choline chloride substituted for NaCl. This value was subtracted from uptake observed using 50 mM NaCl, to give the Na+-gradient-dependent uptake.

[b]Results were averaged from duplicate determinations repeated 2 or 3 times on each preparation. Variation was ± 10–25% among duplicates.

[c]M.E.F. – mouse embryo fibroblasts.

Since K_m values for iso-Abu transport, which vary as a function of Na+ concentration as shown above, were not altered in these preparations, it seems likely that the interaction of Na+ with the carrier is not altered.

As a preliminary indication of differences in Na+ permeability, vesicles from these cell types were loaded with iso-Abu in the presence of a standard initial Na+ gradient, and then efflux was measured after dilution, maintaining a constant external Na+ concentration, as shown in Fig. 10. The observed efflux would depend on the levels and mobility of iso-Abu carriers as well as the internal iso-Abu and Na+ concentrations at the time of dilution. Efflux from vesicles from confluent normal cells was faster than from those prepared from subconfluent normal cells and slower than efflux from SV 40-transformed cell membranes. As expected, addition of monensin at the time of dilution increased iso-Abu efflux from all types of preparations. Interestingly, monensin eliminated differences in efflux between confluent and subconfluent nontransformed cells, an indication that their efflux differences were mainly due to a relatively higher membrane Na+ permeability in confluent cell membranes. The increased efflux maintained in vesicles from SV 40-transformed cells after addition of monensin indicates that iso-Abu carriers are also increased in these cells relative to nontransformed cells.

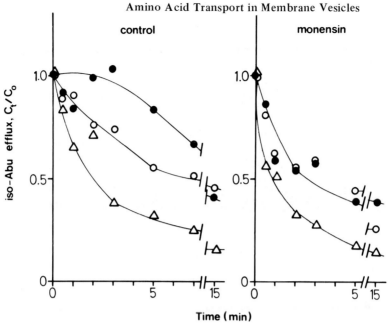

Fig. 10. Efflux of iso-Abu accumulated in vesicles from nontransformed and SV 40-transformed mouse fibroblasts. Vesicles from subconfluent (●) or confluent (○) cultures of nontransformed mouse embryo fibroblasts or SV 40-transformed cultures (△) were incubated 15 min with 50 mM NaCl and 0.17 mM ^{14}C-iso-Abu. Then efflux was measured after addition of A) 2% Me_2SO or B) 30 μM monensin in 2% Me_2SO and 10-fold dilution with respect to ionophore and labeled substrate. C_t/C_O represents the ratio of solute at the indicated times relative to that observed at zero-time. 100% was 1.2 nmol/mg (●), 0.42 nmol/mg (○), and 1.2 nmol/mg (△).

DISCUSSION

The properties of these neutral amino acid transport systems of isolated membrane vesicles from mouse fibroblast cell lines provide experimental confirmation of some of the general predictions derived from the alkali-ion gradient hypothesis for active transport of several organic solutes in animal cells as formulated by Christensen (1, 2), Crane (3, 9), Heinz (4) and others. It could be clearly demonstrated that the accumulation of several neutral amino acids across the plasma membrane is unequivocally coupled to the direction and magnitude of a driving force specifically and directly provided by a Na^+ gradient imposed across the membrane. The nature of this noncovalent energy coupling dictated that amino acids translocate by movement of a positively-charged complex and accumulate electrogenically *trans* to the face of the membrane exposed to the higher Na^+ concentration. Thus, it was further established that the ability to utilize electrical potential differences across the menbrane as an additional driving force for transport of these amino acids required Na^+ and is an integral feature of this mechanism. The molecular basis of the coupling process appeared to be an increased affinity of the carrier for the amino acid substrate directly induced by interaction of Na^+ with the membrane, concurrent with the formation of a positively charged translocating complex.

The experimental support on which these conclusions are based depends upon: 1) the ability to rapidly collect vesicles under conditions where efflux of intravesicular solutes and nonspecific binding are minimal, 2) the demonstration that transport carrier activity in membrane vesicles fulfils standard criteria which serve to identify specific transport systems in the cell, such as countertransport stimulation of uptake and efflux, saturation kinetics of rates of uptake, competitive interactions, and pH profiles, 3) the ability of vesicles to maintain a maximal accumulation of an amino acid above its external concentration which reflects the quasi-steady-state response of the carrier to the magnitude and direction of a specific ion-gradient driving force, this accumulation representing a specific response since a variety of experimental conditions designed to collapse the driving force also reduced the apparent internal amino acid concentration to its known external concentration, and 4) interpretation of the effects of ionophore probes based on their known actions characterized in artificial and biological membranes (32).

In addition to the relationship between electrical currents, gradients of several ions, and amino acid transport in vesicles, studies were directed to several additional questions concerning the translocation mechanism beyond the general predictions of the alkali-ion gradient hypothesis. Do other ions interact with the carrier to directly influence this mechanism? How is the utilization of driving force for transport apportioned between electrical and chemical components? Are these components utilized by separate or interdependent processes? Does the uncomplexed carrier bear a net charge? Does Na^+ induce a conformational change in the carrier?

Although a chemical role for other physiological ions has been proposed (2, 53), Na^+ was the only ion which could be demonstrated to make a chemical contribution to the driving force for neutral amino acid transport in vesicles. The Na^+ did not make an appreciable electrical contribution under these conditions, and in fact caused a small depolarization. The K^+ could only contribute electrically to the driving force. Efflux of internal K^+ by endogenous pathways or catalyzed by the nonelectrogenic ionophore nigericin did not affect amino acid uptake but electrogenic efflux of K^+ mediated by valinomycin markedly enhanced influx and accumulation of amino acids by specific generation of a measurable interior-negative membrane potential as confirmed by suitable control experiments. Also, external K^+ did not stimulate amino acid transport in the absence or presence of Na^+. This observation can be rationalized because cellular chemical K^+ gradients directionally oppose Na^+ gradients and would tend to diminish the driving force from Na^+ gradients if K^+ interacted directly with the carrier. Proton gradients could not be demonstrated to be obligatorily coupled to this process as shown by the lack of inhibition by proton conductors such as nigericin, 2,4-dinitrophenol, and FCCP and the relative insensitivity of quasi-steady-state neutral amino acid accumulation to external pH. The initial rate of Na^+-dependent uptake showed a distinctive pH profile with maximal uptake at external pH 7.4 but Na^+-independent uptake was relatively pH-independent in this range (20). Since this is the same pH as the buffer enclosed within the vesicles, no pH gradient would be expected when maximal uptake was observed. Predictably, H^+ gradients could contribute electrically to the transport process and also make a secondary, non-obligatory chemical contribution as identified by the pH profile of Na^+-dependent uptake in vesicles.

An important observation was that stimulation of amino acid uptake by interior-negative membrane potential, and conversely, dissipation of membrane potential by amino acid transport, required the presence of Na^+. The Na^+-dependent utilization of energy derived from charge asymmetry did not require a gradient of Na^+ although energy from a Na^+ gradient could also be utilized in an additive manner with that from the electrical contribution. This intimate connection between Na^+ and the response of the carrier to membrane potential implies that each contributes energetically through a common mechanism: Na^+ is required to convert translocation to an electrogenic process represented by movement of a positively-charged "ternary complex" of Na^+, substrate, and membrane component, stoichiometry and molecular disposition unspecified. From this observation that amino acid translocation is nonelectrogenic in the absence of Na^+, plus previous evidence (53) that the zwitterion seems to be the predominant translocated species, it may be deduced indirectly that the unbound carrier is uncharged at this pH. Since the preferred orientation of an electroneutral carrier would not be influenced by membrane potential, this possible contribution to the total driving force is eliminated. From the fact that the loaded carrier responds to membrane potential in the presence of Na^+, it may be concluded that binding and release of substrate on either side of the membrane are not rate-limiting, as discussed for several theoretical possibilities (54). Probably, the translocation step is rate-limiting, and its mobility increased by Na^+, as predicted from effects of Na^+ on K_m, according to a cotransport kinetic model (55).

Many important details of this mechanism remain to be established, notably the stoichiometry and disposition of interactions between ion, substrate, charge, and membrane component. Studies of the regulation of this transport system indicate that direct hormonal regulation of the active Na^+ pump and also changes in membrane Na^+ permeability may be primary targets to rapidly modulate influx of amino acids, in addition to possible changes in levels of carrier molecules and alterations in membrane fluidity. Since the system responds directly to electrical differences, other electrogenic processes, membrane depolarization, or electrical coupling between different cells could also make important contributions. The relatively small observed fluctuations in Na^+-dependent amino acid transport in cell populations could merely represent an indirect monitor of the primary interactions of certain hormones with the cell membrane. It remains to be established whether this transport regulation is obligatory, permissive, or dissociable from the regulation of cell proliferation.

ACKNOWLEDGMENTS

I thank Dr. Renato Dulbecco for generous support and encouragement, Mrs. Patricia Pettican for excellent technical assistance, the Cell Production Unit for providing cell cultures, and Dr. H. R. Kaback for a gift of ^{3}H-triphenylmethylphosphonium. This work was supported by National Cancer Institute, Department of Health, Education, and Welfare grant 5P32CA05174-02.

REFERENCES

1. Christensen HN, Riggs TR: J Biol Chem 194:57, 1952.
2. Riggs TR, Walker LM, Christensen HN: J Biol Chem 233:1497, 1958.
3. Crane RK: Physiol Rev 40:789, 1960.
4. Kromphardt H, Gobecker H, Ring K, Heinz E: Biochim Biophys Acta 74:549, 1963.
5. Mitchell P: In Kleinzeller A (ed): "Membrane Transport and Metabolism." New York and London: Academic Press, 1961, pp 22–34.
6. Schultz SG, Curran PF: Physiol Rev 50:637, 1970.
7. Mitchell P: Bioenergetics 3:63, 1973.
8. Harold FM: Ann NY Acad Sci 227:297, 1974.
9. Crane RK: Rev Physiol Biochem Pharmacol 78:99, 1977.
10. Kaback HR: Science 186:882, 1974.
11. Illiano G, Cuatrecasas P: J Biol Chem 246:2472, 1971.

12. Hopfer U, Nelson K, Perrotto J, Isselbacher KJ: J Biol Chem 248:25, 1973.
13. Lee JW, Beygu-Farber S, Vidaver GA: Biochim Biophys Acta 298:446, 1973.
14. Colombini M, Johnstone RM: J Membr Biol 15:261, 1974.
15. Hochstadt J, Quinlan DC, Roder RL, Li CC, Dowd D: In Korn W (ed): "Methods in Membrane Biology." New York: Plenum Press, 1974, vol 5, pp 117–162.
16. Lever JE: Proc Natl Acad Sci USA 73:2614, 1976.
17. Kamat VB, Wallach DFH: Science 148:1343, 1965.
18. Quinlan DC, Parnes JR, Shalom R, Garvey TQ, Isselbacher KJ, Hochstadt J: Proc Natl Acad Sci USA 73:1631, 1976.
19. Lever JE: J Cell Physiol 89:779, 1976.
20. Lever JE: J Biol Chem 252:1990, 1977.
21. Quinlan DC, Hochstadt J: J Biol Chem 251:344, 1976.
22. Essner E, Novikoff AB, Masek BJ: Biophys Biochem Cytol 4:711, 1958.
23. Sottocasa GL, Kuylenstierna B, Ernster L, Bergstrand A: J Cell Biol 32:415, 1967.
24. de Duve C, Pressman BC, Gianetto R, Wattiaux R, Appelmans F: Biochem J 60:604, 1955.
25. Lowry OH, Rosebrough NJ, Farr AL, Randall RJ: J Biol Chem 193:265, 1951.
26. Shuldiner S, Kaback HR: Biochemistry 14:5451, 1975.
27. Rosenburg T, Wilbrandt W: J Gen Physiol 41:289, 1957.
28. Oxender DL, Christensen HN: J Biol Chem 238:3686 1963.
29. Christensen HN: Adv Enzymol Relat Areas Mol Biol 32:1, 1969.
30. Isselbacher KJ: Proc Natl Acad Sci USA 69:585, 1972.
31. Garcia-Sancho J, Sanchez A, Christensen HN: Biochim Biophys Acta 464:295, 1977.
32. Pressman BC: Ann Rev Biochem 45:501, 1976.
33. Mitchell P, Moyle J: Biochem J 105:1147, 1967.
34. Goldman DE: J Gen Physiol 27:37, 1943.
35. Hodgkin AL: Proc R Soc London Ser B 148:1, 1958.
36. Kessler RJ, Tyson CA, Green DE: Proc Natl Acad Sci USA 73:3141, 1976.
37. Murer H, Hopfer U, Kinne R: Biochem J 154:597, 1976.
38. Ramos S, Kaback HR: Biochemistry 16:848, 1977.
39. Dulbecco R: Proc R Soc London Ser B 189:1, 1975.
40. Jimenez de Asua L, Rozengurt E: Nature 251:624, 1974.
41. Holley RW: Proc Natl Acad Sci USA 69:2840, 1972.
42. Isselbacher K: Proc Natl Acad Sci USA 69:585, 1972.
43. Peters JH, Hausen P: Eur J Biochem 19:509, 1971.
44. Quastel MR, Kaplan JG: Exp Cell Res 63:230, 1970.
45. Glynn IM, Karlish SJD: Ann Rev Physiol 37:13, 1975.
46. Heinz E, Geck P, Pietrzyk C: Ann NY Acad Sci 1264:428, 1975.
47. Rozengurt E, Heppel L: Proc Natl Acad Sci USA 72:4492, 1975.
48. Lever JE, Clingan D, Jimenez de Asua L: Biochem Biophys Res Commun 71:136, 1976.
49. Jimenez de Asua L, Clingan D, Rudland PS: Proc Natl Acad Sci USA 72: 2724, 1975.
50. Kimelberg HK, Mayhew E: J Biol Chem 250:100, 1975.
51. Cunningham DD, Pardee AB: Proc Natl Acad Sci USA 64:1049, 1969.
52. Quinlan DC Hochstadt J: Proc Natl Acad Sci USA 71:5000, 1974.
53. Christensen HN, Handlogten ME: Proc Natl Acad Sci USA 72:23, 1975.
54. Geck P, Heinz E: Biochim Biophys Acta 443:49, 1976.
55. Heinz E, Geck P, Wilbrandt W: Biochim Biophys Acta 255:442, 1972.

Journal of Supramolecular Structure 6:125–133 (1977)
Molecular Aspects of Membrane Transport 65–73

Energy Sources for Amino Acid Transport in Animal Cells

Erich Heinz, Peter Geck, Christian Pietrzyk, Gerhard Burckhardt, and Barbara Pfeiffer

Abteilung Physikalische Biochemie, Gustav-Embden-Zentrum der Biologischem Chemie, J. W. Goethe Universität, D-6000 Frankfurt a.M. 70, Theodor-Stern-Kai 7, Federal Republic of Germany

The existence of an electrogenic Na^+ pump in Ehrlich cells which substantially contributes to the membrane potential, previously derived from the distribution of the lipid soluble cation tetraphenylphosphonium (TPP^+), could be confirmed by an independent method based on the quenching of fluorescence of a cyanine dye derivative, after the mitochondrial respiration had been suppressed by appropriate inhibitors. The mitochondrial membrane potential, by adding to the overall potential as measured in this way is likely to cause an overestimation of the membrane potential difference (p.d.). But since this error tends to diminish with increasing pump activity, the true p.d. of the plasma membrane should easily account for the driving force to drive the active accumulation of amino acids in the absence of an adequate Na^+ concentration gradient. Accordingly, the F_2-aminoisobutyric acid (AIB) uptake rises linearly with the distribution of TPP^+ at constant Na^+ concentrations, suggesting that each responds directly to membrane potential. There is evidence that the electrogenic (free) movement of Cl^- is slow, at least at normal p.d., whereas a major part of the Cl^- movement across the cellular membrane appears to occur by an electrically silent Cl^--base exchange mechanism. By such a mode Cl^-, together with an almost stoichiometric amount of K^+, may under certain conditions move into the cell against a high adverse electrical potential difference. This "paradoxical" movement of K^+Cl^- contributing to the deviation of the Cl^- distribution from the electrochemical equilibrium distribution, is not completely understood. It is insensitive towards ouabain but can almost specifically be inhibited by furosemide. As a likely explanation a H^+-K^+ exchange pump was previously offered, even though unequivocal evidence of such a pump is so far lacking. According to available evidence the electrogenic movement of free Cl^- is too small, at least at normal orientation of the p.d., to significantly shunt the electrogenic pump potential so that the establishment of such a potential is plausible. The evidence presented is considered strong in favor of the gradient hypothesis since even in the absence of an adequate Na^+ concentration gradient, the electrogenic Na^+ pump will contribute sufficient extra driving force to actively transport amino acid into the cells.

Key words: amino acid transport, gradient hypothesis, electrogenic cation pump, electrolyte movements, ouabain, furosemide

The active uptake of neutral amino acids by Ehrlich cells, as in other animal cells and tissues, is according to many investigators secondary active transport, i.e., driven by the electrochemical potential gradient of Na^+ via cotransport (gradient hypothesis),

Received March 14, 1977; accepted April 1, 1977

rather than directly by the affinity of a chemical reaction, as in primary active transport. There are other investigators, however, who reject this gradient hypothesis, and, indeed, the evidence available, though strongly in its favor seems not to have been sufficient to fully exclude some contribution of primary active transport to the overall process. The controversy is illustrated by the following 2 findings. It has been shown by Eddy and others that active uptake of neutral amino acids may take place during complete inhibition of metabolism, as long as appropriate gradients of Na^+ and K^+ can be maintained (1). This observation appears to indicate that only the ion gradients are required to drive the transport, but not directly the metabolic energy supply. On the other hand, it has been shown that metabolically fully active cells may also actively accumulate amino acids while gradients of both Na^+ and K^+ are absent or even moderately inverted (2). This observation in contrast to the first one seems to indicate that only metabolic activity, but not the ion gradients, is required for transport. To reconcile these 2 seemingly contradictory observations some researchers have assumed that there are 2 separate pathways for active amino acid transport, one coupled to the Na^+ flow and the other directly coupled to a metabolic reaction. It has never been possible, however, to separate 2 such pathways, even though neither of the 2 mentioned alternatives alone gives optimal transport, which is obtained only in the presence of both Na^+ gradient and respiration. The main arguments raised against the gradient hypothesis nowadays are energetic ones, i.e., the question whether the electrochemical potential gradient of Na^+ is adequate under all conditions (3). The energetic adequacy of any driving device depends on 2 conditions. Firstly, that enough energy is available, and secondly, that the coupling between the driver process and the transport process is tight enough to warrant sufficient transfer of this energy. As to the latter condition, it could be shown by the methods of irreversible thermodynamics that coupling between amino acid influx and Na^+ influx is indeed fairly tight (4) whereas a direct coupling between the same amino acid influx and the hydrolysis of ATP is barely detectable (5). Hence the second condition appears to be fulfilled. The actual amount of energy available from the Na^+ electrochemical potential gradient is still a matter of controversy, especially in the above mentioned case of inverted Na^+ and K^+ gradients (2). More recently, however, we have been questioning whether electrochemical potential difference of Na^+ as it is derived from separate estimates of concentrations and electrical potential differences, may not be greatly underestimated so that its energetic inadequacy, whenever it is observed, may not be real. In particular, the electrical component of this potential difference (p.d.) may be much higher than has hitherto been assumed, especially in view of the evidence that an electrogenic Na^+ transport system (pump) appears to operate in these cells (6). If the p.d. were exclusively due to a disequilibrium of ion distribution, the electrochemical potential difference of Na^+ could not exceed a ceiling value

$$\frac{RT}{F} \ln \frac{[Na^+]' \cdot [K^+]''}{[Na^+]'' \cdot [K^+]'} \, {}^* .$$

During inversion of both Na^+ and K^+ distribution the electrochemical p.d. of Na^+ should then clearly be in the wrong direction. An electrogenic pump, however, has a contribution to the overall electrical potential of its own, which in the nonsteady state may raise the electrochemical p.d. of Na^+ well above the ceiling value so that in spite of inverted gradients the electrochemical p.d. of Na^+ would be in the right direction to drive the amino acid into the cell. There is experimental evidence that the uptake of the neutral amino acid AIB is accelerated by raising the electrical p.d., even in the absence of a Na^+

*The superscripts ' and '' refer to extracellular and intracellular space, respectively.

concentration gradient (Fig. 1). It can also be shown that the cells are depolarized by neutral amino acids, which, acting as Na$^+$ ionophores, tend to shunt the electrical p.d. (7–9).

Somewhat uncertain still, however, is the actual magnitude of the contribution to the overall potential by this pump. All methods applied to estimate the electrical p.d. appear to involve some error: The p.d. estimated on the basis of the Cl$^-$ distribution is by far too small. As will be discussed below, the distribution of this ion is unsuitable for a p.d. estimate. The p.d. measured by microelectrodes (10) is higher but probably not high enough. Apart from the fact that this method has not yet been applied under the condition of inverted alkali ion distribution, the puncture of the cell by microelectrodes has been found to damage the cellular membrane so that owing to a rapid decay, the values obtained by this method are likely to be underestimated. On the other hand, p.d. estimates based on the distribution of passively permeant ions, such as lipophilic tetraphenylphosphonium ions (TPP$^+$) or fluorescent dyes, give much higher values, which would easily suffice to explain active amino acid transport entirely on the basis of the gradient hypothesis. They are, however, likely to give overestimated values because these presumably also contain the electrical p.d. of the mitochondrial membrane. With some such cations, the interference by mitochondrial p.d. may be so strong that the contribution

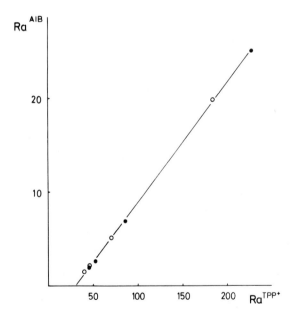

Fig. 1. Effect of electrical potential on the uptake of AIB. The ordinate shows the AIB distribution ratio (R_a) after 4 (○) and 6 (●) min incubation of K$^+$-depleted Ehrlich cells in the presence of various ouabain concentrations in the extracellular medium. The abscissa shows the corresponding ratios of the distribution of tetraphenylphosphonium (TPP$^+$) ions, also after 4 and 6 min incubation. The K$^+$-depleted cells were preincubated with TPP$^+$ (10 μM) and K$^+$-free Krebs-Ringer-phosphate buffer, pH 7.4, for 10 min at 37°C before buffers containing AIB (0.1 mM), K$^+$ (18 mM), and ouabain (0–0.75 mM) were added. It is seen that beyond a certain value of TPP$^+$ distribution the transient accumulation ration of AIB linearly rises with the ration of TPP$^+$. The Na$^+$ concentrations inside and outside the cell are approximately the same in all experiments. To the extent that the TPP$^+$ distribution ratio indicates the electrical potential, the experiment shows that the transient accumulation ratio of AIB is strongly influenced by the electrical potential or by the difference in electrochemical activity of Na$^+$.

to the p.d. by the electrogenic pump can be measured only after selectively blocking the mitochondrial respiration and supplying energy to the Na^+ pump from glycolysis. Figure 2 shows the measurements of the p.d. using a method based on the quenching of the fluorescence of a cyanine derivative. This method seems to depend more strongly on mitochondrial interference than does the TPP^+ method. An increase in electrical potential after stimulating the electrogenic Na^+ pump by extracellular K^+ can with this method be observed only after mitochondrial metabolism has been selectively blocked by some respiratory inhibitor whereas the energy source of the electrogenic Na^+ pump is glycolysis. The difference in behavior between lipophilic cations and the fluorescent dyes in this respect might be attributed to a different permeability of the mitochondria membrane to these substances. A p.d. estimate obtained under such precautions appears to provide an independent confirmation of our previous estimates with TPP^+ (Fig. 3). It can be predicted that the effect of extracellular K^+ on the mitochondrial p.d. is opposite to that on the

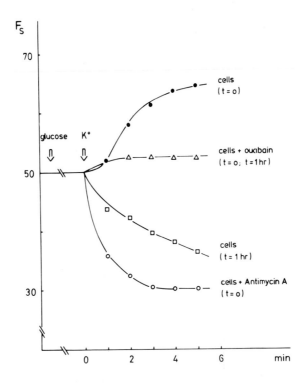

Fig. 2. Electrogenic pump potential as shown by fluorescence change. The fluorescence (F_s) of 3 μm 3, 3′-dipropylthiodicarbocyanine iodide (DiS-C_2-(5)] in a 0.3% suspension of K^+-depleted cells is plotted vs the incubation time. Cells were equilibrated with dye for 10 min. Glucose (2.5 mM), antimycin A (0.5 μM), and ouabain (1 mM) were added 5 min before the addition of K^+. The pump was stimulated by a single dose of KCl (36 mM). The immediate reaction (t = 0) is an increase in fluorescence indicating a decrease in the electrical potential (●). This p.d. decrease is presumably of mitochondrial origin and due to an increase in oxidative phosphorylation. If mitochondrial metabolism is inhibited by antimycin A the fluorescence is quenched by the addition of K^+ indicating an increase in potential. This increase in potential presumably refers to the plasma membrane potential and is due to the electrogenic pump (○). Both effects can be inhibited by the addition of ouabain (△). Also in the absence of antimycin A quenching is observed after 1 hr presumably because of partial inhibition of mitochondrial respiration due to depletion of O_2 (□).

p.d. increase attributable to the electrogenic pump p.d. This appears to be borne out by the behavior of the fluorescence with and without mitochondrial respiration: while the electrogenic pump is stimulated the mitochondrial p.d. goes down, presumably owing to the stimulation of oxidative phosphorylation. Hence our overestimation of the electrical p.d. of the plasma membrane should be smallest with maximum activity of the electrogenic pump, and therefore probably close enough to the real value to justify the assumption that the electrochemical p.d. of Na^+ is high enough to drive the active amino acid transport even during the inversion of the Na^+ and K^+ gradients.

Two questions come up in this context: Why is the electrical p.d. not shown by the Cl^- distribution, and why does Cl^- not shunt the electrogenic pump potential to a greater extent than it appears to do? To answer these questions, we have studied the behavior of Cl^-.

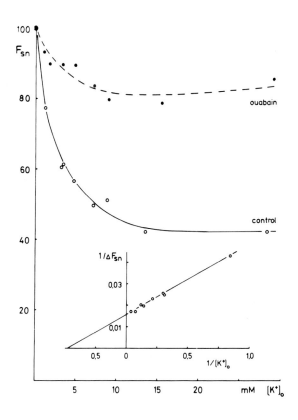

Fig. 3. Electrogenic pump potential as a function of extracellular potassium concentration, as demonstrated by the fluorescence levels of the supernatants (F_{sn}) of 0.09% cell suspensions. In K^+-depleted cells the electrogenic pump was stimulated by the addition of increasing amounts of K^+ to the extracellular medium in the presence of glucose (2.5 mM) and of antimycin A (0.5 μM) to block mitochondrial potential formation. Incubation time with K^+: 5 min. It is seen that the maximum quenching of fluorescence, indicating the maximum development of a potential, increases with extracellular K^+, reaching a maximum at about 10 mM. The shape of this curve is similar to that obtained with the distribution of TPP^+ under similar conditions. In the presence of 1 mM ouabain the potential generation is largely suppressed. In the insert the reciprocal fluorescence change ($\triangle F_{sn} = 100 - F_{sn}$) is plotted vs the reciprocal K^+ concentration suggesting a Michaelis-Menten relationship with respect to K^+.

The magnitude of the contribution of an electrogenic pump to the electrical membrane potential depends on concentration and permeabilities of the other ions, in particular of Cl^-. If these ions penetrate the cellular membrane very fast, they will shunt the pump potential and keep it to low values, as appears to be the case in some other systems. Hence it is of crucial importance in the present context to show that this mobility is indeed low enough to allow the generation of the postulated pump potential. Measurement of the actual rate of Cl^- permeability turned out to be very difficult since under different conditions the permeability coefficient for Cl^- appears to vary over wide ranges. Apparently there are several pathways of different velocities for Cl^- movement of which the more rapid ones are probably due to an electrically silent exchange between Cl^- and some base anions, most likely OH^-. In addition, the permeability coefficient of the cellular membrane for net movement of Cl^- appears to be influenced by the electrical potential of the membrane as if there were some rectification. It seems that with a normal orientation of the potential, i.e., positive outside, the movement of Cl^- is indeed very slow, but it becomes accelerated if the potential is reversed (11). Hence it seems that the electrogenic pump, by tending to raise the potential in the outward direction tends to decrease the permeability for Cl^-, thus providing the very condition necessary to avoid shunting by Cl^-. The outward net movement at least appears to be slow enough to allow the generation of a sufficiently high pump potential. For the same reason, the Cl^- distribution may in transient states depend more strongly on the pH difference than on the electrical p.d. but in the end it should nonetheless tend toward equilibrium with the electrical p.d. This, however, is not observed. On the contrary, Cl^- may under certain conditions even move into the cell steadily away from electrochemical equilibrium. (Fig. 4). This "paradoxical" Cl^- uptake is associated with an almost stoichiometric movement of K^+ as is usually observed with K^+-depleted cells if the extracellular K^+ rises above 5–10 mM (12). Since under these conditions the electrical p.d. is still strongly positive outside, the distribution of Cl^- at any time during this process should give highly erroneous p.d. estimations if the simple Nernst equation is applied. The mechanism of this paradoxical KCl uptake, which is inhibited by metabolic inhibitors but not by ouabain, is not fully understood. One explanation has been based on a hypothetical H^+-K^+ exchange pump which is stimulated by extracellular K^+ and which by raising the cellular OH^- level should favor the uptake of Cl^- via the mentioned Cl^--OH^- exchange mechanism. This pump would thus lead to a net uptake of KCl in exchange for a stoichiometric amount of water molecules. Since in the present system H^+ extrusion would be readily neutralized by OH^- via the rapid Cl^--OH^- exchange mechanism, such a proton pump cannot easily be detected. We therefore tried to catch a H^+ extrusion before neutralization using a sensitive, rapid-response pH-meter after triggering the hypothetical proton pump by a K^+ pulse. To delay neutralization via the Cl^--OH^- exchange mechanism the Cl^- in the suspending medium was replaced by gluconate. Ouabain was also added to prevent interference of the Na^+-K^+ pump with the proton pump. No evidence of a rapid H^+ extrusion, however, could be detected under these conditions. There still remains the possibility that the H^+ are not pumped directly into the medium but first into some intracellular compartment which communicates with the outside medium to let the H^+ leak out in exchange for K^+. Thus the appearance of OH^- in the cytoplasm may initially precede that of H^+ in the medium. The existence of an intracellular compartment with an acidity higher than that of the cytoplasm would be in agreement with the previously reported evidence of an intracellular inhomogeneity with respect to pH (3). It had been shown by the applica-

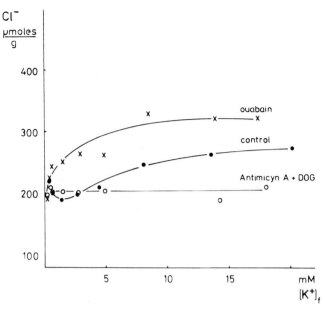

Fig. 4. Effect of extracellular K^+ on the chloride movement. The Cl^- content (μmoles/gram dry weight) of K^+-depleted Ehrlich cells incubated for 3 min in Krebs-Ringer-phosphate buffer, pH 7.4, containing different inhibitors is plotted against the extracellular K^+ concentration (mM). To K^+-depleted cells K^+-free buffers were added with and without antimycin A (0.4 μM) and 2-desoxyglucose (DOG, 5 mM) and preincubated for 5 min at 37°C before buffers with different K^+ content with and without ouabain (1.25 mM) were added for the final 3 min of incubation. In agreement with previous findings the increase of extracellular K^+ in a suspension of K^+ depleted cells causes an uptake of chloride if the extracellular K^+ exceeds about 5 mM. This uptake is strongly enhanced in the presence of ouabain. The chloride uptake is completely abolished after metabolic inhibition by antimycin A and 2-desoxyglucose.

tion of both an acidic and a basic pH probe that the distribution ratio was far from reciprocity, which should have been expected for a homogeneous compartment, whereas reciprocity is produced by metabolic inhibition (3). Further investigation is required to test this hypothesis.

A suitable tool to study this special KCl uptake may be given by the inhibitor furosemide which at 2 mM not only strongly inhibits Cl^- permeability, but also almost specifically eliminates the mentioned paradoxical Cl^- uptake and also the concomitant K^+ uptake (Fig. 5). On the other hand, furosemide does not appear to effect the ouabain-sensitive electrogenic Na^+ pump (Fig. 6).

Whatever its mechanism, the paradoxical Cl^- movement would account for inability of the Cl^- distribution to indicate the true electrical p.d. It also gives an explanation why the p.d. of the electrogenic pump is not shunted to a greater extent by electrogenic Cl^- movement.

The presented observations are all in agreement with the postulates of the gradient hypothesis, and moreover provide a rational explanation for observations seemingly arguing against this hypothesis. They are therefore regarded as a strong support of the gradient hypothesis.

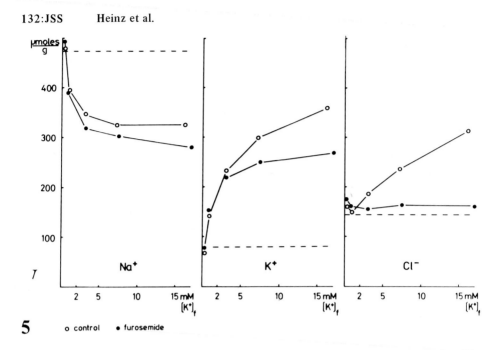

Fig. 5. The effect of 2 mM furosemide on ion transport in Ehrlich cells. The cellular content of Na$^+$, K$^+$, and Cl$^-$ (μmoles/gram dry weight) after 5 min incubation in presence ($\bullet$) and absence ($\circ$) of furosemide is plotted against the final K$^+$ concentration of the medium. The dotted lines give the corresponding levels at zero time. To K$^+$-depleted cells in K$^+$-free medium buffers of varied K$^+$ content with and without furosemide (2 mM) were added. The Na$^+$ extrusion, stimulated by extracellular K$^+$, is only slightly increased in the presence of furosemide. In agreement with previous experiments, Cl$^-$ uptake occurs if the extracellular K$^+$ exceeds about 5 mM, in spite of the pump potential, which is outside positive. This "paradoxical" Cl$^-$ uptake is almost completely inhibited by furosemide. The K$^+$ uptake rises with increasing extracellular K$^+$. As Fig. 6 shows, this K$^+$ uptake is composed of 2 components, one related to the Na$^+$-K$^+$ pump and an additional one which is ouabain insensitive and is about equivalent to the "paradoxical" Cl$^-$ uptake. In the presence of furosemide this extra K$^+$ uptake, equivalent to the paradoxical Cl$^-$ uptake is inhibited.

Fig. 6. Effect of ouabain and furosemide on ion transport in Ehrlich cells. The cellular content of Na$^+$, K$^+$, and Cl$^-$ (μmoles/gram dry weight) is plotted against the incubation time. To K$^+$-depleted Ehrlich cells in K$^+$-free medium K$^+$-containing (15 mM) buffers with and withoug furosemide (2 mM) and ouabain (1 mM) were added. The extrusion of Na$^+$ is triggered and this extrusion appears not to be changed significantly by the presence of furosemide. The presence of ouabain, however, blocks this Na$^+$ extrusion in the presence and absence of furosemide. The addition of K$^+$ causes a rapid uptake of K ions, which is significantly inhibited by 2 mM furosemide. 1 mM ouabain reduces the K$^+$ uptake considerably without abolishing it. This ouabain insensitive residual K$^+$ uptake appears to be strongly inhibited by furosemide. The addition of K$^+$ causes first an extrusion and then at higher K$^+$ concentrations an uptake of Cl$^-$. This uptake is significantly increased in the presence of ouabain. This increase presumably presents the real, ouabain insensitive Cl$^-$ uptake, which in the absence of ouabain is counteracted by the pump potential. In the presence of furosemide each curve appears to be displaced by about the same amount indicating that under these conditions the Cl$^-$ movement appears to follow in the direction of the potential being outward in the absence of ouabain and remains on the same level in the presence of ouabain. $\circ$) Control, $\triangle$) ouabain, $\bullet$) furosemide, $\blacktriangle$) ouabain + furosemide.

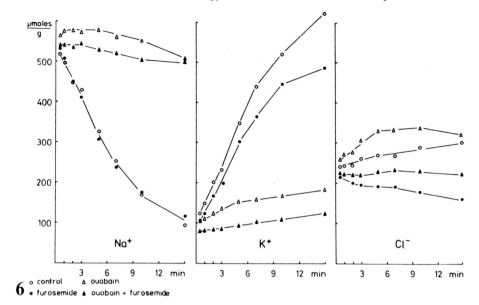

6 o control   Δ ouabain
   • furosemide ▲ ouabain + furosemide

ACKNOWLEDGMENTS

The authors wish to thank Mrs. E. Heyne, Miss G. Werner, and Miss C. Henschel for competent technical assistance, and Mrs. E. Kemsley for preparing the manuscript.

The work described in this paper has been supported by a grant of the Deutsche Forschungsgemeinschaft (He 102/14). Part of the cyanine preparations used in this work was kindly given to us by Dr. A. S. Waggoner, Amherst, Mass.

REFERENCES

1. Eddy AA: Biochem J 108:195, 1968.
2. Schafer JA, Heinz E: Biochim Biophys Acta 249:15, 1971.
3. Heinz E, Pietrzyk C, Geck P, Pfeiffer B: In Gárdos G, Szász I (eds): "Energetic Coupling of Amino Acid Transport." Amsterdam:North Holland, 1975, pp 107–115.
4. Heinz E, Geck P: Biochim Biophys Acta 339:426, 1974.
5. Geck P, Heinz E, Pfeiffer B: Biochim Biophys Acta 339:419, 1974.
6. Heinz E, Geck P, Pietrzyk C: Ann NY Acad Sci 264:428, 1975.
7. Geck P, Heinz E, Pietrzyk C: In Silbernagl S, Lang F, Greger R (eds): "The Membrane Potential of Ehrlich Cells and Its Significance for Active Amino Acid Transport." Stuttgart:Thieme, 1976, pp 33–39.
8. Laris PC, Pershadsingh HA, Johnstone RM: Biochim Biophys Acta 436:475, 1976.
9. Philo RD, Eddy AA: Biochem Soc Trans 3:904, 1975.
10. Lassen UV, Nielsen A-MT, Pape L, Simonsen LO: J Membr Biol 6:269, 1971.
11. Heinz E, Geck P, Pietrzyk C: "Energy Sources of Active Transport." Amsterdam:Excerpta Medica, 1976, pp 313–319.
12. Heinz E, Geck P, Pietrzyk C, Pfeiffer B: In Semenza G, Carafoli E (eds): "Electrogenic Ion Pump As an Energy Source for Active Amino Acid Transport in Ehrlich Cells." Berlin, Heidelberg:Springer Verlag, 1977, pp 236–249.

Journal of Supramolecular Structure 6:135–153 (1977)
Molecular Aspects of Membrane Transport 75–93

Characterization of a Periplasmic Protein Related to sn-Glycerol-3-Phosphate Transport in Escherichia Coli

Manfred Argast, Günter Schumacher, and Winfried Boos

Department of Biology, University of Konstanz, Postfach 7730, D-7750 Konstanz, West Germany

The cold osmotic shock procedure releases a protein (GLPT) from the cell envelope of Escherichia coli that is related to the transport of sn-glycerol-3-phosphate in this organism. The evidence for this correlation is as follows: 1) GLPT is under the regulatory control of the glpR gene. 2) Some glpT mutants that were isolated as phosphonomycin resistant clones do not synthesize GLPT. Revertants of these mutants (growth on sn-glycerol 3-phosphate) again synthesize GLPT. 3) Some amber mutations in glpT reduce the amount of GLPT while suppressed strains produce normal amounts. 4) Transfer of a plasmid carrying the glpT genes into a strain lacking GLPT and sn-glycerol-3-phosphate transport restores both functions in the recipient. Transport and GLPT synthesis in the plasmid carrying strain are increased 2- to 3-fold over a fully induced wild-type strain, but appear to be constitutive. GLPT is a soluble protein of molecular weight 160,000 composed of 4 identical subunits. The 160,000 molecular weight complex is stable in 1% sodium dodecylsulfate at room temperature. Upon boiling in 1% sodium dodecylsulfate GLPT dissociates into its subunits. Likewise, 8 M urea at room temperature dissociates GLPT into its subunits. Dialysis of dissociated GLPT against phosphate or Tris-HCl buffer, pH 7.0, allows renaturation to the tetrameric form. The protein is acidic in nature (isoelectric point 4.4).

In contrast to the typical transport-related periplasmic-binding proteins, no conditions could be found where pure GLPT exhibited binding activity toward its supposed substrate, sn-glycerol-3-phosphate.

In vivo new appearance of transport activity for sn-glycerol-3-phosphate transport occurs only shortly before cell division. However, GLPT synthesis does not fluctuate during the cell cycle. The available evidence indicates a cell-division-dependent processing of GLPT in the cell envelope as a reason for the alteration in transport activity.

Transport in whole cells is sensitive to the cold osmotic shock procedure, demonstrating the participation of an essential periplasmic component. However, isolated membrane vesicles that are devoid of periplasmic components, including GLPT, are fully active in sn-glycerol-3-phosphate transport. Therefore, we conclude that GLPT is essential in overcoming a diffusion barrier for sn-glycerol-3-phosphate established by the outer membrane. Attempts to isolate mutants that are transport negative in whole

Received March 14, 1977; accepted March 16, 1977

cells due to a defect in GLPT but are active in isolated membrane vesicles have failed so far. All GLPT mutants tested, whether or not they synthesize GLPT, are not active in isolated membrane vesicles.

Iodination of whole cells with [^{125}I] followed by osmotic shock reveals that several shock-releasable proteins including GLPT become radioactively labeled. This indicates that some portions of GLPT are accessible to the external medium.

Key words: periplasmic proteins, transport, precursor

Recently, the application of a sensitive 2-dimensional polyacrylamide gel electrophoresis technique for the analysis of shock-releasable proteins from E. coli revealed the existence of a protein (GLPT) that was closely related to the transport system of sn-glycerol-3-phosphate (G3P) (Silhavy et al., 1976a).

This transport system previously has been studied mainly by Lin and his collaborators (Lin, 1976) with the following results: 1) Transport of G3P occurs against the concentration gradient without chemical alteration of the substrate. 2) The apparent K_m of the transport system is 12 μM and in E. coli this system is the only one that transports G3P. 3) The transport system is under negative control of the gene product of glpR, the regulator gene of the glp regulon. 4) The glpT mutants that exhibit a phenotypically negative transport behavior for G3P were mapped to be located at min 48 on the linkage map of E. coli, cotransducible with the nalA marker. From these studies it was not clear whether or not the glpT operon consists of more than one gene.

In the biochemical studies of different bacterial active transport systems, so far 3 types of proteins have been implicated.

A. The M-Protein, the Gene Product of the lacY Gene

The M-protein is tightly bound to the inner membrane of E. coli and is most likely the only protein sufficient to account for all the phenomena observed with the lactose transport system (Kennedy, 1970). It is characteristic for this system that it still operates in isolated membrane vesicles (Kaback, 1972) and derives its energy from the proton motive force as defined by Mitchell (Ramos et al., 1976). Several other systems for sugars and amino acids have been found to operate in membrane vesicles. However, their responsible carriers have not been identified.

B. Numerous Soluble Periplasmic Substrate-Binding Proteins

Even though not proven for all binding proteins the following generalization can be made about the soluble periplasmic substrate-binding proteins (Boos, 1974): 1) They establish the recognition site of the respective active transport system. 2) They are not the only component of the system. 3) Binding-protein-mediated transport systems do not operate in membrane vesicles, and their energy coupling may involve the direct participation of cellular ATP (Berger and Heppel, 1974).

C. Outer Membrane Components

These proteins have been identified as transport components that facilitate the diffusion of substrate through the outer membrane. These outer membrane components usually have additional functions as colicin or phage receptors. For instance, the outer membrane component for the maltose transport system is the receptor for phage λ (Szmelcman and Hofnung, 1975).

The present publication is concerned with the characterization of a shock releasable protein that belongs to the *glpT* operon and may be part of the G3P transport system.

MATERIAL AND METHODS

Bacterial strains used are listed in Table I together with their source. The isolation of phosphonomycin resistant strains was done according to Venkateswasan and Wu (1972). Strains carrying the *glpT*$^+$ operon on the Col E1 plasmid were obtained by cross-streaking strain JA 200/3-46 with a *glpT* strain on an agar plate containing sn-glycerol-3-phosphate as the sole carbon source. Minimal medium A (Miller 1972) with 0.4% glycerol was used as growth medium. For BUG-6 low-phosphate medium (Garen and Levinthal, 1960) was used with 0.4% glycerol as carbon source. The osmotic shock procedure was done according to Neu and Heppel (1965). Two-dimensional polyacrylamide gel electrophoresis was done according to Johnson et al. (1975) with the modification described by Silhavy et al. (1976a). Purification of GLPT, enzymatic tests, polyacrylmide gel electrophoresis in sodium dodecylsulfate (SDS), and cross-linking of GLPT were described previously (Boos et al., 1977). Preparation of membrane vesicles was done according to Hirata et al. (1974), and transport in these vesicles was measured as described by Boos et al. (1977). Iodination of intact bacteria was done according to Mueller and Morrison (1974). The iodinated cells were subjected to the osmotic shock procedure of Neu and Heppel (1965) and the periplasmic proteins subjected to 2-dimensional polyacrylamide gel electrophoresis. The gels were stained with Coomassie Blue. All spots were cut out, and the gel pieces were solubilized by the addition of 0.5 ml 30% H_2O_2 containing 10% concentrated ammonia. These mixtures were incubated overnight at 42°C. The radioactivity was measured in a liquid scintillation counter after addition of 10 ml dioxan-based scintillation fluid.

To isolate and solubilize the proteins of the cytoplasmic membrane, the exclusive solubilization of these proteins in Triton X-100 in the presence of $MgCl_2$ was used. The procedure was essentially that described by Schnaitman (1971). The solubilized proteins were precipitated in 80% ethanol at 4°C. The precipitate was washed once with 80% ethanol, solubilized in guanidinium thiocyanate, and subjected to 2-dimensional polyacrylamide gel electrophoresis as described by Johnson et al. (1975).

Synchronization technique: Strain 72 was synchronized by a stationary-phase method (Cutler and Evans, 1966) which takes advantage of the tendency of cells to synchronize themselves when they enter the stationary phase of growth.

Transport of sn-glycerol-3-phosphate during synchronous growth: To 100 μl of cells 10 μl of [^{14}C] G3P was added to a final concentration of 1 μM. After 30 sec 100 μl were removed and filtered through a Millipore filter of 0.65 μm pore size. The filter was washed with 10 ml growth medium. This test is linear for bacterial cultures with optical densities (ODs) (at 576 nm) from 0.1 to 1.0. The activities during synchronous growth were expressed as the amount of G3P uptake during the initial 30 sec incubation period.

GLPT pulse labeling: To aliquots of 5 ml of the growing culture [U-^{14}C]-leucine (0.75 · 10^6 cpm) and [U-^{14}C]-alanine (0.75 · 10^6 cpm) were added at a final concentration of 0.1 μM. Labeling was allowed to proceed for 10 minutes. It was stopped by the addition of an 1,000-fold excess of unlabeled amino acids. The culture was allowed to grow for 10 more minutes. Then, the cells were harvested by centrifugation and washed

TABLE I. Strains of E. Coli Used

Strain no.	Parent	Sex	Isolation procedure	Genotype	Reference
72		Hfr		Δ (glpR-malA), phoA	Cozzarelli et al. (1968)
LA 3400	72	Hfr	P1 transduction to mal⁺	glpT⁺, mal⁺, phoA	Silhavy et al. (1976a)
LA 3404	72	Hfr	Resistance against phosphonomycin, spontaneous mutants	Δ (glpR-malA), phoA, glpT	Silhavy et al. (1976a)
LA 108	BUG-6 temperature sensitive for cell division, i.e., septum formation	F⁻	Spontaneous; growth on β-glycerolphosphate	strA, gal, xgl, mtl, phoR	Reeve et al. (1970) Shen and Boos (1973)
JA 200/3-46	–	F⁺		Δ trpE5, recA; carrying plasmid colE1, pLCNr. 3–46	Clarke and Carbon (1976)
G 810	MC 4/00	F⁻	Mu cts: :malT; P1 transduction to mal⁺, glpR; Mu cts: :glpT, phosphonomycin resistant	araD, Δlac, strA, glpR, Mu cts::glpT	Silhavy et al. (1976b)
G 810/3	G 810	F⁻	Growth at 42°C	araD, Δlac, strA, glpR, glpT	this study
G 810/3 3-46	G 810/3 and JA 200/3-46	F⁺	Cross-streaking on sn-glycerol-3-phosphate	araD, Δlac, strA, glpR, glpT⁺	this study
165	–	–		glpR, glpD, glpA⁺, thr, leu, thi	Miki, Harvard Medical school, personal communication

once with growth medium minus carbon source. The cells were then transformed into spheroplasts by the EDTA-lysozyme treatment in the presence of 20% sucrose (Neu and Heppel, 1964). The supernatant of these spheroplasts that contained the proteins of the periplasmic space was analyzed by SDS polyacrylamide gel electrophoresis together with an unlabeled sample of the osmotic shock proteins. The Coomassie-Blue stained band identified on these gels with GLPT was cut from the gel, dissolved in 0.5 ml 30% H_2O_2 containing 50 μl conc. NH_3 and counted in the liquid scintillation counter with 5 ml dioxan-naphtaline based scintillation fluid.

RESULTS

1. Evidence for GLPT Being a Gene Product of the *glpT* Operon

   A. Genetics and osmotic shock. Figure 1 shows the 2-dimensional electrophoresis analysis of the shock-releasable proteins from a strain with a fully induced G3P transport

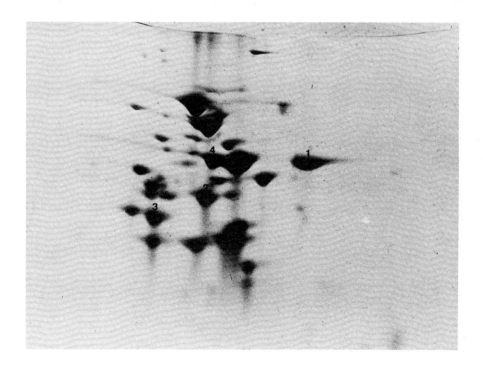

Fig. 1. Two-dimensional polyacrylamide gel electrophoresis of shock fluid of strain LA 3400 (wild-type) grown in the presence of glycerol. The first dimension consists of electrophoresis in 8 M urea (pH 8.4), followed by electrophoresis in 0.2% sodium dodecylsulfate (pH 6.48). About 300 μg of protein was applied. The numbers and molecular weights correspond to the following proteins: 1) GLPT: 2) galactose-binding protein (Boos and Gordon, 1971); 3) ribose-binding protein (Willis and Furlong, 1974); 4) maltose-binding protein (Kellerman and Szmelcman, 1974). With modifications taken from Silhavy et al. (1976a) (with permission from the authors and publisher).

system. Using this technique we previously demonstrated that one spot designated GLPT only appeared in glycerol-grown wild-type cells, or independent of inducer, in *glpR* strains that are constitutive for all *glp* operons. Moreover, some *glpT* mutants that are unable to accumulate G3P lack GLPT, while their revertants again produce it. Treatment of the cells with the cold osmotic shock procedure that releases GLPT results in a drastic reduction of the ability of shocked cells to accumulate G3P (Shilhavy et al., 1976a).

B. Nonsense mutants in *glpT*. The shock-releasable proteins of 3 characterized mutants that carry nonsense mutations in *glpT* and their respective suppressed derivatives (obtained from Dr. Weissenbach, Harvard Medical School, Boston) gave the following picture: 2 mutants synthesized a sharply reduced amount of GLPT, while the suppressed strain again synthesized it. Transport activity for G3P corresponded to the amount of GLPT seen in these mutants. One mutant showed normal GLPT levels, even though it was transport negative. Thus, in 2 of these mutants the nonsense mutation must have occurred in a gene proximal to the gene that codes for GLPT, affecting the amount but not the structure of GLPT. This demonstrates that the *glpT* operon must consist of more than one gene.

C. Insertions of phage Mu into *glpT*. Insertion of phage Mu into any gene of the E. coli chromosome destroys the gene product of this gene and has a strong polar effect on the distal genes of the same operon. Mutants carrying a temperature-sensitive Mu phage in *glpT* were obtained by Dr. Silhavy (Harvard Medical School, Boston) and analyzed for their content of shock-releasable GLPT. Out of 8 transport negative mutants 7 lacked GLPT while 1 contained the normal amount (not shown). This again suggests that GLPT

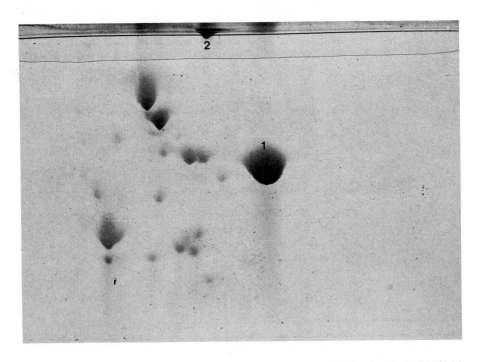

Fig. 2. Two-dimensional polyacrylamide gel electrophoresis of shock fluids of strain JA 200/3-46 grown on glycerol. Spot No. 1 is monomeric GLPT, and No. 2 tetrameric GLPT. Conditions as described in the legend to Fig. 1.

is coded for by a gene that is contained in the *glpT* operon.

D. The *glpT* plasmid. From a bank of colE1 plasmid carrying different genes of the E. coli chromosome (Clarke and Carbon, 1976) 3 strains were obtained containing the *glpT* gene. The electrophoretic analysis of the periplasmic shock proteins of such a strain grown on glycerol is shown in Fig. 2. As can be seen this strain contains a large amount of GLPT. One of the *glpT* carrying plasmids was then transferred into a strain that was transport negative for G3P and did not contain GLPT. Figure 3 shows the transport activity of the recipient strain with and without the plasmid. As a control, the transport activity of a fully-induced normal wild-type strain is included. The introduction of the *glpT* carrying protein results in a 2- to 3-fold increase of transport activity for G3P over a fully-induced wild-type strain.

However, transport activity as well as the amount of GLPT synthesized is not any longer inducible by G3P as is the normal wild-type strain. The GLPT operon of the plasmid is apparently constitutive even though the recipient strain contains a wild-type *glpR* gene product, the regulator protein for the *glp* operon. Figure 4 shows the electrophoretic analysis of the shock proteins of the receptor strain with and without the plasmid. As can be seen the only new spot in the protein pattern is GLPT. The increased transport activity for G3P correlates with the increased amount of GLPT produced by these strains. The size of the *glpT* carrying plasmid was measured as $9-10 \cdot 10^6$ daltons, while the original plasmid DNA is in the order of $4 \cdot 10^6$ daltons. Thus, a DNA segment of $5 \cdot 10^6$ daltons of the E. coli chromosome corresponding to no more than 5 genes is responsible for the transport activity of G3P and the synthesis of GLPT.

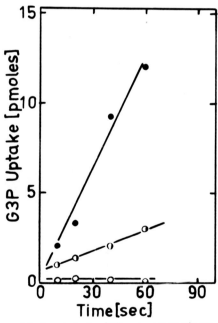

Fig. 3. Transport of sn-glycerol-3-phosphate in *glpT* mutant G 810/3 (○), its *glpT*⁺ (plasmid) carrying derivative (●), and a normal wild-type strain, LA-3400 (◐). The bacterial cultures were resuspended in 10 mM Tris-HCl, 150 mM sodium chloride (pH 6.5) to an optical density (OD) of 1.0 at 576 nm. [^{14}C]-G3P was added at an initial concentration of 0.3 μM. The data are expressed in terms of amount substrate taken up per 50 μl sample. All operations were done at room temperature.

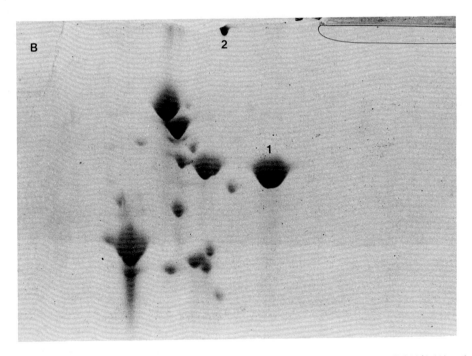

Fig. 4. Two-dimensional polyacrylamide gel electrophoresis of shock fluids of strain G 810/3 (A) and G 810/3, 3-46 (B). Spot No. 1 is monomeric GLPT, and No. 2 is tetrameric GLPT. Conditions as described in the legend to Fig. 1.

2. Properties of GLPT

A. Quaternary structure. GLPT was purified from the osmotic shock fluid (Boos et al., 1977). The rational of the purification procedure was based on the properties of GLPT during 2-dimensional polyacrylamide gel electrophoresis, i.e., a molecular weight of 40,000 and a rather acidic isoelectric point. Therefore, GLPT was purified by Sephadex G-100 chromatography followed by preparative isoelectric focusing (pI 4.4).

In the absence of an adequate enzymatic test, 2-dimensional polyacrylamide electro-phoresis was used to monitor the purification. Two problems were encountered: 1) the presence of a proteolytic activity that splits GLPT into a peptide of 35,000 daltons, and 2) the fact that GLPT in the native form exhibits quaternary structure and is composed of

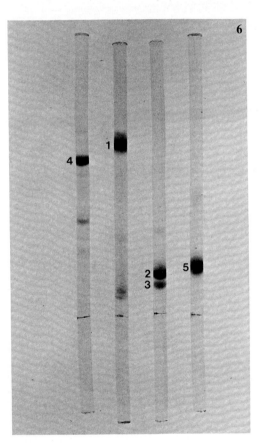

Fig. 5. Polyacrylamide gel electrophoresis in SDS of a purified preparation of GLPT. The sample was boiled for 10 min in 1% SDS, 1% mercaptoethanol. About 15μg protein were applied. The gels contained 5% acrylamide. The ink mark indicates the migration of bromphenol blue. Electrophoresis was performed at 8 mA per tube for 5 h. The direction of migration is from top to bottom. Taken from Boos et al. (1977) with the permission of authors and publisher.

Fig. 6. Polyacrylamide gel electrophoresis in SDS of purified preparations of GLPT with and without boiling in SDS. The samples were boiled for 10 min in 1% SDS, 1% mercaptoethanol. About 20 μg protein were applied. 1) Protein GLPT, unboiled sample; 2, 3) protein GLPT, boiled sample; 4) E. coli β-galactosidase; 5) chicken serum albumin. Experimental conditions as described in the legend to Fig. 5. Taken from Boos et al. (1977) with the permission of authors and publisher.

4 identical subunits of approximately 40,000 molecular weight. The purest preparation analyzed by the usual SDS electrophoresis is shown in Fig. 5. In this case the sample had been boiled in 1% SDS prior to electrophoresis. In contrast, merely incubating the sample in SDS without boiling prior to electrophoresis reveals that GLPT occurs as a complex with molecular weight of 160,000. Figure 6 demonstrates this phenomenon and compares the 2 different species of GLPT with marker proteins of molecular weight 135,000 (E. coli β-galactosidase) and 45,000 (chicken serum albumin). In contrast to the GLPT preparation seen in Fig. 5, the preparation used here was stored at 4°C for several days. As a consequence proteolytic degradation of GLPT occurred which was revealed by the additional presence of a polypeptide of 35,000 molecular weight after boiling of the 160,000 molecular weight complex. Since this proteolytic digestion occurred in the 160,000 molecular weight complex and does not result in an apparent reduction in the molecular weight of the complex its quaternary structure remains intact despite the proteolytic degradation.

Dissociation of the GLPT complex can easily be followed by SDS-polyacrylamide gel electrophoresis in crude shock fluid since the tetrameric form of GLPT is by far the slowest moving band in this technique. This is demonstrated in Fig. 7. Dissociation of the

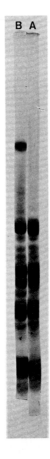

Fig. 7. Polyacrylamide gel electrophoresis in SDS of crude shock fluid of strain G 810/3, 3-46.
A) Sample boiled in SDS for 10 min prior to electrophoresis, B) untreated sample.

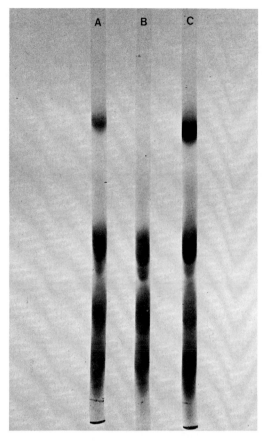

Fig. 8. Polyacrylamide gel electrophoresis in SDS of crude shock fluids of strain G 810/3, 3-46.
A) Treated for 30 min with 8 M urea, followed by dialysis against 100 mM Tris-HCl, pH 7.0; B) treated
for 30 min in 8 M urea; C) untreated sample. About 150 μg protein were applied on the gels. Experimental conditions as described in the legend to Fig. 5.

complex by 8 M urea or boiling SDS is reversible. Dialysis overnight of the treated sample
against 10 mM Tris-HCl buffer, pH 7, results in reformation of the complex. This is shown
in Fig. 8.

Cross-linking of the GLPT complex by a series of diimidates of varying chain length
and subsequent analysis of the cross-linked products after boiling in SDS reveals that only
a species of twice the molecular weight of the monomer can be linked together (not
shown).

This dimeric species cannot be seen in experiments where the GLPT complex is
exposed for 30 min to 1% SDS at different temperatures. While dissociation is nearly
complete at temperatures above 40°C, at temperatures below 40°C only the monomeric
and tetrameric but not the dimeric form can be seen. Due to the analysis of shock proteins
by 2-dimensional electrophoresis involving initial separation of the proteins in 8 M urea,
we at first had not realized the tetrameric nature of GLPT. However, reexamination of
these gels, particularly of shock proteins derived from a *glpT* plasmid carrying strain, revealed the existence of the tetrameric form of GLPT on 2-dimensional gels. This can easily
be seen on Figs. 2 and 4 (spots No. 2) To avoid artifactual appearance of multiple spots

caused by long exposure to urea (carbamylation by cyanate) we routinely added urea to the protein sample only shortly before electrophoresis in the first dimension. Dissociation of GLPT under these conditions is apparently slow and is not completed during the time of the electrophoretic run. Dissociation is completely stopped after the electrophoresis in the second dimension is begun and the protein freed from urea. This is obvious since no trailing of spot No. 2 (GLPT complex) occurs in the second dimension. This again demonstrates the surprising resistance of the GLPT complex against SDS at room temperature.

B. Tests for possible enzymatic activity. From the genetic correlation concerning the map positions of the different enzymes contained in the *glp* regulon, the identity of GLPT with glycerolkinase or with either of the catabolic flavin-linked and membrane-bound glycerolphosphate dehydrogenases was excluded. Tests performed with GLPT to detect any enzymatic activity of glycerolphosphatase or glycerolkinase activity were negative. Similarly, equilibrium dialysis using [^{14}C]-labeled G3P as ligand did not reveal any binding activity of GLPT towards G3P under a variety of different conditions of ionic strength and metal ions. The experimental setup was such that a dissociation constant of 10^{-4} M or less could be detected. For these experiments purified GLPT had to be used since crude preparations contain a phosphatase activity that hydrolyzed G3P to glycerol and inorganic phosphate.

3. Cell-Division-Dependent Transport Activity for G3P; Precursor or Positioning Mechanism for GLPT

Studies by Ohki (1972) demonstrated that transport activity of G3P in E. coli was subject to a cell-cycle-dependent regulation. Similar phenomena have been found with the transport activity of the β-methylgalactoside system (Shen and Boos, 1973). In the latter system the corresponding periplasmic substrate recognition site, the galactose-binding protein, was similarly affected in its synthesis. Synthesis of material cross-reacting with anti-galactose-binding protein antibodies occurred during cell division but not during cell elongation. Similar observations were made in BUG-6, a mutant which is temperature sensitive for cell division (Reeve et al., 1970). This strain synthesized galactose-binding protein only at the permissive temperature (Shen and Boos, 1973). Therefore, it was of interest to determine if GLPT synthesis would exhibit the same phenomena. Figure 9 shows an experiment where several parameters were measured during synchronized cell growth: 1) optical density, 2) cell number, 3) transport of G3P, 4) rate of total protein synthesis (TCA-precipitable material), 5) rate of GLPT synthesis (EDTA-lysozyme-sucrose extractable [^{14}C]-labeled material that comigrates with tetrameric GLPT on SDS gels).

This experiment clearly showed the cell-division-dependent alteration of transport activity for G3P but did not reveal a corresponding alteration in the rate of synthesis for the tetrameric form of GLPT as recovered by the lysozyme-EDTA-sucrose method.

This would mean that it cannot be the periodically altering synthesis of GLPT that evokes the cell-division-dependent alteration in G3P transport activity. However, it might be caused by a cell-division-dependent processing of a GLPT precursor, or its cell-division-dependent positioning into a functional state in the cell envelope. Experiments with BUG-6, the temperature-sensitive cell division mutant, are relevant in this respect. This strain grows and divides normally at 35°C. When shifted to 42°C, division stops immediately and long filaments form. The mutation does not affect DNA replication but prevents septum formation (Reeve et al., 1970).

The BUG-6 mutant when grown overnight at 42°C on glycerol does not exhibit any transport activity for G3P, while cultures grown at 35°C transport normally. Analysis of

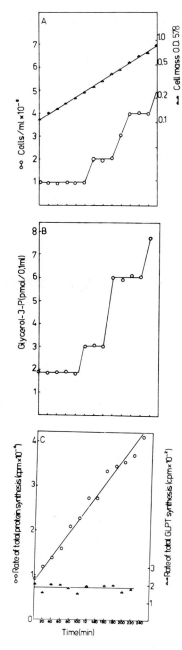

Fig. 9. Net transport activity and GLPT synthesis in synchronized cells of strain 72. A) Growth was followed by measuring the optical density at 578 nm. Cell number was measured by plating for single cell colonies on nutrient broth agar plates. B) Transport activity (initial rate of uptake) is given in pmol of $[^{14}C]$-G3P taken up per 30 sec per 0.1 ml of growth culture. C) Pulse labeling of the GLPT was done for 10 min with $[U-^{14}C]$-leucine, $[U-^{14}C]$-valine and $[U-^{14}C]$-alanine followed by a 10 min chase with excess of unlabeled amino acids. Incorporation of radioactivity is given in cpm per 0.1 ml EDTA-lysozyme-20% sucrose extractable $[^{14}C]$-labeled material that comigrates with tetrameric GLPT on SDS gels. Rate of total protein synthesis is followed by the incorporation of radioactivity in TCA precipitable material.

the periplasmic shock proteins released from both cultures revealed the following: Both preparations contained large amounts of material that in the immunodiffusion technique cross-reacts with anti-GLPT antibodies. However, only the preparation grown at the permissive temperature contained GLPT as judged by 2-dimensional gel electrophoresis. Therefore, the cross-reactivity to GLPT must be caused by a GLPT precursor.

When a culture of BUG-6 that had been growing for 50 min at the nonpermissive temperature is shifted to the permissive temperature under simultaneous addition of chloramphenicol, increase of transport activity is observed (Fig. 10). Since cell division can be initiated under these conditions (Reeve and Clark, 1972) the cell-cycle-dependent processing of GLPT is very likely responsible for the observed alteration in transport activity in normal strains. It also suggests that the primary temperature-sensitive defect in BUG-6 is contained in a proteolytic enzyme.

At the present time it is not clear why BUG-6 releases the precursor of GLPT upon osmotic shock. It also is not clear at what location in the cell envelope the processing occurs. From the shock releasability of the precursor one would think that processing occurs in the periplasm. However, from the cell-division-dependent correlation of this phenomenon, processing on or within the cytoplasmic membrane is a more likely picture. In this respect it is relevant to refer to the shock releasability of the elongation factor Tu, which is pictured to be connected to the inner site of the cytoplasmic membrane (Jacobson and Rosenbusch, 1976).

4. Transport of G3P in Membrane Vesicles, the G3P Carrier

Previous experiments by other workers have shown that G3P can be transported actively in membrane vesicles. This is surprising in view of our finding of the periplasmic nature of GLPT, namely, that it is one of the gene products of the glpT operon. If GLPT is an essential component of the G3P transport system, in analogy to many periplasmic substrate-binding proteins (Boos, 1974), membrane vesicles should be unable to transport G3P. However, this is not the case. Figure 11 shows the transport activity of 3 membrane vesicle preparations: from a fully induced wild-type strain, from a constitutive strain, and from a mutant that does not transport G3P and lacks GLPT. As can be seen the transport activity observed in whole cells is analogous to the transport activity measured in their isolated membrane vesicles. Therefore, either transport in membrane vesicles is mediated by residual GLPT in these membranes, or GLPT is only essential for transport of G3P in whole cells but not in membrane vesicles. Two-dimensional polyacrylamide gel electrophoresis of the proteins contained in these membrane vesicles did not reveal any residual amounts of GLPT (not shown). Therefore, it is clear that GLPT is not essential for the energy dependent active transport of G3P in membrane vesicles, and its only function must reside in overcoming a possible diffusion barrier for G3P in the outer envelope structure of E. coli.

As a consequence it is clear that membrane vesicles must contain a protein, different from GLPT, that is coded for by the glpT operon and that is responsible for the active transport of G3P, the "G3P carrier." The above-mentioned electrophoretic analysis of the proteins contained in membrane vesicles did not show any difference in the protein pattern between a wild-type strain and one that should not contain the G3P carrier. However, the availability of strains that contained glpT carrying plasmids (and thus must produce more of the G3P carrier) together with the use of an extraction procedure specific for proteins of the inner (cytoplasmic) membrane (Schnaitman, 1971), made possible the identification of a protein that is a very likely candidate for the G3P carrier protein. Figure 12 shows

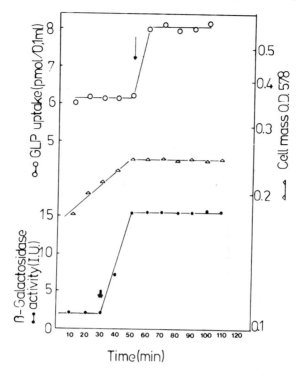

Fig. 10. Effect of chloramphenicol on G3P transport activity of E. coli BUG-6. The culture was grown for 50 min at 42°C. At the time indicated by the arrow the temperature of the culture was shifted to 35°C with simultaneous addition of chloramphenicol (150 μg/ml final concentration). Growth and transport assay as described in the legend to Fig. 9. β-galactosidase activity, induced at the time indicated by the double arrow, was assayed as described by Ullman et al. (1968).

the electrophoretic analysis of proteins contained in the cytoplasmic membrane of a transport-negative strain (it does not contain GLPT) in the presence and absence of the *glpT* carrying plasmid. As can be seen, there are minor quantitative differences in the amounts of several proteins. However, the plasmid-carrying strains contain one spot (x) that is absent in the *glpT* strain without plasmid. Therefore, this protein must be coded for by one of the plasmid genes. The purification of this protein and its test for binding activity toward G3P will be necessary to establish it as the as yet unknown G3P carrier.

5. Some Periplasmic Proteins Are Accessible From the Outside Medium

Since GLPT does not participate in the energy dependent translocation step of G3P in the cytoplasmic membrane it is likely to be connected at least functionally to the outer envelope structure of the Gram negative E. coli cell.

If by some structural arrangement in the outer membrane GLPT would facilitate the diffusion of G3P it is not unlikely that at least some portion of its polypeptide chain should be extended to or be in contact with the external medium. To test this possibility lactoperoxidase mediated iodination with [^{125}I] was performed with whole cells of E. coli. Figure 13 shows the 2-dimensional polyacrylamide gel electrophoresis pattern of the periplasmic shock proteins released after the iodination procedure. After staining, all spots were cut from the gel, dissolved in H_2O_2-ammonia, and counted in the liquid

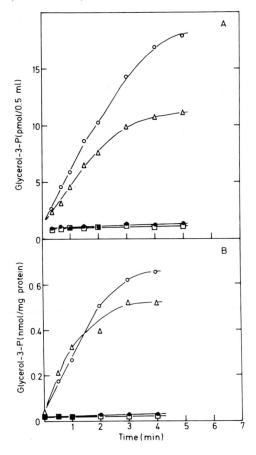

Fig. 11. Transport of G3P in whole cells and membrane vesicles of wild-type and *glpT* mutants of
E. coli. A) Uptake in whole cells. The bacterial cultures were resuspended in 100 mM Tris-HCl, pH 7.0,
to an absorbance of 1.0 at 578 nm. [^{14}C]-G3P was added at an initial concentration of 80 nM. The
data are expressed in terms of amounts G3P taken up per 0.5 ml samples. All operations were done at
room temperature. B) Uptake in membrane vesicles. The uptake mixture contained per filtered
aliquot 70 μg membrane protein; initial [^{14}C]-G3P concentration was 0.8 μM, and D-lactate was added
at a final concentration of 20 mM. ○) Constitutive strain 72; △) inducible strain LA 3400; ●) *glpT*
mutant LA 3404; ▢) *glpT* mutant LA 3404. All strains were grown with glycerol as carbon source.
Taken from Boos et al. (1977) with the permission of authors and publisher.

scintillation counter. The counts that were obtained are indicated on Fig. 13. Surprisingly, many periplasmic proteins were labeled. Although GLPT is among the labeled proteins, its radioactivity is only 15% of the most heavily labeled proteins. The fact that not
all proteins were labeled throughout demonstrates that this method is specific for surface
located proteins and indicates such a position also for GLPT.

DISCUSSION

In this paper we discussed the different aspects of GLPT, a protein that is closely
related to the transport system of G3P in E. coli. All available evidence points to the
conclusion that GLPT is coded for by one of the genes contained in the *glpT* operon.

Mutants that lack GLPT map within the *glpT* region. Suppressible nonsense mutations in *glpT* or Mu insertions in *glpT* affect the amount of GLPT produced. Finally, plasmids that contain the *glpT* region produce large amounts of GLPT in a strain that otherwise lacks this protein. At the present time we do not have a *glpT* mutant carrying a nonsense muta- tion in GLPT that we could identify as such on polyacrylamide gels. Therefore, we cannot yet exclude the possibility that GLPT production might be dependent on a hypo- thetical positive regulator that is coded for by one of the *glpT* genes. If such is the case, the structural gene for GLPT could be located anywhere on the E. coli chromosome.

In comparison to other periplasmic proteins GLPT has some unusual properties. It consists of 4 identical subunits of 40,000 molecular weight. This GLPT complex is stable in 1% SDS at room temperature. Treatment with 8 M urea or with SDS at temperatures above 40°C dissociates the complex. This denaturation is reversible by simple dialysis against Tris or phosphate buffers at pH 7.0.

Until now the relationship of GLPT to G3P transport has been indirect, based on the genetic correlation that has been discussed. Clearly, GLPT is not similar to the numer- ous periplasmic substrate-binding proteins that have been identified as the recognition sites for active transport systems. In contrast to these proteins GLPT does not exhibit binding activity towards its supposed substrate nor are membrane vesicles devoid of the corresponding transport activity. Therefore, the only remaining function for GLPT in G3P transport is to act as a facilitator for the diffusion of G3P through the outer mem- brane of E. coli. The observation that GLPT can be iodinated by lactoperoxidase in intact cells indicates that GLPT could have access to the outer surface of the cell.

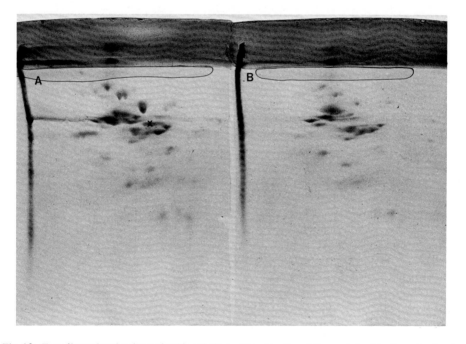

Fig. 12. Two-dimensional polyacrylamide gel electrophoresis of proteins contained in the cytoplasmic membrane of the *glpT* strain G 810/3 (B), and its *glpT*⁺ plasmid carrying derivative (A). Both strains were grown with glycerol as carbon source. The proteins were extracted from the cell envelope by Triton X-100 in the presence of MgCl₂. Two hundred micrograms protein were applied on the gels.

No information as to the in vivo location or structure of GLPT is as yet available. In particular there is no evidence for the pore-like structure which one might propose for a protein that facilitates the diffusion of a substrate through an otherwise hydrophobic membrane. However, one cannot fail to note the highly acidic nature of GLPT. Using isoelectric focusing, a pI of 4.4 was found for GLPT. This is reflected in the fast electro-phoretic mobility of GLPT in gels containing urea. Possibly the abundance of negative charges in GLPT at neutral pH is connected to its function in G3P transport.

As with other periplasmic proteins very little is known concerning either its bio-synthesis or its translocation through the cytoplasmic membrane. The electrophoretic

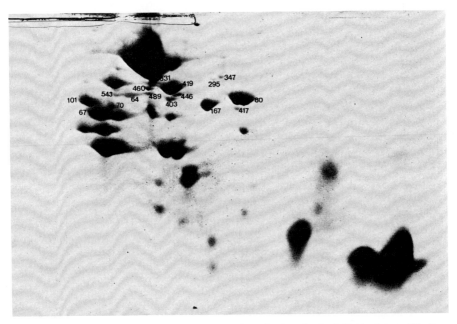

Fig. 13. Two-dimensional polyacrylamide gel electrophoresis of crude shock fluid of strain 72 after lactoperoxidase catalyzed iodination with [^{125}I] of intact cells. The Coomassie Blue stained spots were cut out from the gel, dissolved in H_2O_2-ammonia, and counted. The numbers depicted on the spots represent their counts per minute. Background activity is 15–20 cpm.

analysis of the shock proteins from a mutant that is temperature sensitive for cell division demonstrates, at least for this strain, the existence of a GLPT precursor. This precursor undergoes proteolytic processing before it attains its final form. The picture is even further complicated by the observation that newly formed capacity for G3P transport appears to be correlated to events occurring only at certain times during the cell cycle. If this phenomenon is related to GLPT at all, it must be the cycle-dependent processing of this protein, but not its synthesis, that is causing the effect. At present it is not clear where in the cell envelope this processing takes place or whether it is related to the translocation of GLPT through the cytoplasmic membrane.

Since GLPT is not necessary for active transport in membrane vesicles the *glpT* operon must code for at least one more protein, the G3P carrier. Indeed, the electro-phoretic analysis of proteins contained in the cytoplasmic membrane did reveal a poly-peptide that might be the as yet unknown G3P carrier. This indication is based on the observation that this protein is produced in a *glpT* strain only after the introduction of the *glpT* carrying plasmid. The biochemical characterization and identification of this protein as the G3P carrier will have to await its certainly very challenging purification.

REFERENCES

1. Berger EA, Heppel LA: J Biol Chem 249:7747, 1974.
2. Boos W, Gordon AS: J Biol Chem 246:621, 1971.
3. Boos W: Ann Rev Biochem 43:123, 1974.
4. Boos W, Hartig-Beecken I, Altendorf K: Eur J Biochem 72:571, 1977.
5. Clarke L, Carbon J: Cell 9:91, 1976.
6. Cozzarelli NR Freedberg WB, Lin ECC: J Mol Biol 31:371, 1968.
7. Cutler RG, Evans JE: J Bacteriol 91:469, 1966.
8. Garen A, Levinthal C: Biochim Biophys Acta 38:470, 1960.
9. Hirata H, Altendorf K, Harold FM: J Biol Chem 249:2939, 1974.
10. Jacobson GR, Rosenbusch JB: Nature 261:23, 1976.
11. Johnson WC, Silhavy TJ, Boos W: Appl Microbiol 29:405, 1975.
12. Kaback HR: Biochim Biophys Acta 265:367, 1972.
13. Kellerman O, Szmelcman S: Eur J Biochem 47:139, 1974.
14. Kennedy EP: In Beckwith JR, Lipser D (eds): "The Lactose Operon." New York: Cold Spring Harbor Laboratory, 1970.
15. Lin ECC: Annu Rev Microbiol 30:535, 1976.
16. Miller IH (ed): "Experiments in Molecular Genetics." New York: Cold Spring Harbor Laboratory, 1972.
17. Mueller TH, Morrison M: J Biol Chem 249:7568, 1974.
18. Neu HC, Heppel LA: Biochem Biophys Res Commun 14:109, 1964.
19. Neu HC, Heppel LA: J Biol Chem 240:3685, 1965.
20. Ohki VM: J Mol Biol 68:249, 1972.
21. Ramos S, Schuldiner S, Kaback HR: Proc Natl Acad Sci USA 72:1892, 1976.
22. Reeve JN, Groves DJ, Clark DJ: J Bacteriol 104:1052, 1970.
23. Reeve JN, Clark DJ: J Bacteriol 110:117, 1972.
24. Schnaitman CA: J Bacteriol 108:545, 1971.
25. Shen BHP, Boos W: Proc Natl Acad Sci USA 70:1481, 1973.
26. Silhavy TJ, Hartig-Beecken I, Boos W: J Bacteriol 126:951, 1976a.
27. Silhavy TJ, Casadaban JM, Shuman HA, Beckwith JR: Proc Natl Acad Sci USA 73:3423, 1976b.
28. Szmelcman S, Hofnung M: J Bacteriol 124:112, 1973.
29. Ullmann A, Jacob F, Monod J: J Mol Biol 32:1, 1968.
30. Venkateswasan PS, Wu HCP: J Bacteriol 110:935, 1972.
31. Willis RC, Furlong CE: J Biol Chem 249:6926, 1974.

Journal of Supramolecular Structure 6:169–177 (1977)
Molecular Aspects of Membrane Transport 95–103

Transport in Halobacterium Halobium: Light-Induced Cation-Gradients, Amino Acid Transport Kinetics, and Properties of Transport Carriers

Janos K. Lanyi

Ames Research Center, National Aeronautics and Space Administration, Moffett Field, California 94035

Cell envelope vesicles prepared from H. halobium contain bacteriorhodopsin and upon illumination protons are ejected. Coupled to the proton motive force is the efflux of Na^+. Measurements of ^{22}Na flux, exterior pH change, and membrane potential, $\Delta\psi$ (with the dye 3,3′-dipentyloxadicarbocyanine) indicate that the means of Na^+ transport is sodium/proton exchange. The kinetics of the pH changes and other evidence suggests that the antiport is electrogenic ($H^+/Na^+ > 1$). The resulting large chemical gradient for Na^+ (outside $\gg$ inside), as well as the membrane potential, will drive the transport of 18 amino acids. The 19th, glutamate, is unique in that its accumulation is indifferent to $\Delta\psi$: this amino acid is transported only when a chemical gradient for Na^+ is present. Thus, when more and more NaCl is included in the vesicles glutamate transport proceeds with longer and longer lags. After illumination the gradient of H^+ collapses within 1 min, while the large Na^+ gradient and glutamate transporting activity persists for 10–15 min, indicating that proton motive force is not necessary for transport. A chemical gradient of Na^+, arranged by suspending vesicles loaded with KCl in NaCl, drives glutamate transport in the dark without other sources of energy, with V_{max} and K_m comparable to light-induced transport. These and other lines of evidence suggest that the transport of glutamate is facilitated by symport with Na^+, in an electrically neutral fashion, so that only the chemical component of the Na^+ gradient is a driving force.

The transport of all amino acids but glutamate is bidirectional. Actively driven efflux can be obtained with reversed Na^+ gradients (inside > outside), and passive efflux is considerably enhanced by intravesicle Na^+. These results suggest that the transport carriers are functionally symmetrical. On the other hand, noncompetitive inhibition of transport by cysteine (a specific inhibitor of several of the carriers) is only obtained from the vesicle exterior and only for influx: these results suggest that in some respects the carriers are asymmetrical.

A protein fraction which binds glutamate has been found in cholate-solubilized H. halobium membranes, with an apparent molecular weight of 50,000. When this fraction (but not the others eluted from an Agarose column) is reconstituted with soybean lipids to yield lipoprotein vesicles, facilitated transport activity is regained. Neither binding nor reconstituted transport depend on the presence of Na^+. The kinetics of the transport and of the competitive inhibition by glutamate analogs suggest that the protein fraction responsible is derived from the intact transport system.

Key words: Halobacterium halobium, amino acid transport, sodium-proton exchange, asymmetry of transport system, reconstitution of glutamate transport

Received March 14, 1977; accepted April 26, 1977

INTRODUCTION

In intact bacteria the active transport of various amino acids and sugars is energized either by terminal oxidation or by hydrolysis of ATP (1, 2). When the source of energy is terminal oxidation the means of energy coupling has been shown to be the H^+ gradient generated. In such transport it is assumed that the translocation of the substrates is coupled to the movement of H^+ down its electrochemical gradient — symport (3). Transport of this type is obtained also in cell envelope vesicles (4, 5), which are devoid of soluble components. The transport of those substrates which depend on ATP hydrolysis but not proton motive force is more complex. In such transport soluble factors have also been implicated, the "shockable proteins" (2), many of which have been isolated and show binding of the substrates of transport.

The energetics of transport in the first category has occupied many students of transport, and the relationship of various cation gradients to substrate accumulation is still under intense scrutiny. Another, more recent approach to the study of transport is the isolation and characterization of membrane proteins which accomplish the translocation of substrates across the membranes. Such studies (6—10) will be undoubtedly followed in the future by attempts to describe the molecular details of the mechanisms responsible for the movement of the substrate and for the coupling between the substrate and cation fluxes.

With the discovery of bacteriorhodopsin (11, 12) Halobacterium halobium became a prime candidate for transport studies, since the function of this pigment is to generate a gradient for H^+ across the cell membrane upon illumination (12, 13). Membranes of this organism are thus easily energized by illumination and deenergized by witholding illumination. Cell envelope vesicles from H. halobium, used for these studies, have inside-in membrane orientation (13—15). These preparations are inert unless ion gradients are established, either by preloading with the appropriate salts, or by illumination or substrate oxidation. Since the envelopes contain both the purple membrane and the rest of the cytoplasmic membrane, the components which couple proton motive force to energy-utilizing processes are present. The secondary energized events which follow H^+ extrusion include the transport of Na^+, K^+, and amino acids. The relationship of these to one another and to the gradient of H^+ has been examined in detail.

Although the nature and mechanism of the membrane components which catalyze the translocation of the amino acids remain obscure, the existence of a protein fraction which binds glutamate and which can be reconstituted with soybean lipids to yield specific transport for this amino acid has been demonstrated.

LIGHT-INDUCED GRADIENTS FOR H^+ AND Na^+

In H. halobium envelope vesicles light-induced proton release can be conveniently followed by pH measurements in the extravesicle medium (14, 16—18). The existence of a gradient of protons is inferred from such measurements on the basis of the fact that the number of protons which appear in the medium is many times the number of pigment molecules (18). In addition, heavily buffered vesicles release 2—3 times more H^+ during illumination (18), suggesting that H^+ release is normally accompanied by a pH rise inside the vesicles. Within a minute after the beginning of the illumination an equilibrium is reached, where the H^+ extrusion is equal to H^+ influx, due to various dissipative mechanisms. The number of protons translocated at equilibrium is greatly dependent on pH; optimal H^+ release is near pH 3.4—4, and is greatly diminished at and above pH 6

(19). The pH dependence may be caused by direct pH effects on bacteriorhodopsin, by mass-action effect of H^+ concentration inside the vesicles, or by possible lack of counterion permeability at higher pH, which would allow the development of higher electrical potentials but lower pH gradients. The last does not seem to be the case since the electrical potentials are also less at higher pH (17), and passive cation permeability rises rather than falls with pH (20). Drachev et al. (21) reported that in a planar membrane containing bacteriorhodopsin the photopotentials were diminished when H^+ translocation was against a pH gradient.

The electrical potentials which arise during illumination have been measured using the fluorescent dye 3,3'-dipentyloxadicarbocyanine (di-0-C_5) (16–18). As described earlier by Sims et al. (22), membrane potential (negative inside) will cause this cationic dye to accumulate in the vesicle interior resulting in aggregation and fluorescence quenching. Diffusion potentials for K^+ in the presence of valinomycin were used to calibrate the fluorescence changes, and yielded a slope of -0.33 percent/mV up to about -110 mV (17).

The pH changes and electrical potential during illumination depend greatly on the presence of Na^+ inside the envelope vesicles, while the composition of the exterior solution has much less influence. When the vesicles contain only KCl the pH difference which develops across the membrane amounts to about 2 pH-units and the potential, $\Delta\psi$, is low, about -30 mV (17). When the vesicles contain only NaCl, ΔpH is smaller and $\Delta\psi$ is increased to as much as -100 mV (17). Vesicles containing both NaCl and KCl show complex time-dependent changes during illumination for both ΔpH and $\Delta\psi$ (18, 23). Dependent on the amount of Na^+ inside the vesicles, there is an initial period of smaller pH change and larger potential, similar to the results obtained with NaCl-loaded vesicles. After this initial period ΔpH increases and $\Delta\psi$ decreases until the gradients resemble those obtained with KCl-loaded vesicles. The lengths of the initial periods are roughly proportional to the concentration of NaCl in the vesicles.

The significance of the above observations became clear once it was discovered that during illumination Na^+ is extruded against its electrochemical potential (16, 18). Light-induced Na^+ efflux is thus accompanied by decreased ΔpH and increased $\Delta\psi$, until ^{22}Na is depleted from the vesicles. Such a result is consistent with H^+/Na^+ exchange, or in Mitchell's terminology, antiport (3). Protons returned by the antiporter are reejected by bacteriorhodopsin, and the net result is the removal of Na^+ (and positive charges), leading to the increased membrane potential observed.

If the stoichiometry of H^+ to Na^+ exchanged were 1:1, there would be no net transport of charges and the antiport would be driven only by ΔpH. However, when the light-induced pH changes in NaCl-containing vesicles are measured at a higher pH, at pH 7–7.5, the initial H^+ extrusion is seen to be followed by H^+ influx, leading to a transient reversal of ΔpH (18). This reversal suggests that the H^+/Na^+ antiport cannot proceed with a stoichiometry of 1:1, because such a process would be abolished as ΔpH approaches zero. On the other hand, at higher H^+/Na^+ stoichiometries the exchange would be electrogenic and would be driven also by the electrical potential. Potential in this model would serve to balance the system at a reversed pH gradient. The electrogenicity of the H^+/Na^+ exchange is suggested also by results from a more direct approach (MacDonald and Lanyi, unpublished experiments), where Na^+-gradient (outside $>$ inside) dependent [^{3}H] triphenylmethyl phosphonium ion uptake was observed in KCl-loaded, nonenergized vesicles.

The characteristic pH changes during illumination, which reveal the operation of H^+/Na^+ antiport, are not seen during a second illumination unless an interval of several hours is allowed between the 2 illuminations (18). The slowness of the return of the effect with the incubation in the dark is consistent with the depletion of the vesicles of Na^+ during the first illumination and the slowness of the exchange of K^+ and Na^+ in the dark (20).

The efflux of Na^+ could not proceed extensively unless the inward movement of another ion decreased the magnitude of the electrical potential which develops. Earlier results by Kanner and Racker (19) indicate that the membranes are somewhat permeable to K^+, and illumination will drive the uptake of K^+ even in the absence of valinomycin. It seems reasonable that the influx of K^+ would compensate for the efflux of Na^+ in these vesicles. The scheme of ionic fluxes in the model proposed above is given in Fig. 1.

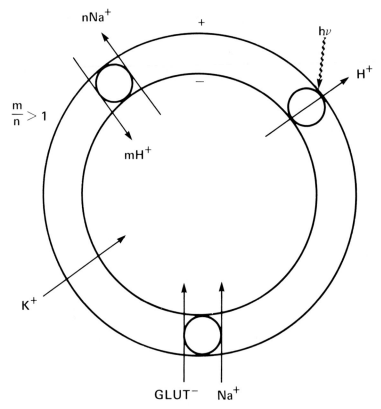

Fig. 1. Scheme of energy transduction in H. halobium cell envelope vesicles. As explained in the text, 3 active components are proposed: bacteriorhodopsin, activated by light, the proton-sodium antiporter, and the glutamate carrier, which functions as a symporter for glutamate and Na^+. Unlike the transport of glutamate, the transport of other amino acids responds to membrane potential, in addition to a chemical gradient for Na^+, implying that those translocations involve net transport of charges. The accumulation of K^+ may be facilitated by a specific ionophore, but such has not been found yet.

AMINO ACID TRANSPORT KINETICS

According to the principle of chemiosmotic energy coupling the ion gradients, which arise as a result of metabolic activity, will drive the accumulation of transported substrates (3). The relationship of the cation gradients in H. halobium cell envelopes to substrate transport is therefore of interest. Findings which identify such gradients as driving forces for transport should be considered indirect evidence for the existence of specific membrane components which couple ion and substrate movements to one another.

Cell envelopes from H. halobium actively transport 19 amino acids when illuminated (14–18, 24–26). The energetics of such transport have been described in some detail, and the results suggest that the driving force for most amino acids includes both chemical (ΔNa^+) and electrical ($\Delta\psi$) components of the Na^+ gradient which arises during illumination. Glutamate transport, which is somewhat different in this respect from the others, has proved particularly interesting. The transport of this amino acid can be driven either by illumination (16, 24) or by substrate oxidation (27). The results in both cases are similar: the transport kinetics are dependent on the concentration of Na^+ inside the vesicles. Analogously to the light-induced pH changes, described in the previous section, and on a similar time scale, glutamate transport during illumination shows lags whose duration is determined by the intravesicle Na^+ concentration. Lags are also observed when the light intensity is lowered. On the other hand, when the intravesicle concentration of Na^+ is set at a very low value and the external medium contains Na^+, transport of glutamate occurs even in the dark. At a very high initial Na^+-concentration difference (outside/inside $\geqslant 500$) the rates of transport (and the K_m as well, $\sim 1.3 \times 10^{-7}$M) were very similar to transport induced by illumination. These results strongly suggest that the sodium gradient developed during illumination is a driving force for glutamate transport, and imply that the mechanism involves glutamate-Na^+ symport.

Lack of glutamate transport during the illumination of the vesicles, until a large chemical gradient for Na^+ is developed, argues that ΔNa^+ is an obligatory ingredient of the driving force, particularly since under these conditions large electrical potentials are observed (18). The fact that transport of glutamate can be induced with an artificially set Na^+ gradient argues that the Na^+ gradient alone is sufficient as a driving force. The latter question was approached also by measuring transport activity following illumination, after various lengths of incubation in the dark. Both components of the proton motive force (ΔpH and $\Delta\psi$) decay within 1 min under these conditions (17, 18), but ΔNa^+ and glutamate transport persists for 10–15 min. It may be concluded that a chemical gradient for Na^+ alone will drive glutamate transport.

Agents which discharge one or both components of the proton motive force inhibit light-induced glutamate transport but have much less effect on transport driven in the dark by a chemical Na^+ gradient in KCl-loaded vesicles. It has been concluded from such results that proton motive force is also a driving force for glutamate uptake, but only insofar as it produces a chemical gradient for Na^+ (25). Thus, as indicated in Fig. 1, energy conservation in this system involves first the photochemical reactions of bacteriorhodopsin, second a gradient for protons, produced by the light-induced H^+ translocation, third a gradient for Na^+, generated by proton/sodium antiport, and fourth, gradients for amino acids, such as glutamate, achieved by symport with Na^+.

As noted above, all the amino acids can be transported in H. halobium with a Na^+ gradient. This is probably a consequence of the fact that the organisms have adapted to living at high concentrations of NaCl (28–30). There exist Na^+-dependent transport systems in other microorganisms, however, and recent results indicate that some of these, e.g., proline transport in Mycobacterium phlei and glutamate transport in Escherichia coli B, are driven by a sodium gradient generated by respiratory activity (MacDonald and Lanyi, manuscripts in preparation). Thus, it is likely that parts of the scheme presented in Fig. 1 will apply to nonhalophilic bacteria as well.

SYMMETRY OF AMINO ACID TRANSPORT CARRIERS

Many different lines of evidence point to the existence of proteins which are responsible for substrate translocation. One of the questions relating to the molecular details of such transport carriers, which can be studied with a kinetic approach, is the symmetry of the protein involved with respect to transport in the 2 directions. In H. halobium envelope vesicles all the amino acids but glutamate show efflux when illumination is stopped or when excess unlabeled amino acid is added (14, 25, MacDonald, unpublished experiments). Such efflux, which has been studied also in vesicles preloaded with radioactive amino acids, has a large Na^+-dependent component, suggesting that the translocation involves the carrier in this direction also. Half-saturating concentration for Na^+ is about 150 mM for exchange diffusion of a number of amino acids in either direction (MacDonald, unpublished experiments). Moreover, prearranged Na^+ gradients in the reverse direction (inside > outside) accelerate amino acid efflux, indicating that both active and passive transport are bidirectional. These results are consistent with either symmetrical carriers for the amino acids or carriers with mixed orientation in the membranes. Investigations of the noncompetitive inhibition of transport by cysteine have resolved this question. Cysteine, which seems to inactivate some of the transport carriers in H. halobium (but not via a redox reaction), acts in an asymmetrical manner: both passive and active influx of methionine are inhibited when cysteine is added on the outside of the vesicles, but remain unaffected when the inhibitor is added to the inside only (Helgerson and Lanyi, manuscript in preparation). On the other hand, efflux of methionine, passive or actively driven, is not inhibited by cysteine when present on either (or both) side of the membrane. Thus, the transport carrier for methionine a) must be uniformly oriented in the envelope vesicles and b) appears as symmetrical in transport functions but asymmetrical in the cysteine effect.

The glutamate transport system is clearly different from the others in that the efflux of glutamate does not occur upon adding unlabeled glutamate to either energized or nonenergized envelopes (24). Since glutamate is chemically unchanged during transport, the irreversibility observed must reflect the properties of the transport carrier. Although glutamate would not exit passively, it can be driven from preloaded vesicles by a reverse Na^+ gradient (Na^+ inside, K^+ outside) (24). The functional asymmetry of this carrier may therefore reflect specific requirements for different cations on either side of the membrane. At this time it is difficult to visualize a molecular mechanism which would generate such properties.

RECONSTITUTION OF TRANSPORT

Ultimately, the questions relating to the functioning of transport carriers will be solved only after the proteins involved are isolated in a functional form. Since these proteins are membrane-bound and their assay requires the measurement of substrate fluxes across membranes, attempts to isolate them have succeeded only recently, after techniques to solubilize and reconstitute membrane proteins became available (31–33).

When H. halobium cell membranes are disrupted with cholate, the soluble fraction binds [³H] glutamate. Figure 2 shows a Scatchard plot of the binding data, indicating a single kind of binding site, with the dissociation constant $K_{diss} = 6 \times 10^{-8}$ M (Lanyi, Yearwood-Drayton, and MacDonald, manuscript in preparation). The proteins(s) involved in the binding elute from an Agarose column, equilibrated with detergent-buffer, at an approximate molecular weight of 50,000. This fraction, but not the others eluted, can be reconstituted with soybean lipids to yield lipoprotein vesicles permeable to glutamate. Reconstituted transport is shown in Fig. 3. No uptake of aspartate is observed, as expected, since in H. halobium this amino acid does not share its carrier with glutamate (25). The transport of glutamate in the liposomes appears to be facilitated equilibration

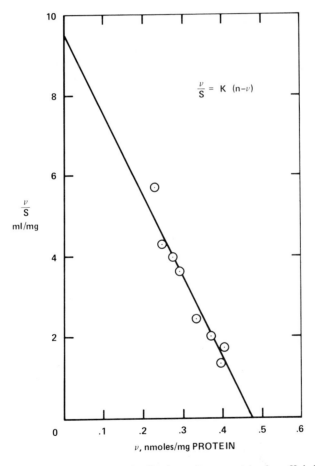

Fig. 2. Glutamate binding by cholate-solubilized membrane proteins from H. halobium. The binding data (Lanyi, Yearwood-Drayton, and MacDonald, manuscript in preparation) are plotted according to the modified Scatchard equation shown: v) bound [³H] glutamate, nmol/mg protein; S) total glutamate concentration (essentially equal to free glutamate), nmol/ml; and n) specific glutamate binding activity, nmol/mg protein. From the intercept on the ordinate a dissociation constant of 6×10^{-8} is calculated.

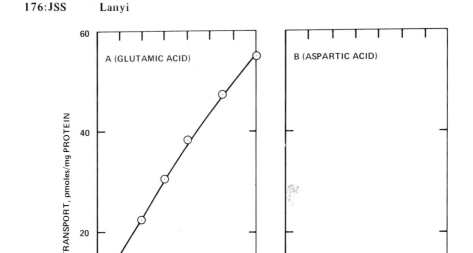

Fig. 3. Transport of glutamate and aspartate in liposomes reconstituted with a membrane protein fraction from H. halobium. Reconstitution by the cholate dialysis method (31). At zero time the lipoprotein vesicles were added to buffer containing [^{3}H]-labeled amino acid and at the indicated times samples were withdrawn. Retained radioactivity was determined by rapid elution of the liposomes from small gel columns (Lanyi, Yearwood-Drayton, and MacDonald, manuscript in preparation). ⊙) Complete system; △) protein omitted. No transport was seen when the lipids were omitted.

rather than active accumulation. Neither binding nor reconstituted transport requires Na$^+$.

Specificity of reconstituted transport is often ascertained by the use of inhibitors (6, 8–10). In the glutamate transport system of H. halobium only structural analogs of glutamic acid are available as inhibitors. Results with these show that the kinetic constants for competitive inhibition by kainic acid, α-methylglutamic acid, and N-methylglutamic acid form a consistent pattern in 3 systems: transport in intact envelopes, binding in cholate-buffer, and transport in the reconstituted liposomes. It was concluded from these studies (Lanyi, Yearwood-Drayton, and MacDonald, manuscript in preparation) that the recognition site for glutamate is the same in the 3 measurements. The protein fraction isolated thus appears to be derived from the intact transport system, but has lost the ability to couple the translocation of glutamate with that of Na$^+$. Binding and transport of glutamate are therefore not obligatorily linked to the coupling function. The protein fraction is presently about 12- to 15-fold purified. Future studies with further purified material may reveal more properties of the protein(s) involved and shed some light on the mechanism of the substrate translocation.

REFERENCES

1. Harold FM: Bacteriol Rev 36:172, 1972.
2. Berger EA, Heppel LA: J Biol Chem 249:7747, 1974.
3. Mitchell P: Symp Soc Gen Microbiol 20:121, 1970.
4. Kaback HR: Biochim Biophys Acta 265:367, 1972.
5. Kaback HR: Science 186:882, 1974.
6. Kasahara M, Hinkle PC: Proc Natl Acad Sci USA 73:396, 1976.
7. Hirata H, Sone N, Yoshida M, Kagawa Y: Biochem Biophys Res Commun 69:665, 1976.
8. Crane RK, Malathi P, Preiser H: FEBS Lett 67:214, 1976.
9. Zala CA, Kahlenberg A: Biochem Biophys Res Commun 72:866, 1976.
10. Amanuma H, Motojima K, Yamaguchi A, Anraku J: Biochem Biophys Res Commun 74:366, 1977.
11. Oesterhelt D, Stoeckenius W: Nature (London) New Biol 233:149, 1971.
12. Oesterhelt D, Stoeckenius W: Proc Natl Acad Sci USA 70:2853, 1973.
13. Lozier RH, Niederberger W, Bogomolni RA, Hwang SB, Stoeckenius W: Biochim Biophys Acta 440:545, 1976.
14. MacDonald RE, Lanyi JK: Biochemistry 14:2882, 1975.
15. Lanyi JK, MacDonald RE: In Fleischer S, Packer L (Eds): "Methods in Enzymology." New York: (In press).
16. Lanyi JK, Renthal R, MacDonald RE: Biochemistry 15:1603, 1976.
17. Renthal R, Lanyi JK: Biochemistry 15:2136, 1976.
18. Lanyi JK, MacDonald RE: Biochemistry 15:4608, 1976.
19. Kanner BI, Racker E: Biochem Biophys Res Commun 64:1054, 1975.
20. Lanyi JK, Hilliker K: Biochim Biophys Acta 448:181, 1976.
21. Drachev LA, Frolov VN, Kaulen AD, Liberman EZ, Ostroumov SA, Plakunova VG, Semenov AYu, Skulachev VP: J Biol Chem 251:7059, 1976.
22. Sims PJ, Waggoner AS, Wang C-H, Hoffman JF: Biochemistry 13:3315, 1974.
23. Lanyi JK, MacDonald RE: Fed Proc Fed Am Soc Exp Biol 36:1824, 1977.
24. Lanyi JK, Yearwood-Drayton V, MacDonald RE: Biochemistry 15:1595, 1976.
25. MacDonald RE, Lanyi JK: Fed Proc Fed Am Soc Exp Biol 36:1828, 1977.
26. Lanyi JK: In Capaldi RA (Ed): "Membrane Proteins in Energy Transduction." New York: Marcel Dekker (In press).
27. Belliveau JW, Lanyi JK: Arch Biochem Biophys 178:308, 1977.
28. Larsen H: In Gunsalus IC, Stanier RY (Eds): "The Bacteria." New York: Academic Press, 1963, vol 4, pp 297–342.
29. Larsen H: Adv Microb Physiol 1:97, 1967.
30. Lanyi JK: Bacteriol Rev 38:272, 1974.
31. Kagawa Y, Racker E: J Biol Chem 246:5477, 1971.
32. Racker E: J Biol Chem 247:8198, 1972.
33. Racker E: FEBS (Fed Eur Biochem Soc) Proc Meet 10:25, 1975.

Journal of Supramolecular Structure 6:179—189 (1977)
Molecular Aspects of Membrane Transport 105—115

Role of the Membrane Potential in Serum-Stimulated Uptake of Amino Acid in a Diploid Human Fibroblast

Mitchel L. Villereal and John S. Cook

The University of Tennessee-Oak Ridge Graduate School of Biomedical Sciences, and Cancer and Toxicology Program, Biology Division, Oak Ridge National Laboratory, Oak Ridge, Tennessee 37830

The Na^+-dependent accumulation of α-aminoisobutyric acid (AIB), measured in normal growing and quiescent (serum-deprived) HSWP cells (human diploid fibroblast), was found to be twofold higher (AIB_{in}/AIB_{out} = 20—25) under the normal growing conditions. Serum stimulation of quiescent cells increases their AIB concentrating capacity by approximately 70% within 1 hr. These observations suggest that the driving forces for AIB accumulation may be reversibly influenced by the serum concentration of the growth medium. Addition of valinomycin (Val) to cells preequilibrated with AIB causes an enhanced accumulation of AIB, suggesting that the membrane potential can serve as a driving force for AIB accumulation. After pre-equilibration with AIB in 6 mM K^+, transfer to 94 mM K^+ with Val results in a marked and rapid net loss of AIB. The effect of Val on the accumulation of AIB is greatest in quiescent cells, with the intracellular AIB concentrations reaching those seen both in Val-stimulated normal cells and in Val-stimulated serum-stimulated cells. By adjusting $[K^+]_0$, in the presence of Val, the membrane potential of growing cells can be matched to that of quiescent cells or vice versa. When this is done, the two accumulate AIB to the same extent. Hence the AIB accumulating capacity is characteristic of the membrane potential rather than of the growth state. In summary, these data suggest that the accumulation of AIB in HSWP cells is influenced by changes in membrane potential and that a serum-associated membrane hyperpolarization could be responsible for the increased capacity for AIB accumulation in serum-stimulated cells.

Key words: valinomycin, human fibroblast, amino acid transport, serum stimulation, membrane potential

In recent years, many investigators have sought to determine whether alterations in membrane permeability can be correlated with changes in cell growth state. Several studies have demonstrated that a decrease in the transport of several low-molecular-weight nutrients (inorganic phosphate, uridine, amino acids) occurs when cells go from a state of growth to one of quiescence. These observations have led to speculation that alterations in membrane permeability to metabolites may, by controlling their intracellular concentrations, play a role in regulating cell growth and transformation (1, 2).

The transport of amino acids, in particular, has been extensively studied as a possible growth regulatory mechanism. The well-documented dependence of cell growth on amino

Received March 14, 1977; accepted March 18, 1977

acid supply (3), as well as the small margin of safety observed in some cells between maximal transport rate of and metabolic demands for amino acids (4), suggest that this pathway is well suited for such a regulatory function. Investigations of amino acid transport indicate that the rate of amino acid uptake in logarithmically growing, nontransformed cells is substantially higher than in either confluent [3T3 cells (1)], hyperconfluent [chick embryo cells (5)] or serum-deprived, quiescent cells [chick embryo cells (5), human diploid cells (6)]. The enhanced amino acid uptake seen in growing cells suggests the presence of an increased number of transport sites in the cell membrane. However, in some cases, changes in transport are too rapid to be due to the synthesis and incorporation of new transport sites into the membrane. Also, observations indicate that the level of amino acid accumulation is elevated in growing cells (1, 5). Although an increase in the number of transporters could increase the rate of transport, it would not increase the steady-state level of amino acid concentrating capacity of the cell. It therefore appears that when quiescent cells enter a growth state there must be either an increase in the driving force for amino acid uptake or a more efficient coupling to existing forces.

The present study attempts to identify the source of energy for the elevated capacity for accumulation of amino acids in growing cells. Amino acid transport is studied in a skin-derived, human diploid fibroblast strain (HSWP), using the nonmetabolizable amino acid analog α-aminoisobutyric acid (AIB) as a model substrate for the Na^+-dependent amino acid transport system. We show that AIB accumulation is greater in growing cells than in serum-deprived, quiescent cells and that the effect is partially reversed within 1 hr after serum-stimulation of quiescent cells. The possible role of the membrane potential in driving the enhanced amino acid accumulation in growing cells is tested by using valinomycin to alter the membrane potential. Evidence is presented which suggests that the enhanced level of AIB accumulation in growing cells is the result of a growth-related membrane hyperpolarization.

MATERIALS AND METHODS

Cells and Growth

HSWP cells, human diploid fibroblasts, were derived from human foreskin by J. D. Regan (ORNL). They were cultured in Eagle's minimum essential media (KC Biological Inc.) containing 10% fetal calf serum (KC Biological Inc.), and 25 μg/ml gentamicin (Schering Corp.). Stock cultures were maintained at confluence in the same medium with 1% fetal calf serum. Cells were grown at $37°C$ in a 95% air-5% CO_2 atmosphere, and were used between the 10th and 25th passage. Cells were removed from stock flasks by trypsinization and were seeded onto individual coverslips (11 $\times$ 25 mm) in petri dishes as originally described by Foster and Pardee (1) and modified by Salter and Cook (7). Twenty-four hours after plating, the cells were fed with growth media containing either 10% fetal calf serum (growing cells) or 0.1% fetal calf serum (serum-deprived cells). Transport measurements were performed on growing cells at subconfluent densities and on serum-deprived cells following 2–4 days on 0.1% fetal calf serum. Serum-deprived cells are considered to be quiescent when [^{3}H] thymidine incorporation into the acid insoluble fraction drops below 10% of growing controls. Approximately 18–22 hr after serum stimulation (20% fetal calf serum) of serum-deprived cells, there is a burst of incorporation of [^{3}H] thymidine into acid-soluble material.

Transport Studies

Amino acid transport was measured using α-aminoisobutyric acid (AIB), a non-metabolizable substrate of the Na^+-dependent, amino acid concentrating, transport system [A system (8)]. The assay medium consisted of amino acid-free Eagle's minimum essential medium (EMEM) with Hanks' salts, 20 mM HEPES (N-2-hydroxyethylpiperazine-N'-2-ethanesulfonic acid), 0.1 mM AIB, and ~ 1 μCi/ml ^{3}H-AIB (ICN). In some experiments the Na^+ and K^+ concentrations of the assay media were altered by equimolar substitution of either potassium chloride or choline chloride for sodium chloride in the Hanks' salts. In all media [NaCl + KCl + ChCl] = 144 mM. Valinomycin (Val; Sigma Chemical Co.) was added to some assay media as an ethanol solution with the appropriate amount of ethanol also being added to control media ($<$ 0.25% final concentration).

Four-milliliter aliquots of assay medium were placed in 15 $\times$ 45-mm shell vials (Kimble-Products) and equilibrated at $37°$C. Coverslips were removed from their growth media, drained by touching to an absorbent paper towel and placed in shell vials for the required times. To terminate AIB uptake, the coverslips were removed from the assay media, and rapidly washed 3 times in cold Tris-buffered saline (pH 7.4). They were then drained and placed in a glass scintillation vial containing 1 ml of 0.1 N NaOH to lyse the cells. Following lysis, 0.1 ml of 1 N HCl and 10 ml of Triton X-100/toluene counting solution, containing 5.5 g/liter Permablend I (Packard), were added. Radioactivity was determined in a Nuclear-Chicago Mark II Spectrometer.

Protein Measurement

Protein was measured by the intrinsic fluorescence of tryptophan residues (9). Protein was solubilized by placing each coverslip in a shell vial containing 4 ml of 0.2% sodium dodecyl sulfate (SDS). The emission of the SDS extract was measured at 338 nm, using an excitation wavelength of 286 nm (Perkin Elmer model 204 fluorescence spectrophotometer). Bovine serum albumin dissolved in 0.2% SDS was utilized as a standard.

Water and Electrolyte Measurements

Cell water content was measured as the distribution volume of ^{3}H-3-O-methyl glucose, which is passively distributed in HSWP cells (10). The distribution of L-glucose, which is not transported in these cells, was used to measure extracellular medium not removed by washing. The water-to-protein ratio (μg H_2O/μg protein) was 4.4 $\pm$ 0.10 ($\eta = 5$) for growing cells in the 6 mM K^+, 138 mM Na^+ assay medium. This ratio was not appreciably altered by any of the experimental manipulations (i.e., serum deprivation, valinomycin addition, decrease in $[Na^+]_0$, increase in $[K^+]_0$).

Na^+ and K^+ concentrations were determined flame photometrically using Li^+ as an internal standard.

RESULTS

HSWP cells cultured in 10% fetal calf serum grow to confluent densities of 120 $\pm$ 15 μg protein per coverslip within 6–8 days after plating. In contrast, cells which are deprived of serum 24 hours after plating, become quiescent within 3–5 days. The serum-deprived cells grow to densities of 30 $\pm$ 5 μg protein per coverslip and can achieve confluent densities only upon restoration of serum to the growth medium.

AIB uptake was measured in both growing and serum-deprived cells 3–5 days subsequent to plating on coverslips, which corresponds to a time of logarithmic growth for the serum-sufficient cells and of quiescence for the serum-deprived cells. Although AIB uptake in growing cells is relatively slow, reaching steady state in about 60 min, the AIB concentration ratios (AIB_i/AIB_0) which can be maintained are 20–25 (Fig. 1). The AIB accumulation in quiescent, serum-deprived cells is considerably lower than that in growing cells, with AIB ratios of only 8–10. However, the AIB concentrating capacity of the quiescent cells can be at least partially restored by serum stimulation (20% fetal calf serum for 1 hour) prior to measurement of AIB uptake. Serum-stimulated cells can accumulate AIB to concentration ratios of 12–15.

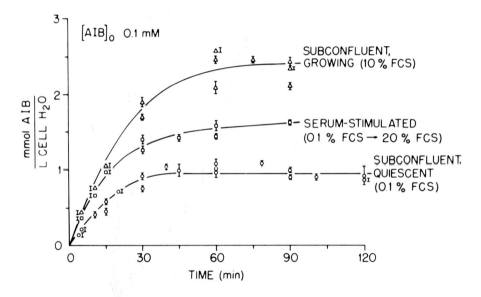

Fig. 1. Effect of serum deprivation and serum stimulation on AIB uptake in subconfluent HSWP cells. For assay, cells were removed from the indicated growth medium and placed in a serum-free, amino acid-free, Eagle's minimum essential medium (6 K^+, 138 Na^+) containing 0.1 mM AIB. AIB uptake is plotted vs time of accumulation for data compiled from 4 experiments. Mean $\pm$ SE of at least 3 determinations at each time point is shown.

The data in Fig. 2 demonstrate that the uptake of AIB in growing HSWP cells is Na^+-dependent. When extracellular sodium concentrations are decreased by equimolar replacement with choline, the accumulation of AIB is also decreased. For example, when the environmental sodium concentration is lowered from 138 mM to 10 mM, the accumulation ratio drops from 20 to 4.

The capacity for concentrating amino acids in growing cells is dependent on the membrane potential as well as on the Na^+ concentration gradient. If cells are equilibrated with 0.1 mM AIB for 75 min, in a control medium of 6 mM K^+ and 50 mM Na^+, and then transferred to a 50 mM Na^+, 0.1 mM AIB environment containing the K^+-ionophore valinomycin (Val), the final level of AIB accumulation is markedly influenced by the external concentration of K^+. In the absence of Val, growing cells incubated in this control environment can achieve an AIB accumulation ratio of 8 (Fig. 3). Membrane hyperpolarization by the addition of Val to cells in this medium enables an accumulation ratio of at least 10

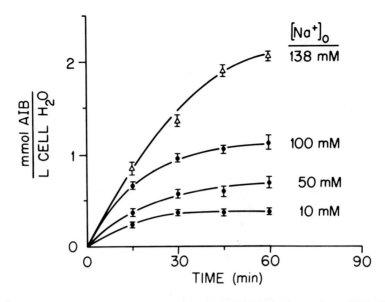

Fig. 2. Na⁺ sensitivity of AIB uptake in subconfluent, growing HSWP cells. Cells were removed from their normal growth medium and placed in a serum-free, amino acid-free medium containing 0.1 mM AIB, 6 mM K⁺, and Na⁺ + Ch⁺ = 138 mM. AIB uptake vs time of accumulation is plotted for a representative experiment. Mean ± SE of at least 3 determinations is shown.

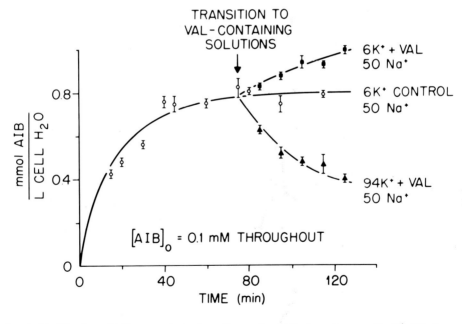

Fig. 3. Modification of AIB accumulation by valinomycin and its dependence on the K⁺ diffusion gradient. Cells were removed from their normal growth medium and incubated in a serum-free, amino acid-free, 6 mM K⁺, 50 mM Na⁺ medium for 75 min. They were then transferred to either i) the same medium, ii) a 6 K⁺, 50 mM Na⁺ medium containing Val (10 μg/ml), or iii) a 94 mM K⁺, 50 mM Na⁺ medium containing Val. The intracellular AIB concentration is plotted vs time of accumulation for data from 3 experiments. Mean ± SE of 9 determinations is shown.

to be achieved. However, if cells preequilibrated in the control environment are transferred to a Val-containing medium with 94 mM K⁺, causing a membrane depolarization, there is a marked and rapid net loss of intracellular AIB.

The experiments depicted in Fig. 4 were designed to compare the effects of membrane hyperpolarization on the AIB-accumulating capacity of cells in 3 different growth states. Either growing, quiescent, or serum-stimulated cells were incubated in a medium containing 0.1 mM AIB, 138 mM Na⁺, and 6 mM K⁺ for 90 min, at which time they were transferred to an identical medium containing Val (10 μg/ml). The presence of Val enabled growing cells to increase their accumulation ratio of AIB from 25 to 34 (Fig. 4). The effect of Val stimulation is more dramatic in quiescent cells, causing a three- to four-fold increase in the previously depressed intracellular AIB concentration, to an AIB accumulation ratio of 34. This corresponds to the same concentrating capacity observed in Val-stimulated, growing cells. It is important to emphasize that the effects of serum stimulation and Val stimulation on AIB accumulation in quiescent cells are not additive. The combined treatment with Val and serum drives the internal AIB concentration to the same value observed in quiescent cells treated with Val alone. Thus, Val stimulation overrides the differences in AIB accumulation associated with the growing, quiescent, and serum-stimulated growth states.

Since a membrane hyperpolarization was observed to negate the growth associated

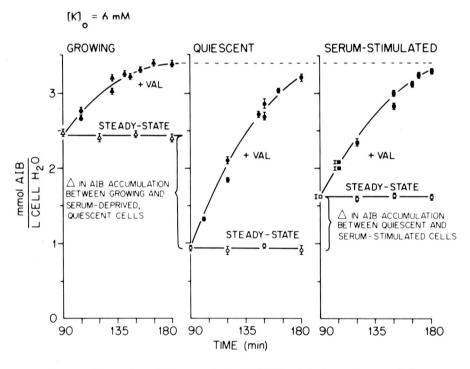

Fig. 4. Effect of valinomycin on AIB accumulation by HSWP cells in 3 growth states. Cells were removed from their growth medium and incubated in a serum-free, amino acid-free EMEM (6 K⁺, 138 Na⁺) containing 0.1 mM AIB for 90 min. Valinomycin (10 μg/ml) was added to half the population of cells at 90 min, while the other half served as controls. AIB uptake is plotted vs time of accumulation for data compiled from 3 experiments. Mean ± SE of at least 3 determinations is shown.

differences in AIB accumulation, the next series of experiments investigated whether the increased concentrating capacity in growing cells might indeed be due to a higher membrane potential. In this series of experiments, the Na^+ concentration of the assay medium was maintained at 50 mM to allow flexibility in manipulating K^+ concentration. Membrane potentials were estimated by varying the external K^+ concentration in the presence of Val, and determining at what value of $[K^+]_0$ the AIB accumulation equaled the accumulation in the control medium (6 mM K^+, no Val). This method is based on the "null point" technique used with the fluorescent dyes (11). After AIB accumulation has reached steady state in the controls (6 mM K^+, no Val), $[K^+]_0$ is adjusted to a series of values in the presence of Val and the change in AIB is determined for each value of $[K^+]_0$. The value of $[K^+]_0$ in the presence of Val at which there is no change in AIB concentration ratio is determined and the membrane potential calculated from the equation for a K^+ electrode

$$(E_m = \frac{RT}{F} \ln \frac{[K^+]_0}{[K^+]_i}).$$

Since cells lose about 10% of their internal K^+ content after 60 min in the presence of Val, the potential was calculated based on the value of $[K]_i$ measured at the end of the experiment. This potential is taken to be the same as the preexisting potential of the control cells in 6 mM K^+ without Val. The complete data for the determination of the "null point" are presented elsewhere (12).

The data in Fig. 5 demonstrate that in the control environment (6 mM K^+, no Val) the cells accumulate AIB to about an eightfold concentration ratio, and experience little or no change of AIB when cells are transferred to a 21 mM K^+ environment containing Val. Thus, the "null point" value for $[K^+]_0$ is approximately 21 mM, when $[K^+]_i = 125$ mM (measured at the end of the experiment). This corresponds to a membrane potential of

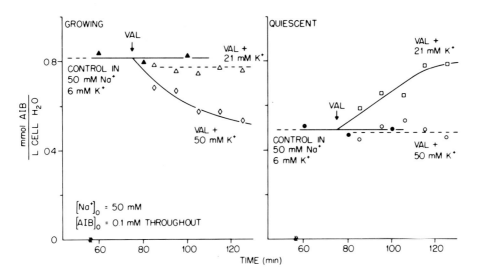

Fig. 5. AIB accumulation in quiescent and growing cells: the effect of matching their membrane potentials. Cells were removed from their growth medium and incubated in a serum-free, amino acid-free, 6 mM K^+, 50 mM Na^+, 0.1 mM AIB medium for 75 min. They were then transferred to either a 21 mM K^+, 50 mM Na^+, Val-containing medium or a 50 mM K^+, Val-containing medium. Intracellular AIB concentration is plotted vs time of accumulation for a representative experiment. Mean of at least 3 determinations is shown.

−47 mV. On the other hand, in the control environment quiescent cells accumulate AIB to a concentration ratio of approximately 5 and undergo no change in internal AIB concentration when transferred to a 50 mM K⁺ environment containing Val (Fig. 5). Thus, quiescent cells have a "null point" of 50 mM K⁺, again with $[K^+]_i = 125$ mM (measured at the end of the experiment), which corresponds to a membrane potential of −24 mV.

Quiescent cells in a 21 mM K⁺ + Val medium, where their membrane potential is matched to that determined for growing cells, accumulate AIB to the same level as do growing cells in a control environment (Fig. 5). Conversely, growing cells in a 50 mM K⁺ + Val medium, where their membrane potential is matched to that determined for quiescent cells, decrease their AIB concentration ratio to the same level found in control quiescent cells,

DISCUSSION

Studies of growth-related alterations in amino acid transport have dealt primarily with the kinetics of initial transport rates. The observation that in some cells the maximum amino acid transport rate is higher during growth than quiescence suggests a larger number of transport sites in growing cells. However, since the V_{max} for AIB uptake can be a function of more than the number of transport sites, and under some conditions may even reflect primarily a change in the membrane potential (13), then the growth-associated alterations in transport may be more complex than a simple increase in the number of transport sites. Steady state transport experiments can help clarify this point by determining whether increased transport rates during growth are accompanied by an increased amino acid accumulating capacity. An increase in concentrating capacity would be independent of the number of transport sites, but would instead be dependent on an additional source of energy.

The observations that AIB accumulation in confluent 3T3 cells is 30% lower than in nonconfluent cells (1) and that valine equilibrium uptake (expressed only as cpm/mg protein) into the acid-soluble fraction of chick embryo cells is about fourfold higher in growing than in hyperconfluent cells (5) suggest that more than an increase in the number of transport sites is required to explain growth-associated alterations in amino acid transport. The above mentioned findings, coupled to the observations that growing HSWP cells can accumulate AIB to a twofold higher concentration ratio than can quiescent cells (Fig. 1) indicate that in growing cells there is either an increase in the driving force for amino acid transport or a more efficient coupling to existing forces.

One possible source of energy for an enhanced amino acid-concentrating capacity in growing cells, is the electrochemical potential energy stored in the cation gradients. A growth-related increase in the Na⁺ electrochemical potential could provide an increased driving force for the accumulation of Na⁺-dependent amino acids. The response of AIB uptake in growing cells to alterations in the Na⁺ concentration gradient (Fig. 2) is consistent with this idea. However, not only a change in the Na⁺ concentration gradient, but also an alteration in the membrane potential affects the Na⁺ electrochemical gradient. Thus, a growth-associated increase in the membrane potential could provide the necessary energy for the enhanced AIB accumulation observed in growing cells.

Experiments to test this hypothesis were based on the assumption that in the presence of Val the cell membrane behaves like a K⁺ electrode so that

$$E_m = \frac{RT}{F} \ln \frac{[K^+]_o}{[K^+]_i}.$$

It has been determined that Val specifically increases the K^+ conductance of Amphiuma red cell membranes (11), artificial lipid membranes (14, 15), and many other systems. Although no comparable electrical measurements are available in HSWP cells, we have found that Val induces a fourfold increase in K^+ exchange (measured with ^{86}Rb) when the normal K^+ exchange components are inhibited by ouabain and furosemide (unpublished observations). This occurs presumably by electrical coupling of influx and efflux through a conductive pathway.

The accumulation of AIB in growing cells was demonstrated to be sensitive to Val-mediated alterations in the membrane potential (Fig. 3). When the membrane is hyper-polarized by adding Val to cells in 6 mM K^+ medium there is a 36% increase in the AIB concentrating capacity. In contrast, a dramatic decrease in AIB accumulation is observed when the membrane is depolarized by transferring cells to a 94 mM K^+ medium containing Val. Thus, the response of AIB accumulation to changes in membrane potential indicates that a growth-associated alteration in membrane potential could provide the energy required for an enhanced AIB accumulation in growing HSWP cells.

If the enhanced AIB concentrating capacity in growing cells is due solely to an increase in membrane potential, then one would predict that AIB accumulation could be driven to the same level in quiescent and growing cells by hyperpolarizing both to the same membrane potential. However, if the energy were provided by some other source (i.e., an increase in the Na^+ concentration gradient) or if the enhancement were due to tighter coupling to existing energy sources, then one would expect that a hyperpolarization would increase the concentrating capacity in both quiescent and growing cells, but that accumulation would still be higher in the growing cell. Therefore, the observation that a membrane hyperpolarization drives the AIB accumulation ratio to 34 in both quiescent and growing cells suggests that a growth-associated difference in membrane potential does exist and is responsible for the higher AIB accumulation observed in growing cells (Fig. 4). Also, the observation that Val stimulation of serum-stimulated cells again drives the AIB accumulation ratio to 34 is consistent with this hypothesis, and suggests that the membrane potential in serum-stimulated cells is somewhere intermediate between that of growing and quiescent cells.

An attempt was made to measure indirectly the membrane potential of growing and quiescent cells by varying the external K^+ concentration in the presence of Val, and determining at what value of $[K^+]_0$ the AIB accumulation corresponded to control accumulation (6 mM K^+ environment, without Val). The "null point" value of $[K^+]_0$ for growing cells is approximately 21 mM which predicts a membrane potential of -47 mV (Fig. 5). In contrast, the "null point" value of $[K^+]_0$ for quiescent cells is about 50 mM, which corresponds to a membrane potential of -24 mV.

It must be pointed out that, although this method is adequate for measuring differences in membrane potential between the two growth states, the actual values of the potentials measured in a 50 mM Na^+ medium may deviate from those that exist in a normal growth environment. Since we have no data on how the actual cytoplasmic Na^+ concentrations (16, 17) vary with the reduction of extracellular Na^+ concentration, we cannot estimate the contribution of the Na^+ gradient to the potential. However, since the membrane potentials of both growing and quiescent cells were measured in the same environment it appears that a real difference in potential does exist. If anything, the difference in potentials may be underestimated in a 50 mM Na^+ medium. In a 138 mM Na^+ environment, growing cells accumulate more than twice as much AIB as do quiescent cells (Fig. 1), while in a 50 mM Na^+ environment, AIB accumulation is only 1.5 times higher in growing cells (Fig. 5). A lower potential in quiescent cells implies that the Goldman equation (18), describing the potential in quiescent cells, may be weighted more in favor of the Na^+ potential than in growing cells. Thus, reducing the external Na^+ concentration might cause a larger increase in the potential in quiescent cells than in growing cells, thereby decreasing the difference between them.

The most convincing evidence that an increased membrane potential is responsible for the enhanced AIB accumulation in growing cells is the observation that accumulation in quiescent cells can be matched to that of growing cells by increasing the membrane potential of the quiescent cells to −47 mV. Conversely, the accumulation of AIB in growing cells can be matched to that in quiescent cells by decreasing the membrane potential of growing cells to −24 mV. Thus, by adjusting their membrane potentials one can mimic the growth-associated alterations of AIB accumulation seen in HSWP cells.

It is of interest to note that cell cycle-dependent variations in AIB accumulation ratios have been observed in Ehrlich ascites tumor cells (19). Minimum AIB accumulation is seen in early M phase with maximum AIB ratios, representing a threefold increase, occurring in S phase. Since it has been demonstrated that Na^+-dependent amino acid accumulation in ascites cells responds to Val-mediated alterations in membrane potential (20, 21), one can speculate that cell cycle-dependent variations in membrane potential could be responsible for the observed changes in AIB accumulation ratios. Although no measurements of membrane potential throughout the ascites cell cycle have been reported, comparable measurements in cultured Chinese hamster cells indicate that their membrane potential is low after collection of the cells in mitosis, reaches a maximum in S, and falls off again in G2 (22). In general, these measurements roughly parallel the rates of AIB uptake (allowing for an increasing cell surface area throughout the cycle) observed in synchronized CHO cells by Sander and Pardee (23). Also, the ouabain-insensitive, furosemide-insensitive component of K^+ flux, presumably the diffusional K^+ flux, in the ascites cell approximately triples during S phase (24). This finding is consistent with a hyperpolarization during S phase in the ascites cell.

One can speculate that changes in amino acid accumulation capacity in response to growth-associated alterations in membrane potential could serve to regulate cell growth. Others have proposed a role for the membrane potential in the regulation of mitotic activity (25–27). At present we have no information concerning the mechanism for growth-associated alterations of membrane potential in the HSWP cell. The possibility that changes in ionic fluxes, comparable to those seen in the ascites cell and in serum-stimulated 3T3 cells (28) may occur is currently being investigated. Also under investigation is the possibility that alterations in membrane potential are responsible for the differences in amino acid transport observed between transformed and nontransformed cells.

ACKNOWLEDGMENTS

This research was sponsored jointly by the National Cancer Institute and The Energy Research and Development Administration under contract with the Union Carbide Corporation. MLV is a Postdoctoral Investigator on Carcinogenesis Training Grant CA 05296 from the NCI.

REFERENCES

1. Foster DO, Pardee AB: J Biol Chem 244:2675, 1969.
2. Holley RW: Proc Natl Acad Sci USA 69:2840, 1972.

3. Eagle H: Science 148:42, 1965.
4. Hempling HG: In Harris EJ (ed): "Transport and Accumulation in Biological Systems." Baltimore: University Park Press, 1972, p 271
5. Bhargava PM, Vigier P: J Membr Biol 26:19, 1976.
6. Birckbichler PJ, Whittle WL, Dell'orco RT: Proc Soc Exp Biol Med 149:530, 1975.
7. Salter DW, Cook JS: J Cell Physiol 89:143, 1976.
8. Christensen HN: In Heinz E (ed): "Na^+-Linked Transport of Organic Solutes." New York: Springer-Verlag, 1972, p 39.
9. Avruch J, Wallach DFH: Biochim Biophys Acta 233:334, 1971.
10. Salter DW: Personal communication, 1976.
11. Hoffman JF, Laris PC: J Physiol 239:519, 1974.
12. Villereal ML, Cook JS: (In preparation).
13. Geck P, Heinz E: Biochim Biophys Acta 443:49, 1976.
14. Henderson PJF: Annu Rev Microbiol 25:393, 1971.
15. Haydon DA, Hladky SB: Quart Rev Biophys 5:187, 1972.
16. Pietrzyk C, Heinz E: Biochim Biophys Acta 352:397, 1974.
17. Smith TC, Adams R: J Membr Biol (In press).
18. Goldman DE: J Gen Physiol 27:37, 1943.
19. Tupper JT, Mills B, Zorgniotti F: J Cell Physiol 88:77, 1976.
20. Gibb LE, Eddy AA: Biochem J 129:979, 1972.
21. Reid M, Gibb LE, Eddy AA: Biochem J 140:383, 1974.
22. Sachs HG, Stambrook PJ, Ebert JT: Exp Cell Res 83:362, 1974.
23. Sander G, Pardee AB: J Cell Physiol 80:267, 1972.
24. Mills B, Tupper JT: J Cell Physiol 89:123, 1976.
25. Cone CD: J Theor Biol 30:151, 1971.
26. Cone CD, Tongier M: J Cell Physiol 82:373, 1973.
27. McDonald TF, Sachs HG, Orr CW, Ebert JD: Dev Biol 28:290, 1972.
28. Rozengurt E, Heppel L: Proc Natl Acad Sci USA 72:4492, 1975.

Journal of Supramolecular Structure 6:191—204 (1977)
Molecular Aspects of Membrane Transport 117—130

Hormonal Regulation of Hepatic Amino Acid Transport

Michael S. Kilberg and Otto W. Neuhaus

*Division of Biochemistry, Physiology and Pharmacology, Section on Biochemistry,
The University of South Dakota, Vermillion, South Dakota 57069*

The transport of 2-aminoisobutyric acid (AIB) into liver tissue was increased by both insulin and glucagon. We have now shown that these hormones do not stimulate the same
transport system. Glucagon, possibly via cAMP, increased the hepatic uptake of AIB by
a mechanism which resembled system A. This glucagon-sensitive system could be monitored by the use of the model amino acid MeAIB. In contrast, the insulin-stimulated
system exhibited little or no affinity for MeAIB and will be referred to as system B. On
the basis of other reports that the hepatic transport of AIB is almost entirely Na^+ dependent and the present finding that the uptake of 2-aminobicyclo [2,2,1] heptane-2-carb-
oxylic acid (BCH) was not stimulated by either hormone, we conclude that system B is
Na^+ dependent. Furthermore, insulin added to the perfusate of livers from glucagon-pretreated donors suppressed the increase in AIB or MeAIB uptake. Depending upon the
specificities of systems A and B, both of which are unknown for liver tissue, the insulin/
glucagon ratio may alter the composition of the intracellular pool of amino acids.

Key words: insulin, glucagon, transport, amino acids, diabetes

Amino acid transport in liver cells, controlled by various hormones, may be an
important factor in regulating hepatic metabolism by altering the composition of the
amino acid pools. Thus, it is known that substrate supply can regulate gluconeogenesis (1).
Using α-aminoisobutyric acid (AIB) as a monitor of transport, numerous investigations
have shown that its accumulation in the liver is increased in response to such hormones as
glucocorticoids (2), growth hormone (3), insulin (4, 5), and glucagon (6, 7). Furthermore,
the hormonal stimulation of hepatic amino acid uptake has been demonstrated both in
vivo (8, 9) and in vitro (5, 10). Of the hormones used, insulin and glucagon have been the
most rigorously investigated, perhaps because of their importance in the regulation of
carbohydrate metabolism.

These studies were taken from a dissertation presented by Mr. Michael S. Kilberg to the Graduate
School, The University of South Dakota in partial fulfillment of the degree of Doctor of Philosophy.
Present address: Department of Biological Chemistry, The University of Michigan, Ann Arbor,
Michigan 48104.

Received March 14, 1977; accepted April 26, 1977

Both insulin (4, 5, 10) and glucagon (6, 7) stimulate the transport of AIB yet they are recognized as metabolically antagonistic hormones. The purpose of the present study was to distinguish between the stimulatory effects of insulin and glucagon using model amino acids and to establish whether or not their antagonistic effects also could be described in terms of hepatic amino acid transport.

MATERIALS AND METHODS

Determination of Amino Acid Uptake in Vivo

One hour prior to sacrifice, 1.0 μCi (10 μCi per ml 0.9% saline) of radioactively labeled amino acid was injected via the tail vein. Injections were spaced 5 min apart to allow time for removal of the liver exactly 60 min later. The animals were sacrificed by decapitation; the livers were removed, rinsed in 0.9% saline, and weighed. The entire liver was immersed in 15.0 ml of 0.9% saline. After homogenization in a Sorvall Omnimixer, a 0.5 ml aliquot of the homogenate was mixed with 10.0 ml of Bray's fluid and the radioactivity was determined in a scintillation spectrometer. The dose of radioactivity injected was determined daily by counting 0.5 ml of a 1:1000 dilution of the labeled amino acid. Results are reported as percent of the injected radioactivity recovered in the liver, i.e., (cpm per liver/total cpm injected) $\times$ 100.

Isolated Liver Perfusion

Liver perfusion was performed in a manner similar to that of Miller (11), with a few modifications. Human red cells were isolated from whole blood by centrifugation at 4°C. The cells were then washed 3 times by centrifugation in chilled Ringer solution. The packed cells, 30 ml, were stored in an Erlenmeyer flask containing 1 ml of a solution of 38 mg of glucose, 300 units of penicillin G, and 3.0 g of streptomycin sulfate dissolved in Ringer solution. The flasks were covered with parafilm, stored between 2°C and 4°C, and the cell suspensions were used during the next 72 h. The cells were diluted on the day of perfusion with 70 ml of Krebs-Ringer bicarbonate buffer, pH 7.4, containing 2.1 g of fraction V BSA, 1,000 units of heparin, and 250 mg of glucose. The portal vein and the thoracic vena cava cannulas were made from Intramedic polyethylene tubing. The flow rate of the perfusate averaged 1–3 ml/g liver/min. A preliminary perfusion of 30 min was performed before the labeled amino acid and hormones (where indicated), were added to the perfusate. The first sample was collected following a 3-min mixing time and was recorded as the zero time point for the subsequent calculations.

Samples of perfusate, 1.0 ml, were removed at 0, 10, 20, 30, 45, 60, 90, and 120 min; each was deproteinized by the addition of 4.0 ml of absolute ethanol. After centrifugation, a 1.0 ml aliquot of supernatant was added to 10.0 ml of Bray's scintillation fluid and the mixture was counted in a scintillation spectrometer (Packard). The percent of the amino acid accumulated by the liver was calculated from the decrease in radioactivity in the perfusate; i.e., [(cpm/total perfusate at t_x)/(cpm/total perfusate at t_o)] $\times$ 100 gives the percent uptake per total liver. The total perfusate volume used in the calculation was adjusted at each time to account for the removal of the 1.0 ml sample. The percent uptake per liver was then multiplied by 10.0/liver weight so that the data were normalized and reported as the percent uptake per 10 g of liver tissue. Samples (1 g) of liver, taken at the termination of randomly selected experiments, were homogenized, and the radioactivity was determined. In all cases, 95–100% of the radioactivity removed from the perfusate was recovered in the liver; less than 1.0% was found in the bile.

The final amino acid concentration in the perfusate was 1.2 mM, representing approximately 2×10^6 cpm. When glucagon-pretreatment was used, 10.0 μg of the hormone per 100 g body weight were injected subcutaneously 1 h prior to the removal of the liver. Hormones given in vitro were added at the same time as the labeled amino acid and in the concentration indicated in the text. "Glucagon-free" insulin and "insulin-free" glucagon were the generous gifts of Dr. Mary Root of Eli Lilly. The cAMP content of the perfused livers was determined by the use of a radioimmunoassay kit purchased from New England Nuclear Corporation, Boston, Massachusetts. Instructions for tissue preparation and the assay protocol were supplied with the kit.

Experimental Diabetes

Experimental diabetes was induced by an intravenous injection of alloxan monohydrate (50 mg per 100 g body weight) in 0.9% saline. The animals, which had been starved 48 h before the injection, were refed immediately after treatment and given a 1% glucose solution instead of water for the first 24 h. For those rats receiving insulin, 4 units of protamine zinc insulin (Lilly) were given subcutaneously each day. Control rats received an equal volume of 0.9% saline. Blood glucose levels were estimated by the method of Feteris (12). Rats were considered diabetic when the plasma glucose content was greater than 300 mg per 100 ml of plasma 48 h after the treatment with alloxan. Plasma was obtained by centrifugation of blood samples in the presence of 0.25 ml of anticoagulant solution containing potassium oxalate (4.0 g/100 ml) and sodium fluoride (4.0 g/100 ml) in glass distilled water. To 0.1 ml of plasma, 5.0 ml of o-toluidine reagent were added and the mixture was heated for 10 min in a boiling water bath. A 0.1 ml aliquot of a standard containing 100 mg of glucose per 100 ml was included during each set of incubations. The cuvettes were then removed and allowed to cool to room temperature by placing them in tap water for 3 min. The absorbance at 635 nm was determined for each tube during the following 30 min. All samples were run in duplicate while the standards were in triplicate. The data were expressed as mg glucose per 100 ml of plasma.

Reagents

The radioactive amino acids, 1-^{14}C-BCH, 1-^{14}C-AIB, 1-^{14}C-cycloleucine, and 1-^{14}C-MeAIB were purchased from New England Nuclear Corporation, Boston, Massachusetts. The unlabeled BCH was the kind donation of Dr. Halvor N. Christensen, The University of Michigan. The o-toluidine reagent, heparin, Fraction V BSA, alloxan monohydrate, and unlabeled amino acids were obtained from Sigma Chemical Company. Fisher Scientific Company was the source for the components of the liquid scintillation fluid. All other reagents and chemicals were of the highest quality obtainable.

RESULTS

Hepatic Uptake of AIB Stimulated by Glucagon and Insulin

Figure 1 shows the stimulation of AIB accumulation in livers perfused with a single, 1.0 μg dose of glucagon. The livers treated with glucagon accumulated 46.7 ± 6.1% of the added AIB per 10 g of tissue, compared with a control of 20.4 ± 0.8%. Although the AIB levels in the control livers approached a steady state after 60 min, the stimulated uptake was nearly linear during this time and continued to increase even after 120 min. The elevated uptake could be observed as early as 10 min following the addition of glucagon.

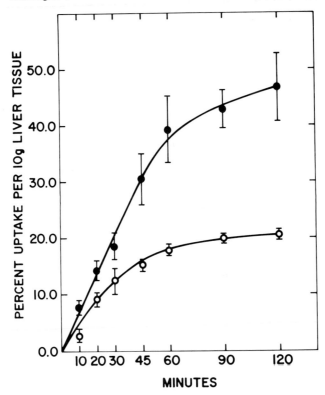

Fig. 1. Accumulation of AIB in isolated, perfused livers. Livers were taken from normal rats and perfused with (●—●) or without (○—○) 1.0 μg of glucagon added in vitro. The hormone and the labeled amino acid were added 3 min prior to time zero. The results are the average ± SD of the data obtained from 3–6 livers.

As seen in Fig. 2, the addition of 1.0 μg of insulin to the perfusate increased the final accumulation of AIB to 29.7 ± 0.4% while 5 μg increased it to 46.9 ± 1.8%. These stimulations of AIB uptake by the perfused liver following the in vitro administration of glucagon and insulin are in accord with the earlier work of Chambers et al. (2).

Interrelation of Glucagon- and Insulin-Regulated Transport

The stimulation of AIB uptake in livers exposed individually to glucagon and insulin was confirmed in the preceding section. However, these 2 hormones also have antagonistic actions which may be manifested as opposing effects on amino acid transport. Glucagon exerts its effect on AIB uptake via cAMP production (6, 13) whereas insulin is known to exert its antagonistic influence on metabolism either by diminishing the cAMP levels (14) or by counteracting the actions of the nucleotide (15). Stimulation by both hormones could be explained by the existence of separate hormone-sensitive systems. The antagonistic effect of insulin may be explained by an inhibition of the glucagon-sensitive system. If this is indeed the case, it is conceivable that a dose of insulin might be found which could suppress the glucagon-stimulated AIB uptake without simultaneously stimulating the insulin-responsive system. Figure 3 shows typical results for the addition of insulin in 2 doses (1 or 5 μg). It is apparent that the livers obtained from donor rats, pretreated with 10 μg of glucagon/100 g body weight, exhibited an accumulation after 3 h of 41.4 ± 3.4%

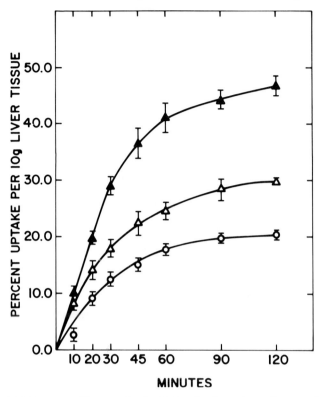

Fig. 2. Uptake of AIB by normal livers perfused with insulin. Controls are the open circles (○——○); livers treated with 1.0 or 5.0 μg of insulin are designated by open (△——△) and closed (▲——▲) triangles, respectively. Data are the average ± SD of 3–6 determinations.

compared with the control of 20.4 ± 0.8%. When livers from glucagon-pretreated rats were subsequently perfused with 1 μg of insulin, the AIB accumulation was reduced to 32.0 ± 0.9% after 2 h. Doses of 2 and 3 μg also suppressed the effect of glucagon but to lesser degrees (data not shown). A 5 μg dose of insulin, however, enhanced the uptake to 50.7 ± 4.2% of the AIB. Thus it appears that doses of 3 μg or less of insulin suppressed the glucagon effect more than they stimulated the simultaneous uptake. The 5 μg dose, however, yielded an overriding stimulation compared with its concomitant inhibition of the glucagon effect. These data suggest the existence of 2 hormonally responsive processes. Both glucagon and insulin stimulate AIB uptake when used individually. On the other hand when used together insulin has the capability of suppressing the effect of glucagon.

Antagonistic Effect of Insulin on the Glucagon-Stimulated Amino Acid Transport

In the preceding sections, the existence of 2 separate hormonally sensitive processes was suggested as well as an inhibitory effect of insulin on the glucagon-stimulated process. The distinction of these 2 phenomena was complicated by their simultaneous occurrence. Glucagon-stimulated uptake is thought to have the specificity of the A- or alanine-preferring transport as described for Ehrlich ascites cells (7, 16). Since Christensen et al. (17) have shown that monomethylation of the α-amino nitrogen of amino acids restricts their transport to the A system, it seemed likely that N-methyl AIB (MeAIB)

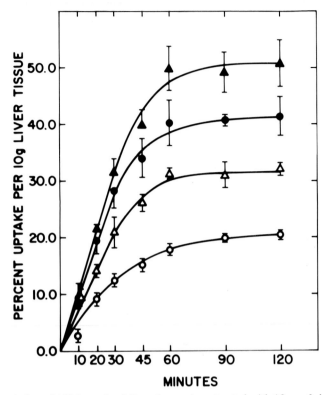

Fig. 3. Accumulation of AIB in perfused livers from rats pretreated with 10 μg of glucagon per 100 g body weight plus insulin added in vitro. Livers from pretreated rats (●—●) were exposed to 1.0 μg (△—△) or 5.0 μg (▲—▲) of insulin. Control livers are shown by the open circles (○—○). Insulin was added at the same time as the labeled amino acid. Data are the average ± SD of 3–6 perfusions.

would selectively respond to glucagon but not to insulin. This amino acid might be used to distinguish the inhibitory effect of insulin from its stimulatory action.

Pretreatment in vivo with glucagon markedly stimulated the hepatic transport of MeAIB. Figure 4 shows that the maximal uptake by livers from the treated animals reached 45.9 ± 2.8% after 60 min while the controls were 24.6 ± 0.9% of the MeAIB. As anticipated, perfusion of normal livers with 5 μg of insulin had no effect on the uptake of MeAIB (Fig. 5). Thus the accumulation of MeAIB is stimulated by glucagon but not by insulin.

Since MeAIB responded well to the stimulatory effect of glucagon but did not respond to insulin, it appears that the latter hormone does not influence the A transport system. It is conceivable, however, that the insulin might exert its antagonistic effect by suppressing the glucagon-stimulated uptake of MeAIB. Both 1 and 5 μg of insulin added to the perfusate of a liver from a glucagon-pretreated rat suppressed the uptake of MeAIB (Fig. 4). These data substantiate the existence of an antagonistic effect of insulin on the glucagon-stimulated uptake of amino acids by the liver.

Absence of Na+-Independent L Transport in Hormone-Stimulated Amino Acid Transport

Although the stimulated transport of AIB induced by glucagon is known to require Na^π (7), the Na^+ dependence of the newly recognized component of AIB uptake needed

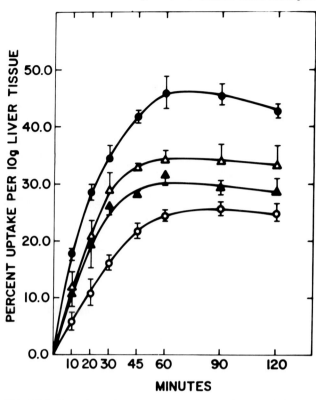

Fig. 4. Uptake of MeAIB in livers from glucagon-pretreated rats (10 μg per 100 g body weight) (●—●) and the effect of adding 1.0 μg (△—△) or 5.0 μg (▲—▲) insulin to the perfusate. Control livers are shown by the open circles (○—○). Data are the average ± SD of 3–6 determinations.

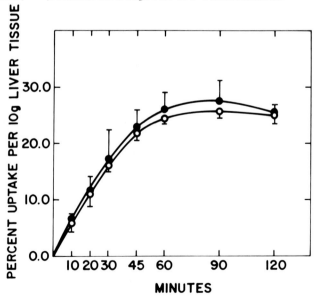

Fig. 5. Uptake of MeAIB in normal and insulin-treated livers. Effect of perfusion with (●—●) or without (○—○) the addition of 5.0 μg of insulin is shown. Data are the average ± SD of 3–6 perfusions.

TABLE I. Effect of Insulin and Glucagon on BCH Uptake in Isolated Perfused Liver*

Hormone added	Time of perfusion (min)						
	10	20	30	45	60	90	120
Control	5.4	9.2	11.7	11.7	12.0	14.0	13.5
5.0 μg insulin[a]	4.5	6.9	10.7	13.4	14.4	13.5	13.5
Glucagon[b] 10 μg/100 g body weight	4.0	7.9	9.5	12.0	13.6	12.9	13.8

*Data are the averages of 2 perfusions for each condition and are expressed as the percent of the added BCH accumulated per 10 g liver tissue.

[a]5.0 μg of insulin were added to the perfusate at the same time as the labelled amino acid.

[b]Glucagon was injected subcutaneously 1 hr prior to liver isolation.

TABLE II. Effect of Insulin and Glucagon on Hepatic cAMP Levels in Isolated Perfused Liver*

Hormone added	cAMP content after 120 min
Control	0.92 ± 0.20
1 μg glucagon	1.40 ± 0.20[b]
1 μg insulin	0.72 ± 0.02[c]
5 μg insulin	0.90 ± 0.08[c]
Glucagon pretreatment[a]	2.17 ± 0.30[d]
Glucagon pretreatment[a] plus 1 μg insulin	1.50 ± 0.30[e]
Glucagon pretreatment[a] plus 5 μg insulin	1.23 ± 0.20[f]

*Livers were perfused with glucagon or insulin (see Materials and Methods) for 120 min, then immediately submersed in liquid nitrogen and stored frozen until assayed. Data are expressed as nmoles cAMP per g liver and are the average of 3–6 livers.

[a]Glucagon (10 μg/100 g body weight) was injected subcutaneously 1 h prior to liver isolation. Insulin, when given, was added to the perfusate of the livers taken from glucagon-pretreated donors.

[b]$p < 0.02$ when compared to the control.

[c]not statistically different when compared to the control.

[d]$p < 0.001$ when compared to the control.

[e]$p < 0.05$ when compared to glucagon pretreatment.

[f]$p < 0.01$ when compared to glucagon pretreatment.

to be determined. Evidence for Na^+ dependence was obtained indirectly by determining the absence of any stimulation on the uptake of 2-aminobicyclo[2,2,1] heptane-2-carboxylic acid (BCH), a model amino acid carried only by the Na^+-independent L (leucine-preferring) system (18). When ^{14}C-BCH was introduced into the perfusate, control livers accumulated approximately 14% of the added amino acid per gram of liver tissue (Table I); neither the addition of 5 μg of insulin nor pretreatment of the donor rat with glucagon had any effect on the uptake of BCH. Thus both the insulin- and the glucagon-stimulated transport of the liver are believed to occur by Na^+-dependent systems.

Effect of Glucagon and Insulin on Hepatic cAMP Levels

It is believed that the effect of glucagon on amino acid uptake is mediated by cAMP. Insulin can oppose the action of glucagon either by suppressing cAMP levels (14) or by counteracting the metabolic effects of this nucleotide (15). To demonstrate that the antagonism of glucagon and insulin on the uptake of AIB was accompanied by alterations of cAMP levels, measurements were made of this nucleotide in all the foregoing experiments following their termination. As seen from Table II, glucagon administered either in vitro or in vivo, increased the hepatic cAMP. Glucagon perfused in vitro increased the cAMP from a normal of 0.92 to 1.40 nmoles/g liver whereas pretreatment of the donor rat yielded a level of 2.17 nmoles/g. Tews et al. (6) observed a maximal stimulation of AIB transport when the cAMP levels were approximately 3 nmoles/g liver. Insulin, on the other hand, did not affect the normal content of cAMP as already observed by Park et al. (14).

Consistent with the amino acid transport data, 1 or 5 μg of insulin depressed the cAMP levels to 1.50 ± 0.30 and 1.23 ± 0.20 nmoles per g liver, respectively, when introduced into the perfusate of livers from glucagon-pretreated donor rats. Therefore, it may be concluded that insulin lowered the glucagon-stimulated levels of cAMP. It may also be concluded that glucagon and insulin each stimulated AIB uptake by a different mechanism, the former via a cAMP-mediated process as suggested by Tews et al. (19) and the latter via a system not involving this cyclic nucleotide. Although insulin decreased the glucagon-stimulated transport and also partially lowered the level of cAMP, it may not be assumed that these processes are necessarily related.

Table II shows that a cAMP level of 1.40 nmoles/g liver was associated with a stimulated AIB uptake (Fig. 1) whereas 1 μg of insulin, which reduced the cAMP from 2.17 to 1.50 nmoles/g, suppressed the accumulation of AIB (Fig. 4). The reason for the divergent effects on AIB uptake for the same apparent cAMP concentration is presently unknown.

AIB Transport Studied in the Diabetic Rat

Although it is known that experimental alloxan diabetes is accompanied by a hyper-glucagonemia (20), no effort has been made to demonstrate that this phenomenon results in a stimulation of hepatic amino acid transport. Based on the preceding studies it would be expected that AIB accumulation should be elevated in the diabetic state and that this increased uptake could be counteracted by insulin. Rats were injected with alloxan mono-hydrate and the uptake of AIB measured at the times indicated for a total of 14 days (Fig. 6). Before treatment with alloxan, approximately 6% of the injected AIB was re-covered in the liver. In contrast, 48 h after injecting the alloxan, the accumulation was 17.4 ± 0.7%. On the 8th day the animals were divided into 2 groups; 1 received 4 units of protamine zinc insulin daily while the controls received saline. Within 24 h, treatment with insulin reversed the diabetes-induced stimulation of AIB transport (Fig. 6). Those rats receiving insulin accumulated 5% of the injected amino acid per liver, while the AIB uptake in the animals given saline remained elevated at 20%. To support the hypothesis that the insulin had reversed an action of glucagon upon the liver, the plasma glucose levels were determined in the same animals. As expected, plasma glucose levels exceeded 300 mg per 100 ml 48 h after the administration of alloxan and continued to increase for the duration of the experiment (Data not shown). Treatment of these rats with daily injections of insulin decreased the plasma glucose to control values resulting in a general profile which was similar to that of the AIB transport. This reversal of the elevated glucose levels may be caused by a decrease in the circulating glucagon (21, 22).

Hepatic Accumulation of Other Model Amino Acids in Diabetic Rats

As already described, MeAIB serves as a monitor for the stimulation of the glucagon-sensitive system A but not of the insulin-sensitive transport. To prove that the stimulation of AIB uptake in diabetes is the result of the hyperglucagonemia, MeAIB uptake was monitored in livers from control and diabetic rats. Figure 7 shows that diabetes resulted

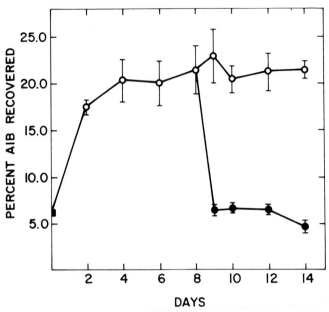

Fig. 6. Percent of the total AIB injected recovered in the livers of alloxan diabetic rats. On day 8, one half of the remaining animals received insulin (●—●) as described in the section on Materials and Methods; controls (○—○) were given an equal volume of saline. Each point is the average ± SD of 4 rats

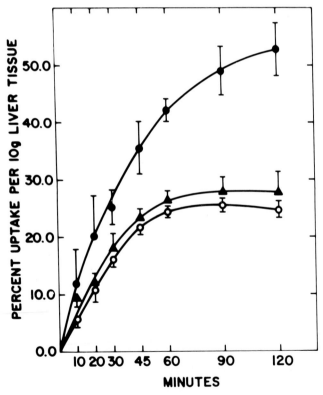

Fig. 7. Uptake of MeAIB in livers from diabetic rats perfused with (▲—▲) or without (●—●) 5.0 µg of insulin. Control livers are shown as the open circles (○—○). Insulin was added in vitro at the same time as the labeled amino acid. Each curve is the average ± SD of 3–6 livers.

TABLE III. Effect of Insulin on Hepatic cAMP Levels in Isolated, Perfused Livers From Alloxan-Diabetic Rats*

Treatment	cAMP content after 120 min
Control	0.92 ± 0.20
Diabetic	2.69 ± 0.20[b]
Diabetic plus 5.0 μg insulin[a]	1.10 ± 0.08[c]

*Diabetes was induced by injecting alloxan as described in Materials and Methods. The same livers as those used for Fig. 7 were employed for these determinations. After perfusion for 120 min, the livers were submersed in liquid nitrogen and stored frozen until assayed. The data are the average $\pm$ SD of 3–6 livers and are expressed as nmoles cAMP per g liver tissue.

[a]Insulin was added to the perfusate at the beginning of the experimental period as described in Materials and Methods. [b]$p < 0.001$ when compared to the control. [c]$p < 0.001$ when compared to the diabetic.

TABLE IV. In Vivo Uptake of BCH and Cycloleucine in Livers of Alloxan-Diabetic Rats*

	BCH	Cycloleucine
Control	2.08 ± 0.14 (5)	2.19 ± 0.15 (4)
Diabetic	2.53 ± 0.18 (3)[a]	6.38 ± 0.32 (3)[b]

*Determination of uptake is described in Materials and Methods. Data are reported as percent of injected amino acid recovered in the liver after 60 min. The number of rats is shown in parenthesis.
[a]$p < 0.02$ when compared to the control
[b]$p < 0.001$ when compared to the control

in a twofold stimulation in accumulation of MeAIB, following 2 h of perfusion. When 5 μg of insulin were introduced into the perfusate, the stimulated accumulation of MeAIB was abolished (Fig. 7). These results confirm those previously described for the glucagon-pretreated rats.

Because of the known hyperglucagonemia, hepatic cAMP levels are also elevated in alloxan diabetes (23). Treatment with insulin decreases the cAMP content and reverses the glucagon-induced changes in carbohydrate metabolism (14). In accord with these observations the data of Table III show the increased levels of cAMP in perfused livers from alloxan-diabetic rats. Perfusion of these livers with 5 μg of insulin reduced the cAMP content to near control values. Thus the interrelation of glucagon and insulin as studied in alloxan-diabetes support the evidence provided in the preceding section using only exogenous hormones.

The limitation of uptake to Na^+ dependent systems was also tested by determining the uptake of BCH and cycloleucine (Table IV). The uptake of BCH which monitors the Na^+-independent L system (18) was not affected to the same degree as cycloleucine, a model amino acid transported by both Na^+-dependent A and the Na^+-independent L systems (16).

DISCUSSION

Although both insulin (4, 5, 10) and glucagon (6, 7) stimulate the uptake of AIB by the liver, the present report demonstrates that they do not enhance a common transport mechanism. Glucagon stimulated system A, mediated by cAMP (6, 7, 13), as suggested by

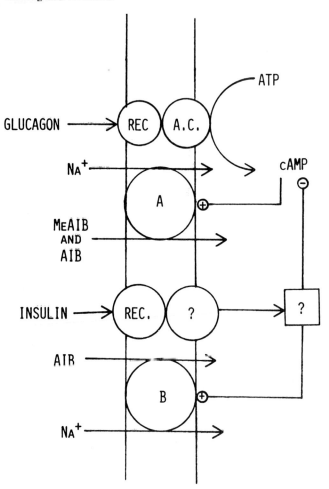

Fig. 8. A schematic diagram showing the relationship between the pancreatic hormones insulin and glucagon and Na⁺-dependent amino acid transport at the liver plasma membrane.

the response of MeAIB to the hormone whereas insulin did not affect the methylated model amino acid. For the time being, we suggest that the insulin-sensitive transport simply be called system B. Thus, the hepatic uptake of AIB can be assumed to be the result of a combination of systems A and B. Previous investigations have suggested the existence of 2 systems for AIB uptake in the liver (24, 25). Furthermore, Flory and Neuhaus (24) have shown that the transport of AIB is almost entirely Na^+ dependent, thus system B is probably also Na^+ dependent. In support of this concept is the fact that neither insulin nor glucagon affected the hepatic accumulation of BCH, a model amino acid carried entirely by the Na^+-independent L-process (18).

According to LeCam and Freychet (7, 26) the systems for neutral amino acid transport, present in the Ehrlich ascites tumor cell, are also found in the liver. These authors showed that AIB uptake was only partially inhibited by MeAIB (7). The uninhibited component was attributed to system ASC (26), despite the fact that Christensen has shown that ASC exhibits no affinity for AIB (27). In any event it appears that the liver possesses 2 AIB transporting systems, 1 inhibited by MeAIB, the other not. It is conceivable that the latter may be the same as the insulin-sensitive process described in the present report. For the time being the provisional designation of system B is preferred to distinguish the insulin-stimulated process from the glucagon-sensitive system A.

Exposure of rats to whole-body irradiation (γ rays) has been used to study the role of amino acid transport in controlling hepatic metabolism (28, 29). The radiation-induced uptake of amino acids by the liver occurs via system A and results from an excessive release of glucagon from the pancreas (29, 30). A stimulation of amino acid uptake caused by exposure to γ rays was accompanied by both gluconeogenic and glycogenic states (29). Since the effect of glucagon on hepatic transport appears to involve the A system, which transports many of the gluconeogenic amino acids, it appears likely that the specificity of the A system is an important factor in regulating the gluconeogenic state of the liver. In contrast, the insulin-sensitive system B may present the liver with an entirely different spectrum of amino acids. What effect this may have on hepatic metabolism must await a determination of the specificity of system B.

It is apparent that insulin has the ability to regulate hepatic, neutral amino acid transport in 3 distinct ways. First, insulin can act on the pancreas to decrease the circulating levels of glucagon and, therefore, can suppress the stimulation of system A by glucagon. Secondly, the data reported here show that insulin can modulate the effect of glucagon on system A at the liver cell. Finally, insulin can cause the elevation of an additional Na^+-dependent system, namely, system B. Unger has proposed that the insulin/glucagon molar ratio is more important in homeostasis than is the absolute concentration of either hormone (20). It is possible that the I/G ratio is reflected as an A/B transport ratio at the liver membrane (Fig. 8) and thereby controls the composition of the intracellular amino acid pool.

ACKNOWLEDGMENTS

Special thanks are expressed to Mr. Alan Conroy for assistance with the liver perfusion experiments. This work was supported by U.S. Atomic Energy Commission Grant AT (11-1)-1754 and U.S. Public Health Service Grant AM 11146.

REFERENCES

1. Exton JH, Mallette LE, Jefferson LS, Wong EHA, Friedmann N, Miller TB Jr, Park CR: Rec Prog Horm Res 26:411, 1970.
2. Chambers JW, Georg RH, Bass AD: Mol Pharmacol 1:66, 1965.
3. Jefferson LS, Schworer CM, Tolman EL: J Biol Chem 250:197, 1975.
4. Krawitt EL, Baril EF, Becker JE, Potter VR: Science 169:294, 1970.
5. Miller LL, Griffin EE: In Bartosek I, Guaitani A, Miller LL (eds): "Isolated Liver Perfusion and Its Applications." New York: Raven Press, 1973, p 139.
6. Tews JK, Woodcock Colosi N, Harper AE: Life Sci 16:739, 1975.
7. LeCam A, Freychet P: Biochem Biophys Res Commun 72:893, 1976.
8. Sanders RB, Riggs TR: Endocrinology 80:29, 1967.
9. Harrison LI, Christensen HN: Biochem Biophys Res Commun 43:119, 1971.
10. Kletzien RF, Pariza MW, Becker JE, Potter VR, Butcher FR: J Biol Chem 251:3014, 1976.
11. Miller LL: In Bartosek I, Guaitani A, Miller LL (eds): "Isolated Liver Perfusion and Its Application." New York: Raven Press, 1973, p 11.
12. Feteris WA: Am J Med Technol 31:17, 1965.
13. Tews JK, Woodcock Colosi N, Harper AE: Am J Physiol 228:1606, 1975.
14. Park CR, Lewis SB, Exton JH: In Fritz IB (ed): "Insulin Action." New York: Academic Press, 1972, p 509.

15. Goldberg ND: In Weissmann G, Claiborne R (eds): "Cell Membranes." New York: Hospital Practice Publishing Co, 1975, p 185.

16. Oxender DL, Christensen HN: J Biol Chem 238:3686, 1963.

17. Christensen HN, Oxender DL, Liang M, Vatz KA: J Biol Chem 240:3609, 1965.

18. Christensen HN, Handlogten ME, Lam I, Tager HS, Zand R: J Biol Chem 244:1510, 1969.

19. Tews JK, Woodcock NA, Harper AE: J Biol Chem 245:3026, 1970.

20. Unger RH: N Engl J Med 285:443, 1971.

21. Unger RH: Metabolism 23:581, 1974.

22. Unger RH: Diabetes 25:136, 1976.

23. Pilkis SJ, Exton JH, Johnson RA, Park CR: Biochim Biophys Acta 343:250, 1974.

24. Flory W, Neuhaus OW: Radiat Res 68:138, 1976.

25. Grimm J, Manchester KL: Biochim Biophys Acta 444:223, 1976.

26. LeCam A, Freychet P: J Biol Chem 252:148, 1977.

27. Christensen HN, Liang M, Archer EG: J Biol Chem 242:5237, 1967.

28. Kilberg MS, Neuhaus OW: Radiat Res 64:546, 1975.

29. Kilberg MS, Neuhaus OW: Radiat Res 66:597, 1976.

30. Kilberg MS, Neuhaus OW: Fed Proc Fed Am Soc Exp Biol 36:910, 1977.

Journal of Supramolecular Structure 6:205—213 (1977)
Molecular Aspects of Membrane Transport 131—139

Amino Acid Transport Systems in Animal Cells: Interrelations and Energization

Halvor N. Christensen

Department of Biological Chemistry, The University of Michigan, Ann Arbor, Michigan 48109

After summarizing the discrimination of the several transport systems for neutral amino acids in the cell of the higher animal, I discuss here the ways in which 2 dissimilar transport systems interact, so that one tends to run forward for net entry and the other backwards for net exodus. An evaluation of the proposals for energization shows that uphill transport continues when neither alkali-ion gradients nor ATP levels are favorable. Evidence is presented that under these conditions a major contribution is made by another mode of energization, which may depend on the fueling of an oxidoreductase in the plasma membrane. This fueling may involve the export by the mitochondrion of the reducing equivalents of NADH by one of the known shuttles, e.g., the malate-aspartate shuttle. After depletion of the energy reserves in the Ehrlich cell by treating it with dinitrophenol plus iodoacetate concentrative uptake of test amino acids is restored by pyruvate, but in poor correlation with the restoration of alkali-ion gradients and ATP levels. This restoration by pyruvate but not by glucose is highly sensitive to rotenone. A combination of phenazine methosulfate and ascorbate will also produce transport restoration, before either the alkali-ion gradients or ATP levels have begun to rise. The restoration of transport applies to a model amino acid entering by the Na^+-independent system, as well as to one entering by the principal Na^+-dependent system, restoration being blocked by ouabain, despite the weak effect of ouabain on the alkali-ion gradients in the Ehrlich cell. Quinacrine terminates very quickly the uptake of model amino acids, before the alkali-ion gradients have begun to fall and before the ATP level has been halved. Quinacrine is also effective in blocking restoration of uphill transport by either pyruvate or the phenazine reagent. Preliminary results show that vesicles prepared from the plasma membrane of the Ehrlich cell quickly reduce cytochrome c or ferricyanide in the presence of NADH, and that the distribution of a test amino acid between the vesicle and its environment is influenced by NADH, quinacrine, and an uncoupling agent in ways consistent with the above proposal, assuming that a majority of the vesicles are everted.

Key words: amino acid transport in animal cells, energization of transport systems, discrimination of transport systems, reverse operation of transport systems, Ehrlich cell, NADH dehydrogenase, alkali-ion gradients, phenazine methosulfate, ouabain

TRANSPORT SYSTEMS AND THEIR INTERACTION

The transport systems for neutral amino acids in the Ehrlich cell are taken to be approximately representative of those of the cells of the higher animal in general, with more and more evidence supporting that interpretation. Most conspicuous are a broad-

Received March 13, 1977; accepted April 3, 1977

range Na$^+$-dependent system called A and a broad-range Na$^+$-independent system called
L (1). Another Na$^+$-dependent system almost completely specific to glycine is not seen
in the Ehrlich cell but has been described for nucleated and reticulated red blood cells,
and may occur in a variant form with broader specificity in the intestine and kidney
(Summaries, Refs. 2, 3). A third Na$^+$-dependent system called ASC has an intermediate
range of specificity embracing 3- to 5-carbon straight-chain amino acids and their hydroxy
and sulfhydryl derivatives, also asparagine and glutamine, and the prolines. Its differences
from System A leave no doubt of the independence of the 2: System ASC is less sensitive to H$^+$ and will not accept Li$^+$ as a substitute for Na$^+$. The receptor site of System
ASC binds the alkali ion at a closely specified point in juxtaposition to the hydroxyl group
of ordinary (trans) 4-hydroxyproline (4), whereas System A binds Na$^+$ or Li$^+$ at a different,
less precisely localized point rather nearer the β-carbon atom of the bound amino acid
substrate (5). The ASC system is conspicuous in immature and nucleated red cells (6–8),
leukemia cells (9), and lymphocytes (10); it participates in placental transport (11), and
in general it appears to be ubiquitous.

The relation between Systems A and L is the most interesting one, because essentially all the neutral amino acids are to some degree transported by both of these systems,
although in different proportions. Because System A characteristically is strongly concentrative, and System L more weakly so, it can be shown that steady states are set up in
which net uptake for various amino acids takes place by System A, and net exodus by
System L (12). Amino acids with weak reactivity with System A and much stronger
affinity for System L will therefore maintain only moderate gradients across the membrane,
whereas those with the opposite pattern of preference will maintain rather high cellular
levels relative to the extracellular fluid. The comparison is complicated somewhat, however, by effects of amino acid structure on the intensity of accumulation by System L.

My comments may be running contrary to an assumption about System L that has
gained unfortunate currency, namely that it produces exchange but not net uptake. This
idea is easily refuted by experiments with model substrates specific to System L, using
rather high concentrations to minimize the contribution of exchange (13). Net uptake and
net exodus of phenylalanine by it also can be shown (14). I do not mean by this caveat,
however, to question the importance of the heteroexchange activity of System L.

A specialization of one transport system in net uptake and another (by running
backward) in net exodus of the same substrate seems highly useful in establishing the
transport asymmetry of epithelial cells from one pole to the other, so that transcellular
migration of nutrients can be produced without threatening the nutrition of the epithelial
cell as in the small intestine, for example (15). But in nonepithelial cells, this specialization
may also be advantageous in enhancing the regulatability of transport. If the energy
sources of Systems A and L are distinct, as most evidence suggests, the duality of transport allows us to generalize the Mitchell hypothesis to include amino acid transport across
the plasma membrane. The 2 energizing reactions can be coupled via the 2 transports.

ENERGIZATION

The more steeply uphill System A in the cells of the higher animal can be driven by
linked, down-gradient flows of the alkali ions, particularly the inward flow of Na$^+$ (16–
18). This source of energy can account to a large extent, but not completely (19, 20) for
the uphill transport of test amino acids when the cellular ATP is largely depleted. The
energy calculated to be thus made available might perhaps become sufficient if hypotheti-

cal increases in the transmembrane potential could be taken into account (21, 22). We have observed, however, conditions under which steeply uphill, Na^+-dependent uptake of 2-aminoisobutyric acid (AIB) or its N-methyl derivative (MeAIB) continues with little if any associated uptake of Na^+ (13). Unless an ion actually moves in cotransport with the neutral amino acid, the value of the transmembrane potential appears to contribute nothing to an explanation. Hence the alkali-ion gradient hypothesis remains inadequate to explain Na^+-dependent amino acid uptake.

Furthermore, respiratory poisons may largely fail to stop the uphill entry of the amino acids, whether Na^+-dependent or Na^+-independent. Eddy observed distinctly stronger accumulation of glycine for a given alkali gradient when respiratory metabolism took place than when it was prevented (19). Schafer and Williams have particularly emphasized the failure of the concentrative uptake of AIB to be eliminated when the ATP of the Ehrlich cell is sharply lowered by respiratory poisons, even when the electrochemical gradient of Na^+ is unfavorable (23). We find that after ATP depletion by simultaneous treatment of this cell with 2,4-dinitrophenol (0.1 mM) and iodoacetate (1 mM), the concentrative uptake of various amino acids is restored by supplying 10 mM pyruvate before either the ATP level or the alkali-ion gradients are restored. Sensitivity of this effect to inhibition by rotenone suggests a mainly mitochondrial origin for the energy under these conditions (24). We raise the question, does the energy flow from the mitochondrion in a form other than ATP?

In the bacterial cell, electron transport and oxidative phosphorylation occur in the plasma membrane. Furthermore, large gradients of H^+ and of the electrical potential can be maintained across the plasma membrane, and nutrient molecules can be concentrated by their electrophoretic cotransport with H^+ made possible by these gradients. In the ascites tumor cell, in contrast, as in many other animal cells, the pH gradient and the transmembrane potential gradients tend to be quite small. Do the plasma membranes of animal cells differ from bacterial cells in not using proton gradients at all to intermediate between metabolic energy release and nutrient transport? Is amino acid transport energized on totally different principles in the animal cell? We have proposed that the mediating H^+ gradients in the animal cell may be generated in and largely restricted to the membrane interior (13, 25, 26). The present question is, however, a different one; not how the energy transduction occurs, but in what form is the energy brought to the membrane.

ATP BREAKDOWN AND ENERGIZATION OF AMINO ACID TRANSPORT

The conventional assumption has been that the flow of ATP to the membrane may serve to the extent that preexisting gradients, e.g., of the alkali ions, fall short. An amino acid stimulation of the Na^+- and K^+-dependent ATPase activity associated with the plasma membrane has been reported by Forte et al. (27). This activity applied, however, to both D and L isomers, and is shared with unnatural chelating agents. We have observed a stimulation of Mg^{2+}-dependent ATPase activity in a preparation from the plasma membrane of the Ehrlich cell, in the absence of both Na^+ and K^+ and in the presence of ouabain (28). These properties might well correspond to Na^+-independent System L which the Na^+ gradient apparently does not energize at all. The stimulating amino acids inappropriately include L-ornithine (not D-ornithine), however; furthermore the norbornane amino acid is not stimulatory, even though it is a model substrate for System L; nor does it block the stimulatory effect of L-ornithine. We conclude that either ATP does not directly energize System L, or that we have not yet detected the ATPase corresponding to that system, or else perhaps that the ATPase activity has undergone alteration of its specificity during separation of the membrane fraction.

AN OXIDOREDUCTASE SYSTEM ENERGIZING TRANSPORT IN THE PLASMA MEMBRANE?

The question which now presents itself is, does the plasma membrane of the animal cell unexpectedly retain a redox system which allows it to energize transport? Do reducing equivalents flow from the mitochondrion to such a system? Although dehydrogenase activity for NADH has been detected in the plasma membrane of various cells, including nonnucleated red blood cells (29), hepatocytes, and adipocytes (30–32), these observations have been mainly incidental to the study of marker enzymes and until recently (31, 32) attracted little interest. It has occurred to us that NADH might reach the dehydrogenase of the plasma membrane by the transfer of its reducing equivalents from the plasma membrane via one of the shunts.

For example, the malate-aspartate shunt might serve, although the direction of that particular shunt has been seen as favoring movement of reducing equivalents into the mitochondrion. Participation by that shunt could explain some earlier findings, as follows: Because transamination on both sides of the mitochondrial membrane is obligatory, a deficiency of vitamin B_6 might interfere with the maintenance of gradients of AIB by liver and muscle of the rat, with respect to the blood plasma (33), and vitamin B_6 analogs might interfere with intestinal amino acid transport, as has been repeatedly reported (34–36). Observation of these effects was at one time held to support an older hypothesis that the aldehyde group of pyridoxal phosphate might serve to take hold of the amino acid for transport. Other shuttles, described or undescribed, should also be considered.

HORMONAL SENSITIVITY OF PLASMA MEMBRANE NADH DEHYDROGENASE

The observation that NADH stimulates the adenylcyclase activity of plasma membranes of the hepatocyte and adipocyte (30) led to a search for the membrane-borne NADH sensor, with the result that responsive NADH dehydrogenase activity was discovered. Crane and Löw have recently reported that the NADH dehydrogenase of these 2 cells is characteristically sensitive to quinacrine (atebrin), azide, and triiodothyronine (31). The activity of the adipocyte membrane is stimulated by ACTH or glucagon, that of the liver cell by glucagon, at just the concentrations at which these hormones stimulate the adenylcyclase activity (32). Löw and Crane consider that the membrane NADH dehydrogenase has a monitoring function, i.e., in serving to regulate membrane activities to correspond to the metabolic state of the cell. Our present proposal adds the idea that this system may also serve to drive transport of amino acids and possibly other substances, as a significant biological alternative and complement to energization by cotransport with Na^+ and by ATP breakdown.

PRELIMINARY FINDINGS

We have now found informative a comparison between the rate of restoration of the following parameters after 30 min of depletion of energy reserves of the Ehrlich cell by 0.1 mM dinitrophenol plus 1 mM iodoacetate, or by 100 μg of rotenone per liter: 1) Concentrative uptake of model amino acids; 2) Cellular ATP level; 3) Gradients of Na^+ and K^+ across the membrane. When a HEPES-buffered Krebs-Ringer medium containing 10 mM pyruvate is substituted for the poisoning solution, the uptake of [^{14}C] MeAIB is restored in 3 min, during which time the ATP level rises, although the alkali-ion gradient (whether expressed as $([Na^+]_{out} \times [K^+]_{in})/([Na^+]_{in} \times [K^+]_{out})$ or as $[Na^+]_{out}/[Na^+]_{in}$) remains below 50% of normal. This early restoration of MeAIB transport by pyruvate is highly sensitive to rotenone, whereas that by glucose is not, a result which leads us provisionally to assign the restorative effect of pyruvate to mitochondrial oxidation.

ARTIFICIAL HYDROGEN DONOR

We have been able to replace pyruvate with the combination 0.1 mM phenazine methosulfate plus 20 mM sodium ascorbate for restoring MeAIB transport. This restorative effect differs from that of pyruvate, however, in that it is insensitive to rotenone inhibition. Figure 1 in our recent report (37) compares these restorative effects on (dinitrophenol + iodoacetate)-treated cells (left-hand section) and on rotenone-treated cells (right-hand section). Note that restoration of the MeAIB gradient had already progressed in the first min, and was already half complete in 4 or 5 min, whereas the restoration of the ATP level and of the alkali-ion gradients had not yet begun. Transport restoration by pyruvate was not obtained in the right-hand section of the cited figure, where rotenone had been used as the metabolic poison. Figure 2 in the same paper showed that restoration of the fully Na^+-independent uptake of the norbornane amino acid by phenazine-ascorbate was quite parallel. The amino acid gradients in this case typically continued to decline for 15 min after the dinitrophenol-iodoacetate had been removed, before application of the restorative reducing agent.

Figure 1 of the present paper shows a similar triplet of parallel experiments, in all of which, however, a 15-min delay was introduced after the dinitrophenol-iodoacetate or rotenone treatment before adding the phenazine-ascorbate. This figure also shows that restoration of amino acid gradients proceeds without rise in the severely depressed ATP levels or alkali-ion gradients. We argue that ATP of mitochondrial origin should pass through the cytoplasmic pool, to increase the cellular content. ATP of glycolytic origin might instead be introduced into a relatively small membrane-associated pool, as has been proposed for the transfer of ATP in the human red blood cell to the $(Na^+ + K^+)$-ATPase (38, 39). Such a compartmentation seems unlikely, however, for ATP exported by the mitochondrion. Note that the restoration occurred under the same conditions for the Na^+-independent uptake of the norbornane amino acid and the fully Na^+-dependent uptake of MeAIB.

ACTION OF OUABAIN ON PMS-ASCORBATE RESTORATION

Figure 1 also shows that ouabain blocks the restorative effect of phenazine-ascorbate, not only for the Na^+-dependent uptake of MeAIB, but apparently also for the Na^+-independent uptake of the norbornane amino acid (Fig. 1). This effect is all the more remarkable because in this cell ouabain acts only sluggishly to decrease the alkali-ion gradients; also because we lack any obvious basis for blockage by ouabain of Na^+-independent amino acid uptake. Perhaps you will suppose I have been wrong to stress in my first section the importance of discriminating among the transport systems, because the test of Fig. 1 seems to indicate a common basis of energization of 2 of them at least under selected conditions. This strong effect of ouabain, seen when the alkali-ion gradients are already unfavorable, and even for a model amino acid whose concentrative uptake is Na^+-independent, suggests however, that we do not yet fully understand the full action of ouabain on membrane energetics. Kimmich has suggested that this agent acts on an ATPase serving for both the Na^+-dependent transport of an organic metabolite and for the transport of the alkali ions (40). This proposal still does not seem broad enough to cover the present effect. A close

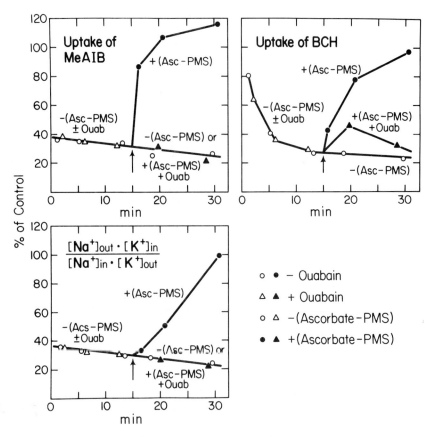

Fig. 1. Time course of restoration by phenazine methosulfate (0.1 mM) plus sodium ascorbate (20 mM) (ASC-PMS) of the 30-sec uptake of MeAIB (upper left) and of 2-aminonorbornane-2-carboxylic acid (upper right). Comparison with restoration of the alkali-ion gradients (lower left) $([Na^+]_{out} \cdot [K^+]_{in})/([Na^+]_{in} \cdot [K^+]_{out})$. The Ehrlich cells had been treated for the 30 min preceding the interval shown here with a 0.1 mM dinitrophenol and 1 mM iodoacetate, but these agents were absent from the medium for 15 min before the phenazine-ascorbate-reagent was added at the point indicated by the arrows. The $[^{14}C]$ amino acids were set at 20 μM. Note that the rate of amino acid accumulation into the cell responds immediately, whereas the alkali-ion gradient responds sluggishly. As shown on Fig. 1 of reference (37), the cellular ATP levels in the meantime do not recover perceptibly in the presence of the phenazine-ascorbate reagent. Note that the presence of ouabain at 2 mM largely prevents the recovery of amino acid uptake, even where it is Na$^+$-independent.

interaction among the modes of membrane energization is indicated also by the continued influence of the transmembrane potential on amino acid uptake by cells treated with dinitrophenol and iodoacetate: The presence of valinomycin or thiocyanate produced a 40–60% stimulation of MeAIB uptake without any increase in the ATP concentration and despite the low alkali-ion gradients.

QUINACRINE

Crane and Löw considered characteristic the quinacrine sensitivity of the NADH dehydrogenase of the plasma membrane of the hepatocyte and the adipocyte (31). We find that quinacrine at 2–3 mM terminates the uptake of MeAIB by the Ehrlich cell more quickly and more completely than any other inhibitor heretofore tested. After 30 sec of contact, the uptake of MeAIB measured during that 30 sec was decreased by 75–85%, before the cellular ATP level had been halved and before the alkali-ion gradients had even begun to fall. We reason that in this short time interval quinacrine is likely to have inhibited mainly dehydrogenase action in the plasma membrane, and rather less than in the mitochondrion. A further delay may be expected in the communication of the consequences of mitochondrial inhibition to the plasma membrane. To strengthen this argument, quinacrine should be tested on mitochondria-free plasma membrane vesicles. Quinacrine is as effective as ouabain in eliminating the restoration by pyruvate or by phenazine of MeAIB transport in the energetically depleted cells (see Fig. 1).

PRELIMINARY EXPERIMENTS WITH VESICLES PREPARED FROM THE PLASMA MEMBRANE OF THE EHRLICH CELL (28)

These vesicles were incubated for 30 sec in Krebs-Ringer phosphate medium containing 0.2 mM [^{14}C] MeAIB, at pH 7.4 and 37°C. The vesicles were then separated by filtration on a glass-fiber filter, and the radioactivity of the vesicles referred to the protein content. The following tabulation compares the ^{14}C taken up into the vesicles under these conditions with the amount taken up when the indicated agents were present during the 30 sec:

	nmoles MeAIB/mg protein
Control	0.148
0.4 mM quinacrine	0.231
0.33 ng FCCP/ml	0.224
1.8 mM NADH	0.071

These effects correspond to the acceptance of NADH by the plasma membrane to produce transport, if we suppose that more of our vesicles were everted than right-side-out. Extrusion of the entering MeAIB in that case would appear to have been inhibited to the extent of 51 and 56%, respectively, by quinacrine and by trifluoromethoxy-carbonyl-cyanide phenylhydrazone.

DISCUSSION

On the basis of the above results we propose provisionally that amino acid transport by the Ehrlich cell can be energized either by the alkali-ion gradients, by cellular ATP, or by reducing equivalents that may reach the plasma membrane from the mitochondrion by way of an unidentified shuttle. The natural electron acceptor which presumably accounts for the effectiveness of the reducing equivalents remains unidentified. In the case of the heretofore studied NADH dehydrogenase of plasma membranes, various artificial acceptors have been effective, including ferricyanide, glyoxylate, and dichlorophenol indophenol.

The malate-asparate shunt is known to be operative in the Ehrlich cell (41). Its contribution should be recognizable by a sensitivity to inhibition by the pyridoxal phosphate binding reagent, aminooxylate. A preliminary test failed to show inhibition of phenazine-ascorbate restoration of MeAIB uptake by the Ehrlich cell by this reagent at 0.5 mM.

The conditions of our experiments may be such as to maximize the proposed contribution of reducing equivalents from the mitochondrion to transport. It is known that the malate-asparate shunt tends to transfer reducing equivalents inwardly when ATP levels are high, and outwardly when ATP production is restricted (42). At the same time it does not appear to be fully proved that the breakdown of ATP per se drives amino acid transport; hence it would be premature to ascribe limits to the contribution of the mitochondrion by other of its products than ATP.

The effect of ouabain to block transport restoration may well have an origin other than its diminution of the alkali-ion gradients. It is known that external K^+ at 0–2 mM concentrations regulates oxygen consumption by this cell (43). These are K^+ levels that also govern alkali-ion transport by the plasma membrane; furthermore these effects of K^+ can be blocked by ouabain (Ref. 43 and references therein). The present sensitivity of Na^+-independent transport to ouabain suggests that amino acid transport is energized in part by a component of respiratory metabolism controlled by K^+-binding sites at the external surface of the cell.

Finally, we should note that an ability of the phenazine-ascorbate mixture or of NADH to energize transport by the plasma membrane has quite different implications than it does for vesicles of E. coli (44) or B. subtilis (45). In these bacterial organisms the plasma membrane is known to contain the respiratory chain, and energization of transport appears to occur through reactions within that chain. In cells of the higher animal, in contrast, the respiratory chain is considered to have taken its place in the inner mitochondrial membrane.

ACKNOWLEDGMENTS

I acknowledge research support from the Institute of Child Health and Human Development, Grant HD01233, National Institutes of Health, U.S. Public Health Service, and the important collaboration of the coauthors named in the list of references.

REFERENCES

1. Oxender DL, Christensen HN: J Biol Chem 238:3686, 1963.
2. Christensen HN: Adv Enzymol 32:1, 1969.
3. Christensen HN: Curr Top Membr Transp 6:227, 1975.
4. Thomas, EL, Christensen HN: Biochem Biophys Res Commun 40:277, 1970.
5. Christensen HN, Handlogten ME: J Membr Biol, in press.
6. Vidaver G: Biochemistry 3:662, 1964.
7. Thomas EL, Christensen HN: J Biol Chem 246:1682, 1971.
8. Eavenson E, Christensen HN: J Biol Chem 242:5386, 1967.
9. Wise WC: J Cell Physiol 87:199, 1976.
10. Wise WC: Fed Proc Fed Am Soc Exp Biol 35:605, 1976.
11. Enders RH, Judd RM, Donohue TM, Smith CH: Am J Physiol 230:706, 1976.
12. Christensen HN: In Levi G, Battistin L, Lajtha A (eds): "Transport Phenomena in The Nervous System." New York: Plenum Press, 1973, p 3.
13. Christensen HN, deCespedes C, Handlogten ME, Ronquist G: Biochim Biophys Acta 300:487, 1973.
14. Christensen HN, Handlogten ME: J Biol Chem 243:5428, 1968.
15. Christensen HN: In Proc 6th Int Cong Nephrol, Florence. Basel: Karger, 1975, p 134.
16. Christensen HN, Riggs TR: J Biol Chem 194:57, 1952.
17. Riggs TR, Walker LM, Christensen HN: J Biol Chem 233:1479, 1958.
18. Schultz SG, Curran PF: Physiol Rev 50:637, 1970.
19. Eddy AA: Biochem J 108:489, 1968.

20. Schafer JA, Heinz E: Biochem Biophys Acta 249:15, 1971.
21. Gibb LE, Eddy AA: Biochem J 129:979, 1972.
22. Heinz E, Geck P, Pietrzyk C, Pfeiffer B: In Semenza G, Carafoli E (eds): "Proc of FEBS Symp 42, Biochemistry of Membrane Transport." Berlin: Springer, 1977, p 236.
23. Schafer JA, Williams AE: In Silbernagel G, Lang F, Greger R (eds): "Amino Acid Transport and Uric Acid Transport." Stuttgart: Georg Thieme, 1976, p 20.
24. Christensen HN, Garcia-Sancho J, Sanchez A: J Supramol Struct, Suppl 1:154, 1977.
25. Christensen HN, deCespedes C, Handlogten ME, Ronquist G: Ann NY Acad Sci 227:335, 1974.
26. Christensen HN, Handlogten ME: Proc Natl Acad Sci USA 72:23, 1975.
27. Forte JG, Forte TM, Heinz E: Biochim Biophys Acta 298:827, 1973.
28. Im WB, Christensen HN, Sportés B: Biochim Biophys Acta 436:424, 1976.
29. Samudio I, Canessa M: Biochim Biophys Acta 120:165, 1966.
30. Löw H, Werner S: FEBS Lett 65:96, 1976.
31. Crane FL, Löw H: FEBS Lett 68:153, 1976.
32. Löw H, Crane FL: FEBS Lett 68:157, 1976.
33. Riggs TR, Walker LM: J Biol Chem 233:132, 1958.
34. Jacobs FA, Hillman RSL: J Biol Chem 232:445, 1958.
35. Akedo H, Sagawa T, Yoshikawa S, Suda M: J Biochem (Tokyo) 47:124, 1960.
36. Ueda K, Akedo H, Suda M: J Biochem (Tokyo) 48:584, 1960.
37. Garcia-Sancho J, Sanchez A, Handlogten ME, Christensen HN: Proc Natl Acad Sci USA 73, vol 74, p 1488.
38. Parker JC, Hoffman JF: J Gen Physiol 50:893, 1967.
39. Proverbio F, Hoffman JF: Ann NY Acad Sci 242:459, 1974.
40. Kimmich GA: Biochemistry 9:3669, 1970.
41. Greenhouse WVV, Lehninger AL: Cancer Res 36:1392, 1976.
42. Bremer J, Davies EJ: Biochim Biophys Acta 376:387, 1975.
43. Levinson C, Hempling HG: Biochim Biophys Acta 135:307, 1967.
44. Konings WN, Barnes EM, Kaback HR: J Biol Chem 246:5857, 1971.
45. Hayakawa K, Veda T, Kasaka I, Fukui E: Biochem Biophys Res Commun 72:1548, 1976.

Journal of Supramolecular Structure 6:215–228 (1977)
Molecular Aspects of Membrane Transport 141–154

Perspectives and Limitations of Resolutions — Reconstitution Experiments

Efraim Racker

Section of Biochemistry, Molecular and Cell Biology, Cornell University, Ithaca, New York 14853

Reconstitutions of membranous activities can tell us how many components are required and what their functions are. The mitochondrial proton pump is used as an example. Moreover, the biological activity, such as P_i transport, can be used in reconstituted vesicles as an assay during the isolation of the transporter.

Reconstitution experiments reveal the importance of membrane asymmetry and allow us to study conditions of vectorial assembly.

The mechanism of action of ion pumps has been successfully analyzed in reconstituted liposomes. We can study the movement of ions and the electrogenicity of the system without interference by other unrelated processes.

Based on studies with the resolved Ca^{2+}-ATPase of sarcoplasmic reticulum, we propose a novel formulation of the mechanism of ATP-driven ion pumps in which cyclic binding of Mg^{2+} plays a key role.

Key words: reconstitutions of ion pumps, coupling factors of oxidative phosphorylation, phospholipids, role in ion pump activity, mechanism of ATP-driven Ca^{2+} pump, oxidative phosphorylation, a new hypothesis, ATPases of membranes

It is apparent from this conference that resolution and reconstitution of membrane-linked functions in artificial liposomes have become fashionable. It seems appropriate to take stock and to evaluate what we can expect from such experiments and to point out their limitations. It should be obvious that reconstituted systems do not tell us exactly what happened in the natural membrane. Any biological system separated from the multitude of other intersecting functions is, strictly speaking, an artifact. The primary purpose of resolutions and reconstitutions is indeed the simplification of the system to the minimal number of components required for functions. The price we pay for getting away from the turbulence of metabolic events that take place in natural membranes seems worthwhile to anyone who has been exposed to the frustrations associated with analysis of events in cells or in organelles.

What can we learn from reconstitution experiments? I have outlined in Table I the subjects I shall discuss. In each case I shall stress limitations and deficiencies of the approach of resolution and reconstitution.

Abbreviations: DCCD – N,N′-dicyclohexylcarbodiimide; F_1, F_2, F_6 – coupling factors 1 (ATPase), 2 and 6 respectively; OSCP – oligomycin sensitivity conferring protein.
Received March 4, 1977; accepted April 19, 1977

TABLE I. What Can We Learn From Reconstitutions?

1. What are the parts required and what are their functions?
2. What are the functions of the phospholipids?
3. What is the role of asymmetry and how do we achieve asymmetrical orientation in reconstitution?
4. What is the mechanism of action of ion pumps?
   a) What ions are moving?
   b) Is it a carrier or a channel?
   c) Is the system electrogenic or electrically silent?
   d) How can we study the molecular mechanism of ion pumps?

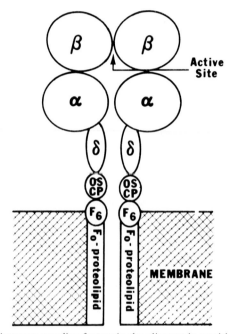

Fig. 1. Relationship between coupling factors in the oligomycin-sensitive ATPase complex.

WHAT ARE THE PARTS REQUIRED AND WHAT ARE THEIR FUNCTIONS?

I have chosen the mitochondrial proton pump to illustrate the reconstitution approach. Almost 20 years ago we resolved from submitochondrial particles the first coupling factor (F_1) for oxidative phosphorylation (1, 2). We learned 2 significant facts: The coupling factor was an ATPase, and it was resistant to oligomycin. At first sight both these facts were confusing. How can an ATPase function as a coupling factor without hydrolyzing the ATP it helps to generate? This puzzle was solved by the discovery of a mitochondrial protein which inhibits ATPase activity without interfering with ATP generation (3). The second puzzle, the resistance of the ATPase activity to oligomycin led to the discovery of a membranous complex that confers oligomycin sensitivity to added F_1 (4). This feature of conferral proved most valuable as an assay during subsequent fractionations and isolations of several additional coupling factors (cf 5).

The scheme of the oligomycin-sensitive ATPase shown in Fig. 1 is incomplete. It is not meant to convey either the number or the positions of the subunits of F_1. It is meant to convey our present knowledge of the broad relationship between the headpiece (F_1) and the other components of the complex. It shows the factors needed for the attachment of F_1 to the membrane and their relationship to the proton channel (F_0). It is partly based on studies on ATPase from E. coli (6) and from spinach chloroplasts (7) which have identified the δ subunit of F_1 as the peptide chain responsible for the binding of F_1 to the membrane. Oligomycin sensitivity conferring protein and F_6 (8) are the membrane components that interact with the subunit of F_1, with OSCP probably located between F_6 and the δ subunit.

What happened to F_2? Its fate exposes one of the limitations of reconstitution experiments. We can readily show that F_2 (factor B) is required for oxidative phorphorylation in defective submitochondrial particles (9). However, in our hands reconstituted liposomes have thus far failed to display a dependency on this factor. For this and other reasons we proposed that F_2 serves as a "sealing factor" which is required to lower the proton permeability of the inner mitochondrial membrane (5). The artificial liposomes made with a large excess of phospholipids appear to be sufficiently impermeable to protons without added F_2.

The function of F_0 as a proton channel was first proposed by Mitchell (10). The reactivity of F_0 with oligomycin and DCCD has been a major aid in the course of its isolation. The highly purified preparations from mitochondria (11), chloroplasts (12), and bacteria (13), which can be used for the reconstitution of liposomes that catalyze DCCD-sensitive proton translocation, contain 2–3 protein bands in addition to the subunits of F_1 and the other known coupling factors. A great deal of attention is now being paid to the proteolipid which was isolated from mitochondria (14, 15) and bacteria (16) based on its reactivity with ^{14}C-labeled DCCD. After extraction with chloroform-methanol, a water soluble proteolipid apoprotein, virtually free of phospholipids, was isolated from mitochondria and shown to enter readily into preformed liposomes suspended in aqueous media (17). The proteolipid apoprotein formed a proton channel which however was not blocked by the addition of DCCD or oligomycin. More gentle procedures (12, 18) are required for the preservation of reactivity with DCCD and for the reconstitution of a DCCD-sensitive proton channel. Further purification and demonstration of capacity to interact with coupling factors is needed for the evaluation of the role of the DCCD-sensitive proteolipid and of other hydrophobic components in the formation of the proton channel.

The reconstitution of the oligomycin-sensitive ATPase into liposomes and the demonstration that these vesicles catalyze oligomycin-sensitive ATP-driven proton translocation (19) was a decisive factor in our acceptance of the chemiosmotic hypothesis and stimulated further attempts to reconstitute oxidative phosphorylation.

The second example which illustrates the value of the reconstitution approach is the respiratory chain of mitochondria. Several complexes of the oxidation chain have been isolated (20) and incorporated with the native mitochondrial orientation into liposomes (21–24). As will be shown later these vesicles exhibit the phenomenon of respiratory control and they generate during oxidation a proton motive force which facilitates the electrogenic inward transport of K^+ in the presence of valinomycin (21). Although these experiments clearly demonstrate that the complexes catalyze proton translocation, the isolation and identification of the individual components which participate in electron transport and in proton translocation, is still a problem for the future.

The reconstitution of the oxidation complexes in the submitochondrial orientation together with the mitochondrial ATPase yielded vesicles which catalyzed oxidative phos-

TABLE II. Reconstitutions of Oxidative Phosphorylation*

Phospholipids	P:2e ratio		
	Site 1	2 + 3	3
Phosphatidylethanolamine + Phosphatidylcholine (4:1)	0.5	0.63	0.4 (0.35)
Phosphatidylethanolamine + Phosphatidylcholine (1:1)	0.8	—	0.2

*The experimental procedures were as described (34, 35).

phorylation. As shown in Table II, when the cholate dialysis procedure was used for reconstitution a phosphatidylethanolamine:phosphatidylcholine ratio of 4 was optimal. The reconstitution of site 2 + 3 yielded with succinate a P:O ratio of 0.63, which was about twice as high as the P:O ratio of 0.35 obtained with the same vesicles with ascorbate-phenazine methosulfate as substrate. The reconstitution of site 3 of oxidative phosphorylation in submitochondrial orientation posed difficulties which illustrate the limitations of our current reconstitution methods. I shall return to this problem when I discuss problems associated with asymmetric reconstitution.

The demonstration (25) that bacteriorhodopsin, which catalyzes a light-driven translocation of protons, can substitute for the mitochondrial respiratory enzymes in the above described experiments, proved that the chemiosmotic mechanism described by Mitchell (10) is operative in a reconstituted system. Since no oxidation-reductions accompany the translocation of protons via bacteriorhodopsin, it was unnecessary to assume that a direct contact takes place between members of the respiratory chain and the ATPase as proposed in the original conformational hypothesis (26). Moreover, the fact that the purple patches of halobacteria contain bacteriorhodopsin as the only protein component excludes the possibility of a direct contact between the ATPase and rhodopsin during light-driven ATP generation in the intact bacteria (27).

In spite of this persuasive evidence we have to agree that mechanisms that operate in model systems or even in bacteria need not be the same in membranes of higher organisms that have evolved and have become more complex and more efficient.

Recent experiments with buffered chloroplasts (28) suggest that the proton flux during light-driven electron transport may not be in equilibrium with the bulk phase of the internal chloroplast water. It is appealing to think in terms of a thin layer of "structured water" associated with the membrane surface, a concept which would eliminate some of the difficulties posed by thermodynamic calculations based on the measured pH differences between the inside of chloroplasts and the external medium. It may be of interest in the future to conduct similar experiments with reconstituted systems since it should be possible to control the surface:volume ratio by varying the size of vesicles, e.g., by fusion (29).

WHAT ARE THE FUNCTIONS OF THE PHOSPHOLIPIDS?

The primary function of the phospholipids is to provide a compartment which is impermeable to ions and other solutes that must be kept within the cell at appropriate concentrations. The phospholipids must also serve as a suitable matrix for the embedding

TABLE III. Restoration of Activities to Delipidated Ca^{2+}-ATPase*

Additions to delipidated enzyme	ATPase μmoles/min/mg	Phosphoenzyme nmoles/mg	Transport nmoles/min/mg
None	0	0.3	0
Phosphatidylcholine	4.0	1.9	0
Phosphatidylethanolamine	2.6	0.8	183
Acetyldilaurylphospha-tidylethanolamine	0.12	2.3	(no vesicles)
Phosphatidylethanolamine/Phosphatidylcholine (4:1)	3.3	1.5	344[a]

*The experimental procedure was as described (30).
[a]Phosphatidylethanolamine/phosphatidylcholine (3:1)

of proteins which facilitate the entry of desirable and the exit of undesirable solutes. It seems likely that an asymmetric assembly of phospholipids participates in the asymmetric orientation of transport proteins although experimental evidence for such a relationship is not as yet available. Finally, there are indications that the phospholipids specifically influence the individual catalytic steps involved in the transport process

It can be seen from Table III that a preparation of Ca^{2+}-ATPase from sarcoplasmic reticulum that has been stripped of phospholipids no longer catalyzed the catalytic functions associated with its enzymatic activity (30). Phosphatidylcholine restored to the protein the ability to form phosphoenzyme and to hydrolyze ATP, but did not allow for efficient Ca^{2+} transport. Phosphatidylethanolamine on the other hand was less effective in the first 2 functions, but formed vesicles that catalyzed Ca^{2+} transport. Both phospholipids added together during reconstitution gave the highest values for Ca^{2+} transport. An interesting example is acetyldilaurylphosphatidylethanolamine. It was completely unsuitable for the formation of transport vesicles, it supported very low ATPase activity, but it was the most effective lipid for phosphoenzyme formation. It is apparent from these data that by the appropriate choice of phospholipids we can disect the first step of the catalytic process and study it independently of subsequent events.

The chemical modification of the amino group of phosphatidylethanolamine was explored in experiments conducted in collaboration with Dr. H. G. Khorana (31). As shown in Table IV, acetylation of phosphatidylethanolamine did not impair the ability of the phospholipid to form vesicles with bacteriorhodopsin. Although the kinetics of H$^+$ pumping were somewhat altered, the overall pump function was not significantly altered by the introduction of the acetyl group. On the other hand, the Ca^{2+} transport activity of reconstituted vesicles was lost completely. Significantly, it could be restored by incorporation of an hydrophobic alkylamine such as stearylamine or oleolylamine. Amines with shorter chain lengths were less effective. Of particular significance is the observation that the ATPase activity was not well supported by acetylated phospholipid as already indicated by the data shown in Table III. Here again, incorporation of stearylamine sustained the catalytic function. Since dicetylphosphate inhibited and could be quantitatively titrated against the stearylamine, it appears that the catalytic function of the ATPase is greatly influenced by the surface charge of the membrane in which it is embedded. Acetylphosphatidylethanolamine was also ineffective for the reconstitution of the mitochondrial ATPase which catalyzes a ^{32}P$_i$-ATP exchange. Stearylamine again sustained the activity.

TABLE IV. Acetylphosphatidylethanolamine in Reconstituted Systems*

	Phosphatidyl- ethanolamine (PE)	Acetyl PE	Acetyl PE plus stearylamine
Proton pump	2,360	2,400	—
Calcium pump	114	0	109
Calcium ATPase	890	200	740
$^{32}P_i$-ATP exchange[a]	133	0	112

*The activity of the proton pump is expressed as the extent of n atoms H^+ pumped per mg of bacteriorhodopsin; the other activities are expressed as nmoles/min/mg protein.
[a]Reconstituted with phosphatidylcholine plus the other indicated lipids. The experimental procedure was as described (31).

However, in this case we could show that the acetylphosphatidylethanolamine phospholipid was required in addition to phosphatidylcholine, although the latter was capable by itself of forming impermeable compartments as illustrated by studies with the bacteriorhodopsin proton pump. These studies show how reconstitution can shed light on the contribution of individual phospholipids and allow us to modify the chemistry of the participating components.

Among the limitations of this approach, at least at the present time, is our lack of understanding of the significance of the large variety of lipids present in natural membranes. Some of these components may play a role that we do not appreciate. Thus, reconstitutions with highly purified or even synthetic phospholipids that do not contain the minor components may lead us astray. For example, the role of large quantities of alkylated phospholipids present in many natural membranes is unknown. Reconstitution experiments have thus far not revealed any specific functions for these phospholipids (32), but I have little doubt that there is a good reason for their presence.

Another drawback of reconstitutions is that phospholipid requirements vary with the method used for the formation of liposomes. With the cholate dialysis procedure a phosphatidylethanolamine:phosphatidylcholine ratio of 4 is optimal (see Table II) for most of the systems studied thus far. With the cholate dilution and particularly with the sonication procedure, the lipid requirements are quite different. For example, for the reconstitution of the Ca^{2+} pump by sonication, phosphatidylethanolamine alone was rather ineffective. Transport activity was much more dependent on the presence of phosphatidylcholine than in the case of the dialysis procedure. Differences in procedure must be invoked also in the discrepancies of observations recorded in the literature, such as the effectiveness of phosphatidylcholine as the sole lipid in the reconstitution of the Ca^{2+} pump (33) which is contrary to the data shown in Table III.

WHAT IS THE ROLE OF ASYMMETRY AND HOW CAN WE ACHIEVE IT?

One of the lessons we learned during studies of oxidative phosphorylation is that the orientation of proteins in reconstituted vesicles varies greatly depending on the experimental procedure. It is important in such experiments to devise assays which determine both the percent incorporation and the relative orientations of the protein in the artificial

TABLE V. Uncoupling of Oxidative Phosphorylation by Cytochrome c*

	ATP formation (nmoles/min)
Complete reconstituted systems	9.3
Minus polylysine	5.3
Minus polylysine plus cytochrome c (10 μg)	2.7
Minus polylysine plus cytochrome c (25 μg)	1.8
Plus ETP$_H$ particles	18.8
Plus ETP$_H$ particles plus cytochrome c (25 μg)	19.4

*The reconstituted vesicles (350 μg) were assayed as described (34) with and without 50 μg of polylysine. The ETP$_H$ particles were assayed only without polylysine.

membrane. This is particularly essential in systems that must be oriented unidirectionally to permit measurements of functions as in the case of the third site of oxidative phosphorylation. With all methods of reconstitution used thus far, cytochrome oxidase is preferentially assembled in the mitochondrial orientation with the cytochrome c side of the enzyme facing the medium. Our reconstitution of site 3 of oxidative phosphorylation required however the opposite, namely the inside-out orientation of submitochondrial particles, since under these conditions the system was independent of the transport of P_i and ADP which are not permeant without the aid of specific transporters. With luck and the help of cytochrome c we achieved some submitochondrial orientation of cytochrome oxidase which was masked, however, by an excess of the enzyme in the right-side-out orientation (34). Only when we learned to displace all of the residual cytochrome c from the outside surface of the vesicles, could we demonstrate phosphorylation coupled to electron transport. When cytochrome oxidase was allowed to function in both directions, the membrane potential and the ΔpH was collapsed and no phosphorylation took place. In contrast to natural submitochondrial particles the reconstituted vesicles were thus uncoupled in the presence of small amounts of cytochrome c as shown in Table V.

This story reveals one of the major limitations of reconstitution: our inability to guide the orientation of a protein in a desired manner. Although the incorporation procedure, which I shall describe below, allows for unidirectional orientation, we have no control over its direction: it is right-side out in the case of cytochrome oxidase and inside out in the case of the oligomycin sensitive ATPase. Whether the final outcome suits the experimenter is mainly a matter of luck. In the case of bacteriorhodopsin the preferred inside-out orientation of reconstituted vesicles was a convenient feature which permitted the light-driven formation of ATP by the mitochondrial ATPase which also selects the inside-out orientation during reconstitution by current procedures.

At least in some instances assays can be devised that tell us accurately the quantitative distribution of inside-out and right-side-out orientations. It is also often important to establish the percentage of total protein that is incorporated into the liposomes. Such methods are available for some systems and will be described later for cytochrome oxidase vesicles. In this case a detergent (Tween 80) can be used to open the vesicles without inhibiting enzymatic activity. In the case of reconstituted vesicles catalyzing site 1 of oxidative phosphorylation, spectral change with impermeant reactants were used (35). Specific inhibitors which react with one side of an asymmetric transporter allow us in some instances to determine accurately the inside-out and right-side-out distribution of recon-

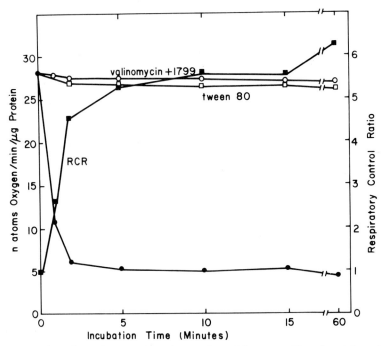

Fig. 2. Incorporation of cytochrome oxidase into performed liposomes. Experimental conditions were as described (39). ●———● oxygen uptake; ■———■ respiratory control ratio (RCR).

stituted vesicles. For example, atractyloside was used for the evaluation of the adenine nucleotide transporter (36), and N-ethylmaleimide for phosphate transport (37). It helps when hydrophobic counterparts of the inhibitors are available such as bongkrekic acid for the nucleotide transporter and N-benzylmaleimide for the P_i transporter. They serve as controls assuring proper operation of the transport system. Moreover, impermeant inhibitors such as atractyloside or ouabain can be included during reconstitution to allow interaction with the sensitive site of the transport system. Such impermeant inhibitors can be useful in establishing the amount of enzyme that has escaped reconstitution, which may be in large excess of the portion that can be incorporated into liposomes. An example for this is the reconstituted Na^+, K^+-ATPase from electric eel (38). Since both ouabain and ATP do not readily permeate through liposomes, all the ATPase activity which is ouabain-sensitive represents the fraction of enzyme that has not been reconstituted either inside out or right-side out.

RECONSTITUTION BY THE INCORPORATION PROCEDURE

We have recently described a method of reconstitution which avoids scrambling of orientation (39). The procedure is simple and consists of preparing liposomes which contain phosphatidylethanolamine and about 30% of an acidic phospholipid. When a hydrophobic protein is added to such vesicles it is rapidly incorporated in an asymmetric and unidirectional orientation. As illustrated in Fig. 2 the oxidation of reduced cyto-chrome c by cytochrome oxidase takes place at a rapid rate (zero incubation time). When the enzyme is added to phosphatidylserine containing liposomes, the uptake of oxygen

TABLE VI. Sequential Incorporation of Proteins Into Liposomes*

	Respiratory control	
	Expected	Observed
Exp. 1		
Cytochrome oxidase added to liposomes		10
F_0 added to COV plus liposomes	8.2	5.1
Exp. 2		
Cytochrome oxidase added to liposomes		4.7
Cytochrome oxidase added to F_0 liposomes plus liposomes	1.5	4.4

*Cytochrome oxidase or the hydrophobic protein of the ATPase complex (F_0) were added to an equal mixture of protein-free liposomes and protein-containing liposomes as indicated. The expected results were calculated based on random incorporation. Experimental details were as described (42).

drops within a few minutes to 10–20% of the original rate. On addition of valinomycin and 1799 (or nigericin) the original rate of oxidation is restored. This phenomenon, which we look upon as a form of respiratory control, is characteristic for cytochrome oxidase vesicles reconstituted by other procedures (22). The unambiguous interpretation of these observations is that the protein is incorporated asymmetrically into the proton-impermeable liposomes. The collapse of both the ΔpH (e.g., by nigericin) and of the membrane potential by valinomycin is required to release the control mechanism imposed by the proton motive force. When 3% Tween 80 was added to open the vesicles (40) the rate of oxidation was the same as in the presence of the ionophores. As mentioned earlier, if cytochrome c was present during reconstitution, both inside-out and right-side-out orientation of cytochrome oxidase was observed. Stimulation of respiration by ionophores was then only one half or less of the stimulation observed in the presence of Tween 80 (40).

The incorporation procedure illustrated in Fig. 2 was effective with several other membrane proteins, e.g., mitochondrial ATPase, QH_2 cytochrome c reductase, and Ca^{2+}-ATPase (39, 41). The method lends itself to interesting explorations. What effect has one protein already present in the membrane on the incorporation of a second protein? We can give a protein which is to be incorporated a choice between liposomes which are protein-free and liposomes which already contain a protein. We can calculate the results expected from a random distribution provided we have an assay that tells us what proportion of vesicles contain both proteins in the same membrane. The rather unexpected results that were observed in such experiments, opened a field which I facetiously refer to as "molecular psychology."

As shown in Table VI the data show marked deviation from random distribution. The assay used in these experiments is based on the observation that cytochrome oxidase vesicles catalyze a low rate of respiration in the presence of reduced cytochrome c. When a proton channel is present in the same vesicle (22) the rate of oxygen uptake is greatly increased. In these experiments the hydrophobic protein of the oligomycin-sensitive ATPase was used as a proton channel.

The conclusions that can be drawn from these experiments are: 1) Cytochrome oxidase avoids liposomes that contain the proton channel. The enzyme is preferentially incorporated into protein-free liposomes. On the other hand, cytochrome oxidase chooses vesicles that contain QH_2 cytochrome c reductase over protein-free liposomes (41). 2) The hydrophobic proteins of the oligomycin-sensitive ATPase are incorporated into liposomes that contain cytochrome oxidase in preference to liposomes which contain no protein. Thus, there appears to be considerably specificity in the selection process that takes place during the incorporation of proteins into the liposomes.

Once more I should like to point to the limitations of such reconstitution procedures. Although these experiments may indeed have physiological significance and may shed light on the mechanism of protein incorporation into membranes, as well as on trans-membranous protein movements during secretion, it is clear that the assembly of cyto-chrome oxidase in natural membrane is a much more complex and integrated process of subunit assembly which may take place from both sides of the membrane (42).

WHAT IS THE MECHANISM OF ACTION OF ION PUMPS?

a. What ions are moving? The complexity of natural membranes, such as the mito-chondrial inner membrane in which many translocation processes take place simultaneous-ly, makes it difficult to assess the individual translocation events and the compulsory movements of solutes associated with them. Even in much simpler systems, such as in the sarcoplasmic reticulum, the permeability of the membrane to phosphate has precluded the evaluation of a hypothesis (43) that the terminal phosphate of ATP moves together with the Ca^{2+}. This mechanism could be eliminated in experiments with ATP^{32} and recon-stituted Ca^{2+} vesicles which are quite impermeable to phosphates (44). Although we admit once more that the reconstituted model does not necessarily represent the events in the natural membrane, the above mentioned experiments demonstrated that a translocation of the terminal phosphate of ATP is not essential for the mechanism. Thus the burden of proof rests on the proponent of the hypothesis who has to show that the mechanism operative in the natural membrane is fundamentally different from that in the reconstituted system.

b. Is it a carrier or a channel? An example for an experimental approach to this question has been described for bacteriorhodopsin which was reconstituted with phospho-lipids with known transition temperatures. The experiments showed that the pump was operative well below transition temperatures eliminating a mobile carrier mechanism. Rather than describing these experiments in detail I refer the reader to the original paper (45). I want to point out here some of the limitations of this approach. For example. we have attempted similar experiments with the Ca^{2+}-ATPase of sarcoplasmic reticulum. This enzyme would have been particularly suitable for such experiments since complete delipi-dation with retention of reconstitutive activity has been achieved (30). Such delipidation experiments have yet to be performed with bacteriorhodopsin. Unfortunately in the case of the Ca^{2+}-ATPase we do not have available phospholipids with appropriate transition temperatures that are suitable for Ca^{2+} transport and Ca^{2+}-ATPase activity. The available phospholipids have transition temperatures which are too low, and the rates of Ca^{2+} translo-cation at these temperatures are too slow to permit an unambiguous interpretation of the data.

c. Is the system electrogenic or electrically silent? The best illustration for the difficulties encountered in the evaluation of the electrogenicity of ion movements in natural membrane is the controversy about the nature of the transport of adenine nucleo-tides and phosphate in the mitochondrial membrane. Peaks of controversy are reached when various proponents disagree with each other as well as with themselves. In the case of the nucleotide transporter, estimates of electrogenicity range from 0 to 85% (cf 46, 47).

TABLE VII. Reconstitution of P_i Transporter*

Additions	P_i/OH	P_iP_i exchange
	nmoles/min/mg	
Reconstituted vesicles	32	87
" plus valinomycin	64	—
" plus nigericin	58	—
" plus valinomycin plus nigericin	117	99

*The experimental procedure was as described (48).

TABLE VIII. Reconstitution of Nucleotide Transporter*

Reconstitution	ATP/ADP exchange nmoles/min/mg	
	Without ionophores	Plus valinomycin plus nigericin
Nucleotide transporter	102	210
P_i transporter	0	< 20
Nucleotide transporter plus P_i transporter	233	237

*The experimental procedure was as described (49).

Experiments with reconstituted vesicles allow analysis of the transport in isolation. We have reported previously (36) that adenine nucleotide transport is stimulated by iono- phores which collapse the membrane potential indicating electrogenicity. Recently we have obtained similar results with reconstituted P_i transporter (48, 49). As shown in Table VII, in reconstituted vesicles the P_i/OH exchange is markedly stimulated by addition of both valinomycin and nigericin. The P_i/P_i exchange is rapid and not stimulated by the ionophores. We proposed that valinomycin is required to collapse the membrane potential when negatively charged phosphate ions move into the vesicles and are electrically not compensated by the charge of OH$^-$ moving out (or H$^+$ moving in). Nigericin is required to collapse the resulting ΔpH. In the experiment shown in Table VIII the nucleotide trans- porter was reconstituted alone as well as together with the P_i transporter. The latter stimulated the ATP/ADP exchange in the absence of the ionophores but not in their presence. We therefore suggest that in the reconstituted system both P_i and adenine nucleo- tide transporter are electrogenic, and that the overall process of P_i and ADP moving in and ATP moving out is essentially electrically neutral. Thus these 2 functionally related pro- cesses may be electrically coupled.

These suggestions are based on reconstitution experiments and cannot be applied without reservations to the events in mitochondria where a membrane potential is imposed by the respiratory chain. Moreover, there are data suggesting that in mitochondria P_i trans- port is electrically neutral (50, 51). It should be remembered, however, that it is virtually impossible to eliminate side reactions in intact mitochondria. Thus H$^+$/K$^+$ and H$^+$/Na$^+$ antiporter activities were invoked to explain differences in P_i transport depending on the monovalent cations that were used (51). Moreover, the possibility that valinomycin might stimulate the swelling of mitochondria in the presence of ammonium phosphate was not explored (or not recorded). Finally, the possibility that the transporter does indeed

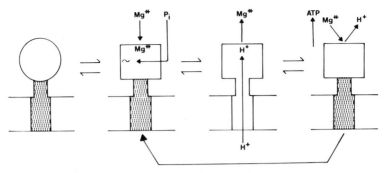

Fig. 3. Proposal for the mechanisms of ATP generation by the proton translocating mitochondrial ATPase.

operate in artificial liposomes by a mechanism that is different from that in natural membranes cannot be ruled out, even though such a notion may be distasteful to some.

　　　d. **How can we study the molecular mechanism of ion pumps?** Since I have recently reviewed the extensive work that has been performed in numerous laboratories on the mechanism of action of ATP-driven ion pumps (52), I shall restrict the discussion here to our recent calorimetric measurements with various ATPases (53, 54). We have shown that the isolated Ca^{2+}-ATPase catalyzes the net formation of ATP from P_i and ADP in a 2-step reaction (55). The ATP formed was not firmly bound by the enzyme but released into solution. Moreover, the process of ATP formation could be repeated if the protein was precipitated in the presence of EDTA. Finally, formation of a Ca^{2+} gradient across a membrane could be ruled out by performing experiments in the presence of a Ca^{2+} ionophore such as A-23187. All these experiments pointed to an interaction between inorganic ions and the proteins as the driving force for ATP formation. Calorimetric measurements with the Na^+,K^+-ATPase from electric eel (53) or with the Ca^{2+}-ATPase from sarcoplasmic reticulum (54) revealed large enthalphy changes during the interaction of the protein with either Mg^{2+} or P_i (about 40 Kcal/mole of enzyme). These and other findings suggest that major conformational changes of the protein take place during interaction with the ions.

　　　A novel formulation of the mechanism of oxidative and photophosphorylation was proposed based on these experiments (56). Briefly, the following sequence of events is visualized to take place during ATP formation by the mitochondrial ATPase (Fig. 3). The interaction of Mg^{2+} alters the conformation of membranous F_1 in such a manner that the protein can now accept a phosphate by either forming a covalent intermediate as in the case of the Na^+, K^+- or Ca^{2+}-ATPases or by forming an activated ionic complex as suggested by Jenks (57). A 2-stage model of 2 conformational states of the enzyme in which the equilibrium is pulled toward the low energy form by addition of Mg^{2+} was suggested in our first publication of the calorimetric measurements (53). In the next step the proton flux generated by the respiratory chain reaches the Mg^{2+} bound to the enzyme and pushes it into the medium. With the discharge of the phosphoryl group, which is transferred to ADP, the enzyme returns to the closed form which excludes the proton flux from the Mg^{2+} site. Now the Mg^{2+} can return to its binding site displacing the proton into the medium and the process recycles. According to this formulation the major function of the proton flux is the displacement of Mg^{2+}, and the major function of the conformational changes is the closing and opening of the proton channel which communicates with the Mg^{2+} site as well as changes in the affinity of the protein to Mg^{2+} and Ca^{2+}.

I need not advertise the limitations associated with conclusions based on experiments with ATPase proteins that are removed from their natural membrane. Yet these experiments have led to the formulation of a new hypothesis which explains at the molecular level how an ion gradient can be utilized to generate the formation of ATP from ADP and P_i. Moreover, this hypothesis suggests experiments that can be conducted with functional membranes which might tell us whether the predicted replacements of Mg^{2+} by H^+ take place during the process of oxidative and photophosphorylation.

ACKNOWLEDGMENTS

This investigation was supported by Grant number CA-08964, awarded by the National Cancer Institute, DHEW; Grant number BC-156 from the American Cancer Society; Grant number BMS-75-17887 from the National Science Foundation.

REFERENCES

1. Pullman ME, Penefsky HS, Datta A, Racker E: J Biol Chem 235:3322, 1960.
2. Penefsky HS, Pullman ME, Datta A, Racker E: J Biol Chem 235:3330, 1960.
3. Pullman ME, Monroy GC: J Biol Chem 238:3762, 1963.
4. Racker E: Biochem Biophys Res Commun 10:435, 1963.
5. Racker E: "A New Look at Mechanisms in Bioenergetics." New York: Academic Press, 1976.
6. Smith JB, Sternweiss PC, Heppel LA: J Supramol Struct 3:248, 1975.
7. Younis HM, Winget GD, Racker E: J Biol Chem 252:1814, 1977.
8. Knowles AF, Guillory RJ, Racker E: J Biol Chem 246:2672, 1971.
9. Racker E, Fessenden-Raden JM, Kandrach MA, Lam KW, Sanadi DR: Biochem Biophys Res Commun 41:1474, 1970.
10. Mitchell P: Biol Rev Cambridge Philos Soc 41:445, 1966.
11. Serrano R, Kanner BI, Racker E: J Biol Chem 251:2453, 1976.
12. Winget GD, Kanner N, Racker E: Biochim Biophys Acta 460:490, 1977.
13. Sone N, Yoshida M, Hirata H, Kagawa Y: J Biol Chem 250:7917, 1975.
14. Cattell KJ, Knight IG, Lindop CR, Beechey, RB: Biochem J 125:169, 1970.
15. Stekhoven FS, Waitkus RF, van Moerkerk HTB: Biochemistry 11:1144, 1972.
16. Filingame RH: J Biol Chem 251:6630, 1976.
17. Racker E: In Quagliariello E, Papa S, Palmieri F, Slater EC, Siliprandi N (eds): "Proceeding of International Symposium on Electron Chains and Oxidative Phosphorylation." Amsterdam: North-Holland Publishing Company, 1975, p 401.
18. Nelson N, Eytan E, Notsani B-E, Sigrist H, Sigrist-Nelson K, Gitler C: Proc Natl Acad Sci USA (In press).
19. Kagawa Y, Racker E: J Biol Chem 246:5477, 1971.
20. Hatefi Y, Haavik AG, Fowler LR, Griffiths DE: J Biol Chem 237:2661, 1962.
21. Hinkle PC, Kim J-J, Racker E: J Biol Chem 247:1338, 1972.
22. Racker E: J Membr Biol 10:221, 1972.
23. Ragan CI, Hinkle PC: J Biol Chem 250:8472, 1975.
24. Leung KH, Hinkle PC: J Biol Chem 250:8467, 1975.
25. Racker E, Stoeckenius W: J Biol Chem 249:662, 1974.
26. Boyer PC: In King TE, Mason HS, Morrison M (eds): "Oxidases and Related Redox Systems." New York: John Wiley & Sons, vol 2, p 994, 1965.
27. Danon A, Stoeckenius W: Proc Natl Acad Sci USA 71:1234, 1974.
28. Ort DR, Dilley RA, Good NE: Biochim Biophys Acta 449:108, 1976.
29. Miller C, Racker E: J Membr Biol 26:319, 1976.
30. Knowles AF, Eytan E, Racker E: J Biol Chem 251:5161, 1976.
31. Knowles AF, Kandrach A, Racker E, Khorana HG, J Biol Chem 250:1809, 1975.
32. LaBelle EF, Racker E: J Membr Biol 31:301, 1977.
33. Warren GB, Toon DA, Birdsall NJM, Lee AG, Metcalfe JC: Proc Natl Acad Sci USA 71:622, 1974.

34. Racker E, Kandrach A: J Biol Chem 248:5841, 1973.
35. Ragan CI, Racker E: J Biol Chem 248:2563, 1973.
36. Shertzer H, Racker E: J Biol Chem 251:2446, 1976.
37. Rhodin TR, Racker E: Biochem Biophys Res Commun 61:1207, 1974.
38. Racker E, Fisher LW: Biochem Biophys Res Commun 67:1144, 1975.
39. Eytan GD, Matheson MJ, Racker E: J Biol Chem 251:6831, 1976.
40. Carroll R, Racker E: J Biol Chem (In press).
41. Eytan GD, Schatz G, Racker E: In Abrahamsson S, Pascher I (eds): "Structure of Biological Membranes." New York: Plenum Press, Nobel Symposium 34, Sweden, June 1976, p 373, 1977.
41. Schatz G, Mason TL: Annu Rev Biochem 43:51, 1974.
43. Martonosi A: In "Current Topics in Membranes and Transport." New York:Academic Press, vol 3, p 83, 1973.
44. Knowles AF, Racker E: J Biol Chem 250:3538, 1975.
45. Racker E, Hinkle PC: J Membr Biol 17:181, 1974.
46. Klingenberg M, Rottenberg H: Eur J Biochem 73:125, 1977.
47. Thayer WS, Hinkle PC: J Biol Chem 248:5395, 1973.
48. Banerjee RK, Shertzer HG, Kanner BI, Racker E: Biochem Biophys Res Commun 75:772, 1977.
49. Shertzer HG, Kanner BI, Banerjee RK, Racker E: Biochem Biophys Res Commun 75:779, 1977.
50. Chappell JB: Br Med Bull 24:150, 1968.
51. Mitchell P, Moyle J: Eur J Biochem 9:149 (1969).
52. Racker E: Trends in Biochemical Sciences 1:244, 1976.
53. Kuriki Y, Halsey J, Biltonen R, Racker E: Biochemistry 15:4956, 1976.
54. Epstein M, Kuriki Y, Biltonen R, Racker E: International Symposium on Calcium Binding Proteins, (abstract) June 5–10, Cornell Univ, Ithaca, NY, 1977.
55. Knowles AF, Racker E: J Biol Chem 250:1949, 1975.
56. Racker E: Annu Rev Biochem (In press).
57. Jenks WP: Adv Enzym 43:219, 1975.

Journal of Supramolecular Structure 6:229–238 (1977)
Molecular Aspects of Membrane Transport 155–164

Isolation and Purification of Bacterial Membrane Proteins by the Use of Organic Solvents: The Lactose Permease and the Carbodiimide-Reactive Protein of the Adenosinetriphosphatase Complex of Escherichia Coli

Karlheinz Altendorf, Margot Lukas

Lehrstuhl für Mikrobiologie II, Universität Tübingen D-7400 Tübingen, W.-Germany

Brigitte Kohl, Clemens R. Müller and Heinrich Sandermann, Jr.

Lehrstuhl für Biochemie der Pflanzen, Biologisches Institut II, Universität Freiburg, D-7800 Freiburg, W.-Germany

Techniques for the solubilization and fractionation of integral membrane proteins have been developed in recent years. A small portion of membrane protein (about 2%, proteolipid fraction) will partition into chloroform or 1-butanol, and, in several cases, these proteins retain functional activity. A virtually complete solubilization can be achieved at neutral pH by use of aprotic solvents, like hexamethylphosphoric triamide or N-methylpyrrolidone.

At relatively low concentrations (< 3 M) aprotic solvents inhibited β-D-galactoside transport by whole cells and the derivative membrane vesicles of Escherichia coli, but this inhibition could be largely reversed by a simple washing procedure. At higher concentrations of aprotic solvent (5–6 M), 50–80% of the total protein of lactose transport-positive membrane vesicles was solubilized. When these extracts were added to intact lactose transport-negative membrane vesicles, lactose transport was reconstituted, the required energy being provided by either respiration (e.g., addition of D-lactate) or by a K^+ diffusion potential established with the aid of valinomycin.

The dicyclohexylcarbodiimide (DCCD)-reactive subunit of the E. coli ATPase complex was found to partition into chloroform, and to be amenable to further purification in organic solvent. Ether precipitation and chromatography on DEAE-cellulose and hydroxypropyl-Sephadex G-50 yielded an homogeneous polypeptide of an apparent molecular weight of 9,000.

The purified and unlabeled DCCD-reactive protein was incorporated into K^+-loaded liposomes, and a membrane potential was generated by the addition of valinomycin. There are indications that the DCCD-reactive protein alone made the membrane specifically permeable for protons.

Key words: Escherichia coli, lactose permease, carbodiimide-reactive protein, Ca^{2+}, Mg^{2+}-ATPase, aprotic solvents, organic solvents, integral membrane proteins, bioenergetics

Received March 13, 1977; accepted April 10, 1977

INTRODUCTION

Many peripheral membrane proteins (1) of bacterial membranes, like the periplasmic binding proteins of various transport systems (2) and the BF_1 portion of the ATPase complex (3), have been purified to homogeneity. However, great difficulties still remain in the solubilization and purification of integral membrane proteins (1) in their active forms. The solubilization of these proteins by detergents has, in a number of cases (4, 5), been an important step in the purification procedure. However, in other cases detergents are highly inhibitory. For example, binding of a high-affinity β-D-galactoside to the E. coli lactose permease is completely abolished in the presence of 0.1% Trition X-100 (6). This protein has therefore not yet been solubilized in a functional form.

Two alternative methods for the solubilization of integral membrane proteins also have been used. Chaotropic agents effectively solubilize E. coli membrane proteins (7), and the D-lactate dehydrogenase has been solubilized in active form by 0.75 M guanidine hydrochloride (8). The present paper will describe, however, a further method for the solubilization of internal membrane proteins using organic solvents.

RESULTS AND DISCUSSION

Solubilization by Organic Solvents

Isolation of proteolipids. The solubilization of certain membrane proteins like the enzyme II of the phosphotransferase transport system (9) and phospholipase A_1 (10) of E. coli, requires the addition of 5 and 15% (vol/vol) 1-butanol, respectively. When more than about 20% (vol/vol) 1-butanol is added to aqueous buffer, a 2-phase system results. Proteolipids can therefore be defined as polypeptides which, in a water/organic solvent 2-phase system, will partition into the organic phase. This definition leaves open the question of whether partitioning is due to the hydrophobicity of the polypeptide proper, or to complex formation with lipids.

Protein fractions soluble in organic solvent (chloroform) were initially discovered in a number of tissues (11), and the myelin proteolipid has been particularly well studied (12). Recent reviews are available (13, 14).

The proportion of the protein of various biological membranes which, at pH values greater than 4, will partition into a chloroform or 1-butanol phase is typically around 2%. Functional activity may be retained under such conditions. Solubilization with 1-butanol, and a number of newly-developed methods applicable to organic solvents, were used for the purification to homogeneity of the extremely hydrophobic C_{55}-isoprenoid-alcohol kinase of Staphylococcus aureus (15, 16). Similar techniques were also successfully applied to the following catalytic membrane proteins: the glycosyl-transferases (17) and C_{55}-isoprenoid-alcohol kinase (18) of Klebsiella aerogenes, the proline transport system of E. coli (19) and the dicyclohexylcarbodiimide (DCCD)-reactive subunit of the E. coli ATPase complex (see below). In addition, the diglyceride kinase activity of E. coli membranes (20) has recently been found to partition in high yield into the 1-butanol phase when the standard procedure (16) for 1-butanol extraction (at pH 4.2) was used (E. Bohnenberger and H. Sandermann Jr., unpublished results). Proteolipids important in cellular energetics may also be extracted using organic solvents. For example, extremely hydrophobic polypeptides which partition from water into chloroform or 1-butanol phases have been isolated from the mitochondrial ATPase (21) and sarcoplasmic Ca^{2+}-ATPase (22), and the latter proteolipid appears to act as an ionophore (23).

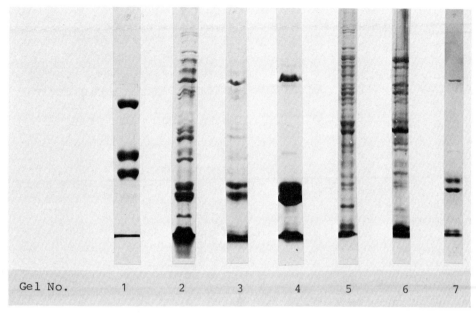

Gel No.   1   2   3   4   5   6   7

Fig. 1. Sodium dodecyl sulfate-polyacrylamide gel electrophoresis of polypeptides extracted with aprotic solvents from E. coli membrane preparations. Polyacrylamide gels were run according to Laemmli (54). Staining was by Coomassie Blue. Gel No. 1: Standard proteins (from top: bovine serum albumin, mol wt 67,000; ovalbumin, mol wt 43,000; aldolase, mol wt 40,000; chymotrypsinogen, mol wt 25,000). Gel No. 2: Total membrane preparation of E. coli prepared according to Osborn and Mason (53). This material was used for preparing the extracts of gels no. 3 and 4. Gel No. 3: Extract from total membrane (53) prepared using 90% (vol/vol) HMPT at pH 7. The extracts were in all cases the supernatants after ultracentrifugation (100,000 × g, 120 min). Gel No. 4: Extract of total membrane (53) prepared using 90% (vol/vol) MP at pH 7. Gel No. 5: Membrane vesicle protein prepared according to Kaback (35). Gel No. 6: Extract of membrane vesicles prepared using 90% (vol/vol) HMPT at pH 7. Gel No. 7: Extract of membrane vesicles prepared using 90% (vol/vol) MP at pH 7.

A virtually complete solubilization of membrane proteins in solvents like chloroform or 1-butanol can be achieved at strongly acidic pH values (pH < 2), but under such conditions functional activities are lost, and membrane reconstitution appears to lead to artifacts (24).

Aprotic solvents. Aprotic solvents, in particular hexamethylphosphoric triamide (HMPT), N-methylpyrrolidone (MP), and dimethylsulfoxide, bring about extensive solubilization (up to 80%) of E. coli membrane proteins under mild conditions (25, 26). These solvents are all freely miscible with water, and solubilized a number of polypeptides (Fig. 1). Depending on the pH and solvent concentration, some selectivity of solubilized polypeptides can be achieved. For example, the prominent outer membrane polypeptides of molecular weight (mol wt) 30,000–40,000 (27) appear to be enriched under certain conditions (gel no. 4, and perhaps no. 7, of Fig. 1).

The retention of certain enzyme activities in aprotic solvent has been known for some time (28), and a recent application of aprotic solvents is in the field of cryoenzymology (29, 30). Aprotic solvents are also known to induce the differentiation of certain mammalian cells, and to produce phase separation phenomena with artificial phospholipid bilayers (31).

The Lactose Permease System of E. coli

Characterization under inactivating conditions. Fox and Kennedy were able to label the lactose permease protein (y-gene product of the lac-operon) by an indirect affinity labeling procedure (32). The protein, labeled at an essential cysteine-residue by radioactive N-ethyl-maleimide, was subsequently solubilized by the detergents, Triton X-100 (32) or sodium dodecyl sulfate (SDS) (33), and its mol wt was determined as about 30,000 (33).

Reversible inhibition by aprotic solvents. At relatively low concentrations (< 3 M), aprotic solvents like N-methylpyrrolidone (Fig. 2) or dimethyl sulfoxide inhibited β-D-galactoside transport by whole cells of E. coli. The inhibition was not caused by membrane leakiness since even at high solvent concentrations the control values for nonspecific hydrolysis by intracellular β-galactosidase were not increased.

In contrast to detergents like SDS or Triton X-100, aprotic solvents were easily removed by washing with buffer, and cellular transport activity was fully restored in this way (25, 34; Fig. 2). Lactose uptake by isolated membrane vesicles (35) was more sensitive to aprotic solvents, but again the inhibition could be largely reversed by a simple washing procedure (34; Fig. 3).

Reconstitution of lactose-transport negative membrane vesicles. At high concentrations of aprotic solvent, 50–80% of the total protein of transport-positive vesicles from E. coli ML 308–225 were solubilized. For solubilization of membrane proteins the vesicles were washed once with 100 mM potassium phosphate buffer (pH 8.0), collected by centrifugation, and solubilized at 25°C for 5 min by incubation in 90% hexamethylphosphoric triamide containing 100 mM LiCl, 50 mM Tris-SO$_4$ (pH 7.5), and 0.3 mM dithiothreitol.

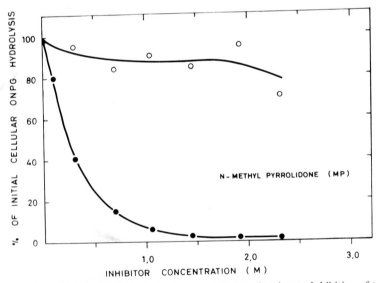

Fig. 2. Reversible inhibition of β-D-galactoside transport by aprotic solvents. Inhibition of the cellular hydrolysis of o-nitrophenyl-β-D-galactopyranoside in the presence of N-methylpyrrolidon (●——●), and after removal of the aprotic solvent by pelleting the cells and resuspending in buffer (reactivation, ○——○). Experimental details are given in Ref. 34.

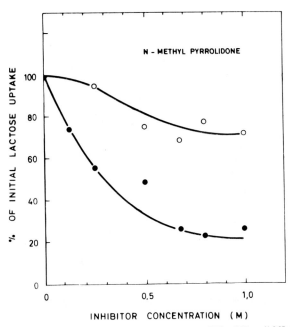

Fig. 3. Inhibition of [^{14}C] lactose uptake of membrane vesicles (35) of E. coli ML 308-225 by N-methylpyrrolidone (●——●), and after removal of the aprotic solvent by pelleting the vesicles and resuspending in buffer (reactivation, ○——○). Experimental details are given in Ref. 34.

After ultracentrifugation (140,000 × g, 2 hr) this extract (R-extract) was used in the reconstitution experiments. For that purpose membrane vesicles of the transport-negative strain E. coli ML 35 were suspended in 50 mM Tris-SO$_4$ (pH 7.5), 2 mM MgSO$_4$, and 0.3 mM dithiothreitol (0.7 mg of protein in 0.5 ml). The mixture was sonicated (Branson sonifier, model S 125, microtip attachment) in an ice-bath for 10–15 sec, during which time 50 μl R-extract was added (0.46 M final solvent concentration). The vesicles were washed once with the above buffer and assayed for [^{14}C] lactose uptake (42). Lactose accumulation was observed in the presence of D-lactate, whereas no uptake could be detected in the presence of uncoupler (1 μM CCCP) or of high-affinity inhibitor (5 mM thio-βD-digalactoside) (Fig. 4). It should be emphasized that the steady-state uptake of lactose (12 nmol/mg vesicle protein) by the reconstituted vesicles was significantly above the 1:1 binding stoichiometry expected if the transport protein added had been completely reconstituted (2 nmol/mg protein, e.g., Ref. 36). Lactose uptake could also be reconstituted by using a potassium diffusion potential, which was established with the aid of valinomycin (34, 42; Fig. 5). This technique of energization is, in principle, also applicable to reconstitution in artificial liposomes which are free of the respiratory enzymes of E. coli ML 35 membrane vesicles. Such simplified systems may yield answers to as yet open questions, such as the energy requirement for binding rather than transport of D-galactosides (36, 37), or the mechanism of symport of protons and β-D-galactosides (38).

It has been possible to fractionate the lipid and polypeptide components of the aprotic solvent extracts employed for reconstitution on columns of Sephadex LH-20, Sephacryl S-200, or Controlled Pore Glass CPG-10. Phospholipids appeared to occur in monomeric form in these extracts and were clearly separated from the eluted polypeptides.

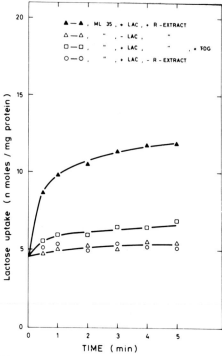

Fig. 4. Reconstitution of $[^{14}C]$ lactose transport in membrane vesicles (35) prepared from cells of the lactose-transport negative strain, E. coli ML 35. The following method was applied for the preparation of the extract used for the reconstitution. Membrane vesicles prepared from E. coli ML 308-225 (transport positive, 12 mg protein) were solubilized in 1 ml of 90% (vol/vol) hexamethyl phosphoric triamide, containing 100 mM LiCl, 50 mM Tris-SO_4 (pH 7.5), 0.3 mM dithiothreitol. For the reconstitution, the above extract (50 μl, after ultracentrifugation, 140,000 × g, 2 hr) was added with constant sonication (10–15 sec, Branson Sonifier, model S 125; temperature ~ 2°C) to a suspension (0.5 ml) of membrane vesicles (0.7 mg protein) of E. coli ML 35 in 50 mM Tris-SO_4 (pH 7.5), 2 mM $MgSO_4$, 0.3 mM dithiothreitol. The membrane vesicles were reisolated by centrifugation, and assayed for $[^{14}C]$ lactose uptake in the presence (▲———▲) or absence (△———△) of D-lactate, or in the presence of D-lactate with the addition of 5 mM thio-β-digalactoside (□———□). No uptake was observed when the addition of R-extract was not included in the above procedure (○———○). Further experimental details are given in Ref. 34.

Hopefully, these studies will finally allow the definition and characterization of the minimum number of components required for lactose permease activity.

The Carbodiimide-Reactive Protein of E. coli

The membrane-bound Ca^{2+}, Mg^{2+} stimulated adenosine triphosphatase complex from bacteria plays a crucial role in energy-conserving reactions (39). It has been demonstrated that this complex is composed of 2 structurally distinct entities. One component is the ATPase (BF_1) which actually catalyzes hydrolysis of ATP and which belongs to the category of peripheral proteins. The other component (BF_0) is buried within the cytoplasmic membrane and thus belongs to the category of integral proteins. The hydrolysis of ATP by the bacterial ATPase complex is coupled to the translocation of protons (39–41). Therefore, it was reasonable to assign the BF_0 component a role in the translo-

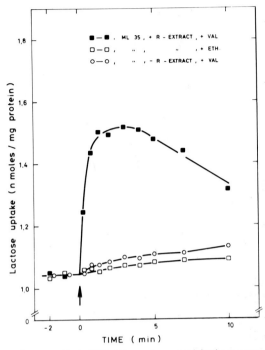

Fig. 5. Lactose transport in reconstituted ML-35 membrane vesicles in response to an artificial membrane potential. Membrane vesicles (35), prepared from cells of E. coli ML 35 were loaded with K^+ as previously described (42), and, in a control experiment, assayed for $[^{14}C]$ lactose uptake after addition of valinomycin (o ——o). The transport assay was also performed after prior addition of aprotic solvent extract (■——■), and a control, in this experiment, was obtained by omission of valinomycin (□——□). Experimental details are given in Ref. 34.

cation of protons. This view was supported recently by several lines of experimental evidence (42–46).

Energy-transducing reactions carried out by the ATPase complex from bacteria are inhibited by DCCD, as is also the case in mitochondria and chloroplasts. It is now well established that the inhibitor exerts its irreversible effect on the BF_0 part.

Since the DCCD-reactive protein is the only component so far which has been implicated in the proton translocation process directly, it was conceivable that this protein alone might be required for the translocation of protons. Therefore, the characterization and purification of the DCCD-reactive protein seemed warranted.

Characterization and purification of the DCCD-reactive protein. For the identification of the DCCD-reactive protein it was necessary to demonstrate that the reaction of DCCD with that protein was related to inhibition of the ATPase activity. This was accomplished by using mutants where the ATPase activity was no longer inhibited by DCCD (47, 48). After treatment of E. coli membranes derived from wild type and mutant with $[^{14}C]$ DCCD, the proteolipids were extracted with chloroform/methanol (2:1). After precipitation of the proteolipids with diethyl ether from the washed chloroform/methanol extract, it could be demonstrated by SDS-gel electrophoresis, that one protein was labeled by $[^{14}C]$ DCCD in membranes of the wild type, but not in the mutants (Fig. 6; for further details see Refs. 47 and 48). A direct correlation between the reaction of $[^{14}C]$ DCCD

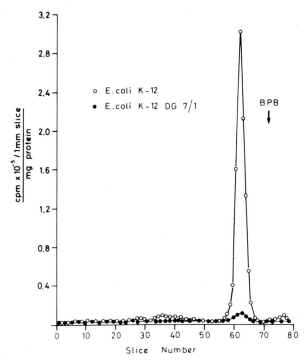

Fig. 6. Sodium dodecyl sulfate-acrylamide gel electrophoresis of [^{14}C]-labeled proteolipid preparations of E. coli K 12 (wild type) and E. coli K 12 DG 7/1 (mutant strain). The preparation of membranes from both wild type and mutant strain, the labeling of the membranes with [^{14}C] DCCD (14.5 mCi/mmol), and the extraction of the membranes with chloroform/methanol (2:1) were performed as already described (51). The proteolipids were precipitated from the chloroform/methanol extract with diethyl ether and subsequently taken up in chloroform/methanol (2:1). Samples (~ 5 μg of protein) from the wild type and the mutant strain were electrophoresed on gels (13% acrylamide, 0.65% N,N'-diallyltartardiamide, for further details see Ref. 51) containing 0.2% sodium dodecyl sulfate, and 1 mm slices were dissolved in 2% periodic acid and counted. The position of the tracking dye, bromophenol blue (BPB) is marked by an arrow.

with that protein and the inhibition of the ATPase activity was thus established. Less than 10% of the membrane protein was soluble in chloroform/methanol, achieving in this first step a greater than 10-fold purification of the DCCD-reactive protein. It should be emphasized that this extraction procedure was performed without the addition of acid.

Despite the fact that about 80–85% of radioactivity from [^{14}C] DCCD incorporated into membranes was due to unspecific labeling, further purification of the DCCD-reactive protein could be followed by monitoring radioactivity, since this protein was the only radioactively labeled component extracted into chloroform/methanol and subsequently precipitated with diethyl ether (47, 48).

The crude proteolipid fraction still contained several protein components and a considerable amount of phospholipids. Since DEAE-cellulose column chromatography has proven to be a powerful separating system for proteins (16) and lipids (49) in the presence of organic solvents, the crude proteolipid fraction was applied to such a column (50, 51). Sodium dodecyl sulfate-gel electrophoresis and determination of the phosphorus content revealed that DEAE-cellulose chromatography separated the DCCD-reactive protein from other contaminating proteins as well as from phospholipids. The final purification of this protein was achieved by chromatography on hydroxypropyl-Sephadex G-50 in the presence of chloroform/methanol containing ammonium acetate (50, 51).

Reconstitution of the unlabeled DCCD-reactive protein. For the reconstitution experiments, it was necessary to apply the DCCD-reactive protein in an unlabeled form. The unlabeled protein was purified in the same way as the labeled one (50, 51). To demonstrate that the purified, unlabeled DCCD-reactive protein had retained at least some of its biological activity, the following method was used: The protein, together with a phospholipid mixture in chloroform from a thermophilic bacterium (52), were evaporated to dryness by rotary evaporation. The residue was dispersed in buffer and sonicated. The liposomes were loaded with potassium by heat treatment in the presence of high concentrations of potassium phosphate (53). These K^+-loaded liposomes containing the DCCD-reactive protein were washed with sucrose to remove the potassium outside. Addition of valinomycin resulted in an efflux of K^+. Since K^+ ions carry a positive charge, an electrical potential difference was thus generated across the membrane of the liposomes. Following the pH changes in the medium there are indications that the DCCD-reactive protein makes the membrane of the liposomes specifically leaky for protons and treatment with DCCD reduces the high permeability. This indicates that the DCCD-reactive protein purified by this method retains at least part of its biological activity.

ACKNOWLEDGMENTS

This work was supported by the Deutsche Forschungsgemeinschaft (SFB 76, Sa 180/9).

We thank Dr. H. Hirata for providing us with a sample of phospholipids from a thermophilic bacterium.

REFERENCES

1. Singer SJ, Nicolson GL: Science 175:720, 1972.
2. Boos W: Annu Rev Biochem 43:123, 1974.
3. Vogel G, Steinhart R: Biochemistry 15: 208, 1976.
4. Machtiger NA, Fox CF: Annu Rev Biochem 42:575, 1973.
5. Tanford C, Reynolds JA: Biochim Biophys Acta 457:133, 1976.
6. Kennedy EP, Rumley MK, Armstrong JB: J Biol Chem 249:33, 1974.
7. Moldow J, Robertson J, Rothfield L: J Membr Biol 10:137, 1972.
8. Reeves JP, Hong J-S, Kaback HR: Proc Natl Acad Sci USA 70:1917, 1973.
9. Kundig W, Roseman S: J Biol Chem 246:1407, 1971.
10. Scandella CJ, Kornberg A: Biochemistry 10:4447, 1971.
11. Folch J, Lees M: J Biol Chem 191:807, 1951.
12. Folch-Pi J, Sakura JD: Biochim Biophys Acta 427:410, 1976.
13. Folch-Pi J, Stoffyn PJ: Ann NY Acad Sci 195:86, 1972.
14. Montal M: Annu Rev Biophys Bioeng 5:119, 1976.
15. Sandermann H, Strominger JL: Proc Natl Acad Sci USA 68:2441, 1971.
16. Sandermann H, Strominger JL: J Biol Chem 247:5123, 1972.
17. Lomax JA, Poxton IR, Sutherland IW: FEBS Lett 34:232, 1973.
18. Poxton IR, Sutherland IW: Microbios 15:93, 1976.
19. Amanuma H, Motojima K, Yamaguchi A, Anraku Y: Biochem Biophys Res Commun 74:366, 1977.
20. Schneider EG, Kennedy EP: Biochim Biophys Acta 441:201, 1976.
21. Tzagoloff A, Rubin MS, Sierra MF: Biochim Biophys Acta 301:71, 1973.

22. MacLennan DH: Can J Biochem 53:251, 1975.
23. Racker E, Eytan E: J Biol Chem 250:7533, 1975.
24. Wehrli E, Moser S, Zahler P: Biochim Biophys Acta 426:271, 1976.
25. Sandermann H: Hoppe-Seyler's Z Physiol Chem 355:1246, 1974.
26. Kohl B, Sandermann H: Abstr 10th IUB Congress, Hamburg, p 267, abstr no 05-1-299.
27. Hindennach I, Henning U: Eur J Biochem 59:207, 1975.
28. Singer SJ: Adv Protein Chem 17:1, 1962.
29. Douzou P: Mol Cell Biochem 1:15, 1973.
30. Fink AL: J Theor Biol 61:419, 1976.
31. Lyman GH, Papahadjopoulos D, Preisler HD: Biochim Biophys Acta 448:460, 1976.
32. Fox CF, Kennedy EP: Proc Natl Acad Sci USA 54:891, 1965.
33. Jonas THD, Kennedy EP: J Biol Chem 244:5981, 1969.
34. Altendorf K, Müller CR, Sandermann H: Eur J Biochem 73:545, 1977.
35. Kaback HR: Methods Enzymol 22:99–120, 1972.
36. Schuldiner S, Kerwar GK, Kaback HR, Weil R: J Biol Chem 250:1361, 1975.
37. Belaich A, Simonpietri P, Belaich J-P: J Biol Chem 251:6735, 1976.
38. West I, Mitchell P: Bioenergetics 3:445, 1972.
39. Harold FM: Curr Top Bioenerg 6:83, 1976.
40. Mitchell P: FEBS Lett 33:267, 1973.
41. Mitchell P: FEBS Lett 50:95, 1975.
42. Altendorf K, Harold FM, Simoni RD: J Biol Chem 249:4587, 1974.
43. Rosen BP: Biochem Biophys Res Commun 53:1289, 1973.
44. Rosen BP, Adler LW: Biochim Biophys Acta 387:23, 1975.
45. Patel L, Schuldiner S, Kabach HR: Proc Natl Acad Sci USA 72:3387, 1975.
46. Patel L, Kaback HR: Biochemistry 15:2741, 1976.
47. Fillingame RH: J Bacteriol 124:870, 1975.
48. Altendorf K, Zitzmann W: FEBS Lett 59:268, 1975.
49. Rouser G, Kritchevsky G, Yamamoto A, Simon G, Galli C, Baumann AJ: Methods Enzymol 14:272, 1969.
50. Fillingame RH: J Biol Chem 251:6630, 1976.
51. Altendorf K: FEBS Lett 73:271, 1977.
52. Hirata H, Sone N, Yoshida M, Kagawa Y: Biochim Biophys Res Commun 69:655, 1976.
53. Osborn MJ, Manson R: Methods Enzymol 31:642, 1974.
54. Laemmli UK: Nature (London) 227:680, 1970.

Journal of Supramolecular Structure 6:239–247 (1977)
Molecular Aspects of Membrane Transport 165–173

The Folate and Thiamine Transport Proteins of Lactobacillus Casei

Gary B. Henderson, Edward M. Zevely, Robert J. Kadner, and F. M. Huennekens

Department of Biochemistry, Scripps Clinic and Research Foundation, La Jolla, California 92037

Two separate binding proteins, one specific for folate and the other for thiamine, have been isolated from membrane fragments of Lactobacillus casei. Purification to homogeneity was achieved by fractionation of the Triton-solubilized proteins with microgranular silica (Quso G-32) and Sephadex G-150. Amino acid analyses revealed that the folate (M_r = 25,000) and thiamine (M_r = 29,000) binders have unusually low polarity constants, 0.32 and 0.26, respectively. Evidence obtained with intact cells has established a direct role for these binding proteins in transport of the corresponding vitamins: A) In each case, the processes of binding and transport showed similarities in substrate affinities and repression by excess vitamin in the growth medium. B) Competition studies employing amethopterin, 5-formyl tetrahydrofolate, and 5-methyl tetrahydrofolate (for folate) and thiamine monophosphate and thiamine pyrophosphate (for thiamine) have shown that the ability of these compounds to inhibit the transport of the corresponding vitamins is paralleled by their ability to inhibit binding. C) Amethopterin-resistant mutants which are defective in folate transport have a comparable defect in ability to bind folate. D) Amethopterin-resistant cells which (compared with the parent cell line) contain folate transport systems with altered affinities for amethopterin also contain binding proteins whose affinities for amethopterin have changed by equivalent amounts. E) Both the transport and binding of folate by one of the mutants were stimulated (approximately 3-fold) in parallel by the addition of mercaptoethanol.

Key words: folate, thiamine, transport, binding proteins, Triton X-100

INTRODUCTION

The active transport of folate (1–4) and thiamine (4, 5) into L. casei proceeds via 2 separate systems readily distinguishable by their substrate specificity. The uptake processes are similar in other respects, however, such as their energy requirements, rates of vitamin uptake, dependence on pH and temperature, and regulation by amount of vitamin in the growth medium. In conjunction with the transport processes, the cells also have the common ability to bind appreciable amounts ($\sim 2 \times 10^4$ molecules/cell) of both folate

R.J. Kadner contributed to this investigation while on leave from the Department of Microbiology, University of Virginia School of Medicine, Charlottesville.
Received March 14, 1977; accepted March 28, 1977

(4, 6, 7) and thiamine (4, 5). The components responsible for this binding are expressed only in cells propagated under conditions of vitamin limitation and have high affinities for their respective substrates. The present report summarizes evidence for a relationship between vitamin-binding activity and vitamin transport. The binding components have been solubilized from membrane preparations of L. casei, purified to homogeneity, and shown to be extremely hydrophobic, water-insoluble proteins. Folate transport mutants have also been isolated and employed in conjunction with the parent cells to demonstrate kinetic similarities between the binding and transport processes.

MATERIALS AND METHODS

Lactobacillus casei var. rhamnosis (ATCC 7469) were grown according to the general procedure described previously (7) in the medium of Flynn et al. (8) containing either 5 nM folate plus 5 μM thiamine (for studies on the folate transport system) or 5 μM folate and no added thiamine (for studies on the thiamine transport system). Folate transport mutants (RX-1 through RX-21) were isolated by selecting for cells with resistance to amethopterin. Details of the procedures employed for this purpose will be described elsewhere. The minimum concentrations of folate (values in parentheses) required to give full growth of the individual mutants were as follows: RX-2 and RX-3 (500 nM); RX-1 (100 nM); RX-4 and RX-5 (10 nM); and RX-6 through RX-21 (5 nM).

Measurement of the binding of folate or thiamine by intact cells was determined in assay mixtures consisting of 0.05 ml of [G-^{3}H] folate (120,000 dpm/nmole) or [thiazole-2-^{14}C] thiamine (35,000 dpm/nmole), 0.05 ml of a desired addition, and 0.9 ml of a washed cell suspension (6 $\times$ 10^8/ml in 0.05 M potassium phosphate, pH 6.8). After 5 min at 4°C, the cells were collected on a Millipore filter (0.22 μm) and washed with 2 1-ml portions of ice-cold phosphate buffer. The filters were placed in 5 ml of a dioxane-based scintillation fluid (1), and the radioactivity was determined. Control values (usually 5–10%) were calculated as described previously (7). Vitamin transport was determined by a procedure similar to that described above for binding. In this case, cell suspensions were preincubated with glucose (5 mM) for 5 min at 37°C prior to their addition to the assay mixture. The samples were then incubated for an appropriate interval at 37°C. The amount of folate or thiamine bound by cells (see above) served as the control for the corresponding transport process. A general procedure for measuring vitamin-binding activity in membrane or supernatant fractions derived from intact cells has been described previously (7).

Protein in intact cells (after sonication for 1 min at 23°C) or membrane preparations was determined by the method of Lowry et al. (9). Protein in samples containing Triton was determined by the biuret reaction (10). Bovine serum albumin served as the standard.

RESULTS AND DISCUSSION

I. Characterization of the Binding Proteins

Cellular location. Information on the intracellular location of the folate- and thiamine-binding components of L. casei was obtained by fractionation experiments. Intact cells (having a binding capacity of 0.15 nmole folate or 0.26 nmole thiamine per 20 mg protein) were disrupted either osmotically following treatment with lysozyme or by passage through a homogenizer (7); the supernatant and particulate (membrane)

maximal in cells grown with 1—10 nM folate, declines progressively at higher concentrations (50% loss at 50 nM), and is totally repressed at 1 μM folate (Fig. 1). Folate-binding capacity showed an identical response (50% loss at 55 nM) to these growth conditions. In a similar fashion, thiamine is transported and bound optimally in cells propagated in medium containing up to 10 nM thiamine (Fig. 2). At higher concentrations, a decline in transport and binding was observed although at slightly different rates; reductions of 50% occurred at 37 nM and 17 nM, respectively, for the 2 processes. An exact coincidence in the latter curves was not obtained, and the reason for this difference is not yet clear. Thiamine added to the growth medium may interfere with these measurements since separate experiments have shown that unlabeled thiamine, once bound to the cells, does not readily exchange at 4°C with the [^{14}C] thiamine of the assay mixture. It is clear, however, that the rates of both folate and thiamine transport are closely related to the amounts of the corresponding binding proteins within the cell.

Binding affinities. Analyses of binding parameters provided an alternative means for comparing the binding and transport processes. For investigation of the folate system, affinity constants were determined for several of the compounds (folate, amethopterin, 5-methyl tetrahydrofolate, and 5-formyl tetrahydrofolate) previously shown to compete with folate for transport (1). The results show that amethopterin was bound with the highest affinity to both the receptor protein and the transport system (Table III). In this respect 5-formyl tetrahydrofolate and folate were intermediate while 5-methyl tetrahydrofolate was bound with the lowest affinity. The actual values for the affinity constants showed considerable variation (16—210 nM), yet for each compound tested, the dissociation constant for the binding protein was 3-fold lower than the Michaelis constant for the transport system (Table III). Thus, a linear relationship could be demonstrated when the affinity constants for transport were plotted against the corresponding values for the binding protein (cf. Fig. 3).

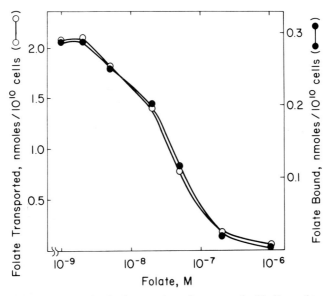

Fig. 1. Effect of folate concentration in the growth medium upon the binding and transport of folate by L. casei. Assay mixtures contained 1.0 μM [^{3}H] folate and were incubated for 10 min at 37°C or 5 min at 4°C for the measurement of transport and binding activity, respectively.

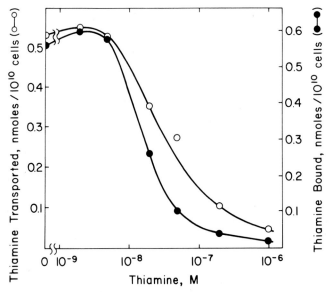

Fig. 2. Effect of thiamine concentration in the growth medium upon the binding and transport of thiamine by L. casei. Assay mixtures contained 1.0 μM [^{14}C] thiamine and were incubated for 2 min at 37°C or 5 min at 4°C for the measurement of transport and binding activity, respectively.

TABLE III. Comparative Affinity of Various Compounds for the Folate-Binding Protein and Folate Transport System of L. casei*

| | Affinity constant | |
| Folate compound | Binding protein | Transport system |
	nM	nM
Amethopterin	16	50
5-Formyl tetrahydrofolate	26	60
Folate	36	95
5-Methyl tetrahydrofolate	85	210

*The K_m for [^{3}H] folate transport (1) and the K_D for [^{3}H] folate binding (6) were determined from double-reciprocal plots of folate taken up by cells vs free folate. The corresponding affinity constants for amethopterin, 5-formyl tetrahydrofolate, and 5-methyl tetrahydrofolate were determined as K_i values by the method of Dixon and Webb (14); assay mixtures contained 0.1 μM or 0.4 μM [^{3}H] folate and variable concentrations (0.05–1.0 μM) of the indicated folate derivatives. Samples were incubated for 2 min at 37°C and 5 min at 4°C for measurement of transport and binding, respectively.

A similar relationship could also be established between the transport system and binding protein for thiamine. In this case, percent inhibition, rather than an inhibition constant, was employed for the comparison, since the affinity constants for thiamine of both the transport system ($K_m \cong 10$ nM) and the binding protein ($K_D \cong 10$ nM) were too low to measure conveniently. The competition of unlabeled thiamine, thiamine phosphate, and thiamine pyrophosphate with ^{14}C-labeled thiamine for binding and transport is illustrated in Table IV. The results show that thiamine phosphate and thiamine pyrophosphate are good inhibitors of thiamine binding and transport, and that the sequential addition of

TABLE IV. Inhibition of [^{14}C] Thiamine Binding and Transport by Thiamine Compounds*

Addition	Concentration	[^{14}C] Thiamine bound	Inhibition	[^{14}C] Thiamine transported	Inhibition
	μM	nmoles/10^{10} cells	%	nmoles/10^{10} cells	%
None	–	0.50	–	0.49	–
Thiamine	0.05	0.36	28	0.36	26
	0.2	0.22	56	0.19	61
Thiamine phosphate	0.1	0.43	14	0.42	14
	0.4	0.31	38	0.32	35
	1.0	0.20	60	0.21	57
Thiamine pyrophosphate	0.4	0.45	10	0.43	12
	1.0	0.35	30	0.39	20
	4.0	0.30	40	0.32	35

*Assay samples were prepared (at 4°C) by combining 0.1 ml of 1.0 μM [^{14}C] thiamine and 0.1 ml of the indicated (unlabeled) thiamine compound followed by the addition of 0.8 ml of cells. The binding and transport of [^{14}C] thiamine was then determined following incubation for 5 min at 4°C or 2 min at 37°C, respectively.

phosphate groups onto the vitamin progressively lowers the ability of the analogs to compete with the parent compound. It was also observed that as the concentrations of thiamine and its phosphorylated derivatives were varied, the binding and transport of thiamine were each inhibited to the same degree. Thus, the transport system and binding protein have the same relative affinity for each of these thiamine compounds.

   Folate transport mutants. A series (RX-1 through RX-21) of amethopterin-resistant cell lines which contain a defective folate transport system has been isolated. In each case, a reduction in the transport of folate was accompanied by a comparable loss in ability to bind folate. In several of the cell lines, the ability to transport and bind folate was essentially absent, whereas other mutant cells retained up to 50% of the transport and binding activity of the parent cells.

   Since the folate transport mutants were selected for resistance to a folate analog (amethopterin), it seemed possible that mutant cells might have arisen which retained an effective means for the transport of folate but not amethopterin. In cells of this type, the binding protein and transport system might be altered in their affinity for folate compounds. To test this possibility, affinity constants for folate and amethopterin were determined in several of the cell lines. One of the mutants selected for this study was RX-13 since it retained the highest capacity to transport folate (~50% of wild-type cells.) In these cells, the K_M for folate transport (110 nM) and the K_D for folate binding (25 nM) were found to be comparable to the corresponding values obtained in the parent cells (cf. Table III), while the affinities of the transport system (K_i = 3,000 nM and the binding protein (K_i = 800 nM) for amethopterin were both 60-fold lower than in the wild-type cells. A similar result was obtained with mutant RX-21. When measured in the presence of mercaptoethanol (see below), the interaction of amethopterin with the transport system (K_i = 650 nM) and the binding protein (K_i = 170 nM) was 10-fold less efficient in this mutant than in the parent cells, while the parameters relative to folate binding were, again, virtually unchanged. When the affinity constants for mutants RX-13 and RX-21 were plotted in Fig. 3, the following relationships were

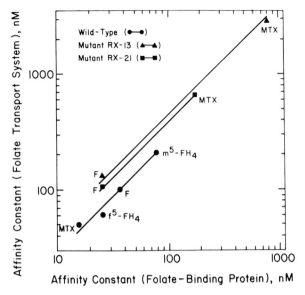

Fig. 3. Relationship between the affinity of various compounds for the folate transport system and the folate-binding protein of wild-type and mutant cells of L. casei. The affinity constants were determined as described in the legend to Table III. Cells of mutant RX-21 were incubated with 25 mM mercaptoethanol for 5 min at 37°C prior to the addition of folate compounds.

apparent: A) The plot of the affinity constants for each mutant was parallel to and nearly coincident with that for the wild-type cells; B) the 3-fold difference between the affinity constants of the binding protein and transport system in the wild-type cells (cf. Table III) was maintained in the mutant cell lines.

The transport system and binding activity of mutant RX-21 were both dependent on added mercaptoethanol. When measured under standard conditions, binding and transport of folate by these cells were only one-third of the levels characteristic of the parent cell line. Mercaptoethanol added to the assay mixtures increased both the binding and transport of folate by 3-fold (Table V). Thus, analyses of the mutant cell lines were able to establish correlations between the transport and binding processes in 3 separate areas, i.e., relative amounts, binding affinities, and responses to an external agent.

III. Concluding Remarks

Evidence has been presented which shows that the membrane-associated folate- and thiamine-binding proteins of L. casei participate in vitamin transport. These conclusions are based primarily upon the cellular location and amino acid compositions of the binding proteins, a parallel relationship between the interaction of specific ligands with the binding proteins and transport system, and, in the case of folate, the properties of transport mutants. The results do not establish whether the binding proteins are merely receptor sites or the actual carriers of the vitamins across the cell membrane, although the extreme hydrophobicities of these proteins argue in favor of the latter possibility. Reconstitution of a vitamin-transporting system by insertion of the purified proteins into liposomes should be able to resolve this question. Other aspects of the transport process, such as the

TABLE V. Effect of Mercaptoethanol on the Binding and Transport of Folate by Mutant RX-21*

Process	No addition	Mercaptoethanol	Stimulation
		25 mM	−fold
Folate binding (nmoles/10^{10} cells)	0.09	0.29	3.2
Folate transport (nmoles/10^{10} cells)	0.8	2.2	2.8

*Cells were preincubated with mercaptoethanol for 5 min at 37°C prior to the addition of labeled folate. The binding and transport of folate was then determined following incubation for 5 min at 4°C or 10 min at 37°C, respectively.

mechanism for energy-coupling, are not yet clear. Evidence has been obtained, however, that folate transport is directly linked to glycolysis and that both oxidoreduction equivalents and ATP may be required as sources of energy (3).

ACKNOWLEDGMENTS

G. B. Henderson is a recipient of a Public Health Service Postdoctoral Fellowship (CA00616) from the National Cancer Institute, National Institutes of Health.

This investigation was supported by grants from the National Cancer Institute, National Institutes of Health (CA 6522), and the American Cancer Society (CH-31). The authors are indebted to Karin Vitols for assistance in the preparation of this manuscript and to Dr. Irving Crawford of the Department of Microbiology for the determination of the amino acid compositions of the binding proteins.

REFERENCES

1. Henderson GB, Huennekens FM: Arch Biochem Biophys 164:722, 1974.
2. Shane B, Stokstad ELR: J Biol Chem 250:2243, 1975.
3. Huennekens FM, Henderson GB: In Pfleiderer W (Ed): "Chemistry and Biology of Pteridines." Berlin: Walter de Gruyter, pp 179–196, 1976.
4. Henderson GB, Zevely EM: Fed Proc Fed Am Soc Exp Biol 35:1357, 1976.
5. Henderson GB, Zevely EM, Huennekens FM: (Submitted to J Bacteriol).
6. Henderson GB, Zevely EM, Huennekens FM: Biochem Biophys Res Commun 68:712, 1976.
7. Henderson GB, Zevely EM, Huennekens FM: J Biol Chem (In press).
8. Flynn LM, Williams VB, O'Dell BD, Hogan AG: Anal Chem 23:180, 1951.
9. Lowry OH, Rosebrough MJ, Farr AL, Randall RJ: J Biol Chem 193:265, 1951.
10. Gornall AG, Bardawill CS, David MM: J Biol Chem 177:751, 1949.
11. Hirs CHW: Methods Enzymol 11:59, 1967.
12. Goodwin TW, Morton RA: Biochem J 40: 628, 1946.
13. Capaldi RA, Vanderkooi G: Proc Natl Acad Sci USA 69:930, 1972.
14. Dixon M, Webb EC: "Enzymes." 2nd Ed. New York: Academic Press, pp 328–330, 1964.

Journal of Supramolecular Structure 6:313–323 (1977)
Molecular Aspects of Membrane Transport 175–185

Thermodynamics, the Structure of Integral Membrane Proteins, and Transport

S. J. Singer

Department of Biology, University of California at San Diego, La Jolla, California 92093

Membranes are structures whose lipid and protein components are at, or close to, equilibrium in the plane of the membrane, but are not at equilibrium across the membrane. The thermodynamic tendency of ionic and highly polar molecules to be in contact with water rather than with nonpolar media (hydrophilic interactions) is important in determining these equilibrium and nonequilibrium states. In this paper, we speculate about the structures and orientations of integral proteins in a membrane, and about how the equilibrium and nonequilibrium features of such structures and orientations might be influenced by the special mechanisms of biosynthesis, processing, and membrane insertion of these proteins. The relevance of these speculations to the mechanisms of the translocation event in membrane transport is discussed, and specific protein models of transport that have been proposed are analyzed.

Key words: peripheral and integral proteins, membrane biosynthesis, hydrophobic and hydrophilic
         interactions

It has become clear in recent years that over long distances in the plane of the membrane each membrane surface is close to equilibrium with the aqueous phase bathing it, whereas over the much shorter distance across the membrane the 2 surfaces are normally far from equilibrium. The near equilibrium state in the plane of the membrane is reflected in part by the rapid intermixing of components that can occur over the entire membrane surface (8, 34). By contrast, the nonequilibrium state across the membrane is reflected in the normally very slow or negligible rates of mixing of membrane components from one surface to the other (see below). This nonequilibrium state is not simply due to the fact that the 2 aqueous compartments that are separated by the membrane of a living cell are themselves not in equilibrium with one another, but is rather an intrinsic property of the membrane-water system itself, because the asymmetry of the membrane persists when the cell is lyzed and the membranes are isolated.

Some years ago, we discussed semiquantitatively some of the thermodynamic factors influencing membrane structure (29). In particular, the important roles played by hydrophobic and hydrophilic interactions were stressed. Hydrophobic interactions have become

Received April 15, 1977; accepted May 5, 1977

well-appreciated in molecular biology since the classic paper by Kauzmann (15), and reflect, crudely speaking, the thermodynamic tendency of nonpolar structures to sequester themselves from contact with water. Hydrophilic interactions have received less attention. As a class of interactions, they are responsible for the strong thermodynamic tendency of highly polar and ionic structures to remain in contact with water if given the choice between an aqueous and a nonpolar environment.

The near equilibrium and nonequilibrium features of membrane structure discussed above follow from these hydrophobic and hydrophilic interactions. The well-known bilayer structure of phospholipids in membranes, for example, is one result of these interactions. At equilibrium the nonpolar fatty acyl chains are sequestered from contact with water, thereby maximizing hydrophobic interactions, while the polar head groups are exposed to the aqueous phase, maximizing hydrophilic interactions. On the other hand the very slow rates of trans-bilayer rotations (flip rates) of phospholipids [with an estimated minimum half-time of 80 days in synthetic bilayer vesicles (24)], very likely reflect hydrophilic interactions, namely, the large free energy of activation required to transfer the polar headgroup of a phospholipid molecule through the nonpolar interior of the membrane (29–31).

The structures of membrane proteins must also result from the interplay of these thermodynamic factors. Lenard and I (18) and Wallach and Zahler (37) suggested independently that membrane proteins [or, rather, those that we now call integral proteins (29)] would generally exhibit amphipathic molecular structures, with predominantly hydrophilic segments (containing all the ionic and highly polar residues, such as saccharides) exposed to the aqueous phase, and predominantly hydrophobic segments embedded in the membrane interior. At the time that this proposal was made, no integral membrane proteins had been studied in sufficient structural detail to test it, but since then the existence of a large number of such amphipathic structures has been demonstrated.

We further suggested (31) that integral membrane proteins would exhibit negligibly slow trans-membrane rotations, again because of the very large free energies of activation involved in moving their hydrophilic segments across the nonpolar membrane interior. This accords with the facts: so far as is known, each integral protein is present in a unique orientation in its membrane. This orientation is most likely not an equilibrium condition, but rather must reflect the specific mechanisms by which the integral proteins are inserted into the membrane during or after biosynthesis, as discussed below.

It seems appropriate, now that the crude first approximation suggestions about the structure of integral proteins have been borne out, and a considerable number of such proteins is currently under investigation, to examine in somewhat greater detail than previously what factors might influence the structures of integral proteins, in particular what might be the equilibrium and nonequilibrium effects on such structures, and how these structures bear on the problem of membrane transport. Some of the ideas about membrane protein biosynthesis discussed in this paper have been recently put forward by others as well (25, 27).

THE STRUCTURES OF INTEGRAL PROTEINS

At this early stage in our knowledge of the structures of integral proteins, at least 4 classes of such structures have to be considered. These are depicted in Fig. 1.

In what follows, we will emphasize the biosynthetic sites of these proteins with respect to the membrane. Reference to the 2 faces of the generalized membrane in Fig. 1 as cis and trans reflects this emphasis. Thus, in single membrane procaryotes, the cyto-

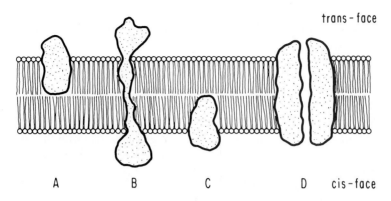

Fig. 1. At least 4 classes of membrane integral proteins need to be considered, depicted schematically in this figure. The cis and trans faces are defined in the text, the cis side being related to that surface to which membrane-bound ribosomes are attached. A and C represent proteins that are only part-way embedded in the bilayer, from the trans and cis sides, respectively. Whether proteins of type A actually exist, however, is not clear. Proteins of type B are trans-membrane proteins with hydrophilic segments protruding from both membrane faces, and a hydrophobic segment in between which is embedded in the hydrophobic interior of the bilayer. Transport proteins, it is suggested, are all type D, subunit aggregates with aqueous channels running down the axis of the aggregate. See text for further details.

plasmic face of the plasma membrane is the cis face, the one to which membrane-bound ribosomes become attached. For the plasma membranes of eucaryotic cells, however, the corresponding cis face is that facing outside, the cytoplasmic surface being the trans face. This is because protein biosynthesis on membrane-bound ribosomes occurs in the endoplasmic reticulum of eucaryotic cells; new plasma membrane most likely arises by a process of vesiculation of the reticulum and fusion of the vesicles with already existing plasma membrane. This process results in an inversion of the surfaces of the vesicle and plasma membranes (22, 12). It is therefore more useful to define the membrane face in terms of biosynthetic origins rather than final membrane orientation.

Returning to the protein structures of Fig. 1, cytochrome b_5 of endoplasmic reticulum is a well-known example of type 1C (32). It has one hydrophilic domain containing the NH_2 terminus, protruding from the cis face of the membrane, and a hydrophobic domain, containing the COOH terminus, embedded in the membrane. A distinctive feature of such a type 1C protein is its spontaneous binding to lipid bilayers and membranes (33) suggesting the absence of a significant second hydrophilic domain attached to the hydrophobic one, as in type 1B. On the other hand, it is not really known how deeply the hydrophobic domain of cytochrome b_5 is embedded in the membrane. It is certainly large enough to span the thickness of the membrane, but probably does not.

Glycophorin is the current paradigm for proteins of type 1B, where a single linear hydrophobic portion of the polypeptide chain is embedded within the membrane, connecting 2 hydrophilic domains exposed on either side (35). It is of interest that the NH_2 terminus of glycophorin is exposed at the trans face of the membrane, and the COOH terminus at the cis face. This feature is discussed in the following section.

The integral proteins depicted in Fig. 1A and C appear to be closely similar in structure. They are considered to be distinct for our purposes, however, since the former protrudes from the membrane surface opposite to the side where protein biosynthesis occurs (the trans face), whereas the latter is located on the cis face. Bretscher (6) suggested that integral proteins of the type shown in Fig. 1A do not exist, i.e., that integral proteins which project a hydrophilic segment from the trans face of a membrane must all be proteins that span the membrane (type 1B). This suggestion was based on the dispositions of only the 2 major polypeptides of the human erythrocyte membrane. I think, however, that it is premature at this time to exclude the existence of membrane proteins of type 1A. It is difficult to obtain positive evidence for their existence; one can only infer this from a relatively loose hydrophobic association of the protein with the membrane (a criterion which does not discriminate it clearly from a peripheral protein) and from the inability to label the protein from the cis-face side. On such grounds, the IgM-like receptor on B lymphocytes is a good candidate for a type IA protein (36).[1]

Integral proteins of type 1D have not been widely discussed as yet. On thermodynamic grounds, we proposed (29, 30) that all proteins involved in specific transport through membranes are of this type, unaware that for other reasons Jardetzky (14) had made a similar proposal earlier. Proteins of type 1D are molecular aggregates of some small number (2, 3, or 4) of identical or similar subunits which span the membrane. Such aggregates would have (two-, three-, or fourfold) rotation axes perpendicular to the plane of the membrane, i.e., each chain of the aggregate would have the same orientation in the membrane so that the aggregate was asymmetrically positioned in the membrane.

An important feature of such aggregates is that they may generate a narrow water-filled channel down the central axis of the aggregate. The surfaces of the polypeptide chains lining the channel could contain ionic and polar groups, because they could interact with the water in the channel and thereby satisfy their hydrophilic interactions.

Structures similar to type 1D are very common among soluble proteins (20). Of those soluble aggregates with only a single rotation axis, dimers are the most prevalent kind. Cytoplasmic malate dehydrogenase is an example of such an homotropic dimer; its x-ray crystallographic structure is known to a resolution of 3.0 Å (11). A water-filled channel traverses the entire length of the molecule down the twofold axis. In principle, the only structural change required to convert cytoplasmic malate dehydrogenase into a type 1D integral membrane protein is to produce an hydrophobic outer surface on the aggregate where it would come in contact with the hydrophobic interior of the membrane.

At the time this proposal was first made, no integral membrane proteins with such a structure were known. Since then evidence has been obtained with 2 transport proteins that is at least consistent with their having type ID structures: the Na^+, K^+-ATPase (17), and the erythrocyte band 3 anion transport protein (38). In each case, chemical crosslinking and other studies have shown that these 2 proteins are both noncovalently bound homotropic dimers within the membrane; and each monomer chain spans the membrane since it can be labeled from both sides of the membrane (26, 5). What is not yet known is whether these dimers have a continuous aqueous channel down their twofold rotation axes, and whether their respective specific ion binding sites are located within the channel.

In addition to these 2 proteins bacteriorhodopsin has been shown to exist as a trimeric aggregate in the purple membrane (10), and functions as a H^+ ion transport protein (19). However, such proteins may be structurally different from other transport

[1] Parenthetically, one reason that it is very important to know whether type IA proteins do in fact exist is that if they do not, then all integral proteins exposed at the trans face, being trans-membrane, are at least potentially capable of direct linkage to cytoskeletal proteins (3).

proteins, because H$^+$ ion transport need not occur by way of a specific binding site within an aqueous channel; it could conceivably occur by transfer of the proton along a chain of different proton-accepting groups within a relatively hydrophobic matrix.

Another important structural feature of integral proteins is the number of trans-membrane folds made by a polypeptide chain that is of type IB or ID. A single bacteriorhodopsin molecule, for example, has about 80% of its chain folded into 7 helical segments spanning the membrane. It is not known how these segments are interconnected, nor what is the disposition of all of the remaining 20% of the chain, but if the helical segments are connected by hydrophilic bends in the chain, then the insertion of such molecules into the membrane raises thermodynamic problems.

THE MECHANISM OF TRANSLOCATION IN TRANSPORT

A few years ago, the "rotating carrier" mechanism for the translocation event in transport was popular. In this scheme, a transport protein translocated its ligand-bound active site from one membrane surface to the other by rotating about an axis parallel to the plane of the membrane. For thermodynamic reasons, however, we considered that such protein rotations or trans-membrane diffusions were unlikely. Furthermore, several investigators in recent years have shown (16, 13, 7) that the attachment of large proteins (antibodies or lectins) to transport proteins in intact membranes does not alter their transport or transport-coupled enzyme activities, a result that is very difficult to reconcile with a rotating carrier mechanism.

Jardetzky (14) and I (29, 30) have suggested an alternative mechanism. Specific transport proteins are all proposed to be type ID proteins, capable of existing in at least 2 conformationally distinct states each of which retains the same basic orientation in the membrane. In one state, however, the ligand-binding site is accessible to the aqueous phase on one side of the membrane; in the other state, the site is accessible to the other side (Fig. 2). The affinity of the site for the ligand is different in the 2 states. Conversion from one state to the other involves an energy-requiring rearrangement of the subunits, driven by the concentration gradient of the ligand in the case of facilitated diffusion, or by some other energy source (ATP hydrolysis, membrane potential, etc.) in the case of active transport.

With a homotropic dimer of type ID, for example, 2 ligand binding sites would be present, related by the twofold rotation axis. These sites might exhibit cooperativity in binding and transport of either the positive or negative type (28).

With minor modifications, the same basic mechanism could be extended to account for exchange diffusion and for group translocation types of transport.

An important feature of a subunit aggregate is that it is a structure which allows large changes in the spatial arrangement of the aggregate to occur with a relatively small input of energy. This allows an active site within the channel to be exposed alternatively to 2 different aqueous compartments bathing the membrane without much change in the disposition of the site in the direction perpendicular to the membrane (Fig. 2).

The role of the periplasmic binding proteins in a large number of cases of bacterial transport has been mystifying. On the one hand, the evidence is very extensive that they are critically involved in their respective transport processes. On the other hand, their localization in the periplasmic space and their solubility properties have been difficult to reconcile with a membrane-mediated role in transport. The possibility that they shuttled back and forth across the inner bacterial membrane was often considered in the past. In

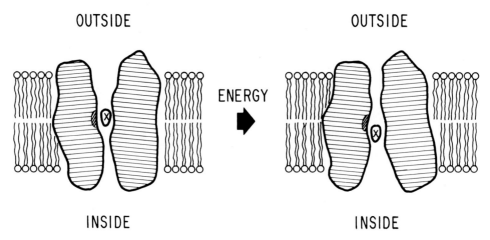

Fig. 2. The aggregate rearrangement mechanism for the translocation event in active transport. See text for details. [Reprinted from (30), with permission.]

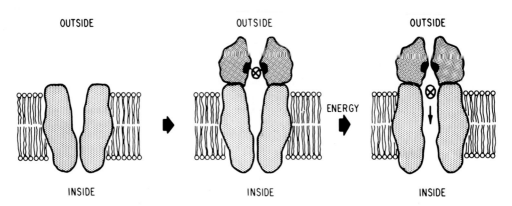

Fig. 3. A schematic diagram of a proposed mechanism for the involvement of periplasmic binding proteins in transport. The binding protein (shaded), with an active site for the ligand X, is considered to be a peripheral protein attached to a so-called portal protein (stippled) which is an integral protein of type D, Fig. 1. The mechanism of translocation of X is then projected to be similar to that depicted in Fig. 2. [Reprinted from (30), with permission.]

view of the thermodynamic considerations mentioned earlier, and the model of transport represented in Fig. 2, we suggested (30) a general mechanism for their action, depicted in Fig. 3. The significant features of this mechanism are: a) the periplasmic binding proteins are proposed to be peripheral proteins (29) which, when functional in transport, are attached to the trans face of the inner membrane; b) their attachment is (noncovalently) to exposed sites on specific integral proteins (which might be called portal proteins) spanning the membrane; c) these putative portal proteins exist as subunit aggregates in the membrane, forming channels much like those in Fig. 2, except that they do not possess the active sites to bind the transported ligand. These active sites are on the periplasmic binding protein; and d) the mechanism of translocation of the ligand across the membrane during transport involves a coordinated subunit rearrangement of the periplasmic and portal proteins much like that depicted in Fig. 2.

At the time we first considered this proposal, we sought for evidence in the literature to support it. We came upon the elegant studies of Ames and Lever (1) which showed that the high-affinity histidine transport system in S. typhimurium required the products of 2 genes in order to function: one gene, his J, coded for a periplasmic binding protein; the other, his P, coded for an unknown product. We suggested that the his P gene product was the portal protein. Although the his P gene product has not yet been identified, and it is not known whether it is an integral protein of the membrane, the suggestion that the his J-his P protein interaction is a peripheral-intergral protein interaction has been adopted by Ames and Spudich (2), and important evidence consistent with this suggestion has been adduced by these authors. There is also preliminary evidence that such a system of 2 components may function in other cases involving periplasmic binding proteins (Hogg, this volume, p 411 [MAMT, p 273]).

While the basic scheme of translocation is the same for the 2 types of transport systems depicted in Figs. 2 and 3, the advantage of the latter is that, in principle, several different periplasmic binding proteins could use the same portal protein channel, as required; quite possibly, even other types of periplasmic proteins, such as those involved in chemotaxis (9), can use the same portal proteins.

It should be obvious that the schemes depicted in Figs. 2 and 3 were originally meant to convey general features of the translocation mechanisms proposed. When they were put forward, it was hoped that they would provide a point of view about how these systems work, rather than precise structural prescriptions of the mechanisms. The x-ray crystallographic studies of Hogg and Quiocho (this volume) have shown, for example, that the single chain of the arabinose binding protein is itself a pseudo-dimer: it has 2 similar but not identical domains related by a pseudo twofold axis of rotation, with an aqueous channel down the twofold axis. Only one of the domains has a sugar-binding site, which is located on the face of the channel. Clearly, such a single pseudo-dimer molecule is structurally closely analogous to the true dimer of periplasmic binding protein depicted in Fig. 3. The aqueous channel of a single arabinose binding protein, with its single active site, might then occupy the position of the aqueous channel of the dimer of periplasmic protein shown in that figure. The important features of the mechanism proposed in Fig. 3 are essentially unaltered by such a substitution. Many more specific changes can be accommodated into the mechanism as new information is acquired, assuming that the mechanism is basically correct.

THE BIOSYNTHESIS OF INTEGRAL PROTEINS

The biosynthesis of membrane proteins is a subject of intense speculation and little experimental information at present. Our purpose in discussing it is to stress our opinion that the insertion of at least certain types of integral proteins in membranes cannot occur spontaneously, and must therefore occur by suitable mechanisms, and that, therefore, the biosyntheses and the structures of integral proteins are intimately connected to one another.

The propositions of this section are that: i) membranes are not made de novo, but grow by the synthesis of lipids, and the insertion of integral proteins, within preexisting membranes acting as templates; ii) all integral proteins are originally synthesized on membrane-bound ribosomes; iii) various forms of posttranslational processing of integral proteins may occur; and iv) the final structures of integral proteins in the membrane, in particular whether they are equilibrium or nonequilibrium structures, reflect these biosynthetic and processing events.

It is not our object here to try to document these propositions fully. The first 2 are discussed in the interesting review article by Sabatini and Kreibich (1976). The second is at present a matter of controversy, some of which arises from a failure to discriminate between peripheral and integral proteins of membranes. Peripheral proteins are indeed very likely made on free ribosomes, as are cytoplasmic proteins. Our own indirect evidence regarding the site of biosynthesis of cytochrome b_5 is briefly discussed below.

It is our conviction on thermodynamic grounds that proteins of type IA and IB, with hydrophilic segments protruding from the trans face of the membrane, cannot find their proper orientation in the membrane by a spontaneous process (i.e., they are not equilibrium structures). It has been proposed (4, 27) that type IB proteins are synthesized and inserted into membranes by a mechanism analogous to the synthesis and disposition of secretory proteins. It is proposed that at the NH_2 terminus of the polypeptide chain, as it is translated off the messenger RNA of the membrane-bound ribosome, there is a short segment of a hydrophobic peptide; this so-called signal peptide (4) directs the nascent polypeptide chain into the lipid interior of the attached endoplasmic reticulum membrane. After passage of the signal peptide through the lipid, a hydrophilic segment of the chain sequence is translated and passes through the membrane via a newly-generated hydrophilic protein channel in the membrane, and the signal peptide is removed by proteolysis. If the protein is a secretory protein, the entire sequence is ultimately transferred across the membrane in this fashion, to fold up in its native soluble conformation in the cisternal space on the trans side of the membrane. If the protein is an integral membrane protein of type IB, however, the first hydrophilic part of the sequence, after traversing the membrane, is followed by a hydrophobic sequence which, it is presumed, remains embedded in the lipid interior of the membrane. The next hydrophilic sequence, after synthesis, is then retained on the cis side of the membrane. The first and second hydrophilic segments, after release of the nascent chain, then fold up independently on their respective sides of the membrane.

Proteins of type IA (if they exist) might be synthesized similarly and either i) have as their COOH termini the hydrophobic segment that followed, in sequence, the signal peptide and the hydrophilic segment of the chain; or ii) be secreted entirely into the cisternal space, to become attached at a later stage by some suitable hydrophobic processing (see below). If the former occurred, the NH_2 terminus would be exposed at the trans face, but the COOH terminus would be embedded in the membrane interior; if the latter occurred, the position along the chain where the hydrophobic processing happened would determine the orientation of the chain termini.

Cytochrome b_5, as mentioned above, is considered to be an integral protein of type IC. It has been suggested that such proteins are synthesized on free ribosomes and are then detached into the cytoplasm, to find their way to the cis-face side of their appropriate membrane (6). If this were so, however, what would be the explanation of the finding that cytochrome b_5 is present on the cis side of the endoplasmic reticulum, and not, for example, on the cis side of the plasma membrane to which it would also have access (23)? José Remacle and I, in unpublished studies, have shown by ferritin-antibody labeling experiments that, in vitro, purified cytochrome b_5 can indeed attach spontaneously to the cis side of plasma membrane fragments of liver cells. This argues that the absence of cytochrome b_5 from the plasma membrane in vivo is not due to some thermodynamic barrier

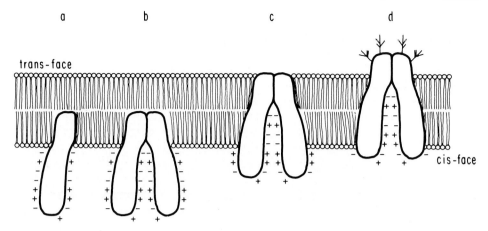

Fig. 4. A hypothetical process for the generation of a protein of type D, Fig. 1, starting from a protein of type C, Fig. 1 (a), which dimerizes to form an extended aqueous channel exterior to the membrane (b), and which after hydrophobic processing of its exposed cis-face hydrophilic surfaces, is embedded more deeply into the membrane (c), and finally, after hydrophilic processing on the trans-face side, becomes a trans-membrane aggregate (d). See text for further details.

to binding, but rather suggests that cytochrome b_5 is directed to the endoplasmic reticulum membrane because it is synthesized on ribosomes attached to that membrane.

It is of interest that the NH_2 terminus of the cytochrome b_2 molecule is part of the hydrophilic segment exposed on the cis side of the membrane. This suggests that if the molecule is made on membrane-bound ribosomes, the hydrophilic NH_2-terminal segment never enters the membrane but remains in the aqueous phase on the cis side of the membrane during translation of the nascent chain. Only after the hydrophobic COOH-terminal segment is synthesized and released, does that portion of the molecule insert spontaneously into the lipid bilayer. It is of great importance to know whether these chain termini positions are generally found with all type IC proteins. If so, then the different positions in the membrane that are taken up by type IB and IA proteins, on the one hand, and type IC proteins, on the other, may simply depend upon whether they are, or are not, initiated by a signal peptide on the NH_2 terminus of their nascent chains.

Of primary interest in connection with transport is the biosynthesis of proteins of type ID, for reasons given above. The monomer subunits of such proteins are presumed to have hydrophilic residues lining the aqueous channels that are formed by their oligomers. It is thermodynamically unreasonable for such trans-membrane monomers to be inserted individually into the membrane because the channel hydrophilic residues would then be in contact with the lipid interior of the membrane. It seems likely, therefore, that type ID and IB proteins, although both trans membrane, are synthesized and inserted into membranes by entirely different mechanisms.

One can simplify matters somewhat by proposing that the synthesis of type ID and IC proteins is related. Two monomers of a type IC protein, after they were individually synthesized and were bound to the cis side of the membrane (Fig. 4a), might spontaneously dimerize (Fig. 4b). The dimer might form an extended aqueous channel that was initially entirely exterior to the membrane. Specific processing or modification of the dimer might then occur that displaced the dimer more deeply into the membrane (Fig. 4c), and eventually caused it to protrude from the other side (Fig. 4d). As a consequence, the aqueous channel in the dimer might finally extend entirely across the membrane.

We are already familiar with certain specific posttranslational processing of membrane proteins, such as glycosylation by membrane-bound glycosyl transferases. As the highly polar oligosaccharides of membrane glycoproteins are found exclusively on the trans face of eukaryotic cell membranes (21), these residues must be attached only after the protein has spanned the membrane and become exposed at the trans face. Glycosylation of suitable residues of a protein close to the trans-side water-membrane interface may serve to "pull" a larger volume fraction of the protein through the membrane into the aqueous phase on the trans side. Other processing reactions of the protein on the cis face, such as the acylation of polar and ionic residues situated close to the water-membrane interface, or the formation of nonionic amide linkages between a glutaminyl-residue and an ionic ϵ-NH_3^+ of a lysyl residue (transpeptidation), to mention just a few possibilities, may likewise serve to "push" deeper into the membrane those regions of an integral protein that were exposed on the cis side of the membrane. The sequential operation of such cis side and then trans side processing reactions could therefore result in the proper positioning and conformation of a type ID protein in the membrane (Fig. 4).

It is not intended that these fanciful speculations about biosynthesis be taken very seriously in their details. They are presented primarily to illustrate the problems that arise when thermodynamic principles are applied to integral membrane proteins and to emphasize the likely connection between the biosynthesis and structures of these proteins. It is certain that there are still many surprises ahead of us as the experimental analysis of these proteins continues to develop.

ACKNOWLEDGMENTS

S. J. S. is an American Cancer Society Research Professor. Original studies discussed in this paper were supported by United States Public Health Service Grants GM-15971 and AI-06659.

REFERENCES

1. Ames GF-L, Lever J: Proc Natl Acad Sci USA 66:1096, 1970.
2. Ames GF-L, Spudich EN: Proc Natl Acad Sci USA 73:1877, 1976.
3. Ash JF, Singer SJ: Proc Natl Acad Sci USA 73:4575, 1976.
4. Blobel G, Sabatini DD: In Manson LA (ed): "Biomembranes." New York: Plenum Press, vol 2, p 193, 1971.
5. Bretscher MS: J Mol Biol 59:351, 1971.
6. Bretscher MS: Science 181:622, 1973.
7. Dutton A, Rees ED, Singer SJ: Proc Natl Acad Sci USA 73:1532, 1976.
8. Frye CD, Edidin M: J Cell Sci 7:313, 1970.
9. Hazelbauer GL, Adler J: Nature (London) New Biol 230:101, 1971.
10. Henderson R, Unwin PNT: Nature (London) 257:28, 1975.
11. Hill E, Tsernoglou D, Webb L, Banaszak LJ: J Mol Biol 73:577, 1972.
12. Hirano H, Parkhouse B, Nicolson GL, Lennox ES, Singer SJ: Proc Natl Acad Sci USA 69:2945, 1972.
13. Ho MK, Guidotti G: J Biol Chem 250:675, 1975.
14. Jardetzky O: Nature (London) 211:969, 1966.
15. Kauzmann W: Adv Protein Chem 14:1, 1959.
16. Kyte J: J Biol Chem 249:3652, 1974.
17. Kyte J: J Biol Chem 250:7443, 1975.

18. Lenard J, Singer SJ: Proc Natl Acad Sci USA 56:1828, 1966.
19. Lozier RH, Bogomolni RA, Stoeckenius W: Biophys J 15:955, 1975.
20. Matthews BW, Bernhard SA: Annu Rev Biophys Bioeng 2:257, 1974.
21. Nicolson GL, Singer SJ: J Cell Biol 60:236, 1974.
22. Palade GE: In Hayashi T (ed): "Subcellular Particles." New York: Ronald Press, p 64, 1959.
23. Remacle J, Fowler S, Beaufay H, Berthet J: J Cell Biol 61:237, 1974.
24. Roseman M, Litman BJ, Thompson TE: Biochemistry 14:4826, 1975.
25. Rothman JE, Lenard J: Science 195:743, 1977.
26. Ruoho A, Kyte J: Proc Natl Acad Sci USA 71:2352, 1974.
27. Sabatini DD, Kreibich G: In Martonosi A (ed): "The Enzymes of Biological Membranes." New York: Plenum Press, vol 2, p 531, 1976.
28. Seydoux F, Malhotra OP, Bernhard SA: CRC Crit Rev Biochem 2:227, 1974.
29. Singer SJ: In Rothfield LI (ed): "Structure and Function of Biological Membranes." New York: Academic Press, p 145, 1971.
30. Singer SJ: Annu Rev Biochem 43:805, 1974.
31. Singer SJ, Nicolson GL: Science 175:720, 1972.
32. Spatz L, Strittmatter P: Proc Natl Acad Sci USA 68:1042, 1971.
33. Strittmatter P, Rogers MJ, Spatz L: J Biol Chem 247:7188, 1972.
34. Taylor RB, Duffus WPH, Raff MC, de Petris S: Nature (London) New Biol 233:225, 1971.
35. Tomita M, Marchesi VT: Proc Natl Acad Sci USA 72:2964, 1975.
36. Vitetta ES, Uhr JW: Biochem Biophys Acta 415:253, 1975.
37. Wallach DFH, Zahler PH: Proc Natl Acad Sci USA 56:1552, 1966.
38. Yu J, Steck TL: J Biol Chem 250:9176, 1975.

Journal of Supramolecular Structure 6:325–331 (1977)
Molecular Aspects of Membrane Transport 187–193

Hormonal Regulation of Membrane Phenotype

Sarah A. Carlson and Thomas D. Gelehrter

Departments of Human Genetics and Internal Medicine, University of Michigan Medical School, Ann Arbor, Michigan 48109

Incubation of rat hepatoma cells (HTC) in tissue culture with glucocorticoids alters several membrane properties characteristic of transformed cells, without affecting the growth rate of these cells. Variant cell lines resistant to dexamethasone inhibition of plasminogen activator production have been isolated using an agar-fibrin overlay technique to detect plasminogen activator production by individual colonies of HTC cells. The resistance to dexamethasone is not secondary to abnormal or absent glucocorticoid receptors, but due to a lesion in a later step in hormone action specific for plasminogen activator. These variants should prove useful for the study of the mechanism of steroid action as well as for the analysis of the role of proteases in the hormonal regulation of membrane function.

Key words: plasminogen activator, hepatoma cells, transformed membrane phenotype, glucocorticoid-resistant variants, glucocorticoids, amino acid transport, hormonal regulation, glucocorticoid regulation

Neoplastic transformation is associated with structural and functional alterations in the cell membrane which may have an important role in growth regulation. Hormones can also affect membrane phenotype in a manner similar or opposite to neoplastic transformation (1, 2). In HTC[1] cells, an established line of rat hepatoma cells in tissue culture, glucocorticoids alter several membrane properties characteristic of transformed cells, but do not affect the growth rate of these cells. Dexamethasone, a synthetic glucocorticoid, a) rapidly and reversibly inhibits the rate of influx of selected amino acids by a process requiring concomitant protein synthesis[2] (3); b) decreases the number of microvilli on the surface of HTC cells in suspension culture, as assessed by scanning electron microscopy; and c) increases the adhesiveness of HTC cells (4). Dexamethasone also decreases the production of plasminogen activator (5), an intracellular protease which may significantly modulate various membrane properties. Using an agar-fibrin overlay technique to detect plasminogen activator production by individual colonies of HTC cells (6), we have selected lines of HTC cells resistant to the inhibitory effect of dexamethasone. Combined genetic and biochemical analysis of such dexamethasone-resistant variants should facilitate study of hormonal regulation of specific membrane phenotypes.

[1]Abbreviations: HTC – hepatoma tissue culture; AIB – α-aminoisobutyric acid; BSA – bovine serum albumin.
[2]McDonald, R. A. and Gelehrter, T. D., submitted for publication.

Received April 1, 1977; accepted April 27, 1977.

AGAR-FIBRIN OVERLAY TECHNIQUE

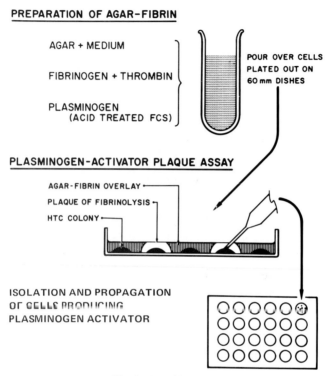

Fig. 1. Agar-fibrin overlay.

METHODS

HTC cells are an established line of rat hepatoma cells which have been in continuous culture for the past 12 years (7). The cells were grown in spinner culture without antibiotics in Eagle's minimal essential medium (MEM) for suspension culture, modified to contain 0.05 M Tricine, 0.5 g/liter $NaHCO_3$, 2 mM glutamine, 5% calf serum, and 5% fetal calf serum.

An agar-fibrin overlay technique (6) was used to detect the production of plasminogen activator by individual colonies (Fig. 1). HTC cells (100–200 cells per dish) were plated in 60-mm tissue culture dishes in alpha-MEM (Flow Laboratories) without added hormones, supplemented with 10% fetal calf serum, and allowed to grow into colonies for 5–7 days. The cells were then incubated overnight in serum-free MEM with or without 0.1 μM dexamethasone. The plates were washed with serum-free medium, covered with 1.4 ml of 0.9% agar containing 0.7 × medium, 2.8 mg/ml fibrinogen, 1 U/ml thrombin, 3.6% acid-treated fetal calf serum (as a source of plasminogen), and with or without 1 μM dexamethasone, and incubated at 37°C for 24 h. The thrombin activates the fibrinogen to form fibrin, and an opalescent, fibrin meshwork is formed in the agar. Colonies producing plasminogen activator activate the plasminogen to plasmin, which in turn lyses the fibrin, and a clear area, or plaque, is formed. This technique is not destructive and allows recovery of selected colonies through the agar overlay, and their propagation.

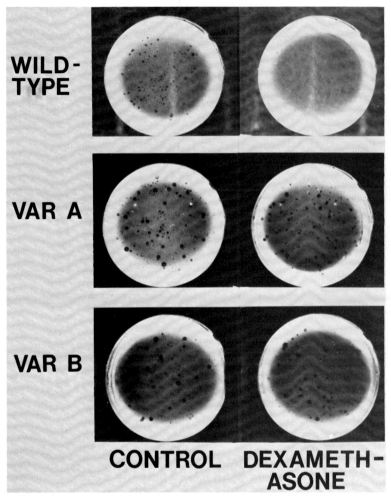

Fig. 2. HTC cells were plated in dilute suspension and allowed to grow into colonies. The cells were incubated for 22 h in serum-free medium with or without 0.1 μM dexamethasone. The plates were then covered with the agar-fibrin overlay in the continued presence or absence of dexamethasone (1 μM) and photographed after 24 h incubation at 37°C.

Transport of α-aminoisobutyric acid (AIB) was measured as described by Heaton and Gelehrter (8). Tyrosine aminotransferase was assayed by the method of Spencer and Gelehrter (9).

Fibrinogen (77% clotable) was purchased from Calbiochem and purified as described by Strickland and Beers (10). Bovine thrombin was purchased from Parke-Davis. Dexamethasone was a gift from Merck and Company. All other compounds were of reagent grade.

RESULTS

The effect of dexamethasone on plasminogen activator production in HTC cells is shown in the upper portion of Fig. 2. In the absence of dexamethasone, 100% of the colonies produced fibrinolytic plaques. After 18–24-h incubation in the presence of 0.1

DEXAMETHASONE INDUCTION OF TYROSINE AMINOTRANSFERASE

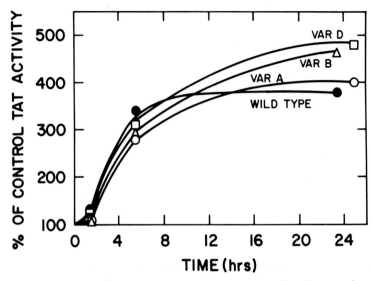

Fig. 3. Wild-type and variant HTC cells were incubated in suspension culture in serum-free medium with 0.1% BSA with or without 0.1 μM dexamethasone. At the times indicated, samples were taken for assay of tyrosine aminotransferase (9).

μM dexamethasone, less than 1% of the colonies produced plaques and most of these tended to be smaller than those produced in the absence of dexamethasone.

Isolation of Dexamethasone-Resistant Variants

Because dexamethasone inhibits plasminogen activator production, colonies which are resistant to dexamethasone inhibition should continue to produce plasminogen activator in the presence of the hormone. Hormone-resistant colonies will thus form plaques in the agar overlay in the presence of dexamethasone. Recovery of such plaque-forming colonies and retesting by the overlay technique has resulted in the isolation of populations of cells partially resistant to the inhibitory effects of dexamethasone on plasminogen activator production. By repeating (3—5 times) the process of isolation of plaque-forming colonies, serial propagation in the absence of dexamethasone, and retesting by the overlay technique, we have isolated a number of HTC cell lines fully resistant to the inhibitory effect of dexamethasone on protease production. The effect of dexamethasone on plasminogen activator production in 2 of these variants is shown in the lower portion of Fig. 2. Essentially all of the colonies of variants A and B produce plaques in the presence as well as the absence of dexamethasone, and the size of the plaques under the 2 conditions is the same.

Analysis of Dexamethasone-Resistant Variants

The basis for the hormonal resistance in these variant cell lines has been examined by testing for the presence of normal dexamethasone receptor function. This has been accomplished by the simple expedient of testing whether these lines have dexamethasone-

DEXAMETHASONE INHIBITION OF AIB TRANSPORT

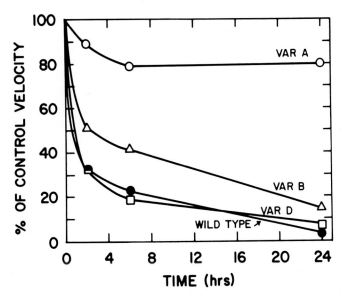

Fig. 4. Wild-type and variant HTC cells were incubated in suspension culture in serum-free medium with 0.1% BSA with or without 0.1 μM dexamethasone. At the times indicated, samples were taken for measurement of AIB transport (8).

inducible tyrosine aminotransferase, a well-characterized, non-membrane-associated response to glucocorticoids in HTC cells. Transaminase induction requires binding of the glucocorticoid to a specific cytoplasmic receptor, and translocation of the steroid-receptor complex to the nucleus where it interacts with nuclear chromatin, resulting in accumulation of specific messenger RNA (11). Furthermore, it is generally accepted that all glucocorticoid actions in a given cell are mediated by a single class of receptors (11, 12). As shown on Fig. 3, all variants tested (including 3 other lines not shown) show normal wild-type induction of tyrosine aminotransferase, unequivocally demonstrating normal glucocorticoid receptor function.

The specificity of the hormonal resistance in these variant lines has been tested by examining the ability of dexamethasone to alter other aspects of the transformed membrane phenotype. Figure 4 shows the effect of dexamethasone on AIB transport in wild-type and hormone-resistant HTC cells. Variants B and D, and all but 1 of the other lines tested, show wild-type inhibition of AIB transport. In contrast, variant A appears to be pleiotropic, in that it shows at least partial resistance to the inhibitory effects of dexamethasone on AIB transport.

DISCUSSION

HTC cells provide a useful model for studying the hormonal regulation of membrane phenotype in neoplastic cells. Several membrane properties are modulated by glucocorticoids, without changing the growth rate of these cells. The hormonal responsiveness of this cell line has been extensively characterized. In order to study the mechanisms by which hormones regulate membrane phenotype, it would be useful to have variants resistant to specific hormone-mediated events. We report here the isolation of several, spontaneously occurring variant cell lines which are resistant to the dexamethasone inhibition of plasminogen activator production. The resistant phenotype of these variants has remained stable for 6–9 months (more than 100 generations) in culture in the absence of dexamethasone.

The resistance to dexamethasone is not secondary to abnormal or absent glucocorticoid receptors. All variants tested show wild-type induction of tyrosine aminotransferase, unequivocally demonstrating the presence of normal glucocorticoid receptor function. The lesion in these variants is presumably in some more distal step in hormone action, after the binding of hormone-receptor complexes to nuclear chromatin. This finding is in contrast to the great majority of glucocorticoid-resistant variants described previously. In mouse lymphoma lines, for example, essentially all variants which have been analyzed show deficient or defective glucocorticoid receptor (12). A probable exception to this situation is the recent report by Thompson et al. (13) describing the isolation of variant HTC cells in which tyrosine aminotrasferase is not inducible by dexamethasone, and in which glucocorticoid receptor function appears to be normal.

HTC cells resistant to the dexamethasone inhibition of plasminogen activator production should prove useful in the analysis of the role of proteases in the hormonal regulation of membrane phenotype. It is possible that dexamethasone regulation of membrane characteristics is secondary to a proteolytic removal or alteration of membrane proteins. It seems clear however that the inhibition of AIB transport is not secondary to an inhibition of plasminogen activator production since variants resistant to the latter effect retain normal inhibition of transport. On the other hand, it is quite possible that the dexamethasone induction of adhesiveness in HTC cells is secondary to the decrease in protease activity which in turn would allow the accumulation of specific glycoproteins involved in adhesion (14). This hypothesis is directly testable using the dexamethasone-resistant variants described in this report. The use of such variants should provide a powerful alternative to the use of protease inhibitors in the analysis of the role of proteases in regulation of membrane function; the problems associated with the latter approach have been critically reviewed (15). Finally, these variants should also provide a useful tool for the study of the mechanism by which dexamethasone regulates plasminogen activator production. Isolation of other variant cell lines resistant to specific hormone-mediated functions affecting different membrane properties should prove similarly useful.

ACKNOWLEDGMENTS

We thank Dr. Sidney Strickland for assistance with establishing the agar-overlay technique. This research was supported by National Institutes of Health Grant GM 15419. S.A.C. is supported by a predoctoral training grant from the National Institutes of Health, 5-T01-GM00071. T.D.G. is the recipient of a Faculty Research Award from the American Cancer Society.

REFERENCES

1. Holley RW: Nature 258:487, 1975.
2. Hsie AW, Jones C, Puck TT: Proc Natl Acad Sci USA 68:1648, 1971.
3. Risser WL, Gelehrter TD: J Biol Chem 248:1248, 1973.

4. Ballard P, Tomkins GM: J Cell Biol 47:222, 1970.
5. Wigler M, Ford JP, Weinstein IB: In Reich E, Rifkin D, Shaw E (eds): "Proteases and Biological Control." New York: Cold Spring Harbor Laboratory, 1975, p 849.
6. Jones P, Benedict W, Strickland S, et al: Cell 5:323, 1975.
7. Thompson EB, Tomkins GM, Curran JF: Proc Natl Acad Sci USA 56:296, 1966.
8. Heaton JH, Gelehrter TD: J Biol Chem 252:2900, 1977.
9. Spencer CJ, Gelehrter TD: J Biol Chem 249:577, 1974.
10. Strickland S, Beers WH: J Biol Chem 251:5694, 1976.
11. Rousseau GG: J Steroid Biochem 6:75, 1975.
12. Yamamoto KR, Gehring U, Stampfer MR, et al: Rec Prog Horm Res 32:3, 1976.
13. Thompson EB, Aviv D, Lippman ME: Endocrinology 100:406, 1977.
14. Pouyssegur JM, Pastan I: Proc Natl Acad Sci USA 73:544, 1976.
15. Roblin R, Chou I-N, Black PH: Adv Cancer Res 22:203, 1975.

Journal of Supramolecular Structure 6:333—344 (1977)
Molecular Aspects of Membrane Transport 195—206

Isolation of Membrane Vesicles With Inverted Topology by Osmotic Lysis of Azotobacter vinelandii Spheroplasts

Eugene M. Barnes, Jr., and Pinakilal Bhattacharyya

Marrs McLean Department of Biochemistry, Baylor College of Medicine, Houston, Texas 77030

Membrane vesicles were prepared from Azotobacter vinelandii spheroplasts by lysis in either potassium phosphate (pH 7.0) or Tris[1]-acetate (pH 7.8) buffers. These 2 types of preparations differ considerably in their properties: 1) Examination by scanning electron microscopy reveals that the Pi vesicles consist primarily of closed structures 0.6—0.8 μm in diameter with a rough or particulate surface similar to that of spheroplasts. The Tris vesicles are significantly smaller, 0.1—0.3 μm in diameter, and have a much smoother surface structure. 2) Antisera from rabbits immunized with A. vinelandii lipopolysaccharide antigen will agglutinate Pi vesicles but not Tris vesicles. 3) Tris vesicles have a fourfold higher specific activity of latent H^+-ATPase than Pi vesicles. After exposure to Triton X-100 similar ATPase activities are observed for both types of vesicles. 4) Pi vesicles transport calcium in the presence of ATP or lactate at less than 30% of the rates observed for Tris vesicles. 5) Tris vesicles have less than 22% of the transport capacity of Pi vesicles for accumulation of labeled sucrose and less than 3% of the capacity for valinomycin-induced uptake of rubidium observed during respiration. 6) Quinacrine fluorescence intensity is reduced by 30% during lactate oxidation and 20% during ATP hydrolysis by Tris vesicles. Under similar conditions, fluorescence in Pi vesicles is quenched by only 7% and less than 2%, respectively. These findings suggest that Pi vesicles have the normal orientation of the intact cell whereas Tris vesicles have an inverted topology.

Key words: calcium transport, quinacrine fluorescence, rubidium transport, sucrose transport, lipopolysaccharide antibody, scanning electron microscopy, topology, membrane vesicles, Azotobacter vinelandii

Cytoplasmic membrane vesicles isolated from bacteria according to the techniques pioneered by Kaback (1) have proven to be a powerful system for elucidation of active transport mechanisms. The topological orientation or sidedness of such preparations and the homogeneity of topology in the vesicle population are key factors in interpretation of solute transport experiments, especially if results are to be evaluated in chemiosmotic terms. Kaback has cited considerable evidence (1—3) in support of his claim that Escherichia coli membrane vesicles prepared by the established technique have a normal or right-side-out orientation, i.e., identical to that of the cell. Perhaps most convincing are freeze-cleave electron microscopic studies, repeated in several laboratories (1, 4, 5), which

[1] Abbreviations: CCP — Chlorocarbonylcyanide phenylhydrazone; LPS — Lipopolysaccharide; Tricine — N-tris-(hydroxymethyl)-methylglycine; Tris — Tris-(hydroxymethyl)-aminomethane.

Received April 29, 1977; accepted May 11, 1977

confirm that preparations of the Kaback type are almost completely right-side out. However, extensive examination of certain membrane markers known to be located at the internal face of the cytoplasmic membrane in intact cells and spheroplasts is not consistent with this interpretation. For example, it has been shown that ATPase, NADH dehydrogenase, succinic-ferricyanide reductase, and α-glycerol-P ferricyanide reductase are located on the inner surface of the plasma membrane of E. coli spheroplasts (6–9). These enzyme systems in spheroplasts are inaccessible to impermeant substrates (ATP, NADH, ferricyanide) or to specific antibodies, but become accessible when the permeability barrier is destroyed. In spheroplast vesicles (kabackosomes), however, about 50% of the total activity is accessible to impermeant substrates for these marker enzymes (8, 9) and to the ATPase antibody (6, 7). But other markers, such as D-lactic dehydrogenase, remain inaccessible in spheroplast vesicles (3). The most widely accepted interpretation of these conflicting findings is that the E. coli spheroplast vesicles have a normal gross topology, but that certain internal components become externalized during the isolation procedure. The topology of such preparations would thus be functionally heterogenous.

In the course of our studies with Azotobacter vinelandii we observed that vesicles prepared by lysis of spheroplasts in Tris acetate buffer transport calcium in the presence of oxidizable substrates (10, 11) or ATP (12) but show very low transport activity for respiration-coupled uptake of glucose or sucrose. These latter solutes are accumulated by A. vinelandii membrane vesicles (13, 14) prepared by lysis of spheroplasts in the potassium phosphate buffer employed by Kaback. Since most bacterial cells actively extrude calcium (15) and inverted vesicles from E. coli accumulate calcium (16), it appeared that our Tris preparations everted during isolation. In light of the foregoing discussion, however, it was not clear whether only selected components (i.e., ATPase and dehydrogenase) became inverted or the entire surface became everted. In this paper we report further studies which support the latter conclusion.

METHODS

Membrane vesicles were isolated from Azotobacter vinelandii strain O after growth on 1% glucose or 1% sucrose by a procedure described earlier (12, 13).

Samples for scanning electron microscopy were immediately fixed in 2% glutaraldehyde in 0.1 M cacodylate buffer (pH 7.4) containing 0.1 M sucrose at $4°C$ for at least 1 h. The samples were then rinsed with cold buffer and dehydrated in a graded series of alcohol followed by dehydration in amyl acetate. The samples were further processed by critical point drying in a Denton DCP1 apparatus using oil-free liquid CO_2 as an intermediate fluid. Specimens were immediately coated with 50–70 Å of a 60:40 alloy of gold palladium with a DC sputtering module in a Kinney KSM-2 evaporator at 2×10^{-5} torr. Specimens were examined with a 100C JEOL electron microscope equipped with ASID at 40 kV using 30–50° tilting angle in a side entry goniometer. Images were recorded on Polaroid 105 N/P film from a 2000 line CRT. Photographic enlargement did not exceed 1.6 diameters.

Antigenic lipopolysaccharide was isolated from Azotobacter vinlandii cells by the method of Boivin and Mesrobeanu as described by Staub (17). Aqueous solutions of LPS were sterilized by filtration and stored at $4°C$. Rabbits were injected with an aliquot of this solution equivalent to 250 μg dry weight of LPS and then received a booster injection of 500 μg on day 20 and 2.5 mg on day 30. Blood was collected on day 35, incubated at $0°C$ for 2 h, and then centrifuged at 48,000 $\times$ g for 30 min. The supernate was adjusted to 50% saturation of ammonium sulfate and the precipitate collected by centrifugation. The pellet was dissolved in 0.1 M sodium phosphate buffer (pH 7.0) and stored at $-20°C$. Precipitating antibody was detected by discontinuous counterimmunoelectrophoresis (18). A weak but distinct precipitin line was observed between immune globulin and LPS antigen but no reaction occurred with normal rabbit globulin. For antibody agglutination experiments, membrane vesicles were pelleted by centrifugation and resuspended in 0.15 M sodium chloride buffered with 10 mM potassium phosphate (pH 7.0). Membrane vesicles (500 μg protein) were incubated in a final volume of 60 μl with 100 μg globulin at 0°C for 1 h. These incubations were inspected by phase contrast microscopy using a Leitz Orthoplan microscope and photographed with a Leitz Orthomat camera on KB-14 film (ASA 100).

ATPase activity of membrane vesicles was activated with trypsin and assayed as previously described (12). The methods for measuring transport of calcium (12), sucrose (14), and rubidium (19) were also reported earlier.

For fluorescence measurements, vesicles were pelleted by centrifugation and resuspended in a buffer prepared by titration of 0.10 M Tricine with Tris base to pH 8.0. The assay mixture (2 ml) contained this buffer solution plus 1 μM quinacrine hydrochloride, 1 mM $MgSO_4$, and 0.2 mg/ml vesicle protein. Quinacrine fluorescence was recorded as described (19) at excitation and emission wavelengths of 425 nm and 500 nm, respectively.

RESULTS

Membrane vesicles were prepared by osmotic lysis of lysozyme-EDTA spheroplasts from Azotobacter vinelandii strain O. Two different buffers were employed as the lysis medium and in subsequent washing of the vesicles. Preparations referred to as Pi vesicles were lysed and washed in potassium phosphate (pH 7.0), whereas Tris vesicles were prepared in Tris-acetate (pH 7.8). Both the spheroplasts and derived vesicles were examined by scanning electron microscopy as shown on Figs. 1 and 2. The Pi vesicles consist primarily of closed structures 0.6–0.8 μm in diameter which have a rough or particulate surface topology similar to that of the spheroplasts. In contrast, the Tris preparations contain a much higher content of small vesicles, 0.1–0.3 μm in diameter. These latter preparations have a much smoother surface structure than is observed for spheroplasts or Pi vesicles. Although extensive aggregation of vesicles is observed in the electron photomicrographs this probably represents an artifact due to fixation in glutaraldehyde. Little aggregation of vesicles is observed by phase contrast light microscopy in untreated preparations but extensive aggregation is noted after incubation in 4% glutaraldehyde (not shown).

In order to further examine the surface components of these vesicles, lipopolysaccharide antigen was extracted from A. vinelandii and injected into rabbits. Immune globulin gave a precipitin line with purified LPS in discontinuous counterimmunoelectrophoresis but none was observed using preimmune globulin. The LPS antibody produced extensive agglutination of Pi vesicles as shown in the phase contrast micrographs of Fig. 3. This effect was blocked by addition of excess LPS; no agglutination was produced by preimmune globulin. However, incubation of LPS antibody with Tris vesicles failed to produce significant agglutination.

Previous studies from this laboratory established that Tris vesicles isolated from A. vinelandii contain a trypsin-activated, dicyclohexylcarbodiimide-sensitive ATPase which catalyzed proton uptake (12). The F_1 subunit of this enzyme is thought to be attached to

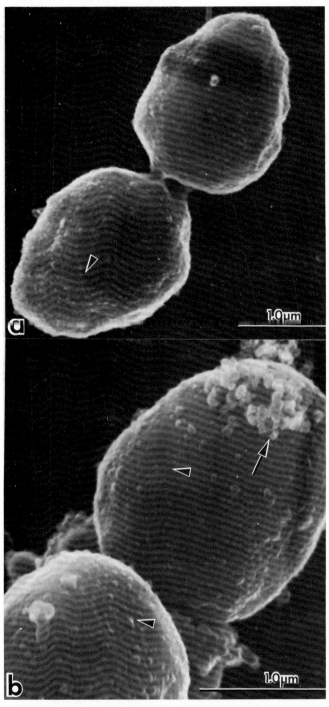

Fig. 1. Scanning electron photomicrographs of A. vinelandii cells (a) and spheroplasts (b). Specimens were prepared and examined as described under Methods.

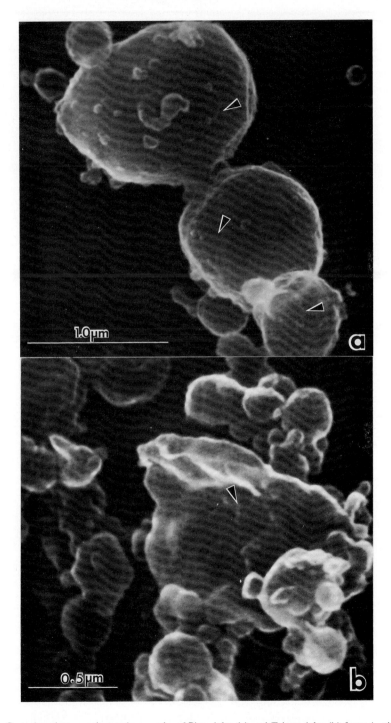

Fig. 2. Scanning electron photomicrographs of Pi vesicles (a) and Tris vesicles (b) from A. vinelandii. Specimens were prepared and examined as described under Methods.

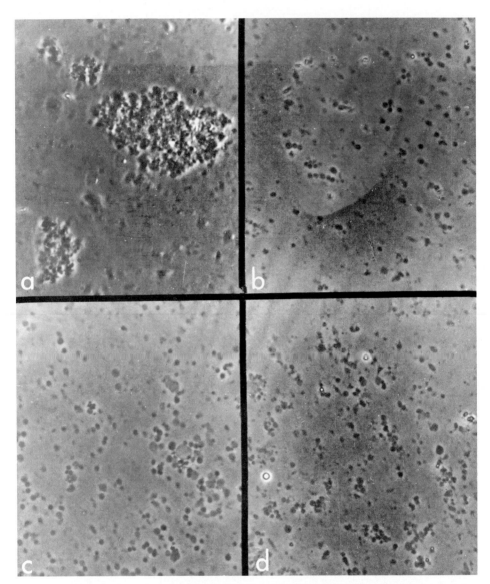

Fig. 3. Agglutination of membrane vesicles by lipopolysaccharide antibody. Vesicles were incubated with antibody and phase contrast photomicrographs obtained as described under Methods. a) Pi vesicles (500 μg protein) incubated with 100 μg LPS antibody; b) Pi vesicles (500 μg) incubated with 100 μg control (preimmune) globulin; d) Tris vesicles (500 μg) incubated with 100 μg LPS antibody.

the cytoplasmic face of the plasma membrane in bacteria (6, 9). Since ATP is not permeant, we sought to employ ATPase activity, unmasked by trypsin treatment, as an index of membrane sidedness in both Tris and Pi vesicles. These activities are shown in Table I. Both Tris and Pi vesicles have relatively low ATPase activity until exposed to a brief trypsin treatment. Trypsinization of Tris vesicles results in nearly a sevenfold increase in ATPase activity. This is presumably due to destruction of a polypeptide inhibitor as

TABLE I. ATPase Activity of A. vinelandii Membrane Vesicles*

| Carbon source | Lysis buffer | Vesicle incubation | | ATPase activity |
		Trypsin	Triton	
				nmol/min/mg
Sucrose	Tris	−	−	17
		+	−	116
		+	+	134
Sucrose	Pi	−	−	16
		+	−	31
		+	+	118
Glucose	Pi	−	−	8
		+	−	19
		+	+	59

*Vesicles were isolated from cells grown on sucrose or glucose by spheroplast lysis in phosphate or Tris buffer (as indicated). Where indicated, latent ATPase was activated by exposure of vesicles to trypsin (62 μg/mg membrane protein at 0°C) in 0.1 M Tris-acetate (pH 7.8) for 2 min. Where indicated, the incubation medium for trypsinization contained 0.1% Triton X-100. After addition of soybean trypsin inhibitor, aliquots were removed for ATPase assay as described in Methods.

described for the oligomycin-sensitive ATPase in heart mitochondria (20). Trypsin treatment under these conditions does not make vesicles grossly leaky since they are able to retain accumulated calcium (cf., Table II). Exposure of Tris vesicles to trypsin in the presence of 0.1% Triton X-100 results in only a 13% further increase in ATPase activity.

On the other hand, exposure of Pi vesicles to trypsin yields only a twofold increase in ATPase activity. But the presence of Triton increases ATPase dramatically, i.e., up to a sevenfold overall increase. Values for ATPase activity are also reported for Pi vesicles from cells grown on glucose as sole carbon source. These latter preparations have been shown to extrude protons during respiration (19) and accumulate glucose actively (13). The same pattern is observed for Pi vesicles isolated from sucrose-grown cells although the ultimate ATPase levels are about twofold higher in preparations from the sucrose culture. It seems clear that intact Tris vesicles have much greater trypsin-activated ATPase activity (about

TABLE II. Rates of Energy-dependent Calcium Uptake by A. vinelandii Membrane Vesicles*

| Carbon source | Lysis buffer | Calcium uptake rate | |
		D-Lactate	ATP
		nmol/min/mg	nmol/min/mg
Sucrose	Tris	15.6	.9.2
Sucrose	Pi	4.6	< 0.2
Glucose	Pi	2.3	< 0.2

*Vesicles were isolated from cells grown on sucrose or glucose by lysis in Pi or Tris buffer, as indicated. The vesicles were washed in 50 mM Tris-acetate (pH 7.8) and assayed in this buffer containing 2 mM MgSO$_4$, 50 μg membrane protein, 40 μM ^{45}Ca (36 Ci/g-atom), and either 20 mM D-lactate or 2 mM ATP, as described under Methods. Vesicles were trypsin treated (Table I) for assays of ATP-dependent uptake and untreated preparations were used for lactate-dependent uptake. All values for Ca uptake were corrected for the uptake by controls that lacked energy source.

fourfold) than the comparable Pi vesicles. However, after destruction of the permeability barrier with Triton, similar activities are observed.

Several species of bacteria have been shown to actively extrude calcium (15). Thus the proton-coupled accumulation of calcium by bacterial membrane vesicles can be an index of sidedness as well as the functional state of the ATP- and respiration-driven proton pumps. A comparison of rates for energy-linked calcium transport by A. vinelandii membrane vesicles is given in Table II. As indicated, Tris vesicles accumulate calcium at appreciable rates in the presence of D-lactate or ATP (after trypsin treatment). The rates shown have been corrected for the energy-independent uptake which is typically less than 5% of the coupled rate. In contrast, Pi vesicles in the presence of lactate transport calcium at less than 30% of the rate observed with Tris vesicles. In the presence of ATP, Pi vesicles do not accumulate significant amounts of calcium either before or after the usual trypsin treatment.

It was previously established that Pi vesicles isolated from A. vinelandii grown on glucose develop a membrane potential (inside-negative) and concentrate glucose by a proton-coupled mechanism (19). Since the Tris vesicles used in this study were derived from cells grown on sucrose, we examined the active accumulation of sucrose and Rb^+ (in the presence of valinomycin) in both vesicle types. Pi vesicles actively accumulate [^{14}C] sucrose from the medium during the oxidation of L-malate (Table III). Under similar conditions, Rb^+ is also accumulated but only after addition of valinomycin. This latter indicates that a transmembrane electric potential develops across the vesicles during respiration. The Tris vesicles, on the other hand, accumulate sucrose at only 20% of the level achieved by Pi vesicles. Furthermore, Tris vesicles fail to take up significant amounts of Rb^+ in the presence of valinomycin. These latter observations are not due to a gross defect in respiration since malate is rapidly oxidized by these preparations and drives calcium accumulation (11).

In order to study the development of an energized state during respiration or ATP hydrolysis by these preparations, the fluorescence of quinacrine hydrochloride was also examined. As shown on Fig. 4, lactate oxidation by Tris vesicles results in a 30% reduction in quinacrine fluorescence intensity. This quenching was reversed by addition of the uncoupler CCP. Quinacrine fluorescence was not significantly quenched by lactate addition to Pi vesicles derived from glucose-grown cells and was reduced only by 7% in Pi vesicles

TABLE III. L-Malate-dependent Accumulation of Sucrose and Rubidium (in the presence of valinomycin) by Vesicles*

Lysis buffer	Sucrose uptake	Rb^+ Uptake
	nmol/mg/10 min	nmol/mg/10 min
Pi	6.5	25.2
Tris	1.4	0.6

*For sucrose uptake vesicles were washed in 50 mM KPi buffer (pH 7.0) and assayed in that buffer containing 2 mM $MgSO_4$, 50 μM FAD, 20 mM L-malate, and 40 μM [^{14}C]-sucrose as described under Methods. For Rb uptake vesicles were washed in 50 mM NaPi (pH 7.0) and assayed in that buffer containing the above except that 40 μM $^{86}Rb^+$ and 1 μM valinomycin replaced sucrose.

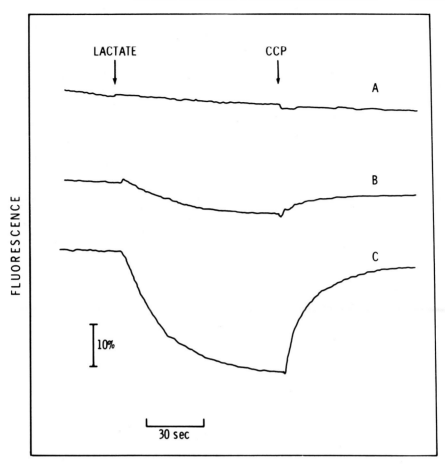

Fig. 4. Effect of lactate oxidation by membrane vesicles on quinacrine fluorescence. The assay mixture contained 0.10 M Tris-tricinate (pH 8.0), 1 mM $MgSO_4$, 0.2 mg/ml membrane protein, and 1 μM quinacrine hydrochloride. Fluorescence was recorded as described under Methods; the vertical bar indicates scale callibration equivalent to 10% of the initial fluorescence intensity. As indicated by the first arrow, DL-lactate (Tris salt) was added at a final concentration of 1 mM. At the second arrow, CCP was added at a final concentration of 1 μM. Trace A) Pi vesicles isolated from glucose-grown cells; B) Pi vesicles from sucrose-grown cells; C) Tris vesicles from sucrose-grown cells.

from sucrose-grown cells. Since quenching of quinacrine fluorescence reflects development of a proton gradient, inside-acid (21), these observations are consistent with the relative rates of calcium uptake given in Table II. Similar results were obtained on addition of ATP to trypsinized vesicles as shown in Fig. 5. ATP hydrolysis by Tris vesicles results in a 20% reduction of quinacrine fluorescence which was reversible by uncoupler. Addition of ATP to Tris vesicles before trypsin treatment gave no fluoresence quenching (data not shown). However, ATP hydrolysis by Pi vesicles, previously treated with trypsin, failed to produce any significant effect on quinacrine fluorescence intensity. This again is consistent with the effects of ATP on calcium transport.

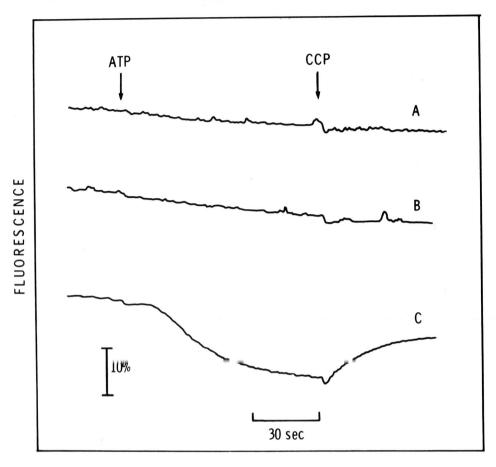

Fig. 5. Effect of ATP hydrolysis by membrane vesicles on quinacrine fluorescence. The assay conditions were identical to those described in the legend of Fig. 4 except that trypsin-treated vesicles were employed (Methods) and 0.3 mM ATP (Mg salt) replaced lactate. Trace A) Pi vesicles isolated from glucose-grown cells; B) Pi vesicles from sucrose-grown cells; C) Tris vesicles from sucrose-grown cells.

DISCUSSION

The results reported here clearly indicate topological differences in Pi and Tris vesicles isolated from A. vinelandii. The Pi vesicles, prepared by a method similar to Kaback's, appear to have properties (proton extrusion, sugar accumulation) similar to the intact cell (19). On the other hand, Tris vesicles clearly possess properties usually associated with an inverted (inside-out) topology, i.e., proton and calcium accumulation (13). The major point at issue, however, is whether: i) the Tris lysis of spheroplasts brings about inversion of selected membrane components leaving the gross topology normal or ii) Tris lysis causes a true everting of the membrane in a significant population of the vesicles. The findings presented here support the second interpretation; this rests on 7 lines of evidence.

1) Scanning electron micrographs indicate that the Tris vesicles are significantly smaller in diameter than Pi vesicles.

2) Pi vesicles have a rough or particulate topology, revealed by the scanning electron

microscope, which is similar to that of spheroplasts. Tris vesicles appear to have a much smoother surface.

3) Pi vesicles are readily agglutinated by LPS antibody, but Tris vesicles are not agglutinated. Using ferritin-conjugated LPS antibody, Shands (22) demonstrated the presence of LPS antigen on the outer surface of Salmonella typhimurium spheroplasts and plasma membranes obtained by osmotic lysis.

4) Tris vesicles have fourfold higher specific activity of trypsin-activated H^+-ATPase than intact Pi vesicles. After dissolution of the permeability barrier with Triton, similar ATPase activity is observed in both types of vesicles.

5) Tris vesicles rapidly accumulate calcium during oxidation of D-lactate or hydrolysis of ATP. Pi vesicles transport calcium at less than 30% of the rates observed for Tris vesicles during lactate oxidation and at negligible rates in the presence of ATP.

6) Tris vesicles have less than 22% of the transport capacity of Pi vesicles for the respiration-coupled accumulation of $[^{14}C]$ sucrose, and less than 3% of the capacity for accumulating rubidium (in the presence of valinomycin) during respiration.

7) The fluorescence intensity of quinacrine is markedly reduced when lactate or ATP is added to Tris vesicles. Much less quenching is observed with Pi vesicles under identical conditions.

These findings support the interpretation that the Pi vesicles consist primarily of structures with the same topology as the intact cell, whereas the Tris preparations are principally composed of everted vesicles. It is difficult, however, to quantitate the extent to which each fraction is contaminated with vesicles of opposite topology. The electron photomicrographs, LPS-antibody studies, and fluorescence measurements cannot be stringently applied for quantitative purposes. Using ATPase and calcium transport activity as markers for inverted vesicles, we can estimate that Pi vesicles contain at most 29% and 27%, respectively, of inverted structures. Using sucrose and rubidium uptake as indices of right-side-out vesicles, we estimate the maximum contamination of Tris vesicles by right-side-out structures at 21% and 3%, respectively. Of course, these approximations rest on the assumptions that: i) these markers do not undergo transmembrane flip-flop and ii) the lysis and washing procedures do not cause marker activation. These possibilities cannot be rigorously excluded.

The mechanism of inversion is of great interest to us. Although lysis of A. vinelandii spheroplasts in Tris acetate buffer clearly promotes inversion, similar treatment of E. coli spheroplasts apparently does not. Such vesicles from E. coli do not accumulate significant amounts of calcium in the presence of D-lactate (H. R. Kaback, unpublished observations). A significant factor may be the network of internal membrane thought to be formed by invagination of the A. vinelandii plasma membrane (23, 24). Osmotic lysis of spheroplasts in a relatively permeant buffer such as Tris acetate may induce sealing and/or release of vesicles with inverted topology. Such structures may be present but trapped within the internal space of Pi vesicles. Further investigation is required to clarify this issue.

ACKNOWLEDGMENTS

This work was supported by NIH Grant GM 18962 and NSF Grant PCM 75-13591. E.M.B. is a recipient of NIH research career development award AM 00052. The authors acknowledge the collaboration of Dr. I. Daskal with the scanning electron microscopy, Dr. G. Dreesman with immunoelectrophoresis, and Dr. W. Wray with phase contrast photomicrographs.

REFERENCES

1. Kaback HR; Science 186:882, 1974.
2. Short SA, Kaback HR, Kaczorowski G, Fisher J, Walsh CT, Silverstein SC: Proc Natl Acad Sci USA 71:5032, 1974.
3. Short SA, Kaback HR, Kohn LD: J Biol Chem 250:4291, 1975.
4. Altendorf KH, Staehelin LA: J Bacteriol 117:888, 1974.
5. Konings WN, Bisschop A, Voenhuis M, Vermeulen CA: J Bacteriol 116:1465, 1973.
6. Van Thienen G, Postma PW: Biochim Biophys Acta 323:429, 1973.
7. Hare JB, Olden K, Kennedy EP: Proc Natl Acad Sci USA 71:4843, 1974.
8. Weiner JH: J Membr Biol 15:1, 1974.
9. Futai M: J Membr Biol 15:15, 1974.
10. Barnes EM Jr: Fed Proc Fed Am Soc Exp Biol 33:1457, 1974.
11. Barnes EM Jr, Roberts RR, Bhattacharyya P: Membr Biochem 1 (In press).
12. Bhattacharyya P, Barnes EM Jr: J Biol Chem 251:5614, 1976.
13. Barnes EM Jr: J Biol Chem 248:8120, 1973.
14. Barnes EM Jr: Arch Biochem Biophys 163:416, 1974.
15. Silver S: In Weinberg ED (ed): "Microorganisms and Minerals." New York: Marcel Dekker, Microbiology Series, (1976).
16. Rosen BP, McClees JS: Proc Natl Acad Sci USA 71:5042, 1974.
17. Staub AM: Methods Carbohydr Chem 5:92, 1965.
18. Dreesman GR, Hollinger FB, Melnick JL: Appl Microbiol 24:100, 1972.
19. Bhattacharyya P, Shapiro SA, Barnes EM Jr: J Bacteriol 129:756, 1977.
20. Pullman ME, Monroy GC: J Biol Chem 238:3762, 1963.
21. Schuldiner S, Rottenberg H, Avron M: Eur J Biochem 25:64, 1972.
22. Shands JW: Ann NY Acad Sci 133:292, 1966.
23. Oppenheim J, Marcus L: J Bacteriol 101:286, 1970.
24. Pate JL, Shah VK, Brill WJ: J Bacteriol 114:1346, 1973.

Journal of Supramolecular Structure 6:345–353 (1977)
Molecular Aspects of Membrane Transport 207–215

Active Calcium Transport Via Coupling Between the Enzymatic and the Ionophoric Sites of Ca^{2+} + Mg^{2+}-ATPase

Adil E. Shamoo, Terrence L. Scott, and Thomas E. Ryan

Department of Radiation Biology and Biophysics, University of Rochester School of Medicine and Dentistry, Rochester, New York 14642

The 20K dalton fragment of Ca^{2+} + Mg^{2+}-ATPase obtained from the tryptically digested sarcoplasmic reticulum has been further purified using Bio-Gel P-100. This removed low-molecular-weight UV-absorbing and positive Lowry-reacting contaminants. The ionophoric activity of the 20K fragment in both oxidized cholesterol and phosphatidylcholine:cholesterol membranes is unaltered by this further purification. The 20K selectivity sequence in phosphatidylcholine:cholesterol membranes is Ba^{2+} > Ca^{2+} > Sr^{2+} > Mn^{2+} Mg^{2+}.

Digestion of intact sarcoplasmic reticulum vesicles with trypsin, which results in the dissection of the hydrolytic site (30K) from the ionophoric site (20K), is shown to disrupt energy transduction between ATP hydrolysis and calcium transport. This further implicates the 20K dalton fragment as a calcium transport site.

These data and previous evidence are discussed in terms of a proposed model for the ATPase molecular structure and the mechanism of cation transport in sarcoplasmic reticulum.

Key words: Ca^{2+} + Mg^{2+} ATPase, transport, inophore, energy coupling

The nature of the mechanism for directing the energy released by ATP hydrolysis to the vectorial work of ion transport is a fundamental problem of biological research. The customary approach in dealing with this problem has been to study the activation and kinetics of ATP hydrolyzing membrane-bound enzymes. More recently reconstitution of membrane proteins into vesicles has allowed a direct measure of transport.

Our approach has been to identify the ion-transporting site (ionophoric site) as an entity separate from the enzymatic site of the transport ATPases. This seems feasible for a number of reasons. The high-molecular-weight Na$^+$ + K$^+$-ATPase from excitable tissue and the Ca^{2+} + Mg^{2+}-ATPase from muscle sarcoplasmic reticulum have been shown to contain the entire mechanism necessary for active transport (1–4). The ion translocating site is probably contained within a small part of the much larger membrane-bound enzyme. Under conditions of assay, the isolated ion-bearing site must have an inherent affinity for the transported ion. Separation of this ion transporting or ionophoric site from the parent molecule may result in the isolation of a valinomycin-type ionophore (5, 6) with a hydro-

Received March 14, 1977; accepted June 1, 1977

phobic exterior but in any case must have a specific affinity for the transported ion. This approach has led to our identification of several ionophores isolated as part of membrane-bound transport proteins (7–12).

In our definition, ionophoric activity is the ability of a substance to increase black lipid membrane conductance. Criteria which link this ionophoric activity to transport have recently been reviewed by us (5, 6). Briefly, these are: the ability of the ionophore to move the ion transported by the parent protein across a black lipid membrane; this phenomenon must show dependence and/or selectivity to that ion; evidence showing that this activity is isolated from definite parts of the parent protein and inhibitor work demonstrating that the transport is affected in both the native enzyme system and an ionophore-doped black lipid membrane (BLM).

Rabbit white skeletal muscle sarcoplasmic reticulum has been shown to contain an ATP-dependent pump for Ca^{2+} (13). Vesicles assembled from this sarcoplasmic reticulum purified $Ca^{2+} + Mg^{2+}$-ATPase are able to pump Ca^{2+}. The pure $Ca^{2+} + Mg^{2+}$-ATPase has also been shown to exhibit Ca^{2+}-dependent and -selective ionophoric activity in the BLM (8). The 100K dalton $Ca^{2+} + Mg^{2+}$-ATPase as part of the sarcoplasmic reticulum vesicles is cleaved into peptides of 45K and 55K daltons. The enzymatic and ionophoric activities are retained on the 55K dalton fragment.

Subsequent cleavage of the 55K dalton fragment to 30K and 20K dalton fragments separates the enzymatic and ionophoric sites. The 30K dalton fragment contains the site of ATP hydrolysis and the 20K dalton fragment the ionophoric site (10–12). Ruthenium red and mercuric chloride, inhibitors of transport in the intact system, inhibit the ionophoric activity of the 55K and 20K dalton fragments. Methylmercury, an inhibitor of the hydrolytic site of the enzyme, does not inhibit the ionophoric activity (14).

This report presents evidence indicating that the 20K dalton fragment purified on Bio-Gel A-1.5m contains smaller protein pieces not detectable using sodium dodecyl sulfate (SDS)-polyacrylamide gel electrophoresis. The contaminating proteins were removable by SDS exclusion chromatography on Bio-Gel P-100. Evidence is presented that cleavage of the 55K dalton fragment to 30K and 20K dalton fragments in sarcoplasmic reticulum vesicles causes an interruption of Ca^{2+} transport but not ATPase activity. Data is also presented on the ionophoric properties of the 20K dalton fragment in phosphatidylcholine:cholesterol (5:1 mg/mg) membranes.

MATERIALS AND METHODS

Preparation

Sarcoplasmic reticulum and the $Ca^{2+} + Mg^{2+}$-ATPase were prepared from rabbit white skeletal muscle by the method of MacLennan (15). The tryptic fragments of the ATPase of 55K, 45K, 30K, and 20K daltons were prepared and purified by the method of Stewart et al. (11). At times a Sephacryl S-200 column was substituted for Bio-Gel A-1.5m.

Purification of the 20K Dalton Fragment on Bio-Gel P-100 Column

A Bio-Gel P-100 column (180 × 2.5 cm) was equilibrated with 0.5% SDS, 50 mM Tris-HCl, pH 7.0, 1 mM dithiothreitol, 0.02% NaN_3. The column was eluted with the same column buffer and 2-ml fractions were collected every 20 min.

Thin Layer Chromatography of 20K

Fractions presumed to be the 20K dalton fragment from Bio-Gel A-1.5m were concentrated with the aid of an Amicon concentrator. Then $1-2$ μl of 1 mg/ml was spotted on a silica gel G glass plate. The plate was subjected to ascending solvent of 4:1:1.67 (butyl alcohol:acetic acid:H_2O). Fluorescamine staining was performed according to the method of Udenfriend et al. (16). Plates were also stained in an iodine tank.

Analytical Methods

Protein was determined by the method of Lowry (17). SDS-polyacrylamide gel electrophoresis was carried out according to the methods of Weber and Osborn (18); Laemmli (19); and Swank and Munkres (20).

$Ca^{2+} + Mg^{2+}$-ATPase activity was assayed in 25 mM Tris-Cl, pH 7.50, 100 mM KCl, 5 mM $MgCl_2$, 5 mM [^{32}P] ATP, and 10 μM added $CaCl_2$ and $^{32}P_i$ determined by the method of Martin and Doty (21). Calcium uptake was measured in 25 mM HEPES pH 7.0, 120 mM KCl, 5 mM $MgCl_2$, 5 mM ATP, and 50 μM free ^{45}Ca by Millipore filtration.

Conductance Measurement

The lipid bilayer was formed from egg phosphatidylcholine:cholesterol (5:1 mg/mg in decane). Conductance, capacitance, and ionic selectivities were all measured according to published methods (5, 12).

RESULTS

Purification of 20K Dalton Fragment of $Ca^{2+} + Mg^{2+}$-ATPase

The tryptically digested sarcoplasmic reticulum is treated with potassium deoxycholate (KDOC, pH 8.0) to remove calsequestrin and high affinity calcium binding protein as described previously (11, 12). The digested sarcoplasmic reticulum is solubilized with 10% SDS and passed through an SDS-equilibrated Bio-Gel A-1.5m column. The column is eluted with the previously described buffer (12). Figure 1 represents the elution pattern of the column. Fractions 167–200 are UV-absorbing material which show no protein bands on SDS-gel electrophoresis. Fractions 220–241 contain the 20K dalton fragment as tested on SDS-gel electrophoresis. Fractions 220–232 show the 20K dalton fragment on SDS-gel electrophoresis and contain no impurities as tested on the thin-layer chromatography (TLC) plate (Fig. 2). Fractions 232–243 contain 2 extra fluorescamine stained spots. An 8 M urea SDS-gel electrophoresis of fractions 220–241 shows only one protein band. The 8 M urea gel is able to separate proteins of molecular weight as low as 1,000 daltons (e.g., bacitracin). Therefore, the fluorescamine sensitive spots on silica gel may represent small peptidic fragments below 1,000 daltons. The small fragments may be the result of the tryptic digestion of sarcoplasmic reticulum.

In controlled thin-layer chromatography experiments, we have shown that the fluorescamine sensitive spot with R_f = 0.26 on the silica gel is due to Tris-HCl. However, fractions 6–9 show I_2-sensitive staining indicative of other contaminants. The spot at the origin represents the 20K dalton fragment. The spot with R_f = 0.6 is only slightly I_2 sensitive, and may represent slight lipid contaminants. The spots with R_f values of 0.36 and 0.47 are fluorescamine sensitive and may represent the small peptidic fragments.

In order to purify the 20K dalton fragment eluted from Bio-Gel A-1.5m (fractions 220–241) further, the fractions were concentrated with an Amicon P-10 membrane and passed through an SDS-equilibrated Bio-Gel P-100 column. Figure 3 represents the elution pattern of this column. Fractions 40–56 represent purified 20K dalton fragment as shown by both SDS-gel electrophoresis (Fig. 4) and silica gel TLC patterns as in spots numbered

FRACTIONATION OF SR-TRYPIC FRAGMENTS ON BIO-GEL A-1.5m COLUMN

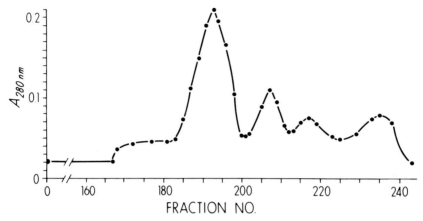

Fig. 1. Fractions of sarcoplasmic reticulum tryptic fragments from a Bio-Gel A-1.5m column (180 × 5 cm) equilibrated with 0.5% SDS, 50 mM Tris-HCl, pH 7.0, 1 mM DTT, and 0.02% NaN_3. The tryptic digest of sarcoplasmic reticulum was run through the column similar to the published method, Stewart et al. (11).

1–5 in Fig. 2. Fractions 60–80 represent the contaminants described earlier. The 20K dalton fragment purified with Bio-Gel P-100 is considered our standard purified 20K dalton fragment for subsequent work described in this paper.

Ionophoric Activity Associated With the 20K Dalton Fragment

The purified 20K dalton fragment just described was dialyzed against 8 M urea for 5 days followed by water for several days as described previously (12). In certain experiments, in order to insure complete removal of SDS, the 20K dalton fragment was treated further by solubilization in 1% potassium cholate and dialysis against the same for several days followed by dialysis against water. The ionophoric data presented here will be those of the cholate-treated 20K dalton fragment. We found no noticeable changes in the ionophoric properties of the 20K dalton fragment after cholate treatment.

Measurement of the diffusion potential in the presence of the 20K dalton fragment and a gradient of calcium chloride yields a permeability ratio $P_{Ca}^{2+}:P_{Cl}^{-}$ of 2.3:1, consistent with previous data in oxidized cholesterol bilayers (10, 12).

Table I presents the biionic potential in mV in the presence of calcium ion on one side of a phosphatidylcholine:cholesterol (5:1 mg/mg) BLM and another divalent cation on the other side. The permeability ratios P_{Ca}^{2+}/P_x, x denoting the other cation, were calculated as detailed in our review (5). The selectivity ratios seen here are consistent with the published selectivity sequence of the 20K dalton fragment in oxidized cholesterol membranes (12) and those of the intact enzyme (8, 9).

The fluorescamine-sensitive contaminant of the 20K dalton fragment when tested alone on oxidized cholesterol or phosphatidylcholine:cholesterol BLMs, shows non-ion-dependent, nonselective ionophoric activity. A collection of the small contaminants from fractions further away from the 20K dalton fragment peak has no ionophoric activity.

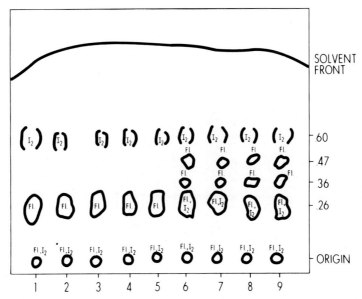

Fig. 2. Silica gel G thin-layer chromatography of 20K dalton fractions from Bio-Gel A-1.5m column. Solvent 4:1:1.67 (butyl alcohol: acetic acid:H$_2$O). Spots 1–9 represents fractions taken in order of elution from the Bio-Gel A-1.5m column. Fl. and I$_2$ represent fluorescamine and iodine-stained spots, respectively.

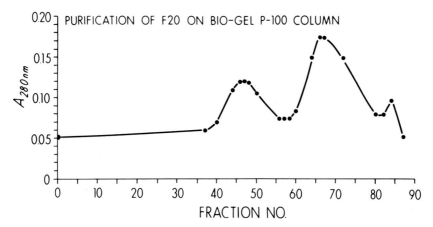

Fig. 3. Purification of 20K dalton fragment. The 20K dalton peak fractions from Bio-Gel A-1.5m were concentrated with an Amicon P-10 membrane and passed through a Bio-Gel P-100 column (180 × 5 cm) equilibrated with 0.5% SDS, 50 mM Tris-HCl, pH 7.0, 1 mM DTT, and 0.2% NaN$_3$.

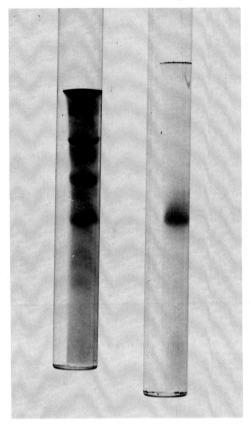

Fig. 4. 10% SDS-polyacrylamide gels of DOC-treated tryptically digested sarcoplasmic reticulum (left) and the 20K dalton fragment after further purification on Bio-Gel P-100 (right).

Uncoupling of Hydrolysis and Transport by Tryptic Digestion of Sarcoplasmic Reticulum Vesicles

Sarcoplasmic reticulum vesicles were digested as described in Methods but with a range of trypsin:protein ratios from 1:5 to 1:1,400. The control was treated with trypsin which had been previously inhibited with trypsin inhibitor. ATPase activity is substantially unaffected by tryptic digestion throughout, but while calcium uptake is unimpaired by cleavage of the 100K dalton molecule to 55K and 45K dalton fragments, it is abolished upon digestion of the 55K dalton fragment to 30K and 20K dalton fragments. Thus, the cleavage of an essential bond between the 30K and 20K dalton fragments uncouples ATP hydrolysis and calcium transport (22).

DISCUSSION

The extreme sensitivity of the bilayer membrane assay requires the highest purity of the material to be tested in order to provide meaningful results. In view of the discrepancies in the amino acid analyses of the 20K dalton fragment prepared in the laboratories of MacLennan (11) and Green (23) by similar digestion procedures, we have further purified

TABLE I. Selectivity of 20K Dalton Fragment in PC:Cholesterol (5:1) Membranes

	Ba^{2+} > Ca^{2+} > Sr^{2+} > Mn^{2+} > Mg^{2+}				
Biionic potential (mV)	−2.1 > 0.0 > 1.4 > 5.8 > 8.8				
P$_{Ca^{2+}}$/P$_X$	0.82 > 1.00 > 1.14 > 1.80 > 2.53				

Biionic potential in mV in the presence of the 20K dalton fragment and 5 mM CaCl$_2$ on one side of a phosphatidylcholine:cholesterol (5:1, mg:mg) bilayer and 5 mM of the other divalent cation on the opposite side.

TABLE II. Uncoupling of Transport From Hydrolysis by Tryptic Digestion of Sarcoplasmic Reticulum

Activity (% control)		Protein (% maximum)	
Ca-dependent ATPase	Ca-uptake	100K dalton fragment	55K dalton fragment
100	100	100	0
106	103	39	95
85	3	0	5
102	0	0	0

Calcium-dependent ATP hydrolysis and ATP-dependent calcium uptake assayed as described in Methods, presented as percent of control activities. Amount of fragments determined by measurement of the area under the peaks in gel scans, expressed as percent of the maximum amounts attained before or during tryptic digestion.

the 20K dalton ionophoric polypeptide whose properties are described here. The 20K dalton fragment purified by the standard procedure contained 2 fluorescamine-sensitive spots in addition to the 20K dalton fragment. This material did not appear on SDS gels which are sensitive down to 1,000 daltons. The isolated contaminants removed by further purification of the 20K dalton fragment with SDS-gel chromatography showed non-specific ionophoric activity or no activity, while the properties of the further purified 20K dalton fragment were the same as before.

Phosphatidylcholine:cholesterol bilayers provide a lipid environment which is more consistent with the composition of the sarcoplasmic reticulum membrane than that of oxidized cholesterol bilayers used previously. In bilayers of either composition, the iono-phoric activity of the 20K dalton fragment possesses the same ion dependence and the same selectivity for divalent cations. This is further evidence that the ionophoric activity of the 20K dalton fragment is not the result of a nonspecific protein-lipid interaction. The 20K dalton fragment requires Ca^{2+} ions for the expression of its ionophoric properties and has a selectivity sequence for divalent cations which is consistent with the selectivity of transport in intact sarcoplasmic reticulum. The ionophoric activity of the intact ATPase (100K daltons), and the 55K and the 20K dalton fragments exhibit the same selectivity sequence for divalent cations (25). Further, specific inhibition of the ionophoric function by mercuric chloride, ruthenium red, etc., has been shown to be the same for the intact ATPase, the 55K dalton cleavage product and the 20K dalton fragment (14). This is con-

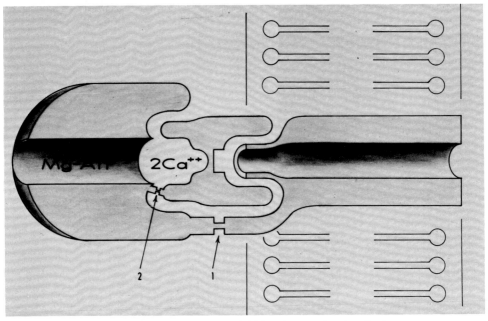

Fig. 5. Model for active transport of calcium in sarcoplasmic reticulum.

sistent with inhibition of transport in intact sarcoplasmic reticulum.

The disruption of calcium transport upon the cleavage of a bond connecting the 30K dalton fragment phosphorylation site to the 20K dalton fragment in intact sarcoplasmic reticulum vesicles is consistent with the identification of the 20K dalton polypeptide as a Ca^{2+}-ionophoric site of the ATPase molecule and is further evidence for localization of the hydrolytic and ionophoric functions in separate portions of the ATPase molecule. ATP hydrolytic activity is unimpaired while Ca transport is uncoupled from hydrolysis by the disruption of this essential bond.

These studies and our previously published work (5–12, 14, 22) lead then to a schematic model of the ATPase molecular structure (Fig. 5) which is consistent with all of the evidence for coupling between hydrolysis and transport in sarcoplasmic reticulum. The 100K dalton ATPase has a hydrophobic portion which may be a channel spanning the membrane, this being the 45K dalton fragment produced by trypsin digestion (9, 24, 25). The 55K dalton portion exists on the exterior, cytoplasmic face and contains both the 30K dalton phosphorylation site and the 20K dalton ionophoric site which is a gate partially buried in the hydrophobic interior at the mouth of the 45K dalton fragment channel. The 20K dalton fragment contains the site of specific Ca^{2+} interaction as demonstrated by the ionophoric properties in this study and previous work. Thus, while both ATP hydrolysis and Ca^{2+} transport remain coupled despite tryptic cleavage of the bond denoted 1, disruption of energy transduction ensues with the breaking of the bond denoted 2, hydrolysis being still functional while transport is abolished since the gate can no longer interact with the hydrolytic site.

ACKNOWLEDGMENTS

This paper is based on work performed under contract with the U.S. Energy Research and Development Administration at the University of Rochester Biomedical and Environmental Research Project and has been assigned Report No. UR-3490-1109. This paper was also supported in part by NIH Grants 1 RO1 AM17571 and 1 RO1 AM18892; a Center Grant ESO-1247 and Program Project Grant ES-10248 from the NIEHS; the Muscular Dystrophy Association (USA); and the Upjohn Company. A.E.S. is an Established Investigator of the American Heart Association.

REFERENCES

1. Racker E: J Biol Chem 247:8198, 1972.
2. Warren GB, Toon PA, Birdsall NJM, Lee AG, Metcalfe JC: Proc Natl Acad Sci USA 71:622, 1974.
3. Goldin SM, Tong SW: J Biol Chem 249:5907, 1974
4. Hilden S, Hokin LE: J Biol Chem 250:6296, 1975.
5. Shamoo AE, Goldstein DA: Biochim Biophys Acta (In press).
6. Blumenthal RP, Shamoo AE: In O'Brien RD (ed): "The Receptors: A Comprehensive Treatise," New York: Plenum Press, Vol 1 (In press).
7. Shamoo AE: Ann NY Acad Sci 242:389, 1974.
8. Shamoo AE, MacLennan DH: Proc Natl Acad Sci USA 71:3522, 1974.
9. Shamoo AE, Ryan TE: Ann NY Acad Sci 264:83, 1975.
10. Shamoo AE, Ryan TE, Stewart PS, MacLennan DH: Biophys J 16:190a, 1976.
11. Stewart PS, MacLennan DH, Shamoo AE: J Biol Chem 251:712, 1976.
12. Shamoo AE, Ryan TE, Stewart PS, MacLennan DH: J Biol Chem 251:4147, 1976.
13. Hasselbach W, Makinose M: Biochem Z 333:518, 1961.
14. Shamoo AE, MacLennan DH: J Membr Biol 25:65, 1975.
15. MacLennan DH: J Biol Chem 245:4508, 1970.
16. Udenfriend S, Stein S, Bohlen P, Dairman W, Leimgruber W, Weigele M: Science 178:871, 1972.
17. Lowry OH, Rosebrough NJ, Farr AL, Randall RJ: J Biol Chem 193:265, 1951.
18. Weber K, Osborn M: J Biol Chem 244:4406, 1969.
19. Laemmli UK: Nature (London) 227:680, 1970.
20. Swank RT, Munkres RD: Anal Biochem 39:462, 1971.
21. Martin JB, Doty DM: Anal Chem 21:965, 1949.
22. Scott TL, Shamoo AE: Biophys J 17:185a, 1977.
23. Thorley-Lawson DA, Green NM: Eur J Biochem 59:193, 1975.
24. Abramson J, Shamoo AE: Biophys J 17:185a, 1977.
25. Shamoo AE, Abramson J: In Wasserman RH et al (eds): "Calcium Binding Proteins and Calcium Function in Health and Disease." Amsterdam: Elsevier, 1977.

Journal of Supramolecular Structure 6:355–362 (1977)
Molecular Aspects of Membrane Transport 217–224

Control of Amino Acid Transport in the Mammary Gland of the Pregnant Mouse

Cinda J. Lobitz and Margaret C. Neville

Department of Physiology, University of Colorado, Medical Center, Denver, Colorado 80262

The regulation of the uptake of the amino acid analog α-aminoisobutyric acid was studied in diced mammary glands from pregnant mice. Stimulation of uptake by insulin was not prevented by inhibitors of protein synthesis; protein synthesis inhibitors decreased uptake by 20%; this response occurred more promptly in insulin-treated tissues. Elimination of extracellular amino acids led to a substantial increase in transport which was not abolished by inhibitors of protein synthesis. These results indicate that insulin does not increase amino acid transport in this system by altering synthesis and degradation of transport protein. They are consistent with a model in which the activity of the existing amino acid transport protein is subject to negative feedback regulation from the intracellular amino acid pool.

Key words: amino acid transport, mammary gland, cell proliferation, feedback regulation

Increases in amino acid transport have been observed as an early event following the stimulation of cell proliferation in many systems (1–4). Sander and Pardee (5) and Topper et al. (6) found cell-cycle-related changes in amino acid transport. Recent studies by Oxender and co-workers (7) demonstrated reciprocal changes in the activity of the A and L systems for neutral amino acid transport when Balb/3T3 cells reach confluency (see also 8, 9). Neither the mechanism nor the significance of these changes is well-understood, although both Holley (10) and Pardee (11) have proposed that alterations in the membrane transport of small molecules may play a role in the control of cell proliferation.

Oka, Topper, and their co-workers (12–14) demonstrated that at least 1 round of cell division ensues within 24 h when explants of mammary glands from pregnant mice are treated with insulin in vitro. A significant increase in the transport of nonmetabolizable amino acid analog, α-aminoisobutyric acid (AIB), occurred within 4 h of insulin treatment (13, 14). We undertook the present study of the mechanism of the insulin stimulation of AIB transport in mammary gland in vitro with the hope of gaining further insight into the mechanisms by which amino acid transport is controlled in proliferating tissue. Our results indicate that AIB transport is regulated in response to changes in the intracellular amino acid pool by a mechanism which does not involve alterations in the synthesis and degradation of amino acid transport protein. A preliminary report of this work has appeared (15).

Received April 14, 1977; accepted April 27, 1977

MATERIALS AND METHODS

Solutions

Unless otherwise indicated all incubations were carried out in TC199 medium modified to increase the buffer capacity by addition of 10 ml of a solution containing 50 mM HEPES buffer, 1.2 mM KCl, 26.4 mM NaCl, 4.9 mM $CaCl_2$, and 0.7 mM KH_2PO_4 to 25 ml of reconstituted TCl99. Solid glucose and bovine serum albumin were added to concentrations of 27 mM and 2%, respectively, and the pH adjusted to 7.4 with NaOH. For the amino acid-free medium we used a modified Ringer bicarbonate containing 104 mM NaCl, 5 mM $NaHCO_3$, 6 mM Na_2HPO_4, 2 mM $CaCl_2$, 1 mM $MgCl_2$, 30 mM glucose, 20 mM HEPES, 2 mg/100 ml phenol red and 2% bovine serum albumin with a final pH of 7.4. Insulin (bovine) was obtained from Eli Lilly (Iletin®). Isotopes were obtained from New England Nuclear Corporation. All chemicals were reagent grade. Budget-solv (RPI) was used for all scintillation counting.

Incubation

All experiments were carried out on tissues acutely isolated from the mammary glands of pregnant Balb-C mice. Our breeding colony was originally started with mice obtained from the colony maintained at the American Medical Center in Denver. For all reported experiments, the mammary gland from a single 16–18-day pregnant mouse was dissected free from excess connective tissue and lymph nodes, diced into 1 mm^2 pieces with crossed razor blades and preincubated in TC199 medium, modified as described above, for 1 h at 37°C to stabilize the tissue. The tissue was then further incubated in the presence or absence of insulin and protein synthesis inhibitors for 0.5–6 h. To measure the amino acid transport capacity 100 mg of tissue was placed in 1 ml of solution containing 0.5 mM α-aminoisobutyric acid (AIB) and 0.5 μCi/ml [^{14}C] AIB for 10 minutes. Experiments to be reported elsewhere indicate that this time of incubation allows estimation of the initial rate of AIB entry into the tissue. Experiments using dissociated alveoli gave quantitatively similar results, suggesting that AIB transport into diced mammary tissue represents transport by the alveolar cells. Methods for determining [^{14}C] alanine incorporation into lipids as well as [^{14}C] tryptophane into protein are given in the legends.

By using relatively large volumes of medium for preincubation (30 ml for 1 g of tissue), pH changes attributable to the high lactate production by this tissue were avoided. When present, insulin was used at 0.4 μU/ml and cycloheximide at 0.5 mg/ml. The sucrose space of the diced tissue averaged 25%, a value used to correct the data for extracellular amino acid.

RESULTS

Figure 1 shows the effects of insulin and protein synthesis inhibitors on AIB uptake in diced mammary glands from pregnant mice. Note that in the control tissue the rate of AIB uptake remains constant over a 6-h period. After 3 h of preincubation insulin had a marked stimulatory effect on the initial rate of AIB uptake. A concentration of cycloheximide sufficient to inhibit protein synthesis by 90% (data not shown) depressed AIB uptake by 20% after 2 h of preincubation. These effects, although quantitatively small, were quite consistent, including the single step decrease of AIB uptake to a new steady level in the presence of cycloheximide. When insulin and cycloheximide were added together, AIB uptake did not differ significantly from the control. This experiment did not

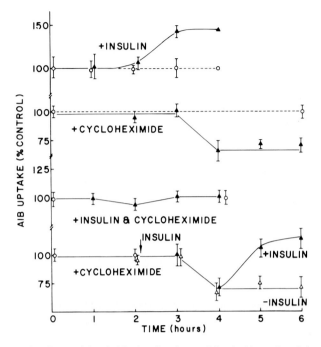

Fig. 1. The effect of insulin, cycloheximide, insulin plus cycloheximide, and cycloheximide followed by insulin on AIB uptake by diced mammary gland from a pregnant mouse. Tissues were preincubated with the indicated agent for varying periods of time (abscissa), followed by incubation with [^{14}C] AIB and 0.5 mM AIB for 10 min. After blotting and weighing, the radioactive amino acid was extracted with 5% trichloroacetic acid and counted in a liquid β-scintillation counter. Here and in Figs. 2–4 all points are the average of quadruplicate determinations. The distance between bars represents 2 standard errors of the means. Control shown as ($\circ$) with a dotted line.

allow us to decide whether cycloheximide inhibited the insulin action or whether the effects of the 2 agents canceled each other out. To clarify this problem, insulin was added to cycloheximide-treated tissues (Fig. 1, bottom) after 2 h of preincubation. Two hours after insulin addition, there was an increase in AIB uptake similar to that seen in the absence of cycloheximide. The finding that the effect of insulin on AIB uptake is not inhibited by inhibitors of protein synthesis suggests that insulin does not act by stimulating the synthesis of transport protein.

We next considered the possibility that insulin prevents amino acid carrier degradation in mammary tissue as has been reported for muscle (16). In Fig. 2, we show the results of the addition of cycloheximide to insulin-stimulated tissue. If insulin were acting by inhibiting carrier degradation, we would expect no effect or a delayed effect of cycloheximide in the presence of insulin. Instead, cycloheximide addition is followed by a prompt decrease in AIB uptake. During the same period cycloheximide had no effect on AIB uptake by the control tissue. Therefore insulin does not appear to act by decreasing carrier degradation in mammary tissue.

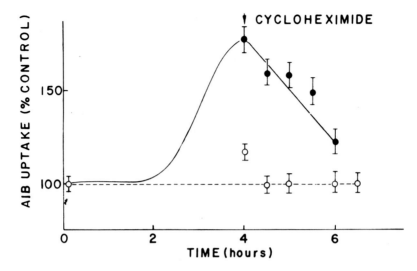

Fig. 2. The effect of cycloheximide on AIB uptake in insulin-treated tissue. Tissues were preincubated in the presence (●) and absence (○) of insulin for the time shown on the abscissa after which AIB transport was determined as detailed in the legend to Fig. 1. Cycloheximide was added to both tissues at 4 h.

Finally we tested the hypothesis that insulin alters the size of the intracellular amino acid pool, regulating amino acid transport through a negative feedback mechanism. If this hypothesis is correct, it should be possible to demonstrate that insulin increases amino acid utilization. To do this we measured alanine incorporation into lipid and tryptophane incorporation into protein as a function of preincubation time in the presence of insulin. Figure 3 shows that both functions are markedly increased over controls, indicating that insulin does increase the rate of cellular processes which utilize intracellular amino acids.

If insulin works through alteration in amino acid pool size, other treatments which increase or decrease these pools should have predictable effects on amino acid transport. The effect of the protein synthesis inhibitor cycloheximide, has already been shown on Fig. 1. The observed decrease in AIB uptake is the result one would expect from decreased amino acid utilization. Puromycin was also studied and found to have similar effects. Figure 2 illustrates the effect of cycloheximide on insulin stimulated tissue; here the decrease in AIB uptake occurs more rapidly than in nonstimulated tissue. This finding suggests that, when the flow of amino acids through the pool was increased in response to insulin, blockage of protein synthesis led to a rapid build-up of the intracellular amino acid concentration necessary to decrease amino acid uptake.

Removal of all amino acids from the bathing medium should decrease intracellular amino acid pools. The results of such an experiment are shown on Fig. 4. When the tissue is removed from TC199 and placed in a Ringer solution containing no amino acids there is a prompt increase in AIB uptake which begins to level off at 2 h. This increase cannot simply be the result of removing competing amino acids from the medium since the extracellular space in this tissue equilibrates within 5 min (experiment not shown). A

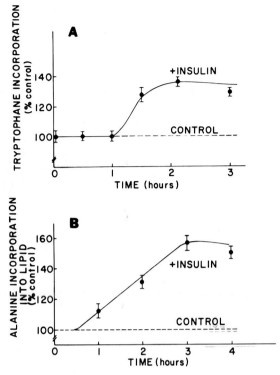

Fig. 3. The effect of insulin on amino acid utilization by diced mammary gland from a pregnant mouse. A) the effect of insulin on tryptophane incorporation into protein. After preincubating, the tissue was incubated in the presence of 5 μCi/ml [^{14}C] tryptophane for 10 min. After blotting and weighing, free radioactivity was extracted with 3 changes of trichloroacetic acid and the protein dissolved in 1 N NaOH and counted. No precipitation of protein occurred during the counting interval using this technique.   B) The effect of insulin on alanine incorporation into lipid. After preincubation with or without insulin for the time indicated on the abscissa, the tissues were incubated in the presence of [^{14}C] alanine for 30 min. After blotting and weighing, the tissues were placed in a glass scintillation vial with 10 ml of a toluene based fluid and shaken at room temperature overnight to extract the lipid soluble components. Prior to counting, 1 ml of water was added to the vials to dissolve the water soluble radioactivity and partition it away from the fluor. A separate experiment shows that no more than 11% of the water soluble activity is counted by this technique.

similar but quantitatively smaller result is seen in the presence of cycloheximide (Fig. 4), again suggesting that synthesis of new protein is not involved in the response to decreased amino acid supply.

DISCUSSION

In studies to be reported elsewhere on diced mammary glands from pregnant mice, we found that insulin increases V_{max} without altering the K_m of transport. From the studies reported here, this alteration in transport capacity is related neither to synthesis of

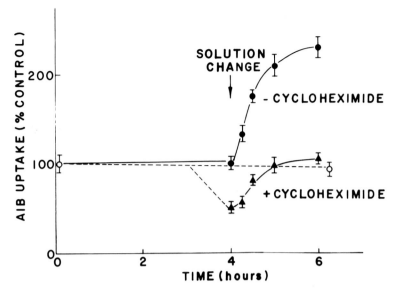

Fig. 4. The effect of removal of external amino acids on AIB uptake by diced mammary gland from a pregnant mouse. Tissues were incubated in TC199 with (▲) or without (●) cycloheximide for 2 h after which the solution was changed to Ringer bicarbonate. AIB uptake was assayed as described in the legend to Fig. 1.

new carrier protein nor to alteration of the rate of carrier degradation. However, the effects on AIB uptake of insulin, protein synthesis inhibitors, and removal of external amino acids are all consistent with the model for regulation of amino acid transport shown on Fig. 5. In this model, negative feedback from the intracellular amino acid pool regulates the rate of transport by preexisting membrane transport elements.

This conclusion differs from that reached by workers studying regulation of amino acid transport in other tissues. For example, Guidotti and his co-workers (16) concluded that insulin acted on the A system of muscle both by protecting membrane sites from degradation and by increasing the rate of synthesis of transport protein (see also 17). A substantial increase in amino acid transport when tissues are preincubated in amino acid-free media has been noted in uterus (18), newborn rat kidney cortex (19), placenta (20, 21), and a variety of avian and mammalian tissues (22, 23). This response to amino acid starvation, termed "adaptive regulation" (19, 22), often appears to be abolished by cycloheximide or puromycin (18, 19, 21, 22). This observation suggested to several workers that synthesis of new transport protein is involved (18, 19, 22). In this regard the results of an early study on yeast by Grensen et al. are instructive (24). These workers observed an increase in arginine uptake during nitrogen starvation and a decrease in arginine uptake in the presence of cycloheximide; because cycloheximide had no effect in nitrogen-starved yeast, they concluded that the alterations in the transport of the amino acid were the result of changes in the level of free intracellular amino acids rather than alterations in the

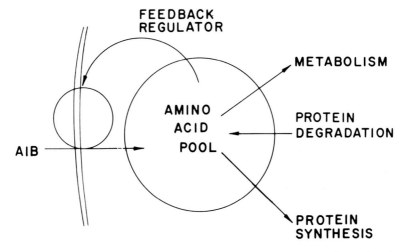

Fig. 5. Model for regulation of amino acid transport in diced mammary gland. For explanation, see text.

amount of transport protein. They suggested that the transinhibition of amino acid transport by intracellular amino acids first proposed by Ring and Heinz (25) might constitute a type of feedback control in this system. Neither the experiments of Guidotti and co-workers (22, 23) nor those of Reynolds et al. (19) on adaptive regulation were designed in such a way as to eliminate feedback effects of intracellular amino acids. The studies of Smith (22) suggest that pool size does play a role in placental tissue.

Recently Oxender and co-workers (7) measured the actual amino acid pools in Balb/3T3 mouse cells under depressed growth conditions. They found that a 2- to 3-fold increase in the amino acid pools was accompanied by a decrease in transport by the A system. These results again suggest a central role for the amino acid pool in the regulation of amino acid transport.

The mechanism of the apparent local feedback regulation in mammary tissue is not yet clear. It could involve direct transinhibition of the type suggested by Ring and Heinz (25). Changes in the membrane potential which alter the electrochemical gradient for Na^+, the apparent driving force for transport by the A system (26, 27), could be involved. A more complex feedback loop involving metabolites of the amino acids in question cannot at this time be ruled out. Although there is no evidence for transcriptional or translational control in the present study, it is important to be aware that regulation of the synthesis of transport protein could be superimposed on local control mechanisms. For example, Pall (28) has observed both transcriptional and local control of methionine transport in yeast. Further understanding of the regulation of amino acid transport in eukaryotes will only come from a detailed correlation of amino acid pools with transport activity under a variety of conditions.

ACKNOWLEDGMENTS

This investigation was supported by Grant number 1R01CA19389, awarded by The National Cancer Institute, DAEW. MCN is the recipient of Research Career Development Award 5 KO4 AM 00.038.

REFERENCES

1. Costlow M, Baserga R: J Cell Physiol 82:411, 1973.
2. Vaheri A, Ruoslahti E, Hovi T, Nordling S: J Cell Physiol 81:355, 1973.
3. Foster DO, Pardee AB: J Biol Chem 244:2675, 1969.
4. Isselbacher KJ: Proc Natl Acad Sci USA 69:585, 1972.
5. Sander G, Pardee AB: J Cell Physiol 80:267, 1972.
6. Topper JT, Mills B, Zorgniotti F: J Cell Physiol 88:77, 1976.
7. Oxender DL, Lee M, Cecchini G: J Biol Chem (In press).
8. Robinson JH: J Cell Physiol 89:101, 1976.
9. Robinson JH, Smith JA: J Cell Physiol 89:111, 1976.
10. Holley RW: Proc Natl Acad Sci USA 69:2840, 1972.
11. Pardee AB: Natl Cancer Inst Monogr 14:7, 1964.
12. Stockdale FE, Juergens WG, Topper YJ: Dev Biol 13:266, 1966.
13. Friedberg SH, Oka T, Topper YJ: Proc Natl Acad Sci USA 67:1493, 1970.
14. Oka T, Perry JW, Topper YJ: J Cell Biol 62:550, 1974.
15. Neville MC, Lobitz CJ: J Supramol Struct (Suppl) 1:699, 1977.
16. Guidotti GG, Franchi-Gazzola R, Gazzola GC, Ronchi P: Biochim Biophys Acta 356:219, 1974.
17. Riggs TR, McKirahan KJ: J Biol Chem 248:6450, 1973.
18. Riggs TR, Pan MW: Biochem J 128:19, 1972.
19. Reynolds R, Rea C, Segal S: Science 184:68, 1974.
20. Smith CH, Adcock EW III, Teasdale F, Meschia G, Battaglia FC: Am J Physiol 224:558, 1973.
21. Smith CH: Pediatr Res 8:697, 1974.
22. Franchi-Gazzola R, Gazzola GC, Ronchi P, Saibene V, Guidotti GG: Biochim Biophys Acta 291:545, 1973.
23. Guidotti GG, Gazzola GC, Borghetti AF, Franchi-Gazzola R: Biochim Biophys Acta 406:264, 1975.
24. Grensen M, Crabeel M, Wiame JM, Bechet J: Biochem Biophys Res Commun 30:414, 1968.
25. Ring K, Heinz E: Biochem Z 344:446, 1966.
26. Heinz E: Ann NY Acad Sci 264:428, 1975.
27. Villereal MJ, Cook JS: J Supramol Struct 6:179, 1977.
28. Pall ML: Biochim Biophys Acta 233:201, 1971.

Journal of Supramolecular Structure 6:363–374 (1977)
Molecular Aspects of Membrane Transport 225–236

Defective Transport of Thymidine by Cultured Cells Resistant to 5-Bromodeoxyuridine

Thomas P. Lynch, Carol E. Cass, and Alan R. P. Paterson

Cancer Research Unit (McEachern Laboratory) and Department of Biochemistry, University of Alberta, Edmonton, Alberta, Canada, T6G 2H7

A line of HeLa cells resistant to 5-bromo-2'-deoxyuridine (BUdR) was established by continuous culture in growth medium containing BUdR; during the selection period, BUdR concentrations, initially 15 μM, were gradually increased to 100 μM. Cells of a clone (HeLa/B5) established from this line were also resistant to 5-fluoro-2'-deoxyuridine (FUdR), but not to the free base, 5-fluorouracil. Although extracts of HeLa/B5 cells exhibited levels of thymidine kinase activity comparable to those of parental cells, rates of uptake of BUdR, FUdR, and thymidine into intact cells were much reduced. The kinetics of uptake of uridine and adenosine, nucleosides which appear to be transported independently of thymidine in HeLa cells, were similar for HeLa/B5 and the parental line (HeLa/0). Relative to thymidine uptake by HeLa/0 cells, that by HeLa/B5 cells was distinctly less sensitive to nitrobenzylthioinosine (NBMPR), a specific inhibitor of nucleoside transport in various types of animal cells. Despite this difference in NBMPR sensitivity, both cell lines possessed the same number of high affinity NBMPR binding sites per mg cell protein. The altered kinetics of thymidine uptake and the NBMPR insensitivity of that function in HeLa/B5 cells suggest that resistance to BUdR is due to an altered thymidine transport mechanism.

Key words: **thymidine transport, nitrobenzylthioinosine, bromodeoxyuridine resistances, HeLa cells, thymidine kinase**

It is generally held that the entry of nucleosides into animal cells is mediated by nucleoside-specific transport mechanisms (1–4). This is clearly evident in the case of human erythrocytes which, because of their inability to phosphorylate or cleave either uridine or thymidine, afforded an opportunity to study the transport of these compounds in the absence of cellular metabolism (4). The participation of a classical facilitated diffusion process in the movement of these permeants across the erythrocyte plasma membrane was recognized by rate saturability, competition between permeants, nonconcentrative permeation, and accelerative exchange diffusion (3–5). Under circumstances in which nucleoside permeants are subjected to cellular metabolism, it has not been possible to

Abbreviations: NBMPR – nitrobenzylthioinosine, 6-[(4-nitrobenzyl)-thio]-9-β-D-riboturanosylpurine FU – 5-fluorouracil; BUdR – 5-bromo-2'-deoxyuridine; FUdR – 5-fluoro-2'-deoxyuridine; HEPES – N-2-hydroxyethylpiperazine-N'-2-ethanesulfonic acid; MEM – Minimal Essential Medium

Received April 18, 1977; accepted May 10, 1977

clearly determine whether transport or the metabolic events involved in the uptake process determine the kinetics of that process. In this report, "uptake" refers to the total quantity of a permeant found to be cell-associated after incubation of those cells with the permeant under defined conditions; the term includes both unaltered permeant and all metabolites thereof.

Kinetic studies (1, 2) have shown the uptake process to have the following characteristics in cultured cells of several types: a) rate saturability, b) competition between permeants, and c) inhibition of uptake by compounds that do not affect nucleoside phosphorylation. These characteristics indicate that a rate-limiting step in nucleoside uptake is mediated; it has been generally assumed that transport is that rate-limiting step and, accordingly, initial rate kinetics have been interpreted in terms of transport.

After brief uptake intervals, nucleosides are found in cells primarily in the form of nucleotide metabolites; however, it is not clear whether transport, or a subsequent metabolic step such as phosphorylation, which might even be coupled to the transport event, is rate-limiting in the uptake of particular nucleosides (6–13). For example, studies of the kinetics of thymidine metabolism and uptake in HeLa cells, using a rapid sampling procedure, demonstrated the approximate equivalence of early rates of uptake and thymidine phosphorylation (6). In Novikoff hepatoma cells (1) and HeLa cells (6–9), several nucleoside uptake mechanisms are distinguishable in terms of permeant specificity. It is not known whether this specificity is imparted to the nucleoside uptake process at the level of transport or at that of nucleoside metabolism. Recent studies of thymidine transport in ATP-depleted Novikoff hepatoma cells (10) have identified a low affinity mechanism which appears to resemble the nucleoside transport mechanism of human erythrocytes (3–5). Plagemann et al. (10) have suggested that in Novikoff cells thymidine uptake may proceed a) by a process in which the rate-limiting step in thymidine uptake is phosphorylation rather than transport, or b) by way of 2 systems, a facilitated diffusion system with low substrate specificity, together with a second system which may involve substrate phosphorylation.

In the present work, HeLa cells were selected for BUdR resistance in an attempt to obtain variants with defects in the membrane transport mechanism for thymidine. A number of examples of BUdR resistance in cultured cells have been attributed to decreased thymidine kinase activity (14–17); alteration in thymidine transport has also been suggested as a basis for BUdR resistance (18, 19).

In this report, we compare the properties of a BUdR-resistant clone (HeLa/B5) with those of the parental cell line (HeLa/0) in respect to kinetics of nucleoside uptake, thymidine kinase activity, and sensitivty to nitrobenzylthioinosine (NBMPR), a potent and specific inhibitor of nucleoside transport (20). Previous studies with human erythrocytes demonstrated that NBMPR occupancy of high affinity binding sites decreased uridine transport in proportion to the number of sites occupied (21, 22). HeLa cells possess similar high affinity binding sites for NBMPR, occupancy of which results in a general inhibition of nucleoside uptake (6–9, 23). We have recently shown that a) the thymidine kinase activity of HeLa cell extracts was not inhibited by NBMPR at concentrations greatly in excess of those which inhibited thymidine uptake in intact HeLa cells, and b) the relative proportions of thymidine anabolites in HeLa cells were similar in the absence and presence of NBMPR, although in the latter instance, total anabolism of thymidine was drastically reduced (6). These results are consistent with the view that NBMPR inhibits the transport step which initiates the anabolism of thymidine in HeLa cells.

MATERIALS AND METHODS

Chemicals

[5-^{3}H] BUdR was obtained from New England Nuclear Corporation, Boston, Massachusetts; other labeled compounds were purchased from Amersham/Searle, Oakville, Ontario. Dr. S. R. Naik prepared NBMPR by an established method (24) from 6-thioinosine generously provided by Developmental Therapeutics Program, National Cancer Institute, Bethesda, Maryland. [^{35}S] NBMPR was prepared by G. J. Lauzon of this laboratory (23).

Cell Culture

A cell line resistant to BUdR was selected by serial passage of HeLa S3 cells in monolayer culture in medium containing BUdR, initially at 15 μM; as subculturing proceeded, the BUdR concentration was increased progressively to 100 μM. From the resistant line, a clone (HeLa/B5) was established; characterization of BUdR resistance in HeLa/B5 cells will be described elsewhere (25). The HeLa/B5 clone was chosen for transport studies because of an apparent decrease in thymidine uptake activity relative to the parental cells (HeLa/0). HeLa/B5 cells had a slightly slower proliferation rate and were somewhat larger than HeLa/0 cells; values for cell volume, DNA, and protein content were 1.5–1.7 times larger in the resistant cells (25).

HeLa/0 and HeLa/B5 cells were maintained as monolayer cultures in antibiotic-free Eagle's minimal essential medium (MEM) supplemented with 10% calf serum. After 6 to 8 weekly transfers, cultures were restarted from frozen stocks; the latter were consistently free of Mycoplasma (Dr. J. Robertson, Department of Medical Bacteriology, University of Alberta).

Measurements of cell proliferation rates, nucleoside uptake rates, and [^{35}S]NBMPR binding employed replicate monolayer cultures prepared by the following procedure (6). Inocula from stock monolayer cultures were expanded into suspension cultures in spinner flasks which subsequently became inocula for suspension cultures in 2-liter round bottom flasks kept under continuous agitation with a vibrating mixer (Vibro-Mixer, Model E1, Chemapec, Hoboken, New Jersey); the latter cultures were single-cell suspensions. Throughout this procedure, cultures were maintained in logarithmic growth by dilution with growth medium to keep cell concentrations between 0.5 and 1.0 $\times$ 10^5 cells/ml. The medium for suspension cultures (MEM-S) consisted of calcium-free MEM supplemented with 5% calf serum, penicillin (100 units/ml), streptomycin (100 μg/ml), and 2 mM HEPES (pH 7.4) Replicate monolayer cultures (50 or more) were established in 2-oz prescription bottles (Brockway Glass, Brockway, Pennsylvania) with cells from Vibro-Mixer suspension cultures in attachment medium (MEM with 10% calf serum, HEPES, and antibiotics) and were incubated at 37°C in 5% CO$_2$ and air. Cell culture materials were purchased from Grand Island Biological Company, Calgary, Alberta.

Nucleoside Uptake Assay

Uptake of ^{3}H-nucleosides by replicate monolayers was measured at 20–21°C using a rapid sampling procedure described previously (6); the uptake medium (MEM-T) consisted of bicarbonate-free MEM supplemented with 12 mM NaCl, 20 mM HEPES buffer (pH 7.4, 20°C), and ^{3}H-nucleosides. Uptake assays using 24-h monolayer cultures (approximately 10^6 cells/bottle) were performed in triplicate, and each culture was processed individually as follows. After removing growth medium, the culture bottle was placed horizontally with cells uppermost, 4 ml of MEM-T was added, and the assay was initiated by rapid rotation of the bottle 180° about its long axis to immerse cells. Five seconds before the uptake interval ended, MEM-T was removed by suction and, to terminate the assay, the cell sheet was rapidly flooded with 60 ml of ice-cold 0.154 M NaCl solution (saline). Fifteen seconds later, the saline was removed by aspiration and the bottle was drained thoroughly. After digestion of cell sheets in 1.5-ml portions of NCS tissue solubilizer (Amersham/Searle), 10.0 ml of Bray's fluor (26) was added to each bottle and a sample from each was taken for liquid scintillation counting. For determination of protein content (27), replicate cultures were processed as described above, as far as completion of the draining step.

Thymidine Kinase Activity

Thymidine kinase (ATP:thymidine 5'-phosphotransferase, EC 2.7.1.75) was determined in extracts of replicate monolayer cultures, prepared as follows. After 1 wash with cold saline, to each monolayer was added 1.0 ml of extraction buffer (10 mM NaF; 0.01 M KCl; 2 mM dithioerythritol; 25 mM 6-aminocaproic acid; 20 μM thymidine; 0.01 M Tris-HCl, pH 7.4; 10% glycerol), and after successive freezing (dry ice-ethanol) and thawing steps (repeated 4 times, with mixing after thawing), extracts were centrifuged (4 min, 13,000 $\times$ g). Supernatant fractions were reserved for determination of protein content (27), and thymidine kinase activity using the method of Lee and Chang (28). In the latter, cell extracts were incubated for timed intervals in a medium containing 30 μM [^{3}H] thymidine and the thymidine phosphates formed were isolated on DEAE cellulose paper (29); the ^{3}H-content of such samples was determined by liquid scintillation counting after combustion in a Packard Model 306 Sample Oxidizer.

Affinity Chromatography and Polyacrylamide Gel Electrophoresis

Thymidine kinase affinity gels were prepared by the methods of Kowal and Markus (30) as modified by Lee and Cheng (28). p-Aminophenylthymidine-3'-phosphate was linked in the presence of 1-ethyl-3(3-dimethylaminopropyl)-carbodiimide to a Sepharose 4B (Pharmacia, Montreal, Quebec) derivative prepared by CNBr activation followed by reaction with 6-aminocaproic acid (31). Suspensions of HeLa cells (10^7 cells/ml) in extraction buffer (see above) were disrupted with 15-sec bursts of 20-kc ultrasound, and after precipitation with streptomycin sulfate, ammonium sulfate fractions were prepared (28). The resulting precipitates were dissolved in buffer (0.01 M Tris-HCl (pH 7.4), 10% glycerol, 2 mM dithioerythritol, 0.5 mM EDTA) and dialyzed overnight at 4°C against 500 ml of the same buffer. Portions (1.0 ml) of such dialysates were applied to thymidine kinase affinity gel columns (7.0 $\times$ 0.5 cm) which were eluted with appropriate buffers (28); eluate fractions (2.0 ml) were assayed for thymidine kinase activity. Crude extracts and fractionated, dialyzed extracts were subjected to electrophoresis on 5% polyacrylamide gels (28).

[^{35}S] NBMPR Binding Assay

To quantitate the binding of NBMPR to HeLa cells, a modification of a procedure described previously (23) was employed. Replicate monolayer cultures were exposed to 4-ml portions of medium (MEM-T) containing [^{35}S] NBMPR at various concentrations for analysis; cell sheets were then rinsed with 60 ml of cold saline and dissolved in 1.5-ml portions of NCS tissues solubilizer. After addition of Bray's fluor solution to the latter, samples were assayed for ^{35}S activity (as in the nucleoside uptake assay) to determine cell-associated [^{35}S]NBMPR. Medium samples (1.0 ml), after addition to 13.5 ml of Bray's solution, were assayed for ^{35}S activity to determine the concentration of free NBMPR.

RESULTS

Cells of the BUdR-resistant line from which the HeLa/B5 clone was isolated possessed levels of thymidine kinase activity comparable to those of HeLa/0 cells; however, thymidine uptake rates were significantly lower in the former than in the latter under comparable conditions (25). These and other related characteristics were found in cells of the HeLa/B5 clone. The experiments of Table I demonstrated that HeLa/B5 cells were resistant to BUdR and FUdR. Cells of the parental line, HeLa/0, ceased proliferation after 48 h of exposure to BUdR (100 μM), or to FUdR (0.01 μM), whereas proliferation of HeLa/B5 cells was unaffected. Both lines were equally sensitive to FU, suggesting that the resistant character of HeLa/B5 cells was due to altered uptake or metabolism of the nucleoside.

That BUdR/FUdR resistance was a consequence of impaired uptake of these agents is indicated by a comparison of the uptake of radioactive BUdR, FUdR, and thymidine by HeLa/0 and HeLa/B5 cells. Representative time courses of uptake are presented in Fig. 1 for thymidine uptake by both cell types, and reciprocal plots of uptake rates and substrate concentrations (thymidine, BUdR, and FUdR) are presented in Fig. 2. Uptake data have been expressed in terms of cellular protein content rather than cell number be-

TABLE I. Effect of FUdR, BUdR, and FU on Proliferation of HeLa/0 and HeLa/B5 Cells*

Agent (μM)	HeLa/0		HeLa/B5	
	Cell number ($\times 10^{-6}$)	Percent	Cell number ($\times 10^{-6}$)	Percent
Expt. I: FUdR				
0	2.27	100	1.05	100
0.01	0.39	17	1.00	95
0.05	0.10	4	1.02	97
0.1	0.07	3	0.78	74
1.0	0.07	3	0.15	14
Expt. II: BUdR				
0	2.83	100	1.16	100
50	0.42	15	0.98	85
100	0.33	12	0.96	83
Expt. III: FU				
0	3.51	100	1.42	100
1	1.76	50	1.16	82
5	0.38	11	0.39	28

*Replicate monolayer cultures were started with 1×10^5 cells in 3.5 ml of MEM-A medium containing 10% calf serum; 4 h later, 0.5 ml of medium without additives (control) or analog-containing medium was added. Growth curves for the 2 cell types in the presence and absence of the analogs were determined from cell numbers measured after particular intervals of growth at 37°C; 3 bottles for each condition (agent and concentration) were withdrawn for determination of cell number. Only those data for cell numbers after 96 h of growth are presented above (25).

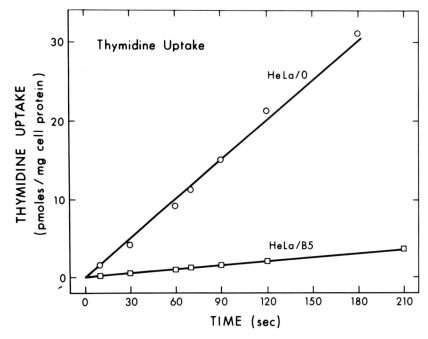

Fig. 1. Time course of thymidine uptake by HeLa/0 and HeLa/B5 cells. Replicate monolayer cultures were exposed at 20°C to MEM-T medium containing 0.05 μM [^{3}H-methyl] thymidine for the intervals indicated. Cellular content of labeled permeant was determined as specified in Materials and Methods.

cause HeLa/B5 cells are larger than HeLa/0 cells (25). It should be noted that uptake of thymidine and thymidine analogs by HeLa/B5 cells is a mediated process, as indicated by the linear reciprocal plots of Fig. 2 and the sensitivity of thymidine uptake to NBMPR (see below).

Because phosphorylation of BUdR and FUdR by thymidine kinase is an essential step in the manifestation of the cytotoxicity of these agents, the properties of the thymidine kinases of HeLa/0 and HeLa/B5 cells were compared. Kinetic studies of the thymidine kinase activity of unfractionated "freeze-thaw" extracts from HeLa/0 and HeLa/B5 cells showed little difference in the apparent kinetic constants for this enzyme activity; with thymidine as the substrate, these values were obtained: K_m, 8–10 μM; V_{max}, 400–450 pmoles thymidine phosphorylated/min/mg protein. When either crude or partially purified extracts from HeLa/B5 and HeLa/0 cells were chromatographed on the thymidine kinase affinity gel, elution profiles for thymidine kinase activity were essentially the same for extracts from either cell type (Fig. 3). Electrophoresis of crude cell extracts on 5% polyacrylamide gels revealed similar profiles of thymidine kinase activity for both cell types (Fig. 4); evidently the slower component (approximately 90% of the total activity) was the cytosol enzyme, while the faster component was of mitochondrial origin (16, 17, 28). These experiments indicated that a) the thymidine kinase content of the variant and parental cells was similar, and b) the kinetic characteristics and the physico-chemical properties that determine chromatographic and electrophoretic mobility of the thymidine kinases were similar in both cell types. Thus, the resistance of HeLa/B5 cells

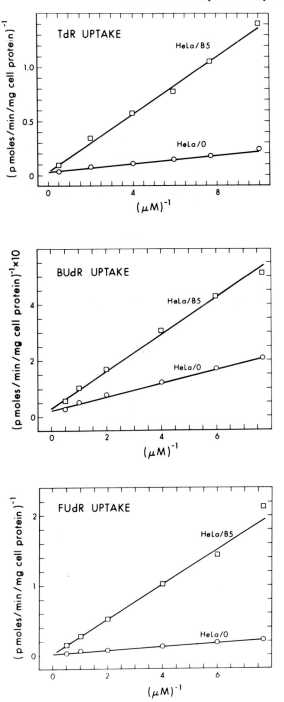

Fig. 2. Kinetics of uptake of thymidine, BUdR, and FUdR by HeLa/0 and HeLa/B5 cells. Replicate monolayer cultures were exposed at 20°C for 60 sec to MEM-T medium containing the indicated concentrations of ^{3}H-labeled permeant and the cellular content of permeant was then determined as given in Materials and Methods. Redrawn from Lynch et al. (25).

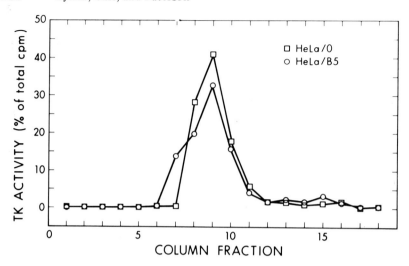

Fig. 3. Elution profiles from thymidine kinase affinity chromatography of extracts from HeLa/0 and HeLa/B5 cells. "Freeze-thaw" extracts, prepared from cells proliferating exponentially in suspension culture, were fractionated with ammonium sulfate and dialyzed overnight (28). Samples (1 ml, 4–6 mg protein) of the fractionated extracts were applied to 0.5 × 7 cm columns of thymidine 3'-phosphate-linked sepharose gel which were eluted with Tris-HCl buffer (pH 7.5) containing 10% glycerol, 5 mM dithioerythritol, and graded concentrations of thymidine. Samples (20 μl) from each eluate fraction (2.0 ml) were assayed for thymidine kinase activity as in Fig. 4.

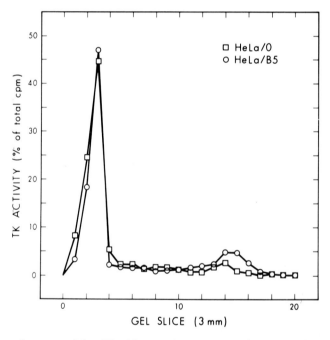

Fig. 4. Electropherogram of thymidine kinase activity from HeLa/0 and HeLa/B5 cells. Samples (50 μl) of "freeze-thaw" extracts, prepared from cells proliferating exponentially in suspension culture, were subjected to electrophoresis on 5% polyacrylamide gels. Gel slices (3 mm) were assayed for thymidine kinase activity by incubation for 1 h at 37°C with 150 μl of the [^{3}H-methyl] thymidine-containing assay mixture as described in Materials and Methods, and the reaction product was isolated on DEAE-cellulose paper for assay of ^{3}H activity.

to the thymidine analogs did not appear to be attributable to changes in thymidine kinase activity. The possibility that the substrate specificity of the thymidine kinase in the variant cells might be altered was not investigated.

We considered that BUdR/FUdR resistance might result from defects in the operation of a membrane transport mechanism; to test this possibility, several aspects of nucleoside uptake were compared in both cell types. Uptake data have been expressed in terms of cellular protein content because HeLa/0 and HeLa/B5 cells differ in size; apparent kinetic constants for uptake of thymidine, thymidine analogs, and for adenosine and uridine are listed in Table II. The 2 cell types did not differ in respect to kinetic constants for uptake of adenosine and of uridine; however, K_m values for uptake of thymidine, BUdR, and FUdR by HeLa/B5 cells were significantly higher than those found with HeLa/0 cells, whereas V_{max} values were essentially identical for uptake of all 3 permeants.

TABLE II. Apparent Kinetic "Constants" for Nucleoside Uptake by HeLa/0 and HeLa/B5 Cells*

Permeant	HeLa/0		HeLa/B5	
	K_m	V_{max}	K_m	V_{max}
BUdR	1.0	43	2.8	42
Thymidine	0.6	32	4.0	30
FUdR	1.3	47	11.3	45
Uridine	6.0	117	4.5	110
Adenosine	3.6	285	2.0	333

*Nucleoside uptake rates were derived from the amounts of ^{3}H-nucleoside taken up by replicate monolayer cultures during 60 sec of incubation at 20°C; rates were measured at various nucleoside concentrations and constants were estimated from reciprocal plots of rates and concentrations (25): K_m, μM, V_{max}, pmoles/min/mg protein.

HeLa/B5 cells were less responsive than HeLa/0 cells to inhibition of thymidine uptake by NBMPR, a potent inhibitor of nucleoside transport. The concentration of NBMPR that reduced thymidine uptake to 50% of control values (IC_{50}) in HeLa/B5 cells was 4 μM; this IC_{50} value is about 800 times higher than that reported previously (6) for HeLa/0 cells (NBMPR inhibition of thymidine uptake, $IC_{50} = 0.05$ μM). Because NBMPR inhibits transport of thymidine without effect on thymidine metabolism in HeLa cells, the reduced sensitivity of HeLa/B5 cells to NBMPR suggests altered function of the thymidine transporter. The IC_{50} values for NBMPR inhibition of uptake of adenosine and guanosine by HeLa/B5 and HeLa/0 cells were approximately 0.05 μM, indicating that transport of these nucleosides was unaffected in the variant cells (25).

To determine if changes in the number of NBMPR binding sites, or in properties of the latter, could explain reduced NBMPR sensitivity of the thymidine uptake mechanism in HeLa/B5 cells, binding of [^{35}S]NBMPR to monolayer cultures was examined (Fig. 5). Mass law analysis of these binding data by the method of Scatchard (32) indicated that HeLa/0 and HeLa/B5 cells possess similar numbers of NBMPR binding sites (per mg cellular protein) and that the affinity of these sites for NBMPR is similar in both cell types. The mass law plot indicates that binding at low NBMPR concentrations was due to a single type of high affinity binding site; assuming that 1 molecule of NBMPR was bound to each such site, HeLa/B5 and HeLa/0 cells bound 1.24 and 1.25 pmoles NBMPR/mg total

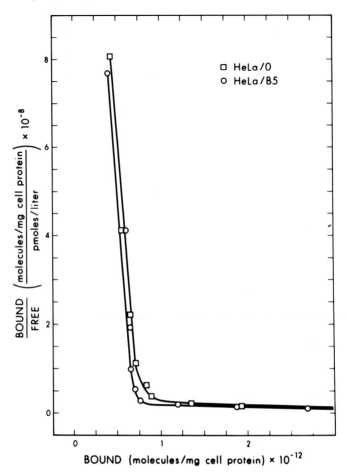

Fig. 5. Scatchard plots of NBMPR binding by HeLa/0 and HeLa/B5 cells. Replicate monolayer cultures were exposed at 20° for 5 min to 4 ml MEM-T medium containing various concentrations of [^{35}S]NBMPR; the [^{35}S]NBMPR-content of medium samples and of monolayers (after 1 wash with cold, buffered saline) was then determined.

protein*, respectively. Dissociation constants calculated from these data were 4.2×10^{-10} and 4.4×10^{-10}M for HeLa/0 and HeLa/B5 cells, respectively. These results indicate that the decreased responsiveness of HeLa/B5 cells to inhibition of thymidine uptake by NBMPR was not due to alterations in number of NBMPR binding sites, nor in the affinity of such sites for transport inhibitor.

DISCUSSION

Two types of BUdR resistance arising from decreased cellular uptake of BUdR have been found in animal cells: a) resistance attributable to a deficiency in thymidine kinase

---

*Recent studies in this laboratory (A.R.P. Paterson and C.E. Cass) have shown that the number of NBMR binding sites per HeLa cell is higher in S phase than in other cell cycle phases. Thus, values reported here for asynchronous cell populations represent the average of the NBMPR binding capacities of cells in the different cell cycle states.

activity has been the more frequently studied (14–17) and b) resistance due to impairment of the transport mechanism by which BUdR enters the cell has also been reported (18, 19). In the present study, the latter mechanism appears to account for BUdR/FUdR resistance in HeLa/B5 cells; we suggest that a reduced affinity of the thymidine transport mechanism for thymidine and thymidine analogs is responsible for low rates of uptake of the latter by HeLa/B5 cells, relative to those of HeLa/0 cells.

Two mechanisms for transport of nucleosides by animal cells have been described: a) the low affinity, low specificity systems found in erythrocytes (3–5) and in polymorphonuclear leucocytes (33) and b) the high affinity systems seen in various lines of cultured animal cells [for example, Novikoff hepatoma (1), HeLa (6–9), 3T3 (34)] which appeared to have distinct permeant specificities. Recent studies of thymidine uptake by ATP-depleted Novikoff hepatoma cells (10) have shown that these cells possess a low affinity transport mechanism with low specificity for nucleosides resembling the facilitated diffusion mechanism of human erythrocytes. In addition, studies from this laboratory of NBMPR inhibition of nucleoside uptake in HeLa cells have demonstrated the existence of NBMPR binding sites of a single class, occupancy of which results in inhibition of 4 functionally distinct nucleoside uptake mechanisms (6–9, 23).

BUdR and FUdR evidently enter HeLa cells by means of the thymidine transport mechanism and are phosphorylated by thymidine kinase. That the decreased uptake of these agents by HeLa/B5 cells was not a consequence of changes in thymidine kinase activity was indicated by the following: a) the cellular content of thymidine kinase activity was similar in HeLa/B5 cells and HeLa/0 cells, b) similar kinetic constants for thymidine kinase were obtained for both cell types, and c) substrate affinities for this enzyme were apparently comparable in both cell types because thymidine kinase elution profiles were similar when extracts of either cell type were chromatographed on thymidine kinase affinity gels.

Because NBMPR was known to interact in a highly specific manner with the nucleoside transport mechanism(s) of HeLa cells (6–9, 23), NBMPR was employed as a probe of the apparent alteration in thymidine transport in HeLa/B5 cells. Although thymidine uptake in both HeLa/0 and HeLa/B5 cells was less than 10% of control values in the presence of 10–12 μM NBMPR, at lower NBMPR concentrations thymidine uptake by HeLa/B5 cells was markedly less sensitive than that of HeLa/0 cells to inhibition by NBMPR. Sensitivity to NBMPR of adenosine, guanosine, and uridine uptake was comparable in both cell types. These results, together with those of the kinetic studies of thymidine uptake here reported, indicate that the activity of the thymidine transporter is altered in the variant cells. The relationship between the NBMPR binding site and the thymidine transporter is not understood, apart from the fact that NBMPR occupancy of the former prevents operation of the latter. The comparison of NBMPR binding by HeLa/0 and HeLa/B5 cells reported here indicates that the numbers of cellular binding sites and the dissociation constants for bound NBMPR were similar in both cell types. These findings suggest that HeLa/B5 cells are resistant to BUdR and FUdR because some aspect of the thymidine transporter function is defective.

ACKNOWLEDGMENTS

This work was supported by the Medical Research Council of Canada and the National Cancer Institute of Canada. We acknowledge the competent technical assistance of E. Howell, J. Paran, and E. Lau; we are grateful to G. J. Lauzon for his participation in the NBMPR binding experiments.

REFERENCES

1. Plagemann PGW, Richey DP: Biochim Biophys Acta 344:263, 1974.
2. Berlin RD, Oliver JM: Int Rev Cytol 42:287, 1975.
3. Cass CE, Paterson ARP: J Biol Chem 247:3314, 1972.
4. Oliver JM, Paterson ARP: Can J Biochem 49:262, 1971.
5. Cass CE, Paterson ARP: Biochim Biophys Acta 291:734, 1973.
6. Cass CE, Paterson ARP: Exp Cell Res 105:427, 1977.
7. Paterson ARP, Kim SC, Bernard O, Cass CE: Ann NY Acad Sci 255:402, 1975.
8. Paterson ARP, Naik SR, Cass CE: Mol Pharmacol 1977, in press.
9. Paterson ARP, Babb LR, Paran JH, Cass CE: Mol Pharmacol 1977, in press.
10. Plagemann PGW, Marz R, Erbe J: J Cell Physiol 89:1, 1976.
11. Steck TL, Nakata Y, Bader JP: Biochim Biophys Acta 190:237, 1969.
12. Shuster GS, Hare JD: In Vitro 6:427, 1971.
13. Hare JD: Cancer Res 30:684, 1970.
14. Kit S, Dubbs DR, Piekarski LJ, Hsu TC: Exp Cell Res 31:297, 1963.
15. Littlefield JW: Biochim Biophys Acta 95:14, 1965.
16. Kit S, Leung W-C, Kaplan LA: Eur J Biochem 39:43, 1973.
17. Kit S, Leung W-C, Trkula D: Arch Biochem Biophys 158:513, 1973.
18. Breslow RE, Goldsby RA: Exp Cell Res 55:339, 1969.
19. Freed JJ, Mezger-Freed L: J Cell Physiol 82:199, 1973.
20. Paterson ARP, Oliver JM: Can J Biochem 49:271, 1971.
21. Cass CE, Gaudette LA, Paterson ARP: Biochem Biophys Acta 345:1, 1974.
22. Cass CE, Paterson ARP: Biochem Biophys Acta 419:285, 1976.
23. Lauzon GJ, Paterson ARP: Mol Pharmacol 1977, in press.
24. Montgomery JA, Johnston TP, Gallagher A, Stringfellow CR, Schabel FM Jr: J Med Pharm Chem 3:265, 1961.
25. Lynch TP, Cass CE, Paterson ARP: Cancer Res (Submitted).
26. Bray GA: Anal Biochem 1:279, 1960.
27. Hartree EF: Anal Biochem 48:422, 1972.
28. Lee L-S, Cheng Y-C: J Biol Chem 251:2600, 1976.
29. Ives DH, Durham JP, Tucker VS: Anal Biochem 28:192, 1969.
30. Kowal EP, Markus G: Fec Proc Fed Am Soc Exp Biol 34:700, 1975.
31. Nishikawa AH, Bailon P: Anal Biochem 64:268, 1975.
32. Edsall JT, Wyman J: In "Biophysical Chemistry." New York: Academic Press, 1958.
33. Taube RA, Berlin RD: Biochim Biophys Acta 255:6, 1972.
34. Cunningham DD, Remo RA: J Biol Chem 248:6282, 1973.

Journal of Supramolecular Structure 6:375—381 (1977)
Molecular Aspects of Membrane Transport 237—243

Association of (Ca+Mg)-ATPase Activity With ATP-Dependent Ca Uptake in Vesicles Prepared From Human Erythrocytes

Eugene E. Quist and Basil D. Roufogalis

Laboratory of Molecular Pharmacology, Faculty of Pharmaceutical Sciences, University of British Columbia, Vancouver, British Columbia, Canada, V6T 1W5

Ghost membranes prepared from human erythrocytes exhibit 2 distinct (Ca+Mg)-ATPase[1] activities (Quist and Roufogalis, Arch Biochem Biophys 168:240, 1975). (Ca+Mg)-ATPase activity dependent on a water soluble protein fraction is selectively lost from ghost membranes during preparation of vesicles under low ionic strength, slightly alkaline conditions. In this study, the Ca^{2+} dependence of the remaining membrane bound (Ca+Mg)-ATPase activity and ATP-dependent Ca uptake in vesicles were compared. The Ca^{2+} activation curves for (Ca+Mg)-ATPase activity and Ca uptake into vesicles were parallel over a Ca^{2+} range of 0.3—330 μM, and both curves have 2 apparent K_A values for Ca^{2+} of 0.45 and 100 μM. Addition of a concentrated soluble protein fraction containing predominantly spectrin to the vesicles increased (Ca+Mg)-ATPase activity over twofold but did not affect the rate of Ca uptake. These findings suggest that the (Ca+Mg)-ATPase activity remaining in vesicles after extraction of the water soluble proteins is associated with the Ca pump whereas (Ca+Mg)-ATPase activity dependent on the soluble protein fraction is associated with some other function.

Key words: human erythrocytes; ATP-dependent Ca uptake; (Ca+Mg)-ATPase; spectrin; inside-out vesicles

While it is generally accepted that (Ca+Mg)-ATPase activity in human erythrocyte membranes is associated with Ca transport (1, 2), others have suggested that at least part of the (Ca+Mg)-ATPase activity may be associated with contractile-like proteins in these membranes (3, 4, 5). Recently, it was demonstrated by using $LaCl_3$ as a selective inhibitor of transport (Ca+Mg)-ATPase that only 50% of the total (Ca+Mg)-ATPase activity in re-sealed ghosts (6) and intact erythrocytes (7) is associated with Ca transport.

Further evidence for the presence of at least 2 separate (Ca+Mg)-ATPase activities was recently obtained in erythrocyte ghost membranes (5). Incubation of ghosts under conditions producing formation of vesicles and a loss of water soluble membrane proteins (8) resulted in a selective loss of (Ca+Mg)-ATPase activity having a K_A for Ca^{2+} of 2 μM

[1] Abbreviations: (Ca+Mg)-ATPase — Ca^{2+}-activated, Mg^{2+}-dependent adenosine triphosphate; EGTA — ethylene glycol bis(β-aminoethyl ether)-N,N,N',N'-tetraacetic acid, Tris — Tris (hydroxymethyl)-methylamine; SDS — sodium dodecyl sulfate.

Received April 1, 1977; accepted June 2, 1977

(5). This activity previously referred to as high affinity (Ca+Mg)-ATPase activity is maximally activated by 5–10 μM Ca^{2+} (5, 9, 10). Addition of the water soluble protein fraction (5) containing predominantly spectrin (11) to extracted vesicles under isotonic conditions restores this (Ca+Mg)-ATPase activity. This component of (Ca+Mg)-ATPase activity is thought to be involved in the regulation of erythrocyte membrane deformability (12).

The (Ca+Mg)-ATPase activity remaining in vesicles prepared from ghost membranes has 2 K_A values for Ca^{2+} of 0.5 and 100 μM (5). This so-called low-affinity (Ca+Mg)-ATPase activity is maximally activated by 200–300 μM Ca^{2+} (5, 9, 13). In this study, we show that (Ca+Mg)-ATPase activity and ATP-dependent Ca uptake in resealed erythrocyte vesicles have similar affinities for Ca^{2+}. We also show that addition of the soluble protein fraction to vesicles increases (Ca+Mg)-ATPase activity over twofold but does not affect the rate of Ca uptake.

MATERIALS AND METHODS

Preparation of Vesicles

Whole human blood preserved in acid-citrate-dextrose was obtained from the Red Cross blood bank and was used within 20 days of collection. Ghosts were prepared from washed erythrocytes by stepwise hemolysis as previously described (6). Vesicles were prepared from only freshly prepared ghosts by a modification of the method of Steck et al. (8). One volume of ghosts (3.5 – 4.0 mg protein/ml) was diluted with 5 volumes of 0.5 mM NaCl or 0.1 mM EGTA and 1.0 mM Tris-maleate (pH 8.0) and incubated for 30 min at 37°C. Both procedures result in a complete disruption of ghosts to vesicles (diameter < 1 μm) as determined by phase contrast microscopy. The suspension was centrifuged for 15 min at 20,000 × g at 4°C. The supernatant containing predominantly bands I, II, and V polypeptides as determined by SDS-polyacrylamide gel electrophoresis (5, 11) was collected and concentrated sixfold by ultrafiltration and is referred to as the soluble protein fraction. The pellet consisting of vesicles was suspended to the original volume of ghosts used in 15 mM NaCl and 5 mM Tris-maleate, pH 7.1. To facilitate resealing, 1 volume of the vesicle preparation was suspended in 3 volumes of medium containing 10 mM Tris-maleate, pH 7.1, 4 mM $MgCl_2$, and 0.5 mM $CaCl_2$. After equilibration for 5 min at 4°C, isotonicity was restored with 2.9 M NaCl (4 ml to 80 ml of suspension). The suspension was incubated for 10 min at 25°C and centrifuged at 7,000 × g for 10 min at 4°C. The vesicles were washed twice with 30 volumes of 55 mM Tris-maleate, pH 7.2, 6.4 mM $MgCl_2$, and 66 mM NaCl to remove extravesicular Ca^{2+} at 7,000 × g at 4°C and suspended with the same solution to the original volume of ghosts used. One volume of this vesicle preparation was suspended in 2 volumes of 55 mM Tris-maleate, pH 7.2, at 37°C, 6.4 mM $MgCl_2$, 66 mM NaCl, and 0.15 mM EGTA. The concentration of $CaCl_2$ was varied in this mediium. In some experiments the soluble protein fraction was included to a final concentration of 0.5 mg of protein/ml.

Determination of Ca Uptake and ATPase Activity

To study Ca uptake and ATPase activity, the vesicle preparation was preincubated for 10 min at 37°C and the reaction was started by the addition of Na_2ATP (2 mM final concentration). At appropriate time intervals, 2 ml and 1 ml aliquots were removed for either inorganic phosphate (P_i) or Ca analysis, respectively. The concentration of Ca^{2+}

was calculated by equations given by Katz et al. (14) using a Ca-EGTA stability constant of of $10^{10.65}$ (15).

The Ca content of the vesicles was determined by suspending 1 ml of vesicle suspension in 8 ml of cold 110 mM NaCl and 0.5 mM $LaCl_3$. $LaCl_3$ was included to block Ca uptake (6) and to displace loosely bound extravesicular Ca (16). The tubes were centrifuged at 15,000 X g for 15 min and the supernatant was completely removed by aspiration. Further washing of the vesicles with the $LaCl_3$ solution did not lower the Ca content of the vesicles. Ca was extracted from the pelleted vesicles by the method of Sparrow and Johnstone (17) as previously described for resealed ghosts (6). Ca was determined by atomic absorption spectrophotometry (Techtron AA-5) in the presence of 30 mM $LaCl_3$.

ATPase activity in the vesicles was stopped by mixing 2 ml of suspension with 1 ml of 20% trichloroacetic acid. Inorganic phosphate (P_i) was determined by the method of Fiske and SubbaRow (18). (Ca+Mg)-ATPase activity was corrected for Mg-ATPase activity obtained in the absence of added Ca by subtraction.

Characterization of Vesicles

The proportion of resealed inside-out vesicles present was determined by measuring the fraction of latent acetylcholinesterase activity (19) in the presence and absence of 0.1% Triton X-100 as described by Steck and Kant (20). Similarly, the proportion of resealed right-side-out vesicles was found by determining the amount of latent glyceraldehyde 3-phosphate dehydrogenase activity (21). Protein concentration was determined by the method of Lowry et al. (22).

RESULTS

In the final vesicle preparation, 44–48% of the total membranes consisted of resealed inside-out vesicles (i.e., 46% of the total acetylcholinesterase activity was latent) and 39% resealed right-side-out vesicles (39% of the total glyceraldehyde-3-phosphate dehydrogenase activity was latent). The remaining activity suggests that approximately 15% of the membranes were unsealed. Inside-out vesicles were not routinely separated from right-side-out vesicles as (Ca+Mg)-ATPase activity of this vesicle preparation was lost after separation on a dextran T-110 gradient according to Steck and Kant (20).

The presence of latent acetylcholinesterase and glyceraldehyde-3-phosphate dehydrogenase activities indicates that the vesicles are relatively impermeable to their respective substrates. Other observations indicate that the vesicles are impermeable to Ca and ATP. In the absence of extravesicular $CaCl_2$, there was no change in vesicular Ca content of vesicles incubated for 25 min at $37°C$. EGTA (0.1 mM) in the extravesicular medium would presumably extract Ca from leaky vesicles. In the absence of added $CaCl_2$ and in the presence of 2 mM ATP in the extravesicular medium, the Ca content also does not change under the above conditions. Therefore, the right-side-out vesicles are impermeable to ATP as Ca is rapidly lost from vesicles resealed in the presence of AFP when incubated at $37°C$ (not shown). In agreement with Steck and Kant (20), the vesicles retain their relative impermeability as determined by latent marker enzyme activities for several days when stored at $4°C$ although (Ca+Mg)-ATPase activity is greatly reduced.

The Ca^{2+} dependence of ATP dependent Ca uptake and (Ca+Mg)-ATPase activity in the vesicles is compared on Fig. 1. The Ca^{2+} activation curves for Ca uptake and (Ca+Mg)-ATPase activity are essentially parallel over a Ca^{2+} concentration range of 0.1–330 μM. A plot of this data according to Eadie (23) yields curves with 2 distinct slopes with K_A

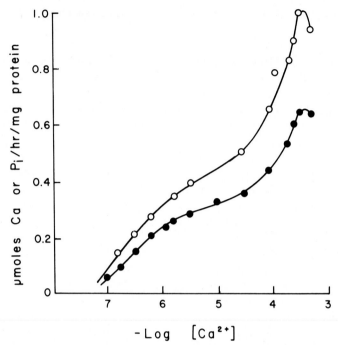

Fig. 1. Ca^{2+} dependence of (Ca+Mg)-ATPase activity (●) and ATP-dependent Ca uptake (○) in vesicles prepared from human erythrocyte ghosts.

values for Ca^{2+} of 0.45 and 100 μM. Both Ca uptake and (Ca+Mg)-ATPase activity are saturated at 330 μM Ca^{2+}. In this study, Ca uptake was dependent on ATP at all Ca^{2+} concentrations studied, and Ca uptake and (Ca+Mg)-ATPase activity were linear for at least 10 min (see Fig. 2A and 2B).

The rate of Ca uptake is approximately 1.5-fold greater than (Ca+Mg)-ATPase activity over the Ca^{2+} concentration range studied (Fig. 1). The presence of unsealed membranes (15% of total membranes) may yield an apparent lower stoichiometry of 1.5 (μmoles Ca taken up/μmoles ATP hydrolyzed). Unsealed membranes can be readily calculated to contribute to 25% of the total (Ca+Mg)-ATPase activity in this vesicle preparation based on the following 2 assumptions. First, only inside-out vesicles (46%) and unsealed membranes (15%) would contribute to (Ca+Mg)-ATPase activity due to the inaccessibility of ATP to the cytoplasmic side of right-side-out vesicles (39%). Second, unsealed membranes would not likely contribute to Ca uptake as the membranes are washed with 0.5 mM LaCl$_3$ solution prior to Ca analysis (see Methods). By subtracting (Ca+Mg)-ATPase activity due to unsealed membranes from total (Ca+Mg)-ATPase activity in Fig. 1, the stoichiometry of Ca uptake becomes 2.0.

Addition of the soluble protein fraction to the vesicles increased (Ca+Mg)-ATPase activity 2.5- and 2.1-fold at Ca^{2+} concentrations of 29 and 230 μM, respectively (Fig. 2A). Under the same conditions, addition of the soluble protein fraction to the vesicles was found not to have any effect on the rate of ATP-dependent Ca uptake.

DISCUSSION

In order to determine which component of total (Ca+Mg)-ATPase activity in erythrocyte membranes is associated with Ca transport, ATP-dependent Ca uptake and

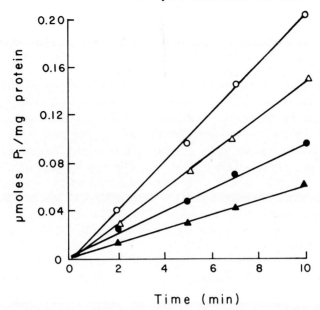

A

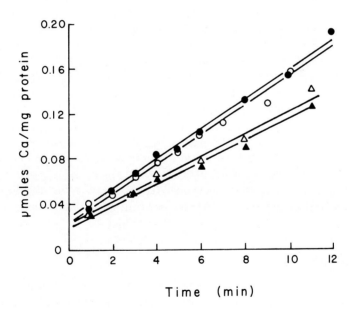

B

Fig. 2. Effect of the addition of the soluble protein fraction to vesicles on (Ca+Mg)-ATPase activity and ATP-dependent Ca uptake. A) Time course of (Ca+Mg)-ATPase activity in vesicles in the presence of 29 μM Ca^{2+} ($\blacktriangle$) and 230 μM Ca^{2+} ($\bullet$). Time course of (Ca+Mg)-ATPase activity after the addition of the soluble protein fraction to the vesicles in the presence of 29 μM Ca^{2+} ($\triangle$) and 230 μM Ca^{2+} ($\circ$). B) Time course of Ca uptake in vesicles in the presence of 29 μM Ca^{2+} ($\blacktriangle$) and 230 μM Ca^{2+} ($\bullet$). Time course of Ca uptake after the addition of the soluble protein fraction to the vesicles in the presence of 29 μM Ca^{2+} ($\triangle$) and 230 μM Ca^{2+} ($\circ$).

(Ca+Mg)-ATPase activity were compared in inside-out vesicles prepared from erythrocyte ghosts. Vesiculation requires incubation of ghosts in low ionic strength, slightly alkaline medium (8) in which the (Ca+Mg)-ATPase activity previously referred to as high-affinity activity is lost (5). The parallel nature of the Ca^{2+} activation curves for Ca uptake and (Ca+Mg)-ATPase activity and similarity of their Ca^{2+} affinities suggests that Ca transport is supported by (Ca+Mg)-ATPase activity remaining bound to the membrane after low ionic strength extraction of soluble proteins (Fig. 1). The apparent stoichiometry of 1.5 for Ca uptake in vesicles is lower than 2.0 obtained in resealed ghost (6) or intact erythrocytes (7, 24). However, correction for (Ca+Mg)-ATPase activity contributed by unsealed membranes to total (Ca+Mg)-ATPase activity raises the stoichiometry to 2.0 (see Results). The previously determined stoichiometry of less than 1 for ATP-dependent ^{45}Ca uptake into inside-out vesicles (26) might be due to leakage of Ca out of the vesicles, as the membranes were frozen prior to the ^{45}Ca-uptake studies. In our study, only freshly prepared vesicles were used and were resealed in the presence of $CaCl_2$ under conditions optimal for resealing of ghosts (6).

To determine if the soluble proteins extracted from ghosts under low ionic strength conditions contribute to ATP-dependent Ca uptake in vesicles, the concentrated soluble protein fraction was reconstituted with the vesicles. Recombination of the soluble protein fraction with the vesicles increased (Ca+Mg)-ATPase activity over twofold but did not alter the rate of Ca uptake into these vesicles (Fig. 2A and 2B). Therefore, (Ca+Mg)-ATPase activity dependent on the soluble protein fraction (5) or highly purified spectrin (25), may not be associated with the Ca pump. This finding is further evidence that high affinity or more appropriately spectrin-dependent (Ca+Mg)-ATPase actitivy is involved in some other function such as regulation of erythrocyte membrane deformability (12).

In human erythrocyte vesicles, low-affinity or spectrin-independent (Ca+Mg)-ATPase activity is characterized by 2 apparent dissociation constants of 0.45 and 100 μM (Fig. 1). The existence of a Ca^{2+} binding site with a dissociation constant of 0.45 μM seems compatible for a Ca transport system which functions to maintain the intracellular concentration of Ca^{2+} below 1 μM (15). The physiological significance of the low Ca^{2+} affinity site (K_A of 100 μM) is less clear. This site is probably not an artifact of membrane preparation as Ca transport does not saturate until an intracellular Ca^{2+} concentration of 200 μM in intact human erythrocytes loaded with Ca^{2+} using the ionophore A-23187 (27). However, Scharff has found that the low affinity site is not apparent in ghost membranes prepared in the presence of 1.5 mM $CaCl_2$ (13). Scharff has suggested that Ca^{2+} may control the properties of membrane bound (Ca+Mg)-ATPase and that the different Ca^{2+} affinities for the pump (Ca+Mg)-ATPase may correspond to different states of the Ca pump (13).

ACKNOWLEDGMENTS

This research was supported by the Medical Research Council of Canada Grant Ma-4078. EEQ is a recipient of a Postdoctoral Fellowship from the Medical Research Council of Canada.

REFERENCES

1. Schatzmann HJ, Vincenzi FF: J Physiol (London) 231:369, 1969.
2. Lee KS, Shin BC: J Gen Physiol 54:713, 1969.

3. Ohnishi T: J Biochem 52:307, 1962.
4. Palek J, Stewart G, Lionetti FJ: Blood 44:583, 1974.
5. Quist EE, Roufogalis BD: Arch Biochem Biophys 168:240, 1975.
6. Quist EE, Roufogalis BD: FEBS Lett 50:135, 1975.
7. Sarkadi B, Szasz I, Gerloczy A, Gardos G: Biochim Biophys Acta 464:93, 1977.
8. Steck TL, Weinstein RS, Strauss JH, Wallach DFH: Science 168:255, 1970.
9. Schatzmann HJ, Rossi GL: Biochim Biophys Acta 241:379, 1971.
10. Wolf HU: Biochem J 130:311, 1972.
11. Fairbanks G, Steck TL, Wallach DFH: Biochemistry 10:2606, 1971.
12. Quist EE, Roufogalis BD: Biochem Biophys Res Commun 72:673, 1976.
13. Scharff O: Biochim Biophys Acta 443:206, 1976.
14. Katz AM, Repke DI, Upshaw JE, Polascik MA: Biochim Biophys Acta 205:473, 1970.
15. Schatzmann HJ: J Physiol (London) 235:551, 1973.
16. Van Breeman C: Arch Int Physiol Biochim 77:710, 1969.
17. Sparrow MP, Johnstone BM: Biochim Biophys Acta 90:426, 1964.
18. Fiske CH, SubbaRow YJ: J Biol Chem 66:375, 1925.
19. Ellman GL, Courtney KD, Andres V Jr, Featherstone RM: Biochem Pharmacol 7:88, 1961.
20. Steck TL, Kant JA: In Fleischer S, Packer L (eds): "Methods in Enzymology." New York: Academic Press, 1974, vol 31, part A, p 172.
21. Cori GT, Slein MW, Cori CF: J Biol Chem 173:605, 1948.
22. Lowry OH, Rosebrough NJ, Farr AL, Randall RJ: J Biol Chem 193:265, 1951.
23. Eadie GS: J Biol Chem 146:83, 1942.
24. Ferreira HG, Lew VL: Nature (London) 259:47, 1976.
25. Quist EE: PhD Thesis, University of British Columbia, Vancouver, BC, Canada, 1975.
26. Weiner ML, Lee KS: J Gen Physiol 59:462, 1972.
27. Sarkadi B, Szasz I, Gardos G: J Membr Biol 26:357, 1976.

Journal of Supramolecular Structure 6:383–388 (1977)
Molecular Aspects of Membrane Transport 245–250

Effects of Sodium Ions on the Electrical and pH Gradients Across the Membrane of Streptococcus lactis Cells

Susan L. Barker and Eva R. Kashket

Department of Microbiology, Boston University School of Medicine, Boston, Massachusetts 02118

Energized cells of Streptococcus lactis conserve and transduce energy at the plasma membrane in the form of an electrochemical gradient of hydrogen ions (Δp). An increase in energy-consuming processes, such as cation transport, would be expected to result in a change in the steady state Δp. We determined the electrical gradient ($\Delta\psi$) from the fluorescence of a membrane potential-sensitive cyanine dye, and the chemical H^+ gradient (ΔpH) from the distribution of a weak acid. In glycolyzing cells incubated at pH 5 the addition of NaCl to 200 mM partially dissipated the Δp by decreasing $\Delta\psi$, while the ΔpH was constant. The Δp was also determined independently from the accumulation levels of thiomethyl-β-galactoside. The Δp values decreased in cells fermenting glucose at pH 5 or pH 7 when NaCl was added, while the ΔpH values were unaffected; cells fermenting arginine at pH 7 showed similar effects. Thus, these nongrowing cells cannot fully compensate for the energy demand of cation transport.

Key words: protonmotive force, sodium ions, Streptococcus lactis

The work entailed in the transport of solutes against electrochemical gradients requires the input of metabolic energy. However, what portion of the cells' energy budget is expended for this work is not known. To approach this problem, we have begun our studies with resting-cell suspensions of the homolactic fermenter Streptococcus lactis. The rationale was based on the findings, that, in general, bacterial catabolism does not appear to be finely regulated. A number of workers has concluded that bacteria lack control mechanisms for the coupling of energy production to processes involving energy utilization (1–6). For example, Thomas and Batt (7) showed that nongrowing S. lactis cells ferment glucose at rates governed by the rate of sugar supply. Thus, it would not be surprising to encounter situations where the rate of energy supply, e.g., by glycolysis, has not been increased to meet an increased demand by energy-consuming processes. Thus, solutes which are massively transported, such as K^+ and Na^+, could be expected to utilize sufficient metabolic energy to cause measurable changes, even when the supply of energy-yielding substrates is not limiting.

The index of metabolic energy chosen for study is the protonmotive force (Δp), since bacteria transduce and conserve energy at the plasma membrane in for the form of an electrochemical gradient of hydrogen ions, according to the chemiosmotic theory of Mitchell (8, 9). The gradient (Δp) is generated by the proton-translocating membrane-bound ATPase complex which uses the ATP formed only by fermentative pathways in S. lactis cells. The Δp consists of a membrane potential ($\Delta\psi$) and a pH gradient (ΔpH) across the bacterial plasma membrane. These parameters bear the relationship

$$\Delta p = \Delta\psi - 59\ \Delta pH$$

where ΔpH equals the pH_{out} (pH of the bulk medium) minus the pH_{in} (pH of the cytosol). The value 59 is a combination of constants for expression of ΔpH in mV at 25°C.

In a previous communication we reported that addition of K^+ to fermenting cells of S. lactis increased the ΔpH and decreased the $\Delta\psi$ (10). Thus, when the increase in ΔpH could not compensate for the decrease in $\Delta\psi$, the Δp was partially dissipated. In the present communication we report that the addition of Na^+ causes no change in ΔpH, but decreases the $\Delta\psi$, again leading to partial dissipation of Δp. More Na^+ than K^+ is required to effect a measurable change in Δp in these fermenting cells. A preliminary report has appeared (11).

METHODS

Streptococcus lactis ATCC 7962 cells were grown and prepared as described previously (10). Stock suspensions of cells (1.4 mg dry weight/ml) plus (5×10^{-5} M) 1, 1'-dipropyl-2,2'-thiodicarbocyanine dye were prepared as before (10).

The transmembrane electric potential ($\Delta\psi$) was calculated from the fluorescence quenching of the cyanine dye. The fluorescence intensity was related to the K^+ diffusion potential obtained in valinomycin-treated cells suspended in media of various K^+ concentrations (10). Under the conditions of the present experiments, addition of NaCl to 100 mM or 200 mM had no effect on the fluorescence intensity, with K^+ diffusion potentials ranging from 0 to 190 mV.

The pH gradient across the membrane (ΔpH) was estimated from the distribution of [^{14}C]benzoic acid (12, 10). The Δp was also determined from the accumulation levels of [^{14}C]thiomethyl-β-galactoside (TMG) (13, 14). The sources of the chemicals used are listed in Ref. 10.

RESULTS

The experiments consisted of incubating S. lactis cells in media of varying NaCl concentrations and measuring the Δp. The energy was supplied by the fermentation of either glucose or arginine. The Δp was measured by 2 methods: a) from the sum of the values for $\Delta\psi$ (determined from the fluorescence quenching of the cyanine dye) plus those for ΔpH, and b) from the accumulation levels of TMG.

Effects of NaCl on $\Delta\psi$

The fluorescence intensity of suspensions of cells plus dye incubated at pH 5 decreased on addition of glucose, indicating an increase in $\Delta\psi$ (Fig. 1). The fluorescence intensity level reached and maintained for at least 3 min was higher as the concentration of

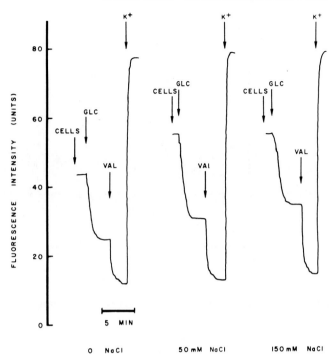

Fig. 1. Effect of NaCl on fluorescence intensity of fermenting cells: 3 representative tracings of chart recordings. Stock cells (1.4 mg dry weight/ml) and dye (5 × 10^{-5}M) (10) were diluted 1:10 with 0.1 M citrate-Tris buffer, pH 5.0, and NaCl was added to the concentrations indicated under each tracing. The reaction mixtures were added to the cuvettes at the time indicated "CELLS"; glucose was added at time "GLC." The values corresponding to the fluorescence intensity at 2 min after glucose addition were read off a calibration curve as described previously (10). At time "VAL," 2 μl of 1 × 10^{-2} M valinomycin was added; at "K$^+$," KCl was added to 230 mM final concentration.

NaCl in the medium was increased, indicating a progressively lower Δψ. With these cells the maximum Δψ had not been reached (Fig. 1). When valinomycin was added, the fluorescence intensity decreased further, indicating an increase in Δψ. This Δψ was attributed to a diffusion potential for K$^+$ ions, since the ionophore had rendered the membrane specifically permeable to that cation, and since the concentration ratio in/out of potassium was very large (> 1,000). Indeed, when K$^+$ was added to concentrations greater than 200 mM, the fluorescence rose, as expected from a decrease in the Δψ due to a decrease in the potassium diffusion potential. Cells incubated with NaCl were capable of showing a potassium diffusion potential almost as great as that of cells incubated without NaCl, since valinomycin addition lowered the fluorescence almost to the same value.

The chemical gradients of H$^+$ (ΔpH) were determined in parallel reaction mixtures. To calculate Δp, the values obtained for Δψ and ΔpH were added. As shown on Fig. 2, increasing the NaCl concentration from 0 to 200 mM resulted in a decrease in the poise of Δp from about 150 to 90 mV. The ΔpH was found to remain constant at about 1 pH unit (inside alkaline), while the Δψ decreased with increasing NaCl in the medium. It should be noted, however, that the standard errors of the means were large, ranging from 5 to 22% of the mean values, suggesting that quantitative results obtained by this technique should be considered with reservation.

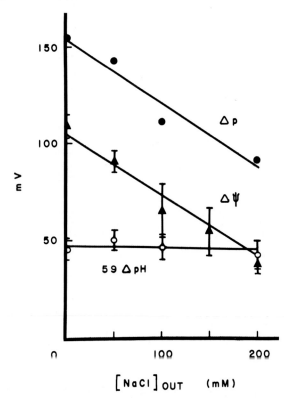

Fig. 2. Effect of NaCl on the electrochemical gradient of hydrogen ions. The $\Delta\psi$ was calculated from the fluorescence intensity and the ΔpH was measured from the distribution of [^{14}C] benzoic acid (see Methods). The Δp was calculated by adding $\Delta\psi$ and 59 ΔpH (see Introduction). Each point represents the average of 5 experiments ± the standard error of the mean. The lines were derived by the method of least squares.

Effects of NaCl on the Accumulation of TMG

The accumulation levels of TMG, which have been shown to correlate directly with Δp (13, 14) were also used to measure the Δp at various NaCl concentrations. This method, in addition to confirming in general the results obtained by the cyanine dye technique, also enabled extension of the experiments to incubation conditions (pH > 5) where the fluorescence intensity reaches levels beyond those in the linear portion of the calibration curve (10). We also established, by performing counterflow experiments (15, 16), that Na$^+$ (or K$^+$) had no effect on the TMG carrier per se.

When the effects of NaCl on Δp were tested, as calculated from TMG levels, there was a decrease in Δp with increasing amounts of NaCl, up to 200 mM NaCl (data not shown). At pH 5, with glucose as energy source, the addition of 200 mM NaCl decreased the Δp from 145.4 to 125.9 mV (Table I). This decrease is less than that calculated from the cyanine dye data, where the Δp decreased to 91.4 mV. However, the latter method has been noted to be inaccurate. At pH 7, the Δp decreased from 140.7 to 119.4 mV, with glucose as energy source.

When arginine was used as energy source, there was no significant decrease in Δp at pH 5. At pH 7, however, the Δp was lowered from 131.0 to 106.1 mV.

Since the ΔpH values were not affected by increasing NaCl in the medium, the decreases in Δp are assigned to decreases in $\Delta\psi$.

TABLE I. Effects of NaCl on TMG Accumulation and Transmembrane pH Gradient

Energy source	pH$_{out}$	$\Delta p (59 \log [TMG]_{in}/[TMG]_{out})$		59 ΔpH (mV)	
		None	200 mM NaCl	None	200 mM NaCl
Glucose	5	145.4 ± 0.8	125.9 ± 1.4	88.8 ± 5.4[a]	66.2 ± 7.6[a]
Glucose	7	140.7 ± 1.7	119.4 ± 1.5	25.5 ± 4.4	23.5 ± 2.5
Arginine	5	99.3 ± 4.1	93.3 ± 5.6	58.9 ± 2.4	59.4 ± 1.9
Arginine	7	131.0 ± 2.7	106.1 ± 2.5	0.0	0.0

The accumulation of TMG and the ΔpH were measured as described previously (10). The values are expressed in equivalent units ± the standard error of the mean. Each value is the average of 6 to 32 determinations, using 2 to 5 batches of cells. The Δp values obtained with NaCl were significantly different from those without NaCl (p < 0.02 by Student's t test), except those for arginine-energized cells at pH 5.

[a] The ΔpH values with or without NaCl were not significantly different.

DISCUSSION

Experiments with other membrane systems have suggested that the rate of H$^+$ extrusion may be rate limiting. For example, Mitchell and Moyle (17) showed that Δp in mammalian mitochondria was about 230 mV when respiration was limited by insufficient inorganic phosphate acceptor (state 4 respiration). When ADP was added to the system (state 3 respiration) the Δp decreased to about 200 mV. Similarly, Pick et al. (18) showed a ΔpH of 3.7 units in illuminated chloroplast preparations deficient in phosphate acceptor; when ADP was added to the ΔpH decreased by about 0.3 units (there is no $\Delta\psi$ component in chloroplast preparations). In intact illuminated cells of Halobacterium halobium Michel and Oesterhelt (19) have measured $\Delta\psi$ from the distribution of the permeant cation triphenylmethyl phosphonium and the ΔpH with the dimethyloxazolidine method. Treatment with dicyclohexylcarbodiimide, an inhibitor of the membrane ATPase, resulted in an increase in Δp, the result of increases of both $\Delta\psi$ and ΔpH. Presumably this increase occured because the synthesis of ATP was abolished in the presence of the inhibitor, and thus influx of H$^+$ was no longer effected by this system. Thus the utilization of Δp for ATP synthesis appeared to lower the poise of Δp in these 3 systems.

The results reported here can be interpreted as follows: Na$^+$ ions enter the cell down their electrochemical gradient, thus depressing the $\Delta\psi$. As result, the membrane ATPase can pump more H$^+$ out of the cell, which would give rise to an increase in ΔpH, were it not for the Na$^+$/H$^+$ antiporter. Such a carrier is presumed to be analogous to that described for S. faecalis (20, 21) or Escherichia coli cells (22). The activity of this carrier results in the electroneutral exchange of Na$^+$ for H$^+$, thus preventing the expected increase in ΔpH and, at the same time, extruding Na$^+$.

The effects of Na$^+$ addition on Δp and its components are different from those seen with K$^+$ addition. Concentrations of 10 mM KCl or less resulted in decrease of $\Delta\psi$ and increase of ΔpH (10). In contrast, NaCl up to 200 mM effected no change in ΔpH, but decreased the $\Delta\psi$. Addition of 50 mM NaCl did not alter the effects seen with 50 mM KCl (data not shown), thus supporting the view that one cation is transported independently of the other.

In these nongrowing, fermenting cells of S. lactis, therefore, the poise of the proton electrochemical gradient can be altered by the addition of cations that the cells actively transport. While these experiments yield no information on cation flux rates, they suggest that these cells cannot fully compensate for the energy demand of cation transport.

ACKNOWLEDGMENTS

This work was funded by Grant BMS75-10955 from the National Science Foundation. We wish to thank Dr. Alan Waggoner for the cyanine dye.

REFERENCES

1. Forrest WW: Symp Soc Gen Microbiol 19:65, 1969.
2. Forrest WW, Walker DJ: Adv Microb Physiol 5:213, 1971.
3. Gunsalus IC, Schuster CW: In Gunsalus IC, Stanier RY (eds): "The Bacteria." New York: Academic Press, 1961, vol 2, p 1.
4. Holms WH, Hamilton ID, Robertson AG: Arch Microbiol 83:95, 1972.
5. Rosenberger F, Elsden SR: J Gen Microbiol 22:726, 1960.
6. Senez JC: Bacteriol Rev 26:95, 1962.
7. Thomas TD, Batt RD: J Gen Microbiol 58:371, 1969.
8. Mitchell P: Biochem Soc Symp 22:142, 1963.
9. Harold FM: Bacteriol Rev 36:172, 1972.
10. Kashket ER, Barker SL: J Bacteriol 130:1017, 1977.
11. Kashket ER, Barker SL: J Supramol Struct (Suppl) 6:1, 1977.
12. Maloney PC, Kashket ER, Wilson TH: In Korn E (ed): "Methods in Membrane Biology." New York: Plenum Press, 1975, vol 5, p 1.
13. Kashket ER, Wilson TH: Proc Natl Acad Sci USA 70:2866, 1973.
14. Kashket ER, Wilson TH: Biochem Biophys Res Commun 59:879, 1974.
15. Wong PTS, Wilson TH: Biochim Biophys Acta 196:336, 1970.
16. Kashket ER, Wilson TH: J Bacteriol 109:784, 1972.
17. Mitchell P, Moyle J: Eur J Biochem 1:471, 1969.
18. Pick U, Rottenberg H, Avron M: FEBS Lett 32:91, 1973.
19. Michel H, Oesterhelt D: FEBS Lett 65:175, 1976.
20. Harold FM, Papineau D: J Membr Biol 8:45, 1972.
21. Harold FM, Altendorf K: In Bonner F, Kleinzeller A (eds): "Current Topics in Membranes and Transport." New York: Academic Press, 1974, vol 5, p 1.
22. West IC, Mitchell P: Biochem J 144:87, 1974.

Journal of Supramolecular Structure 6:389–398 (1977)
Molecular Aspects of Membrane Transport 251–260

Energetics of Galactose, Proline, and Glutamine Transport in a Cytochrome-Deficient Mutant of Salmonella typhimurium

A. P. Singh and P. D. Bragg

Department of Biochemistry, University of British Columbia, Vancouver, British Columbia, Canada V6T 1W5

The effect of inhibitors and uncouplers on the osmotic shock-sensitive transport systems for glutamine and galactose (by the β-methyl galactoside permease) was compared to their effect on the osmotic shock-resistant proline and galactose permease systems in cytochrome-deficient cells of Salmonella typhimurium SASY28. Both osmotic shock-sensitive and -resistant systems were sensitive to uncouplers and to inhibitors of the membrane-bound Ca^{2+}, Mg^{2+}-activated adenosine triphosphatase. This suggests that uptake by both types of systems is energized in these cells by an electrochemical gradient of protons formed by ATP hydrolysis through the ATPase.

Key words: amino acid transport, transport energetics, cytochrome-deficient mutant, shock-sensitive transport, shock-resistant transport, Salmonella typhimurium

We have previously utilized a strain of Escherichia coli not able to form cytochromes in the absence of 5-aminolevulinic acid in order to study the energization of membrane-dependent processes under conditions in which the contribution of substrate oxidation through the respiratory chain is minimized (1–3). Moreover, these cells have low levels of endogenous substrates so that it is not necessary to pretreat the cells with uncouplers to deplete these energy reserves. In the present paper we have used a similar strain of Salmonella typhimurium to investigate the energization of the transport of galactose, proline, and glutamine.

The energization of the transport of proline in E. coli is believed to be fundamentally different from that of glutamine. Thus, transport of proline is driven by an electrochemical gradient of protons across the membrane ("energized state") generated by substrate oxidation through the respiratory chain or by hydrolysis of ATP by the membrane-bound Ca^{2+}, Mg^{2+}-activated ATPase[1] (4), whereas transport of glutamine does not involve the energized state but appears to require the direct involvement of ATP by a mechanism not utilizing the ATPase (5). Berger and Heppel (6) have extended this hypothesis to include certain other transport systems present in E. coli. They suggest that all transport systems which

[1] Abbreviations: ATPase – adenosine triphosphatase; CCCP – carbonylcyanide-m-chlorophenylhydrazone; DCCD – N, N′-dicyclohexylcarbodiimide.

Received May 4, 1977; accepted May 4, 1977

like glutamine have a binding-protein which can be released by osmotic shock treatment of intact cells are energized directly by ATP. Shock-resistant systems, like that of proline, are driven by the energized state.

S. typhimurium possesses both shock-sensitive and shock-resistant transport systems for D-galactose (7). In the cells used in our experiments the methyl β-thio-D-galactoside systems I and II are absent and galactose is transported by the binding-protein dependent β-methyl galactoside permease or by the binding-protein independent galactose permease system. Although Wilson (8, 9) has suggested that the β-methyl galactoside and galactose permease systems are both driven by ATP directly, the results of Henderson et al. (10) which show that the uptake of galactose through the galactose permease, but not that of β-methyl galactoside, is accompanied by symport of protons suggest that the galactose permease system may be driven by the energized state.

In this paper we have examined the action of compounds which effect the energization of transport in cytochrome-deficient cells of S. typhimurium. We have found that there is essentially no difference between the response to these compounds of the binding-protein-dependent and -independent transport systems. Moreover, the sensitivity of glutamine and β-methyl galactoside permease systems to uncouplers and to inhibitors of the Ca^{2+}, Mg^{2+}-dependent ATPase suggests that their uptake, like that of proline and of galactose through the galactose permease system, is driven by the energized state generated on ATP hydrolysis by the ATPase.

METHODS

Growth of Organism

S. typhimurium SASY28 (SA1889) (hem A$^-$, pro$^-$, pur E$^-$), originally isolated by Dr. A Sasarman (University of Montreal, Canada), was obtained from Dr. K. E. Sanderson (University of Calgary, Canada).

For studies of proline and glutamine transport the cells were grown aerobically at 37°C in the absence of 5-aminolevulinic acid in trypticase soy broth (Difco) containing 10 mg/liter adenine. In studies of galactose transport the cells were grown in a minimal salts medium (2) lacking sodium citrate and containing 0.5% (wt/vol) D-fructose, 0.5% (wt/vol) bactotryptone (Difco), 50 mg/liter proline and 10 mg/liter adenine. Galactose and β-methyl galactoside transport was induced by inclusion of 10 mM D-galactose or 2 mM D-fucose, respectively. The cells were harvested in the mid-exponential phase of growth, washed twice at 22–23°C with minimal salts medium or 50 mM Tris-HCl buffer, pH 7.3, containing 5 mM MgCl$_2$, and resuspended in the respective medium at a cell density giving 2.8 mg protein/ml.

Measurement of Transport

Uptake of [^{14}C] proline and [^{14}C] glutamine was measured as previously described (2). The incubation mixture to measure the uptake of [^{14}C] galactose and [^{14}C] β-methyl galactoside contained in a total volume of 0.1 ml, 0.01 ml cell suspension, 0.08 ml of either minimal salts medium or 50 mM Tris-HCl buffer, pH 7.3, containing 5 mM MgCl$_2$, and 100 μg/ml chloramphenicol. These buffers were used interchangeably since the rate of uptake was not noticeably affected by the nature of the buffer used. The energy source was added at the concentration indicated. The cells were preincubated without the energy source for 10 min at 23°C. The energy source was then added followed after 5 min by 10 μl of the radioactive transport solute diluted with unlabeled solute to give a final concentration of 150 μM (31 μM with β-methyl galactoside). In some experiments designed to measure the β-methyl galactoside permease system [^{14}C] galactose was added at a final concentration of 0.6 μM. Uptake was terminated by the addition of 3 ml 150 mM NaCl immediately followed by filtration of the mixture through a Millipore membrane filter (25 mm diameter; pore size, 0.45 μm). The cells were washed twice with 3 ml 150 mM NaCl. After drying, the filters were dissolved in Bray's scintillation fluid and the radioactivity measured with a Packard Tri-Carb liquid scintillation spectrophotometer, model 2425. L-[U-^{14}C] proline (290 Ci/mol) and D-[1-^{14}C] galactose (95 Ci/mol) were obtained from Amersham-Searle Corporation. L-[U-^{14}C] glutamine (235 Ci/mol) and [^{14}C] methyl β-D-galactopyranoside (8.12 Ci/mol) were purchased from New England Nuclear Corporation.

For inhibition studies the cells were preincubated at 23°C for 10 min in the minimal salts or Tris-MgCl$_2$ buffer with the indicated concentrations of inhibitor prior to the addition of the energy source. The 50 mM Tris-HCl buffer, pH 7.3, containing 5 mM MgCl$_2$ was always used for experiments with arsenate.

Determination of Intracellular ATP Concentration

These experiments were based on the conditions used in the transport assays except that the experiments were scaled up 10-fold. To a final volume of 1 ml of cells suspended at 0.24–0.34 mg protein/ml of the buffer various concentrations of the inhibitors were added, and the samples preincubated at 23°C for 10 min prior to the addition of the energy source (glucose or fructose). After 5 min incubation at 23°C the ATP was extracted from the cell suspension by the addition of 0.5 ml 1.4 M perchloric acid. In the experiments with 2,4-dinitrophenol, the cells were pelleted by centrifugation and the supernatant discarded. The cells were resuspended in 1 ml buffer and 0.5 ml perchloric acid was added as above. The mixture was agitated for 10 sec with a Vortex mixer and then kept at 0°C for 15 min. The cells were sedimented by centrifugation at 8,000 $\times$ g for 5 min and 1.0 ml of the supernatant removed for neutralization with 0.5 ml 0.72 M KOH in 0.15 M KHCO$_3$. The mixture was kept at 0°C for 30 min and then centrifuged to remove the precipitate of potassium perchlorate. From the supernatant 1.0 ml was removed and frozen in an ethanol-dry ice bath. The sample was allowed to thaw at room temperature and the freezing-thawing process repeated. This process resulted in the precipitation of further potassium perchlorate which was sedimented by centrifugation. Samples (10 μl) of the supernatant were assayed for ATP by the luciferin-luciferase method (11).

Measurement of Proton Translocation

The cells were harvested in the mid-exponential phase of growth, washed twice with 2 mM glycylglycine buffer, pH 6.8, containing 100 mM KCl and 5 mM MgCl$_2$, and suspended in the same buffer at a cell concentration of 2.3 mg protein/ml. Proton translocation was measured using 5 ml samples of the cell suspension as described before (3).

RESULTS

Transport in Cytochrome-Deficient Cells of S. Typhimurium SASY28

As shown on Fig. 1 cytochromes were not formed in S. typhimurium SASY28 unless the medium was supplemented with 5-aminolevulinic acid. The rate of oxidation of NADH and D-lactate by membrane particles prepared from cytochrome-deficient cells was negligible. Thus, in contrast to the results with cytochrome-containing cells, the uptake of D-galactose and of β-methyl D-galactoside (data not shown) could not be energized by

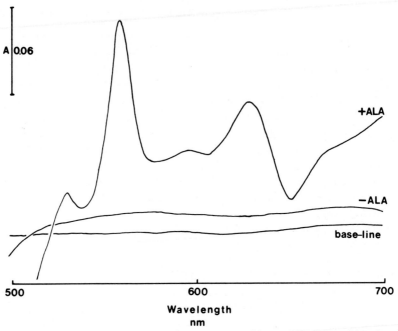

Fig. 1. Dithionite reduced minus oxidized difference spectra of intact cells of S. typhimurium SASY28 grown in the presence and absence of 5-aminolevulinic acid (ALA) (50 mg/liter). Concentration of cells, 0.15 g wet weight/ml.

2

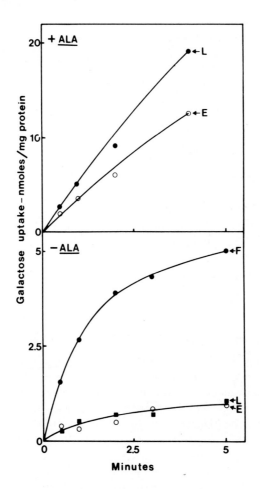

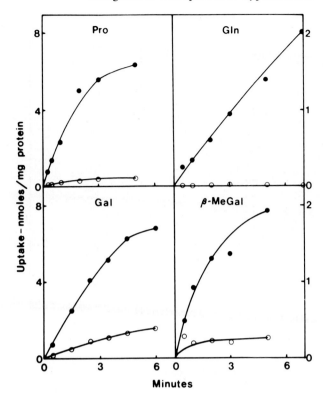

Fig. 3. Uptake of [^{14}C] proline (Pro), glutamine (Gln), galactose (Gal) and β-methyl galactoside (β-MeGal) by cytochrome-deficient cells in the presence of 10 mM glucose (Pro, Gln) or 20 mM fructose (Gal, β-MeGal) (•) or in the absence of exogenous substrates (○). Galactose and β-methyl galactoside transport was induced by growth in the presence of 2 mM D-fucose.

D-lactate in cytochrome-deficient cells (Fig. 2). However, ATP generated by glycolysis of glucose and fructose could support the uptake of proline, glutamine, galactose, and β-methyl D-galactoside in cytochrome-deficient cells (Figs. 2 and 3). Uptake of these solutes was linear with time for the first minute following addition to the cell suspension.

Effect of Inhibitors and Uncouplers on Uptake of Proline, Glutamine, and Galactose

The effect of inhibitors and uncouplers on the uptake of proline, glutamine, and galactose by cytochrome-deficient cells is shown on Figs. 4–6. Uptake of galactose was measured at 2 concentrations (0.6 μM and 150 μM). Kinetic studies with these cells showed that 2 transport systems for galactose with K_m values of 0.6 μM and 27 μM were present. These values can be compared to the K_m values of 1 μM and 50 μM determined for galactose uptake by the β-methyl galactoside and galactose permease systems, respectively, in transport mutants of S. typhimurium (7).

Fig. 2. Uptake of [^{14}C] galactose in cytochrome-containing (+ALA) and cytochrome-deficient cells (–ALA). Galactose transport was induced by growth in the presence of 10 mM D-galactose. The concentration of 5-aminolevulinic acid, where present, was 50 mg/liter. Uptake was energized by endogenous substrate (E), 20 mM D-lactate (L) or 20 mM fructose (F). Concentration of galactose, 150 μM.

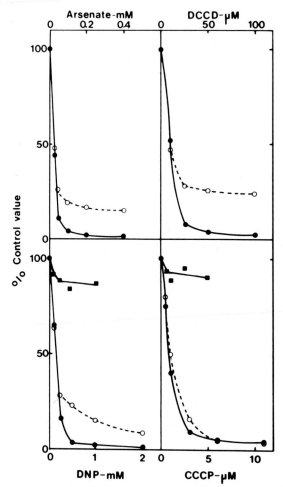

Fig. 4. Inhibition of the uptake of proline (●) and glutamine (○) in cytochrome-deficient cells by arsenate, DCCD, 2,4-dinitrophenol (DNP)and CCCP, and the effect of inhibitors on the level of cellular ATP (■). Uptake was energized by 10 mM glucose. The 100% control values for glutamine and proline uptake were 0.24 and 1.1 nmol/min/mg protein, respectively. The 100% ATP level was 2.63 nmol/mg protein.

Fig. 5. Inhibition of the uptake of 0.6 μM (○) and 150 μM (●) D-galactose in cytochrome-deficient cells by arsenate, DCCD, 2,4-dinitrophenol (DNP), and CCCP, and the effect of inhibitors on the level of cellular ATP (■). Uptake was energized by 20 mM fructose. The 100% control values at 0.6 μM and 150 μM galactose were 1.3–2.0 and 160 nmol/min/mg protein, respectively. The 100% ATP level was 5.2–5.6 nmol/mg protein.

Fig. 6. Effect of azide on the uptake of D-galactose (left) (○, 0.6 μM; ●, 150 μM), proline (●) and glutamine (○) (right), and on the levels of cellular ATP (■) in cytochrome-deficient cells. The uptake of galactose was energized by 20 mM fructose and that of proline and glutamine by 10 mM glucose. The 100% control values for 0.6 μM galactose, 150 μM galactose, proline, and glutamine were 1.4, 156, 2.7, and 0.61 nmol/min/mg protein, respectively. The 100% ATP levels were 6.3 (left) and 3.9 nmol/mg protein (right), respectively.

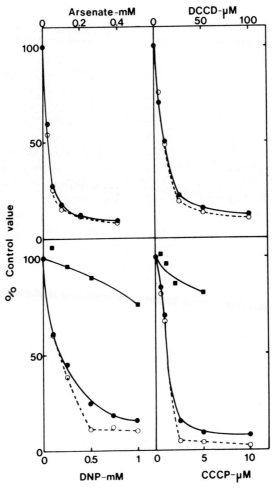

5

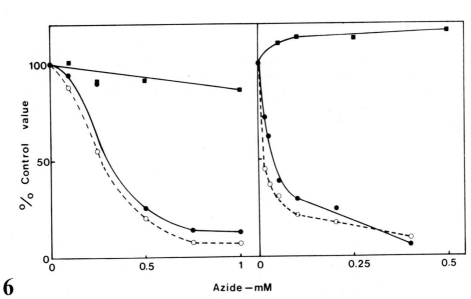

6

Azide – mM

Arsenate inhibits the formation of ATP by glycolysis. Since the energy source for transport in cytochrome-deficient cells is ATP generated by glycolysis, the inhibition of proline, glutamine, and galactose uptake by arsenate confirmed the dependency of the uptake of these solutes on the availability of ATP. To distinguish between the use of ATP through the ATPase or its use by an as yet unknown mechanism independent of the ATPase, the effects of azide and DCCD were examined (Figs. 4–6). These compounds inhibit the membrane-bound Ca^{2+}, Mg^{2+}-activated ATPase (12). Uptake of galactose, proline, and glutamine was inhibited by azide and DCCD suggesting that the ATPase was involved in the transport of these solutes. Azide had little effect on the cellular level of ATP in the cytochrome-deficient cells with fructose and glucose as energy sources (Fig. 6). The small increase in the level of ATP found with the latter substrate is consistent with the blocking by azide of the hydrolysis of ATP by the ATPase.

The uncouplers 2,4-dinitrophenol and CCCP inhibited the uptake of proline, glutamine, and galactose (Figs. 4 and 5). These uncouplers are believed to inhibit transport by dissipating the electrochemical gradient of protons (energized state) (4). However, Wilson (8) and Curtis (13) in interpreting the inhibitory effect of these compounds on shock-sensitive transport systems in wild-type E. coli have suggested that they affect these systems by causing dissipation of the ATP pool of the cell and not because they cause breakdown of the proton gradient. As can be seen from Figs. 4 and 5 concentrations of uncoupler which cause maximal inhibition of the uptake of proline, glutamine, and galactose produced only small changes (5–15%) in the level of cellular ATP. The effect of 2,4-dibromophenol was similar to that of 2,4-dinitrophenol. These results are consistent with the findings of Pavlasova and Harold (14) on the effect of uncouplers on the ATP pools of wild-type E. coli.

Proton Symport During Galactose Uptake

According to the chemiosmotic hypothesis (4) uptake of neutral solutes, such as galactose, driven by an electrochemical gradient of protons should occur with symport of protons. A proton gradient was generated by cytochrome-deficient cells of S. typhimurium SASY28 metabolizing endogenous substrates or following addition of fructose. Addition of D-galactose or D-fucose, which is taken up by both the β-galactoside permease and galactose permease systems (7), resulted in uptake of protons (Fig. 7). The uncoupler CCCP abolished proton uptake following addition of galactose (data not shown) or fucose. There was no consumption of protons following addition of β-methyl D-thiogalactoside since the permease for this glycoside had not been induced. Although the β-methyl galactoside permease was present, uptake of protons did not occur on addition of this galactoside. A similar result was recently reported for E. coli (10).

DISCUSSION

The use of cytochrome-deficient cells of S. typhimurium has enabled us to investigate the energization by ATP of the uptake of certain solutes without interference from respiratory-chain mediated energization. Moreover, these cells have low endogenous activity so that it is not necessary to pretreat the cells with an uncoupler to deplete the endogenous energy reserves. This procedure, which has been used by some workers, is obviously undesirable if the effects of uncoupling agents on transport are to be studied subsequently.

Cytochrome-deficient cells of S. typhimurium SASY28 will take up proline, glutamine, galactose, and β-methyl galactoside in the presence of ATP generated by the glycoly-

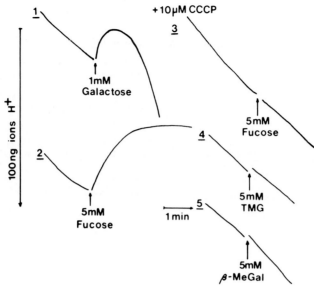

Fig. 7. Proton uptake following addition of solute (at the arrow) to cytochrome-deficient cells metabolizing endogenous substrates (1, 5) or 20 μM fructose (2–4). TMG: β-Methyl thiogalactoside; β-MeGal: β-methyl galactoside.

sis of glucose and fructose. Our principal finding is that the uptake of proline, glutamine, and of galactose used at 2 different concentrations to measure the β-methyl galactoside permease and the galactose permease system (8, 16), is inhibited by the uncouplers 2,4-dinitrophenol and CCCP, and by azide and DCCD, inhibitors of the Ca^{2+}, Mg^{2+}-activated ATPase. These results suggest that the uptake of all of these solutes in the cytochrome-deficient cells is driven by an electrochemical gradient of protons which is formed by hydrolysis of glycolytically generated ATP by the membrane-bound Ca^{2+}, Mg^{2+}-activated ATPase. Since the level of intracellular ATP is little affected by the inhibitors and uncouplers over a concentration range which drastically inhibits uptake, it is not likely that inhibition is caused by an effect on the ATP pool.

Our results do not agree with those obtained with cytochrome-containing E. coli. Thus, Berger and Heppel (5, 6) found that although proline transport was driven by the energized state generated by ATP hydrolysis through the Ca^{2+}, Mg^{2+}-activated ATP, glutamine transport was energized by a different, as yet unknown, mechanism. In contrast to the findings of Kerwar et al. (15), Wilson has suggested that both the galactose permease and the β-methyl galactoside permease systems are driven by a mechanism similar to that used for glutamine transport. However, Parnes and Boos (16) have suggested that uptake by the β-methyl galactoside permease in E. coli is driven by the energized state. The recent results of Henderson et al. (10) suggest that the proton gradient is involved in the galactose permease system although uptake of protons was not found with the transport of β-methyl galactoside.

Other workers have obtained results which are difficult to reconcile with the Berger-Heppel hypothesis. Thus, Plate et al. (17) found that colicin K, which appears to act by deenergization of the energized state of the membrane, inhibited the transport of both proline and glutamine in an ATPase mutant of E. coli. This suggests that glutamine and proline transport systems share a common element sensitive to colicin K. Bradbeer and

Woodrow (18) found that the shock-sensitive transport of vitamin B_{12} in E. coli was driven by the energized state. Lieberman and Hong (19) have isolated a temperature-sensitive mutant in the common element of shock-sensitive and shock-resistant systems. They have suggested that ATP might function as a regulatory effector which could direct the use of the energized state to the transport of solutes belonging to the shock-sensitive systems. This would explain the apparent requirement of these systems for ATP. Recently, Rhoads and Epstein (20) have found that the Trk A system for potassium transport in E. coli shows a need for both the energized state and ATP. Thus, it can be concluded from the results presented in this paper, and from those of other workers, that there is no clear-cut distinction between shock-sensitive and shock-resistant transport systems in their mechanism of energization.

ACKNOWLEDGMENTS

We wish to thank Dr. K. E. Sanderson for the gift of the organism used in this work and the Medical Research Council of Canada for financial support.

REFERENCES

1. Singh AP, Bragg PD: Biochim Biophys Acta 396:229, 1975.
2. Singh AP, Bragg PD: Biochim Biophys Acta 423:450, 1976.
3. Singh AP, Bragg PD: Biochim Biophys Acta 464:562, 1977.
4. Simoni RD, Postma PW: Annu Rev Biochem 44:523, 1975.
5. Berger EA: Proc Natl Acad Sci USA 70:1514, 1973.
6. Berger EA, Heppel LA: J Biol Chem 249:7747, 1974.
7. Postma PW: J Bacteriol 129:630, 1977.
8. Wilson DB: J Bacteriol 126:1156, 1976.
10. Henderson PJF, Giddens RA, Jones-Mortimer MC: Biochem J 162:309, 1977.
11. Stanley PE, Williams SG: Anal Biochem 29:381, 1969.
12. Roisin MP, Kepes A: Biochim Biophys Acta 275:333, 1972.
13. Curtis SJ: J Bacteriol 120:295, 1974.
14. Pavlasova E, Harold FM: J Bacteriol 98:198, 1969.
15. Kerwar GK, Gordon AS, Kaback HR: J Biol Chem 247:291, 1972.
16. Parnes JR, Boos W: J Biol Chem 248:4429, 1973.
17. Plate CA, Suit JL, Jetten AM, Luria SE: J Biol Chem 249:6138, 1974.
18. Bradbeer C, Woodrow ML: J Bacteriol 128:99, 1976.
19. Lieberman MA, Hong J-S: Arch Biochem Biophys 172:312, 1976.
20. Rhoads DB, Epstein W: J Biol Chem 252:1394, 1977.

Journal of Supramolecular Structure 6:399–409 (1977)
Molecular Aspects of Membrane Transport 261–271

Methods for Rapidly Altering the Permeability of Mammalian Cells

Leon A. Heppel and Nizar Makan

Section of Biochemistry, Molecular and Cell Biology, Cornell University, Ithaca, New York 14853

Various agents alter mammalian cells so that they rapidly become nonspecifically permeable to substances that ordinarily do not penetrate intact cells. Thus, toluene renders liver cells permeable to nucleotides and macromolecules. Tween 80 and Tween 60 act in similar fashion, and the effect is reversible. Dextran sulfate reversibly alters the permeability of Ehrlich ascites tumor cells, which offers a tool for studying the control of macromolecular syntheses and other processes. Brief exposure to external ATP alters the permeability of certain transformed mouse cells but not of untransformed cells. The effect of ATP is rapidly reversible.

Key words: permeability, detergents, ATP

The purpose of this review is to call attention to methods which have become available for altering the permeability of mammalian cells within a few minutes. We are particularly concerned with passive permeability and we are interested in agents which render the cell nonspecifically permeable to substances that ordinarily do not penetrate intact cells, such as nucleotides, actinomycin D, etc. Thus we have excluded papers which describe the stimulation or inhibition of normally occurring uptake systems by serum, hormones, or other agents. These matters have been reviewed by others (see, for example, Ref. 1). Only passing reference is made to conditions which affect the uptake of macromolecules; this has become a large and developing field (see, for example, Refs. 2 and 3). We are mainly concerned with the use of agents such as toluene, Tween 80, Tween 60, detergents, dextran sulfate, and ATP in increasing the passive permeability of animal cells. Agents whose effect is reversible are of particular interest to us because they make possible some interesting biological studies of control mechanisms. Normally impermeant cofactors can be inserted at will into living cells with certain agents to be discussed below.

TOLUENE

The bacterial literature contains many publications on the use of toluene to increase cell permeability. As an example, Moses and Richardson (4) were able to demonstrate semiconservative replication of DNA in toluenized E. coli, using labeled deoxyribonucleoside triphosphates, ATP, Mg^{2+} and K^+. Synthesis was temperature sensitive at the restrictive temperature in certain dna_{ts} E. coli mutants.

Received April 12, 1977; accepted May 17, 1977

Replicative synthesis showed a 20-fold stimulation by ATP. If 1% Triton X-100 was present as well, Moses observed that the cells also became permeable to macromolecules (5). The DNA repair reaction in bacteria treated in this way was inhibited by exogenous antibody to DNA polymerase I.

Toluene has been used in Deutscher's laboratory to render isolated individual liver cells permeable to charged molecules and macromolecules (6, 7). Liver cells in culture were treated with 7–9% toluene for 2 min at $0°C$. These cells became permeable to t-RNA and to $[^{14}C]$ATP. Exogenous t-RNA and $[^{14}C]$ATP entered the toluenized cells and reacted with internal tRNA nucleotidyl transferase and aminoacyl-tRNA synthetase. The reaction products then left the cell. Table I shows that at least 75% of the total synthetase activity for several amino acids remained within the cell after toluene treatment, while Table II indicates that the total units of aminoacyl-tRNA synthetase found after toluene treatment of cells was somewhat greater than that found by homogenization of an equivalent weight of intact liver.

These cells could be maintained for periods of up to 1 h and they appeared to be relatively intact upon electron microscopic examination. The toluene treatment did, however, extract a considerable fraction of membranal 5'-nucleotidase and cellular phospholipid. The cells, of course, were not viable.

DETERGENTS: TWEEN 80 AND TWEEN 60

Kay (8) reported that marked alterations in cell permeability were brought about by Tween 80 in Ehrlich-Lettré ascites cells. His work is of particular interest because he demonstrated that growth of these cells was normal after the treatment and they were viable as transplants. Thus, the effect was reversible. Increased permeability was demonstrated by uptake of the dye Lissamine green. Acid soluble nucleotides decreased by 50%, and amino acids were also removed by the Tween treatment. However, while levels of glutamine, aspartic acid, glycine, glutamic acid, leucine, and tyrosine were drastically reduced, the level of cellular serine was not altered. There was a reduction of $[^{14}C]$ formate incorporation into protein and DNA, while RNA synthesis was stimulated.

TABLE I. Location of Aminoacyl-tRNA Synthetases During Assay*

Amino acid	Supernatant	Pellet
	units/g cells	
Arginine	2.48	7.95
Lysine	2.20	5.51
Alanine	1.78	5.14
Threonine	2.62	7.50

*After toluene treatment the cells were suspended in 5 volumes of assay mixture without labeled amino acid and preincubated for 5 min at $37°C$. The cells were then centrifuged at $100 \times g$ for 2 min and the cell pellet was resuspended in 5 volumes of the same medium. Ten microliters of the cell and supernatant fractions were then assayed for aminoacyl-tRNA synthetase activities.

TABLE II. Comparison of Aminoacyl-tRNA Synthetase Activity in Cells Made Permeable by Toluene Treatment or Opened by Other Methods*

Amino acid	Homogenization of intact liver		Sonication of individual liver cells		Nonidet P-40 treatment of individual liver cells		Toluene treatment of individual liver cells	
	Super-natant	Pellet	Super-natant	Pellet	Super-natant	Pellet	Super-natant	Pellet
				units/g of liver				
Arginine	6.51	0.74	14.8	0.63	15.0	0.23	1.30	10.2
Lysine	9.08	0.13	11.3	0.19	13.3	1.60	1.21	10.32
Alanine	4.15	0.75	6.21	0.54	6.46	0.48	0.54	4.3
Threonine	7.63	0.86	11.9	0.49	12.6	0.98	1.19	10.1

*Isolated cells were suspended in 5 volumes of Buffer A containing 20% glycerol. Homogenization of intact liver (0.5 g) was done in a Dounce homogenizer using 10 strokes of a loose fitting pestle. Sonication of individual liver cells (0.5 ml) was performed using a Branson Sonifier cell disruptor at a setting of 1 for a total of 60 sec in 30-sec bursts with a 30-sec cooling period. Nonidet P-40 treatment involved incubating 0.2-ml aliquots of cells with 50 μl of a 10% Nonidet P-40 solution for 10 min in ice. Toluene treatment of individual liver cells was performed on a 0.2-ml aliquot as described earlier. All samples were centrifuged at 100 × g for 2 min, the pellets were resuspended in 0.2 ml of buffer, and both the supernatant and the pellet fractions were assayed for aminoacyl-tRNA synthetase activities.

Incorporation of $^{32}P_i$ into RNA was stimulated in both the nuclear and cytoplasmic fractions.

Kay reported an extraordinary stimulation of $^{32}P_i$ incorporation into phospholipids (two- to fourfold), in both nuclear and cytoplasmic fractions. Curiously enough, only phosphatidyl serine was labeled. He concluded from this that Tween 80 disrupts membrane structure and certain membrane components are reconstituted in the course of recovery. Tween 80 may affect other cell structures as well, since cardiolipins, which are mitochondrial constituents were completely lost as a result of the treatment with Tween 80.

In a very careful study, Malenkov et al. (9) incubated Ehrlich and hepatoma 22 cells with 1% Tween 60 in neutral phosphate buffer for 30—60 min. Tween made the cells susceptible to lysis by trypsin and they stained diffusely with neutral red instead of revealing the usual neutral red granules. When the treated cells were incubated with fluorescent albumin and washed, diffuse fluorescence of cytoplasm and nucleus was observed. The electrophoretic mobility of the cells decreased, as did their electrical resistance.

These changes were reversed within 2 hr at 37°C in the presence of a culture medium supplemented with 20% bovine serum. Thus, the treatment with Tween 60 made animal cells permeable to molecules as large as albumin, and the effect was reversible.

OTHER DETERGENTS

A number of investigators have used other surface active agents to increase the permeability of animal cells to normally impermeant molecules and ions. For example, Hodes et al. (10) found that a series of sodium alkyl sulfates and sulfonates, as well as a series of nonionic phenoxy polyoxyethylene ethanols promoted the penetration of a dye, nigrosin, through the plasma membrane of Ehrlich ascites cells. However, these were relatively dangerous reagents. Increased dye entrance occurred at concentrations only slightly lower than those necessary to produce membrane lysis.

Seufert (11) studied the effect of anionic, cationic, nonionic, and amphoteric surface active substances on synthetic lipid bilayers. All 4 types of detergents lower the membrane resistance and develop potentials to varying degrees, at least transiently. A mechanism for channel formation as a result of detergent-lipid interaction was proposed. This is an example of many publications in which the properties of lipid bilayers have been altered by a vast number of lipid and nonlipid materials. We cannot review this literature here.

GENETIC APPROACHES COUPLED WITH THE USE OF DETERGENTS

Genetic methods have been used to alter the permeability properties of mammalian cells. This does not fall within the province of this review, which is limited to procedures that rapidly alter the permeability to small molecules. However, in some cases the genetic change has been analyzed with the help of detergents and these matters will now be discussed.

A good example is provided by work from a group in Toronto (12–14) on the isolation of mutants resistant to colchicine, a drug that inhibits mitosis. In some cases the basis for resistance was a defect in the system responsible for uptake of the drug by the cells. Such permeability mutants showed reduced ability to take up ^{3}H-labeled colchicine into intact cells with no decrease in intracellular binding activity (12). In their studies, ethylmethanesulfonate was employed as the mutagen and colchicine resistant (CH^R) lines of stable phenotype were isolated from Chinese hamster ovary (CHO) cells. Successive single-step selections were performed for increasing resistance. The CH^R lines were resistant to other drugs including actinomycin D, vinblastine, and colcemid. The degree of cross-resistance was correlated with the degree of colchicine resistance; thus, the mutation appeared to be associated with a generalized decrease in nonspecific permeability.

Ling and Thompson (13) went on to show that Tween 80 potentiated the toxicity of colchicine, both in wild type Chinese hamster ovary cells and in the mutant. This was apparently due to a four- to sixfold stimulation in the rate of colchicine uptake. A similar potentiation had previously been shown by Riehm and Biedler (14) in the case of CHO cell lines resistant to actinomycin D. It should be noted that several workers have described actinomycin resistant lines (for example, Refs. 15, 16).

In a later study, Carlsen, Till, and Ling (17) performed a careful study of the kinetics of colchicine uptake into parental cells and drug-resistant CHO mutants. They concluded that colchicine permeates the cells by an unmediated process for various reasons, including: a) No substrate saturation was observed. b) Sulfhydryl reagents did not inhibit. This is hardly diagnostic, but a good many carrier-mediated systems are known to be sensitive to sulfhydryl reagents. c) No competition was noted for colcemid, a compound of quite similar structure. (This argument also is suggestive, but not conclusive.) d) Colchicine uptake is greatly stimulated by nonionic detergents and local anesthetics.

These authors believe that the detergents and local anesthetics act by causing an increase in membrane fluidity. It is of interest that the maximal rate of colchicine uptake obtainable with Tween 80 in the colchicine-resistant mutant was only half that of the parental line. Apparently part of the alteration associated with the genetic change to colchicine resistance cannot be reversed by the detergent.

USE OF OSMOTIC SHOCK

An interesting paper by Kaltenbach (18) describes the use of a form of osmotic shock to alter the permeability of Ehrlich ascites tumor cells. His procedure was as follows: The ascites tumor was collected in cold buffered saline, and washed by centrifuging twice at 200 X g. The ascites tumor cells were modified by suspending them in distilled water for 4 min and then adding an equal volume of double strength saline to restore isotonicity. The fraction of cells "modified" by this treatment was determined by counting those that took up the dye when suspended in 0.2% nigrosin solution, which stained them black. The degree of change was controlled by varying the time of exposure to distilled water.

The modified cells were able to glycolyze glucose; however, the formation of lactic acid became dependent upon the addition of glycolytic cofactors such as ATP and NAD. Presumably these nucleotides had leaked out of the shocked cells so that it was necessary to add them as supplements. Furthermore, the treated cells were able to convert fructose diphosphate to lactic acid. The control cells were unable to accomplish this conversion because normally a permeability barrier exists for fructose diphosphate and other phosphate esters.

DEXTRAN SULFATE

Several publications have appeared from Racker's laboratory on the use of dextran sulfate in order to increase the permeability of Ehrlich ascites tumor cells (19–21). In the paper of Schnolnick et al. (19) the effect of dextran sulfate and other sulfated polysaccharides on glycolysis was investigated. All of them inhibited aerobic glycolysis, probably by an effect on the cell surface. It appeared highly unlikely that materials of such high molecular weight and charge would permeate into the cells. The inhibition of glycolysis was reversed by AMP and P_i added during the assay of glycolysis. Presumably, dextran sulfate caused leakage of nucleotides. Consistent with this, the ATP level was quite low in dextran sulfate-treated cells, but greatly increased when AMP was present during the treatment with dextran sulfate.

In a later study, McCoy et al. (20) noted that dextran sulfate caused enhanced permeability to Rb^+ and increased staining by erythrosin B. Respiration was reduced and thus was reversed by P_i alone. In agreement with the earlier studies, restoration of glycolysis required both P_i and AMP. The damage induced by dextran sulfate was reversible, for the lesion was repaired by injection of dextran sulfate-treated Ehrlich cells into mice. The cells could also be repaired by incubating them with ascites fluid that had been heated for 7 min in a boiling water bath and then centrifuged to remove coagulated proteins. The repaired cells were able to carry out glycolysis without a supplement of AMP and were not stained by erythrosin B. The uptake of Rb^+ was also restored. Furthermore, while the dextran sulfate-treated cells showed stimulation of Ca^{2+} uptake by external ATP, this effect was abolished after repair with ascites fluid. The repair activity was composed of both dialyzable and nondialyzable factors (see below).

Exposure of Ehrlich ascites tumor cells to dextran sulfate also impaired the active transport of α-aminoisobutyric acid, and it rendered the cells permeable to sorbitol, (unpublished experiments carried out by M. Kasahara, P. C. Hinkle, and L. A. Heppel). Subsequent incubation with ascites fluid restored active transport and repaired the permeability barrier in dextran sulfate-treated cells. Ascites fluid was ineffective after dialysis or addition of ethylene glycol bis(β-aminoethylether)-N,N′-tetraacetic acid, suggesting the involvement of Ca^{2+} ions in repair. Treatment with $CaCl_2$ and glucose under specific conditions restored the transport activity of dextran sulfate-treated cells as effectively as was the case for ascites fluid. McCoy et al. (20) were able to confirm the observation that ethylene glycol bis(β-aminoethylether)-N,N'- tetraacetic acid abolished the repair activity of ascites fluid and agreed that Ca^{2+} is one of the dialyzable components required for the repair process. However, they observed that Ca^{2+} alone was not effective, Ca^{2+} plus dialyzed ascites fluid gave partial repair, and undialyzed ascites fluid was optimal for the repair process.

A major purpose in developing methods for changing permeability to small molecules was to study the regulation of metabolic processes under conditions where general cell organization was not greatly disrupted. Accordingly, McCoy and Racker (21) examined protein synthesis in dextran sulfate-treated ascites tumor cells. They determined that a suitable K^+/Na^+ ratio was needed for incorporation of [^{14}C] valine into protein in the treated cells. This was expected in view of the known requirement for K^+ observed in cell-free systems. The process was inhibited by rotenone, which blocks mitochondrial respiration and ATP synthesis, which was another expected finding. However to overcome the rotenone inhibition both inorganic phosphate and either glucose or glucose-6-phosphate were required. This was unexpected, since there is sufficient residual phosphate present in dextran sulfate-treated cells to support respiration. The virtually complete dependency on added P_i was attributed to the accumulation of phosphorylated intermediates in the presence of glucose, which depletes residual intracellular P_i and nucleoside triphosphates.

Quercetin, an inhibitor of the Na^+, K^+-ATPase, reduced the incorporation of [^{14}C] valine into protein in intact ascites tumor cells in a high Na^+ medium, but had little or no effect in dextran sulfate-treated cells in a high K^+-low Na^+ medium. These findings suggest a relationship between protein synthesis and operation of the Na^+,K^+-ATPase. The Na^+,K^+-ATPase is held to control the intracellular concentration of P_i and adenine nucleotides.

Experiments of this sort demonstrate the usefulness of treatments that alter permeability in the analysis of biosynthetic processes.

TREATMENT OF ANIMAL CELLS WITH EXTERNAL ATP

A very substantial literature has accumulated in which various effects of exogenous ATP on animal cells are described. In most cases these effects are ascribed to changes in permeability induced by ATP. The effects are usually highly specific and are not obtained by nonspecific chelating agents nor by other nucleoside triphosphates.

Effect of External ATP on Suspensions of TA_3 Ascites Tumor Cells

Stewart, Gasic, and Hempling (22, 23) reported that when exogenous ATP was added to suspensions of TA_3 ascites tumor cells in Ca^{2+} and Mg^{2+} free media, a significant increase in cell volume could be measured, which was reversible on addition of divalent cations. The effect was temperature sensitive and highly specific for ATP. An effect was seen with 0.2 mM ATP and it saturated at about 1 mM. In a later paper, Hempling, Stewart, and Gasic (24) studied this phenomenon further and observed that 1 mM ATP caused a dramatic loss of K^+ and Na^+ down their respective concentration gradients.

These authors devised a very specific theory to explain their data. They believed that ATP produced a major change in the passive permeability of the membrane for these ions and suggested that the effect was due to a response of a contractile protein in the membrane to ATP, creating a hydrophilic passage for ions. Other strains of Ehrlich cells did not show this response to ATP. They suggested that differences in the outer coat, in the mucopolysaccharide layer, might provide a barrier for ATP in some strains and not in others.

Effect of External ATP on HeLa Cells

Aiton and Lamb (25) reported that exogenous ATP caused a 15-fold increase in efflux and influx of K^+. The ATP-stimulated K^+ efflux was transient, reaching a maximum within 1–2 min and declining with a half time of about 4 min. It is interesting that only 5×10^{-6} M ATP was required to produce half maximal stimulation of both influx and efflux, whereas concentrations near 1 mM were used in studies with mouse cell lines. Removal of external K^+ reduced the ATP-stimulated efflux by 60–70%, suggesting a large $K^+:K^+$ exchange component. If a second application of ATP was made, the response was less marked; after 6 h it was only 10% of the initial response. However, after 24 h the cells recovered their sensitivity to ATP so that the K^+ efflux was 80% as great as the initial response.

Trams (26) observed that 6×10^{-5} M ATP in the medium caused an immediate fivefold increase in efflux of $^{42}K^+$ from NN astrocytes. At suitable doses this effect of ATP was also seen in the N-18 neuroblastoma, HeLa cells, KB cells, glioblastoma GL-26, and L-929 fibroblasts.

Effect of External ATP on Ca^{2+} Accumulation by Chick Embryo Fibroblasts

Perdue (27) reported that cultured chick embryo fibroblasts accumulate Ca^{2+} in the presence of Mg^{2+} and ATP. The effect of ATP saturated at 3 mM and the optimal concentration of Mg^{2+} was 6 mM. The stimulation of uptake was highly specific: other nucleoside triphosphates were inactive. The uptake of Ca^{2+} was inhibited by Mn^{2+}, mersalate, oligomycin, and hydroxylamine. Perdue suggested that the function of this Ca^{2+} uptake may be to control motility in ameboid-like cells.

Miscellaneous Effects of External ATP

Many other effects of external ATP have been reported and a few of these will now be mentioned. Rorive and Kleinzeller (28) found that exogenous ATP affects the water and ionic contents of the cells of kidney tubules by an interaction with the cell membrane and Ca^{2+} ions. Exogenous ATP alters cell adhesion, aggregation, and movement in fibroblasts (29, 30). It changes ionic fluxes in mast cells (31) and it inhibits (H^+) secretion in the frog gastric mucosa (32).

Chang and Cuatrecasas observed that addition of only 5–50 μM ATP to isolated fat cells inhibits insulin-stimulated glucose oxidation under conditions that measure glucose transport (33). ATP inhibits the insulin-stimulated rate of 3-O-methylglocose transport. The authors speculate that their effect involves selective membrane phosphorylation.

Interesting studies have been carried out with dog red blood cells, which undergo a rapid increase in Na^+-K^+ permeability and an alteration in physical properties when exposed to at least 0.2 mM ATP (34). The effect is reversible on washing the cells. Other nucleotides and chelators are inactive, and the effect is prevented by Ca^{2+} and Mg^{2+} ions. Another study (35) reports that ATP causes a greater than eightfold increase in Na^+ influx.

It is worth mentioning that exogenous ATP has been postulated as a neurotransmitter (36).

Effect of Exogenous ATP on Untransformed and Transformed Mouse Cells

Rozengurt and Heppel (37) observed that external ATP caused a rapid and many-fold increase in the permeability of transformed cells to p-nitrophenylphosphate. The effect was specific for ATP and was not found for other nucleoside triphosphates, nor was it obtained by the use of EDTA, inorganic pyrophosphate, or other chelating agents. This great increase in permeability was observed for 3T6, SV3T3, and PY3T3 cells, but not for 3T3 cells nor for mouse embryo secondary cultures. It was rapidly reversed when the cell cultures in petri dishes were transferred to serum-free growth medium. The effect of ATP was noted in Tris-saline or HEPES-saline mixtures buffered at pH 7.7–8.2. Magnesium ions did not interfere as long as the ratio of Mg^{2+} to ATP did not greatly exceed 1:1.

In a later study (38) it was found that external ATP produced an extensive and non-specific permeability change in transformed cells leading to a depletion of intracellular pools labeled with $[^3H]$ uridine, $[^3H]$ adenosine, ^{86}Rb, or $[^3H]$-2-deoxyglucose. These pools consisted of over 90% phosphate esters which ordinarily do not penetrate the membrane barrier of mouse cells. When 3T6, SV3T3, and PY3T3 cells were exposed to as little as 0.2–0.3 mM ATP in Tris-saline, HEPES-saline, or phosphate-saline for several minutes, there resulted a 20–30-fold increase in rate of efflux of the acid-soluble uridine nucleotide pool. Little or no stimulation was seen with resting or growing, untransformed 3T3 cells or with secondary cultures of mouse embryo fibroblasts. Again, the presence of Mg^{2+} was allowed as long as the ratio of Mg^{2+} to ATP did not greatly exceed 1:1. The increase in permeability produced by ATP was highly specific; it was not obtained with GTP, CTP, UTP, ADP, inorganic pyrophosphate, EDTA, and other chelating agents, cAMP, and various analogues of ATP. The effect showed a pH optimum at about 8.2 and was temperature dependent. Stimulation of efflux persisted after removal of ATP. This fact made it possible to determine that the enhanced rate of efflux was as rapid at $20°C$ as at $37.5°C$, in contrast to the effect of temperature on ATP activation itself. This abnormal change in permeability was readily reversed, within 3 min, by incubating the cultures with serum-free growth medium or with neutral Tris-saline containing Ca^{2+} and Mg^{2+}. In complete growth medium the acid-soluble pools were restored within 1 hr, and the treatment with ATP could be repeated. When cells that had lost their acid-soluble pools were incubated once more in Dulbecco's serum-containing medium, they grew and divided at a normal rate.

Makan and Heppel, in unpublished work, have demonstrated the usefulness of this technique in studies of the control of glycolysis and of the hexose monophosphate shunt in transformed cells. The acid-soluble pools were nearly completely depleted so that glycolysis became almost totally dependent on the addition to the medium of glucose, P_i, ADP, and NAD^+ in order to obtain a rate equal to that of untreated cells. Phosphorylated intermediates, such as glucose-6-phosphate and fructose-1,6-diphosphate served as sub-strates for lactic acid formation in ATP-treated cultures of transformed cells but were in-active when ATP was omitted. Table III describes experiments with citrate, a well-known inhibitor of glycolysis in extracts. It has no effect on control 3T6 cells, but after treat-ment with ATP, citrate inhibits glycolysis. In the case of the nontransformed, 3T3 cells, citrate has no effect on glycolysis in control as well as in ATP-treated cultures. Table IV demonstrates that in ATP-treated, transformed, 3T6 cultures both NADP and glucose-6-phosphate are able to enter the cells and form NADPH. The reaction is catalyzed by in-tracellular glucose-6-phosphate dehydrogenase; release of enzyme into the supernatant solution was excluded.

Effect of ATP on Ca^{2+} Uptake by Ehrlich Ascites Cells

Landry and Lehninger (39) observed that extracellular ATP supported Ca^{2+} uptake by Ehrlich ascites cells. They considered the possibility that ATP is effective only because it leaks through damaged cells. However, in view of the many reported effects of ATP on intact cell preparations they also presented the alternative possibility, namely "that extracellular ATP may have two effects: one to increase membrane permeability of the cell to both Ca^{2+} and ATP, and the other to supply energy during its oligomycin-sensitive hydrolysis in the mitochondria for the transport of cytosolic Ca^{2+} into the mitochondrial matrix."

TABLE III. Effect of Citrate on Rate of Glycolysis of ATP-treated Normal and Transformed Cells*

Additions	Normal cells (3T3)		Transformed cells (3T6)	
	−ATP	ATP treated	−ATP	ATP treated
	(lactic acid formed, nmoles/ml/10 min)			
(a) glucose, P_i, ADP	104	110	370	252
(b) glucose, P_i, ADP + 5 mM citrate	116	92	308	181
(c) glucose, P_i, ADP + 10 mM citrate	98	106	349	78

*Swiss mouse 3T3 (normal) or 3T6 (transformed) cells were subcultured into 33-mm Nunc dishes in Dulbecco's modified Eagle's medium containing 10% fetal calf serum. In the late log or early confluent phase, the cell cultures were washed and treated with 1 ml of Medium A plus or minus 0.5 mM ATP. After 10 min at 37.5°C, the medium was replaced by Medium A minus ATP but containing the appropriate additions as shown. Further incubation was carried out for 10 min at 37.5°C. The supernatant fluid was then removed and assayed for lactate. Medium A contains: Tris-HCl, 0.1 M; NaCl, 0.05 M; $CaCl_2$, 50 μM; Dextran 500, 5 mg per ml; pH (at 23°C) 8.2.

TABLE IV. ATP-Induced Permeability to Glucose-6-Phosphate and $NADP^+$ and Subsequent Effect on the Hexose Monophosphate Shunt*

	Treatment	OD_{340} of supernatant
(A)	Controls: (did not receive ATP treatment in Stage I)	
	Stage II: Buffer A + 50 mM glucose-6-phosphate + 0.4 mM $NADP^+$	0.161
(B)	Cultures permeabilized with ATP in Stage I	
	Stage II: (a) Buffer A + 50 mM glucose-6-phosphate + 0.4 mM $NADP^+$	2.581
	(b) Buffer A only	0.055
	(c) Buffer A + 50 mM glucose-6-phosphate	0.083
	(d) Buffer A + 0.4 mM $NADP^+$	0.136
(C)	Supernatant of ATP-treated cultures: Stage I only	0.031
(D)	Reagent Blank (Buffer A + 50 mM glucose-6-phosphate + 0.4 mM $NADP^+$)	0.048

*Swiss mouse 3T6 cells were subcultured into 33-mm Nunc dishes in Dulbecco's modified Eagle's medium containing 10% fetal calf serum. In the late log or early confluent phase, the cell cultures were washed and treated with 1 ml of medium A plus 0.5 mM ATP. After 10 min at 37.5°C (Stage I) the medium was replaced by 1 ml medium A minus ATP but containing the appropriate additions as shown. Further incubation was carried out for 10 min at 37.5°C (Stage II). The supernatant fluid was then removed and the amount of NADPH that was formed was measured at 340 nm.

Resch (40) carried out similar experiments, and she prefers the first possibility, namely that ATP enters damaged cells, as mentioned by Landry and Lehninger. She also observed stimulation of Ca^{2+} uptake by ATP; CTP, UTP, and ADP were ineffective. However, she noted that the ATP-stimulated uptake, as well as basal uptake, increased with storage of the cells over one to several days. The stimulation of freshly harvested cells by ATP was often very small. The effect of aging for 1–3 days was reproduced by heating them for a short while at $30°C$. In view of these observations one must be careful in interpreting the experiments just cited. Another point to keep in mind is that different laboratories have used different strains of Ehrlich ascites cells; Stewart, Gasic, and Hempling (21) reported that ATP caused a dramatic loss of K^+ and gain of Na^+ only in one out of several strains of Ehrlich cells.

Effect of External ATP on Isolated Chromaffin Granules

Pollard et al. (41) observed that isolated secretory vesicles from the adrenal medulla, known as chromaffin granules, release their contents when exposed to Ca^{2+},Mg^{2+}-ATP and high levels of chloride ion. They postulate that the mechanism involves a chloride gradient, and anion entry resulted from Mg^{2+}-ATP-mediated changes in transmembrane potential. At $37°C$, this anion permeation step resulted in release of vesicle contents. The effect was specific for ATP, although AMPPNP was also active. Reagents such as SITS and pyridoxal phosphate, which block anion movement across red cell membranes, also blocked release of epinephrine and protein from these granules. They concluded that release from the granules occurred as a result of exposure to Ca^{2+}, followed by Mg^{2+}-ATP, evoked anion entry through a specific anion channel.

How Might ATP Become Available at the External Surface of Cells?

Since only small concentrations of ATP occur in the extracellular space the question arises whether any of the effects of exogenous ATP described in this review have physiological significance. Several possibilities for providing extracellular ATP have been advanced. Agren and Ronquist (42) have described the formation of extracellular ATP by tumor cells. Evidence also exists that ATP may be translocated from the cytosol to the exterior of the cell (26, 28). The existence of nucleotide pyrophosphatases (43), ectoATPase (44), and ATP requiring protein kinases (45) on the surface of animal cells almost makes it obligatory that some mechanism for providing external ATP should exist.

GENERAL CONCLUSIONS

This short review makes no pretense of completeness. Our purpose is simply to call attention to some interesting papers which are concerned with relatively rapid alterations in permeability of animal cells. There are suggestions in some instances that these rapid changes in permeability may operate physiologically as important control mechanisms. In other cases the main emphasis has been to demonstrate a technique for temporarily altering cell permeability so that small molecule pools can be depleted and impermeant molecules in any desired concentration can be inserted into the cell. In several cases the procedure for increasing permeability does not alter cell viability or subsequent rate of growth. Thus, the opportunity arises for taking a fresh look at control mechanisms for glycolysis, macromolecule syntheses, and other processes. One may compare the results of studies with reconstituted systems with investigations in which internal cell organization is disrupted as little as possible.

REFERENCES

1. Pardee AB, Rozengurt E: In Fox CF (ed): "Biochemistry of Cell Walls and Membranes." London: Butterworths (1975) pp 155-185.
2. Ryser HJ-P: Science 159:390, 1968.
3. Petitpierre-Gabathuler M-P, Ryser HJ-P: J Cell Sci 19:141, 1975.
4. Moses RE, Richardson CC: Proc Natl Acad Sci USA 67:674, 1970.
5. Moses RE: J Biol Chem 247:6031, 1972.
6. Hilderman RH, Deutscher MP: J Biol Chem 249:5346, 1974.
7. Hilderman RH, Goldblatt PJ, Deutscher MP: J Biol Chem 250:4796, 1975.
8. Kay ERM: Cancer Res 25:764, 1965.
9. Malenkov AG, Bogatyveva SA, Bozhkova VP, Modjanova EA, Vasiliev Ju M, Exp Cell Res 48:307, 1967.
10. Hodes ME, Palmer CG, Warren A: Exp Cell Res 21:164, 1960.
11. Seufert WD: Nature (London) 207:174, 1965.
12. Till JE, Baker RM, Brunette DM, Ling V, Thompson LH, Wright JA: Fed Proc Fed Am Soc Exp Biol 32:29, 1973.
13. Ling V, Thompson LH: J Cell Physiol 83:103, 1974.
14. Riehm H, Biedler JL: Cancer Res 32:1195, 1972.
15. Goldstein MN, Hamm K, Amrod E: Science 151:1555, 1966.
16. Kessel D, Bosmann HB: Cancer Res 30:2695, 1970.
17. Carlsen SA, Till JE, Ling V: Biochim Biophys Acta 455:900, 1976.
18. Kaltenbach JP: Arch Biochem Biophys 114:336, 1966.
19. Scholnick P, Lang D, Racker E: J Biol Chem 248:5175, 1973.
20. McCoy GD, Resch RC, Racker E: Cancer Res 36:3339, 1976.
21. McCoy GD, Racker E: Cancer Res 36:3346, 1976.
22. Gasic G, Stewart C: J Cell Physiol 71:239, 1969.
23. Stewart CC, Gasic G, Hempling HG: J Cell Physiol 73:125, 1969.
24. Hempling HG, Stewart CC, Gasic G: J Cell Physiol 73:133, 1969.
25. Aiton JF, Lamb JF: J Physiol 14P, 1975.
26. Trams EG: Nature (London) 252:480, 1974.
27. Perdue JF: J Biol Chem 246:6750, 1971.
28. Rorive G, Kleinzeller A: Biochim Biophys Acta 274:226, 1972.
29. Jones BM: Nature (London) 212:362, 1966.
30. Knight VA, Jones BM, Jones PCT: Nature (London) 210:1008, 1966.
31. Dahlquist R, Diamont B, Kruger PG: Int Arch Allergy 46:655, 1974.
32. Sanders SS, Butler CF, O'Callaghan J, Rehm WS: Am J Physiol 230:1688, 1976.
33. Chang K-J, Cuatrecasas P: J Biol Chem 249:3170, 1974.
34. Parker JC, Snow RL: Am J Physiol 223:888, 1972.
35. Romualdez A, Volpi M, Sha'afi RI: J Cell Physiol 87:297, 1976.
36. Burnstock G: In Iverson LL, Iverson SD, Snyder SH (eds): "Handbook of Psychopharmacology." New York: Plenum Publishing Corporation, 1974, vol 5.
37. Rozengurt E, Heppel LA: Biochem Biophys Res Cummun 67:1581, 1975.
38. Rozengurt E, Heppel LA: J Biol Chem (In press).
39. Landry Y, Lehninger AL: Biochem J 158:427, 1976.
40. Resch RC: Undergraduate Honors Thesis, Cornell University, 1974.
41. Pollard HB, Zinder O, Hoffman PG, Nikodejevic O: J Biol Chem 251:4544, 1976.
42. Agren G, Ronquist G: Acta Physiol Scand 75:124, 1969.
43. Evans WH: Nature (London) 250:391, 1974.
44. Ronquist G, Agren G: Cancer Res 35:1402, 1975.
45. Mastro AM, Rozengurt E: J Biol Chem 251:7899, 1976.

Journal of Supramolecular Structure 6:411–417 (1977)
Molecular Aspects of Membrane Transport 273–279

L-Arabinose Transport and the L-Arabinose Binding Protein of Escherichia coli

Robert W. Hogg

Department of Microbiology, School of Medicine, Case Western Reserve University, Cleveland, Ohio 44106

The active accumulation of L-arabinose by arabinose induced cultures of Escherichia coli is mediated by 2 independent transport mechanisms. One, specified by the gene locus *araE*, is membrane bound and possesses a relatively "low affinity." The other, specified in part by the genetic locus *araF*, contains as a functional component the L-arabinose binding protein and functions with a "high affinity" for the substrate. The L-arabinose binding protein has been purified, partially characterized, crystallized, and sequenced.

Key words: L-arabinose, transport, binding proteins, sequence

L-arabinose is a pentose which can serve as the sole source of carbon and energy for the enteric bacteria Escherichia coli. The transport proteins and enzymes engaged in the utilization of arabinose are inducible (by arabinose) and regulated in a positive manner by the product of the gene *araC* (Fig. 1, Ref 1).

The active accumulation of arabinose by induced E. coli cells is mediated by 2 genetically distinct transport systems. One, the product of the gene *araE*, was described by Novotny and Englesberg (2) and will be referred to as the "low-affinity" L-arabinose transport system. The second transport system contains as a functional component the L-arabinose binding protein, the product of the gene *araF*, and represents the "high-affinity" L-arabinose transport system. *AraE*, at 61 minutes, and *araF*, tentatively located at 45 minutes on the *E. coli* chromosomal map, are both regulated in a positive manner by the *araC* gene product which is located adjacent to the contiguously linked arabinose operon *araD, A,B,I,O* at 1 minute (3, 1).

"Low-Affinity," *araE*, L-arabinose Transport

Novotny and Englesberg first described the active accumulation of L-arabinose by E. coli (2). Accumulation of arabinose was observed to be an energy dependent, inducible function with a temperature optimum of 25°C. The K^{en} (K_m of entry) for L-arabinose was determined to be 1.25×10^{-4} M and D-fucose, D-xylose, and D-galactose were found to competitively inhibit L-arabinose uptake. The accumulated arabinose was recovered

Received April 1, 1977; accepted June 6, 1977

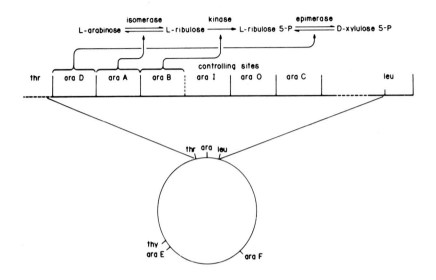

Fig. 1. The L-arabinose operon of Escherichia coli

unmodified from the cytoplasm of the cell. Mutants defective in L-arabinose transport were isolated and described by Isaacson and Englesberg (4). The transport defect was designated *araE* and was observed to cotransduce with the *thyA* locus.

We have since determined that "low-affinity" L-arabinose uptake capacity is retained by membrane vesicles prepared from induced E. coli cells (unpublished). Vesicles prepared from *araE* mutant strains do not demonstrate this capacity. Energization of this transport function appears to be via a chemiosmotic mechanism as described by Henderson (5). "Low-affinity" transport can thus be concluded to represent a membrane-bound *araE* gene product. A partial purification of this protein has been achieved using double isotope techniques and membrane solubilization with Brij 36T (unpublished).

"High-Affinity," *araF*, L-arabinose Transport

The L-arabinose binding protein was isolated from E. coli B by Hogg and Englesberg (6) and from E. coli K12 by Schleif (7). The binding protein was implicated in L-arabinose transport when it was observed that the specificity of ligand binding by the purified protein and the pattern of inhibition of binding by various arabinose analogues were identical to those of the in vivo transport process. Mutations in all genes specifying L-arabinose metabolic enzymes (L-arabinose isomerase, *araA;* L-ribulo kinase, *araB;* and L-ribulose-5-phosphate-4-epimerase, *araD*) did not alter in vivo transport or in vitro binding (6). In addition, Schleif isolated a mutant of E. coli K12 which demonstrated reduced transport capacity and also a lower content of binding protein than the parental strain (7).

The observation that *araE* mutations exerted no apparent effect on the binding protein led to a search for binding protein mutants and a reevaluation of L-arabinose uptake kinetics. Non-*crm*-producing binding protein mutants were obtained using the antibody plate assay described by Hogg (8). This technique permits rapid screening of large numbers

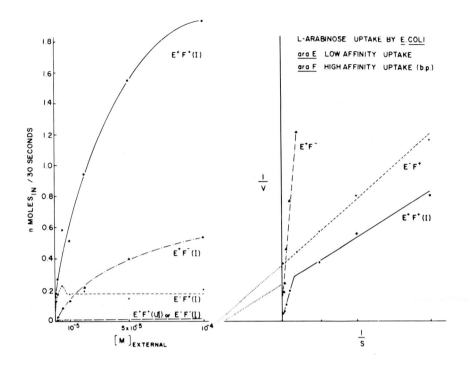

Fig. 2. The kinetics of L-arabinose transport in Escherichia coli. The reciprocal plot on the right panel represents the same data plotted over the same concentration range and units used in the figure on the left.

of suspected mutants by growing such mutants on plates containing arabinose and antibody directed against the binding protein. Subsequent lysis with chloroform and toluene releases cell contents and crm+ or crm− colonies are readily distinguished after 24-h incubation at 4°C. A detailed kinetic analysis of induced cells which had been extensively washed indicated that the kinetics of L-arabinose uptake were biphasic (Fig. 2). Mutant strains defective in araE and/or araF indicate that "low-affinity" L-arabinose uptake is associated with araE and demonstrates a K^{en} of 1×10^{-4} M. "High-affinity" uptake was associated with araF and the K^{en} was determined to be 1×10^{-6} M. Transport defects could be transferred, araE cotransducing with thymine, araF cotransducing with histidine. It was also noted that the strain used by Novotny and Englesberg in their earlier studies possessed only "low-affinity," araE, transport. This condition probably facilitated the later isolation of araE mutants from the same genetic background by Isaacson and Englesberg.

The fact that the uptake of araE+ araF− and araE− araF+ strains is not additive to the the uptake observed in araE+ araF+ strains suggests that although both systems are capable of independent function, a synergistic effect occurs when both are present. Rotman has reported binding-protein-independent galactose uptake by the methylgalactoside system (9, 10); however, this uptake is low and appears restricted to components of a binding-protein-mediated complex. In our case 2 apparently independent systems are capable of interaction.

Transport studies conducted on cells subjected to osmotic shock (11) indicated that "low-affinity" transport capacity was retained by shocked cells and "high-affinity" transport functions were lost.

PURIFICATION AND PROPERTIES OF THE L-ARABINOSE BINDING PROTEIN

Cell-free extracts or osmotic shock supernatants of induced E. coli cells served as an abundant source of the L-arabinose binding protein. E. coli *araA39,* a strain containing a nonsense mutation in the isomerase gene, produces approximately 2 mg of binding protein protein per gram wet weight of cells. The procedure for purification, from cell-free extracts, involves passing the supernatant of a 50% ammonium sulfate fractionation through a 0.01 M DEAE column. The eluant of this column is concentrated and further fractionated by column chromatography on DEAE (0.001 M KPO_4, pH 7.8, gradient from 0 to 0.03 M KCl). This procedure separates the L-arabinose binding protein from the D-galactose binding protein. Arabinose binding material was concentrated and further purified by column chromatography on Sephadex G75 equilibrated and eluted with 0.01 M KPO_4, pH 7.8 (10). The purified protein can be crystallized from 2-methyl-2, 4-pentanediol. The molecular weight of the purified protein is 33,300.

Sedimentation velocity studies and optical rotatory dispersion analysis in the presence and absence of arabinose indicated that no major structural reorientation occurred when arabinose was bound. Arabinose binding did result in minor structural rearrangements as indicated by a 5-nm blue shift in the fluorescence emission spectra suggesting that some tryptophan residues enter a more hydrophobic environment when arabinose is bound Arabinose was also observed to protect some tryptophan residues from N-bromosuccinimide oxidation (12). Sulfhydryl reactive agents were found to eliminate binding capacity; however, this inhibition was completely reversible for those agents which can be inactivated by thiols such as 2-mercaptoethanol (unpublished). Binding of arabinose was uneffected by ionic strengths between 0 and 1.5 M KCl and occurred over a pH range of 6.4–9.5.

Studies of this nature suggest that a tryptophan residue or residues is involved in or near the L-arabinose binding site and the single cysteine is required to maintain an active configuration. The insensitivity of binding to ionic strength and pH suggest a minimum role for charge interactions in the binding process; however, the in vivo environment may provide considerably different conditions.

SEQUENCE ANALYSIS OF THE L-ARABINOSE BINDING PROTEIN

The complete amino acid sequence of the L-arabinose binding protein has been determined (13). The protein contains 306 amino acid residues in a single peptide chain and is presented on Fig. 3. The sequence of the protein contains few distinguishing features. The single cysteine residue is located at position 64 and a highly charged region, -Arg-Arg-Arg-, occurs at residues 149, 150, and 151. The sequence analysis and the crystallographic analysis (18) will provide a detailed molecular model of the L-arabinose binding protein.

BINDING PROTEIN HOMOLOGY

Earlier studies by Parsons and Hogg have indicated that antibody directed against the L-arabinose binding protein of E. coli will form an immunoprecipitate with the D-

```
GLU -  asn -  leu -  lys -  leu -  gly -  phe -  leu -  val -  lys -
gln -  pro -  glu -  glu -  pro -  trp -  phe -  gln -  thr -  glu -  20
TRP -  lys -  phe -  phe -  asp -  lys -  ala -  gly -  lys -  asp -
leu -  gly -  phe -  GLU -  VAL -  ile -  lys -  ile -  ala -  val -  40
pro -  asp -  gly -  GLU -  LYS -  thr -  leu -  asn -  ala -  ile -
asp -  ser -  leu -  ala -  ala -  ser -  gly -  ala -  lys -  gly -  60
phe -  val -  ile -  cys -  thr -  pro -  ASP -  PRO -  lys -  leu -
gly -  ser -  ala -  ile -  val -  ala -  lys -  ala -  ARG -  gly -  80
tyr -  asp -  MET -  lys -  val -  ile -  ala -  val -  asp -  asp -
gln -  phe -  val -  asn -  ala -  lys -  gly -  lys -  pro -  MET - 100
asp -  thr -  val -  pro -  leu -  val -  MET -  MET -  ala -  ala -
thr -  lys -  ile -  gly -  glu -  ARG -  gln -  gly -  gln -  glu - 120
leu -  TYR -  lys -  glu -  MET -  gln -  lys -  ARG -  gly -  TRP -
asp -  val -  lys -  glu -  ser -  ala -  val -  MET -  ala -  ile - 140
thr -  ala -  asn -  glu -  leu -  asp -  thr -  ala -  ARG -  ARG -
ARG -  thr -  thr -  gly -  ser -  MET -  asp -  ala -  leu -  lys - 160
ala -  ala -  gly -  phe -  pro -  glu -  lys -  gln -  ile -  tyr -
gln -  val -  pro -  thr -  lys -  ser -  asn -  asp -  ile -  pro - 180
gly -  ala -  phe -  asp -  ala -  ala -  asn -  ser -  MET -  leu -
val -  gln -  his -  pro -  glu -  val -  lys -  his -  TRP -  leu - 200
ile -  val -  gly -  MET -  asn -  asp -  ser -  thr -  val -  leu -
gly -  gly -  val -  ARG -  ala -  thr -  glu -  gly -  gln -  gly - 220
phe -  lys -  ala -  ala -  asp -  ile -  ile -  gly -  ile -  gly -
ile -  ASN -  GLY -  val -  asp -  ala -  val -  ser -  glu -  leu - 240
ser -  lys -  ala -  gln -  ala -  thr -  gly -  phe -  tyr -  gly -
ser -  leu -  leu -  pro -  ser -  pro -  asp -  val -  his -  gly - 260
tyr -  lys -  ser -  ser -  glu -  MET -  leu -  tyr -  asn -  TRP -
val -  ala -  lys -  asp -  val -  glu -  pro -  pro -  lys -  phe - 280
thr -  glu -  val -  thr -  asp -  val -  val -  leu -  ile -  thr -
ARG -  asp -  asn -  phe -  lys -  glu -  glu -  leu -  glu -  lys - 300
lys -  gly -  leu -  gly -  gly -  lys
```

Fig. 3. The amino acid sequence of the L-arabinose binding protein

galactose binding protein isolated from the same source (14). Using an immunocompeti-tion assay the degree of homology was roughly quantitated and found to be approximately 20%. The fact that binding proteins function in similar environments and carry out similar processes led to the supposition that any existing homology may represent structural requirements for active function. Perhaps there exists a common or similar structure for interaction with the membrane or a membrane-bound translocating protein. Conceivably such units might even specify the periplasmic nature of these proteins.

An extensive consideration of molecular homology must await the completion of additional sequence analysis; however, a comparison of the amino terminal residues of the L-arabinose-, D-galactose-, and D-ribose-binding proteins is possible. The amino terminal sequences of the ribose- and galactose-binding proteins isolated from Salmonella typhimur-ium were determined by Koshland and Bradshaw (manuscript in preparation). Sequence analyses of the arabinose- and galactose-binding proteins isolated from E. coli and S. typhimurium were determined by Hogg and Hermodson (manuscript in preparation).

The data tentatively suggest that some homology exists between the 3 proteins; however, more extensive sequence analysis will be required to substantiate these possibilities.

The possibility that homology exists between the L-arabinose-binding protein and the D-galactose-binding protein (for review of galactose-binding proteins see Ref. 15) is supported by our observation that the gene locus for the arabinose-binding protein is located at approximately 44.5 minutes on the E. coli linkage map and the gene locus for the galactose-binding protein is located at 45.5 minutes (16). The possibility of gene duplication is under consideration.

MUTANTS DEFECTIVE FOR "HIGH-AFFINITY" TRANSPORT OF L-ARABINOSE

A number of mutants defective in the "high-affinity" transport of L-arabinose have been isolated. Initially, mutants which were unable to produce a protein that would cross-react with antibody directed against the L-arabinose binding protein were obtained using the *crm* plate technique described above (8). Such mutants are of interest for genetic mapping and for the demonstration of cotransfer of loss of function and genetic lesion; however, they are of little value for biochemical studies of binding protein structure and function. An extensive search for mutants producing defective L-arabinose binding protein was undertaken using as a parental strain E. coli K12 containing a mutation at the *araE* locus to effectively eliminate "low-affinity" L-arabinose uptake. Following mutagenesis, colonies demonstrating a more arabinose-negative phenotype than the parental strain when grown on EMB arabinose or TPTC arabinose, were isolated. Mutations in the structural genes *araD*, *araA*, and *araB*, the regulatory gene *araC*, and regulatory sites *araI* and *araO* were eliminated by enzymatic assay for *araA* and *araB* gene products and by growth in the presence or absence of L-arabinose. The remaining mutants were screened for "high-affinity" L-arabinose uptake using concentrations of L-arabinose [10^{-5}M] where only binding-protein-mediated uptake would be observed.

Mutants obtained in this manner fall into 2 classes. One class has lost arabinose-binding-protein-mediated uptake though they retain D-galactose uptake (presumably mediated by the methyl galactoside uptake system). The second class appears to be defective for both L-arabinose and D-galactose uptake. D-galactose uptake was originally included to eliminate pleiotrophic mutations; however, the doubly defective mutants are of considerable interest. Amongst the *ara⁻*, *gal⁺* mutations, 3 subclasses exist. The first class appears to bind L-arabinose normally in vitro, however, no in vivo uptake occurs. The second class exibits altered binding properties in vitro and does not transport in vivo. The third class does not bind arabinose in vitro and does not transport in vivo. Since binding proteins are recognized as such because they bind their respective ligands, we assume that mutants in the second and third classes consist of alterations which affect the binding site proper or the molecular configuration which insures the binding site. Some of the mutations in class one effect those regions of the molecule which are required for functions other than the recognition of substrate. These regions of the molecule could be envisioned as determining, membrane interactions, interactions with an energized translocating protein embedded in the membrane, sites involved in oligimer formation, or portions of the molecule which specify the periplasmic nature of an immature variety of the functional protein.

Mutations which result in the loss of both arabinose- and galactose-binding-protein-mediated uptake may possibly represent lesions in a common unit analogous to the *hisP* mutations described by Kustu and Ames (17).

CONCLUSIONS

The L-arabinose-binding protein and the role it plays in the active accumulation of L-arabinose by Escherichia coli is still under consideration.

ACKNOWLEDGMENTS

RWH was supported by United States Public Health Service Grant 1-R01-AM-13791 from the National Institute of Arthritis and Metabolic Diseases.

REFERENCES

1. Englesberg E, Irr J, Power J, Lee N: J Bacteriol 90:946, 1965.
2. Novotny C, Englesberg E: Biochem Biophys Acta 117:217, 1966.
3. Brown C, Hogg RW: J Bacteriol 111:606, 1972.
4. Isaacson D, Englesberg E: Bacteriol Proc 113:114, 1964.
5. Henderson PT, Skinner A: Biochem Soc Trans 2:543, 1974.
6. Hogg R, Englesberg E: J Bacteriol 100:423, 1969.
7. Schleif R: J Mol Biol 46:185, 1969.
8. Hogg RW: J Bacteriol 105:604, 1971.
9. Robbins A, Rotman B: Proc Natl Acad Sci USA 72:423, 1975.
10. Robbins A, Guzman R, Rotman B: J Biol Chem 251:3112, 1976.
11. Neu H, Heppel L: J Biol Chem 240:3685, 1965.
12. Parsons R, Hogg RW: J Biol Chem 249:3602 1974.
13. Hogg RW, Hermodson M: J Biol Chem 252:5135–5141, 1977.
14. Parsons R, Hogg RW: J Biol Chem 249:3608, 1974.
15. Boos W: Annu Rev Biochem 43:123, 1974.
16. Boos W: J Biol Chem 247:5414, 1972.
17. Kustu SG, Ames GF: J Bacteriol 116:107, 1973.
18. Quiocho F: J Supramol Struct 6:503, 1977.

Journal of Supramolecular Structure 6:419—431 (1977)
Molecular Aspects of Membrane Transport 281—293

Leucine Binding Protein and Regulation of Transport in E. coli

Dale L. Oxender, James J. Anderson, Mary M. Mayo, and Steven C. Quay

Department of Biological Chemistry, University of Michigan, Ann Arbor, Michigan 48109

Leucine is transported into E. coli cells by high-affinity transport systems (LIV-I and leucine-specific systems) which are sensitive to osmotic shock and require periplasmic binding proteins. In addition leucine is transported by a low-affinity system (LIV-II) which is membrane bound and retained in membrane vesicle preparations. The LIV-I system serves for threonine and alanine in addition to the 3 branched-chain amino acids. The LIV-II system is more specific for leucine, isoleucine, and valine while the high-affinity leucine-specific system has the greatest specificity.

A regulatory locus, *livR* at minute 22 on the E. coli chromosome produces negatively regulated leucine transport and synthesis of the binding proteins. Valine-resistant strains have been selected to screen for transport mutants. High-affinity leucine transport mutants that have been identified include a LIV-binding protein mutant, *livJ*, a leucine-specific binding protein mutant *livK* and a nonbinding protein component of the LIV-I system, *livH*. A fourth mutant, *livP*, appears to be required only for the low-affinity LIV-II system. The existence of this latter mutant indicates that LIV-I and LIV-II are parallel transport systems. The 4 mutations concerned with high-affinity leucine transport form a closely linked cluster of genes on the E. coli chromosome at minute 74.

The results of recent studies on the regulation of the high-affinity transport systems suggests that an attenuator site may be operative in its regulation. This complex regulation appears to require a modified leucyl-tRNA along with the transcription termination factor rho. Regulation of leucine transport is also defective in relaxed strains.

Among the branched-chain amino acids only leucine produces regulatory changes in LIV-I activity suggesting a special role of this amino acid in the physiology of E. coli. It was shown that the rapid exchange of external leucine for intracellular isoleucine via the LIV-I system could create an isoleucine pseudoauxotrophy and account for the leucine sensitivity of E. coli.

Key words: regulation, amino acid transport, mutants, leucine sensitivity, leucine, isoleucine, valine

E. coli have developed 2 basic types of active transport systems for leucine. One of these transport systems is a low-affinity membrane-bound system (LIV-II) and can be observed in membrane vesicle preparations described by Kaback (1). High-affinity trans-

Received May 27, 1977; accepted June 3, 1977

port systems for leucine (LIV-I and leucine-specific) in E. coli are sensitive to osmotic shock (2) and require periplasmic binding proteins. The leucine binding proteins have been described in various review articles (3–5). The high-affinity LIV-I system has a K_m for leucine of 10^{-7} M and also transports isoleucine, valine, threonine, and alanine. The leucine-specific system comprises about 20% of the high-affinity transport capacity for leucine of wild-type E. coli K-12. The regulation of the high-affinity transport systems which respond to the level of leucine in the medium has been extensively studied in our laboratory (6–10). Mutations resulting in a loss of leucine repression have been mapped and characterized (6, 10). The mutations livR (signifying a derepression of the LIV-I and leucine-specific systems) and lstR (signifying derepression of the leucine-specific system and the LIV-II systems) determine negatively acting regulatory elements and are genetically closely linked. Our current understanding of the regulation of leucine transport suggests that leucine interacts with tRNA[leu] and the leucyl-tRNA synthetase to produce this regulation. The regulation primarily changes the differential rate of synthesis of transport components relative to total cellular proteins (9). We have recently shown that mutations that alter the hisT gene which codes for a tRNA modifying enzyme (11) and the rho allele which codes for a transcription termination factor (12) also produce a derepression of leucine transport (13, 14). These results suggested that an "attenuator" type of regulation obtains similar to that described for the regulation of tryptophan (15), histidine (16), and branched-chain amino acid (11) biosynthetic pathway enzymes.

Analysis of the structural components of the high-affinity and low-affinity transport systems for leucine has been aided by our recent identification (17) of a genetic complex which codes for at least 4 functions involved in binding protein expression and LIV-II transport expression. This complex is linked to malT on the E. coli genetic map, approximately 180° from the regulatory loci livR and lstR.

METHODS

Growth Conditions

Growth supplements for auxotrophic strains when not indicated otherwise were: 50 μg/ml for all L-amino acids except for L-leucine which was used at a concentration of 25 μg/ml; thymine 50 μg/ml; vitamins 1 μg/ml, glucose 0.2%.

For all experiments the medium consisted of a morpholino propane sulfonate-buffered salts solution (MOPS) described by Neidhardt et al. (18). All supplements were either sterilized by filtration through 0.2 μm membrane filters or autoclaved.

Cultures were grown aerobically in 125- or 250-ml erlenmeyer flasks in a shaking water bath (New Brunswick Scientific Company, Model G-76) that maintained constant temperatures between 30 and 41 ± 0.25°C. The platform rotation was approximately 150 rpm. Cell growth was followed by monitoring absorbance at 420 nm at room temperature using a GCA-McPherson model EU 707-12 spectrophotometer.

Isolation of Binding Protein

Binding protein was isolated by the osmotic shock procedure and binding activity determined by equilibrium dialysis as described previously (5).

Transport Assays

Routine transport assays, as well as the rapid transport assay variant for screening large number of prospective transport mutants, were performed as described previously (10).

Enzymes Assays

The following enzymes were assayed as referenced:

Threonine deaminase: (E. C. 4.2.1.6; L-threonine hydrolyase [deaminating] : (threonine dehydratase) (19) except that the absorbance of the 2,4-dinitrophenylhydrazone derivative of α-ketobutyrate was measured at 530 nm in a Zeiss PMQ2 spectrophotometer.

Acetohydroxy acid synthetase: (AHAS; acetolactate synthetase, E.C.4.1.3.18) (20) 3-isopropylmalate dehydrogenase (IPMP; 2-hydroxy-4-methyl-3-carboxyvalerate-nicotinamide adenine dinucleotide oxidoreductase. E. C. 1.1.1.85) (21).

Mu Phage Procedure

Wild type Mu-1 and Mu*cts* 62 lysates were prepared as already described (17).

Mu Mutagenesis

E. coli cultures were grown to a density of 2×10^8 cells/ml in LBT broth supplemented with 0.01 M MgSO$_4$ and mutagenized by adding Mu or Mu*cts* phage to a multiplicity of infection of 1.0. After lysis at the appropriate temperature (37°C for Mu, 30°C for Mu*cts*) the cultures were allowed to grow overnight. Procedures for detecting Mu lysogeny have been previously described (17).

SDS-Polyacrylamide Gel Electrophoresis

Slab gels containing 11% acrylamide and 0.1% SDS (sodium dodecyl sulfate) were prepared by the method of Laemmli (22). The gels were 1.5 mm thick. Samples of concentrated shock fluids were prepared in 1% SDS by heating for 2 min at 100°C. After electrophoresis the gels were fixed, stained, and destained by the method of Fairbanks (23). The gels were dried on filter paper under vacuum and photographed.

RESULTS

Isolation of Mutants in the Branched-Chain Amino Acid Uptake Systems

Regulatory mutants. Among spontaneous mutants selected for the ability to grow on D-leucine were mutants with elevated levels of branched-chain amino acid transport (6). The mutant allele *livR* results in failure of L-leucine to repress the osmotic shock-sensitive high-affinity LIV-I and leucine-specific transport systems and their respective binding proteins, but has no apparent effect on the membrane-bound low-affinity LIV-II system (6, 10). The *lstR* allele, However, results primarily in a derepression of the leucine-specific transport system and in the level of leucine-specific binding protein, with an apparent two-fold increase in the LIV-II system as well (10 and unpublished experiments). The *lstR* mutation permits the cells to utilize lower levels of D-leucine than the *livR* mutation (Table I). Since these 2 distinct patterns of regulation suggested that different functions were specified by the *livR* and *lstR* alleles, a complementation test was carried out between *lstR* and *livR* mutants. Strains were constructed which were diploid for the E. coli chromosome covering the region in which *lstR* and *livR* are located (Fig. 1). Previous work had indicated very close linkage between these 2 alleles (10). Both *lstR* and *livR* are recessive to their respective wild-type alleles (Table II, strains 1, 2, 3, and 5) which therefore indicates that they produce a negatively operating, diffusable factor. The merodiploid strain listed as number 4 in Table II is homozygous for *lstR* and displays the mutant phenotype. Strains homozygous for *livR* also display the mutant phenotype (10). Strain number 6, however, shows that the wild-type alleles of *livR* and *lstR* can complement

TABLE I. Growth Phenotypes of *liv* Mutants

Mutation[a]	Colony formation[d] on medium supplemented with			
	0.03 µM L-leucine[b]	0.4 µM L-valine[c]	850 µM D-leucine[b]	1,700 µM D-leucine[b]
liv+	+	+	−	−
livR	+	−	−	+
1stR	+	−	+	+
livR livJ	+	+	−	+
livR livK	+	+	−	−
livR livH	+	+	−	−
1stR livH livP	−	+	−	−

[a]Complete strain descriptions are published elsewhere; all strains are isogenic and carry the derepressed transport allele *livR* (10) except *livH livP,* which is in a *lstR* strain background.
[b]Determined in *leu* strain background
[c]Determined in *leu+* strain background
[d]Determined after 72-h incubation at 32°C

TABLE II. Complementation Analysis of *lstR* and *livR* Alleles in F-prime Merogenotes

Merogenote genotype[a]	Phenotype	
	L-leucine uptake, nmol/min/mg dry wt.	Colony formation, 1,700 µM D-leucine
1. F (+ +) / (+ +)	0.020	−
2. F (+ +) / (livR +)	0.012	−
3. F (+ +) / (+ lstR)	0.020	−
4. F (+ lstR) / (+ lstR)	0.052	+
5. F (+ lstR) / (+ +)	0.03	−
6. F (+ lstR) / (livR +)	0.018	−

[a]Strains carrying the F147 gal+ F-prime (35) were constructed from recipients which were, in addition to the designated transport alleles, *recA gal leu.* They were grown on galactose minimal medium supplemented with 25 µg L-leucine per ml. The F147 *lstR* episome was isolated from a D-leucine utilizing homogenote of a *recA+* strain similar in genotype to strain No. 3, above.

each other, yielding the wild-type phenotype. We conclude from these results that *livR* and *lstR* are separate genes, each controlling a different pattern of regulation of the branched-chain amino acid uptake systems. The molecular nature of these gene products is under investigation.

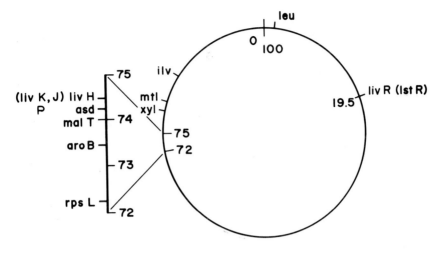

Fig. 1. Simplified genetic map of the E. coli chromosome taken from Bachman, Low, and Taylor (24) showing positions of *liv* mutations discussed within. Alleles in parentheses have not been precisely mapped.

Isolation of Mutants Defective in Branched-Chain Amino Acid Uptake Systems

Mutant strains possessing the *livR* allele have increased sensitivity to L-valine, which inhibits the growth of E. coli K-12 (Table I). Mutations to L-valine resistance were induced by the mutator phage Mu in strains possessing the *livR* allele (Table I). A concentration of L-valine was chosen which would prevent colony formation only in strains with elevated levels of the LIV-I system in order to avoid excessive selective pressure (17). Three biochemically distinct classes of mutants were found which have been designated *livH*, *livJ*, and *livK* (17). These map together by transduction and are closely linked to *malT* (17 and Fig. 1). The transport of L-leucine in these strains has been diminished (Table III) without affecting the transport of other amino acids (17). The mutation *livK* results in a loss of all detectable high-affinity leucine uptake (LIV-I and leucine-specific systems) and a loss of the leucine-specific binding protein, both by in vitro activity (Table I) and by SDS-polyacrylamide gel electrophoresis (Fig. 2). The LIV-binding protein is still present in these strains even though all high-affinity transport is lost. The high-affinity uptake of L-valine and L-isoleucine has also been eliminated (17), as has the ability of the cells to utilize D-leucine (Table 1). The *livJ* mutation eliminates the LIV-I high-affinity uptake but the leucine-specific system and LIV-II system are unaffected. There is a parallel loss of the LIV-binding protein activity and the corresponding band on SDS-polyacrylamide gel slabs (Table III and Fig. 2). D-Leucine utilization, however, is retained (Table I) indicating that the D-leucine utilization phenotype in *livR* and *lstR* strains is presumably due to a derepression of the leucine-specific system. The mutation *livH* eliminates both high-affinity transport systems without producing any detectable change in the mobilities or activities (Table III, Fig. 2 and Ref. 17) of the 2 binding proteins. Evidently *livH* codes for an additional factor of high-affinity uptake, at present unidentified.

TABLE III. Transport Phenotypes of *liv* Mutants

Mutation[a]	Percent of wild-type leucine uptake[b]	L-leucine uptake systems[c]			L-leucine-binding protein activity[d]	
		LIV-I	Leucine-specific	LIV-II	LIV	Leucine-specific
liv+	100	+	+	+	+	+
livK	10	−	−	+	+	−
livJ	55	−	+	+	−	+
livH	15	−	−	+	+	+
livH livP	2	−	−	−	+	+

[a]Strains, same as Table I

[b]L-leucine transport was measured at 5 μM L-leucine.

[c]The leucine-specific and LIV-I systems are defined by their low K_m (0.1−0.4 μM) for L-leucine; the leucine-specific system resists inhibition by L-isoleucine and is responsible for D-leucine uptake. The LIV-II system has a higher K_m (2−4 μM) for L-leucine and also transports L-isoleucine and L-valine. Presence or absence of a given system was determined by kinetic analysis over substrate ranges from 0.02 μM to 20 μM (6).

[d]L-leucine-binding activity was measured on crude osmotic shock fluid; the leucine-specific binding protein activity was determined by resistance of L-leucine binding to L-isoleucine competition (10).

A third group of mutants were obtained which were defective in the low-affinity (LIV-II) transport system. These mutants were obtained by inducing a second mutation in the *livH* mutant strain by mutagenizing the *livH* strain with ethyl methane sulfonate and penicillin-enriching for mutant strains unable to utilize low concentrations of L-leucine (Table I). One of these mutations has been designated *livP*. Conjugational analysis has shown that *livP* is also linked to *malT* (unpublished observations, Fig. 1). Transport of L-leucine in the double mutant *livH livP* has been dramatically lowered over that of the parent *livH* strain (Table III), and kinetic analysis (data not shown) has revealed that the LIV-II system has been reduced to undetectable levels. Thus, in this double mutant strain both high- and low-affinity transport systems for leucine have been mutationally eliminated. Other workers, however (25, 26) have reported additional low-affinity uptake systems for leucine. We are presently characterizing the remaining leucine uptake in the double mutant strain to ascertain the nature of the residual leucine uptake.

A genetic complex is suggested for leucine transport genes in E. coli by these studies since at least 4 different genes specifying components of the known transport systems for leucine map in the same region. We believe that the structural genes for the leucine-binding proteins are part of this complex since a previously reported mutation altering the structure of the leucine-specific binding protein (6) also maps in this area (17). We are presently characterizing other mutants by complementation analysis and fine mapping to better define the genetic organization of the region. We are also attempting to biochemically identify the components coded for by the *livH* and *livP* genes.

Regulation of Leucine Transport

Repression by leucine. Early studies showed that high-affinity leucine transport activity is highly regulated and responds to the level of leucine in the growth medium (6). There is a direct relationship between the level of high-affinity leucine transport and the

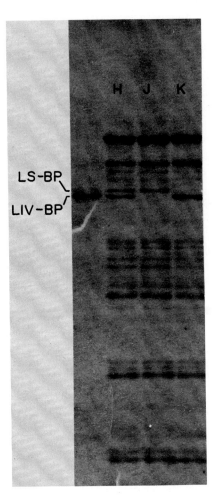

Fig. 2. Sodium dodecyl sulfate-polyacrylamide gel profile of crude osmotic shock fluid proteins. The slab gels were prepared according to the method of Laemmli (22) and were 11% acrylamide. Direction of migration is from top to bottom. A sample of pure LIV-binding protein (LIV-BP) and a trace of the leucine-specific binding protein (LS-BP) are shown on the left; H) shock fluid from a strain carrying the *livH* mutation, J) strain with *livJ* mutation, and K) strain with *livH*. The wild-type profile (not shown) is identical to H.

level of the leucine-binding proteins (17). The correlation between the transport capacity and the amount of binding protein shows that the binding proteins are the rate limiting component in the transport of leucine into E. coli. When E. coli is grown on nutrient medium or minimal medium containing 25–50 μg/ml L-leucine essentially complete repression of the synthesis of both the LIV and the leucine-specific binding proteins occurs and essentially all high-affinity leucine transport is abolished (27).

Separate Regulation of Biosynthesis and Transport of Leucine

The levels of the leucine biosynthetic enzymes are also regulated by the intracellular level of branched-chain amino acids (28). Since both transport and biosynthetic enzymes for leucine were repressed by growth on leucine we attempted to determine whether these 2 processes were regulated in a concerted manner. For these studies we collaborated with Dr. Umbarger of Purdue University. We provided Dr. Umbarger with mutant strains that were derepressed for leucine transport and he gave us several mutants that were derepressed for the biosynthetic enzymes of the branched-chain pathway. We examined the regulation of leucine transport in strains that were derepressed for biosynthesis *(leuABCD, ilvB, ilvADE)* as well as a deletion of the entire leucine biosynthetic operon. Leucine transport activities and leucine-binding protein levels in these strains were regulated in a normal manner. Using the transport mutants from our laboratory, Dr. Umbarger showed that the regulation of the biosynthetic enzymes for the branched-chain amino acids was not altered in mutants with derepressed transport and leucine-binding proteins. The normal regulation of transport in E. coli strains with deletions of the entire biosynthetic operon indicated that none of the gene products of this operon are required for regulation of transport. We concluded from these studies that the transport systems and the biosynthetic enzymes in E. coli are not regulated together by a cis-dominant type of mechanism and although both systems appear to have components in common it is possible to obtain separate regulatory mutations for each system (7, 13).

Role of Leucyl tRNA Synthetase in Transport Regulation

A possible candidate for a common component in the regulation of both transport and biosynthesis of leucine is the requirement for aminoacyl tRNA rather than the free leucine. Extensive studies in other laboratories have indicated aminoacyl-tRNA synthetases are part of the regulatory system for branched-chain amino acid biosynthesis (29). The availability of temperature-sensitive mutants for the leucyl-tRNA synthetase (*leuS1*) permitted us to determine whether this enzyme was also involved in the regulation of the transport of leucine. We examined both the level of transport activity and the level of the periplasmic-binding proteins in the temperature-sensitive mutant strain *(leuS1)* grown at the permissive temperature (36°C) and the nonpermissive temperature (41°C). Strain EB143 has a complete deletion of the leucine biosynthetic operon which avoids the increased endogenous levels of leucine that would be produced in a prototrophic temperature sensitive *leuS* strain grown at 41°C. Strain EB144 served as the isogenic non-temperature-sensitive control strain. The results are presented in Table IV. When mutant strain EB143 was shifted from 36 to 41°C the transport activity for leucine, isoleucine, and valine was greatly increased. Little effect was observed for proline or histidine uptake. Proline is transported by a membrane-bound transport system that derives its energy from the membrane potential while histidine is transported by a periplasmic-binding protein system and derives its energy more directly from ATP. The derepression of threonine deaminase that was observed for strain EB143 when it was shifted to 41°C was the expected response of a branched-chain amino acid biosynthetic enzyme which is subject to multivalent repression. The branched-chain amino acid transport system responds only to changes in the leucine level. Isoleucine and valine tRNA synthetase mutants did not produce regulatory changes in the transport activity. The shock fluid of strains EB143 and EB144 were examined for leucine-binding proteins when the cells were grown at 41°C. There is a fivefold derepression in the synthesis of the binding proteins for leucine when the temperature is shifted to 41°C

TABLE IV. Expression of Transport and a Biosynthetic Enzyme Activity in Strains EB143 and EB144

Strain	Growth conditions[a]	Transport activities[b]			Threonine deaminase[c]
		Leu	Ile	Val	
EB143, *ara-leu△1101, leuS1*	36°C	100	65	87	26
EB144, *ara-leu△1101*	36°C	87	61	70	16
EB143, *ara-leu△1101 leuS1*	41°C	470	773	904	450
EB144, *ara-leu△1101*	41°C	96	70	74	41

[a]Growth in glucose-basal salts medium plus 0.2 mM L-leucine.
[b]Transport was assayed at 1 μM leucine (Leu) or isoleucine (Ile) and 3 μM valine (Val). One hundred percent represents 0.23 mmol leucine taken up per min per kg cells dry weight.
[c]Specific activity represents μmol of α-ketobutyrate formed per min/g of cellular protein. The growth media included 0.4 mM L-leucine and L-isoleucine and 1 mM L-valine.

with little change in the control strain. These results were consistent with a role of the binding proteins in the rate limiting step in leucine transport and further indicated that the synthesis of the binding proteins is regulated by the level of modified leucyl-tRNA or the leucyl-tRNA synthetase. We obtained a mutant strain with a defect in the maturation of leucyl-tRNA to distinguish between these 2 alternatives.

Role of *hisT* Gene Product

The *hisT* locus codes for an enzyme that converts uridine to pseudouridine in the tRNAs for leucine, histidine, and tyrosine. The regulation of biosynthetic enzymes for these 3 amino acids is no longer sensitive to their cognate amino acids in strains containing a *hisT* mutation (11). We obtained E. coli strains containing a *hisT* mutation from Drs. R. P. Lawther and W. Hatfield. An examination of leucine transport activity in the *hisT* strain showed that the regulation was abnormal. The hisT strain showed a higher transport of leucine than the parental strain when grown under repressing conditions and furthermore it failed to derepress leucine transport activity when grown under conditions of leucine limitation. Under these same conditions the biosynthetic pathway for leucine also does not show a repression or derepression regulation (Quay SC, unpublished observations). We reported similar results for a role of the *hisT* gene product in the regulation of leucine transport activity in Salmonella typhimurium (13). The results of these 2 studies with *hisT* strains provide evidence that the repression of transport of leucine requires fully maturated tRNA which is aminoacylated with leucine.

Role of rho in Transport Regulation

Recent reports have shown that several of the biosynthetic operons in bacteria contain an "attenuator" site near the operator region that acts as a barrier to transcription by RNA polymerase (15, 16). The termination factor rho has been implicated in attenuator function. To test if transport for leucine, isoleucine, and valine is under an attenuator-type regulation 2 strains were obtained from Dr. Umbarger of Purdue University, which were leucine auxotrophs with one having a *suA120* allele (14). The *suA120* allele is a missense mutation in the rho factor (30). The initial rate of transport of 8 amino acids was measured in

TABLE V. Effect of a Mutation in Termination Factor *rho* on Amino Acid Transport

| Substrate | Specific transport activity (μmol/min/g cells, dry wt.) | |
	Control strain CU300[a]	SuA mutant CU2054[a]
Arginine (3 μM)	2.92	3.61
Glutamine (1 μM)	1.90	0.63
Histidine (1 μM)	0.20	0.38
Isoleucine (1 μM)	0.12	0.24
Leucine (1 μM)	0.10	0.19
Proline (3 μM)	0.17	0.17
Tryptophan (3 μM)	0.51	0.55
Valine (3 μM)	0.31	0.61

[a]Cells were grown in MOPS-G, 0.2 mM leucine, and 25 mg/liter tryptophan. Cells were harvested and transport assayed as described previously (9).

the *rho* mutant and the isogenic parental strain. The data are presented in Table V. The uptake of leucine, isoleucine, and valine is increased twofold even though leucine is present in the growth medium. Arginine and histidine transport activity is also increased in the *rho* mutant. Tryptophan and proline uptake is unchanged but glutamine is greatly decreased. The kinetics of uptake showed that the K_m values for both the high- and the low-affinity transport systems for leucine in the *rho* mutant were similar to those obtained for the parental strain. The V_{max} value for the high-affinity leucine uptake was increased approximately eightfold in the *rho* mutant. The V_{max} of the low-affinity system was only slightly elevated.

An examination of the leucine-binding activity in the osmotic shock fluid was carried out. The leucine-binding activity showed a fourfold increase in the *rho* mutant. In a similar manner the histidine- and arginine-binding activities in the osmotic shock fluid were increased somewhat while glutamine-binding activity decreased significantly. These results show that *rho*-dependent transcriptional termination is important for leucine-specific repression of branched-chain amino acid transport (14).

Leucine Sensitivity

The *livR* locus, which leads to a trans-recessive derepression of the high-affinity transport system for leucine is responsible for greatly increased sensitivity toward growth inhibition by leucine, valine, serine, and certain analogues such as 4-azaleucine or 5′,5′,5′-trifluoroleucine (31). We recently showed that the ability of the LIV-I transport system to carry out exchange of endogenous amino acids for extracellular leucine is a major factor in leucine sensitivity. When E. coli cells are shifted from nutrient medium to minimal medium containing leucine a long lag time in the resumption of growth is observed. The lag time in the growth represents the time required to derepress certain biosynthetic enzymes. Added isoleucine antagonizes the leucine sensitivity showing that the necessary derepression of the biosynthesis of isoleucine is prevented by leucine. We were able to show that the necessary biosynthetic enzymes for isoleucine cannot by synthesized. The high-affinity transport system serves for rapid exchange of branched-chain amino acids while the low-affinity system does not produce significant exchange of these amino acids.

These results provide a mechanism for leucine sensitivity and explain the increased sensitivity of strains with derepressed high-affinity transport such as that shown by the *livR* mutant strain.

DISCUSSION

The results presented above describe the 2 types of active transport systems in E. coli for the branched-chain amino acids. One type is membrane bound and can be observed in membrane vesicles (LIV-II), and the other type is osmotic-shock sensitive and requires binding proteins to produce active transport of leucine (LIV-I). A regulatory genetic locus, *livR*, at minute 22 on the E. coli chromosome produces negatively regulated leucine transport and synthesis of the leucine-binding proteins. Valine-resistant strains that were still sensitive to the dipeptide glycylvaline were selected as potential LIV-I transport mutants. Among the valine-resistant mutants we identified a LIV-binding protein mutant *(livJ)*, a leucine-specific binding protein mutant (*livK*), and a nonbinding protein component of the LIV-I system *(livH)*. A fourth mutation, *livP*, appears to be a component of the low-affinity LIV-II system. Since a *livP* mutant appears to have normal LIV-I transport we have concluded that the LIV-I and LIV-II systems represent parallel systems even though they may have certain components in common. These mutations concerned with leucine transport form a cluster closely linked to *malT* at minute 74 on the E. coli chromosome. The positions of these mutations are distinctly different from that of *brnQ*, *brnR*, and *brnS* mutations of branched-chain amino acid transport described by Iaccarino and co-workers (26).

Continued studies of the chemical and physical properties of the LIV- and leucine-specific binding proteins are being carried out in collaboration with other laboratories. Dr. Ovchinnikov and co-workers of Moscow, USSR, have recently published the complete amino acid sequence of the LIV-binding protein (32) and are currently sequencing the leucine-specific binding protein which is structurally very similar. Antonov et al. (33) have shown that a concentration-dependent reversible association of the LIV-binding protein may occur producing aggregates with an apparent molecular weight up to 300,000. The presence of these aggregates has been correlated with nonlinear Scatchard plots of leucine-binding activity. Alternative explanations for nonlinear Scatchard plots of leucine binding to the LIV-binding protein have been offered by Anraku and co-workers (34). These studies suggest that binding protein preparations contain bound ligand which could alter the specific activity of added ligand at high protein to ligand ratios.

Figure 3 presents a scheme of the regulatory components of the LIV-I transport system which responds to the level of leucine added to the medium. Modified leucyl-tRNA appears to play a role in the regulation along with the transcription termination factor *rho*. The results of this study suggest an attenuator site may be operative in the regulation of high-affinity leucine transport. This complex regulatory system for transport of leucine is similar to that of its biosynthesis suggesting the important role of the LIV-I transport system to the physiology of the bacterial cell. Since changes in isoleucine or valine levels do not produce regulatory changes in LIV-I a special role of leucine in the physiology of E. coli is implied. The properties of the LIV-I system that give rise to a rapid exchange of leucine for isoleucine, thus creating isoleucine pseudoauxotrophy under certain conditions, have helped to explain the leucine sensitivity of E. coli (31). This complex regulatory system for the transport of leucine which is similar to that of the biosynthetic enzymes provides an example of the important role transport systems can play in amino acid metabolism.

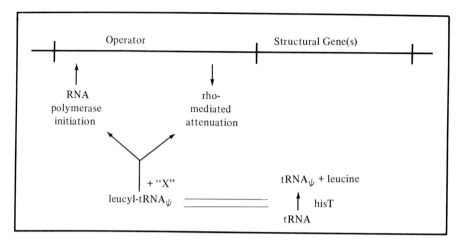

Fig. 3. Model of leucine transport regulation in E. coli.

ACKNOWLEDGMENTS

We wish to thank Dr. H. E. Umbarger of Purdue University and Drs. R. P. Lawther and W. Hatfield of The University of California at Irvine for the E. coli strains containing the *hisT* mutation. This investigation was supported by Public Health Service Grant GM11024 to D.L.O. from the National Institutes of General Medical Sciences.

REFERENCES

1. Kaback HR: Methods Enzymol 22:99, 1971.
2. Neu HC, Heppel LA: J Biol Chem 240:3685, 1965.
3. Oxender DL: Annu Rev Biochem 41:777, 1972.
4. Oxender DL: In Manson LA (ed): "Biomembranes" New York: Plenum Press, 1975, vol 5 p 25.
5. Oxender DL, Quay SC: In Korn ED (ed): "Methods of Membrane Biology." New York: Plenum Press, 1976, vol 6, p 183.
6. Rahmanian M, Claus DR, Oxender DL: J Bacteriol 116:258, 1973.
7. Quay SC, Oxender DL, Tsuyumu S, Umbarger HE: J Bacteriol 122:994, 1975.
8. Quay SC, Kline EL, Oxender DL: Proc Natl Acad Sci USA 72:3921, 1975.
9. Quay SC, Oxender DL: J Bacteriol 127:1225, 1976.
10. Anderson JJ, Quay SC, Oxender DL: J Bacteriol 126:80, 1976.
11. Cortese R, Landsberg RA, Vonder Haar RA, Umbarger HE, Ames BN: Proc Natl Acad Sci USA 71:1857, 1975.
12. Beckwith J, Biochim Biophys Acta 76:162, 1963.
13. Oxender DL, Quay SC: J Cell Physiol 89:517, 1976.
14. Quay SC, Oxender DL: J Bacteriol 130:1024, 1977.
15. Bertrand K, Korn L, Lee F, Platt T, Squires CL, Yanofsky C: Science 189:22, 1975.
16. Artz SW, Broach JR: Proc Natl Acad Sci USA 72:3453, 1975.
17. Anderson JJ, Oxender DL: J Bacteriol 130:384, 1977.
18. Neidhardt FC, Block PL, Smith DF: J Bacteriol 119:736, 1974.
19. Burns RO: In Tabor H, Tabor CW (eds): "Methods in Enzymology." New York: Academic Press, 1971, vol 17, part B, p 555.

20. Stormer FC, Umbarger HE: Biochem Biophys Res Commun 17:587, 1974.
21. Burns RO, Umbarger HE, Gross SR: Biochemistry 2:1053, 1963.
22. Laemli VK: Nature (London) 227:680, 1970.
23. Fairbanks G, Steck TL, Wallach DFH: Biochemistry 10:2606, 1971.
24. Bachman BJ, Low KB, Taylor AL: Bacteriol Rev 40:116, 1976.
25. Wood JM: J Biol Chem 250:4477, 1975.
26. Guardiola J, Iaccarino M: J Bacteriol 117:393, 1974.
27. Peurose WR, Nichoalds GE, Piperno JR, Oxender DL: J Biol Chem 243:5921, 1968.
28. Freundlich M, Burns RO, Umbarger HE: Proc Natl Acad Sci USA 48:1804, 1962.
29. Umbarger HE: In Vogel HJ (ed): "Metabolic Pathways." New York:Academic Press, 1971, vol 5, p 447.
30. Morse DE, Guerlin M: J Mol Biol 63:605, 1972.
31. Quay SC, Dick TE, Oxender DL: J Bacteriol 129:1257, 1977.
32. Ovchinnikov YuA, Aldanova NA, Grinkevich VA, Arzamazova NM, Moroz IN, Nazimov IV: Bioorg Chem (Russian) 3:564, 1977.
33. Antonov VK, Alexandrov SL, Vorotyntseva TI: Adv Enzyme Regul 14:269, 1976.
34. Amanuma H, Itoh J, Anraku Y: J Biochem (Japan) 79:1167, 1976.
35. Low KB: Bacteriol Rev 36:587, 1972.

Journal of Supramolecular Structure 6:433–440 (1977)
Molecular Aspects of Membrane Transport 295–302

Relationship Between Thymidine Transport and Phosphorylation in Novikoff Rat Hepatoma Cells As Analyzed by a Rapid Sampling Technique

Richard Marz, Robert M. Wohlhueter, and Peter G. W. Plagemann

Department of Microbiology, Medical School, University of Minnesota, Minneapolis, Minnesota 55455

Incorporation of thymidine into Novikoff rat hepatoma cells was analyzed with a rapid sampling technique which allowed collection of 12 time points in 20 sec. Transport was studied in the absence of metabolism by using either ATP-depleted cells or a thymidine kinase negative subline. Transport was a rapid, saturable, nonconcentrative process with a K_m of about 85 μM. The intracellular thymidine pool was also rapidly labeled in cells which phosphorylated thymidine, so that a group translocation process involving thymidine kinase can be ruled out. Under all conditions examined, phosphorylation, not the transport, of thymidine was the rate-determining step in its incorporation into the acid-soluble pool. Estimation of transport rates from total incorporation into cells which phosphorylate the substrate is invalid in this cell system and must be questioned in all instances.

Key words: transport; incorporation; uptake; thymidine; nucleoside; Novikoff rat hepatoma cells; rapid sampling technique

Transport rates of substances which are neither metabolized nor accumulated intracellularly (facilitated transport) are difficult to determine accurately. Not only is the total amount of substrate taken up small, but often the process has reached an equilibrium in a matter of seconds, thus making it very difficult to measure initial transport rates. Many investigators, therefore, have estimated the rates of facilitated transport by measuring rates of incorporation[1] of substrates that are rapidly metabolized, mainly phosphorylated, intracellularly, and thereby trapped (see Ref. 1). The phosphorylated products accumulate intracellularly and incorporation is generally linear with time for much longer periods of time (up to 10 min or longer) than in the absence of metabolism. It is obvious, however, that the rates thus obtained can only be considered transport rates if transport, and not metabolism, is the rate-limiting step in the overall process (1, 2). Often investigators have

---

[1] We define "incorporation" as the transfer of radioactivity from extracellular labeled substrate to total cell components which in general involves transport and phosphorylation of the substrate, and its subsequent incorporation into macromolecules. "Transport," on the other hand, refers strictly to the transfer of unaltered substrate by a saturable process from one side of the plasma membrane to the other.

A detailed description of the methodology employed here will be found in volume 20 of "Methods in Cell Biology" (D. Prescott, ed.)

Received April 13, 1977; accepted May 10, 1977.

merely assumed that transport was the rate-limiting step, or that the error they introduced was small and unavoidable. In other cases indirect evidence was offered which suggested that indeed transport was the rate-limiting step, but the validity of these assumptions has been difficult to prove unequivocally (1) and has been questioned (2). Indeed, in recent experiments (3, 4) we have demonstrated that the kinetics of the facilitated transport of thymidine (dThd) as measured in ATP-depleted or dThd kinase-deficient cultured Novikoff rat hepatoma cells, in which dThd is not phosphorylated, differ markedly (both in K_m and V_{max}) from those for dThd incorporation by cells in which dThd is phosphorylated. Measurements of the kinetics of facilitated dThd transport in nonmetabolizing cells was only made possible by the application of a newly developed rapid sampling technique which allows us to obtain up to 12 time points within a 20-sec period (4). We have now applied this methodology to measuring the transport, phosphorylation, and incorporation of dThd into DNA in metabolizing Novikoff cells, and further document that in these cells the rate of dThd incorporation is governed by the rate of its phosphorylation rather than its rate of transport into the cell.

MATERIALS AND METHODS

Wild-type Novikoff rat hepatoma cells (subline N1S1-67) and a dThd kinase-deficient subline thereof (3) were propagated in Swim's medium 67 and enumerated as described previously (5, 6). Cells were collected from exponential phase cultures by centrifugation at $400 \times g$ for about 2 min, and suspended in a basal medium, BM42B (7), to a concentration of $2-4 \times 10^7$ cells/ml for transport measurements or about 2×10^6 cells/ml for incorporation studies. Cells were depleted of ATP by incubation in glucose-free BM42B supplemented with 5 mM KCN and 5 mM iodoacetate (3).

In all incorporation and transport assays the influx of substrate against zero intracellular concentration was measured, i.e., with a "zero-trans" protocol (8). For incorporation measurements at time scales of minutes the cell suspension was supplemented with labeled dThd and incubated on a gyratory shaker at about 200 rpm. Duplicate 1-ml samples of suspension were analyzed for radioactivity in total cell material (9).

For transport and incorporation studies at time scales in seconds, the rapid sampling technique, which has been described in detail elsewhere (4), was employed. Briefly, fixed aliquots of a suspension of cells were rapidly mixed with a solution of radioactive dThd at short time intervals by means of a hand-operated, dual syringe apparatus. Samples emerging from the mixing chamber were dispensed into 12 tubes mounted in an Eppendorf microcentrifuge which contained an oil mixture (density = 1.034 g/ml). Dispensing of samples into sequential centrifuge tubes was paced with a metronome and could be accomplished, comfortably, at 1.5-sec intervals. After the last sample had been mixed, the centrifuge was started, and within an estimated 2 sec the cells had entered the oil phase thus terminating transport. When more than 12 samples were required, cells and substrate were mixed in the same proportions as provided by the mixing apparatus and samples were removed at appropriate times for centrifugation through oil.

After centrifugation the supernatant medium was aspirated. The upper part of the tube was washed once with 1 ml of water which was subsequently removed together with most of the oil. Then 0.2 ml of 0.5 N TCA was added to the tube, and immediately vortexed to disperse the pellet. After 30 min of incubation at 70°C the entire tube and its contents were transferred to a vial containing 8 ml of a modified Bray's solution (5) and analyzed for radioactivity in a liquid scintillation spectrometer.

In order to stop substrate metabolism when cells capable of phosphorylating dThd were used, the centrifuge tubes were set up with 3 phases: the cell substrate mixture (density ~ 1.0 g/ml), above oil (density = 1.034 g/ml), above 0.5 N trichloroacetic acid in 10% (wt/vol) sucrose (density = 1.04 g/ml). The acid phase was separated into 3 fractions: a fraction which was not precipitated by $LaCl_3$ at neutral pH which contained free dThd, an acid-soluble fraction precipitated by $LaCl_3$ at neutral pH, which contained dThd nucleotides, and an acid-insoluble pellet containing DNA.

Total water space and extracellular space in cell pellets obtained by centrifugation through oil were determined in parallel runs in which substrate was replaced by $[^{14}C]$-carboxylinulin in $[^3H]H_2O$. All data were corrected for substrate radioactivity in the extracellular space (generally about 12% of total water space) and normalized to rates per μl cell H_2O.

Initial velocities of transport were computed as zero-time slopes of an integrated rate equation to which data were fit by the method of least squares (4); Michaelis-Menten parameters were computed according to Wilkinson (10).

Chemicals

Radiochemicals were obtained from the following sources: [methyl-^{3}H] dThd (10 Ci/mmole) from ICN (Irvine, California); [carboxyl-^{14}C] carboxyl-inulin (2.6 Ci/g) and $[^3H]$-H_2O (1 mCi/g) from New England Nuclear (Boston, Massachusetts). Thymidine was obtained from Sigma Chemical Co. (St. Louis, Missouri), $LaCl_3$ from Gallard-Schlesinger (Carle Place, New York). Other chemicals were reagent grade from standard suppliers.

RESULTS

The incorporation of dThd (0.2 and 2 μM) into total cell material by untreated wild-type Novikoff cells was approximately linear with time for several minutes both at 25 and 37°C (Fig. 1 A, B). Chromatographic analysis of acid extracts of 6-min labeled cells showed that over 95% of pool radioactivity was associated with thymine nucleotides (data not shown). Initial rates of incorporation, as estimated from the 3-min points, followed normal Michaelis-Menten kinetics (Fig. 2A). The results are similar to those reported previously for Novikoff cells and other cell lines (1).

Quite different results were obtained, when dThd transport per se was measured in ATP-depleted cells. A steady-state intracellular concentration of free dThd, which was about equal to the extracellular concentration, was attained within 20 sec of incubation both at 25 and 37°C (Fig. 1C). The approach to equilibrium was fit to an integrated form of the first-order rate equation: $S_{i,t} = S_0 (1 - e^{-kt})$, where S_i is the intracellular concentration of substrate at time t, S_0 is the extracellular concentration of substrate, and k is a pseudo-first-order rate constant. The rationale for these fits has been described (4). Initial velocities (v_o) were calculated as the slope at zero time. At a dThd concentration of 2μM the v_o of transport was significantly higher than the dThd incorporation rates exhibited by the metabolizing cells (Fig. 1). All values reported in Fig. 1 are directly comparable, since they were obtained in experiments conducted on the same day with the same cell population.

Transport of dThd was a saturable process (Fig. 2B and C) and the "zero-trans" K_m and V_{max} were about 100 times higher than the corresponding values for incorporation into metabolizing cells. The dThd transport K_m at 37°C was significantly higher than that at 25°C. This finding does not imply that the affinity of the carrier for dThd differed

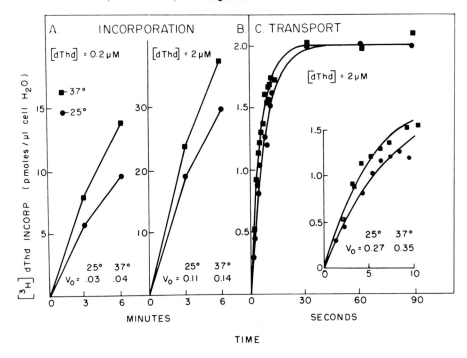

Fig. 1. dThd incorporation into total cell material by untreated NIS1-67 cells (A, B) and dThd transport into ATP-depleted cells (C). The cells were harvested from a single exponential phase culture and suspended to 2×10^6 cells/ml in BM42B (A, B) or to 2×10^7 cells/ml in glucose-free BM42B containing 5 mM KCN and 5 mM iodoacetate (C). The suspensions were incubated at 37°C for about 10 min and then one half of each suspension was equilibrated at 25°. The suspensions were supplemented with 0.2 μM (A) or 2 μM (B) [^{3}H]dThd (680 cpm/pmole) and incubated at the appropriate temperatures. Duplicate 1-ml samples of suspension were analyzed for radioactivity in total cell material (acid-soluble plus acid-insoluble). In (C) cell suspension and substrate solution were mixed in rapid succession (total 509 μl/mixture; 2 μM dThd final concentration, 500 cpm/ pmole) centrifuged through an oil layer and the cell pellet was analyzed for total radioactivity as described in Materials and Methods. The values were corrected for nonspecific trapping. The maximum level of intracellular dThd was assumed to represent equilibrium with the extracellular dThd concentration (2 μM). This assumption agreed in repeated experiments within 15% with results from independent intracellular [^{3}H]H_2O space determinations. The insert in (C) is a replot of some of the data at an expanded time scale. Initial velocities (v_0) of incorporation (A, B) and of transport (C) were estimated from these data (see text) and are expressed in pmoles/μl cell $H_2O \cdot$sec.

correspondingly. According to the kinetic equations developed by Eilam and Stein (8) to describe a simple carrier model, a differential effect of temperature on the mobility of the loaded and unloaded carrier would give the same results.

Kinetic parameters similar to those in Fig. 2 (25°C) were obtained in studies with the dThd kinase-deficient subline of Novikoff cells whether or not the cells were depleted of ATP (data not shown). In 9 experiments with these types of cells the K_m at 25°C ranged from 71 to 118 μM (mean = 88 μM) and the V_{max} values ranged from 11 to 28 pmoles/μl cell $H_2O \cdot$sec, with no obvious connection between the kinetic parameters and the growth stage of cells or the type of cell employed. The observed variability, therefore, may reflect limitations in the methodology, although more detailed studies are needed to determine whether real differences in the cell populations studied may also have been a contributory factor.

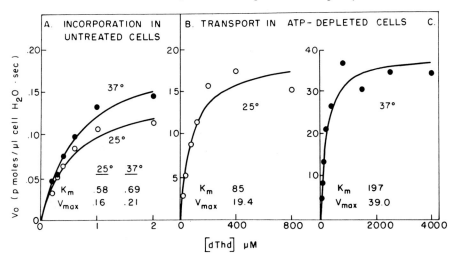

Fig. 2. Initial velocities of dThd incorporation by untreated N1S1-67 cells (A) and of dThd transport into ATP-depleted cells (B, C) as a function of dThd concentration. The experiment was conducted as described in the legend to Fig. 1, except that additional substrate concentrations were employed. Michaelis-Menten parameters were computed using the method of Wilkinson (10). A) Samples of the suspension of 2×10^6 cells/ml were supplemented with 0.2, 0.3, 0.4, 0.7, 1, and 2 μM [^{3}H]dThd (680 cpm/pmole) and the initial rate of incorporation was estimated from duplicate 1-ml samples of suspension analyzed for total cell-associated radioactivity after 3 min of incubation. B) Initial velocities were estimated from complete time courses of intracellular [^{3}H]dThd accumulation (see Fig. 1C). Samples of the suspension of 2×10^7 ATP-depleted cells/ml were mixed with [^{3}H]dThd to a final concentration of 500,000 cpm/sample and unlabeled dThd to final concentrations of 20, 40, 80, 120, 250, 400, and 800 μM. C) As described for B) except that dThd concentrations of 1.5, 2.5, and 4.0 μM were also used.

The discrepancy between the kinetic parameters of transport and incorporation was explored by using the rapid kinetic technique to measure the incorporation of 0.25, 20, and 320 μM dThd into 3 metabolite classes of wild-type cells: free dThd, dThd nucleotides, and DNA.

An intracellular steady-state level of free dThd was achieved in less than 60 sec, at the lower 2 concentrations, and by 200 sec at 320 μM dThd (Fig. 3). These steady states represent the balance between rates of influx, efflux, and phosphorylation of dThd and are determined by, among other things, the exogenous concentration of dThd.

At the lowest dThd concentration, 0.25 μM, the quantity of isotopic dThd incorporated into nucleotides exceeded that of intracellular free dThd after 2 min (Fig. 3A). At this substrate concentration, the incorporation of dThd into the nucleotide pool continued in a linear fashion for a few minutes. At the higher dThd concentrations, net incorporation into nucleotides practically ceased after 1–2 min of incubation (Fig. 3B and C), probably because of feedback inhibition of dThd kinase by TTP and a limited expandibility of the size of the dThd nucleotide pool (11). The amounts of dThd incorporated into DNA during the experimental periods were very small and have thus not been included in Fig. 3; at 23°C macromolecular synthesis in Novikoff cells nearly ceases.

The initial rates of dThd transport and phosphorylation calculated from the data in Fig. 3 are summarized in Table I. The initial rate of transport was calculated by summation of the initial rates of appearance of radioactivity in all intracellular components:

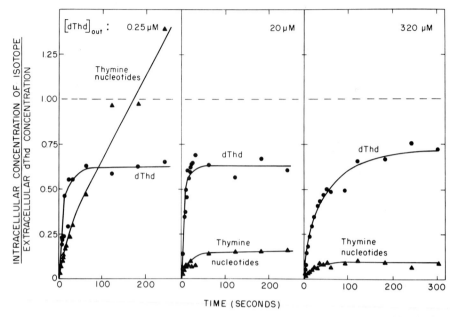

Fig. 3. Accumulation of labeled dThd and thymine nucleotides as a function of time of incubation with various concentrations of $[^3H]$dThd in the medium. Samples of a suspension of 4×10^7 untreated N1S1-67 cells/ml of BM42B were mixed at short intervals with solutions of $[^3H]$dThd to final concentrations of 0.25, 20, and 320 μM and 3,030 cpm/μl. The mixtures were centrifuged through an oil layer into a sucrose layer containing 0.2 N trichloroacetic acid, which was later fractionated into free dThd (●——●), thymine nucleotides (▲——▲), and DNA (not shown) as described in Materials and Methods. The data were normalized by dividing the intracellular concentration of dThd equivalents by the extracellular dThd concentration. The experiment was carried out at ambient temperature (23°C).

TABLE I. Initial Velocities of Thymidine Transport and In Situ Phosphorylation*

External thymidine concentration (μM)	Initial rates (pmoles/μl cell $H_2O \cdot$ sec)	
	Transport	Phosphorylation
0.25	0.0107	0.0025
20	1.61	0.126
320	8.13	1.82

*The experiment was conducted as described in the legend to Fig. 3. The initial rate of transport was calculated by summation of the initial rates of appearance of radioactivity in dThd, dThd nucleotides, and DNA; the initial rate of phosphorylation was similarly the sum of the rates of isotope appearance in dThd nucleotides and DNA.

dThd, dThd nucleotides, and DNA; the initial rate of phosphorylation was similarly the sum of the rates of isotope appearance in dThd nucleotides and DNA. Due to the complexity of the experimental procedure the rates thus obtained must be regarded as estimates. It is obvious, nevertheless, that at all concentrations of dThd tested, the rate of transport was significantly higher than the rate of phosphorylation.

DISCUSSION

Our results clearly show that at all dThd concentrations examined, the rate of dThd transport into the cells occurs at appreciably higher rates than its subsequent intracellular phosphorylation. This conclusion is based on comparisons of the rates of dThd incorporation into total cell material by metabolizing cells with the rates of dThd transport measured directly with cells in which the substrate is not phosphorylated because of lack of ATP or dThd kinase (Figs. 1 and 2), as well as on results from experiments in which both the intracellular accumulation of free dThd and its conversion into nucleotides was analyzed in metabolizing cells (Fig. 3; Table I). When cells are exposed to labeled dThd an intracellular steady-state level of free dThd is attained within 30–200 sec, depending on the extracellular dThd concentration. The fact that even at low extracellular dThd concentrations sizable amounts of free dThd are rapidly accumulated intracellularly seems to exclude a direct involvement of dThd kinase in transport.

At relatively low dThd concentrations (2 μM and below), where the intracellular steady state of dThd is attained quickly, the overall rate of dThd incorporation into cell material (measured at relatively long time scales) must reflect the rate of phosphorylation. Within this range of dThd concentrations, incorporation appears saturable (cf. Fig. 2A; apparent K_m = 0.6 μM). Yet it will be noted in Fig. 3 that, at much higher dThd concentrations, the rate of phosphorylation continues to rise, suggesting that the kinetic behavior of the phosphorylation reaction in situ is more complicated. We do not yet know whether the complication is due to multiple kinases or to an increase in apparent K_m as a function of concentration of the kinase effector dTTP.

In any case, the "zero-trans" K_m for transport in nonmetabolizing cells is about 100 times higher than the K_m apparent in Fig. 2A. Other experiments (12) have shown that the transport system exhibits a broad substrate specificity, and thus resembles the nucleoside transport systems described by Oliver and Patterson, for human erythrocytes (13) and by Taube and Berlin for rabbit polymorphonuclear leukocytes (14). Uridine and dThd are not metabolized in human erythrocytes and initial rates of uptake were estimated from the linear disappearance of substrate from the culture fluid over a 30-sec period. A "zero-trans" K_m for uridine transport of 710 μM was obtained, a value significantly higher than that for dThd transport in Novikoff cells. Uptake data for dThd did not yield a straight line in a Lineweaver-Burk plot, but it was concluded that dThd is transported by the same system as uridine, since it accelerates the efflux of uridine from the cells (13).

A "zero-trans" K_m for dThd transport by polymorphonuclear leukocytes of 50 μM (14) was based on measuring dThd incorporation into metabolizing cells. After 45 sec of incubation with 40 μM [^{3}H] dThd only 17% of the intracellular radioactivity was associated with free dThd, but dThd incorporation was linear for 45 sec; thus the authors assumed that the 45-sec time points used to estimate initial rates were short enough to be valid estimates of the initial rate of dThd transport. It is obvious from our studies (Fig. 3), that in Novikoff cells at least, intracellular dThd concentrations close to the steady-state level are attained in about this time period and thus a 45-sec time point cannot be used to estimate initial rates. The magnitude of the error will depend on the intracellular level of unmetabolized substrate; at concentrations approaching those found in the medium, transport rates will be seriously underestimated because of the high rate of efflux (2).

Our results emphasize that true initial rates of dThd transport are very difficult to obtain in metabolizing cells. In cells incapable of metabolizing dThd, time courses of substrate accumulation in the cells may be fit to an equation describing the approach to intra-extracellular equilibrium and thus time points at which a sizable amount of dThd has accumulated intracellularly still yield information from which initial velocities can be computed. Intracellular accumulation of substrate is linear for an impractically short time because of backtransport of the substrate.

We are not yet equipped with equations to describe the time course of dThd accumulation and phosphorylation in metabolizing cells. Such equations must allow for the kinetics of dThd kinase in situ. From our preliminary data (Fig. 3) it seems that the in situ rates of phosphorylation do not bear a simple hyperbolic relation to dThd concentration. These data also show that at higher concentrations of dThd in the medium (20 μM and above) the dTTP pool becomes maximally labeled within a very short period of time, probably due to feedback inhibition of the dThd kinase and the limited expandability of the dTTP pool (11). Under these conditions the eventual rate of dThd incorporation by the cells will reflect the rate of dTTP incorporation into DNA. Thus our data in Fig. 3 and Table I do not as yet allow calculation of a K_m of the dThd transport process in metabolizing cells. Nevertheless, the time courses of accumulation of dThd in these cells are generally consistent with a value of about 85 μM observed in ATP-depleted or dThd kinase-deficient cells.

ACKNOWLEDGMENTS

This work was supported by USPHS Research Grant CA 16228 and Research Fellowship CA 00800 to R.M.

REFERENCES

1. Plagemann PGW, Richey DP: Biochim Biophys Acta 344:263, 1974.
2. Berlin RD, Oliver JM: Int Rev Cytol 42:287, 1975.
3. Plagemann PGW, Marz R, Erbe J: J Cell Physiol 89:1, 1976.
4. Wohlhueter RM, Marz R, Graff JC, Plagemann PGW: J Cell Physiol 89:605, 1976.
5. Plagemann PGW, Swim HE: J Bacteriol 91:2317, 1966.
6. Graff JC, Plagemann PGW: Cancer Res 36:1428, 1976.
7. Plagemann PGW, Erbe J: J Cell Physiol 83:321, 1974.
8. Eilam Y, Stein WD: Methods Membr Biol 2:283, 1974.
9. Plagemann PGW: J Cell Physiol 77:213, 1971.
10. Wilkinson GN: Biochem J 80:324, 1961.
11. Plagemann PGW, Erbe J: J Cell Biol 55:161, 1972.
12. Wohlhueter RM, Marz R: Fed Proc 36:692, 1977.
13. Oliver JM, Paterson ARP: Can J Biochem 49:262, 1971.
14. Taube RA, Berlin RD: Biochim Biophys Acta 255:6, 1972.

Journal of Supramolecular Structure 6:473–484 (1977)
Molecular Aspects of Membrane Transport 303–314

The Sialoglycoprotein Subunits of Human Placental Brush Border Membranes Characterized by Two-Dimensional Electrophoresis

H. Garrett Wada, Stella Z. Górnicki, and Howard H. Sussman

Department of Pathology, Laboratory of Experimental Oncology, Stanford University Medical School, Stanford, California 94305

A brush border membrane enriched fraction was isolated from human, full-term placenta. This membrane fraction exhibited large membrane fragments with microvilli projecting from the basal membrane in electron micrographs and was enriched tenfold in alkaline phosphatase, a brush border enzyme marker. The sialoglycoproteins associated with this membrane fraction were tritiated by mild periodate oxidation of sialic acid and reduction with tritiated $NaBH_4$. The membranes were solubilized in 8 M urea, 2% Triton X-100, and the tritiated glycoprotein subunits were reduced with β-mercaptoethanol and characterized by 2-dimensional polyacrylamide gel electrophoresis using a method similar to that described by O'Farrell and Bhakdi, Knüferman, and Wallach. The tritiated subunits were detected in the gels by autofluorography. The 2-dimensional subunit "maps" resolved at least 17 major sialoglycoprotein subunits whereas only 10 major periodate-Schiff reagent staining components were resolved by 1-dimensional SDS polyacrylamide gel electrophoresis. Placental alkaline phosphatase (PAP) was identified on the subunit maps by inclusion of ^{32}P-labeled PAP in the tritiated membrane sample. The ^{32}P-labeled PAP corresponded to a major tritiated sialoglycoprotein subunit, which was heterogeneous with respect to charge as demonstrated by 3 closely running spots of the same molecular weight.

Key words: placenta, brush border, sialoglycoprotein, alkaline phosphatase, two-dimensional electrophoresis

The brush border membrane of the syncitial trophoblast of human placenta is in direct contact with the maternal circulation and, therefore, represents the first placental barrier through which numerous solutes are transportable. The human placenta has been demonstrated to actively transport L-amino acids, creatine, Na^+, and K^+, vitamin B_{12}, and acetylcholine (1), and to passively transport sugars, creatinine, p-aminohippurate, urea, tetraethyl ammonium, norepinephrine, and antipyrine (1). Consequently, the brush border membrane of this organ must possess the capabilities for facilitated diffusion and, perhaps, active transport.

The ability to saturate the placental amino acid transport systems and demonstrate

Received March 14, 1977; accepted May 17, 1977

Michaelis-Menten kinetics for transport processes (2, 3) indicate the presence of carrier systems in the placental membranes. Furthermore, the specificity which many transport systems exhibit suggests the presence of carriers at the cell surface which can discriminate between transported and nontransported molecules (4).

Although the mechanisms for solute transport by placental tissue has been investigated by kinetic methods, the molecular components which participate in these mechanisms have not been identified. Since one subunit of the Na^+,K^+-stimulated ATPase has been demonstrated to be a glycoprotein (5) and most proteins on the external cell surface appear to be glycosylated (6, 7), the cell surface glycoproteins of the human placental brush border membranes are prime candidates for transport mediating membrane components. In this paper, we report the characterization of the sialoglycoprotein subunits of the isolated placental brush border by a 2-dimensional electrophoretic technique utilizing isoelectric focusing in the first dimension and sodium dodecyl sulfate (SDS)-polyacrylamide gel electrophoresis in the second. Apparent isoelectric point (pI) and apparent subunit molecular weight (M_r) were calculated from the relative positions of the subunits in two-dimensional maps. The utility of the mapping of membrane subunits was demonstrated by the identification of placental alkaline phosphatase in the maps by covalent labeling of the enzyme with $[^{32}P]$ orthophosphate. Using the physical characterization of the glycoprotein membrane components which the subunit maps provide, further studies may associate specific subunits with transport processes or other cellular functions.

MATERIALS

Substrates for enzyme assays, sodium meta-periodate and E. coli alkaline phosphatase, type III, were obtained from Sigma Chemical Co. Reagents for polyacrylamide gels were from Bio-Rad Laboratories and Canalco, and ampholytes were from LKB Produkter AB. Materials for scintillation spectrophotometry and autofluorography, and scintillation grade Triton X-100 were from New England Nuclear; tritiated sodium borohydride (6 Ci/mmole) was from Schwartz-Mann or Amersham-Searle; and ultrapure urea was obtained from Schwartz-Mann. Other chemicals used were J. T. Baker "analyzed reagent" grade unless otherwise specified.

METHODS

Membrane Preparations

The microvillus membranes of full-term human placentas were prepared by the methods of Carlson, Wada, and Sussman (8) with the following modifications. Minced placental tissue, washed free of maternal blood, was suspended in 5 volumes of isotonic dextrose and stirred in a Waring blender at 4°C for 2.5 min, using a rheostat to reduce the rpm to the lowest possible setting. The connective tissue and major blood vessels remained intact, but soft villus tissue was disrupted. When free nuclei were observed in the suspension by phase contrast microscopy, the undisrupted tissue was filtered off with plastic window screen (1/16 inch mesh), and the purification was carried on from this step as previously described, substituting 5 mM NaEDTA, pH 7.4, for 4 mM $NaHCO_3$, pH 8.1, 1 mM $MgCl_2$ as the buffer in sucrose solutions. The microvillus membranes obtained from this method were found to be of the same purity as those obtained from the earlier methods. The advantage of the modification is the omission of a tedious step in which placental villus tissue is scraped from blood vessels and connective tissue.

Tritiation of Placental Brush Border Membrane Fractions

Placental brush border membranes were labeled with tritium by a modification of the method of Blumenfeld, Gallop, and Liao (9) which utilized oxidation of sialic acid by mild periodic acid oxidation and reduction by $[^3H]NaBH_4$. The conditions for oxidation were that of Van Lenten and Ashwell (10) which were highly specific for sialic acid.

Membranes were suspended in 0.9 ml of cold 0.1 M sodium acetate, 0.15 M NaCl, pH 5.6, and the oxidation was initiated by addition of 0.1 ml of 50 mM sodium meta-periodate. The oxidation was allowed to proceed at $0°C$ in the dark for 10 min and was then terminated by dilution with 2 volumes of cold 50 mM sodium phosphate, 0.15 mM NaCl, pH 7.4 (PBS) and pelleting the placental membranes at $12,000 \times g$ for 10 min. The membranes were washed twice with 3 ml of PBS. The oxidized membranes were suspended in 1 ml of PBS, pH 7.4, and reduced with 1 -2 mCi of $[^3H]NaBH_4$ (6–10 Ci/mmole) for 30 min at $0°C$. Unlabeled $NaBH_4$, approximately 1 mg, was added and reduction was allowed to proceed for 10 min at $0°C$. The membranes were then diluted with 2 volumes of PBS, pelleted, and washed with 3 ml of PBS. The tritiated membranes were stored frozen at $-20°C$.

Labeling of Placental Alkaline Phosphatase

A modification of Milstein's procedure (11) to covalently bind $^{32}PO_4$ to purified E. coli alkaline phosphatase was used to ^{32}P-label placental alkaline phosphatase. The procedure was previously described (8) utilizing an incubation of $^{32}PO_4$ with alkaline phosphatase at pH 5.0, acid denaturation, and acetone precipitation of labeled enzyme. The precipitate was redissolved in 8 M urea, 5% β-mercaptoethanol, 2% Triton X-100, 5 mM $NaPO_4$, pH 8.0, for electrophoretic analysis.

Standards of placental alkaline phosphatase were purified to homogeneity according to the method of Sussman et al. (12). Standards of PAP were tritiated by mild periodate oxidation and reduction by $[^3H]NaBH_4$ according to the method of Van Lenten and Ashwell (10).

Two-Dimensional Electrophoresis

Solubilization of membrane glycoproteins. The tritiated membrane preparations were extracted with 8 M urea, 2% Triton X-100, 5 mM $NaPO_4$, pH 8.0 (saturated with PMSF), at room temperature for 5 min according to Bhakdi, Knüferman, and Wallach (13), chilled to $4°C$ on an ice bath, and centrifuged at $140,000 \times g$ for 60 min at $4°C$. Under these conditions 50–60% of the acid precipitable tritium extracted into the supernatant from placental membranes. The supernatant was treated with 5% β-mercaptoethanol for 10 min at room temperature and utilized for electrophoretic analysis immediately or after concentration by dialysis against a dessicant at $4°C$.

The first dimension: isoelectric focusing. A 2-dimensional electrophoretic system very similar to that described by O'Farrell (14) was utilized to resolve the membrane glycoproteins. The first dimension was isoelectric focusing (IEF) in polyacrylamide gels containing 8 M urea, 0.5% (vol/vol) Triton X-100, 4% acrylamide (10% crosslinker), 10% glycerol, 1% 3–10 pH Ampholine (LKB Produkter) which were cast in 3×100 mm rods. The cathode buffer was 1% NaOH, and the anode buffer was 1% H_3PO_4. The gels were prefocused for 1 h at 150 V. The samples were applied in 100 μl of extraction buffer containing a trace of bromphenol blue. The samples were run in at 150 V for 2 h and focused at 400 V for at least 14 h at $18°C$, constant voltage. After termination of the focusing, the pH gradient was measured in one gel using a membrane electrode (Brinkman Instruments). The other gels were soaked at room temperature in equilibration buffer (2% SDS, 4% β-mercaptoethanol, 10% glycerol, 62.5 mM Tris-HCl, pH 6.8) with gentle agitation for 1 h. Equilibrated gels were either run in the second dimension immediately or stored at −20°C.

IEF gels were sliced into 1-mm slices and counted for ^{3}H and ^{32}P by scintillation spectrophotometry. Each gel slice was dissolved in 0.5 ml of 30% H_2O_2, 0.7% perchloric acid at 37°C for 48 h, and counted with 5 ml of scintillation cocktail (1 vol Triton X-100:2 vol 5 gm Omnifluor dissolved in 1 liter of toluene).

The second dimension: sodium dodecyl sulfate-polyacrylamide gel electrophoresis. The second dimension electrophoresis was run in 0.1% sodium dodecylsulfate (SDS), 8% polyacrylamide gel slabs (1.5 mm thick × 14 cm wide × 10 cm high) according to Laemmli (15). The equilibrated, isoelectric focusing gel rod was sealed to the top of the stacking gel with 0.75% agar solution as described by O'Farrell (14). The sample was electrophoresed into the stacking gel for 20 min at 15 mA/slab with 2% (wt/vol) SDS in the top buffer. The top buffer was replaced with 0.1% SDS buffer containing a trace of bromphenol blue, and the run was continued at 20 mA/slab until the tracking dye reached the bottom of the slab. The gel was then fixed in 50% methanol for 2 h and impregnated with PPO for auto-fluorography according to Bonner and Lasky (16). The PPO impregnated, vacuum dried gels were autofluorographed on Kodak Royal RP X-Omat medical x-ray film at −76°C to detect tritiated glycoprotein subunits. Globular protein standards were used to estimate subunit molecular weights of the glycoprotein spots by the method of Weber and Osborn (17).

RESULTS

The Brush Border Membrane Preparation

The membranes prepared from human, full-term placenta exhibited ultrastructural and enzyme marker characteristics compatible with a brush border membrane. Electron-micrographs demonstrated microvilli projecting from the basal membrane sheets and may represent the terminal web of the brush border (Fig. 1). A brush border enzyme marker, alkaline phosphatase, was enriched tenfold over the tissue homogenate, while other enzyme markers for contaminating organelles remained constant or were decreased (8).

Tritiation of Sialoglycoproteins

The glycoproteins of the placental brush border membranes were heavily labeled by mild periodic acid oxidation and [^{3}H]NaBH$_4$ reduction under conditions reportedly specific for sialic acid (10). The lack of tritium incorporation into a nonglycosylated membrane protein, the 45,000 M_r "actin-like" protein, indicated specific tritiation of glycoproteins. Furthermore, control studies with purified transferrin as a model sialoglycoprotein demonstrated that 85% of the incorporated tritium was on sialic acid which was labile to neuraminidase digestion.

Extraction of 50–60% of the acid precipitable tritiated glycoproteins from the membranes was accomplished by 8 M urea, 2% Triton X-100. Triton X-100 alone would only extract 25% of the acid precipitable tritium.

Two-Dimensional Electrophoresis

The tritiated, extracted glycoprotein subunits were focused in IEF polyacrylamide

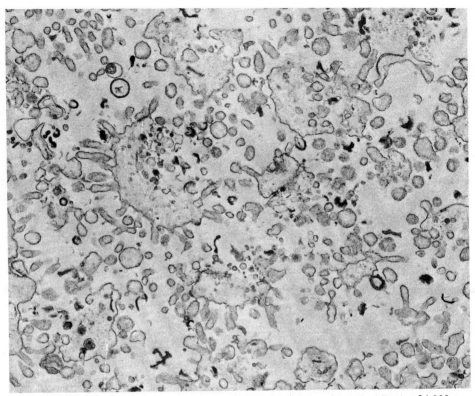

Fig. 1. Electron micrograph of isolated placental brush border membranes. Magnification 24,000 ×
shows representative field of placental brush border membrane preparation.

gels containing 0.5% Triton X-100 and 8 M urea. Lowering of the Triton X-100 gel concentration to 0.1% caused incomplete focusing of the subunits which may be explained by
nonspecific aggregation. Temperature also was critical in attaining complete focusing. At
4°C, focusing of subunits was incomplete in 14 h, while at 18°C focusing was complete
in 14 h. In the second dimensional SDS-polyacrylamide gels, the best definition and resolution of spots was obtained when 2% SDS buffer was used as top tank buffer for the first
20 min of electrophoresis. The higher SDS concentration of the leading ion front may be
necessary to remove excess Triton X-100 associated with the glycoprotein subunits.

   The modified 2-dimensional electrophoretic technique produced subunit maps
resolving the labeled glycoproteins of the placental brush border membranes into 17
major subunits and at least 33 total subunits (Fig. 2a), ranging in M_r from 262,000 to
40,000. Using a one-dimensional SDS-polyacrylamide gel system, only 10 major PAS
positive bands were detected in the same placental brush border preparation (8).

   A control experiment in which placental brush border membranes were reduced with
[^{3}H] NaBH$_4$ prior to meta-periodate oxidation indicated that the labeling of spots identified as sialoglycoproteins was dependent on oxidation. Unoxidized membranes incorporated
only 2% of the tritium incorporated into oxidized membranes as measured by TCA precipitable counts from Triton-urea membrane extracts. Equivalent exposures of 2-dimensional
gels indicated no detectable labeling of sialoglycoprotein spots from unoxidized membranes,

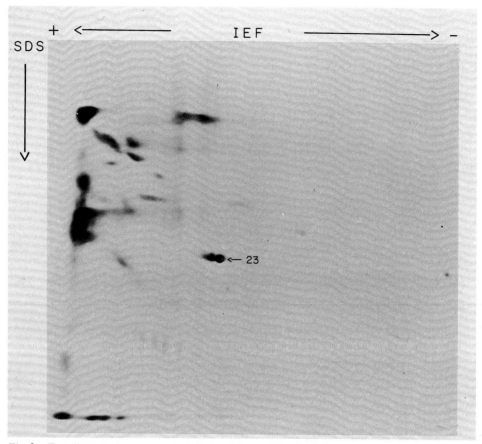

Fig. 2a. Two-dimensional electrophoretic map of placental brush border sialoglycoprotein subunits. 1.4×10^6 dpm of tritiated membrane glycoprotein was focused in a 3–10 pH gradient IEF gel rod for the first dimension and electrophoresed in a 8% acrylamide, 0.1% SDS gel slab for the second dimension. The dried, PPO impregnated gel was autofluorographed at $-76°$C for 18 h.

which are heavily labeled in gels containing equal amounts of protein from oxidized membranes.

The pIs of the subunits resolved in the subunit map were estimated from the co-ordinates (R_f) measured for the center of each spot, using a pH vs R_f standard curve which was determined for each run (Fig. 2b). The glycoprotein subunits ranged in pI from pH 7.2 to pH 4.6 with the majority in the 6–4.6 range (Table I). Most of the glycoprotein subunits were heterogeneous with respect to charge which was demonstrated by close running groups of spots with the same $M_{\overline{r}}$ or single spots with a broad band of pI distribution (Fig. 2).

Identification of a Specific Sialoglycoprotein in Subunit Maps

A specific brush border membrane component, placental alkaline phosphatase (PAP), was identified in the subunit maps by coelectrophoresis of ^{32}P-labeled PAP with tritiated membrane subunits. The ^{32}P-labeled PAP corresponded in the maps with a major sialo-glycoprotein subunit of the placental brush border membrane (Fig. 3). The membrane

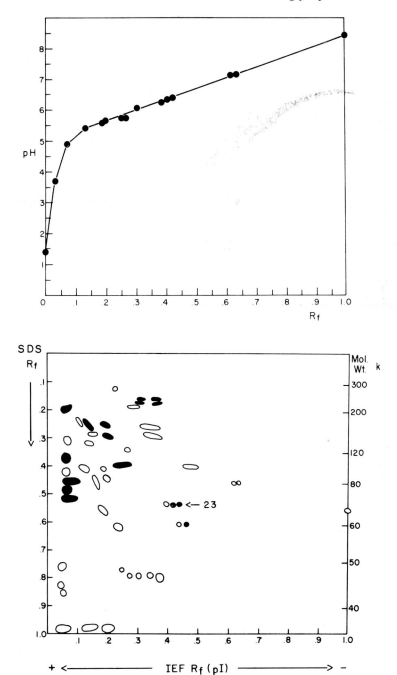

Fig. 2b. Coordinates of sialoglycoprotein subunits in 2-dimensional electrophoretic maps. The top figure shows the pH gradient obtained in the first dimensional IEF gel. The bottom figure illustrates the sialoglycoprotein subunits plotted with respect to R_f in the 2 dimensions. The molecular weight (M_r^-) scale and pI scale were extrapolated from R_f vs M_r^- or pH standard curves. This figure was obtained by tracing the spots directly from the autofluorogram, and the R_f values were assigned from the axis formed at the bottom by the leading ion front of the SDS dimension, and the axis formed at the right side by labeled material which does not penetrate the IEF gel. The left limiting boundary of the map was determined by slicing and counting IEF gels to find the R_f of the tritiated material closest to the anode.

TABLE I. Coordinates of Placental Brush Border Sialoglycoprotein Subunits in 2-Dimensional Maps

Spots		R_f		Apparent pI	Apparent M_r (K)
		X(pI)	Y(M_r)		
1		0.224	0.140	5.72	262.0
2		0.308	0.170	6.09	235.0
3		0.359	0.190	6.19	218.0
4	a	0.060	0.200	4.60	210.0
	b	0.283		5.93	210.0
5		0.368	0.240	5.88	181.0
6	a	0.131	0.258	5.45	170.0
	b	0.194		5.68	170.0
7		0.368	0.272	5.88	161.0
8		0.200	0.294	5.68	148.0
9		0.104	0.250	5.18	175.0
10		0.160	0.286	5.50	153.0
11		0.060	0.307	4.60	142.0
12		0.136	0.318	5.40	136.0
13		0.263	0.344	5.86	123.0
14		0.060	0.381	4.60	107.0
15	a	0.252	0.400	5.82	100.0
	b	0.469		6.57	100.0
16	a	0.131	0.420	5.36	94.3
	b	0.187		5.60	94.3
17		0.197	0.444	5.63	85.2
18		0.202	0.422	5.65	92.4
19		0.162	0.461	5.51	80.0
20		0.060	0.459	4.60	77.8
21		0.060	0.488	4.60	72.5
22	a	0.616	0.463	7.10	74.6
	b	0.631		7.19	74.6
23	a	0.389	0.550	6.22	66.2
	b	0.404		6.30	66.2
	c	0.424		6.39	66.2
24		0.177	0.572	5.60	64.2
25		0.227	0.633	5.73	59.0
26	a	0.429	0.622	6.43	59.9
	b	0.449		6.50	59.9
27		0.066	0.769	4.60	49.0
28	a	0.242	0.794	5.79	47.0
	b	0.272		5.89	47.0
	c	0.303		6.00	47.0
	d	0.338		6.12	47.0
	e	0.369		6.22	47.0
29		0.505	0.844	6.70	44.1
30		0.060	0.871	4.60	42.5
31		0.060	1.00	4.60	< 40
32		0.146	1.00	5.45	< 40
33		0.207	1.00	5.66	< 40

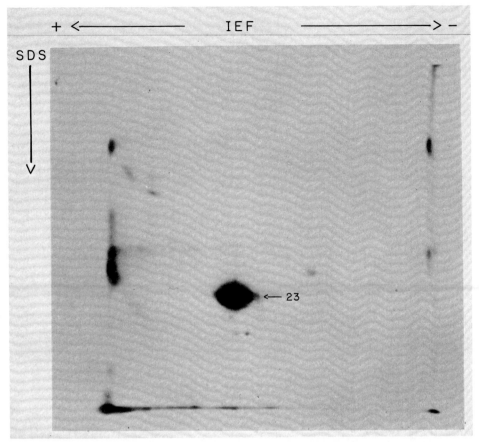

Fig. 3. Identification of placental alkaline phosphatase in 2-dimensional electrophoretic maps. 1.4×10^6 dpm of tritiated membrane glycoprotein and 0.2 μg of ^{32}P-labeled purified PAP were run together in a 2-dimensional gel as described for Fig. 2. The dried, PPO impregnated gel was autofluorographed at $-76°$C for 6 h.

component corresponding to PAP was heterogeneous with respect to charge, giving 3 closely running spots with the same $M_{\overline{r}}$, 66,200 (spot 23, Table I).

A rigorous measurement of the pI of PAP subunits was attained from 1-dimensional IEF gels run on purified PAP. In these experiments a mixture of unlabeled PAP subunits with tritiated PAP subunits showed correspondence between the Coomassie blue stain for protein and counts of tritium. This indicated these subunits had the same pI. However, the active site phosphorylation of PAP caused a 0.1 pH unit, acid shift in pI of the subunits as shown in Fig. 4a using ^{32}P-labeling of the enzyme. IEF gels run on purified E. coli alkaline phosphatase indicated that a similar, 0.1 pH unit, acid shift in pI was present for the phosphorylated subunit relative to unlabeled material in the same gel (Fig. 4b). The occurrence of 2 forms of the unlabeled E. coli phosphatase resolved in the IEF gels was probably due to deamidation of the nascent polypeptide which occurs with long term storage of the purified enzyme (unpublished observation).

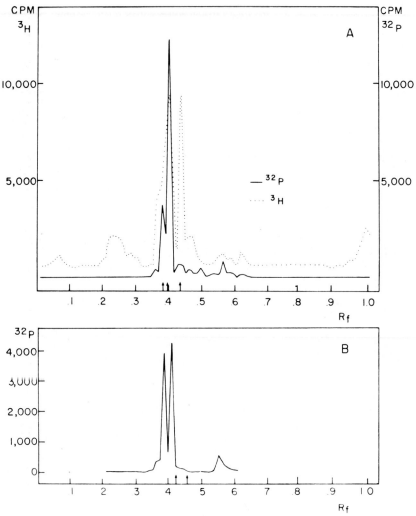

Fig. 4. The effect of phosphorylation on the pI of alkaline phosphatase subunits. A) 1 μg of purified PAP tritiated by the method of Van Lenten and Ashwell (10) and 1 μg of ^{32}P-labeled PAP were focused together in an IEF gel. The gel was fixed in 50% methanol for 2 h, stained with Coomassie blue (.02% Coomassie blue, 5% trichloroacetic acid, 5% sulfosalicylic acid, 18% methanol) for 2 h, destained in 25% ethanol, 5% acetic acid, and stored in 7% acetic acid. Gel was frozen, sliced into 1-mm sections, and solubilized for double label counting of ^{3}H and ^{32}P. The pH gradient in IEF gels were measured on a duplicate gel.

Calculated pI (3 sets of gels)

R_f	0.375	0.397	0.433
Tritiated PAP		6.32 ± 0.11	6.43 ± 0.08
Phosphorylated PAP	6.23 ± 0.07	6.34 ± 0.08	
pI Shift	0.09	0.08	

B) 2 μg of unlabeled E. coli alkaline phosphatase and 0.2 μg of ^{32}P-labeled E. coli enzyme were co-focused in IEF gels, stained, sliced, and counted for ^{32}P. Positions of protein bands are denoted by arrows.

Calculated pI (single set of gels)

R_f	0.389	0.413	0.423	0.456
Unlabeled E. coli AP			6.38	6.50
Phosphorylated E. coli AP	6.30	6.36		
pI shift	0.08	0.14		

312:MAMT

DISCUSSION

The tritiation of placental brush border membranes by a method specific for sialic acid and 2-dimensional electrophoresis of the extracted, labeled membrane components has enabled the characterization of a complex mixture of brush border membrane sialoglycoprotein subunits. Using the parameters of pI and $M_{\overline{r}}$, it was possible to assign coordinates to each subunit resolved. The method was found to be very reproduceable as evidenced by the consistent reproduction of the same pattern of spots for placental membranes from run to run, with duplicate runs yielding identical patterns.

The accuracy of the $M_{\overline{r}}$ assignments from the 2-dimensional maps are similar to those from 1-dimensional SDS-polyacrylamide gel electrophoresis which were reported to be accurate to a standard deviation of 10% (17). It should also be noted that the $M_{\overline{r}}$ for the sialoglycoprotein subunits may not be an accurate estimate of the true subunit molecular weight, due to the anomalous migration in SDS gels of heavily glycosylated subunits (19, 20). The assignment of pI from the focusing dimension of the 2-dimensional maps was found to be more variable than the $M_{\overline{r}}$ assignment. Reproduceability was dependent on complete focusing of subunits, uniformity of IEF gel preparation, and careful pH measurements on the focused gels. The assigned pIs must also be considered to be an apparent value obtained under the particular experimental conditions. For example, native PAP has a mean pI of 4.4 in sucrose gradient IEF (21), while in the present study PAP subunits were found to have a mean pI of 6.30 in IEF polyacrylamide gels containing 8 M urea, 0.5% Triton X-100.

The careful examination of the pI of purified PAP in 1-dimensional IEF gels indicated that tritiation of sialic acid did not alter the pI of the enzyme subunits while phosphorylation caused an acid shift similar to the acid shift observed for the phosphorylated E. coli alkaline phosphatase.

The E. coli enzyme is not glycosylated and is a dimer coded by a single cistron (22, 23), making this enzyme an ideal model for the effect of phosphorylation on pI. The acid shift in pI was expected, since phosphorylation involves the addition of a negatively charged moiety to an uncharged serine residue at the active site of the molecule (18). Based on the evidence that E. coli (24), bovine kidney (25), and intestinal (26) alkaline phosphatases are phosphorylated at a single site per subunit, phosphorylation of PAP would add 1.48 electron charges per subunit.* Since the differences in pI between the multiple molecular forms of unlabeled PAP are similar to the acid shift caused by phosphorylation, the heterogeneity of PAP is due to single charge differences. Peripheral differences, such as deamidation (27) or sialic acid content of the carbohydrate moiety (28) would explain the microheterogeneity of PAP and the other sialoglycoprotein subunits identified in the 2-dimensional, subunit maps. It has been previously shown for the alkaline phosphatase of pig kidney that varying degrees of sialylation was the cause of microheterogeneity, and neuraminidase reduced the different forms of the enzyme to a single, desialylated enzyme (30).

Using PAP as an example, we have shown that 2-dimensional electrophoresis can be used in conjunction with affinity labeling to resolve and identify sialoglycoproteins of the human placental brush border membrane. The identification of other membrane components such as transport carrier molecules and membrane hormone receptors may also be possible utilizing this same approach.

---

*The addition of 1.48 electron charges by phosphorylation is based on pKa of 1.60 and 6.62 for a model organophosphate, ethyl phosphate (29) at pH 6.3.

ACKNOWLEDGMENTS

This work was supported by Public Health Service Research Grant CA13533 from the National Cancer Institute. Dr. Wada is a Postdoctoral Fellow supported by Public Health Service Grant CA05150 from the National Cancer Institute. The authors would like to thank Dr. Klaus Bensch of the Department of Pathology, Stanford University School of Medicine, for electron micrographs of the membrane preparations.

REFERENCES

1. Miller RK, Berndt WO: Life Sci 16:7, 1975.
2. Smith CH: Am J Physiol 224:558, 1973.
3. Miller RK, Berndt WO: Am J Physiol 227:1236, 1974.
4. Oxender DL: Annu Rev Biochem 4:777, 1972.
5. Kyte J: J Biol Chem 246:4157, 1971.
6. Winzler RJ: In Gottschalk A (ed): "Glycoproteins – Their Composition, Structure, and Function." BBA Library, New York: Elsevier Publishing Company, vol 5, pt B, chap 11, section 2, 1972, pp 1268–1293.
7. Guidotti G: Annu Rev Biochem 41:731, 1972.
8. Carlson RW, Wada HG, Sussman HH: J Biol Chem 251:4139, 1976.
9. Blumenfeld OO, Gallop PM, Liao TH: Biochem Biophys Res Commun 48:242, 1972.
10. Van Lenten L, Ashwell G: J Biol Chem 246:1889, 1971.
11. Milstein C: Biochem J 92:410, 1964.
12. Sussman HH, Small PA Jr, Cotlove E: J Biol Chem 243:160, 1968.
13. Bhakdi S, Knüfermann H, Wallach DFH: Biochim Biophys Acta 394:550, 1975.
14. O'Farrell PH: J Biol Chem 250:4007, 1975.
15. Laemmli UK: Nature (London) 277:680, 1970.
16. Bonner W, Lasky RA: Eur J Biochem 46:83, 1974.
17. Weber K, Osborn M: J Biol Chem 244:4406, 1969.
18. Schwartz JH, Crestfield AM, Lipman L: Proc Natl Acad Sci USA 49:722, 1963.
19. Fish WW: Methods Membr Biol 4:254, 1975.
20. Cook GMW, Stoddart RW: "Surface Carbohydrates of the Eukaryotic Cell." New York: Academic Press, 1973, chap 3, p 135.
21. Greene PJ, Sussman HH: Proc Natl Acad Sci USA 70:2936, 1973.
22. Kelley PM, Neumann PA, Scriefer K, Cancedda F, Schlesinger MJ, Bradshaw RA: Biochemistry 12:3499, 1973.
23. Piggot PJ, Sklar MD, Gorini L: J Bacteriol 110:291, 1972.
24. Engström L: Acta Soc Med Ups 64:214, 1959.
25. Cathala G, Brunel C, Chappelet-Tordo D, Lazdunski M: J Biol Chem 250:6046, 1975.
26. Chappelet-Tordo D, Fosset M, Iwatsubo M, Gache C, Lazdunski M: Biochemistry 13:1788, 1974.
27. Robinson AB, McKerrow JH, Cary P: Proc Natl Acad Sci USA 66:753, 1970.
28. Montgomery R: In Gottschalk A (ed): "Glycoproteins – Their Composition, Structure, and Function." BBA Library. New York: Elsevier Publishing Company, vol 5, pt A, chap 4, section 4, 1972, pp 518–528.
29. Sober HA (ed): "Handbook of Biochemistry, Selected Data for Molecular Biology." Chicago: The Chemical Rubber Publishing Company, 1970, p J-190.

Journal of Supramolecular Structure 6:485–494 (1977)
Molecular Aspects of Membrane Transport 315–324

Hexose Transport Regulation in Cultured Hamster Cells

C. William Christopher

John Collins Warren Laboratories, Huntington Memorial Hospital, Harvard University at the Massachusetts General Hospital, Boston, Massachusetts 02114

Hamster (nil) cells maintained overnight in culture medium containing cycloheximide and either glucose or fructose exhibit strikingly different rates of hexose transport and metabolism (i.e., uptake). Pretreatment of cultures with sulfhydryl reagents makes it possible to determine initial transport rates for a physiological sugar such as galactose which is a catabolite in hamster cells. Using galactose transport as a model, hexose uptake enhancements can now be shown to be due almost entirely to increases in the rate of the transport step. The transport regulation can best be accounted for by a model comprised of 2 antagonizing mechanisms. This model involves turnover of transport carriers as well as inhibitory units ("regulators"). The experimental as well as the theoretical model may also apply to the well-known uptake enhancements observed in oncogenically transformed cells.

Key words: cell culture, hexose transport, N-ethylmaleimide, derepression, catabolite inactivation, regulatory factor

The distinction between the transport step for natural (or metabolizable) hexoses and their subsequent metabolism has been difficult to determine experimentally. This is due in part to the uncertain degree to which metabolizable sugars are phosphorylated in vivo by their kinases and in part to the fact that nonmetabolizable analogs such as 3-O-methyl-D-glucose (3-O-meG) may not truly represent their natural counterparts (1, 2). D-Galactose is a natural analog of D-glucose and, relative to glucose metabolism, has a much simpler pattern of metabolic products (3–5). Using D-galactose as a model ligand for hexose transport, it was previously reported that studies of uptake (i.e., transport plus metabolism) enhancements caused by oncogenic virus transformation or glucose deprivation were comparable with similar studies using glucose, 2-deoxy-D-glucose (2-dG) or 3-O-meG in other cell types (3, 6). Using low concentrations of N-ethylmaleimide (NEM), we have been able to completely inhibit the galactokinase (ATP:D-galactose-1-phosphotransferase, E.C. 2.7.1.6) activity of intact hamster cells without interfering with galactose transport. Thus the physiological utilization of galactose by cultured cells can be arrested at the first step after transport, making possible direct studies of hexose transport and transport regulation. As a result of this finding, previously observed changes in galactose uptake caused by extended maintenance of cells in culture media devoid of glycolytic sugars (3, 6–8) can now be unequivocally shown to be due to changes in the rate of the transport step. The changes

Received May 1, 1977; accepted June 3, 1977

in hexose transport rates appear to be due to the synthesis and activation of the hexose carriers counteracted by a catabolite inactivation-like mechanism (9) mediated by a regulatory factor. In addition, de novo synthesis of the regulatory factor is required for inactivation of carriers.

MATERIALS AND METHODS

Cycloheximide and N-ethylmaleimide were purchased from Sigma Chemical Company (St. Louis, Missouri) and cytochalasin B was from Aldrich Chemical Company, Inc. (Milwaukee, Wisconsin). Culture media and sera (fetal calf and dialyzed fetal calf) were obtained from Grand Island Biological Company (Grand Island, New York) and Microbiological Associates (Bethesda, Maryland). Sugar-free culture medium or medium containing fructose contained all components of Dulbecco's modified Eagle's medium (10) except glucose. Radioisotopes of the highest specific activity available were obtained from New England Nuclear Corporation (Boston, Massachusetts).

Culture conditions for the growth and maintenance of hamster fibroblasts (nil strain courtesy of Ms. M. T. Gammon and Dr. K. J. Isselbacher of the Gastrointertinal Unit, Massachusetts General Hospital) were described previously (4, 7, 8). In order to inhibit galactokinase activity, cell monolayers were washed with Dulbecco's phosphate-buffered saline (D-PBS) and preincubated at 37°C for 15 min in D-PBS containing 0.5 mM NEM. Transport of 0.1 mM D-[^{14}C] galactose was for 10 sec at 25°C. Simple diffusion and/or nonspecific hexose adsorption was monitored simultaneously with galactose transport by including 0.1 mM L-[^{3}H] glucose in the assay medium (all transport assays were double label experiments). In 10 to 15 sec immediately following the transport assay, the cells were washed 5 times with 2 ml D-PBS containing 21 μM cytochalasin B and then extracted with 70% ethanol. The ethanol extract was counted as previously described (4, 7, 8), and the data corrected for machine background and channel overlap. Transport of galactose is defined as pmole D-galactose per milligram of cell protein minus pmole L-glucose per milligram cell protein per unit time (usually 10 sec).

RESULTS

Of 3 sulfhydryl reagents tested, only NEM blocked the in vivo activity of galactokinase (Fig. 1, panel II) without affecting transport (Table I). The effective concentration range of NEM was between 0.2 mM and 1.0 mM. Below 0.2 mM NEM, galactokinase activity often contributed to transport measurements and above 1.0 mM NEM cells became progressively more permeable to L-glucose. Although mercuric chloride also inhibited the in vivo activity of galactokinase (Fig. 1, panel IV), it opened the cells to permeation of L-glucose to the extent that there was no net difference between facilitated diffusion and simple diffusion/adsorption (Table I). Treatment of the cells with parahydroxymercuribenzoate (pHMB) resulted in significant inhibitions of galactose transport (Table I). The presence of uridine di-phosphate-hexose (UDP-hexose) and absence of galactose-1-phosphate as well as free galactose in the pHMB-treated cells (Fig. 1, panel III) is like the metabolic pattern of the glucose-fed control (Fig. 1, panel I). These show that the galactose that has entered the cells is immediately processed to UDP-hexose by the activity of at least 2 enzymes (4). The presence, therefore, of UDP-hexose coupled with the absence of free galactose in the extracts

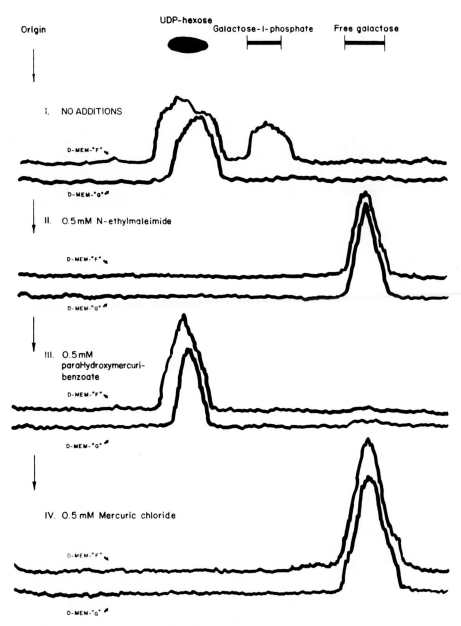

Fig. 1. Confluent monolayers of hamster fibroblasts were washed with sterile Dulbecco's phosphate-buffered saline (PBS) and then maintained in Dulbecco's modified Eagle's minimal essential medium (D-MEM) containing either 22 mM D-glucose (D-MEM-"G") or 22 mM D-fructose (D-MEM-"F") for 18–24 h. The monolayers were then washed with PBS and preincubated for 15 min at $37°C$ in PBS or PBS containing 0.5 mM NEM, 0.5 mM pHMB, or 0.5 mM $HgCl_2$. Following the preincubation the medium was changed to include 0.1 mM D-$[^{14}C]$ galactose with the above reagents and the cells were allowed to take up the galactose for 10 min. The monolayers were then washed with PBS containing 21 μM cytochalasin B and then approximately equivalent amounts of radioactivity were chromatographed as described in detail previously (4).

TABLE I. Effect of Sulfhydryl Reagents on Hexose Transport

Experiment number	Culture condition[a] (Dulbecco's MEM containing 22 mM)	Preincubation[b] (15 min/37°C)	Hexose transport[c] (pmole/mg protein/10 sec)			
			3-O-meG	D-galactose	L-glucose	Net transport (D-[^{14}C]gal-L-[^{3}H]glc)[d]
1	Glucose	no additions	—	8 ± 0	3 ± 1	5 ± 1
		NEM	—	13 ± 2	7 ± 2	5 ± 1
	Fructose	no additions	—	112 ± 11	4 ± 2	109 ± 10
		NEM	—	87 ± 6	5 ± 1	82 ± 6
2	Glucose	no additions	—	8 ± 0	4 ± 0.4	4 ± 0.8
		pHMB	—	6 ± 2	4 ± 0.7	1 ± 1
		HgCl$_2$	—	43 ± 6	44 ± 6	0
	Fructose	no additions	—	102 ± 8	5 ± 0.4	92 ± 11
		pHMB	—	18 ± 3	8 ± 1	10 ± 2
		HgCl$_2$	—	44 ± 1	46 ± 3	0
3	Glucose	no additions	7 ± 1	—	—	—
		NEM	15 ± 1	—	—	—
	Fructose	no additions	110 ± 2	—	—	—
		NEM	110 ± 8	—	—	—

[a]Confluent monolayers of hamster fibroblasts were maintained for 18–24 h in Dulbecco's modified Eagle's medium and then washed with Dulbecco's phosphate-buffered saline (PBS).

[b]Washed cells were preincubated in PBS or PBS containing 0.5 mM sulfhydryl reagent as indicated for 15 min at 37°C.

[c]Preincubated cells were assayed in the presence of 0.5 mM sulfhydryl reagent for transport of 0.1 mM hexose as described in METHODS. Values are the mean ± standard deviation of triplicate samples.

[d]Double label experiments, see METHODS. Experiment 3 was a single label experiment; corrections for simple diffusion adsorption were not made in these cases. In view of corrections for experiment 1, the changes in 3-O-meG transport by glucose-fed cells after treatment with NEM are probably not significant.

indicates that, under the conditions of the experiment, the galactokinase was not noticeably affected by pHMB. Table I also shows that there are no adverse effects of NEM on 3-O-meG transport. Thus these data show that NEM has no effect on the transport of either the metabolizable sugar (galactose) or the nonmetabolizable sugar (3-O-meG) and therefore treatment of whole cells with NEM under carefully controlled conditions can be used to dissociate hexose transport from hexose metabolism.

In NEM-treated cells, D-glucose, 2-dG, and 3-O-meG compete with D-galactose equally well (Fig. 2). The concentration of these sugars that results in 50% inhibition of galactose transport is approximately 5 mM. This value is in close agreement with the esti-mated K_m for D-galactose, 3.8–5.6 mM (Table II). Table II also shows that galactose trans-port changes (enhancements observed after prolonged maintenance of cells in media containing no sugar or fructose and losses of transport activity seen after maintenance in media containing both glucose and cycloheximide) are due to changes in V_{max}.

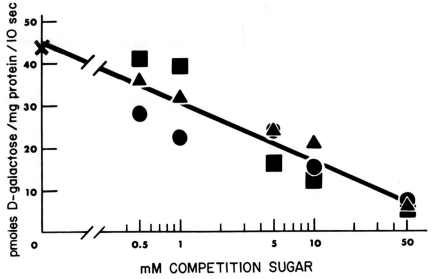

Fig. 2. Monolayers, maintained for 24 h in D-MEM containing fructose, were preincubated for 15 min at 37°C in PBS containing 0.5 mM NEM to inhibit galactokinase activity. Following the NEM treatment, the cells were assayed for 0.1 mM D-galactose transport in the presence of increasing concentrations of D-glucose (●), 2-deoxy-D-glucose (■), or 3-O-methyl-D-glucose (▲).

TABLE II. Galactose Transport Kinetics; K_m and V_{max}

Culture condition	$K_m{}^a$	$V_{max}{}^b$
Glucose	3.8 (1.0)[c]	0.98 (1.0)
Glucose + cycloheximide	5.6 (1.4)	0.27 (0.3)
Fructose	4.4 (1.2)	13.5 (14)

[a] K_m = mM D-galactose
[b] V_{max} = nmole D-galactose transported/mg protein/10 sec
[c] Relative values in parentheses

DISCUSSION

Galactose and Sugar Transport

The use of D-galactose as a probe in studying the regulation of hexose transport gains advantage from the observations that D-galactose is a natural analog of D-glucose, D-galactose supports the growth of animal cells (11), and D-galactose is believed to enter animal cells through the glucose transport system (12). In addition, compared with the multiple isozyme hexokinase (ATP: D-hexose-6-phosphotransferase, E.C. 2.7.1.1) system which phosphorylates a wide range of hexoses without being completely inhibited by sulfhydryl reagents (1), galactokinase restricts its phosphorylation to galactose (or galactosamine) and is completely inhibited by such sulfhydryl reagents as those that can penetrate the plasma membrane of intact cells. Taken together the results of the experiments presented in this paper demonstrate that transport of metabolizable sugars can in fact be measured independently of metabolism. Although other interpretations may apply, the simplest explanation for the results of kinetics measurements at approximate initial rates is that it is the number of active carriers that increases in response to nonglycolytic conditions. Similarly, the decreases in transport in response to inhibitors of protein synthesis during maintenance under glycolytic culture conditions (1, 7, 8) appear to be due to decreases in the number of active carriers.

Since doubts have been expressed concerning the use of 3-O-meG in transport studies (1, 2), this evidence lends validity to the use of 3-O-meG as a reliable indicator of hexose transport. That is, by direct comparison the nonmetabolizable glucose analog, 3-O-meG, is shown to be transported at the same rate as the metabolizable sugar, galactose, and 3-O-meG is as good an inhibitor of galactose transport as glucose and 2-dG. This use of 3-O-meG for hexose transport studies may apply, however, only to the predominant low-affinity transport system for glucose since a high affinity system for glucose does not appear to transport 3-O-meG (1). Nevertheless, changes in hexose uptake detected by the use of glucose, galactose, 2-dG, or 3-O-meG (1, 3, 4, 6–8) can now be attributed to changes in the transport step.

Regulation of Hexose Transport: Derepression/Synthesis vs Inactivation/Turnover, A Model

How is transport of hexose regulated in animal cells? Although the mechanism of regulation is not clear, it is useful to consider a model which accounts for several experimental observations and suggests new experimental approaches. The initial observations of Amos and his co-workers (13) suggesting that derepression of carrier synthesis could account for enhancements of hexose uptake caused by sugar starvation were confirmed by Kletzien and Perdue (14). Based on their recent experimental data Christopher et al. (7, 8) proposed that inactivation and/or turnover of carriers was also involved in regulation and, in addition, a regulatory factor (which promoted the inactivation and/or turnover of the hexose carrier) was itself subject to inactivation (7, 8). Hexose transport control in at least hamster and chick cells appears therefore to be dependent upon 2 opposing forces, namely, synthesis of carriers vs inactivation of carriers. The features of this model are shown in Fig. 3.

During growth under normal culture conditions (Fig. 3A and upper left panel), synthesis and inactivation of the carrier to a functional state are offset by inactivation and/or turnover of the carrier. When protein synthesis in cells maintained in medium containing glucose is blocked by cycloheximide (Fig. 3A and lower left panel), a short period of

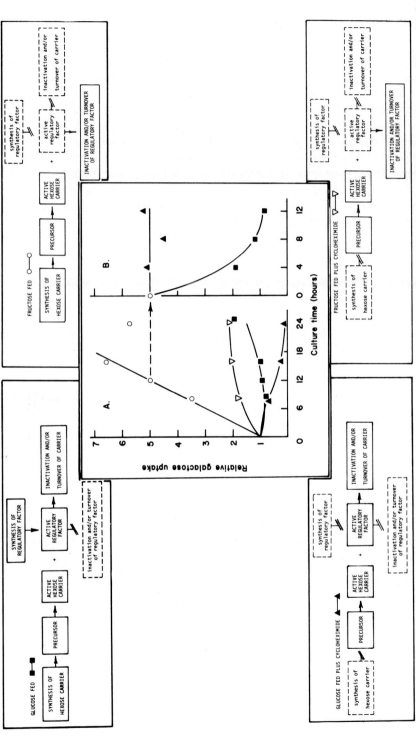

Fig. 3. Parts A and B are modified from Figs. 2 and 4 of Ref. 7. Panels on the periphery of these figures illustrate a scheme of hexose carrier control for each of the experimental observations shown in A. The upper left panel illustrates the hypothetical dynamics of carrier synthesis vs inactivation for cells maintained in medium containing glucose (filled squares in A). The lower left panel illustrates the effects of cycloheximide in medium also containing glucose (filled triangles in A). The upper right panel illustrates regulation in cells maintained in medium containing no sugar or fructose (open circles in A) and the lower right panel illustrates the effects of cycloheximide in the no sugar or fructose-fed cells (open triangles in A). Changes in transport activity depicted in B contribute to the concept of a regulatory factor which appears to become inactive when cells are maintained in media containing no sugar or fructose. The downward regulation of transport by sugar-starved or fructose-fed cells after the addition of glucose to the medium (filled squares in B) is prevented when cycloheximide is added simultaneously with the glucose (filled triangles in B). Cycloheximide concentration was 36 μM; glucose concentration was 11 mM (7).

steady state exists where the apparent rate of activation of the putative precursor carrier is counteracted by the inactivation process. Upon exhaustion of the "precursor pool," further time-dependent inactivation of functional carriers leads to dramatic losses of activity since no new carriers can be synthesized. However, treating the cells with cycloheximide in the absence of glucose or in media containing fructose (Fig. 3A and lower right panel) leads to an increase in transport rates. That the increase under these conditions is immediate and cycloheximide independent further suggests a movement of preformed carriers from an inactive to an active state. This increase in the absence of glucose under otherwise identical culture conditions as those depicted in the lower left panel of Fig. 3 is striking and suggests that glucose (or its catabolites) is required for inactivation/turnover of the carrier. This is supported by the observation that transport rate increases are linear with respect to time when synthesis of carriers continues either in the absence of sugars or in the presence of nonglycolytic sugars such as fructose (Fig. 3A and upper right panel). These observations bear a remarkable resemblance to those described to support the model of the catabolite inactivation mechanism of control in the eukaryotic microorganism, Saccharomyces (9).

Is the Regulatory Factor a Protease?

Some evidence suggests that the regulatory mechanism involved in the inactivation/ turnover may be a proteolytic process. High concentrations of inhibitors of protein synthesis (cycloheximide or puromycin) inhibit the process of inactivation (7, 8). The reactivation of carrier activity requires de novo synthesis of carriers. Following glucose plus cycloheximide treatment, the recovery of activity during maintenance of the cells in medium containing fructose (8) is characterized by an initial 2-h lag period and can be prevented by 7.1 μM cycloheximide (unpublished results). If proteolysis should turn out to be the principal mechanism of hexose transport inactivation, it is all the more interesting that de novo synthesis of a protease (or an unknown cofactor for proteolysis) is also required. This interpretation comes from the observation that the addition of glucose to cells fully "derepressed" by prior maintenance in media containing no sugar results in a rapid inactivation of existing carriers (Fig. 3B). Yet the simultaneous addition of low levels of cycloheximide with the glucose does not lead to inactivation (Fig. 3B). From Fig. 3A it can be seen that cycloheximide, at the concentrations used, does not block inactivation/ turnover in cells that have been continually maintained in media containing glucose. This indicates that the inhibition of the inactivation/turnover process cannot be due to a direct action of cycloheximide (e.g., inhibition of proteolysis) since loss of carrier activity was strikingly evident by 12 h (Fig. 3A).

Is the Regulatory Factor an Allosteric Effector?

Regulation may depend upon the synthesis and action of a regulatory unit of the hexose carrier system. In Fig. 4 a simple model of hexose transport control shows a labile factor interacting with the carrier in a way that does not compete with but impedes the movement of the ligand. In the figure the activity of the triangular shaped regulatory factor is illustrated as being lost when cells are maintained under nonglycolytic conditions (Fig. 4A) and as being present in cells under glycolytic culture conditions (Fig. 4B).

This model might also accomodate the well known effect that transformation has on hexose transport control. In Fig. 4C a defective regulatory unit, induced by a transformation event, is shown only partially blocking transport. In Fig. 4D an alternative model suggests that a transformation induced factor could compete with the regulatory factor for

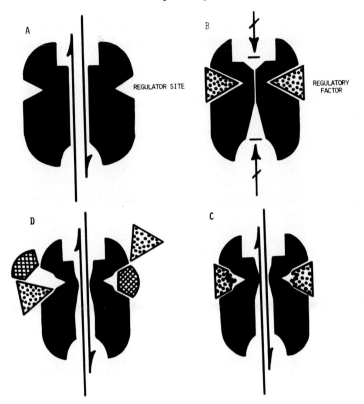

Fig. 4. Simplified, hypothetical model of the interaction of a hexose carrier (filled oval with transport marked by arrows) regulatory factors (stippled). A) Carrier of cells deprived of glycolytic sugars having little or no regulatory factor activity (regulator site is unfilled; transport, influx and efflux, is open). B) Carrier of cells maintained in medium containing glucose (regulatory factor association with regulatory sites of carrier shuts off transport). C) Carrier of transformed cells maintained in media containing glucose and having defective regulatory factors or carriers with defective regulator sites. D) Carrier of transformed cells (glucose-fed) with regulator sites being competed for by normal regulatory factors (stippled) and transformation-specific products (cross hatched). In both C and D transport is only partially regulated.

the regulatory site. These models also could account for further enhancements of hexose transport by sugar-starved, transformed cells (4, 5, 16) by allowing for the inactivation of regulatory activity that is normally caused by nonglycolytic culture conditions. Hence, labile factors and carriers must not be overlooked in efforts to understand the role of transformation in hexose transport control.

ACKNOWLEDGMENTS

My thanks go to Maureen T. Gammon and Dr. Kurt J. Isselbacher for their generous help with cell cultures, to Ms. D. Ullrey and Ms. W. Colby for their expert technical assistance, and to Dr. Herman M. Kalckar for his interest and helpful suggestions throughout all phases of this study. This work was supported by grants from the American Cancer Society (No. BC-120 and BC-120-C), the National Institutes of Health (No. AM-05507-14), and the National Science Foundation (No. BMS71-01291-A03). This is publication No. 1529 of the Cancer Commission of Harvard University.

REFERENCES

1. Christopher CW, Kohlbacher MS, Amos H: Biochem J 158:439, 1976.
2. Romano A: J Cell Physiol 89:737, 1976.
3. Kalckar HM, Ullrey D: Proc Natl Acad Sci USA 70:2502, 1973.
4. Christopher CW, Colby WW, Ullrey D, Kalckar HM: J Cell Physiol 90:387, 1977.
5. Bassham JA, Bissell MJ, White RJ: Anal Biochem 61:479, 1974.
6. Ullrey D, Gammon MT, Kalckar HM: Arch Biochem Biophys 167:410, 1975.
7. Christopher CW, Ullrey D, Colby W, Kalckar HM: Proc Natl Acad Sci USA 73:2429, 1976.
8. Christopher CW, Colby WW, Ullrey D: J Cell Physiol 89:683, 1976.
9. Holzer H: Trends Biochem Sci 1:178, 1976.
10. Dulbecco R, Freeman G: Virology 8:396, 1959.
11. Kalckar HM, Ullrey D, Kijimoto S, Hakomori S: Proc Natl Acad Sci USA 70:839, 1973.
12. Kletzien RF, Perdue JF: J Biol Chem 248:711, 1973.
13. Martineau R, Kohlbacher M, Shaw SN, Amos H: Proc Natl Acad Sci USA 69:3407, 1972.
14. Kletzien RF, Perdue JF: J Biol Chem 250:593, 1975.
15. Kletzien RF, Perdue JF: Cell 6:513, 1975.

Journal of Supramolecular Structure 6:495–502 (1977)
Molecular Aspects of Membrane Transport 325–332

The Mechanism of Sugar-Dependent Repression of Synthesis of Catabolic Enzymes in Escherichia coli

Jose E. Gonzalez and Alan Peterkofsky

Laboratory of Biochemical Genetics, National Heart, Lung, and Blood Institute, National Institutes of Health, Bethesda, Maryland 20014

Previous studies have indicated that the Escherichia coli adenylate cyclase (AC) activity is controlled by an interaction with the phosphoenolpyruvate (PEP):sugar phosphotransferase system (PTS). A model for the regulation of AC involving the phosphorylation state of the PTS is described. Kinetic studies support the concept that the velocity of AC is determined by the opposing contributions of PEP-dependent phosphorylation (V_1) and sugar-dependent dephosphorylation (V_2) of the PTS proteins according to the expression % $V_{AC} = 100/[1 + (Max\ V_2/Max\ V_1)]$. Physiological parameters influencing the rate of the PTS are discussed in the framework of their effects on cAMP metabolism. Factors that increase cellular concentration of PEP (and stimulate V_1) appear to enhance AC activity while increases in extracellular sugar concentration (which stimulate V_2) or internal levels of pyruvate (which inhibit V_1) inhibit the activity of this enzyme.

Key words: adenylate cyclase, catabolite repression, sugar transport

In E. coli, cAMP is required for the synthesis of the mRNA for many catabolic enzymes (1). E. coli growing exponentially on glycerol or succinate have elevated cAMP levels (2) and when exposed to the appropriate inducer rapidly synthesize enzymes for the transport and degradation of that inducer (3). Immediately following the addition of glucose to such cultures, cAMP levels decrease substantially (4, 5); this results in a period of transient repression during which induced enzyme synthesis is inhibited (3). If the level of cAMP is replenished by exogenous addition of this nucleotide, such cells recover from this transient repression (1). In this report, we examine a model for transient repression based on the regulation of cAMP levels by glucose, and other sugar substrates of the phosphoenolpyruvate (PEP):sugar phosphotransferase system (PTS).

The intracellular pool of cAMP is determined by the rates of synthesis, degradation and excretion of this nucleotide. The glucose-dependent decrease in cAMP levels cannot

Address correspondence to: Dr. Alan Peterkofsky, National Institutes of Health, Building 36, Room Room 4c11, Bethesda, MD 20014.

Received May 9, 1977; accepted June 10, 1977

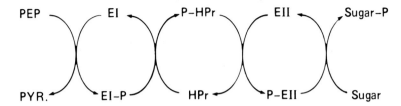

Fig. 1. Mechanism of the PEP:glucose phosphotransferase system in E. coli. The flow of the phosphate group, derived from PEP, to the sugar is thermodynamically favored. PEP can saturate the system with phosphate groups if the concentrations of pyruvate and glucose are low; in the absence of PEP, glucose or pyruvate can convert the proteins to the dephospho forms. EII is composed of 2 proteins, $F^{glc}III$ and $E^{glc}IIB$, specific for glucose. Each different PTS sugar has specific EIIs involved in its transport. The EIIB is a membrane-associated protein. Enzyme I (EI) and HPr are soluble proteins that are required for transport of all PTS sugars.

be explained on the basis of effects on the rates of degradation (6) or excretion (6–8). However, the rate of synthesis of cAMP can be inhibited by glucose (6). The regulatory model outlined below involves an interaction of the PTS with AC.

RESULTS

Figure 1 outlines the PTS, composed of 4 proteins, by which glucose and some other sugars are transported in E. coli. There is a sequential, covalent transfer of the phosphate group derived from PEP across this protein chain to a sugar acceptor (10). This paper is concerned with a model for the regulation of AC activity which is based on the idea that AC interacts with PTS proteins and that when PTS proteins are phosphorylated, AC is active. The model proposes that PEP, which phosphorylates PTS proteins, activates AC, while sugar substrates of the PTS, as well as pyruvate, dephosphorylate PTS proteins and inhibit AC. Since there are opposing effects of PEP compared to pyruvate and sugar in determining the phosphorylation state of the PTS proteins, intermediate levels of AC activity would be expected when both PEP and either pyruvate or sugar are present.

Several lines of evidence indicate that the glucose-dependent inhibition of adenylate cyclase (AC) is coupled to the glucose transport activity. Harwood and Peterkofsky (9) showed that AC activity in intact or permeabilized cells is inhibited by glucose, but not by glucose-6-phosphate, the immediate product of the sugar transport process (10). Magasanik has indicated that transient repression is caused not only by glucose but is also observed with other sugars if their transport systems are present (3). The induction of sugar-specific membrane-associated proteins permit the transport of other hexoses besides glucose. Once sugar transport activity is induced, the sugar is capable of inhibiting AC (11). Figure 2 shows that the sensitivity of AC to inhibition by mannitol parallels the induction of mannitol transport activity. When the mannitol-dependent PTS activity is low, AC activity is relatively insensitive to inhibition by mannitol. Under this condition, the capability of mannitol to dephosphorylate PTS proteins is relatively poor compared to the endogenous PEP-dependent reaction leading to the phosphorylation of PTS proteins. The observation that one PTS substrate (PEP) antagonizes the inhibition of AC by another PTS substrate (sugar) (14) provides further support for the model that AC activity depends on the state of phosphorylation of the PTS proteins.

Figure 3 describes a simplified model for the PTS system. An analysis of this model suggests that, at steady-state, the fraction of the PTS proteins in the phospho form (EP) is

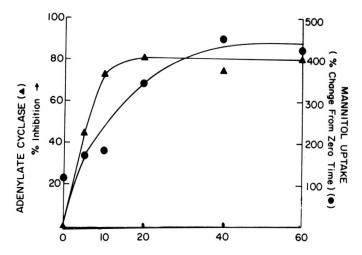

Fig. 2. Kinetics of induction of mannitol inhibition of adenylate cyclase compared to mannitol uptake during exposure of E. coli B to mannitol. E. coli B were grown in minimal medium (12) supplemented with 1% glucose to an A_{650} of 1. At that point, the culture was centrifuged, washed in minimal medium, and resuspended in minimal medium (original volume), adjusted to a mannitol concentration of 15 mM. The cell suspension was shaken at 37°C. At the indicated times, aliquots of the cells were collected and washed on large Millipore membranes (9 cm) then resuspended in their original volume. Aliquots (10 ml) of the washed cells were tested for adenylate cyclase activity in the absence or presence of mannitol (1 mM). The period of pulse-labeling with [^{3}H]adenosine was 1 min. The concentrations of [^{3}H]cAMP were determined as previously described (6). The data are presented as percentage inhibition of the adenylate cyclase activity by mannitol. Measurements of [^{14}C]mannitol uptake were as described by Solomon and Lin (13) using 0.025 ml aliquots of the cell suspension. Assays were done for 0.5, 1.0, 1.5, and 2.0 min. The data are expressed as the percentage change in mannitol uptake rate of the samples compared to the zero time samples. Legend: Effect of mannitol on adenylate cyclase (▲——▲); Mannitol uptake (●——●). Adapted from (11).

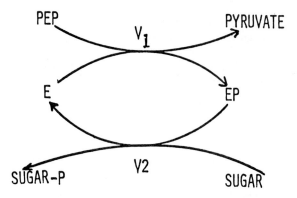

Fig. 3. The interconversion of the PTS proteins between phospho and dephospho forms. E represents the PTS proteins in the dephospho form; EP represents the PTS proteins in the phospho form. V_1 is the PEP-dependent rate of phosphorylation of E; V_2 represents the sugar-dependent rate of dephosphorylation of EP.

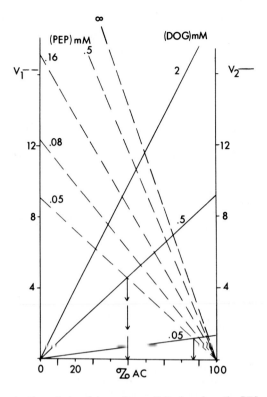

Fig. 4. Graphical determination of adenylate cyclase activity based on the PTS regulatory model. The abscissa labeled % AC also represents the percentage of the PTS proteins in the phospho form. The method used for calculation of maximum V_1 values (Max V_1, where PTS proteins are completely in the dephospho form) and maximum V_2 values (Max V_2, where PTS proteins are completely in the phospho form) at various concentrations of PEP and sugar, respectively, is described in Appendix B. The ordinates labeled V_1 and V_2 represent Max V_1 and Max V_2 values at the intercepts on the ordinates. At any specified concentration of PEP, V_1 is directly proportional to the concentration of the dephospho-PTS, thus justifying the extrapolation (dashed lines) of Max V_1 to zero velocity when all of the PTS is in the phospho form. A similar analysis applies to the construction of the V_2 lines (solid lines). The intersection of any pair of V_1 and V_2 lines corresponds to the steady-state velocity of the PTS reaction at designated concentrations of PEP and DOG (2-deoxyglucose). Since, as indicated in Appendix A, the fractional velocity of AC is equal to the fraction of the PTS in the phospho form, the intersection of the lines also determines the % activity of adenylate cyclase (designated on the abscissa). The arrows indicate the fraction of maximum AC activity at 0.05 mM PEP and either 0.05 mM or 0.5 mM DOG. Values for fractional AC activity determined from this plot are compared with experimental data in Table I.

determined by the maximum V_1 (Max V_1, the velocity of the reaction at any specified concentration of PEP when all of the enzyme is in the dephospho form, E) and by the maximum V_2 (Max V_2, the velocity of the reaction at any specified sugar concentration when all of the enzyme is in the phospho form, EP). We propose that the fractional velocity of AC is equal to the fraction of the PTS in the phospho form, EP. The fractional velocity of AC can be expressed mathematically by the following equation (derived in Appendix A): Fractional Velocity of AC (in %) = 100/ [1 + (Max V_2/Max V_1)]. A graphical solution to this equation is presented in Fig. 4. Steady-state PTS velocities can be read from this graph

TABLE I. Correlation of Experimentally Determined Fractional Adenylate Cyclase Activities With Values Generated by Kinetic Modeling*

PEP (mM)	V_1	2-deoxyglucose (DOG) (mM)									
		0		$0.05(V_2 = 1.4)$				$0.5(V_2 = 9.2)$			
		Experimental		Experimental		Modeling		Experimental		Modeling	
		V_{AC}	FV_{AC}	V_{AC}	FV_{AC}	V_{AC}	FV_{AC}	V_{AC}	FV_{AC}	V_{AC}	FV_{AC}
0.05	9.1	1.14	100	1.00	88	0.99	87	0.57	50	0.57	50
0.08	12.4	1.20	100	1.06	88	1.08	90	0.62	52	0.68	57
0.16	17.2	1.28	100	1.16	91	1.18	92	0.83	65	0.83	65
0.5	23.8	1.32	100	1.21	92	1.24	94	0.95	72	0.95	72

*Adenylate cyclase activity in permeabilized cells of E. coli 1,100 was determined as described (15). Velocity (V_{AC}) is expressed as nmoles cAMP formed per hour per mg of protein. Experimental values for FV_{AC} were calculated as the percentage of the activity obtained at the indicated concentrations of PEP in the absence of DOG. Modeling values for FV_{AC} were calculated from Equation 5 (Appendix B) using the indicated values for V_1 and V_2 obtained as described in Appendix B.

as the ordinate values at the intersection of pairs of V_1 and V_2 lines. These steady-state PTS velocities generate parallel lines in a Lineweaver-Burk plot. Such kinetics are consistent with the expected formation of the stable phosphorylated intermediates previously demonstrated for the PTS and required for the suggested model of PTS-dependent regulation of AC. The intersection point of a particular pair of V_1 and V_2 lines (extrapolated to the abscissa) determines the fractional velocity of AC. Table I summarizes the values of the fractional AC activity obtained graphically and experimentally. The good agreement of the values obtained by the 2 methods supports the notion that the proposed model describes the mechanism of AC regulation.

Other predictions generated by the model equation and fulfilled by preliminary experimental data are as follows:

1) The K_{PEP} and K_{DOG} values in the PTS reaction are the same as the K_{PEP} (for activation) and K_{DOG} (for inhibition) with respect to the AC activity.

2) Inhibition by saturating [DOG] decreases with increasing [PEP]; PEP-dependent activation cannot completely reverse DOG-dependent inhibition.

3) The apparent K_{aPEP} and K_{iDOG} increase as the concentration of the other ligand increases.

4) Subsaturating concentrations of either PEP or DOG are sufficient to completely activate or inhibit AC, respectively, in the absence of the other PTS substrate.

DISCUSSION

The phosphorylation model indicates that extensive inhibition of AC requires both an appropriate concentration in the medium of a sugar transportable by the PTS, as well as a relatively low concentration of PEP in the cell. Addition of glucose to an E. coli culture growing exponentially on glycerol- or succinate-salts medium causes a rapid decrease of the PEP pool (16) thereby enhancing the sensitivity of AC to sugar inhibition. This reduction of intracellular PEP may be the result of sugar phosphate formation via the PTS (16, 17). The high cAMP levels characteristic of growth on glycerol together with the decrease in cAMP after the addition of glucose may be a reflection of the differences in intracellular PEP levels under these conditions. The observed cAMP levels vary in the direction predicted by the model presented here.

The model also predicts that mutants deficient in the production of PEP will be hypersensitive to glucose repression of induced enzyme synthesis. This is the case in PEP synthetase (18) and phosphofructokinase mutants (19).

Mutants leaky in the EI or HPr proteins of the PTS (Fig. 1) are also hypersensitive to glucose repression (20). Such mutants may be unable to effectively phosphorylate PTS proteins (i.e., have low V_1, normal V_2). In contrast, mutants in the membrane-associated sugar-specific proteins of the PTS are unable to dephosphorylate HPr and EI proteins of the PTS (i.e., have low V_2, normal V_1) and thereby, become resistant to sugar inhibition of AC (15). The properties of these mutants are consistent with the model outlined in Fig. 3.

Pyruvate inhibits AC activity (14). Lowry et al. (16) report that the internal concentrations of pyruvate range from 0.04 mM to 0.63 mM according to growth conditions. Since the K_i value for pyruvate inhibition of AC in permeabilized cells is 0.1 mM (data not shown), physiological changes in pyruvate levels may contribute significantly to the regulation of AC activity. The mechanism of AC inhibition by pyruvate may be mediated by the state of phosphorylation of the PTS since, pyruvate, like a sugar substrate, can dephosphorylate the PTS proteins (10) (see Fig. 1).

Pyruvate, like glucose, elicits catabolite repression. Addition of pyruvate to E. coli growing exponentially on glycerol leads to a reduction in the rate of tryptophanase (1, 21, 22, 23) and D-serine deaminase (22) synthesis. Cyclic AMP prevents the pyruvate-dependent repression of enzyme synthesis (21, 23). These observations are consistent with the suggestion (23) that pyruvate decreases cellular cAMP levels.

Cyclic AMP levels are a key factor in regulating induced enzyme synthesis. Various observations in the literature link the expression of induced enzymes to PTS sugars, pyruvate, and PEP. The discussion presented here provides a unifying model compatible with much of these apparently unrelated data. It proposes that phosphorylated PTS substrates increase the steady-state concentration of phospho-PTS proteins thereby activating AC while nonphosphorylated PTS substrates decrease the steady-state concentration of phospho-PTS proteins leading to a deactivation of AC.

REFERENCES

1. Pastan I, Perlman R: Science 169:339, 1970.
2. Peterkofsky A, Gazdar C: Proc Natl Acad Sci USA 68:2794, 1971.
3. Magasanik B: Cold Spring Harbor Symp Quant Biol 26:249, 1951.
4. Makman RS, Sutherland EW: J Biol Chem 240:1309, 1965.
5. Gonzalez JE, Peterkofsky A: Biochem Biophys Res Commun 67:190, 1975.
6. Peterkofsky A, Gazdar C: Proc Natl Acad Sci USA 71:2324, 1974.
7. Epstein W, Rothman-Denes LB, Hesse J: Proc Natl Acad Sci USA 72:2300, 1975.
8. Mawe R, Doore B, McCaman M, Feucht B, Saier M: J Cell Biol 63:211, 1974.
9. Harwood JP, Peterkofsky A: J Biol Chem 250:4656, 1975.
10. Roseman S: In Woessner JF Jr, Huijing F (eds): "The Molecular Basis of Biological Transport." New York: Academic Press, 1972, pp 181–215.
11. Peterkofsky A, Harwood JP, Gazdar C: J Cyclic Nucleotide Res 1:11, 1975.
12. Vogel HJ, Bonner DM: J Biol Chem 218:97, 1956.
13. Solomon E, Lin ECC: J Bacteriol 111:566, 1972.
14. Peterkofsky A, Gazdar C: Proc Natl Acad Sci USA 72:2920, 1975.
15. Harwood JP, Gazdar C, Prasad C, Peterkofsky A, Curtis SJ, Epstein W: J Biol Chem 251:2462, 1976.
16. Lowry OH, Carter J, Ward JB, Glaser L: J Biol Chem 246:6511, 1971.
17. Tyler B, Magasanik B: J Bacteriol 102:411, 1970.

18. Brice CB, Kornberg HL: J Bacteriol 96:2185, 1968.
19. Roehl RA, Vinopal RT: J Bacteriol 126:852, 1976.
20. Pastan I, Perlman RL: J Biol Chem 244:5836, 1969.
21. Botsford JL, DeMoss RD: J Bacteriol 105:303, 1971.
22. McFall E, Mandelstam J: Biochem J 89:391, 1963.
23. Perlman R, Pastan I: Biochem Biophys Res Commun 30:656, 1968.

APPENDIX A

Derivation of the Fractional Velocity Equation for Adenylate Cyclase Activity

From the model in Fig. 3, at steady-state,

$$\frac{\text{Max } V_1}{\text{Max } V_2} = \frac{(EP)}{(E)} \tag{1}$$

Since the amount of PTS proteins is constant,

$$E + EP = 1 \tag{2}$$

From equations 1 and 2,

$$EP = \frac{1}{1 + \dfrac{\text{Max } V_2}{\text{Max } V_1}} \tag{3}$$

We define the fractional velocity of AC (FV_{AC}) as equivalent to the fraction of the PTS that is phosphorylated (EP). Therefore,

$$FV_{AC} = \frac{1}{1 + \dfrac{\text{Max } V_2}{\text{Max } V_1}} \tag{4}$$

APPENDIX B

Method for Calculation of Fractional V_{AC}

As shown in Appendix A the fractional velocity of AC in the presence of the PTS substrates, PEP and DOG is

$$FV_{AC} = \frac{1}{1 + \dfrac{\text{Max } V_2}{\text{Max } V_1}} \text{, where} \tag{5}$$

$$\text{Max } V_1 = \frac{k_1 \, [E] \, [PEP]}{K_{PEP} + [PEP]} \text{, when all of the enzyme is in the E form, and} \tag{6}$$

$$\text{Max } V_2 = \frac{k_2 \, [EP] \, [DOG]}{K_{DOG} + DOG} \text{, when all of the enzyme is in the EP form.} \tag{7}$$

K_{PEP} and K_{DOG} were determined to be 0.11 mM and 0.8 mM, respectively by assays for PTS activity at various concentrations of PEP or DOG.* An experimentally determined value for FV_{AC} at 0.5 mM PEP and 0.5 mM DOG was 0.72. By arbitrarily assigning a value of Max V_2 as 9.2, equation 5 was solved for Max V_1 to give a value of 23.8. Using the values of K_{PEP} and Max V_1 at saturating PEP other values of Max V_1 at different [PEP] may be obtained graphically from the Lineweaver-Burk form of equation (6).

*The assay for phosphotransferase activity used in this study depended on the DOG-dependent conversion of [14C] PEP to [14C] pyruvate. Incubations for PTS activity were carried out as previously described (15), except that [14C] PEP was used as the labeled compound. After a suitable incubation period, an aliquot (50 μl) of the incubation mixture together with 100 μl of pyruvate (1 N) was deposited on a column (0.5 × 2 cm) of Dowex 3-X4 200–400 mesh. [14C] pyruvate was separated from [14C] PEP by elution with 0.2 N HCl. The recovery of pyruvate in the fraction counted was determined by a spectrophotometric measurement at 340 nm. For other details, see the text.

Journal of Supramolecular Structure 6:503–518 (1977)
Molecular Aspects of Membrane Transport 333–348

Crystallographic and Chemical Studies of the L-Arabinose-Binding Protein From E. coli

F. A. Quiocho, G. L. Gilliland, D. M. Miller, and M. E. Newcomer

Department of Biochemistry, Rice University, Houston, Texas 77001

The crystal structure of the L-arabinose-binding protein (ABP), an essential component of the high affinity L-arabinose transport system in E. coli, has been determined at 3.5- and 2.8-Å resolutions. The Fourier maps indicate that the molecule is ellipsoidal with overall dimensions of 70 X 35 X 35 Å (axial ratio $\simeq$ 2:1) and consists of 2 distinct globular domains (designated "P" and "Q"). A tentative trace of the polypeptide backbone is presented. The 2 domains are arranged to create a deep and narrow cleft, the base of which is formed by 3 polypeptide chain segments linking the 2 domains. The arrangements of the secondary structure of the 2 domains are remarkably similar and can be related by a pseudo-twofold axis. Each domain has a pleated sheet core with 2 helices on either side of the plane of the β sheet. This secondary structural arrangement is similar to that found in other proteins, specifically the dehydrogenases and kinases. The structural similarity is particularly intriguing in light of the recent finding in this laboratory that the dye $2',4',5',7'$-tetraiodofluorescein, an adenine analogue which has been shown to bind to several dehydrogenases and kinases, binds to ABP with a dissociation constant of 30 μM.

Experiments performed with protein, modified with the chromophoric probe 2-chloromercuri-4-nitrophenol (MNP), suggest that the binding site is near an essential cysteine residue: modification of the thiol with the mercurial dramatically decreases the ligand-binding affinity of ABP, and conversely, the sugar protects the cysteine from reaction with MNP. The binding of L-arabinose to MNP-labeled protein perturbs the nitrophenol absorbance spectrum. The essential cysteine has been assigned to position 64 in the proposed chain tracing, which is consistent with the amino acid sequence. As an explanation for the failure of the difference Fourier analyses to locate the sugar-binding site, it is postulated that the structure has been solved with the sugar bound. Electron density to which no amino acid residue can be assigned and which could be the sugar molecule is within van der Waals distance of the sulfur atom.

Key words: L-arabinose-binding protein, three-dimensional structure, spectrochemical studies, active transport, chemotaxis

Abbreviations: ABP – L-arabinose-binding protein; MNP – 2-chloromercuri-o-nitrophenol; PCMBS – p-chloromercuribenzenesulfonic acid; TIF – $2',4',5',7'$-tetraiodofluorescein

Received for publication March 14, 1977; accepted May 25, 1977

Following the discovery of the sulfate-binding protein by Pardee in 1966 (1), a number of low-molecular-weight, water-soluble proteins binding a variety of small substrates (sugars, inorganic ions, and amino acids) have been isolated from the osmotic shock fluid of gram-negative bacteria (for reviews see Refs. 2–4). The localization of binding proteins to the peptidoglycan region between the cytoplasmic and outer membranes (5–6) and the correlation of transport-negative mutants with reduced amounts or altered forms of binding proteins (2–4, 7) suggest that binding proteins are essential components of some high-affinity uptake systems. It has also been demonstrated that binding proteins participate in bacterial chemotaxis (8–10). Recent evidence suggests that binding-protein transport (11) and chemotaxis (12–13) systems are linked to other, presumably membrane-bound, protein components which either facilitate the movement of ligand into the cell or signal the flagella to suppress twiddling activity.

One of these binding proteins, the L-arabinose-binding protein (ABP), has been purified from E. coli B/r (14) and crystals suitable for high resolution x-ray analysis have been obtained (15). The L-arabinose-binding protein is an essential component of high-affinity L-arabinose transport (16), and like all binding proteins, is composed of a single polypeptide chain. The amino acid sequence of the L-arabinose-binding protein has recently been determined by Hogg and Hermodson (17), and the molecular weight derived from this analysis is 33,200. The results of 5-Å and 3.5-Å resolution structural analyses, indicating an ellipsoidal and bilobate molecular structure of the binding protein with an axial ratio of 2:1, have recently been reported (18).

In this paper we present results of our crystallographic analysis of the L-arabinose-binding protein and evidence, based upon studies of the protein modified with the chromophoric probe, 2-chloromercuri-4-nitrophenol (MNP), that the ligand binding site is near an essential thiol residue.

METHODS AND MATERIALS

The L-arabinose-binding protein was purified from E. coli B/r strain UP1041 (ara A39) according to the procedure of Parsons and Hogg (14) with the exclusion of the final step (denaturation in 8 M urea to remove residual L-arabinose). (Cells are induced with 20 mM L-arabinose during the exponential phase of growth.) Crystals suitable for x-ray analysis were grown in 2-methyl-2,4-pentanediol and 3 mM potassium phosphate, pH 6.5 (15). These crystals belong to the space group $P2_1 2_1 2_1$ with one molecule per asymmetric unit. The unit cell dimensions recently obtained by least squares analysis of orientation parameters for a number of crystals are: a = 55.44 ± 0.08, b = 71.72 ± 0.15, and c = 77.64 ± 0.23 Å. Diffraction patterns extend to a spacing of approximately 2 Å.

X-ray diffraction intensities were measured with the omega step-scan procedure of Wyckoff et al. (19). A more detailed description of the procedure used for collecting and processing of diffraction data is given elsewhere (18).

ABP utilized for chemical studies was stored at −20°C as a crystalline suspension in 60% 2-methyl-2,4-pentanediol, 0.1% β-mercaptoethanol, and 3 mM potassium phosphate, pH 6.5. Prior to reaction with the thiol-specific reagent, 2-chloromercuri-4-nitrophenol, the protein was dialyzed overnight into a solution of 6 M guanidine-hydrochloride, 5 mM ehtylenediaminetetraacetic acid, 2 mM dithiothreitol, 10 mM Tris-HCl, pH 7.4 and subsequently dialyzed exhaustively against the appropriate buffer.

Assays were routinely performed by equilibrium dialysis in plexiglass microcells modeled after the design of Willis et al. (20). The purified protein exhibits native binding activity (3×10^{-7} M and 4×10^{-7} M for the dissociation constants of L-arabinose and D-galactose, respectively) and migrates as a single band on polyacrylamide gels (14). Protein was determined from either the absorbance at 280 nm ($\epsilon = 0.94$ 1 g^{-1} cm^{-1}) (14) or by the method of Lowry et al. (21) using lyophilized ABP as a standard. All spectrophotometric measurements were carried out with the use of a Cary 118 or Varian 635 Spectrophotometer. Mercury analysis by atomic absorption spectrophotometry was performed with an Instrumentation Lab Inc. Absorption-Emission Spectrophotometer-153.

RESULTS

Structure

The calculation of a much improved 3.5-Å electron density map, followed shortly by a 2.8-Å map, has provided considerable information about the 3-dimensional structure of the L-arabinose-binding protein. The improvement of the initial 3.5-Å resolution map (18) was achieved by the addition of 2 more good heavy-atom derivatives, namely, p-chloromercuribenzenesulfonic acid (PCMBS) and CdI$_2$. A 3.5-Å refinement of these 2 derivatives, together with 2-chloromercuri-4-nitrophenol, K$_2$PtCl$_4$, and K$_2$IrCl$_6$ yielded a figure of merit of 0.76 for the 4,000 reflections phased. This figure of merit is a significant improvement over 0.66, the value obtained in the initial 3.5-Å phase refinement (18). Phases used to calculate electron density maps were obtained by the method of alternate cycles of least-squares refinement of heavy-atom parameters and multiple isomorphous replacement phase determination (22, 23).

The 2.8-Å resolution Fourier map recently calculated was based entirely upon the phases derived by extending the PCMBS and CdI$_2$ derivatives to 2.8-Å resolution. The mean figure of merit was 0.65. The heavy atom parameters (Table I) utilized for the 2.8-Å phasing were derived from a refinement of the final heavy-atom parameters used in the phasing of the 3.5-Å map. A summary of the phasing statistics resulting from the 2.8-Å resolution is presented in Table II. A more detailed description of the 2.8-Å resolution phase refinement will be presented elsewhere.

The complete 2.8-Å electron density map for the L-arabinose-binding protein is viewed up the z-axis in Fig. 1A–E. It will be evident from the description of this map which follows that the molecule is an ellipsoid consisting of 2 distinct globular domains, designated for convenience as the "P" and "Q" domains. The molecule lies inclined to the x-y plane with the P domain visible in the lower left at higher z values and the Q domain in the upper right. The electron density of the protein molecule is quite distinguishable from the low electron density of the mother liquor.

In Fig. 1A the central region of electron density is a portion of the C-terminal helix, and the upper right-hand region is part of the P domain. The regions of electron density around the border are from adjacent ABP molecules.

The protein boundary is clearly delineated in Fig. 1B with electron density from a neighboring molecule appearing at the lower left. The Q domain is fully outlined (upper right) and the P domain is just beginning to emerge (lower left). This section depicts 2 of the 3 chain segments which connect the 2 domains.

Fig. 1C reveals the 2:1 axial ratio of the molecule. The division of the P and Q domains is not clear in this view of the map due to the perspective. There is only one true connection between the 2 domains in this region. Also, there are several helical regions visible in both domains.

TABLE I. Heavy-Atom Parameters Based on the 2.8-Å Resolution Refinement

Site no.	x	y	z	$B(\text{Å})^{2\text{a}}$	Fractional occupancy
		CdI_2 derivative			
1	0.4658	0.9749	0.5988	46.1	1.11
2	−0.0769	1.1326	0.5498	49.6	1.11
3	0.0086	1.1921	0.5457	98.4	1.21
4	0.0618	1.1957	0.7398	76.5	0.77
5	0.3715	1.1388	0.7334	50.8	0.32
6	0.3764	0.2951	0.8578	73.3	0.34
		Hg, PCMBS derivative			
1	0.1078	1.1313	0.7708	31.1	0.77
2	0.3075	0.7365	0.7407	7.8	0.82
3	0.0061	1.1611	0.5286	14.3	0.16
		Hg, MNP derivative			
1	0.1035	1.1275	0.7704	57.3	0.86
2	0.3051	0.7373	0.7409	10.0^{b}	0.27
		$(NH_4)_2PtCl_4$ derivative			
1	0.4521	0.8189	0.6951	28.5	1.02
2	0.3978	0.3113	0.6653	177.1	1.06
3	0.0294	1.1707	0.5287	30.1	0.65
4	0.0634	1.1081	0.7636	10.0^{b}	0.11
5	0.5313	0.8256	0.6677	10.0^{b}	0.15
6	0.4993	0.6735	0.7719	10.0^{b}	0.19
7	0.1912	1.1272	0.6873	17.0	0.18
8	0.3834	0.8418	0.6789	10.0^{b}	0.14
9	0.4917	0.7346	0.8842	74.2	0.31
		K_2IrCl_6 derivative			
1	0.5297	0.7766	0.9345	245.1	0.79
2	0.5326	0.7219	0.8879	153.3	0.42
3	0.4812	0.8744	0.6654	12.8	0.13
4	−0.0537	1.1487	0.5027	216.8	0.28
5	0.0994	1.1287	0.7558	188.1	0.18

[a]Isotropic temperature factor.
[b]Parameters which were not refined.

Several interesting features are seen in Fig. 1D. The molecular boundary is outlined by the dashed line, and the division between the P and Q domains, which is quite pronounced, is indicated by the arrow. Four strands of the P-domain β-sheet are clearly visible. A long segment of helix appears in the Q domain (upper right) and a screw-related helix in another molecule is seen in the upper left. Neighboring molecules protrude into this region of the map at all 4 corners.

TABLE II. Summary of the 2.8-Å Phasing Statistics (Overall Figure of Merit = 65%)

Heavy atom (resolution)	RMS-E[a]	RMS-F_H[b]	RMS-F_H/RMS-E	R factor[c]
Hg, PCMBS (2.8 Å)	53	129	2.43	0.078
Hg, MNP (3.5 Å)	48	81	1.69	0.062
CdI$_2$ (2.8 Å)	54	90	1.67	0.080
(NH$_4$)$_2$PtCl$_4$ (3.5 Å)	104	144	1.38	0.129
K$_2$IrCl$_6$ (5.0 Å)	37	55	1.49	0.051

[a]Root mean square lack-of-closure error $E = (\Sigma_h \epsilon_{hj}^2/n)^{1/2}$, ϵ_{hj} = lack of closure for reflection h of derivative j and n = number of reflections.
[b]Root mean square heavy atom contribution $F_H = (\Sigma_h f_{hj}^2/n)^{1/2}$, f_{hj} = heavy atom scattering factor.
[c]Kraut R factor = $(\Sigma||F_D| - |F_N + f_D||)/\Sigma|F_D|$, F_D = structure factor of the derivative, F_N for the protein and f_D for the heavy atoms.

Figure 1E shows an almost total disappearance of the Q domain with only 2 parallel helical regions of the P domain remaining. Electron density from the 4 neighboring molecules is still visible.

A tentative trace of the course of the polypeptide chain was first derived from an interpretation of the 3.5-Å map. With a few minor modifications, this trace proved to be essentially identical to the one based upon the interpretation of the 2.8-Å map. Attempts were made to fit the complete amino acid sequence (17) to the improved 3.5-Å and later to the 2.8-Å maps. Approximately two-thirds of the amino acid sequence have been successfully fitted to the provisional trace. The identification of the N-terminal peptide (up to residue 95) and the C-terminal peptide (starting from 189) was verified by fitting the known sequence. A stereo drawing of the provisional trace is shown in Fig. 2. Depicted in Fig. 3 is a schematic drawing of the molecule. The molecular dimensions are 70 X 35 X 35 Å. The fact that the L-arabinose-binding protein is ellipsoidal, with an axial ratio of 2:1, could account for the initial overestimation of the molecular weight of 38,000 for the protein by sedimentation and gel filtration techniques (11).

In our proposed trace, approximately the first 35% of the polypeptide chain forms the major part of the P domain and the next 43% forms the entire Q domain. The remainder of the chain (~ 22%) protrudes as a loop or "handle," then extends back into the P domain, and finally forms the C-terminal helix shared by both domains.

The arrangements of the secondary structure ("super secondary structure") of the 2 domains are remarkably similar and both domains can be related by a pseudo-twofold axis (Fig. 3). Each domain has a central pleated sheet core consisting entirely of parallel strands (with the exception of the sixth antiparallel strand in the Q domain), and on either side of the plane of the β sheet lie 2 helices antiparallel to the β sheet. This arrangement of secondary structure is similar to that found in other proteins, specifically the dehydrogenases and kinases, and has been termed the "nucleotide-binding fold" (24). A detailed quantitative comparison of the 2 domains of ABP and of similar domains found in other proteins is presently underway.

This structural similarity is particularly significant in light of a recent finding in this laboratory that the dye 2',4',5',7'-tetraiodofluorescein (TIF) binds to ABP with a dissociation constant of approximately 30 μM. TIF has been used as a spectral and crystallo-

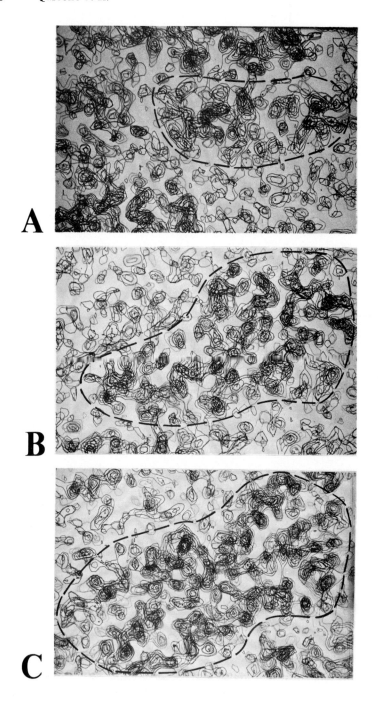

Fig. 1. The complete 2.8-Å electron density map of L-arabinose binding protein viewed up the z axis. Each of the views, A–E, consist of 10 superposed sections and portions of the molecule are enclosed by a dashed line. The bounds of the map are: x = −0.20 to 0.70 (∼ 50 Å), y = 0.25–1.25 (∼ 72 Å), and z = 0.389–0.918 (∼ 41 Å) with the x axis vertical and the y axis horizontal. The upper left corner is at

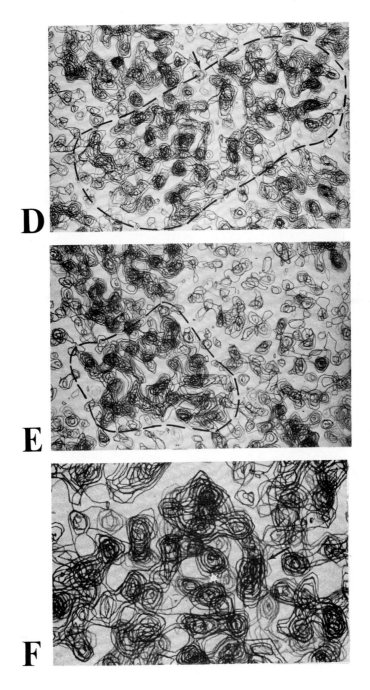

x = −0.20, y = 0.25. A) z = 0.389−0.486. B) z = 0.497−0.594. C) z = 0.605−0.702. D) z = 0.713−0.810. The arrow marks the separation of the 2 domains. E) z = 0.821−0.918. F) An enlargement of the proposed sugar binding site region consisting of 12 superposed sections with z = 0.626−0.745 [y = 0.47−0.93 (∼ 33 Å)]. The arrow indicates the position of an "extraneous" electron density and the asterisk is above the single "essential" cysteine, residue no. 64.

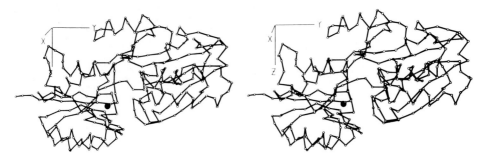

Fig. 2. Stereo drawing of the proposed polypeptide chain trace for L-arabinose-binding protein. The black dot indicates the essential cysteine.

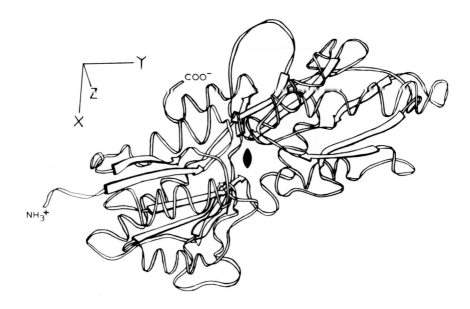

Fig. 3. A schematic representation of L-arabinose-binding protein. The β sheets are indicated by arrows pointing from N to C. A pseudo-twofold axis is located directly in the center of the molecule, perpendicular to the plane of the page.

graphic probe of the nucleotide binding site in a variety of enzymes, e.g., lactate dehydrogenase (25), aspartate transcarbamylase (26), and creatine kinase (27). The difference spectra obtained upon binding of TIF to ABP (Fig. 4) is characteristic of spectra observed for TIF binding to the aforementioned enzymes and suggests the existence of a "super secondary structure" akin to the "nucleotide fold." A similar TIF difference spectrum was also observed for the D-galactose-binding protein. The significance of 2 structurally

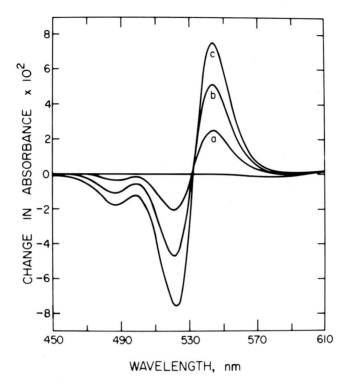

Fig. 4. Difference spectra of free vs tetraiodofluorescein bound to arabinose-binding protein. The sample contained 1.4×10^{-4} M arabinose-binding protein in 0.1 M Tris-hydrochloride, pH 8.2, and the reference cell buffer only. Curves a, b, c are spectra produced by the addition of a small volume of 1 mM tetraiodofluorescein to both cells to give a concentration of 5.0, 13, and 15 μM, respectively.

similar domains in L-arabinose-binding protein and especially the similarity of these domains with those found in unrelated enzymes remains an intriguing mystery. Their existence in this binding protein supports the notion that they can be formed readily and could have evolved independently. We are currently attempting to determine whether the L-arabinose-binding protein may bind nucleotides and various nucleotide analogs.

Essential Thiol Residue and Sugar Binding Site

Several attempts have been made using difference Fourier techniques to locate the sugar-binding site from crystals soaked or cocrystallized in solutions containing L-arabinose or D-galactose. Difference Fourier maps, however, showed no peaks above background. A likely explanation for these failures to locate the sugar-binding site is the possibility that the native protein may have been crystallized with tightly bound L-arabinose (introduced during cell growth).

The demonstration that ABP contains an "essential" cysteine residue provided us with a means to locate the sugar-binding-site region. Both chemical and crystallographic

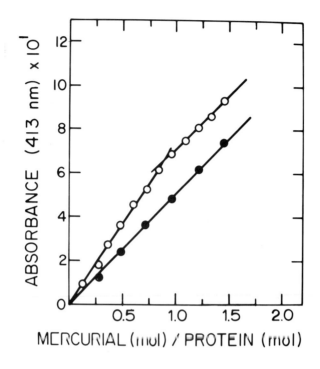

Fig. 5. Titration of arabinose-binding protein with 2-chloromercuri-4-nitrophenol. Protein (4.74 × 10^{-5}M) in 1.5 ml of 5 M guanidine hydrochloride, 5 mM EDTA, 0.1 M triethanolamine hydrochloride buffer, pH 8.0, was titrated with 5-μl aliquots of MNP (1.72 mM). The absorbance at 413 nm of the protein sample and blank was read after each addition. ○) Protein; ●) blank.

analyses indicate that ABP contains one cysteine residue. Titration of the protein in the denatured state with MNP, a thiol-specific chromophoric probe (28), yielded an inflection point at 0.88 ± 0.03 equivalents of added mercurial (Fig. 5) (29). These results were duplicated by a separate titration with Ellman's reagent (30) 5,5'-dithiobis(2-nitrobenzoic acid) (D. M. Miller, manuscript in preparation). Furthermore, mercury analysis of an ABP sample (5.2 mg/ml) incubated with a fivefold excess of MNP for 3 days and separated from unreacted mercurial by passage through a Sephadex G-25 column and exhaustive dialysis gave a ratio of 0.8 moles of mercury per mole of protein.

The observation that thiol-specific reagents dramatically decrease the affinity of ABP for L-arabinose (14, 29) and that L-arabinose inhibits the rate of reaction of MNP and 5,5'-dithiobis(2-nitrobenzoic acid) with the protein suggest that the thiol is near the binding site (D.M. Miller, manuscript in preparation). Figure 6 illustrates the maximal inhibition of binding achieved with the addition of one equivalent of mercurial and the protection afforded ABP by preequilibration with 10^{-5} M L-arabinose.

Furthermore, the addition of L-arabinose to mercurial-labeled ABP perturbs the characteristic nitrophenol absorbance spectrum shown in Fig. 7 suggesting a ligand-

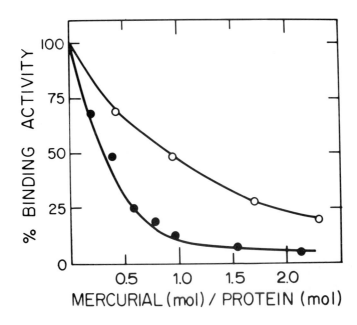

Fig. 6. Binding activity of MNP-modified arabinose-binding protein and protection by ligand. For the protection experiment, to arabinose-binding protein equilibrated (8 h) in the binding cells with L-[1-^{14}C]arabinose were added 5-μl aliquots of MNP. Following incubation overnight at 4°C, 100-μl aliquots were counted for radioactivity. In a second experiment, "sugar-free" protein was modified with MNP by adding 10-μl aliquots of the mercurial at appropriate concentrations to 0.44 ml of arabinose-binding protein and allowing the reaction to proceed overnight at 4°C. For the assay, double chambers separated by a dialysis membrane were filled with 200 μl of MNP-modified protein and 200 μl of L-[1-^{14}C]arabinose, equilibrated and counted as above. The percent of binding activity is expressed as the ratio of the apparent dissociation constant of duplicate samples to that of a control. The buffer was 10 mM Tris-HCl, pH 7.4. Arabinose-binding protein was 0.5 mg/ml and L-[1-^{14}C]arabinose was 10^{-5} M in both experiments. ●) Arabinose-binding protein modified with MNP; ○) arabinose-binding protein reacted with MNP in presence of L-arabinose.

induced conformational change and/or direct interaction between the sugar and the phenolic mercurial.

A nondialyzable mercury-binding site was located by difference Fourier synthesis at position 64 in a tentative trace of the polypeptide chain. Moreover, the native electron density at site 64 is consistent with the assignment of a sulfur atom to this position. Although amino acid analysis had earlier indicated the presence of 2 half-cystine residues (14), the sequence determination showed one cysteine at position 64 (17), consistent with our assignment.

The likelihood that the structure of ABP may have been solved with bound L-arabinose prompted further examination of the 3.5- and 2.8-Å electron density maps, particularly in the region near the single essential cysteine residue. This examination shows in both maps the presence of an "extraneous" density near the thiol group which presently cannot be attributed to the protein molecule, but is of sufficient size and shape to be a

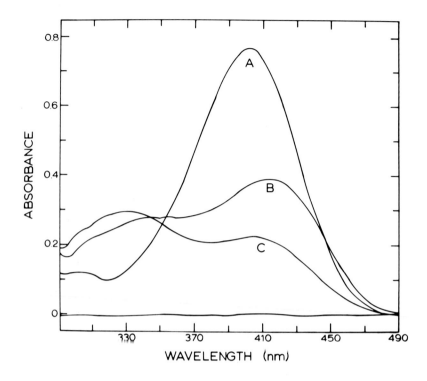

Fig. 7. Comparison of absolute spectra of 2-chloromercuri-4-nitrophenol in buffer (0.1 M Tris-HCl, pH 7.85) and bound to arabinose-binding protein in the absence and presence of L-arabinose. Arabinose-binding protein and MNP were 5×10^{-5} M. A) Absolute spectrum of MNP. B) Absolute spectrum of MNP bound to arabinose-binding protein. C) Absolute spectrum of MNP bound to arabinose-binding protein in the presence of 10^{-2} M L-arabinose.

sugar molecule. The presumed sugar molecule would be bound just to the "right" and within van der Waals distance of the cysteine residue (see Figs. 1F and 2). A native structure with sugar is consistent with the results of the soaking and crystallization experiments.

ABP is fully active in 60% 2-methyl-2, 4-pentanediol, suggesting that the failure of the difference Fourier maps to reveal bound sugar is not a result of the molecule being rendered inactive by its mother liquor. In addition, it can be seen from the packing diagram (Fig. 8) that the protein is readily accessible in the crystal lattice; various organomercurials, no smaller than the sugar substrate, are able to pentrate the crystal lattice and bind the essential thiol. Attempts to obtain "sugar-free" ABP crystals are currently underway.

DISCUSSION

The 3-dimensional structure of the L-arabinose-binding protein is the first structure to be determined in the family of transport proteins. The remarkable features of the structure may prove to be common to all periplasmic binding proteins and may be related to function.

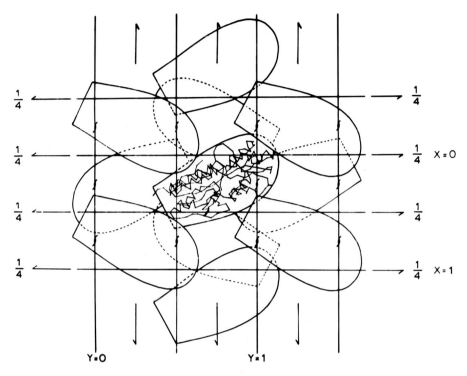

Fig. 8. Packing diagram for L-arabinose-binding protein. The equivalent positions for this space group are x, y, z; ½ −x, −z,½ + z; ½ + x, ½ − y, −z; and −x, ½ + y, ½ −z. The symmetry notation is taken from the International Tables for X-ray Crystallography (32).

All binding proteins share common functions — active transport and bacterial chemotaxis — and are dislodged from bacterial cells by mild osmotic shock as monomeric polypeptides. That binding proteins may be elongated is indicated by the axial ratio of 2:1 for ABP and the preliminary finding that the sulfate-binding protein (MW = 34,000) from S. typhimurium has an axial ratio of 4:1 (33). Furthermore, among subgroups of periplasmic binding proteins (i.e., sugar-binding proteins, amino acid-binding proteins, etc.) a number of observations indicates structural similarity as well. For instance, despite the fact that L-arabinose- and D-galactose-binding protein biosyntheses are controlled by distinctly different genes, *ara*C and *mgl*R respectively (34, 35), the cross-reactivity of antibodies for each shows that the molecules share some regions of similar tertiary structure (36). Furthermore, we have shown that the dye 2′,4′,5′,7′-tetraiodofluorescein, a chromophoric probe for the nucleotide-binding site in a variety of proteins, binds to both L-arabinose-binding protein and D-galactose-binding protein. Circular dichroic measurements for the 2 sugar-binding proteins indicated similar secondary structure content (14, 31). However, the relative amount of helix predicted from circular dichroic measurements (10%) for ABP is incorrect on the basis of the structural results which indicate 35% helix. Perhaps this common molecular conformation has fullfilled a functional and/or structural requirement for binding proteins. It is not unreasonable to suggest that a 3-

dimensional structure is common to a given family of proteins; groups of proteins (i.e., cytochromes, dehydrogenases, and kinases) often have common structural features.

The existence of extensive secondary structure and the packing of this structure ("super secondary structure") is an aspect of the protein that could be relevant to the general properties of binding proteins. Perhaps this "super-structure" could account for the unusual stability of L-arabinose-binding protein and, by extension, the resistance of binding proteins in general to denaturation.

Our structural analysis has further revealed that the 2 distinct globular domains in the L-arabinose-binding protein are very similar. The orientation of the 2 domains is such that both super secondary structures can be related by an approximate local twofold axis (Fig. 3). A pseudo-dyad axis relating 2 domains has similarly been shown in rhodanese (37). A detailed quantitative comparison, using methods developed by Rossmann and coworkers (24), of the 2 domain structures in the binding protein, and with similar domains found in other, unrelated protein structures, is under further investigation. It is of interest to note that the arrangement of the secondary structure in both domains closely resembles the main structural features of the so-called "nucleotide-binding fold" domain found in some protein structures, notably the dehydrogenases and kinases (24), in the sense that all have a central pleated sheet structure consisting mainly of parallel strands and 2 helices on either side of the plane of the sheet, antiparallel to each strand. The significance of the presence of 2 structurally similar domains in L-arabinose-binding protein and especially the similarity of these domains with those found in unrelated enzymes remains a tantalizing mystery. Their occurrence in this binding protein supports the notion that they can be formed readily and could have evolved independently.

Insofar as it is known, the L-arabinose-binding site lies between 2 domains, but is predominantly associated with the P domain. The second domain (Q domain) may be involved to a large extent in binding to other components of the system. The elegant genetic studies of Ames and co-workers (11) on the histidine transport system have shown that the J protein (histidine-binding protein) has 2 distinct binding sites necessary for its function. One site is required for binding histidine, and the other is involved in a direct interaction with another protein component (the P protein) of the system which is presumed to be membrane bound. Nothing is known about the nature and location (relative to the histidine-binding site) of the other site.

The 2 domains in the binding protein could also conceivably facilitate an induced conformational change necessary for transport and bacterial chemotaxis. Since only 3 closely parallel polypeptide chain segments connect the 2 domains (the hinge region), any conformational change(s) could take place in the molecule by a mere shifting of one domain relative to the other. This conformational change could then affect the interaction with other components, the affinity for substrate, or both. The functional relationship between domains is especially important since the sugar-binding site is presumed to be between the domains, near the opening of the cleft.

For efficient transport, the sugar-binding site would be expected to be juxtaposed with other components of the system. This would imply protein:membrane surface interactions encompassing regions on the surface of the 2 domains of the protein and adjacent to the sugar-binding site. We have preliminary observations (currently under further study) which suggest that these specific regions of the binding protein surface may have some local charge properties necessary for hydrophilic interactions with the cytoplasmic membrane surface. This is based on the finding that the binding sites of all the heavy-atom derivatives used so far are unusually concentrated on these regions (or on one side) of the protein molecule (Table I). This side of the molecule contains the presumed sugar-binding site. This remarkable distribution of the heavy-atom binding sites is not a result of the molecular packing of the crystal; Fig. 8 clearly indicates that all sides of the protein are accessible to the solvent. It is noteworthy that all these heavy-atom compounds are anionic, suggesting, in part, local concentrations of positively charged residues on the surface of the protein. These positively charged residues may suggest the possibility of protein inter-actions with either the phosphate moiety of the phospholipids adjacent to the other membrane-bound protein components or the negatively charged residues on the exposed part of the other protein component, or both. Such a hydrophilic interaction would be consistent with the observation that all binding proteins are loosely bound and easily removed by mild osmotic shock treatment.

Currently, the assignment of the sugar-binding site region of the L-arabinose-binding protein depends to some extent upon the identification of the single "essential" cysteine residue. However, the structure of the binding protein may have been solved with bound L-arabinose. To the "right" and within van der Waals distance of the cysteine residue is an "extraneous," nonprotein density peak which we attribute to a bound sugar molecule (Fig. 1F). The presumed sugar-binding site is in the cleft formed by the packing of the 2 domains, abutting the P domain. In addition to the cysteine at least 4 other residues, one of which appears to be a tryptophan, are in the vicinity of the sugar molecule. In most periplasmic binding proteins (e.g., L-arabinose- and D-galactose-binding proteins (14, 31); see also Ref. 2 and 4), the presence of substrates causes changes in the fluorescence of tryptophan residue(s). This fluorescence change may now be attributed more likely to changes in the microenvironment of the tryptophan as a result of direct interaction with the substrate rather than exclusively to conformational change, as has been frequently suggested.

ACKNOWLEDGMENTS

The contributions of former colleagues G. N. Phillips, Jr., V. K. Mahajan, and A. K. Q. Siu are gratefully acknowledged. This investigation was supported by a grant from the Robert A. Welch Foundation (C-581) and by NIH Grant GM-21371.

REFERENCES

1. Pardee AB, Prestidge LS, Whipple MB, Dreyfus J: J Biol Chem 241:3962, 1966.
2. Oxender DL: Annu Rev Biochem 41:777, 1972
3. Rosen BP, Heppel LA: In Lieve L (ed): "Bacterial Membranes and Walls." New York: Marcell Dekker, 1973, p 209.
4. Boos W: Annu Rev Biochem 43:123, 1974.
5. Pardee AB, Watanabe K: J Bacteriol 96:1049, 1968.
6. Nakane PK, Nichoalds GE, Oxender DL: Science 161:182, 1968.
7. Ames GF-L, Lever TL: Proc Natl Acad Sci USA 66:1096, 1970.
8. Adler J: Annu Rev Biochem 44:341, 1975.
9. Koshland DE Jr: In Jaernicke L (ed): "Biochemistry of Sensory Functions." Berlin: Springer-Verlag, 1974, p 133.
10. Adler J: J Gen Microbiol 74:77, 1973.
11. Ames GF-L, Spudich EN: Proc Natl Acad Sci USA 73:1877, 1976.
12. Ordal GW, Adler J: J Bacteriol 117:517, 1974.
13. Strange PG, Koshland DE Jr: Proc Natl Acad Sci USA 73:762, 1976.
14. Parsons RG, Hogg RW: J Biol Chem 249:3602, 1974.

15. Quiocho FA, Phillips GN Jr, Parsons RG, Hogg RW: J Mol Biol 86:491, 1974.
16. Brown CE, Hogg RW: J Bacteriol 111:606, 1972.
17. Hogg RW, Hermodson MA: J Biol Chem 252:5135, 1977.
18. Phillips GN Jr, Mahajan VK, Siu AKQ, Quiocho FA: Proc Natl Acad Sci USA 73:2186, 1976.
19. Wyckoff HW, Doscher M, Tsernoglou D, Allewell NM, Kelly DM, Richards FM: J Mol Biol 27:563, 1967.
20. Willis RC, Morris RG, Cirakoglu C, Schellenberg GD, Gerber NH, Furlong CE: Arch Biochem Biophys 161:64, 1974.
21. Lowry OH, Rosebrough NJ, Farr AL, Randall RJ: J Biol Chem 193:265, 1951.
22. Dickerson RE, Kendrew JC, Strandberg BE: Acta Crystallogr 19:1188, 1961.
23. Lipscomb WN, Coppola JC, Hartsuck JA, Ludwig ML, Muirhead H, Searle J, Steitz TA: J Mol Biol 19:423, 1966.
24. Rossmann MG, Liljas A, Branden C-I, Banazak LJ: In Boyer PD (ed): "The Enzymes," 3rd Ed. New York: Academic Press, 1975, vol II, p 62.
25. Wasserman PM, Lentz PJ: J Mol Biol 60:509, 1971.
26. Jacobsberg LB, Kantrowitz ER, Lipscomb WN: J Biol Chem 250:9238, 1975.
27. Somerville LL, Quiocho FA: Biochim Biophys Acta 481:493, 1977.
28. McMurray CH, Trentham DR: Biochem J 115:913, 1969.
29. Miller DM: Fed Proc Fed Am Soc Exp Biol 35:1586, 1976.
30. Ellman GL: Arch Biochem Biophys 82:70, 1959.
31. Boos W, Gordon AS, Hall RE, Price HD: J Biol Chem 247:917, 1972.
32. Henry NFM, Lonsdale K (eds): "International Tables for X-ray Crystallography." London: Kynock Press, 1969.
33. Langridge R, Shinagawa H, Pardee AB: Science 84:585, 1970.
34. Hogg RW, Englesberg AS: J Bacteriol 100:423, 1969.
35. Lengeler J, Herman KO, Unsold HJ, Boos W: Eur Biochem 19:457, 1971.
36. Parsons RW, Hogg RW: J Biol Chem 294:3608, 1974.
37. Bergsman J, Hol WGJ, Jansonius JN, Kalk KH, Ploegman JH, Smit JDG: J Mol Biol 98:637.

Journal of Supramolecular Structure 6:519–533 (1977)
Molecular Aspects of Membrane Transport 349–363

High-Affinity Phlorizin Binding to Brush Border Membranes From Small Intestine: Identity With (a Part of) the Glucose Transport System, Dependence on the Na+-Gradient, Partial Purification

Carl Tannenbaum, Gerhard Toggenburger, Markus Kessler, Aser Rothstein*, and Giorgio Semenza

Laboratorium für Biochemie, Eidgenössische Technische Hochschule, Universitätstrasse 16, CH 8092 Zürich, Switzerland

In the presence of an NaSCN gradient phlorizin binds with a high affinity ($K_d \simeq$ 4.7 μM) to vesicles derived from brush border membranes of intestinal cells of rabbits. The value for K_d corresponds closely to that of K_i determined from phlorizin inhibition of sugar transport. The apparent affinity for phlorizin is decreased if NaCl is substituted for NaSCN and decreased substantially if the gradient of NaSCN is allowed to dissipate prior to the phlorizin binding. The number of high affinity binding sites is about 11 pmol/mg protein. Additional binding to low affinity sites can amount to as much as 600 pmol/mg protein after prolonged exposure to phlorizin (5 min). The high affinity sites are related to glucose transport based on the similarity of the K_d and K_i values under a variety of conditions and on the inhibition of the binding by D-glucose but not by D-fructose. The transport system and the high affinity phlorizin binding sites can be enriched by a factor of 2–3 by treatment of vesicles with papain, which does not affect the transport system, but considerably hydrolyzes nonrelevant protein.

Key words: phlorizin binding, to intestinal membranes; D-glucose transport; sugar transport; sucrase, small intestinal; small intestine; membrane transport of D-glucose

Phlorizin inhibition of monosaccharide transport across the brush border membrane of small intestine and of renal proximal tubuli has been extensively studied. In particular, the inhibition is known to be fully competitive with transport substrates (1, 2). Phlorizin, on this basis, may be assumed to bind to the sugar binding site of the transport system or to some closely associated site. The binding of phlorizin has been directly demonstrated in brush border membranes from renal tubuli by centrifugation (e.g., Refs. 3, 4) or membrane filtration (e.g., Refs. 5–7) techniques. In this paper we report some of the characteristics of phlorizin binding to vesicles derived from brush border membranes of rabbit small intestine. By using a rapid Millipore filtration technique, we could show that the

*Aser Rothstein is now at the Hospital for Sick Children, Toronto, Ontario, Canada.
Received April 14, 1977; accepted June 1, 1977

binding consists of at least 2 components, a very rapid, low capacity, high affinity, element and a slower, high capacity, low affinity element. The former component is the one that is associated with the inhibition of sugar transport. In the presence of a Na$^+$ gradient its apparent affinity (K_d) is about the same as its apparent affinity as an inhibitor of transport (K_i), but in the absence of a Na$^+$ gradient, the binding affinity is considerably reduced. The high affinity binding is inhibited by D-glucose and the binding site is presumably identical with (a part of) the Na-dependent monosaccharide transport system(s). We also report a partial negative purification of this membrane component.

MATERIALS AND METHODS

Rabbit intestine was collected from freshly killed commercial rabbits at a local slaughterhouse. It was cleaned and frozen before being brought to the laboratory for storage. Vesicles were prepared from the frozen tissue on the day of each experiment according to the procedure of Schmitz et al. (8), as modified by M. Kessler, O. Acuto, C. Storelli, H. Murer, M. Müller, and G. Semenza (Biochim Biophys Acta, in press). Briefly, 20 g of frozen tissue in 1–1.5-g pieces are allowed to thaw in 60 ml of 300 mM mannitol, 12 mM Tris-HCl, pH 7.5. The thawed tissue is mixed in solution with a Chemap Vibro-Mixer for 1.5 min to loosen the mucosal cell fragments from the connective tissue. The mixture is filtered through a Büchner funnel and the connective tissue discarded. The cell fragments are diluted sixfold with water, homogenized in a blender for 3 min at full speed, and then CaCl$_2$ is added to a final concentration of 10 mM. After standing for 20 min, the mixture is centrifuged at 3,000 X g for 15 min and the pellet discarded. The supernatant is then further centrifuged at 27,000 X g for 30 min. The pellet is resuspended in 40 ml of 100 mM mannitol, 10 mM Tris/Hepes, pH 7.5, homogenized in a Teflon Potter Elvehjem Homogenizer and centrifuged at 27,000 X g for 30 min. The pellet, containing the brush border vesicles, is suspended in 1 ml of the appropriate solution.

Uptake experiments were initiated by mixing equal volumes of vesicle suspension and incubation medium containing 100 mM mannitol and sodium and glucose at twice the desired final concentration. This incubation medium also contained the appropriately labeled radioactive compounds. At the desired times, 20-μl aliquots of reaction mixture were withdrawn, quickly pipetted into 1.5 ml ice cold 150 mM NaCl, filtered through a wet Sartorius microfilter (0.6 μm), washed with a further 5 ml of the cold 150 mM NaCl, dissolved in 10 ml of toluene, Triton X-100, butyl PBD, acetic acid scintillation solvent, and counted.

When influx or binding during short (10 sec or less) incubation periods were determined the procedure was slightly different. Ten microliters of vesicles were carefully placed in the bottom of a clear polystyrene test tube fitted into a vibration device controlled by an electric timer. Ten microliters of radioactive incubation medium was placed as a separate drop at the bottom of the tube. At the start of the timer the shaking of the vibrator rapidly mixed the 2 drops together (less than 80 msec). At the chosen time 2 ml of cold 150 mM NaCl were automatically injected into the incubation tube stopping the reaction. The sample was then treated in the usual manner.

The phlorizin binding in the presence of a sodium thiocyanate gradient was determined as described for the uptake studies. The sodium, phlorizin, and either glucose or fructose were all in the incubation medium. The vesicles were exposed to all 3 ligands simultaneously at the start of the incubation. When binding in the absence of a gradient was determined the vesicles were allowed to come to equilibrium (20 min at room temperature, followed by 1 h at 4°C) in 100 mM NaSCN before the phlorizin and either glucose or fructose incubation was begun.

During stopping and washing (total time: 10 sec), undoubtedly some desorption of bound phlorizin did take place (the washout curves will be reported elsewhere). Therefore, the number of binding sites were certainly underestimated. However, since the washing procedure was identical for all test samples (which were incubated either with glucose or fructose present, and with or without a Na^+ gradient across the membrane), the general conclusions concerning the effect of glucose or the effect of a Na^+ gradient on high affinity phlorizin binding sites are valid.

Papain digestion of vesicles, usually with 0.4 or 0.8 units papain/mg protein was carried out at $37°C$ for 20 min. The mixture also contained 1 mM EDTA; 5 mM cysteine, 15 mM Tris/Hepes, pH 7.5, 50 μM $MgCl_2$; 50 μM $CaCl_2$. After the incubation, the mixture was diluted 20-fold with cold 15 mM Tris/Hepes, pH 7.5, and centrifuged at 27,000 × g for 30 min. The pellet then was washed in 10 mM Hepes/Tris, 100 mM mannitol, pH 7.5, and again centrifuged. The treated vesicles were then used for the determination of either glucose transport or phlorizin binding activity.

Protein (9), sucrase (10), and lactase (10) activities were determined by routine procedures.

Sodium dodecyl sulfate-polyacrylamide gel electrophoresis (SDS-PAGE) was carried out in Tris/glycine buffer, pH 8.8, in slab gels, according to Maestracci et al. (11) with minor modifications. D-[1-^{3}H]Glucose was from Amersham, Buks. (5.2 Ci/mmol). [^{3}H(G)] Phlorizin, lot No. 929-221, was obtained from New England Nuclear (Boston, Massachusetts). The radioactive phlorizin yielded with cold phlorizin a single spot by both UV and radioactive detection. It had the same R_f value of Whatman 3MM, using water as the chromatographic solvent at room temperature (J. S. Cook, personal communication, 1977), as authentic recrystallized phlorizin from ICN Pharmaceuticals, Inc., Life Science Group, Plainview, New York. The authenticity of the nonradioactive sample was checked by NMR. Papain was obtained from Boehringer, Mannheim, GFR. All other chemicals were reagent grade.

RESULTS

Kinetics of Phlorizin Inhibition of D-Glucose Uptake

The vesicle preparation yields itself particularly well to the kinetic studies, due to the high and relatively persistant substrate overshoot that occurs when a Na^+ gradient is imposed in the presence of the permeant anion SCN^- [M. Kessler, O. Acuto, C. Storelli, H. Murer, M. Müller, and G. Semenza (Biochim Biophys Acta, in press) and also Fig. 6 below]. The prolonged overshoot is presumably indicative of the dissipation of the Na^+ gradient being slow. Under the conditions employed, the uptake of D-glucose is linear with time up to 2–3 sec, deviates slightly from linearity by 4 sec, and significantly after 8 sec (Fig. 1). Other conditions and lower D-glucose concentrations also yielded essentially linear uptake up to 2 sec, our standard error on the time scale being approximately 2%. We, therefore, used 2-sec incubation periods when studying phlorizin inhibition of D-glucose uptake into these vesicles. The K_i value determined from Dixon plots (Fig. 2) was 7 μM.

Phlorizin Binding

Initial attempts to determine phlorizin binding were carried out by centrifugation techniques after relatively prolonged exposure of the vesicles to phlorizin (up to 30 min). In each case a large low-affinity binding was evident, Scatchard plots providing little or no

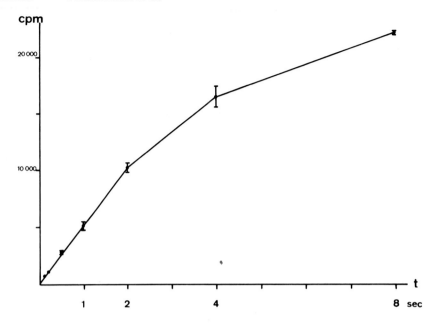

Fig. 1. Time course of D-glucose uptake into membrane vesicles from rabbit small intestinal brush borders. The uptake of D-[^{3}H] glucose (0.1 mM) was measured at room temperature in the presence of a NaSCN gradient, at the start Na$^+_{out}$ = 100 mM, Na$^+_{in}$ = 0 mM. All solutions contained also 100 mM mannitol and 10 mM Hepes/Tris, pH 7.5. The values were obtained from measurements in triplicates, the bars indicate the standard deviations. The figure reports only the initial rising part of the overshooting glucose uptake; after a few minutes the radioactivity in the vesicles declines to approximately 3,000 cpm, corresponding to the equilibration of the inner and outer substrate and Na$^+$ concentrations. Specific radioactivity of the D-glucose used: approximately 250 cpm per pmol glucose.

indication of any high-affinity binding (i.e., no K_d in the range corresponding to the K_i for inhibition). From the time course of phlorizin binding (Fig. 3), it is apparent that a steady state of binding is not reached within the first minute of incubation. Even after 5 min the level of binding still increases. The amount bound at 5 min using 1 μM phlorizin is approximately 5 pmol/mg protein but at 200 μM phlorizin it is over 100 times as high, 600 pmol/mg protein. It is very probable, therefore, that most of the binding at high phlorizin concentrations is to nonspecific sites, unrelated to glucose transport.

The onset of phlorizin inhibition of sugar transport is very rapid both in the small intestine (2) and kidney (12). In the case of brush border membrane vesicles, the inhibitory effect is completed within 2 sec after the addition of phlorizin, with no increase after more prolonged exposure times. It can be concluded, therefore, that the binding to the inhibitory site is virtually complete in 2 sec and that the large binding component after 2 sec in Fig. 3 is to sites that are not associated with inhibition. Furthermore, the inhibition of glucose uptake was determined in the presence of a NaSCN gradient across

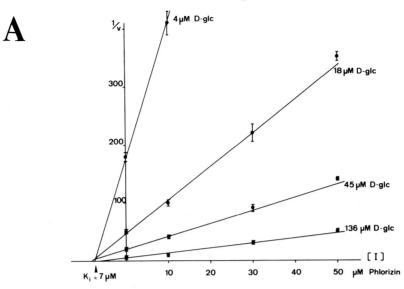

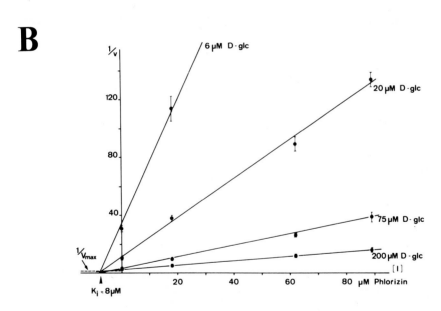

Fig. 2. Dixon plot of phlorizin inhibition of D-glucose uptake into brush border membrane vesicles, at room temperature. A) Incubation time, 2 sec. Phlorizin, D-[³H] glucose and NaSCN were added simultaneously to the membrane vesicles. Final concentrations were: NaSCN (out), 100 mM; mannitol, 100 mM; Hepes/Tris, pH 7.5, 10 mM. The concentrations of D-glucose and phlorizin are indicated in the figure. The uptake values were obtained from measurements in triplicates. The bars indicate the standard deviations, the lines were calculated by linear regression. The velocity v is expressed as nmol of glucose taken up per 176 μg of protein in 2 sec. B) As in A, but with incubations lasting 7 sec. The velocity v is expressed as nmol og glucose taken up per 290 μg of protein in 7 sec. [From G. Toggenburger, M. Kessler, A. Rothstein, G. Semenza, and C. Tannenbaum (in preparation).]

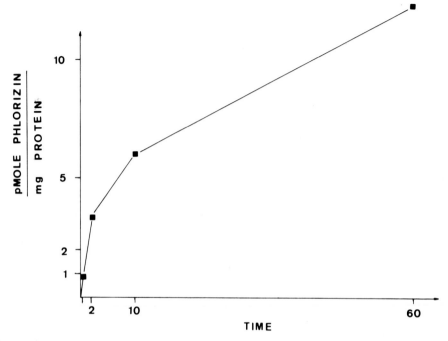

Fig. 3. Time course of phlorizin binding to brush border membrane vesicles. Phlorizin, to a final concentration of 2 μM, was added to membranes which had been previously equilibrated in 150 mM NaCl. Incubation was run at room temperature and was stopped by addition of an excess of ice cold saline, followed by immediate filtration. The time needed for diluting, filtering, and washing was about 10 sec.

the membrane, i.e., NaSCN, glucose, and phlorizin, were all added simultaneously, whereas phlorizin binding mentioned above was measured (Fig. 3) after NaCl had time to equilibrate across the membrane. We therefore remeasured phlorizin binding after 2 sec and under the same conditions as in the case of phlorizin inhibition of glucose transport, i.e., in the presence of a gradient of NaSCN, phlorizin, and glucose (or fructose) added simultaneously. The results of Fig. 4 were obtained. The lower curve (glucose present) represents the glucose nonprotectable phlorizin binding sites. The upper curve (with an equal concentration of fructose as a control) represents the glucose protectable binding sites. When the difference is plotted according to Scatchard we get the results of Fig. 5. Under these conditions (i.e., before the dissipation of the gradients of both NaSCN and glucose, at a time when the glucose uptake is still linear) a glucose-protectable, high-affinity site of phlorizin binding is apparent, with K_d = 4.7 μM and total binding of 11 pmol/mg protein. (based on extrapolation of the high affinity component of the Scatchard plot).

In contrast, if the vesicles are preequilibrated with 100 mM NaSCN prior to the addition of phlorizin, to dissipate the Na^+ and SCN^- gradients, the binding affinity is substantially reduced. In the presence of 25 mM D-glucose, with or without a Na^+ gradient, no high affinity binding is apparent, and indeed the binding is the same in either case.

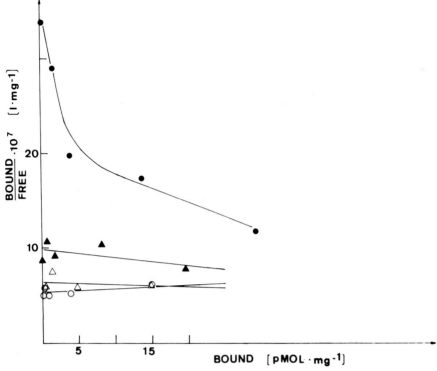

Fig. 4. Phlorizin binding to intestinal brush border membrane vesicles: effect of a NaSCN gradient. Conditions as in the glucose uptake experiments of Fig. 2A (i.e., 2-sec incubations at room temperature in 10 mM Hepes/Tris, pH 7.5, and 100 mM mannitol; phlorizin concentrations from 0.2 to 25 μM). ● and ○) 100 mM NaSCN, phlorizin, and either 25 mM D-fructose (●) or 25 mM D-glucose (○) added simultaneously at time zero. ▲ and △) the vesicles had been equilibrated in 100 mM NaSCN (+ buffer and mannitol) before the addition, at time zero, of phlorizin and either 25 mM D-fructose (▲) or 25 mM D-glucose (△).

Table I compares the K_d values of the D-glucose protectable high-affinity binding sites (as estimated from the steeper, more linear portion of the plot) with the K_i values for inibition of D-glucose uptake obtained under comparable conditions. The K_d and K_i values agree reasonably well and change according to the experimental conditions in the same way: when a NaCl gradient is substituted for the NaSCN gradient, K_d and K_i values both increase (and the initial velocity of D-glucose uptake decreases to approximately one fourth).

Hydrolysis of Phlorizin (13, 14)

Brush border vesicles were incubated for 30 min at room temperature in the standard incubation medium containing phlorizin. They were then extracted with hot ethanol. Thin layer chromatographic analysis of the extract indicated that less than 10% of the phlorizin had been hydrolyzed. The amount of hydrolysis in the 2-sec duration of the binding and inhibition experiments must, therefore, be insignificant.

A

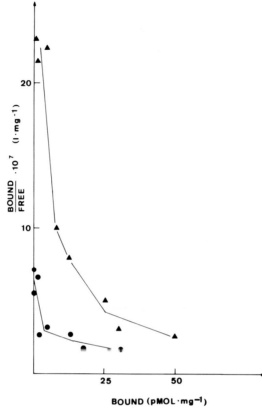

B

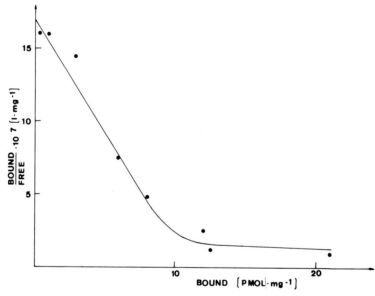

Fig. 5. A) Phlorizin binding to intestinal brush border membrane vesicles: effect of D-glucose. Conditions as in Fig. 4. 100 mM NaSCN, phlorizin (from 0.2 to 100 μM) and either 25 mM D-glucose ($\bullet$) or D-fructose ($\blacktriangle$) were added simultaneously at the start. B) $\bullet$) is a Scatchard plot of the glucose-protectable, NaSCN gradient-dependent phlorizin binding (calculated as the difference between the values in fructose minus the values in glucose of A). [From G. Toggenburger, M. Kessler, A. Rothstein, G. Semenza, and C. Tannenbaum (in preparation).]

TABLE I. Interaction of Phlorizin With Membrane Vesicles From Rabbit Small Intestinal Brush Borders: K_i Values From Phlorizin Inhibition of Initial D-Glucose Uptake and K_d Values for the High-Affinity, D-Glucose-Protectable Phlorizin Binding Sites.

	Phlorizin inhibition of D-glucose uptake	Phlorizin binding[a]	
	---	---	---
	K_i (μM) $\overline{x} \pm$ S.D. (range)	K_d (μM) $\overline{x} \pm$ S.D. (range)	Approximate concentration of high-affinity, glucose-protectable binding sites (pmol per mg membrane protein) (range)
Intact vesicles			
with NaSCN gradient	7.8 ± 1.4 (6–10)	4.7 ± 2.4 (2–7)	9.6 ± 3.2 pmol per mg (5–13)
with NaCl gradient	25–35	23–33	approx. 55–70 pmol per mg
with neither NaSCN nor NaCl gradient	not determined	> 100	
Papain digested vesicles			
with NaSCN gradient	not determined	6.3–16	29–34 pmol per mg

[a]Due to the curvature of some Scatchard plots, quantitative comparision can only be made among experiments carried out under strictly identical conditions.

Partial Negative Purification of the High-Affinity Phlorizin Binding Sites and of the Glucose Transport Agency(ies); the Location of Small Intestinal Sucrase-Isomaltase Complex[1]

Papain treatment has been extensively used to solubilize the sucrase-isomaltase complex and other intestinal hydrolases (e.g., see Refs. 16–19) Since the sucrase-related sugar transport system is not identical with the Na-dependent monosaccharide transport system(s) (20), it seemed appropriate to attempt to enrich the phlorizin high-affinity binding sites and the Na-dependent glucose transport system(s), by controlled papain digestion. In Fig. 6A the data are presented as pmol D-glucose uptake/mg protein, where the protein content is that actually measured in each incubation medium; in B, the results are normalized to the amount of protein present in the vesicles prior to papain solubilization.

It is apparent that papain treatment under the conditions used did not reduce the function of the Na-dependent D-glucose transport system(s) — as shown by the "normal" initial velocity of uptake — nor did it make the vesicles leaky — as shown by the retention of equilibrium concentrations of the substrate in the vesicles after washing with substrate-free media. Thus, the recovery of the transport activity after papain treatment was quantitative or nearly so. However, 30–60% of the proteins — apparently not associated with this transport system(s) — had been removed from the vesicles by papain. This resulted in a twofold increase of specific activity (Fig. 6A and Table II). In addition, the treated vesicles retain the ability to discriminate between D- and L-glucose. There is no indication that papain exposes cryptic transport sites or that it degrades membrane proteins extensively: if SDS-PAGE patterns of the control vesicles, of the treated vesicles, and of the supernatant are compared (Fig. 7), then we see that those proteins that have been removed from the vesicles by papain treatment appear in the supernatant. Only 2 new bands appear in either gel pattern. None of the bands found in the control is absent from the patterns after papain treatment, and there is no increase in bands found in the low molecular weight region of the gel.

[1] Some of these data have already appeared in a preliminary form (15).

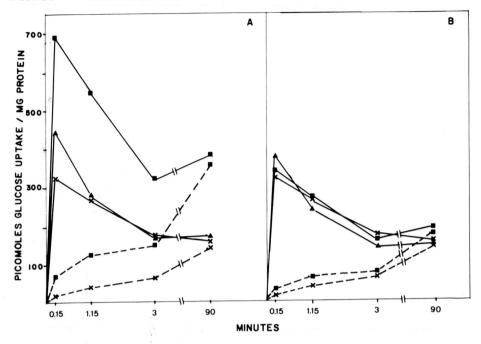

Fig. 6. Uptake of D- and L-glucose by papain (+EDTA) treated brush border membrane vesicles. A) The uptake values are given in terms of amounts of protein remaining in the membrane vesicles after papain treatment. B) in terms of amount of protein present in the original vesicle preparation prior to papain treatment. ———) L-glucose uptake; ———) D-glucose uptake; x———x) control; ■———■) vesicles treated with 0.8 U papain mg^{-1} protein in 1 mM EDTA; ▲———▲) vesicles treated with 1 mM EDTA alone.

TABLE II. Effect of Papain Treatment on Glucose Uptake by Vesicles of Brush Border Membranes

	Uptake[a]		Sucrase[b]
	D	L	
Control	215	11	1.08
After EDTA treatment (1 mM)	295	34	1.11
After EDTA + papain (0.8 U/mg protein)	460	44	< 0.05

[a]Uptake (of D- or L-glucose, respectively) in picomoles · mg^{-1} protein · 10 sec^{-1} at room temp.

[b]Sucrase activity left in the vesicles in units · mg^{-1} protein. The total yield of sucrase activity during the papain treatment (supernatant + pellet) was essentially quantitative.

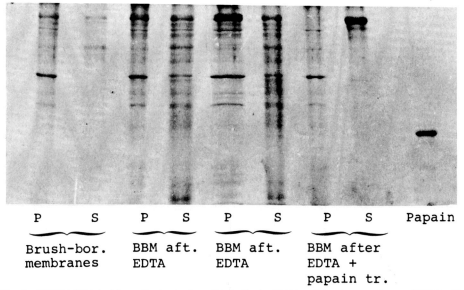

Fig. 7. SDS-PAGE (start from the top) of small intestinal rabbit brush border membranes (BBM) with no treatment (left), after EDTA treatment (center), or after EDTA + papain treatment (right). After each treatment, the membrane vesicles were spun down (P, pellet; S, supernatant).

The vesicles were assayed for sucrase and lactase activity before and after treatment with papain. Treatment of the vesicles with EDTA resulted in the liberation of only about 4% of sucrase activity; treatment with papain liberated 95–98% of this activity (Table II) and also of lactase activity (data not shown). As to the high-affinity phlorizin binding sites, papain treatment of the vesicles resulted in a slight (or no) increase in the apparent K_d value, with a doubling or tripling of the amount of binding sites per mg protein (Fig. 8), as a result of the removal of other proteins from membrane.

It should be emphasized that the increase in specific transport activity reported in Table II was obtained in 3 out of 6 experiments. In the others papain treatment failed to increase the specific transport activity, in spite of differential solubilization of membrane proteins identical to those of the "successful" experiments (as revealed by the SDS-PAGE patterns). We have failed as yet to identify the reason why papain treatment is not quantitatively reproducible. Neither the segment of the intestine, nor the age or the breed of the rabbit nor the contamination by variable amounts of serine-proteases were found to be of relevance.

An important corollary of the results of Fig. 6 and Table II is the information as to the location of small intestinal sucrase. It has already been shown by functional (21, 22), immunologic (23), and morphologic (24, 25) criteria that (at least some) sucrase activity is located at the luminal surface of the brush border membrane. The quantitative solubilization of intestinal sucrase (Table II) under conditions which do not alter the permeability properties of the membrane (Fig. 6), demonstrates that all sucrase is available to a reagent (papain) which was added to the luminal side of the vesicles and which is very unlikely to cross the membrane. Likewise, addition of lytic concentrations of Triton X-100 to native vesicles did not lead to any apparent increase in sucrase activity [M. Kessler, O. Acuto, C. Storelli, H. Murer, M. Müller, and G. Semenza (Biochim Biophys Acta, in press)].

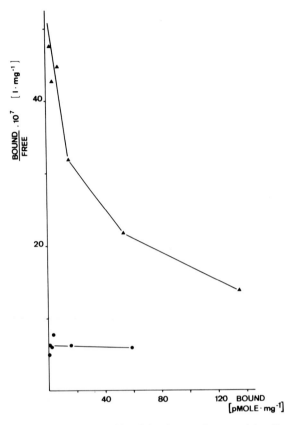

Fig. 8. Phlorizin binding to papain-digested brush border membrane vesicles. Conditions as in Figs. 4 and 5A. ▲) in the presence of 25 mM D-fructose; ●) in the presence of 25 mM D-glucose.

DISCUSSION

Kinetics of D-Glucose Uptake Into Vesicles From Brush Border Membranes and of its Inhibition by Phlorizin

The question may arise as to the validity of measuring parameters of transport systems under nonsteady-state conditions. Specifically, the experiments are performed under conditions of an initial NaSCN gradient, so that during the course of the experiment the gradient dissipates and all parameters dependent on the gradient will change with time. In addition, the rate of Na gradient dissipation, due to the glucose-Na-coupled cotransport, might depend upon the initial external glucose concentration. An effect of glucose on the rate of dissipation of the Na$^+$ gradient is unlikely, however, since the percentage of Na$^+$ that can enter the vesicles through the glucose-dependent pathway is only a small portion of the Na$^+$ entering the vesicles by all pathways [M. Kessler, V. Tannenbaum, and C. Tannenbaum (submitted to Biochim Biophys Acta)]. The complete dissipation of the Na$^+$ gradient (as shown by reversion of glucose flux and by Na$^+$-uptake measurements) takes up to 2 to 3 min, but the measurements of glucose uptake are carried out for only 2 sec. In this short time the flux of Na$^+$ would be relatively small and only a small fraction of the

NaSCN gradient would dissipate, so that it would be a relative constant factor with respect to its effect on transport or binding parameters. This relative constancy is suggested by the finding that the glucose uptake is linear for up to almost 4 sec, and by the finding that the measured "initial" velocity of glucose uptake gives a linear kinetic plot (Lineweaver-Burk, Eadie, Dixon). Substantial dissipation of the Na^+ gradient would be expected to cause some curvature in Fig. 2. In addition, the K_m values for D-glucose uptake measured at 7 sec (i.e., shortly after deviation from linearity) did not differ significantly from the K_m values obtained from 2-sec incubations despite continued dissipation of the gradient. Furthermore the K_i for phlorizin inhibition is also not substantially altered between 2 and 7 sec (Fig. 2). These findings suggest that the reported K_i and K_d values represent minimal values associated with relatively large NaSCN gradients that are not substantially dissipated in the 2 sec required for measurement. A theoretical argument further supports this view. In electrogenic transport systems, Geck and Heinz (26) have shown that kinetic parameters such as K_m and V_{max} may be relatively independent of changes in the membrane potential (such as those that might occur as a result of a partial dissipation of the NaSCN gradient), as long as the membrane potential is very high. Finally, measuring uptake velocities at almost zero-trans concentrations of substrates has the obvious advantage of minimizing possible trans effects.

Although there is uncertainty about the absolute values of the kinetic and binding parameters in the present experiments, the comparison of the K_i for phlorizin inhibition of sugar transport and K_d for phlorizin binding is certainly valid because the size of the gradient and the extent of its dissipation was the same in each case. It is also certain that if the less permeant species, Cl^-, is substituted for SCN^- as the accompanying anion, the apparent affinity of phlorizin measured by K_i or K_d is significantly reduced, and if the gradient is allowed to dissipate completely it is substantially reduced (Table I).

Identity of the High-Affinity Phlorizin Binding Sites With (a part of) the Na-Dependent Glucose Transport Agency(ies).

This identity is borne out by the following observations: i) the K_d values of the high-affinity phlorizin binding sites match the K_i values of phlorizin inhibition of D-glucose transport (Table I); ii) the K_d values and the K_i values change with the experimental conditions in parallel fashion (Table I); iii) high-affinity phlorizin binding is inhibited by D-glucose but not by D-fructose (Figs. 4,5).

From the data in Figs. 4 and 5 there appear to be 10–12 pmol of high-affinity phlorizin binding sites per mg membrane proteins in these vesicles from small intestinal brush borders. Assuming a 1:1 stoichiometric relation with the transport agency(ies) for D-glucose, the "transport turnover number" at 1 mM D-glucose, $20°C$, in buffer pH 7.5 can be calculated to be of the order of 20 sec^{-1} in the presence of a starting NaSCN gradient of 100 mM (out) vs 0 mM (in). The 10–12 pmols of high-affinity phlorizin binding sites found in the vesicles of small intestinal brush border membranes are in very close agreement with the corresponding figure reported for the high-affinity phlorizin binding sites of renal proximal tubulin membranes (3,5). Assuming a molecular weight of 80,000-100,000 daltons for glucose transport agency(ies), this number of high-affinity phlorizin binding sites would indicate that the glucose transport involves only about 0.1% of the membrane proteins.

Our determination of the high-affinity binding sites corresponds rather well to similar values reported for kidney brush border fragments [rat, by Bode et al., (4) and by Glossmann and Neville, (5); dog, by Silverman and Black, (27)]. Others have reported larger numbers for high affinity binding sites [Bode et al., (28), Chesney et al., (7)] in similar preparations from kidney. Due to the different ways of expressing the data, it is more difficult to make a comparison with the results of Diedrich (13) and of Stirling (29, 30) for the small intestine.

At any rate, all quantitative calculations of this sort, including ours, must be taken with caution, due to the critical effect of the incubation conditions (see for example, Table I), the different washing times, the curvature sometimes present in Scatchard plot (see below), and so on.

Elsewhere (M. Kessler, A. Rothstein, G. Semenza, G. Toggenburger, and C. Tannenbaum, in preparation) we present that phlorizin, as generally accepted, is indeed a fully competitive inhibitor (also in the presence of a NaSCN gradient), and not a poorly transported substrate.

Dependence of the High-Affinity Binding Sites Upon the Na^+ Gradient

As pointed out under Results, we were unable to show any high-affinity phlorizin binding in membrane vesicles of rabbit small intestinal brush borders unless a NaSCN (or NaCl) gradient had been imposed on them (Fig. 4). To the best of our knowledge no such dependence upon such a gradient has even been reported for phlorizin binding to kidney tubuli membranes [although the K_d value has been reported to depend on the presence of Na^+ (3)], or to small-intestinal membranes. Since it is not always clear whether the conditions employed by other authors have generated a Na^+ gradient during binding measurements, we do not know whether the need for the Na^+ gradient is limited to the biological preparation used by us.

The dependence of the size of the transient D-glucose overshoot in the presence of a highly permeant anion (31, 32), and the apparent reduction in affinity for phlorizin binding when Cl^- is substituted for SCN^- (Table I), suggests that the membrane potential may modulate the interaction of the carrier with phlorizin as it has been shown with glucose (31, 32). Such findings can be accomodated by a mobile cotransport carrier model in which at least one form of the carrier is charged, for which the kinetics have been developed by Geck and Heinz (26), but a detailed discussion of a mechanism would be inappropriate until more data are collected concerning the individual effects of Na^+ concentrations and of membrane potentials on phlorizin binding.

Scatchard Plots

Another point which is presently being investigated is the curvature of the Scatchard plots of high-affinity, glucose protectable phlorizin binding sites (Figs. 5, 8). The extent of the curvature varied somewhat from preparation to preparation of the vesicles. A number of reasons can be envisaged. One of the possibilities is the existence of two, rather than one, transport agencies for D-glucose, having different K_d values. This would agree with the conclusions reached for hamster small intestine on the basis of kinetic investigations of monosaccharide uptake (33).

ACKNOWLEDGMENTS

We want to thank the skillful cooperation of Miss L. Wahlgren in some of this work. We also thank Dr. J. S. Cook, Oak Ridge National Laboratory, for having drawn our attention to the unreliability of some of the commercial preparations of ^{3}H-labeled phlorizin. This work was partially supported by the Swiss National Science Foundation, Berne.

NOTE ADDED IN PROOF

Subsequent, extensive experimentation has shown that although there is a significant increase in the amount of phlorizin bound by the vesicles when a Na-gradient is imposed across the vesicle membrane, the amount of glucose-protectable phlorizin binding in the absence of a Na-gradient is not insignificant. [See G. Toggenburger, M. Kessler, A. Rothstein, G. Semenza, and C. Tannenbaum (in preparation).]

REFERENCES

1. Alvarado F, Crane RK: Biochim Biophys Acta 56:170, 1962.
2. Diedrich DF: Arch Biochem Biophys 117:248, 1966.
3. Frasch W, Frohnert PP, Bode F, Baumann K, Kinne R: Pflügers Archiv 320:265, 1970.
4. Bode F, Baumann K, Diedrich DF: Biochim Biophys Acta 290:134, 1972.
5. Glossmann H, Neville DM Jr: J Biol Chem 247:7779, 1972.
6. Glossmann H, Neville DM Jr: Biochim Biophys Acta 323:408, 1973.
7. Chesney RW, Sacktor B, Kleinzeller A: Biochim Biophys Acta 332:263, 1974.
8. Schmitz J, Preiser H, Maestracci D, Ghosh BK, Cerda J, Crane RK: Biochim Biophys Acta 323:98, 1973.
9. Lowry OH, Rosebrough NJ, Farr AL, Randall RJ: J Biol Chem 193:265, 1951.
10. Dahlqvist A: Anal Biochem 7:18, 1964.
11. Maestracci D, Preiser H, Hedges T, Schmitz J, Crane RK: Biochim Biophys Acta 382:147, 1975.
12. Lotspeich WD: Harvey Lect 56:63, 1960.
13. Diedrich DF: Arch Biochem Biophys 127:803, 1968.
14. Malathi P, Crane RK: Biochim Biophys Acta 173:245, 1969.
15. Semenza G, Tannenbaum C, Toggenburger G, Wahlgren L: In Semenza G, Carafoli E, (eds): "Proceedings of the FEBS Symposium on the Biochemistry of Membrane Transport (Nr 42)" Heidelberg: Springer Verlag, 1977, p 269.
16. Auricchio S, Dahlqvist A, Semenza G: Biochim Biophys Acta 73:582, 1963.
17. Eichholz A: Biochim Biophys Acta 163:101, 1968.
18. Maestracci D: Biochim Biophys Acta 433:469, 1976.
19. Sigrist H, Ronner P, Semenza G: Biochim Biophys Acta 406:433, 1975.
20. Malathi P, Ramaswamy K, Caspary WF, Crane RK: Biochim Biophys Acta 307:613, 1973.
21. Miller D, Crane RK: Biochim Biophys Acta 52:281, 1961.
22. Ugolev AM, Jesuitova NN, De Laey P: Nature (London) 203:879, 1964.
23. Cummins D, Gitzelmann R, Lindenmann J, Semenza G: Biochim Biophys Acta 160:396, 1968.
24. Johnson CF: Science 155:1670, 1967.
25. Nishi Y, Yoshida O, Takesue Y: J Mol Biol 37:441, 1968.
26. Geck P, Heinz E: Biochim Biophys Acta 443:49, 1976.
27. Silverman M, Black J: Biochim Biophys Acta 394:10, 1975.
28. Bode F, Baumann K, Frasch W, Kinne R: Pflügers Archiv 315:53, 1970.
29. Stirling CE: J Cell Biol 35:605, 1967.
30. Stirling CE, Schneider AJ, Wong MD, Kinter WB: J Clin Invest 51:438, 1972.
31. Murer H, Hopfer H: Proc Natl Acad Sci USA 71:484, 1974.
32. Beck JC, Sacktor B: J Biol Chem 250:8674, 1975.
33. Honegger P, Semenza G: Biochim Biophys Acta 318:390, 1973.

Journal of Supramolecular Structure 7:1–13 (1977)
Molecular Aspects of Membrane Transport 365–377

Kinetics of Na$^+$-Dependent D-Glucose Transport

Ulrich Hopfer

Department of Anatomy and Developmental Biology Center, School of Medicine, Case Western Reserve University, Cleveland, Ohio 44106

The kinetic parameters of the Na$^+$-dependent glucose transport system have been determined in isolated membrane vesicles for D-glucose, Na$^+$, and phlorhizin. The D-glucose flux measurements were carried out by the equilibrium exchange procedure at constant external and internal Na$^+$ concentrations and zero potential. Equations were developed to extract information about K_m and V_{max} from uptake measurements into a vesicle population that is heterogeneous with respect to size (surface to volume ratio). The K_m for D-glucose was 14 mM and independent of the Na$^+$-concentration, while the V_{max} was strongly Na$^+$-dependent and increased 15-fold between 1 and 100 mM Na$^+$. The K_m of Na$^+$ for activation of the V_{max} was 18 mM. The calculated K_I values for phlorhizin were 2.7 and 1.9 μM when determined under active and equilibrating D-glucose flux conditions, respectively.

Key words: microvillus membranes, small intestine, phlorhizin inhibition

In recent years highly purified preparations of plasma membranes have been used to investigate the mechanism of active sugar and amino acid transport. These studies have provided evidence for the existence of transport systems that catalyze coupled flow of the nonelectrolytes with Na$^+$. This cotransport is electrogenic, i.e., the coupled translocation of nonelectrolyte and Na$^+$ across the membrane is associated with the net transfer of a positive charge. In terms of mechanism, the transport reaction can be characterized as a coupled "facilitated diffusion" of 2 substrates. Because of molecular coupling the flow of Na$^+$ down its electrochemical gradient can support nonelectrolyte uptake by cells or isolated membrane vesicles against a gradient, thus accomplishing active nonelectrolyte transport (for reviews see Refs. 1–5). Active nonelectrolyte transport in isolated brush border membrane vesicles is consistent with the studies in intact epithelial sheets which, in addition, have provided the quantitative information of a 1:1 stoichiometry of Na$^+$ and nonelectrolyte (6,7).

Although the overall transport reaction of Na$^+$-dependent nonelectrolyte transport is now reasonably well understood, contradicting results with respect to the relevant kinetic parameters have been reported. For example, in some studies the V_{max}, but not the K_m,

Received May 25, 1977; accepted June 7, 1977

of the Na$^+$-dependent glucose transport system changes with Na$^+$ concentration (6—10) while in others the reverse is observed (11—13). In addition, the values for the K$_m$ of D-glucose range between 80 μM and 10 mM (6—15), an unsatisfactorily high spread which has yet to be explained. Since the kinetic parameters are calculated from measurements of solute fluxes that are functions of the transport system as well as of transmembrane forces, the calculated values may not truely reflect the properties of the carrier in all studies.

It is the purpose of this paper to analyze the kinetics of the Na$^+$-dependent glucose transport system in isolated brush border membrane vesicles under defined transmembrane conditions of potential and Na$^+$ concentrations so that the K$_m$ and V$_{max}$ are functions only of the transport system.

MODEL AND EXPERIMENTAL DESIGN

Figure 1 shows a simple carrier model for Na$^+$-dependent D-glucose transport. Non-electrolyte substrate translocation is accomplished either with or without Na$^+$ by formation of binary or ternary complexes, respectively, with the carrier. This mechanism of cotransport implies that the rate of D-glucose translocation is not only dependent on its own concentration, but also on the Na$^+$ concentration at both membrane faces and the electrical potential difference. The mutual dependence of D-glucose and Na$^+$ fluxes imposes certain restrictions on the experimental conditions by which the kinetic parameters can be determined. Defined K$_m$ and V$_{max}$ values for one substrate can only be obtained when during flux measurements all factors besides the concentration of the particular sub-

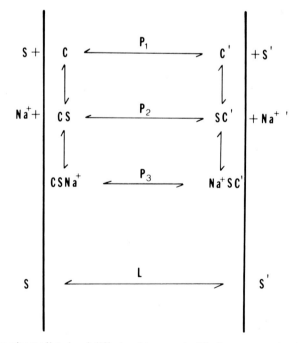

Fig. 1. Model for carrier-mediated and diffusional transport of D-glucose across brush border membrane. C) Carrier, S) nonelectrolyte substrate (D-glucose), P) apparent permeabilities of the various forms of the carrier, L) permeability coefficient of a "leak" pathway.

strate under investigation remain constant within the experimental time span and over the range of substrate concentration. Thus, when measuring the rate of D-glucose transport as a function of its concentration to determine K_m and V_{max}, the Na$^+$ concentration gradient across the membrane and the electrical potential difference should be kept constant. However, an experimental difficulty arises from the D-glucose flux since it is coupled to Na$^+$. The resulting Na$^+$ flux, in turn, changes the electrochemical Na$^+$ gradient across the membrane unless balanced by another Na$^+$ flux in the opposite direction via a glucose carrier-independent pathway in the membrane.

The experimental methods of maintaining a constant electrochemical Na$^+$ gradient across isolated membrane vesicles in spite of changing D-glucose fluxes have not been worked out. Therefore, I have used the equilibrium exchange procedure to circumvent the problem of changing electrochemical Na$^+$ gradient. In this method the substrate concentration is the same at the 2 membrane faces and, hence, there is no net flux. The unidirectional substrate transport is determined after addition of tracer amounts of labeled substrate. K_m and V_{max} are determined from the rate of tracer flux as a function of total substrate concentration (16, 17). The experimental conditions can be further simplified by equal Na$^+$ concentrations on both sides of the membrane and an electrical potential of zero. This condition is easily met in isolated vesicles by sufficiently long preincubations with unlabeled D-glucose and permeant Na$^+$ salts.

THEORETICAL CONSIDERATIONS

In a homogeneous cell or membrane vesicle preparation the uptake of trace amounts of labeled nonelectrolyte under equilibrium exchange conditions is described by a first order rate process which obeys the relationship (17)

$$\ln(1 - a/A) = -Vt/(S + K) \tag{1}$$

whereby a = tracer uptake at time t, A = tracer uptake at equilibrium, S = unlabeled, preequilibrated nonelectrolyte substrate concentration, $K = K_m$ = substrate concentration at half-maximal velocity, and $V = V_{max}$ = maximal velocity.

However, as shown in Figs. 2A and B, uptake of labeled D-glucose by isolated intestinal brush border membranes under these conditions is not a first-order process. An explanation for the absence of a good fit to equation 1 is heterogeneity of vesicle size (see e.g., Ref. 5) and, thereby, the surface to volume ratio, which is determinant for the rate of isotope equilibration.

Equation 1 is still useful for an analysis of the kinetics if it is assumed that the tracer uptake into an individual vesicle is described by a first-order rate process, i.e., if

$$a_i = A_i - A_i \exp[-V_i t/(S + K)]; \tag{2}$$

with i referring to an individual vesicle. The microscopic uptake into an individual vesicle is related to the macroscopically observed uptake of the entire population by the appropriate summation function

$$a = A - \Sigma A_i \exp[-V_i t/(S + K)], \text{ or} \tag{3}$$

$$\ln(1 - a/A) = \ln\Sigma(A_i/A) \exp[-V_i t/(S + K)]. \tag{4}$$

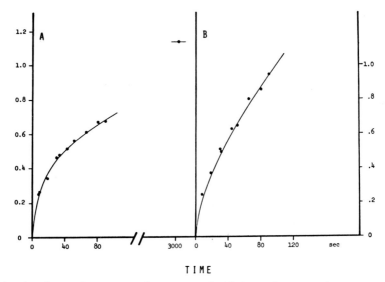

TIME

Fig. 2. Kinetics of tracer D-glucose uptake as measured with the equilibrium exchange procedure. Membrane vesicles were pre-equilibrated with 5 mM unlabeled D-glucose and 50 mM NaSCN for 1 h before labeled D-glucose was added. A) Uptake as a function of time (units of ordinate: nmole × 0.2/mg protein). B) Replot of the data from A in terms of fractional uptake with respect to equilibrium [units of ordinate: $-\ln(1 - a/A)$].

Equation 4 can be simplified if it is assumed that V_i per unit vesicle has a distribution around a macroscopic V such that $V_i = V + \triangle V_i$. Substitution of V_i simplifies equation 4 to

$$\ln(1 - a/A) = -Vt/(S + K) + \ln\Sigma(A_i/A)\exp[-\triangle V_i t/(S + K)] \tag{5}$$

Since $t/(S + K) = -[\ln(1 - a_i/A_i)]/V_i$ (from equation 2) equation 5 can be reduced to

$$\ln(1 - a/A) = -Vt/(S + K) + \ln\ \Sigma i, \tag{6}$$

$$\text{whereby} \quad \Sigma i = \ \Sigma(A_i/A)\,(1 - a_i/A_i)^{\triangle V i/V i}\ .$$

The summation term on the right of equation 6 is a correction factor to account for the heterogeneity of the vesicle population. It is important to note that this term is independent of unlabeled substrate concentration S. In addition, for a given membrane preparation this summation term should be constant if the macroscopic fractional uptake a/A, and thereby a_i/A_i, is kept constant.

Thus, regardless of the time course of the uptake, measurements of the times needed to give the same fractional uptake of the tracer at varying unlabeled, preequilibrated substrate concentration provide the data for calculations of the K_m and a relative V_{max}. For a/A = ½ equation 6 yields (with rearrangements):

$$t_{\frac{1}{2}} = [(S + K)/V]\ [\ln(2\Sigma i)]\,. \tag{7}$$

Hence, a plot of $t_{\frac{1}{2}}$ vs S should yield a straight line whose intercept on the abscissa corresponds to $-K_m$. For V_{max} the following relationship holds:

$$V_{max} = [K/t_{\frac{1}{2}(S=0)}]\ \ln(2\Sigma i). \tag{8}$$

It is obvious from equation 8 that V_{max} cannot be determined exactly in the absence of an evaluation of the summation term. However, as will be demonstrated in the Results section, even a relative V_{max} provides information about properties of the carrier.

Analogous equations can be derived for inhibitors (for derivations of the corresponding flux equations into a homogeneous vesicle population see Ref. 17). The equation relating $t_{1/2}$ to inhibitor concentration reads

$$t_{1/2} = [(K_I + I)/V] \, [K/K_I] \ln(2\Sigma i), \qquad (9)$$

where K_I is the inhibitor concentration exerting half-maximal inhibition. Thus the K_I can be determined from a plot of $t_{1/2}$ vs inhibitor concentration. The experimental requirements for equation 9 to be valid are the absence of an inhibitor flux during the time span of substrate uptake and a substrate concentration well below its K_m.

Equations 7–9 are adequate to analyze transport at low D-glucose concentrations. However, it was found that the observed $t_{1/2}$ became independent of substrate concentration above $\sim$ 50 mM suggesting the existence of another route of translocation for D-glucose across the brush border membrane. For purposes of convenience this pathway will be termed "leak" although it is realized that it may constitute another sugar transport system with a high K_m and low V_{max}. Since the Na^+-dependent uptake is saturated with unlabeled D-glucose at high concentrations, tracer D-glucose uptake proceeds predominantly through the high K_m system under these conditions.

With a second "leak" pathway present, the rate equations have to be modified slightly. Assuming that the rate of tracer uptake through the "leak" is proportional to the difference in tracer concentration and the permeability coefficient (L), the tracer uptake under equilibrium exchange conditions is described by (see also Fig. 1 for the model)

$$a_i = A_i - A_i \exp[-V_i t/(S + K) - L_i t]. \qquad (10)$$

This equation can be related to the macroscopically observed uptake and rearranged in a similar manner as equation 2. An additional assumption must be made with respect to L_i, namely that L_i has a distribution around a macroscopic coefficient L. The rearranged equation reads:

$$\ln(1 - a/A) = -[Vt/(S + K)] - [Lt] + \ln\Sigma(A_i/A)(1 - a_i/A_i)^{\triangle V_i/V_i}\exp[t(L - L_i V/V_i)]. \qquad (11)$$

Equation 11 suggests a procedure by which the carrier-related $t_{1/2}$ can be obtained. If V_i and L_i depend in the same way on vesicle parameters, e.g., surface area, then $(L - VL_i/V_i) = 0$ and equation 11 is reduced to a function in which the summation term is again independent of substrate concentration S. Any error due to the assumption of zero is likely to be small as $\exp[t(L - VL_i/V_i)]$ changes only slowly if $|t(L - VL_i/V_i)|$ remains small (e.g. $\leqslant 0.1$). Setting again a/A = ½ the modified equation 11 can be rearranged to read:

$$(t_{1/2})^{-1} = V/[(S + K) \ln(2\Sigma i)] + L/\ln(2\Sigma i). \qquad (12)$$

The term on the left is the observed $(t_{1/2})^{-1}$. The first term on the right corresponds to the carrier-related $(t_{1/2})^{-1}$ as given by equation 7 while the second can be defined as the $(t_{1/2})^{-1}$ due to the "leak." Since the "leak"-related $t_{1/2}$ can be estimated from the observed $t_{1/2}$ at a high substrate concentration, the carrier-related $t_{1/2}$ can be calculated and plotted as a function of substrate concentration according to equation 7 to yield the kinetic parameters of the relevant transport system. Equation 12 can be rewritten as

$$(t_{1/2})^{-1}{}_{observed} = (t_{1/2})^{-1}{}_{carrier} + (t_{1/2})^{-1}{}_{"leak"}. \qquad (13)$$

EXPERIMENTAL METHODS

Brush border membranes were isolated from intestinal scrapings of Sprague-Dawley rats as described previously (18). The efficiency of the cell fractionation procedure was routinely monitored using sucrase. The enrichment of sucrase in the membrane was on the average 22-fold over the starting homogenate. The final membrane was suspended in a buffer consisting of 0.1 M D-mannitol, 1 mM Tris-Hepes buffer, pH 7.5, and 0.1 mM $MgSO_4$. Measurements of D-glucose uptake were made, after appropriate preincubations, by incubation of the isolated membranes with D-$[1-^3H(N)]$glucose, withdrawal of aliquots after predetermined periods, quenching of the transport by dilution with ice-cold buffer, and collection of the membranes on a filter. Details of the method have been given before (18). To improve the precision of time measurements a microswitch was activated at the time of the mixing of membranes with labeled D-glucose and of the quenching of the aliquots. The signals from the microswitch were recorded continuously with a strip-chart recorder.

Except where noted otherwise, experiments were carried out at $15°C$ in a constant temperature room. When labeled D-glucose was to be measured under equilibrating conditions, i.e., in the absence of a NaSCN gradient, membranes were preincubated with the desired concentration of unlabeled D-glucose and NaSCN as well as monactin ($10 \mu g/ml$) for ~ 1 h at room temperature. Samples were then cooled down to $15°C$ and uptake measurements started by addition of a small volume of labeled D-glucose to the membranes. Sufficient amounts of unlabeled D-glucose and NaSCN had been added to the labeled D-glucose so that no change in medium concentration of either D-glucose or NaSCN occurred upon addition of the tracer. Phlorhizin, when present, was added to the membranes immediately before labeled D-glucose.

$t_{1/2}$ (or occasionally $t_{0.33}$ or $t_{0.4}$) values were estimated from plots of uptake vs time as shown in Fig. 2B. Each curve had 6 time points (from 3 parallel incubations) around the actual $t_{1/2}$ value. The carrier-related $t_{1/2}$ was calculated from equation 13 whereby the "leak"-related $t_{1/2}$ was the value that would give the "best fit" (see below) to the experimental data. The "leak"-related $t_{1/2}$ was equal to or greater than $t_{1/2}$ observed at 100 mM D-glucose or 100 μM phlorhizin, respectively. K_m and relative V_{max} values were calculated using equation 7 and a weighted least-squares analysis according to the guidelines of Cleland (19). The standard deviation of the observed $t_{1/2}$ values was estimated to be proportional to the absolute value of $t_{1/2}$, i.e., $SD(t_{1/2}) \propto t_{1/2}$. The "leak"-related $t_{1/2}$ giving the best fit was found through an iterative procedure in which the "leak"-related $t_{1/2}$ was systematically increased, starting with $t_{1/2}$ observed at 100 mM D-glucose, until the correlation coefficient of the linear function of carrier-$t_{1/2}$ on substrate concentration gave a maximum. Unless otherwise indicated, data are reported as mean ± standard deviation and regression lines are calculated by the weighted least-squares method. All calculations were carried out with a programmable desk calculator (Monroe 1860).

MATERIALS

Phlorhizin was obtained from ICN-Life Science Group, (Cleveland, Ohio) and recrystallized twice from hot water before use. D-$[1-^3H(N)]$glucose was bought from New England Nuclear Corporation (Boston, Massachusetts), and D-$[U-^{14}C]$glucose was obtained from Schwarz-Mann (Orangeburg, New York). Monactin was a gift of Ciba-Geigy, Basel, Switzerland. Other materials were obtained from common commercial sources.

RESULTS

D-Glucose transport into brush border membrane vesicles is much slower in the absence of a Na^+ gradient than in its presence (20). Therefore, the initial experiments were designed to determine whether the observed D-glucose uptake proceeds via the Na^+-dependent glucose transport system under both experimental conditions. Phlorhizin is a relatively specific inhibitor of this system and consequently should be effective whether D-glucose uptake is actively driven by an electrochemical Na^+ gradient or is simply equili-brating. To quantitate the effectiveness of phlorhizin its K_I was determined under condi-tions of 1) active D-glucose transport (in the presence of a NaSCN gradient; Fig. 3), and 2) equilibrating D-glucose transport (in the presence of NaSCN, but absence of a gradient; Fig. 4).

Figure 3 shows the phlorhizin inhibition of active D-glucose uptake by membrane vesicles measured after 12 sec of incubation. D-Glucose accumulation in the vesicles in the absence of phlorhizin was 30-fold above medium concentration. The apparent K_I of $2.7 \pm 0.2 \mu M$ has to be considered an upper limit of the true value for several reasons: 1) The measured uptake at 12 sec is not an initial rate. 2) Increasing phlorhizin concen-trations reduced the D-glucose flux, and thereby also the Na^+ flux, so that the dissipation of the electrochemical Na^+ gradient would be retarded. Consequently, D-glucose uptake at higher phlorhizin concentrations may have been driven by higher forces. 3) The observed uptake was not corrected for "leakage" of accumulated D-glucose from the vesicles. This loss should be highest in vesicles with the highest internal D-glucose concentration, i.e., in

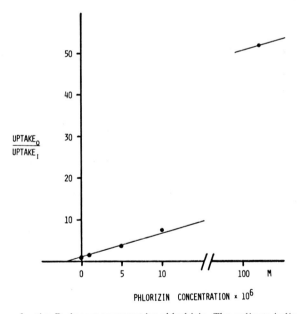

Fig. 3. Inhibition of active D-glucose transport by phlorhizin. The ordinate indicates the ratio of D-glucose uptake in absence (uptake$_0$) and presence of phlorhizin (uptake$_I$). D-Glucose uptake by brush border membranes was measured in the presence of an initial NaSCN gradient (medium: 100 mM; within vesicles: 0 mM). D-glucose concentration: $10 \mu M$. Length of incubation: 12 sec. Temperature: 25°C. The line is calculated by linear regression assuming constant variance.

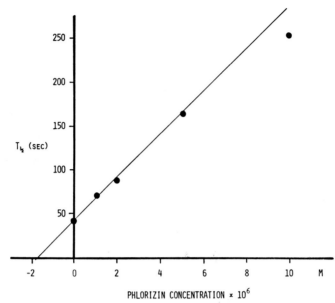

T$_{\frac{1}{2}}$ (SEC)

PHLORIZIN CONCENTRATION × 10^6

Fig. 4. Inhibition equilibrating D-glucose uptake by phlorhizin. D-Glucose uptake was measured in the presence of 100 mM NaSCN on both sides of the membrane, i.e., in the absence of a salt gradient Aliquots of the membranes were preincubated with 100 mM NaSCN, an appropriate amount of phlorhizin added, and D-glucose (10 μM) uptake measured. "Leak"-related $t_{1/4}$ used for correction was 127 sec.

vesicles exposed to the lowest phlorhizin concentrations. In other words, the rate of D-glucose transport per unit driving force may be underestimated in the absence of phlorhizin and overestimated in the presence of phlorhizin, resulting in an overestimate of the K_I.

Figure 4 demonstrates the effectiveness of phlorhizin in inhibiting equilibrating D-glucose uptake, i.e., uptake in the absence of a Na^+ gradient. The K_I under these conditions was determined to be 1.9 ± 0.2 μM which is in excellent aggreement with the above value of 2.7 μM and also those measured in intact epithelia (21). The similarity of the K_I values of phlorhizin, determined from D-glucose flux measurements under active and equilibrating transport conditions, strongly suggests that the translocation route for D-glucose is not dependent on the presence of a Na^+ gradient. This result, therefore, justifies the procedure to be employed in subsequent experiments for K_m and V_{max} determinations of D-glucose which rely on equilibrating tracer uptake and on measurements of the times necessary to achieve constant fractional uptake rather than initial rates.

Figures 5—8 summarize the information that has been obtained about the kinetic parameters of the Na^+-dependent glucose transport system using the equilibrium exchange procedure.[1] Figure 5 shows the carrier-related $t_{1/2}$ as a function of unlabeled D-glucose at

[1] Preliminary analyses of the data were presented at the ICN-UCLA symposium on "Molecular aspects of membrane transport," Keystone, Colorado, March 13—18, 1977 and the American Physiological Society Symposium on "Models for GI transport," FASEB Meetings, Chicago, Illinois, April 4, 1977. Initial calculations had been carried out by the least-squares method without weighing factors. K_m values calculated by this latter method are slightly lower. For example, at 100 mM NaSCN the K_m for D-glucose is 12.2 mM using unweighted least-squares analysis and 14.2 mM with weighing factors. The general conclusions about the effect of Na^+ on K_m and V_{max} are not influenced by the choice of statistical method.

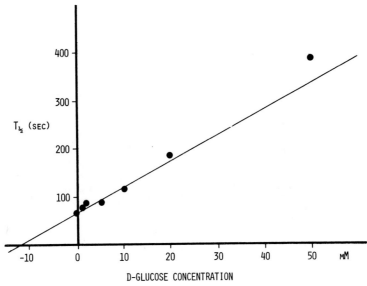

Fig. 5. Kinetics of D-glucose transport by brush border membranes as measured by the equilibrium exchange procedure. Aliquots of the membrane were preincubated with various levels of unlabeled D-glucose and 100 mM NaSCN for 1 h before addition of labeled D-glucose and 100 mM NaSCN for 1 h before addition of labeled D-glucose. "Leak"-related $t_{1/2}$ used for correction was 230 sec.

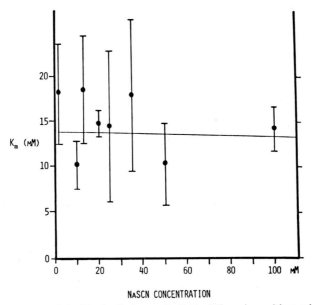

Fig. 6. Na^+ dependence of the K_m for D-glucose. Individual K_m values with standard deviation were obtained from experiments as in Fig. 5 at the various Na^+ concentrations. Bars represent 4 standard deviations. The line was calculated by weighted linear regression.

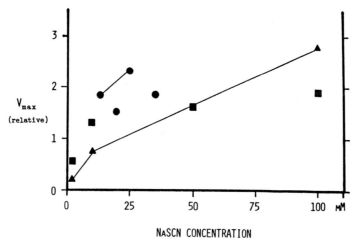

NaSCN CONCENTRATION

Fig. 7. Na+ dependence of the transport rate for D-glucose. Relative V_{max} values $[K/t_{1/2(S=0)}] \times 10^1$ were obtained from experiments as in Fig. 5. Different symbols represent separate batches of animals. The points connected by a line were obtained with the same membrane preparation.

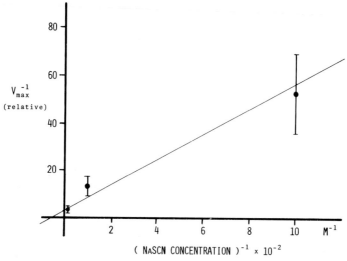

(NaSCN CONCENTRATION)$^{-1}$ x 10^{-2}

Fig. 8. Lineweaver-Burke plot of the Na+ dependence of the transport rate of D-glucose. Data are from a single membrane preparation (replot of data from Fig. 7). Bars represent 4 standard deviations. The line was calculated by weighted linear regression.

a NaSCN concentration of 100 mM for a single experiment. The average K_m value for D-glucose at 100 mM NaSCN was 14.2 ± 1.3 mM.

To obtain some insight into the mechanism of Na+ stimulation of D-glucose transport, K_m and relative V_{max} values were determined at several Na+ concentrations. The effect of Na+ on K_m is summarized in Fig. 6. Within the experimentally accessible range of 1–100 mM Na+ the K_m for D-glucose was constant. In other words, the affinity of D-glucose for the carrier does not appear to be affected by Na+. Uptake experiments could not be carried out at Na^+ concentrations below 1 mM because overall D-glucose transport became dominated by D-glucose influx via a Na^+-independent, apparently non-saturable route.

As shown in Fig. 7 Na^+ increases the maximal velocity of D-glucose uptake. In initial experiments only single values were determined with each membrane preparation. Because of the possibility that the relative V_{max} was dependent on membrane preparation and not on Na^+, the same membrane preparation was used for V_{max} determinations at several Na^+ concentrations at 2 occasions. These experiments (experimental points connected by lines in Fig. 7) demonstrate a significant increase of V_{max} of about 15-fold when Na^+ is raised from 1 to 100 mM.

Figure 8 shows a Lineweaver-Burke plot of one experiment in which V_{max} was determined over a sufficiently wide enough range of Na^+ concentration. The apparent K_m for Na^+ calculated from this plot is 17.9 ± 1.2 mM.

DISCUSSION

The kinetic parameters of D-glucose for the Na^+-dependent glucose transport system have been determined in many different types of epithelial preparations, including intact epithelial tissue in vivo and in vitro isolated cells and brush border membrane vesicles (6–15). The K_m of D-glucose, the K_I of phlorhizin, and the Na^+ dependence of K_m and V_{max} presumably are expressions of certain properties of this transport system. Therefore, one would expect similar K_m and K_I values and similar Na^+ effects in the various studies. The measured K_I for phlorhizin indeed is similar in the various preparations from kidney and small intestine. However, the reported data for the K_m of D-glucose vary by 2 orders of magnitude between 80 μM and 10 mM and Na^+ affects the K_m and not the V_{max} in some studies, while the opposite is found in others. A comparison of results from different laboratories would suggest that there is no correlation of the kinetic parameters of D-glucose transport with tissue (kidney and intestine) or species when analyzed by different techniques.

Part of the reason for the divergency has been pointed out in the earlier part of this paper. A systematic error is introduced into estimates of K_m and V_{max} from net D-glucose flux measurements since corresponding Na^+ fluxes are associated with increasing D-glucose fluxes. Thus, changes in D-glucose concentration influence the electrochemical Na^+ gradient across the membrane, and, therewith, its own driving force. The drop in electrical potential across the brush border membrane due to D-glucose or L-alanine absorption has actually been measured in intact cells with microelectrodes (22–24). A mathematical consequence of this change is an underestimation of the true K_m and V_{max} for a given electrochemical Na^+ gradient. Moreover, if the transport load overwhelms the energy supply, the maximal uptake rate represents a value for transport in the absence of an electrochemical Na^+ gradient. As a consequence the V_{max} becomes independent of the Na^+ concentration over a wide range and the apparent K_m becomes Na^+-dependent.

In isolated membrane vesicles salt gradients have been used to drive active non-electrolyte transport (13, 20). Swamping of this small amount of energy by high non-electrolyte concentrations may account for the observations of Na^+ independence of V_{max} and Na^+ dependence of the K_m of D-glucose and neutral amino acid transport in isolated membrane vesicles by several investigators (13, 25–28). On the other hand, the bias introduced into estimates of K_m and V_{max} of nonelectrolytes, cotransported with Na^+, should be less pronounced in intact cells in vivo where Na^+ influx at the luminal side can be compensated by active Na^+ extrusion at the serosal membrane, provided these studies are not complicated by unstirred layer effects. By this reasoning transport parameters estimated from renal microperfusion studies are least likely to be subject to artifacts. In microperfusion experiments in the rat (14) a K_m of 10.8 mM was found for D-glucose, a value relatively close to 14.2 mM as reported in this paper.

The apparent K_m of D-glucose transport in vitro is much lower than in vivo. Goldner et al. reported an in vitro K_m value of 1.4 mM for rabbit ileum (6). It must represent a lower limit since the electric potential drops about 19% under maximal D-glucose influx conditions (23).

These authors also very thoroughly investigated the effect of mucosal Na^+ concentration on K_m and V_{max} of 3-O-methyl-D-glucose transport in rabbit ileum. In agreement with the results in the isolated vesicles they observed that increasing Na^+ concentrations increased the maximal velocity, but had no effects on the apparent K_m of D-glucose. Since mucosal Na^+ depletion results in a hyperpolarization across the brush border membrane (23) the total electrochemical Na^+ gradient may have been constant in the studies of Goldner et al. (6) in spite of changing Na^+ concentrations. In this case, the agreement of results with respect to the effect of Na^+ would reflect measurements of the same carrier properties.

The main purpose for the transport studies with isolated membrane vesicles was to provide a solid data base of relevant kinetic parameters. To this end, a methodology had to be developed which utilizes the information inherent in the times necessary to obtain constant fractional solute uptake. The $t_{1/2}$ used in these studies depends on uptake of the substrate into most of the vesicles, and the transport data should therefore be less biased than in the more common initial velocity measurements. Initial velocity determinations can be misleading in biological preparations (e.g., the vesicles) that are not homogeneous because the initial velocity is determined by the elements with the fastest rates. These elements may not be representative of the transport properties of the bulk of the preparation.

Using the methodology outlined in this paper, the K_m value and the Na^+ dependence of K_m and V_{max} reflect only properties of the transport system. Application of this method to comparative kinetic studies therefore should resolve the existing discrepancies with respect to the kinetic parameters in renal and intestinal brush border membranes of the various animals.

The reported experiments were not specifically designed to test a specific model of D-glucose translocation, and the ordered binding of nonelectrolyte substrate and Na^+, shown in Fig. 1, represents only 1 of the possibilities. Nevertheless, the Na^+ dependence of K_m and V_{max} can be easily interpreted within the context of that model. The increase in V_{max} with increasing Na^+ concentration can be explained if the translocation steps across the membrane are rate limiting and $P_3 \gg P_2$. In other words, the Na^+-glucose carrier complex is transported much faster across the membrane than the carrier complex without Na^+. The presence of Na^+ shifts the equilibrium between the 2 loaded states of the carrier from the binary D-glucose carrier complex to the ternary Na^+-D-glucose-carrier complex.

Two additional points are noteworthy: 1) Transport as measured by the equilibrium exchange procedure does not necessarily involve translocation of the unloaded form of the carrier across the membrane. Therefore, the K_m and V_{max} may be different for net flux conditions. Such differences have been observed, for example, for the Na^+-independent D-glucose transport in erythrocytes (29). 2) The measured effects of Na^+ in the vesicle system do not take into account the influence of a membrane potential. In particular, it may be possible that the permeability of the Na^+-loaded carrier, if charged, is dependent on the electrical field within the membrane. Thus, the apparent permeability of the Na^+-D-glucose loaded carrier may even be greater in intact cells where a potential of 30–50 mV (inside negative) exists across the brush border membrane (22–24).

ACKNOWLEDGMENTS

I wish to thank Ms. Clara Gulyas and Mr. Rhea Groseclose for valuable help with the experimental part, and Drs. R. Przybylski, P. Will, and K. Neet for valuable comments with regard to the manuscript. The research was supported by NIH Grant AM 18265 and PHS Career Development Award 1KO4 AM00199.

REFERENCES

1. Hopfer U, Sigrist-Nelson K, Murer H. Ann NY Acad Sci 264:414, 1975.
2. Hopfer U. In Quagliariello E, Palmieri F, Singer TP (eds): "Horizons in Biochemistry and Biophysics." Reading, Massachusetts: Addison-Wesley Publishing Company 1976, vol 2, pp 106–133.
3. Silverman M: Biochim Biophys Acta 457:303, 1976.
4. Kinne R. In Bronner F, Kleinzeller A (eds): "Current Topics in Membranes and Transport." New York: Academic Press, 1976, vol 8, pp 209–267.
5. Sacktor B. In Sanadi DR (ed): "Current Topics in Bioenergetics." New York: Academic Press 1977, vol 6, pp 39–81.
6. Goldner AM, Schultz SG, Curran PF. J Gen Physiol 53:362, 1969.
7. Schultz SG, Curran PF. Physiol Rev 50:637, 1970.
8. Kleinzeller A, Kolinska J, Benes I. Biochem J 104:843, 1967.
9. Kleinzeller A, Kolinska J, Benes I. Biochem J 104:852, 1967.
10. Vogel G, Lauterbach F, Kroger W. Arch Ges Physiol 283:151, 1965.
11. Crane RK, Forstner G, Eichholz A. Biochim Biophys Acta 109:467, 1965.
12. Bihler I. Biochim Biophys Acta 183:169, 1969.
13. Aronson PS, Sacktor B. J Biol Chem 250:6032, 1975.
14. von Baeyer H, von Conta C, Haeberle D, Deetjen P: Pfluegers Arch 343:273, 1973.
15. Turner RJ, Silverman M. Fed Proc Fed Am Soc Exp Biol (Abstract) 36:1697, 1977.
16. Eilam Y, Stein WD. In Korn ED (ed): "Methods in Membrane Biology." New York: Plenum Press, 1974, vol 2, pp 283–354.
17. Kotyk A, Janacek K. "Cell Membrane Transport, Principles and Techniques." 2nd Ed. New York: Plenum Press, 1975, pp 64–97, 159–169.
18. Sigrist-Nelson K, Murer H, Hopfer U: J Biol Chem 250:5674, 1975.
19. Cleland WW: Adv Enzymol 29:1, 1967.
20. Murer H, Hopfer U. Proc Natl Acad Sci USA 71:484, 1974.
21. Diedrich DF: Arch Biochem Biophys 117:248, 1966.
22. White JF, Armstrong W McD. Am J Physiol 221:194, 1971.
23. Rose RC, Schultz SG: J Gen Physiol 57:639, 1971.
24. Okada Y, Tsuchiya W, Irimajiri A, Inouye A: J Membr Biol 31:205, 1977.
25. Hamilton TR, Nilsen-Hamilton M. Proc Natl Acad Sci USA 73:1907, 1975.
26. Fass SJ, Hammerman MR, Sacktor B: J Biol Chem 252:583, 1977.
27. Hammerman MR, Sacktor B. J Biol Chem 252:591, 1977.
28. Lever J. J Biol Chem 252:1990, 1977.
29. Kotyk A, Janacek K. In "Cell Membrane Transport, Principles and Techniques." 2nd Ed. New York: Plenum Press, 1975, pp 391–395.

Journal of Supramolecular Structure 7:15—27 (1977)
Molecular Aspects of Membrane Transport 379—391

Orientation of the Protonmotive Force in Membrane Vesicles of Escherichia coli

Lawrence W. Adler, Tomio Ichikawa, Syed M. Hasan, Tomofusa Tsuchiya, and Barry P. Rosen

Department of Biological Chemistry, University of Maryland, School of Medicine, Baltimore, Maryland 21201

Membrane vesicles of Escherichia coli can be produced by 2 different methods: lysis of intact cells by passage through a French pressure cell or by osmotic rupturing of spheroplasts. The membrane of vesicles produced by the former method is everted relative to the orientation of the inner membrane in vivo. Using NADH, D-lactate, reduced phenazine methosulfate, or ATP these vesicles produce protonmotive forces, acid and positive inside, as determined using flow dialysis to measured the distribution of the weak base methylamine and the lipophilic anion thiocyanate. The vesicles accumulate calcium using the same energy sources, most likely by a calcium/proton antiport. Calcium accumulation, therefore, is presumably indicative of a proton gradient, acid inside.

The latter type of vesicle, on the other hand, exhibits D-lactate-dependent proline transport but does not accumulate calcium with D-lactate as an energy source. NADH oxidation or ATP hydrolysis, however, will drive the transport of calcium but not proline in these vesicles. Oxidation of NADH or hydrolysis of ATP simultaneous with oxidation of D-lactate does not result in either calcium or proline transport. These results suggest that the vesicles are a patchwork or mosaic, in which certain enzyme complexes have an orientation opposite to that found in vivo, resulting in the formation of electrochemical proton gradients with an orientation opposite to that found in the intact cell. Other complexes retain their original orientation, making it possible to set up simultaneous proton fluxes in both directions, causing an apparent uncoupling of energy-linked processes. That the vesicles are capable of generating protonmotive forces of the opposite polarity was demonstrated by measurements of the distribution of acetate and methylamine (to measure the ΔpH) and thiocyanate (to measure the $\Delta\psi$).

Key words: protonmotive force, active transport, energy transduction, E. coli

Lawrence W. Adler is presently with the Department of Psychiatry, Sheppard and Enoch Pratt Hospital, Baltimore, MD 21204.

Tomio Ichikawa is presently with the Faculty of Pharmaceutical Sciences, Osaka University, Osaka, Japan.

Tomofusa Tsuchiya is presently with the Department of Physiology, Harvard Medical School, Boston, MA 02115.

Received April 12, 1977; accepted June 8, 1977.

Active transport of solutes in many organisms is a vectorial process in which the solute is concentrated in an unaltered form in 1 compartment of a 2 (or multi-) compartment system. We have been interested in the nature of the biochemical events which determine the directionality of active transport systems. Many bacterial transport systems are secondary in nature, that is, they couple the flow of 1 solute against an apparent electrochemical gradient to a previously established electrochemical gradient of another solute [see the recent reviews by Harold (1) and Rosen and Kashket (2)]. The primary sources of those electrochemical gradients in bacteria are the electron transport chains (3), Mg^{2+}-ATPase (or BF_0F_1) (4–6) and bacteriorhodopsin (7). Each of those is a primary active transport system for protons, converting either chemical or electromagnetic energy into electrochemical potential energy. Escherichia coli does not synthesize bacterio-rhodopsin, but does create electrochemical proton gradients or protonmotive forces (8) presumably with the energy derived from ATP during hydrolysis via the BF_0F_1 or from NADH, D-lactate, or other reduced substrates during their oxidation via flavin-linked dehydrogenases coupled to electron transport chains (1–3, 9, 10).

Are transport carrier proteins symmetrical? Is the only factor which determines the direction of transport by secondary systems the orientation of the electrochemical ion gradient established by the primary system? Or are the carriers "different" on the 2 sides of the membrane, such that they could only work in 1 direction regardless of the orienta-tion of the protonmotive force? In order to answer these question, we began a study of the active transport of calcium in membrane vesicles of E. coli. Our previous results had indi-cated that calcium was actively accumlated by everted membrane vesicles but not by "right-side-out" vesicles when energy was supplied by oxidation of reduced phenazine methosulfate (PMS) (11). Our data further indicated that the calcium transport system works via an antiport mechanism with protons (12), suggesting that an artificially imposed pH gradient (ΔpH), acid inside, should be capable of driving a transient accumulation of calcium in everted vesicles. This was experimentally verified (13). If, however, the carrier is capable of acting symmetrically, with the orientation of the ΔpH determining the direc-tion of calcium transport, then it could be predicted that a reversal of the normal orienta-tion of the ΔpH in "right-side-out" membrane vesicles, from acid outside to acid inside, would result in calcium transport in those vesicles.

We report here that so-called "right-side-out" vesicles do indeed transport calcium inwards in response to an artifically imposed ΔpH.* Burnell et al. (14) have similarly reported that an artifically imposed ΔpH will drive the transport of sulfate in vesicles of Paracoccus denitrificans regardless of the orientation of the vesicles.

However, these results are not necessarily indicative of symmetrical carriers: that interpretation depends on the vesicles having the assumed orientation. There is convincing evidence that vesicles prepared by lysis of intact cells of E. coli with a French press form a single population with membrane having an orientation everted relative to the inner mem-brane of the intact cell (15–18). Vesicles prepared by lysis of spheroplasts according to the method of Kaback (19) have been shown to be a single population (17, 18, 20–22) and have been assumed to have a right-side-out orientation (20, 21, 23). However, evidence from a number of laboratories shows that some proteins which exist solely on the cyto-plasmic surface of the inner membrane in vivo are found on both surfaces in these vesicles

*A preliminary account of a portion of this work was presented at the Annual Meetings of the American Society for Microbiology, Atlantic City, New Jersey, 1975.

(16, 18, 22, 24). This is not consistent with a strictly right-side-out orientation, but suggests that the proteins of the inner membrane become in some way scrambled during lysis of spheroplasts (but not during lysis of intact cells by a French press!). Recent results from our laboratory have suggested that the translocated proteins remain functional components of the larger proton-translocating complexes with which they are associated in vivo (18). In this report we confirm and extend our earlier results with the observation that such vesicles produce protonmotive forces of opposite orientations depending on what energy donor is used during the establishment of the force.

MATERIALS AND METHODS

Preparation of Membrane Vesicles

Escherichia coli K12 strain 7 (25) was grown with shaking at $37°C$ in a basal salts medium (26) supplemented with 68 mM glycerol as a carbon source. Everted membrane vesicles were prepared by lysis of intact cells with a French press as described previously (27). Production of membrane vesicles by osmotic lysis of spheroplasts was performed according to the method of Kaback (19) as modified by Adler and Rosen (18).

Transport Assays

Calcium transport assays were performed at pH 8.0 with 0.5 mM $^{45}CaCl_2$ as described previously (27). Calcium transport driven by an artificially imposed ΔpH was performed as described previously (13). Flow dialysis was performed using a modification (26) of the apparatus described by Colowick and Womack (29). Conditions for measurement of the uptake of weak acids and bases by flow dialysis were as described by Ramos et al. (9).

Assay of NADH Dehydrogenase Activity

NADH dehydrogenase activity was measured during a calcium transport assay in which calcium uptake was supported by 5 mM NADH, as described above. At various times duplicate 0.02 ml samples were withdrawn and diluted into 1 ml of ice water to terminate the reaction. The absorbance at 340 nm of each sample was measured and plotted against time; the slope of the resulting line was taken as a measure of NADH dehydrogenase activity. When the effect of D-lactate on that activity was measured, 20 mM D-lactate was added to the vesicle suspension 2.5 min prior to the addition of NADH.

Protein Determinations

Protein concentrations were determined by a micromodification of the method of Lowry et al. (30).

Chemicals

$^{45}CaCl_2$ (1.3–1.4 Ci/mmol), sodium [^{3}H]acetate, (686 mCi/mmol) and [^{3}H]methylamine (34 mCi/mmol) were purchased from New England Nuclear Corporation (Boston, Massachusetts). Sodium [^{14}C]thiocyanate (8.0 mCi/mmol) was obtained from ICN Pharmaceuticals (Irvine, California). Valinomycin was supplied by Sigma Chemical Company (St. Louis, Missouri). Nigericin was the generous gift of Dr. L. Frank of this department. All other compounds were reagent grade and obtained from commercial sources.

TABLE I. Sources of Energy for Calcium Transport in Everted Membrane Vesicles

Energy source	Calcium uptake
	nmol/30 min/mg membrane protein
None	26
5 mM NADH	175
20 mM Lithium D-lactate	120
20 mM Sodium succinate	85
0.1 mM PMS + 20 mM potassium ascorbate	65
5 mM ATP + 5 mM MgCl$_2$	81
5 mM ADP + 5 mM MgCl$_2$	12

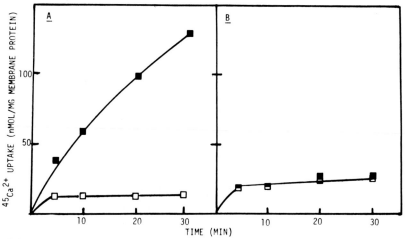

Fig. 1. D-lactate dependent calcium transport in membrane vesicles. Assays were performed as described under Methods. A) everted membrane vesicles prepared by slysis of intact cells with a French press. B) membrane vesicles prepared by osmotic lysis of spheroplasts. Open symbols) +20 mM D-lactate; closed symbols) no exogenous energy source.

RESULTS

Calcium Transport in Everted Membrane Vesicles

As shown in Table I, calcium transport in everted membrane vesicles is energized by the substrates of 3 different dehydrogenases, each of which is a component of the electron transport chain as well as reduced PMS, which couples directly to the cytochrome chain. In addition, ATP also drives calcium uptake (Table I), with energy transduced through the BF_0F_1 (6). We have shown previously that reduced PMS will not energize calcium transport in vesicles prepared by osmotic lysis of the spheroplasts (11), and, as shown in Fig. 1, oxidation of D-lactate likewise does not support calcium transport in such vesicles, although the energy from that oxidation is conserved by transport of calcium in everted vesicles.

Calcium Transport in Membrane Vesicles Driven by an Artificially Imposed ΔpH

Everted vesicles have been shown to transport calcium in response to the imposition of a pH gradient, acid inside (13). Figure 2 demonstrates that vesicles formed by osmotic lysis of spheroplasts respond in a similar manner. Alkalinization of the external medium

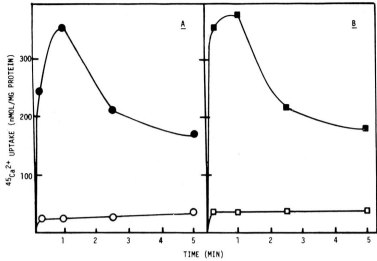

Fig. 2. Calcium transport driven by an artificially imposed ΔpH. Assays were performed as described under Methods. A) everted membrane vesicles; B) vesicles prepared by osmotic lysis of spheroplasts. Open symbols) chemical gradient of protons imposed by a shift in pH from 5.6 to 8.5. Closed symbols) no imposed ΔpH.

produces a transient uptake of calcium which is identical in both types of vesicles. This result would suggest free reversibility of the calcium carrier if the 2 types of vesicles were of the opposite orientation. However, the fact that some inner membrane proteins have been shown to translocate during lysis of spheroplasts (16, 18, 22) obscures the conclusion which could be drawn from this experiment. If a portion of the calcium carriers translocated during the lysis of spheroplasts, then the observed calcium uptake could be due to that fraction of the carriers. It was of interest, therefore, to determine whether other translocated proteins were functional.

Energy-dependent Calcium Transport in Vesicles Prepared by Osmotic Lysis of Spheroplasts

Two enzymes measurable on the external surface of vesicles derived from spheroplasts are NADH dehydrogenase and Mg^{2+}-ATPase. Each is a component of a proton-translocating complex. If the complexes were still functional but now in the opposite direction, it might be possible that a proton gradient, acid inside, would be formed during the oxidation of NADH or hydrolysis of ATP. (Note that the portion of those enzymes located on the inner surface of the membrane would not be active in this in vitro situation, since the membrane of E. coli is relatively impermeable to adenine nucleotides such as ATP or NADH.) Since calcium is transported in the direction of the acid component of the ΔpH, the ability of NADH oxidation of ATP hydrolysis to support calcium uptake could be used as a qualitative measure of the ability to form a ΔpH, acid inside. As shown in Fig. 3, vesicles prepared by osmotic lysis of spheroplasts were capable of coupling energy derived either from oxidation of NADH or hydrolysis of ATP to the uptake of calcium.

D-lactate, which was shown above to be incapable of driving calcium transport in these vesicles (Fig. 1), actually inhibited the utilization of energy derived from NADH oxidation or ATP hydrolysis (Fig. 3). When ATP was used as an energy donor, it could be shown that inhibition of D-lactate oxidation by KCN prevented inhibition by D-lactate

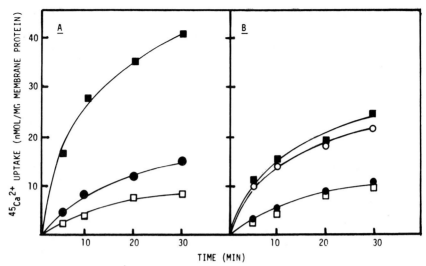

Fig. 3. NADH- and ATP-dependent calcium transport in membrane vesicles prepared by osmotic lysis of spheroplasts. Assays were performed as described under Methods. A) ■—■, +5 mM NADH; ●—●, +5 mM NADH and 20 mM D-lactate; □—□, no exogenous energy source. B) ■—■, +5 mM ATP; ●—●, +5 mM ATP and 20 mM D-lactate, ○—○, +5 mM ATP, 20 mM D-lactate, and 10 mM KCN; □—□, no exogenous energy source. D-lactate was added 2.5 min before the addition of NADH or ATP. KCN was added just prior to D-lactate. In assays utilizing ATP, 5mM $MgCl_2$ was present.

(Fig. 3). The analogous experiment could not be performed with NADH and D-lactate, since KCN inhibits oxidation of both compounds. Compounds such as oxalate and oxamate, which inhibit the D-lactate dehydrogenase, also could not be used, since they form insoluble complexes with calcium. However, the presence of D-lactate during the calcium transport assay did not significantly inhibit the oxidation of NADH (Fig. 4), suggesting that the inhibitory action of D-lactate is the result of an uncoupling of the energy derived from NADH oxidation or ATP hydrolysis, rather than a simple inhibition of the enzymes involved. Similarly, we have shown that D-lactate-dependent uptake of proline by these vesicles is inhibited by the oxidation of NADH or hydrolysis of ATP (18). [It should be pointed out that all of the above results were obtained using a K12 strain of E. coli at pH 8 because of the basic pH optimum of the Mg^{2+}-ATPase (6) and of the calcium transport system (12). Thus, these results may not be directly comparable with those of other laboratories, where proline transport is frequently measured in a ML strain at pH 6.6.]

Measurement of the Orientation of ΔpH and $\Delta\psi$ in Membrane Vesicles

A flow dialysis technique has recently been utilized by Ramos et al.(9) for the measurement of the transmembrane electrochemical gradient of protons. They have concluded that oxidation of D-lactate or reduced PMS results in the generation of a membrane potential, positive outside, which is independent of external pH over a wide range. A pH gradient, acid outside, is established at medium pHs which are neutral to acid, but no pH gradient is established if the external medium is basic.

We have found that everted membrane vesicles accumulate [³H]methylamine or [¹⁴C] thiocyanate during the oxidation of NADH, D-lactate, or reduced PMS or during the hydrolysis of ATP (T. Ichikawa and B. P. Rosen, unpublished results), showing that vesicles

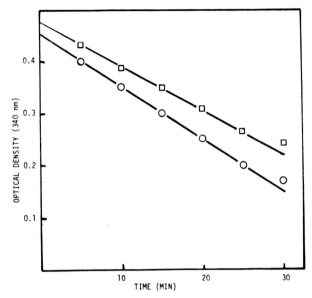

Fig. 4. Effect of D-lactate on NADH oxidation in membrane vesicles prepared by osmotic lysis of spheroplasts. Assays were performed in a manner identical to that utilized for calcium transport, except that nonradioactive CaCl$_2$ was used. At the indicated times samples (0.02 ml) were diluted into 1.0 ml of ice water, and the absorbance at 340 nm was measured. ○—○, +5 mM NADH; □—□, +5 mM NADH and 20 mM D-lactate.

prepared by lysis with a French press form a protonmotive force, acid and positive inside. Vesicles prepared by osmotic lysis of spheroplasts do not accumulate methylamine or thiocyanate during oxidation of D-lactate or reduced PMS, but do accumulate the weak acid acetate when assayed at pH 5.5 (Fig. 5), in confirmation of the results of Ramos et al. (9). At pH 8.0 no acetate uptake was observed, again in confirmation of the results of Ramos et al. (9). On the other hand, methylamine uptake was observed during the oxidation of NADH at pH 8 (Fig. 6) and during the hydrolysis of ATP (data not shown). The experiment shown in Fig. 5 was performed with approximately 3 times as much protein as that shown in Fig. 6. Thus, the pH gradients formed in the 2 experiments are similar in magnitude, but opposite in polarity. Little or no methylamine uptake occurred in the absence of an NADH-regenerating system when that energy source was utilized. Uptake of methylamine was not stimulated by the addition of valinomycin, as would be expected if the uptake of protons by the vesicles were electrogenic (valinomycin would catalyze movement of potassium out of the vesicles, dissipating any membrane potential which would have been formed due to electrogenic uptake of protons). Methylamine uptake was sensitive to nigericin, showing that the formation of a pH gradient had occurred. The reason for the lack of effect of valinomycin is unexplained, especially in light of the fact that NADH oxidation resulted in the uptake of the permeant anion thiocyanate, suggesting the formation of a membrane potential, positive inside. Perhaps the magnitude of the membrane potential is not large enough to create a significant back-pressure on the uptake of protons. It should be pointed out here that these results are being used as a qualitative indication of the orientation of the components of the protonmotive force; future experiments will be concerned with the magnitude of the forces.

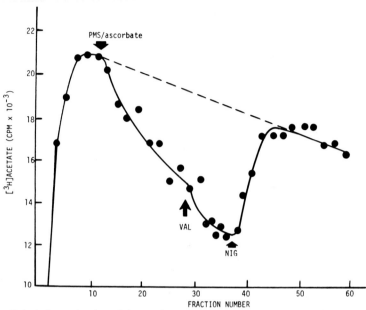

Fig. 5. Flow dialysis determination of the uptake of acetate by membrane vesicles produced by osmotic lysis of spheroplasts. [^{3}H]acetate was added to vesicles (10 mg of protein) at an initial concentration of 28 μM in a total of 1 ml of a buffer consisting of 0.1 M potassium phosphate, pH 5.5, containing 10 mM MgSO$_4$. The same buffer was passed through the lower chamber at a rate of 5 ml per min, with fractions of 1.67 ml collected. Portions (1.0 ml) were mixed with 10 ml of ACS (Amersham/Searle Corp., Arlington Heights, Illinois) and counted in a liquid scintillation counter. At the indicated times were added 0.1 mM PMS and 20 mM potassium ascorbate, pH 5.5, 1.35 μM valinomycin (VAL) and 1.35 μM nigericin (NIG).

DISCUSSION

Two types of E. coli membrane vesicles are currently used for biochemical and physiological studies of energy-linked functions. Everted vesicles, prepared by lysis of intact cells with a French press, have been used for the study of oxidative phosphorylation (3, 15, 33), transport of calcium (6, 11–13, 27), and other questions related to bioenergetics (3, 6, 15, 31–34). These vesicles have been shown to be everted by a number of criteria, including freeze-etch electron microscopy (17) and enzyme localization (16, 18). An everted character is consistent with the observation that substances which are transported outwards by intact cells are transported inwards by these vesicles. For example, primary active transport systems such as the electron transport chain and the BF$_0$F$_1$ normally catalyze the extrusion of protons from intact cells; in everted vesicles they catalyze the uptake of protons (15, 32, 35). This leads to the formation of an electrochemical proton gradient which we have found to be acid and positive inside by measurements of the uptake of the weak base methylamine and the lipophilic anion thiocyanate (T. Ichikawa and B.P. Rosen, unpublished results). Calcium, which is normally extruded from intact cells (36), is transported inwards in everted vesicles (6, 11–13, 27). Since the calcium transport system is a secondary system, most likely a proton/calcium antiport (12, 13), the inwardly acid orientation of the protonmotive force would be a prerequisite for inwardly directed calcium transport. Thus, any reaction which generates a proton-

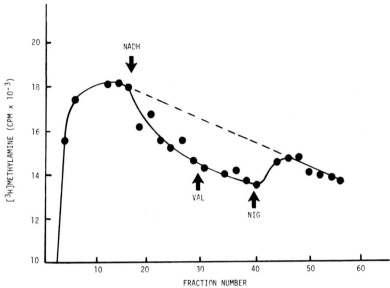

Fig. 6. Flow dialysis determination of the uptake of methylamine in membrane vesicle produced by osmotic lysis of spheroplasts. Conditions were as described in Fig. 5, except that the assay was performed at pH 8.0 with 5 mM NADH as an energy source for the uptake of [3H]methylamine, 0.57 mM, initial concentration. The assay was performed with 3.6 mg of membrane protein.

motive force with that orientation might be expected to drive the accumulation of calcium. In everted vesicles substrates which result in the formation of a pH gradient acid inside include ATP and all tested electron donor substrates of the respiratory chain.

The other type of membrane vesicle derived from E. coli is that described by Kaback (23). Information gained by the use of these vesicles has proven to be enormously valuable for our understanding of the mechanisms of active transport, since this system provided the first in vitro method for the study of vectorial membrane functions. It has been suggested that these vesicles, produced by osmotic lysis of EDTA-lysozyme spheroplasts, on the one hand consist of a homogeneous population and, on the other hand, have an orientation of the membrane which is right-side-out with respect to the inner membrane of the intact cell. The first claim is undoubtedly warranted: freeze-etch electron microscopy (17, 20) and fractionation with antibody directed against membrane proteins (18, 22) have demonstrated only a single class of inner membrane vesicles. Perhaps the most convincing evidence is that each vesicle in the population is capable of transporting the D-lactate analogue 2-hydroxy-3-butenoate (21). However, there has been controversy concerning the second point, the orientation of the membrane of the vesicles. Primarily because the vesicles transport most solutes in the same direction as the intact cell, the claim that the vesicles are right-side-out seemed reasonable. But data concerning the localization of proteins normally associated with only 1 face of the inner membrane in vivo have led to the concept that the vesicles may be mosaics, with certain enzymes translocated from 1 face of the membrane to the other during the lysis event (17, 18, 22). Our data supports that notion, and extends it to include the possibility that translocated enzyme complexes can still be functional. Thus, a complex which establishes an electrochemical gradient of protons acid and positive outside in vivo may produce one of the opposite orientation in the

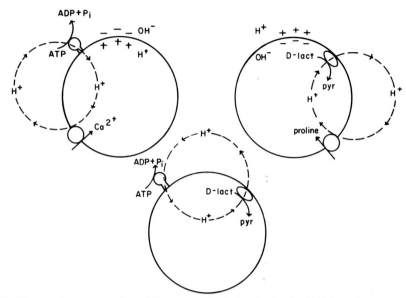

Fig. 7. Diagramatic representation of the proton circulation during the ATP-dependent transport of calcium (top left) and during the D-lactate-dependent transport of proline (top right). During the simultaneous hydrolysis of ATP and oxidation of D-lactate a completed proton circuit is formed resulting in apparent uncoupling (bottom).

in vitro situation following translocation. Again, a secondary solute/H$^+$ antiport which would catalyze extrusion of the solute in vivo would catalyze the uptake of that same solute in this particular in vitro situation. This appears to be the case with the NADH dehydrogenase and BF$_0$F$_1$, which catalyze the formation of protonmotive forces acid and positive inside in both types of membrane vesicles and drive calcium uptake in both types of vesicles. The D-lactate dehydrogenase, on the other hand, does not appear to be translocated during the lysis of spheroplasts (37, 38). Lactate oxidation catalyzes the formation of protonmotive forces of opposite orientation in the 2 types of vesicles, so that D-lactate driven calcium transport occurs in everted vesicles but not in those prepared by the method of Kaback (18).

The simultaneous functioning of 2 primary electrochemical pumps in opposite directions should lead to an apparent uncoupling of either from secondary active transport systems. The calcium transport system, as mentioned above, is linked to a protonmotive force acid and positive inside, while proline transport is linked to a protonmotive force of the opposite orientation. Thus, D-lactate oxidation drives proline but not calcium transport, and NADH oxidation or ATP hydrolysis drives calcium but not proline transport. The oxidation of NADH or hydrolysis of ATP simultaneous with the oxidation of D-lactate allows for neither calcium nor proline uptake, the result of the apparent uncoupling. The proton cycling presumed to occur is illustrated in Fig. 7.

These results in themselves are conceptually unremarkable; they simply reflect the creation of an artifact generated during the preparation of membrane vesicles. However, the fact that the artifact can exist raises certain interesting questions. For example, does the entire NADH oxidase system translocate? If not, how does the translocation of a por-

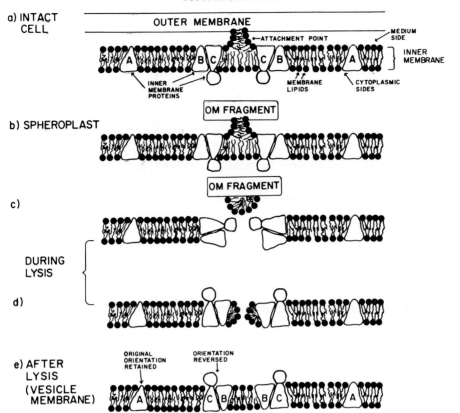

Fig. 8. Model for the translocation of specific inner membrane proteins during the osmotic lysis of spheroplasts. The model is patterned after that described by Alterndorf and Staehelin (17). a) Inner membrane is attached to outer membrane at specific points. Certain proteins, represented by A (an example of which might be the D-lactate dehydrogenase) are located away from the attachment points. Other proteins, represented by B and C (such as the NADH dehydrogenase and BF_0F_1) are localized adjacent to the attachment points. The model does not assume that every molecule of B and C are localized in that region, but that mosaic patches occur more frequently there. b) The outer membrane is degraded during the production of spheroplasts, but fragments still exist which constitute points at which lysis preferentially occurs during an osmotic shock, as in (c). When lysis occurs (c), transient pores are created. Since the inner and outer surfaces of the membrane are topologically identical during the life of the pores, there is no thermodynamic barrier to the lateral diffusion of proteins B and C around the pore as shown in (c) and (d). e) Phospholipids around the sites of the pores reassociate with each other, eliminating the pores, resulting in a reversal of the orientation of a portion of the B and C molecules. Protein A would retain its original orientation, as would the portion of B and C proteins not localized in the vicinity of the transient pore.

tion of the chain allow for the production of a protonmotive force with the opposite orientation? Our knowledge of the arrangement of the components of the electron transport chain and of the proton-translocating ability of the system as a whole is not sufficient at this time to answer this question. While it is obvious that we must know more about the respiratory chain as a whole, it should also be evident that the translocation process itself must be understood. This, then is another and perhaps more important question: how does translocation of enzymes occur; more, how does a specific translocation of certain proteins (such as the BF_0F_1) but not others (such as the D-lactate dehydrogenase) happen? Furthermore, only about half of the BF_0F_1 (and NADH dehydrogenase) activities appear to translocate. What governs this distribution? Related to these questions is the whole problem of how proteins traverse membraneous structures. Altendorf and Staehelin (17) have addressed this question and have postulated a movement of proteins around transient pores created during the lysis of the spheroplasts. Their model, however, did not take into account the apparent specificity of translocation. We have modified their suggestion to include the idea that the transient pores may be created at weak points in the inner membrane, with only specific enzyme complexes being localized near those weak points (Fig. 8). One such location for specific lysis could be points of attachment between the inner and outer membrane, attachments points known to exist in E. coli (39). Flagella are structures which extend through both the inner and outer membrane also (40). A possible structural association between an energy-requiring element such as the flagellar motor and energy-generating systems such as the BF_0F_1 would not be unreasonable from a functional point of view.* Whether such supramolecular structures exist in fact is worthy of further study.

ACKNOWLEDGMENTS

We thank Dr. H. R. Kaback, S. Ramos, and P. Stroobant for their discussions and criticisms. We are especially grateful to Dr. S. Ramos for demonstrating use of flow dialysis as a method for the measurement of the protonmotive force. This work was supported by National Science Foundation Grant PCM-7515896, Public Health Service Grant GM-21648 from the National Institute of General Medical Sciences, and a Basil O'Connor Starter Award from the National Foundation — March of Dimes.

REFERENCES

1. Harold FM: Curr Top Bioenerg (In press).
2. Rosen BP, Kashket ER: In Rosen BP (ed): "Bacterial Transport." New York: Marcel Dekker, (In press).
3. Collins SH, Hamilton WA: J Bacteriol 126:1224, 1976.
4. Schairer HU, Haddock BA: Biochem Biophys Res Commun 48:544, 1972.
5. Berger EA: Proc Natl Acad Sci USA 70:1514, 1973.
6. Tsuchiya T, Rosen BP: J Biol Chem 250:8409, 1975.
7. Lanyi JK: In Capaldi RA (ed): "Membrane Proteins in Energy Transduction." New York: Marcel Dekker, (In press).
8. Mitchell P: Nature (London) 191:144, 1961.
9. Ramos S, Schuldiner S, Kaback HR: Proc Natl Acad Sci USA 73:1892, 1976.
10. Padan E, Zilberstein D, Rottenberg H: Eur J Biochem 63:533, 1976.
11. Rosen BP, McClees JS: Proc Natl Acad Sci USA 71:5042, 1974.
12. Tsuchiya T, Rosen BP: J Biol Chem 250:7687, 1975.
13. Tsuchiya T, Rosen BP: J Biol Chem 251:962, 1976.
14. Burnell JN, John P, Whatley FR: J Biochem 150:527, 1975.
15. Hertzberg EL, Hinkle PC: Biochem Biophys Res Commun 58:178, 1974.
16. Futai M: J Membr Biol 15:15, 1974.
17. Altendorf KH, Staehelin LA: J Bacteriol 117:888, 1974.
18. Adler LW, Rosen BP: J Bacteriol 129:959, 1977.
19. Kaback HR: Methods Enzymol 22:99, 1971.

*DePamphilis and Adler (40) searched for ATPase activity associated with the flagella and with the membrane in the region of the flagellar attachment point. However, their assay conditions (2 mM ATP and 2 mM $MgCl_2$ at pH 7.0) were different from the optima associated with BF_0F_1 (6), so that those results may not be conclusive.

20. Kaback HR: Biochem Biophys Acta 265:367, 1972.
21. Short SA, Kaback HR, Kaczorowski G, Fisher J, Walsh CT, Silverstein SC: Proc Natl Acad Sci USA 71:5032, 1974.
22. Wickner W: J Bacteriol 127:162, 1976.
23. Kaback HR: Science 186:882, 1974.
24. Hare JF, Olden K, Kennedy EP: Proc Natl Acad Sci USA 71:4843, 1974.
25. Hayashi S, Koch JP, Lin ECC: J Biol Chem 239:3098, 1969.
26. Tanaka S, Lerner SA Lin ECC: J Bacteriol 93:642, 1967.
27. Rosen BP, Tsuchiya T: Methods Enzymol (In press).
28. Rosen BP: J Biol Chem 248:1211, 1973.
29. Colowick SP, Womack FC: J Biol Chem 244:774, 1969.
30. Lowry OH, Rosebrough NJ, Farr AL, Randall RJ: J Biol Chem 193:265, 1951.
31. Rosen BP, Adler LW: Biochim Biophys Acta 387:23, 1975.
32. Hasan SM, Rosen BP: Biochim Biophys Acta 459:225, 1977.
33. Schairer HU, Friedl P, Schmid BI, Vogel G: Eur J Biochem 66:257, 1976.
34. Singh AP, Bragg PD: Eur J Biochem 67:177, 1976.
35. West IC, Mitchell P: FEBS Lett 40:1, 1974.
36. Silver S, Kralovic ML: Biochem Biophys Res Commun 34:640, 1969.
37. Short SA, Kaback HR, Kohn LD: J Biol Chem 250:4291, 1975.
38. Futai M, Tanaka Y: J Bacteriol 124:470, 1975.
39. Bayer ME: J Gen Micrbiol 53:395, 1968.
40. DePamphilis ML, Adler J: J Bacteriol 105:396, 1971.

Journal of Supramolecular Structure 7:29—35 (1977)
Molecular Aspects of Membrane Transport 393—399

On the Rate Limiting Step in Downhill Transport Via the LacY Permease of Escherichia coli

Boris Rotman

Division of Biology and Medicine, Brown University, Providence, Rhode Island 02912

Strains of Escherichia coli K12 were constructed for the specific purpose of evaluating the inducibility of the influx mechanism controlled by the *lacY* gene. These strains are heteromerodiploids characterized by a high and relatively constant level of β-D-galactosidase which is not affected significantly by induction of the Lac operon. These properties were obtained by introducing episomal $lacI^+, O^c, Z^+, Y^-$ genes into the cells. In these merodiploids the rate of o-nitrophenyl-β-D-galacto-pyranoside (ONPG) hydrolysis of extracted cells is 50-times that of intact cells. This difference indicates that the rate limiting step in the ONPG hydrolysis by intact cells is influx.

Using a set of merodiploids with and without the LacY transport system, we were able to demonstrate a specific induction of ONPG influx. However, the increase in influx due to induction was only 3.5-fold as compared to the 40-fold increase observed when the LacY permease was measured by intracellular accumulation of $[^{14}C]$TMG.

Key words: transport, induction of influx, LacY permease, β-D-galactosidase, facilitated diffusion

Measurements of carrier-mediated transport across cell membranes may be readily interpreted when nonmetabolizable radioactive substrates are used, since the intracellular radioactivity is a direct function of the difference between influx and efflux (1). Under these conditions, measurements of influx independent of efflux are meaningful if made during the initial substrate uptake when efflux has negligible values. On the other hand, the use of metabolizable substrates for measuring influx offers convenient features while at the same time having serious limitations. This communication is concerned with these limitations and how to circumvent them.

The system chosen for this study is the LacY permease of Escherichia coli which is often measured by the transport of the metabolizable substrate, o-nitrophenyl-β-D-galactopyranoside (ONPG). It has been traditionally assumed that the influx of ONPG via the LacY permease is the rate limiting step in the rate of ONPG hydrolysis by the intact cell (2—4). The hypothesis underlying this assumption is that the excess of β-D-galactosidase present in the cytoplasm assures that practically every molecule of ONPG transported by the permease is immediately hydrolyzed. However, experiments which involve induction of the permease may not fulfill this condition because of the low level of enzyme present at the onset of induction. For this reason the inducibility of the LacY permease measured

Received June 24, 1977; accepted June 27, 1977.

by influx of ONPG must be tested in a system which provides a constant elevated level of β-D-galactosidase under all conditions.

We have fulfilled this criterion by constructing heteromerodiploids of E. coli with $lacI^+,O^c,Z^+,Y^-$ genes in the episomes and $lacI^+,O^+,Z_{del},Y^+$ in the chromosomes. While the LacY permease of these merodiploids may be varied experimentally by induction, the level of β-D-galactosidase is constant since the $lacZ^+$ gene is controlled by the constitutive $lacO^c$ operator in the cis position. In addition, by introducing the $lacI^+,O^c,Z^+,Y^-$ episome in bacteria containing a deletion of all the chromosomal lac genes, it was possible to determine the extent of influx due to passive diffusion or to transport systems other than the LacY permease.

METHODS

Organisms

The bacterial strains used in this study are listed in Table I. Episomal transfers were done according to the methods of Miller (5).

Chemicals

Methyl-1-thio-β-D-[1-^{14}C] galactopyranoside (TMG), 1.85 mC/mmole, was purchased from New England Nuclear Corporation (Boston, Massachusetts), and was purified by paper chromatography (6). Isopropyl-1-thio-β-D-galactopyranoside (IPTG) and o-nitrophenyl-β-D-galactopyranoside (ONPG) were purchased from Nortok Associates (Lexington, Massachusetts).

Growth Conditions.

Cells were grown at 37°C in Davis minimal medium (7) supplemented with 0.4% sodium lactate (Fisher Scientific Company, Pittsburgh, Pennsylvania), and 0.5 μg/ml vitamin B_1. Cultures were aerated on a gyratory shaker. Prior to the experiments, overnight cultures were diluted in fresh, warm medium and allowed to divide for at least 2 generations. For induction, cells were grown in the presence of 5×10^{-4} M IPTG.

Transport Assay

Intracellular accumulation of [^{14}C] TMG was measured using cells harvested during exponential growth. Chloramphenicol at a final concentration of 30 μg/ml was added to the cultures and the cells were centrifuged and washed 3 times with cold growth medium without lactate. The assay mixtures (3 ml final volume) contained 5×10^8 cells suspended in growth medium, 100 μg of chloramphenicol, and 1.17 μM [^{14}C] TMG (10^4 cpm). After

TABLE I. Strains of E. coli K12 Used

Strain No.	Genetic markers[a]
W4680	$lacZ39_{del},strA$
W4980	$lac(Z,Y,A)_{del},strA$
W4980/F'$lacO^c$	$lac(Z,Y,A)_{del},strA$/F'$lacO^c$
S170	$lacZ39_{del},strA$/F'$lacO^c,IacY$
S172	$lac(Z,Y,A)_{del},strA$/F'$lacO^c,lacY$

[a]Genotype abbreviations are given according to Bachmann et al. (13).

incubation at $37°C$ with shaking, the mixtures were filtered through HA millipore membranes. The radioactivity of the dried membranes was determined by liquid scintillation counting with 68% efficiency. Blank values were obtained using formaldehyde-treated cells in the assay (8).

For measurements of ONPG hydrolysis in vivo, the washed cells were diluted with buffer B to a density of about 2×10^8 cells per ml. This buffer consists of 10 mM $MgCl_2$, 100 mM NaCl, and 50 mM 2-mercaptoethanol, adjusted to pH 7.45 $(23°C)$ using acetic acid. The mercaptoethanol was added daily to the buffer. The cell suspension was allowed to equilibrate at $37°C$ for about 5 min, and then 0.2 ml of 3×10^{-2} M ONPG (in water) was added to start the reaction. After development of suitable color, 3 ml of 0.2 M Na_2CO_3 was added to stop the reaction and the mixture was immediately filtered through an HA millipore membrane. The supernatant was collected and its absorbance was measured at 420 nm. A molar extinction coefficient of 4,700 was used to convert absorbance readings to o-nitrophenol concentration.

The β-D-galactosidase activity of disrupted cells was measured using bacterial suspensions treated with deoxycholate and toluene (9, 11). For this assay, 1 ml of a cell suspension was mixed with 0.05 ml of 2% sodium deoxycholate and 3 drops of toluene. The mixture was left at room temperature for 10 min with occassional shaking. At the end of this period, 0.1 ml of the mixture was diluted 1:18 with buffer B and assayed for enzymatic activity as described above.

An enzyme unit for β-D-galactosidase is defined as the amount of enzyme liberating 1 nmole of substrate in 1 min under the indicated conditions.

RESULTS

Construction of a $lacI^+$, O^c, Z^+, Y^- Episome

A number of spontaneous Lac⁻ mutants of $W4980/F'lacI^+O^cZ^+Y^+$ were obtained by penicillin selection in the presence of lactose followed by screening on EMB lactose agar (10). Among these mutants we chose 26 which had high levels of β-D-galactosidase while failing to accumulate [^{14}C] TMG intracellularly. The episome of each presumptive $lacY$ mutant was transfered to W4680 for individual quantitative tests and one of them, termed S170, was selected for further studies. Subsequently the episome of S170 was transfered back to W4980 to yield S172, a strain with a deleted lac operon in the chromosome and the $lacI^+O^cZ^+Y^-$ episome. As shown in Table II, our postulation of the genotypes was confirmed by the phenotypes of the resulting heteromerodiploid strains. Both S170 and S172 had comparable levels of enzyme which were not affected significantly by induction. Strain S170 exhibited an inducible LacY transport system (measured by uptake of [^{14}C] TMG) while S172 was unable to accumulate TMG.

Inducibility of the LacY Transport System in the Merodiploids

The merodiploid S170 was compared to W4680, the parental haploid strain, in terms of inducibility of the LacY transport system. Exponentially growing cultures of each strain were induced in the presence of 5×10^{-4} M IPTG. At intervals cells were harvested, washed, and assayed for intracellular accumulation of [^{14}C] TMG. As shown in Fig. 1, S170 and W4680 exhibited similar patterns of inducibility. In contrast, the uptake of [^{14}C] TMG by either normal or induced cells of strain S172 was indistinguishable from

TABLE II. Phenotype of Parental and Derived Bacterial Strains

Strain	β-D-galactosidase of disrupted cells[a]		Transport	
	Noninduced	Induced	Noninduced	Induced
W4680	0	0	14	681
W4980	0	0	0	0
S170	2,800	3,300	32	502
S172	2,800	2,700	0	0

[a] The β-D-galactosidase activity of cells disrupted by treatment with deoxycholate and toluene was measured. The data is given in enzyme units per 10^9 bacteria. Transport was measured by the intracellular accumulation of $[^{14}C]$ TMG and is expressed in counts per min per 10^9 cells. These values are corrected for a blank of 175–180 cpm; this was obtained when formaldehyde-treated cells were used in the permease assay (8).

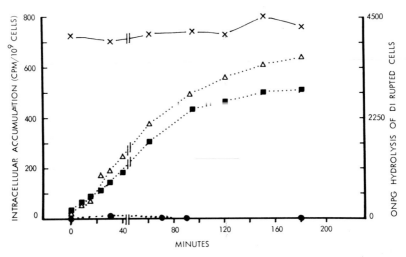

Fig. 1. Induction of the LacY transport system. At zero time, IPTG was added to exponentially growing bacterial cultures. Samples were withdrawn from the cultures at the indicated intervals and the level of LacY permease was determined by the intracellular accumulation of $[^{14}C]$ TMG as indicated in the text. The results are expressed in cpm per 10^9 cells. A blank of 179 cpm was subtracted from each value. Accumulation of W4680 (△··△); S170 (■··■); S172 (●··●). The amount of β-D-galactosidase of S170 (X—X) was measured using cells disrupted by treatment with sodium deoxycholate and toluene (9, 11).

that of negative controls (formaldehyde-treated cells). These experiments also demonstrated that the level of β-D-galactosidase present in either S170 or S172 was not significantly affected by induction since we observed less than a 6% increase in extractable enzymatic activity at the end of a 3.5-h induction period (upper curve of Fig. 1).

Transport Measured by In Vivo Hydrolysis of ONPG

Cells from the same cultures used for determinations of uptake of $[^{14}C]$ TMG were assayed for their ability to transport ONPG using the ONPG hydrolysis of intact cells as the parameter. Noninduced S172 cells (genetically defective in the *lacY* gene) showed about half the rate of ONPG hydrolysis of noninduced cells of S170. Upon induction, the rate of S172 remained relatively constant (less than 20% increase) while that of S170 increased about 3.5-fold (Fig. 2). In both strains intact cells hydrolyzed considerably less than disrupted cells. The ratio of the hydrolysis rates of disrupted cells over that of intact cells was 25 for S170. This ratio decreased by a factor of 10 following an induction period of 70 min. In contrast, S172 exhibited a ratio of 50 which remained constant during induction.

Specificity of Induction of the LacY Transport System Measured by In Vivo Hydrolysis of ONPG

Cultures of S170 were grown in the presence of 10^{-3} M D-galactose, 10^{-3} M D-fucose, 10^{-3} M melibiose, or 5×10^{-4} M IPTG. The hydrolysis of both intact and disrupted cells of each culture was measured after 14.5 h of induction. The results of these experiments (Table III) indicated that among the sugars tested only IPTG is an inducer. We also observed that the level of enzyme in disrupted cells was reduced when cultures were grown in the presence of inducers other than IPTG. This was most noticeable when cells were grown with D-galactose.

DISCUSSION

The inducibility of the LacY transport system has been well documented through studies involving measurements of intracellular accumulation of radioactive, non-metabolizable substrates, i.e., measurements of the difference between influx and efflux. However, no direct evidence has been presented concerning the inducibility of the entry mechanism itself. Previous studies of the LacY permease used ONPG hydrolysis as a measure of influx, relying on the assumption that β-D-galactosidase is always present in

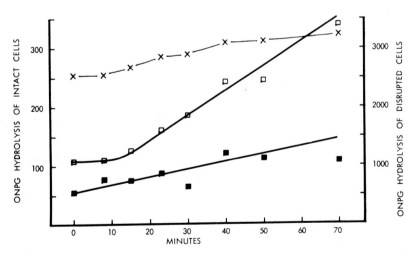

Fig. 2. Induction of the LacY transport system measured by the rate of ONPG hydrolysis of intact cells. We used the same cell suspensions previously assayed for intracellular accumulation of [^{14}C]-TMG in the experiment shown in Fig. 1. Hydrolysis rate of S170 (□–□); S172 (■–■). The rate of hydrolysis of disrupted cells of S170 is shown (X—X).

TABLE III. Induction of the LacY Transport System

Inducer[a]	β-D-galactosidase activity	
	Intact cells	Disrupted cells
IPTG	1,136	4,343
D-fucose	188	2,698
Melibiose	157	2,470
D-galactose	114	1,870
None	108	3,160

[a]Cells were induced overnight. The concentration of inducer was 5×10^{-4} M for IPTG, and 10^{-3} M for the other inducers. The results are given in enzyme units per 10^9 cells.

sufficient excess so that the influx is the rate limiting step. Although there are experimental data supporting this assumption, they are only valid for either constitutive or fully induced bacteria since these were the organisms used in the experiments. For conditions involving induction of the LacY transport system, it is necessary to have a relatively high level of β-D-galactosidase, a condition which is not usually fulfilled in E. coli because the uninduced cell contains only a few molecules of enzyme (11). We have presented here a system specially designed for experiments requiring induction of the LacY transport mechanism. The basis of this system is the use of bacteria containing an episomal *lac* operon with a constitutive operator gene and a defective *Y* gene. These bacteria produce high levels of β-D-galactosidase which are affected only minimally by induction. Because of the *lacY* mutation in the episome, it is possible to study the inducibility of the permease as controlled by the normal chromosomal Y^+ gene.

Using this system, we observed that the rate of ONPG hydrolysis in vivo increases with induction but that the maximum achieved is considerably less than that anticipated on the basis of induction of entry of TMG. Considering that induction causes an increase of 1,000-fold in the β-D-galactosidase level and an increase of 40-fold in the level of TMG accumulation, our finding that there is less than a fourfold increase in ONPG influx is surprising. The possibility that the episome somehow causes a defect in the expression of the normal *lacY* gene of the bacteria is unlikely, since S170 exhibited a normal LacY permease as measured by the intracellular accumulation of radioactive TMG (Fig. 1).

These paradoxical results may be explained by postulating that either ONPG can enter the cell through a system independent of the LacY permease or that induction causes a decrease in the exit rate of TMG. The former idea seems more likely for 2 reasons: i) our results using S172 (a strain with a deletion of the chromosomal *lacY* gene) show that these bacteria exhibit substantial transport of ONPG under either induced or noninduced conditions (Fig. 2); ii) results of previous experiments indicate that the exit mechanism of the LacY permease is induced by IPTG (8, 12).

We also tested the specificity of induction of the LacY influx mechanism using inducers of other transport systems. The results indicate that only inducers of the LacY permease increase the influx.

The heteromerodiploids described here can be of value for future studies on conditions affecting the LacY transport system. A possible application of the merodiploids would be in reexamination of the role of lipid biosynthesis where the use of strains with an inducible β-D-galactosidase may be responsible for the conflicting reports.

ACKNOWLEDGMENTS

I wish to thank Rosario Guzman and Adriana Johnson for their essential contributions. This investigation was supported by Public Health Service grant AM-11194 from the National Institute of Arthritis, Metabolism, and Digestive Diseases.

REFERENCES

1. Rickenberg HV, Cohen GN, Buttin G, Monod J: Ann Inst Pasteur 91:429, 1956.
2. Cohen GN, Monod J: Bacteriol Rev 21:169, 1957.
3. Rotman B: J Bacteriol 76:1, 1958.
4. Herzenberg LA: Biochim Biophys Acta 31:525, 1959.
5. Miller JH: "Experiments in Molecular Genetics." New York: Cold Spring Harbor Laboratories, 1972.
6. Robbins AR, Guzman R, Rotman B: J Biol Chem 251:3112, 1976.
7. Rotman B, Ganesan AK, Guzman R: J Mol Biol 36:247, 1968.
8. Rotman B, Guzman R: Pathol Biol 9:806, 1961.
9. Novick A, Weiner M: Proc Natl Acad Sci USA 43:553, 1957.
10. Cook A, Lederberg J: Genetics 47:1335, 1962.
11. Maloney PC, Rotman B: J Mol Biol 73:77, 1973.
12. Horecker BL, Thomas J, Monod J: J Biol Chem 235:1580, 1960.
13. Bachmann BJ, Low KB, Taylor AL: Bacteriol Rev 40:116, 1976.

Journal of Supramolecular Structure 7:37—48 (1977)
Molecular Aspects of Membrane Transport 401—412

Involvement of Membrane Sulfhydryls in the Activation and Maintenance of Nutrient Transport in Chick Embryo Fibroblasts

H. Smith-Johannsen, J. F. Perdue, M. Ramjeesingh, and A. Kahlenberg

Lady Davis Institute for Medical Research, Jewish General Hospital, Montreal, Quebec, Canada H3T 1E2

At 5 μg/ml, insulin stimulates hexose, A-system amino acid, and nucleoside transport by serum-starved chick embryo fibroblasts (CEF). This stimulation, although variable, is comparable to that induced by 4% serum. The sulfhydryl oxidants diamide (1—20 μM). hydrogen peroxide (500 μM), and methylene blue (50 μM) mimic the effect of insulin in CEF.

PCMB-S,[1] a sulfhydryl-reacting compound which penetrates the membrane slowly, has a complex effect on nutrient transport in serum- and glucose-starved CEF. Hexose uptake is inhibited by 0.1—1 mM PCMB-S in a time- and concentration-dependent manner, whereas A-system amino acid transport is inhibited maximally within 10 min of incubation and approaches control rates after 60 min. A differential sensitivity of CEF transport systems is also seen in cells exposed to membrane-impermeant glutathione-maleimide I, designated GS-Mal. At 2 mM GS-Mal reduces the rate of hexose uptake 80—100% in serum- and glucose-starved CEF; in contrast A-system amino acid uptake is unaffected. D-glucose, but not L-glucose or cytochalasin B, protects against GS-Mal inhibition. These results are consistent with the hypothesis that sulfhydryl groups are involved in nutrient transport and that those sulfhydryls associated with the hexose transport system and essential for its function are located near the exofacial surface of the membrane in CEF.

Key words: transport, sulfhydryl oxidants, p-chloromercuribenzenesulfonate, glutathione maleimide I

When insulin binds to its receptors, the resulting mitogen-receptor complexes generate an array of temporally distinct changes in the cell, commencing with a rapid increase in the flux of ions, sugars, amino acids, and nucleosides across the plasma membrane and culminating in replication. Kinetic analysis of the insulin-induced increase in nutrient transport establishes it as due mostly to an increase in V_{max} rather than to a decrease in K_m (1). This has been interpreted as reflecting the recruitment and/or activation of transport components.

[1]Abbreviations: α-AIB — α-aminoisobutyric acid; α-meAIB — α-methylaminoisobutyric acid; 2-dG — 2-deoxy-D-glucose; DTP — dithiopyridine; NEM — N-ethylmaleimide; PCMB — p-chloromercuribenzene; PCMB-S — p-chloromercuribenzenesulfonate; DMSO — dimethyl sulfoxide.
Received May 31, 1977; accepted June 13, 1977

How the formation of mitogen-receptor complexes triggers the increased activity of transport proteins remains obscure, but membrane sulfhydryls may play a key role in this event. This idea stems from the fact that low concentrations of sulfhydryl oxidants can mimic insulin stimulation of hexose transport. Czech (2) and his colleagues have observed that 20 mM diamide, 4 mM hydrogen peroxide, and 50 μM methylene blue stimulate the uptake of 3-O-methylglucose by isolated brown fat cells to rates comparable to those achieved by 0.1 μg/ml of insulin. Furthermore, compounds which react with sulfhydryl groups inhibit hexose, amino acid, and nucleoside transport. For example, 30 μM PCMB inhibits 2-dG uptake in Novikoff cells (3) and a more hydrophilic derivative, PCMB-S, inhibits 2-dG uptake in mouse embryo cells (4). PCMB (3) and DTP (4) also inhibit uridine transport in cultured cells. Hare (4) observed that PCMB-S inhibits System L amino acid transport to a greater extent than System A in mouse embryo cells. His observations, together with those of Kwock and his colleagues (5, 6) on the effects of γ-ray irradiation on amino acid transport in thymocytes suggest that System L transport components, like those mediating glucose transport, have sulfhydryl groups lying near the exofacial membrane surface whereas the sulfhydryl reactant-sensitive sites of System A transport components lie near the endofacial surface or in hydrophobic regions of the membrane (7).

Recent evidence supporting the idea that some part of the glucose transport system is located in the exterior face of the plasma membrane is based on the observation by Batt et al. (8) that a membrane-impermeant maleimide (GS-Mal) inhibits glucose efflux in human red blood cells. The extent of inhibition was directly related to the binding of GS-Mal to the membrane (9). We have synthesized GS-Mal and have observed that it is a potent inhibitor of glucose transport in CEF while having little or no effect on amino acid transport. These results are communicated in this report along with our observations on the effects of sulfhydryl oxidants and inhibitors on nutrient transport.

MATERIALS AND METHODS

Cell Cultures

The procedures used for culturing chick embryo fibroblasts were as described previously (1). In general, cells were plated at 5×10^5 cells per 60-mm diameter Lux plastic plate or at 1.5×10^5 cells per 35-mm plate, sometimes fed on day 3 and used on day 4 or 5. For serum starvation experiments, CEF were rinsed in either Tris-dextrose or serum-free medium and maintained at 37°C for 6–8 h in serum-free medium prior to the measurement of transport. In some experiments, the serum-free medium was supplemented with 0.15% tryptose-phosphate.

For glucose starvation experiments, cells were rinsed in glucose-free medium and maintained in glucose-free medium containing 4% dialyzed serum for 20–24 h. In certain experiments, cells were starved for both glucose and serum. Under these conditions, although numerous cells often detach from the plate, the majority appear to tolerate glucose and/or serum starvation without incurring ostensible damage.

Experiments With Sulfhydryl Oxidants

Triplicate plates were each rinsed twice with approximately 5 ml warm (37°C) Dulbecco's phosphate-buffered saline (PBS), pH 7.4, and then incubated for 10 min at 37°C in 1 ml PBS containing 5 μg of porcine insulin (Schwarz-Mann, Orangeburg, New York) or various amounts of oxidizing agent. After 10 min, 0.5 ml reaction mixture containing 3 mM α-[1-^{14}C] AIB acid (New England Nuclear Corporation, Boston, Massachusetts; specific activity 0.3 μCi/μmol) and 3 mM [G-^{3}H] 2-dG (New England Nuclear Corporation; specific activity 1.3 μCi/μmol) or 0.75 μM [5,6-^{3}H]-uridine (New England Nuclear Corporation; specific activity 5.2 μCi/μmol) in PBS was added; the final concentration of sugar and of amino acid was 1 mM, while that of uridine was 0.25 μM. Under these conditions, transport was the rate-limiting step for uptake of sugar for at least 10 min. After 5 min incubation at 37°C, uptake of the radiolabeled nutrients by the cells was arrested by rapidly removing the radioactive medium and rinsing each plate 3–5 times with 5 ml ice-cold PBS. The cells were digested in 1 ml 0.2 N NaOH; the digest was transferred to tubes and combined with 0.5 ml H$_2$O rinse. Aliquots were then taken for protein estimation (10) and for determination of radioactivity using dioxane-toluene scintillation fluid and an Intertechnique (Canatech, Incorporated, Montreal, Canada) liquid scintillation counter.

Experiments With PCMB-S

A solution of PCMB-S (Sigma Chemical Co., St. Louis, Missouri) in PBS was prepared just prior to use. Cells grown in 35-mm diameter plates were rinsed twice in warm PBS and incubated in 0.5 ml PBS containing various concentrations of PCMB-S for designated intervals at 37°C. This solution was removed, the cells washed twice in warm PBS, and uptake of 2-dG and α-me AIB was measured over 2 min as in experiments with GS-Mal.

Experiments With GS-Mal

Bismaleimidoethyl ether was prepared by the procedure of Tawney et al. (11) and was used in the synthesis of GS-Mal I according to the method of Abbott and Schachter (9). This species differs from GS-Mal II and III in its relatively short "arm" that joins the reactive maleimide with glutathione. Just before use, GS-Mal was dissolved in cold 50 mM sodium phosphate, pH 7.4, 100 mM NaCl, to a concentration that was twice that of the final concentration designated in the legends to the Figures or Tables. Depending on the concentration of GS-Mal, the pH shifted to 6.8 or was adjusted to this value. Confluent cultures of glucose-starved CEF in 35-mm plates were rinsed twice with warm 50 mM sodium phosphate, pH 6.8, 100 mM NaCl, and then incubated with 0.5 ml phosphate buffer, pH 6.8, containing the substance to be tested for various times at 37°C. Uptake of sugar and amino acid by CEF was measured either directly after this treatment or after 2 rinses in warm PBS and a 30-min incubation in 1 ml PBS. In either case, plates were rinsed with warm PBS and 0.5 ml reaction mixture containing 1 mM α-[1-^{14}C] meAIB (New England Nuclear Corporation; specific activity 0.3 μCi/μmol) and 1 mM [G-^{3}H] 2-dG in PBS (specific activity 1.3 μCi/μmol) was added. Under these conditions, uptake of sugar and amino acid in glucose-starved cells is linear for at least 5 min. The uptake assay was terminated after 2 min by removing the reaction mixture, placing the culture plates on ice, and washing the cells 5 times with ice-cold PBS. The protection against GS-Mal-induced inhibition of 2-dG transport by competing hexoses and competitive inhibitors of glucose transport was examined. Cytochalasin B (Aldrich Chemical Co., Milwaukee, Wisconsin) and phloretin (Sigma) were prepared as stock solutions (2 mg/ml) in DMSO. Controls were treated with equivalent amounts of DMSO during incubation and/or during the transport assay. Cell protein was hydrolyzed in 0.4 ml 0.5 N NaOH, the digest was transferred to tubes and combined with a 0.4 ml H$_2$O rinse. Samples were processed as described above, except Triton X-100:toluene (1:2) was employed as scintillant.

RESULTS

Stimulation of Nutrient Transport by Insulin

Preliminary experiments were conducted to determine the optimal conditions for stimulation of sugar transport in CEF by insulin. In most of the experiments reported here, the fibroblasts were serum-starved for 6–18 h before treatment with various agents and measurement of transport. Raizada and Perdue (12) have shown that although serum-starvation depresses the basal rate of sugar uptake by CEF, the response of these cells to insulin is relatively greater than that of serum-fed CEF.

Confluent cells were incubated in PBS containing 5 μg/ml insulin for varying times up to 60 min and then 2-dG uptake was measured over 5 min. Insulin stimulated CEF to transport sugar at rates averaging 20–30% and up to 70% greater than the rate of sugar uptake in controls. In general, the percent stimulation of transport by insulin was comparable to that induced by 4% serum. Transport was maximally stimulated by 7.5 min exposure of cells to insulin, although in some experiments, this occurred as early as 2 min (data not shown).

The stimulation of sugar and amino acid uptake by insulin was highly variable in that out of 53 experiments, only 50% showed a significant increase in the rates of transport of insulin-treated cells. Similarly, out of 12 experiments involving serum-stimulation of transport, 5 indicated that CEF took up sugar and amino acids at higher rates than controls. The reason for this variability remains obscure.

Stimulation of Nutrient Transport by Sulfhydryl Oxidants

The effect of diamide and hydrogen peroxide on nutrient transport by CEF was studied by exposing cells to various concentrations of these reagents for 10 min and measuring the uptake of labeled 2-dG, α-AIB and/or uridine during 5 min. Figure 1 and 2 show that both of these oxidizing agents stimulate transport and that this effect is concentration-dependent. Maximum response of CEF in these experiments was at 1–20 μM diamide and 500 μM hydrogen peroxide. As seen in Fig. 3, 50 μM methylene blue stimulates sugar transport. In this experiment, methylene blue also stimulated α-AIB uptake, but only by 10%. The magnitude of the stimulation of transport activity induced by diamide and methylene blue was maximum by 10 min treatment of the cells. High concentrations of all 3 reagents inhibit sugar transport whereas they have little or no effect on the basal rate of amino acid transport.

As in studies of the response of CEF to insulin, the stimulation of transport by the chemical oxidants was variable with respect to whether the response occurred, its magnitude and the optimal concentration of reagent required to elicit a response. Approximately half of the 15–19 experiments in which the effect of diamide, hydrogen peroxide, or methylene blue on nutrient transport was tested showed stimulation of uptake by these reagents. In some experiments the sulfhydryl oxidants stimulated transport while insulin failed to do so, and in other experiments the converse was observed.

Effect of PCMB-S

The effect of PCMB-S on nutrient transport in CEF is complex. At very low concentrations (0.01 mM) and short treatment times, PCMB-S causes a slight stimulation of hexose and also amino acid uptake (Table I). At higher concentrations, however, this sulfhydryl reactant inhibits hexose uptake in a concentration- and time-dependent manner.

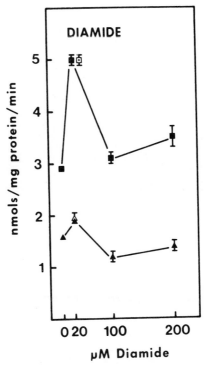

Fig. 1. Stimulation of nutrient transport by diamide. Serum-starved CEF were treated with various concentrations of diamide for 10 min and then uptake measured over 5 min. ▲——▲) uridine uptake; ■——■) α-AIB uptake. Open symbols denote uridine (△) and α-AIB (□) uptake in the presence of 5 μg/ml insulin. Bars represent standard error.

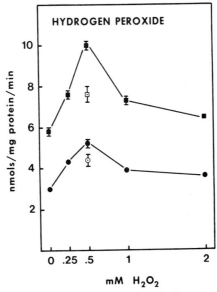

Fig. 2. Stimulation of nutrient transport by hydrogen peroxide. Serum-starved CEF were treated with various concentrations of hydrogen peroxide for 10 min and then uptake measured over 5 min. ●——●) 2-dG uptake; ■——■) α-AIB uptake. Open symbols denote 2-dG (○) and α-AIB (□) uptake in the presence of 5 μg/ml insulin. Bars represent standard error.

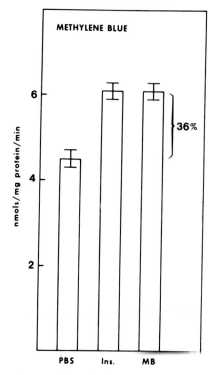

Fig. 3. Stimulation of hexose transport by methylene blue. Serum-fed CEF were treated with 5 μg/ml insulin or 50 μM methylene blue for 10 min and uptake of 2-dG measured over 5 min. Bars represent standard error. The uptake of α-AIB was elevated by insulin and methylene blue 20% and 10% respectively in this experiment.

TABLE I. Effect of PCMB-S on Sugar and Amino Acid Uptake as a Function of Time and Concentration*

PCMB-S (mM)	2-dG uptake (% of control)			α-meAIB uptake (% of control)		
	10 min	30 min	60 min	10 min	30 min	60 min
0.01	129	74,90	–	120	62	–
0.1	88,76	59,80,90	56	50,29	84,77	79
0.5	47	34,47	–	40	77	–
1	50,58	24,30,24	17	32,60	95,85	91
10	–	15	–	–	–	–

*CEF starved for serum and glucose for approximately 24 h were incubated in PBS, pH 7.4, containing various concentrations of PCMB-S for 10, 30, or 60 min. Uptake of sugar and amino acid was measured over 2 min. Data is from 3 separate experiments.

Table I shows that inhibition is rapid; about 50% inhibition is achieved within 10 min at 1 mM PCMB-S. Maximum inhibition is 85% and results from 30 min treatment with 10 mM PCMB-S. Thus 15% of sugar transport activity appears resistant to PCMB-S.

The effect of PCMB-S on amino acid transport is distinct from its effect on sugar transport. Low concentrations (0.01–0.1 mM) or 10 min treatment with higher concentrations (0.1–1 mM) of PCMB-S inhibit α-meAIB uptake greater than 50%. However, this inhibition

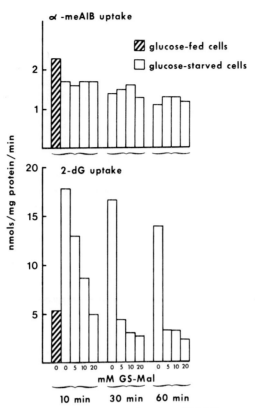

Fig. 4. Effect of GS-Mal on nutrient transport as a function of inhibitor concentration and time of treatment. Glucose-starved (24–27 h) CEF were exposed to 0–20 mM GS-Mal in 50 mM sodium phosphate, pH 6.8, containing 100 mM NaCl for 10, 30, or 60 min. Glucose-fed and glucose-starved controls were treated with pH 6.8 sodium phosphate buffer for comparable intervals. Cells were rinsed twice in PBS and the uptake of α-meAIB and 2-dG was measured over 2 min. The standard error in this experiment averaged less than 5% for sugar uptake and 10% for amino acid uptake.

is transient and when cells are exposed to high concentrations of PCMB-S for 30 or 60 min, amino acid transport is no longer as severely inhibited.

Experiments With GS-Mal

Since GS-Mal breaks down more rapidly at pH 7.4 than in acid conditions (9), treatment of CEF with this compound was carried out in isotonic 50 mM sodium phosphate buffer, pH 6.8. The effect of low pH is to decrease the basal rate of both sugar and amino acid uptake but these transport systems still remain sensitive to the addition of stimulatory and inhibitory agents. To elevate the basal rate of glucose uptake, cells were either glucose-starved or serum- and glucose-starved for approximately 24 h prior to treatment. Under these conditions, glucose uptake by starved CEF is enhanced about fivefold when measured in PBS and twofold after cells are incubated in pH 6.8 buffer. Figure 4 shows the effect of treatment with 5, 10, or 20 mM GS-Mal for 10, 30, or 60 min on 2-dG and α-me AIB uptake in glucose-starved CEF; the latter uptake reflects amino acid transport by System A. A maximum 85% inhibition of 2-dG uptake resulted from exposure of cells to 20 mM GS-Mal for 30 min in this experiment. Experiments were conducted to see if lower con-

TABLE II. Effect of Glucose, Cytochalasin B (CB), and Phloretin on the Inhibition of Hexose Transport by GS-Mal*

Treatment of cells	Uptake 2-dG	
	nmols/mg protein/min	% of control
Expt. 1		
glucose-fed	6.1 ± 0.48	60
glucose-starved	10.2 ± 0.72	100
glucose-starved + 20 mM GS-Mal	3.1 ± 0.13	30
glucose-starved + 50 mM glucose	12.0 ± 1.20	118
glucose-starved + GS-Mal + 50 mM glucose	4.1 ± 0.30	34
glucose-starved + 100 mM glucose	14.1 ± 0.27	138
glucose-starved + GS-Mal + 100 mM glucose	4.1 ± 0.12	29
Expt. 2		
glucose-fed	12 ± 0.22	54
glucose-starved	22.4 ± 0.53	100
glucose-starved + 20 mM GS-Mal	6.0 ± 0.57	27
glucose-starved + 10 μg/ml CB	1.4 ± 0.22	6
glucose-starved + 10 μg/ml CB, reversal	21.6 ± 0.90	96
glucose-starved + GS-Mal + CB, reversal	4.8 ± 0.10	21
"Trapping" control	2.7 ± 0.67	
Expt. 3		
glucose-starved	9.6 ± 0.31	100
glucose-starved + 10 mM GS-Mal	2.7 ± 0.03	28
glucose-starved + 10 μg/ml phloretin	2.0 ± 0.23	14
glucose-starved + 10 μg/ml phloretin, reversal	5.4 ± 0.47	56
glucose-starved + GS-Mal + phloretin, reversal	3.1 ± 0.13	32
"Trapping" control	2.2 ± 0.20	

*Glucose-starved cells (24 h) were incubated in 50 mM sodium phosphate buffer, pH 6.8, containing 100 mM NaCl with or without GS-Mal. Glucose, cytochalasin B, and phloretin were made up double-strength in pH 6.8 buffer and combined with twice-concentrated GS-Mal to give the appropriate final concentration of each compound. In the case of experiments involving cytochalasin B and phloretin, all incubations were carried out in the presence of 0.5% DMSO. After treatment with GS-Mal and/or "protective" agents for 30 min, the cells were rinsed twice with PBS and incubated in PBS for 30 min before a 2 min assay for uptake of 2-dG and α-meAIB. In general, the rate of amino acid transport was not affected by GS-Mal, glucose, cytochalasin B, or phloretin. Values for "trapping" were determined by adding the radioactive assay mixture to rinsed cells on ice and immediately removing this solution.

centrations of GS-Mal might achieve as dramatic an inhibition of 2-dG transport as 20 mM GS-Mal if the duration of treatment were extended. It was found that if cells are exposed to 1 or 2 mM GS-Mal for 60 min, glucose uptake is reduced by greater than 60% or 85%, respectively (see Table III for GS-Mal at 2 mM). Although 20 mM GS-Mal occasionally affects α-meAIB transport, 2 mM GS-Mal never inhibits this transport system. Uridine transport is also unaffected by 2 mM GS-Mal, but preliminary results indicate that leucine, an L-system substrate, may be inhibited.

Experiments were carried out to determine whether substances known to compete with glucose for the hexose-binding site could protect against inhibition by GS-Mal. Since 30 min treatment with 20 mM GS-Mal resulted in maximal inhibition of 2-dG transport,

TABLE III. Effect of D-Glucose and L-Glucose on the Inhibition of Hexose Transport by 2 mM GS-Mal*

Treatment of starved cells	Uptake 2-dG (nmoles/mg protein/min)	% of control
control	4.7 ± 0.20	100
2 mM GS-Mal	2.0 ± 0.06	27
2 mM D-glucose	5.2 ± 0.55	114
GS-Mal + 2 mM D-glucose	4.7 ± 0.10	100
2 mM L-glucose	5.2 ± 0.45	114
GS-Mal + 2 mM L-glucose	2.4 ± 0.60	38
50 mM D-glucose	5.1 ± 0.50	111
GS-Mal + 50 mM D-glucose	5.0 ± 0.10	108
50 mM L-glucose	4.9 ± 0.20	105
GS-Mal + 50 mM L-glucose	1.8 ± 0.25	22
"Trapping" control	1.0 ± 0.05	

*Cells were starved for both serum and glucose for 24 h prior to treatment with GS-Mal in the presence or absence of D- or L-glucose for 60 min. After 2 rinses in PBS, the cells were incubated in PBS for 30 min and then uptake of 2-dG and α-meAIB measured over 2 min. The rate of amino acid transport was constant under all these conditions.

this condition was selected for "protection" experiments. D-Glucose, cytochalasin B, and phloretin were added alone or together with GS-Mal to CEF and incubated for 30 min at 37°C. After 2 washes, the cells were incubated in PBS for another 30 min before the measurement of transport was carried out. The results in Table II show that even 50 or 100 mM D-glucose do not alleviate the inhibitory effect of GS-Mal. Maltose (10 mM) was also tested; it failed to protect against GS-Mal inhibition, but this finding was later explained by the fact that maltose does not compete with 2-dG for uptake by CEF (data not shown). Cytochalasin B (10 μg/ml) inhibits 2-dG uptake in glucose-starved CEF nearly 100%. The inhibition is almost completely reversed 30 min following removal of cytochalasin B from the incubation medium. Although α-meAIB uptake also appears to be reduced in this experiment, in most cases amino acid transport was unaffected in cells treated with cytochalasin B. When cytochalasin B is included in the buffer containing GS-Mal, the rate of sugar transport remains as low as in cells treated with GS-Mal alone after both inhibitors are removed. Similar results were obtained when 10 μg/ml phloretin was added simultaneously with GS-Mal in an attempt to protect the hexose transport component.

The possibility that D-glucose might protect against inhibition of low amounts of GS-Mal was tested and the results are shown in Table III. When D-glucose is present during 60 min treatment of CEF with 2 mM GS-Mal and both glucose and inhibitor are removed for 30 min, the rate of sugar uptake is substantially higher than that in cells exposed to GS-Mal alone. The protective effect of D-glucose ranged from 30 to 100% in several experiments. Equimolar (2 mM) concentrations of D-glucose protect as efficiently as extremely high amounts. Table III also shows that L-glucose fails to protect the hexose transport system against GS-Mal-induced inhibition. The possibility that cytochalasin B might block the inhibition of low amounts of GS-Mal was examined. Unlike D-glucose, 10 μg/ml cytochalasin B does not have any effect on the degree of inhibition of 2-dG uptake by 2 mM GS-Mal.

DISCUSSION

The results of our studies on the effects of sulfhydryl-reacting compounds on nutrient transport in CEF are consistent with the hypothesis that membrane sulfhydryl groups are involved in the activation and maintenance of transport components. In agreement with Czech's observations on isolated fat cells (2), diamide, hydrogen peroxide, and methylene blue stimulate nutrient uptake by CEF to a degree comparable to the increase in transport induced by insulin. Avian fibroblasts appear to be more sensitive than fat cells to diamide and hydrogen peroxide since the concentrations of these sulfhydryl oxidants required to optimally stimulate transport in CEF are much lower than those needed to stimulate hexose uptake in fat cells. In fat cells, it has been shown that diamide, hydrogen peroxide, and methylene blue exert their effect on the transport step, rather than on the metabolism of glucose. Previous work indicates that whereas diamide (13) and hydrogen peroxide (14) primarily oxidize intracellular glutathione, methylene blue oxidizes NADH and NADPH (14) which may secondarily oxidize glutathione. Presumably, reduced glutathione maintains membrane sulfhydryls in the reduced state.

A differential in the sensitivity of nutrient transport systems was also underscored in these studies. At relatively high concentrations of diamide, hydrogen peroxide, and methylene blue, 2-dG uptake, but not α-AIB uptake, was substantially inhibited. The cause for this inhibition is not clear, but the hexose transport component is selectively affected. PCMB-S, a hydrophilic sulfhydryl-reacting compound which does not readily penetrate the membrane (15), has a complex effect on nutrient transport in CEF. At 10 μM PCMB-S stimulates both hexose and amino acid transport slightly. With 10- to 100-fold greater concentrations of PCMB-S and/or longer exposure of cells to this compound, hexose uptake is progressively diminished. However, complete inhibition of 2-dG uptake is not achieved even at 10 mM PCMB-S. These results are consistent with the idea expressed by Hare (4) that sulfhydryl groups lying near the exterior face of the membrane are associated with hexose transport. In contrast to hexose transport, A-system amino acid transport is transiently inhibited by PCMB-S after brief exposure to the sulfhydryl reactant, but with longer treatment, the cells resume close to control rates of α-meAIB transport. This recovery is not due to the inactivation of PCMB-S since it simultaneously inhibits 2-dG uptake. These results contrast with those of Hare (4) who found that uptake of phenylalanine by mouse embryo cells was rapidly and irreversibly inhibited by PCMB-S over a period of 1 h. These conflicting observations may be accounted for by differences in the disposition of A- and L-system amino acid transport components in the membrane or differences between CEF and mouse embryo cells. An explanation is being sought by extending uptake determinations to L-system amino acids as well as to cells grown under different culture conditions.

By using the membrane-impermeant GS-Mal, which reacts irreversibly with free sulfhydryl groups lying on or very close to the exofacial surface of the membrane, we have demonstrated the importance of these groups for hexose transport in CEF. The same sulfhydryl groups which react with maleimide are involved in the binding of D-glucose, since this hexose but not L-glucose protects against GS-Mal inhibition. Two competitive inhibitors of glucose transport in CEF, cytochalasin B and phloretin (16, 17), failed to protect against GS-Mal-induced inhibition and thus presumably bind to a site distinct from that which reacts with maleimide. By inference, this site must also be different from that which binds glucose. Perhaps the cytochalasin B- and phloretin-binding site(s) is close enough to the glucose-binding site to sterically hinder the binding of the sugar and thus display competitive kinetics for hexose uptake. However, their interaction with the site is then not so tight as to block the binding of GS-Mal. On the other hand the apparent competitive inhibition of glucose uptake by compounds whose structure does not resemble that of hexoses must be interpreted with caution. In this regard, Plagemann and Erbe (18) in their studies of the activation energies of 2-dG transport in cultured Novikoff cells suggested that cytochalasin B competitively inhibits hexose uptake by altering the fluidity of the lipid associated with this transport system.

It is evident that the hexose transport system in cultured cells differs from that present in human erythrocytes. For example, maltose which is recognized by the glucose transport system in the red blood cell, although it is not transported, protects against GS-Mal-induced inhibition (8). In CEF, maltose does not compete with 2-dG for uptake and does not protect against the inhibitory effect of GS-Mal in fibroblasts. Whereas in the erythrocyte, both glucose and cytochalasin B protect against inhibition of glucose transport induced by high concentrations of GS-Mal (10–20 mM) (8), only glucose protects against the impermeant-maleimide inhibition in CEF and only when the concentration of the inhibitor is reduced to 1–2 mM. In erythrocytes glucose also inhibits the binding of radiolabeled cytochalasin B, whereas in 3T3 cells glucose fails to compete for the cytochalasin B-binding site (19). Finally, evidence from reconstitution studies of red blood cell membranes strongly suggests that the 55,000–60,000 dalton glycopeptides present in zone 4.5 of sodium dodecyl sulfate-polyacrylamide gel electrophoresis (SDS-PAGE) gels are components of the hexose transport system (20, 21). Recent studies by Banjo and Perdue (22) suggest that polypeptides of 75,000 and 95,000 daltons may comprise the transport system in CEF. This conclusion is based on studies of radiolabeled amino acid incorporation into plasma membrane polypeptides of glucose-starved cells. Future studies will be directed toward the identification and characterization of the hexose transport component(s) in CEF by analyzing extracted membrane peptides from cells exposed to radiolabeled GS-Mal in the presence or absence of glucose.

ACKNOWLEDGMENTS

We wish to thank Ms. Elana Kivity for the culturing of cells. This work was supported by a grant MA-5749 from the Medical Research Council of Canada to J.F.P.

REFERENCES

1. Kletzien RF, Perdue JF: J Biol Chem 249:3383, 1974.
2. Czech MP: J Biol Chem 251:1164, 1976.
3. Plagemann PGW, Richey D: Biochim Biophys Acta 344:263, 1974.
4. Hare JD: Arch Biochem Biophys 170:347, 1975.
5. Kwock L, Wallach DFH: Biochim Biophys Acta 352:135, 1974.
6. Kwock L, Wallach DFH, Hefter K: Biochim Biophys Acta 419:93, 1976.
7. Perdue JF: In Nicolau C (ed): "Virus-transformed Cell Membranes." London: Academic Press, 1977.
8. Batt ER, Abbott RE, Schachter D: J Biol Chem 251:7184, 1976.
9. Abbott RE, Schachter D: J Biol Chem 251:7176, 1976.
10. Lowry OH, Rosebrough NJ, Farr AL, Randall RJ: J Biol Chem 193:265, 1951.
11. Tawney PO, Snyder RH, Conger RP, Leibrand KA, Stiteler CM, Williams AH: J Org Chem 26:15, 1961.
12. Raizada MK, Perdue JF: J Biol Chem 251:6445, 1976.
13. Kosower EM, Correa W, Kinon BJ, Kosower NS: Biochim Biophys Acta 264:39, 1972.

14. Jacob HS, Jandl JH: J Biol Chem 241:4243, 1966.
15. Sutherland RM, Rothstein A, Weed RI: J Cell Physiol 69:185, 1967.
16. Kletzien RF, Perdue JF: J Biol Chem 248:711, 1973.
17. Kletzien RF, Perdue JF: J Cell Biol 63:171a, 1974.
18. Plagemann PGW, Erbe J: J Membr Biol 25:381, 1975.
19. Atlas SJ, Lin S: J Cell Physiol 89:751, 1976.
20. Hinkle PC, Kasahara M: J Supramol Struct (Suppl) 1:143, 1977.
21. Kahlenberg A, Zala CA: J Supramol Struct (Suppl) 1:149, 1977.
22. Banjo BB, Perdue JF: J Cell Biol 70:270a, 1976.

Journal of Supramolecular Structure 7:49—59 (1977)
Molecular Aspects of Membrane Transport 413—423

The Inhibitory Effect of the Artificial Electron Donor System, Phenazine Methosulfate-Ascorbate, on Bacterial Transport Mechanisms

R. G. Eagon, B. D. Gitter, and J. J. Rowe

Department of Microbiology, University of Georgia, Athens, Georgia 30602

The artificial electron donor system, phenazine methosulfate (PMS)-ascorbate, inhibited active transort of solutes in Pseudomonas aeruginosa irrespective of whether the active transport systems were shock sensitive or shock resistant. N,N,N',N'-tetramethylphenylenediamine could be substituted for PMS but a higher concentration was required. PMS-ascorbate also inhibited active transport in several other bacterial species with the exception of Escherichia coli and of a nonpigmented strain of Serratia marcescens. PMS-ascorbate previously has been shown to energize active transport in isolated membrane vesicles, even those prepared from the same bacterial species in whose intact cells active transport was inhibited. The apparent K_m of glucose active transport in untreated cells of P. aeruginosa was 40 μM while the K_m of glucose transport in cells incubated with PMS-ascorbate was 25 mM, and PMS-ascorbate had no effect on efflux of accumulated glucose. These results strongly suggested that facilitated diffusion resulted upon exposure of the cells to PMS-ascorbate. Thus, PMS-ascorbate appeared to have an uncoupler-like effect on cells of P. aeruginosa. The experimental data also pointed out that there are fundamental differences between the response of intact cells and membrane vesicles to exogenous electron donors.

Key words: pseudomonas, transport, phenazinemethosulfate

In attempting to apply the experimental procedure of Berger (1) and Berger and Heppel (2) to Pseudomonas aeruginosa, we noted that active transport of a variety of substrates was inhibited by phenazine methosulfate (PMS)-ascorbate irrespective of whether the cells were normal or starved and irrespective of whether the active transport systems were shock sensitive or shock resistant. This was curious because the artificial electron donor system of PMS-ascorbate energizes active transport by membrane vesicles prepared from P. aeruginosa (3, 4) and from a wide variety of other bacterial species including E. coli (5), and because PMS-ascorbate energizes both shock-sensitive and shock-resistant active transport systems in starved cells of E. coli (2). Thus, the work presented herein was carried out in an effort to characterize and elucidate the inhibitory effect of PMS-ascorbate on active transport systems of intact cells of P. aeruginosa. Evidence will also be presented showing that active transport systems of bacterial species in addition to P. aeruginosa are inhibited similarly by PMS-ascorbate.

Received March 14, 1977; accepted July 1, 1977.

MATERIALS AND METHODS

Cultivation of Organisms

The organisms used for these studies were: P. aeruginosa PAO (formerly Holloway strain 1); P. aeruginosa PAO 57, a mutant derived from PAO which was unable to grow on glucose but which formed the inducible glucose transport system when grown in media containing pyruvate plus glucose (3); P. aeruginosa OSU 64; P. aeruginosa A1466; P. fluorescens ATCC 13525; P. putida ATCC 12633; Escherichia coli ML 308-225; Enterobacter aerogenes UGA; Aeromonas hydrophila (an aerogenic strain); A. hydrophila (an anaerogenic strain); Serratia marcescens Sm7 (a pigmented strain); S. marcescens Sm39 (a nonpigmented strain); Bacillus subtilis SP491; and Staphylococcus aureus 832S-Y.

The strains of S. marcescens and P. aeruginosa A1466 were obtained from Dr. J. J. Farmer, Center for Disease Control, Atlanta, Georgia. The A. hydrophila strains were obtained from Dr. E. B. Shotts, College of Veterinary Medicine, University of Georgia. All other bacterial species and strains were from the culture collection of the Department of Microbiology, University of Georgia.

The Pseudomonas species were cultivated in a chemically defined basal medium and under the cultural conditions previously defined (3, 6). The final concentration of the carbon source was 11 mM glucose for all species except P. aeruginosa PAO 57. The latter was cultivated in 30 mM pyruvate plus 5 mM glucose. All other bacterial species, except S. aureus, were cultivated in the basal salts medium described by Tanaka, Lerner, and Lin (7) to which was added 11 mM glucose in final concentration and, in the case of B. subtilis, S. marcescens, and A. hydrophila, proline was also added to a final concentration of 87 μM. S. aureus was cultivated in tryptic soy broth (Difco Laboratories, Detroit, Michigan).

The basal salts solutions (200 ml) were dispensed in 1-liter Erlenmeyer flasks and sterilized by autoclaving. Concentrated, filter-sterilized solutions of the carbon sources were added aseptically to the sterilized basal salts solutions. In the case of S. aureus, tryptic soy broth was used instead and the flasks were sterilized by autoclaving. Each flask was inoculated with organisms washed from an agar slant composed of the same medium and incubated on a rotary shaker at 30 or 37°C according to the species being cultivated. The cells were allowed to grow to late exponential phase, harvested by centifuging, washed twice in basal salts solution, and then suspended in the basal salts solution to a density of 1 g wet weight per 20 ml. The suspensions were used directly for transport assays.

Transport Assays

The accumulation of radioactivity from [^{14}C]glucose, [^{14}C]gluconate, and [^{14}C]-proline was determined at 30 or 37°C on a reciprocal shaking water bath. Incubation mixtures were held in 10-ml Erlenmeyer flasks and, in final concentration as appropriate, they consisted of: 0.2 ml of cell suspension, 0.1 mM [^{14}C]glucose (7.88 mCi/mmol), 0.1 mM [^{14}C]gluconate (39 mCi/mmol), 10 μM [^{14}C]proline (20 mCi/mmol), 20 mM ascorbate, 140 μM PMS, and basal salts solution to a final volume of 1 ml. The cells were normally preincubated for 10 min in PMS-ascorbate when these reagents were used.

The reactions were started by the addition of substrate. At time intervals, 50-μl samples were withdrawn and delivered over a membrane filter (Millipore Corporation, Bedford, Massachusetts, 25-mm diameter, 0.45-μm pore size) previously overlayed with 1 ml of 0.1 M LiCl, filtered instantaneously, and immediately washed with an additional 5 ml of 0.1 M LiCl.

The filters bearing the cells were removed immediately from the suction apparatus and transferred to vials containing 10 ml of scintillation fluid, and radioactivity was determined as previously described (3, 6).

Osmotic Shock

Cells of A. hydrophila were harvested, washed, and subjected to the cold osmotic shock procedure as described by Nossal and Heppel (8). Cells of P. aeruginosa were subjected to the cold shock and osmotic shock techniques according to the procedure of Gilleland and Murray (9).

Efflux Experiments

The rate efflux of [^{14}C] glucose from P. aeruginosa PAO 57 was measured according to the technique of Lagarde, Pouysségur, and Stoeber (10).

Protein Determination

Protein was measured by the modified buiret procedure as described by King (11).

Chemicals

[U-^{14}C] Glucose and [U-^{14}C] potassium gluconate were purchased from Amersham-Searle (Arlington Heights, Illinois). [U-^{14}C] Proline was obtained from New England Nuclear Corporation (Boston, Massachusetts). PMS and N,N,N',N'-tetramethylphenylenediamine dihydrochloride (TMPD) were purchased from the Sigma Chemical Company (St. Louis, Missouri). All other reagents were purchased from commercial sources in the highest state of purity.

The PMS solutions used in these studies were prepared at frequent intervals. They were stored for short intervals at 0–4°C in flasks wrapped in foil to exclude light.

RESULTS

Inhibitory Effect of PMS-Ascorbate on Active Transport Systems

Transport of proline by P. aeruginosa was extensively inhibited by PMS-ascorbate (Fig. 1). Data from several experiments consistently established that both the rate and extent of proline transport by P. aeruginosa was inhibited 85–90% by PMS-ascorbate. In contrast, PMS-ascorbate stimulated active transport of proline by E. coli (data not shown) and the effect ranged from 10% stimulation of the rate of transport to double the rate.

Active transport of proline by E. coli is a shock-resistant system (2) but it has not been established whether P. aeruginosa transports proline by a shock-resistant or shock-sensitive system. Active transport of glucose by P. aeruginosa, on the other hand, apparently is a shock-sensitive system because a glucose-binding protein has been detected (12). and because membrane vesicles prepared from P. aeruginosa were unable to actively transport glucose but they retained the ability to transport gluconate (3,4) and serine (unpublished observations). On the other hand, gluconate transport by P. aeruginosa apparently is shock resistant because membrane vesicles retained the ability to actively transport gluconate (3, 4). When PMS-ascorbate was tested against these 2 transport systems of P. aeruginosa, both glucose and gluconate transport were found to be as extensively inhibited as proline transport. Results for the PMS-ascorbate inhibition of glucose transport are

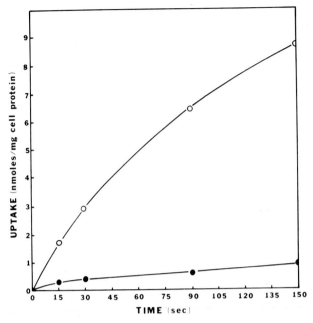

Fig. 1. Uptake of [^{14}C] proline by P. aeruginosa PAO in the absence (open circles) and presence (filled circles) of PMS-ascorbate.

shown in Fig. 2. Thus, both shock-resistant and shock-sensitive transport systems of P. aeruginosa were clearly inhibited by PMS-ascorbate.

Data in Fig. 2. also show that ascorbate had no effect on glucose transport by P. aeruginosa, that PMS alone inhibited glucose transport by about 50%, and that the combination of PMS-ascorbate was the most effective inhibitor. We interpreted these data to indicate that the reduced form of PMS was responsible for the inhibitory effect. We envision that cellular metabolites could reduce PMS when the latter was used alone, but PMS was reduced more effectively when ascorbate was added to the system.

The effect of PMS-ascorbate on the active transport of proline and glucose by a variety of bacterial species is shown in Table I. PMS-ascorbate inhibited proline transport by all species tested except E. coli and S. marcescens Sm39, a nonpigmented isolate. Glucose transport was inhibited in all species tested. These results are especially interesting because PMS-ascorbate has been demonstrated to energize active transport by membrane vesicles prepared from many of these same microorganisms [e.g., P. aeruginosa (3, 4), P. putida, E. coli, S. typhimurium, B. subtilis, and S. aureus (5)].

Inhibitory Concentration of PMS

While maintaining ascorbate at 20 mM, various concentrations of PMS were tested to determine the minimal level required to inhibit active transport of glucose by P. aeruginosa. The experimental results indicated that as little as 7 μM PMS had a pronounced inhibitory effect while the maximal inhibitory effect of PMS was attained at 40 μM (data not shown).

Inhibitory Effect of TMPD

Experiments were done to determine whether PMS could be replaced by TMPD, an alternate artificial electron carrier. It was found that TMPD-ascorbate inhibited glucose transport by P. aeruginosa, but a higher concentration of TMPD than of PMS was required for maximal inhibition (300 μM TMPD vs 40 μM PMS), and TMPD-ascorbate did not inhibit the rate and extent of glucose transport as effectively as PMS-ascorbate (data not shown).

Immediate Inhibitory Effect of PMS-Ascorbate

Incubation mixtures normally were preincubated with PMS-ascorbate for 10 min prior to starting the reaction by the addition of labeled substrate. In order to determine whether the inhibitory effect of PMS-ascorbate was immediate or delayed, we added PMS-ascorbate to incubation mixtures at time zero and after 30 sec of incubation. The results, which are shown in Fig. 3, indicated that the inhibitory effect of PMS-ascorbate was immediate, i.e., no demonstrable lag in inhibition was noted from the time of addition of PMS-ascorbate until the onset of inhibition.

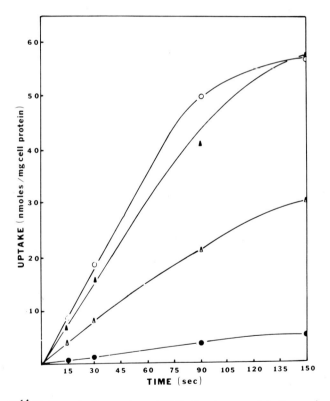

Fig. 2. Uptake of [^{14}C]glucose by P. aeruginosa PAO in the absence and in the presence of PMS, of ascorbate, and of PMS-ascorbate. ○) No additions; ▲) 20 mM ascorbate; △) 140 μM PMS; ●) 140 μM PMS plus 20 mM ascorbate.

TABLE I. Effect of PMS-Ascorbate on Transport of Glucose and Proline in Several Bacterial Species[a]

| | Percent inhibition | | | |
| | Rate of transport | | Total uptake | |
Microorganism	Proline	Glucose	Proline	Glucose
P. aeruginosa PAO	85	90	89	94
P. aeruginosa OSU 64	ND	89	ND	87
P. aeruginosa A1466	ND	82	ND	74
P. putida ATTCC 12633	ND	88	ND	89
P. fluorescens ATCC 13525	ND	65	ND	90
E. coli ML 308-225	Stimulation	50	Stimulation	50
S. marcescens Sm7 (pigmented)	58	ND	42	ND
S. marcescens Sm39 (nonpigmented)	0	ND	15	ND
A. hydrophila (aerogenic)	97	ND	98	ND
A. hydrophila (anaerogenic)	91	ND	98	ND
E. aerogenes UGA	50	ND	52	ND
B. subtilis SP491	57	ND	55	ND
S. aureus 832S-7	43	ND	57	ND

[a]Rate of transport was determined from the linear portion of the uptake curve while total uptake represents the steady state level. ND − not done.

Effect of PMS-Ascorbate on Growth of P. aeruginosa

When PMS and ascorbate (final concentration 14 μM and 20 mM, respectively) were added to the culture medium, a marked increase in the lag phase was observed (Fig. 4). There was also a slightly reduced growth rate after growth was initiated. These data are consistent with the observation that glucose uptake is inhibited by PMS-ascorbate. The results also suggest that growth occurred after the ascorbate had been oxidized and/or after the PMS had been degraded. Moreover, when the concentration of PMS in the culture medium was increased 10-fold, no growth occurred.

Effect of PMS-Ascorbate on Kinetics of Glucose Transport

The apparent K_m for glucose transport by P. aeruginosa PAO 57 in the absence of PMS-ascorbate was 40 μM (Fig. 5). In the presence of PMS-ascorbate, however, the apparent K_m was increased to 25 mM (Fig. 6). The nearly 2,000-fold increase in K_m as a direct effect of PMS-ascorbate suggests that, in the presence of this artificial electron donor system, the active transport system for glucose was changed to a facilitated diffusion system.

Lack of Effect by PMS-Ascorbate on Glucose Efflux

Using P. aeruginosa PAO 57, the rate of efflux of glucose from preloaded cells was measured in the presence and absence of PMS-ascorbate. The rate of efflux of glucose was identical in the 2 systems (data not shown). Thus, since PMS-ascorbate did not effect the rate of efflux of glucose, this indicated that PMS-ascorbate did not alter the permeability of the cytoplasmic membrane to glucose nor change the affinity of the carrier for glucose on the inside of the cytoplasmic membrane. These data are also consistent with the observation that facilitated diffusion resulted on exposure to PMS-ascorbate.

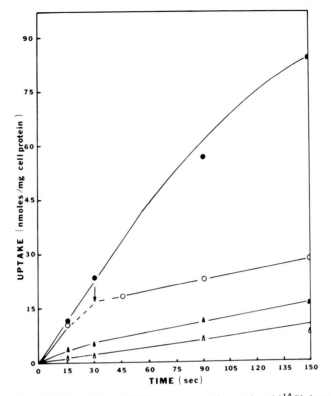

Fig. 3. The immediate inhibitory effect of PMS-ascorbate on the uptake of [^{14}C] glucose by P. aeruginosa PAO. ●) No additions; △) cells were incubated with PMS-ascorbate for 10 min prior to starting the reaction by addition of substrate; ▲) PMS-ascorbate was added to the cells simultaneously with substrate to start the reaction; ○) PMS-ascorbate was added 30 sec after the start of the reaction as indicated by the arrow.

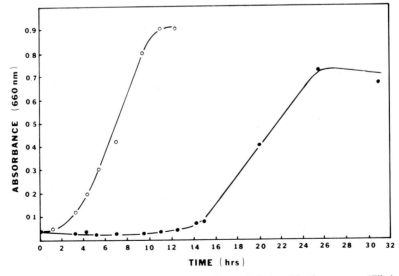

Fig. 4. Growth of P. aeruginosa PAO in the absence (open circles) and in the presence (filled circles) of PMS-ascorbate.

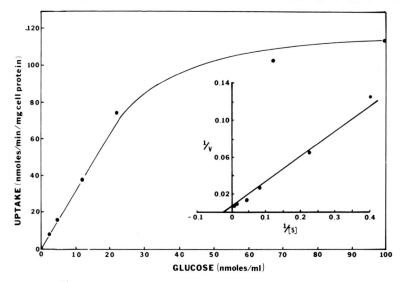

Fig. 5. Kinetics of [^{14}C] glucose uptake in the absence of PMS-ascorbate by P. aeruginosa PAO 57 as a function of exogenous substrate concentration. The reactions were initiated by the addition of cells. Initial rates were approximated by measuring glucose uptake after incubation for 15 sec. The inset is a Lineweaver-Burk plot of the kinetic data.

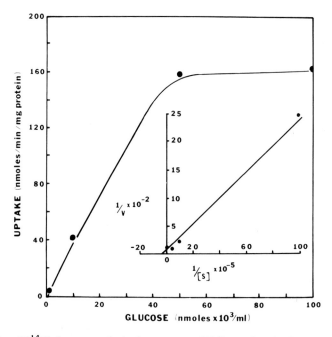

Fig. 6. Kinetics of [^{14}C] glucose uptake in the presence of PMS-ascorbate by P. aeruginosa PAO 57 as a function of exogenous substrate concentration. The reactions were initiated by addition of cells. Initial rates were approximated by measuring glucose uptake after incubation for 15 sec. The inset is a Lineweaver-Burk plot of the kinetic data.

Osmotic Shock Experiments

Active transport by membrane vesicles prepared from P. aeruginosa (3, 4) and from a wide variety of other bacteria (5) is energized by PMS-ascorbate. Thus, in order to determine whether there was a shock-sensitive component responsible for PMS-ascorbate sensitivity, we subjected A. hydrophila to the cold osmotic procedure as described by Nossal and Heppel (8), and P. aeruginosa was subjected to the cold shock and osmotic shock procedures reported by Gilleland and Murray (9). These shocked bacteria exhibited unchanged sensitivity to PMS-ascorbate with respect to proline transport (data not shown), indicating that PMS-ascorbate sensitivity was not due to a shock releasable component.

Effect of PMS-Ascorbate on the Phosphoenopyruvate (PEP) Phosphotransferase System

A curious phenomenon is that PMS-ascorbate inhibited glucose uptake by E. coli (Table I, Fig. 7). Glucose is taken up by E. coli by group translocation via the PEP phosphotransferase system and not by active transport (13). Thus, PMS-ascorbate inhibition of this system was unexpected. We interpreted these data, however, to indicate that PMS-ascorbate stimulated respiration and that the PEP:glucose phosphotransferase system was inhibited by respiration. Other authors similarly have concluded that respiration, or an energized state of the membrane, inhibits glucoside transport by E. coli (14–16).

DISCUSSION

Our experimental results showed that the artificial electron donor system of PMS-ascorbate (and of TMPD-ascorbate as well) inhibited active transport of solutes in a variety of bacterial species. E. coli was a notable exception since PMS-ascorbate stimulated uptake

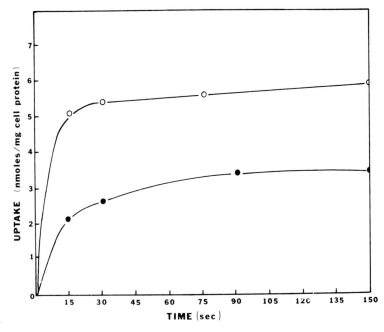

Fig. 7. Uptake of [^{14}C]glucose by E. coli ML 308-225 in the absence (open circles) and presence (filled circles) of PMS-ascorbate. Chloramphenicol (80 μg/ml) was added to the reaction mixtures.

of proline instead. Similarly, other authors have reported that glutamine and serine uptake by cells of E. coli was also stimulated by PMS-ascorbate (2). Why active transport systems in cells of E. coli should respond differently than other bacterial species is not immediately clear.

Another curious phenomenon which became evident from these studies was that intact cells responded differently to the artificial electron donor system, PMS-ascorbate, than membrane vesicles isolated from the same bacterial species (3, 5, 17, 18), and to TMPD-ascorbate, which has been shown to be a satisfactory substitute for PMS (19, 20). Thus, on the one hand, PMS-ascorbate energizes active transport in isolated membrane vesicles while, on the other hand, PMS-ascorbate inhibits active transport by intact cells from which the membrane vesicles were prepared.

Other workers have noted that exogenous electron donors such as pyruvate, succinate, and glycerol (14, 15) and $\underline{\underline{D}}$-lactate (16) inhibited α-methylglucoside (α-MG) uptake by cells of E. coli. α-MG is taken up in E. coli by group translocation via the PEP:glucose phosphotransferase system and not by active transport (13). It was concluded that α-MG uptake was inhibited by the energized membrane state resulting from electron flow (14–16). One group of workers argued against a role of ATP in formation of the energized membrane state (14, 15) while other workers presented evidence that the energized membrane state could also be formed from adenosine triphosphate (ATP) hydrolysis (16). The latter conclusion was based on the observation that galactose inhibited α-MG uptake in E. coli, presumably by formation of ATP via substrate-level phosphorylation. Be that as it may, it is probably that PMS-ascorbate inhibited glucose uptake by E, coli in a man ner analogous to the inhibition of α MG uptake which resulted form the presence of exogenous natural electron donors as reported by other workers (14–16).

By analogy to the inhibition of the E. coli PEP:glucose phosphotransferase system by the energized membrane as discussed above, at first thought it might be suspected that PMS-ascorbate, or TMPD-ascorbate, inhibited active transport of external solutes in P. aeruginosa and other species by formation of an energized membrane state. We think that this cannot be the case, however, because according to current concepts active transport is driven by the energized membrane state (19, 21–24). Thus, it is not likely that the energized membrane could function both to energize and to inhibit active transport.

Similarly, we do not believe that PMS-ascorbate had a specific effect on specific protein carriers because of the generalized effect of electron donors other than PMS ascorbate (i.e., TMPD-ascorbate), and because transport mechanisms in general were inhibited in intact cells by these exogenous electron donors.

It has been shown by Ramos, Schuldiner, and Kaback (23) that at pH 7.5 the electrochemical potential of protons across the membrane of isolated vesicles of E. coli was due almost solely to the electric potential across the membrane and that internal pH was nearly equal to external pH. Thus, since our transport measurements were conducted at pH 7, we presume that the various effects that we noted were due chiefly to the electric potential (or perhaps to its absence in the presence of PMS-ascorbate) across the cytoplasmic membrane and not to a pH gradient.

A distinct possibility in explanation of our observations is that PMS-ascorbate may have fed electrons into the electron transport chain distal to an energy-coupling site for active transport which has been postulated by Kaback and co-workers (24). Electron donors that have been shown to energize active transport in membrane vesicles characteristically donate electrons to the electron transport chain at the cytochrome b or cytochrome c level (3, 17, 19, 20). Thus, if PMS-ascorbate donated electrons distal to the energy-coupling site for active transport, or distal to the terminal energy conservation site, it can be envisioned that the flow of electrons anterior to this site(s) would be impeded. This, in turn, would prevent the formation of the energized membrane and active transport, but not facilitated diffusion, would be inhibited. In support of this hypothesis is our recent observation that PMS-ascorbate has an uncoupler-like effect on P. aeruginosa because, upon exposure to PMS-ascorbate, there was a rapid depletion of intracellular ATP (unpublished preliminary observations).

The difference in response between intact cells and membrane vesicles of P. aeruginosa and other species to PMS-ascorbate cannot be explained at the present time. [The exception is E. coli since intact cells and membrane vesicles appear to respond similarly to exogenous electron donors and membrane vesicles are considered to retain the same orientation as the membrane in the intact cell (23, 24).] PMS-ascorbate, however, energizes active transport in isolated membrane vesicles, and the energy-coupling sites and components of the electron transport chain would be expected to be the same in membrane vesicles as in intact cells. Membrane vesicles, however, are perturbed systems. The possibility exists therefore, that certain electron transport components that are involved in the PMS-ascorbate effect are lost when membrane vesicles are prepared; or, that the electron transport components in isolated membrane vesicles are perturbed in such a manner that the effects of PMS-ascorbate are overcome. An explanation will have to await further experimentation to compare such factors as the electron transport components of intact cells vs membrane vesicles. Similarly, comparative studies between E. coli and other species, such as P. aeruginosa, will also be required.

ACKNOWLEDGMENTS

This investigation was supported in part by Research Grant BMS74-14819 from the National Science Foundation.

REFERENCES

1. Berger EA: Proc Natl Acad Sci USA 70:1514, 1973.
2. Berger EA, Heppel LA: J Biol Chem 249:7747, 1974.
3. Guymon LR, Eagon RG: J Bacteriol 117:1261, 1974.
4. Stinnett JD, Guymon LF, Eagon RG: Biochem Biophys Res Commun 52:284, 1973.
5. Konings WN, Barnes EM, Kaback HR: J Biol Chem 246:5847, 1971.
6. Phibbs PV Jr, Eagon RG: Arch Biochem Biophys 138:470, 1970.
7. Tanaka S, Lerner SA, Lin ECC: J Bacteriol 117:1055, 1967.
8. Nossal WG, Heppel LA: J Biol Chem 241:3055, 1966.
9. Gilleland HE Jr, Murray RGE: J Bacteriol 125:267, 1976.
10. Lagarde AE, Pouysségur JM, Stoeber FR: Eur J Biochem 36:328, 1973.
11. King TE: In Wood WA (ed): "Methods in Enzymology." New York: Academic Press, 1966, vol 9, pp 98–103.
12. Stinson MW, Cohen MA, Merrick JM: J Bacteriol 128:573, 1976.
13. Roseman S: J Gen Physiol 54:138s, 1969.
14. del Campo FF, Hernández-Asenio M, Ramírez JM: Biochem Biophys Res Commun 63:1099, 1975.
15. Hernández-Asenio M, Ramírez JM, del Campo FF: Arch Microbiol 103:155, 1975.
16. Singh AP, Bragg PD: FEBS Lett 64:169, 1976.
17. Barnes EM: Arch Biochem Biophys 152:795, 1972.
18. Johnson CL, Cha Y-A, Stern JR: J Bacteriol 121:682, 1975.
19. Kabach HR: Science 186:882, 1974.
20. Thompson J, MacLeod RA: J Bacteriol 117:1055, 1974.
21. Harold FM: Bacteriol Rev 36:172, 1972.
22. Mitchell P: J Bioenerg 4:63, 1973.
23. Ramos S, Schuldiner S, Kaback HR: Proc Natl Acad Sci USA 73:1892, 1976.
24. Stroobant P, Kaback HR: Proc Natl Acad Sci USA 72:3970, 1975.

Journal of Supramolecular Structure 7:61–77 (1977)
Molecular Aspects of Membrane Transport 425–441

Nutrient Uptake and Control of Animal Cell Proliferation

Gregory S. Barsh and Dennis D. Cunningham

Department of Medical Microbiology, California College of Medicine, University of California at Irvine, Irvine, California 92717

The division of fibroblast-like cells in culture can be regulated by cell density, serum, and various growth factors. This system has been widely utilized as a model to study the regulation of cell proliferation. There are many physiological and metabolic changes that correlate with the proliferative state of the cell. These include changes in morphology, cyclic nucleotide levels, enzyme activities, and certain cell surface properties such as nutrient uptake and chemical composition of the plasma membrane. Of primary concern is determination of which changes might be critical links in the control of cell proliferation and which ones are simply correlated but not causally involved with cell growth. We have discussed evidence which has strongly suggested a fundamental role for uptake of certain nutrients in the regulation of cell growth. In addition, we have presented several methods allowing a critical analysis of a putative cause and effect relationship between nutrient uptake and growth control. One method involves a dose-response study of the effect of a mitogen on uptake and DNA synthesis, while a second method involves search for a particular mitogen that may, under the appropriate conditions, stimulate cell division without stimulating uptake. These two methods are limited, however, since they are not always applicable to any given nutrient or mitogen. A third method which is not limited in its applications involves varying the concentration of a particular nutrient in the medium to control its uptake. In the case of orthophosphate (P_i) or glucose, we have used this "nutrient concentration" method to demonstrate that under normal culture conditions, uptake of these nutrients is not a causal event in the regulation of cell division.

We considered the possibility that intracellular nutrient availability might control cell growth, even under conditions where uptake did not. For P_i and glucose, we assumed intracellular pool size to be an accurate indicator of intracellular nutrient availability and measured these pools under a variety of proliferative conditions. These studies revealed, however, no correlation between pool size and proliferative state of the cells. This clearly demonstrates that for P_i and glucose, intracellular pool sizes are not causally involved in the control of growth. The possibility remains, however, that if these nutrients are compartmentalized within the cell, intracellular pool sizes may not be an accurate indicator of nutrient availability.

For P_i and glucose there are many interesting questions that remain to be answered about the transport mechanisms for these nutrients. For some other nutrients, particularly K^+ and amino acids, in addition to questions dealing with the nature of transport mechanisms, the question of uptake involvement in the control of proliferation remains entirely open. As with P_i and glucose, many observations strongly suggest a fundamental relationship between amino acid or K^+ uptake and control of cell growth. We suggest that the "nutrient concentration" technique used

Received April 20, 1977; accepted June 6, 1977.

in our studies to analyze P_i and glucose uptake is applicable to any nutrient and should, therefore, prove extremely useful for studying the involvement of any uptake change in the regulation of cell proliferation.

Key words: cell culture, growth control, glucose uptake, phosphate uptake

Numerous observations made over the past several years with a variety of different cell types and nutrients have suggested the possibility of a causal relationship between membrane transport[1] and control of cell division. The purpose of the present review is to delineate some of these observations and to point out methods allowing a careful dissection of this relationship.

There is an extensive literature dealing with the general relationship of cell surface properties to neoplastic transformation and regulation of proliferation in so-called normal cells. This subject is treated comprehensively in several recent reviews and conference proceedings (1–9). In this review, we will focus on the putative role of nutrient uptake in the growth of "normal" fibroblast-like cells.

The first part of this article will deal primarily with changes in the uptake of P_i and glucose. The relationship between cell division and changes in uptake of these nutrients is particularly well documented, and experiments have been carried out allowing a careful analysis of this relationship. The second part of this article will summarize the correlations observed between cell division and changes in uptake of nutrients other than P_i or glucose.

BACKGROUND

One line of reasoning which has led investigators to consider nutrient uptake as a key regulator of cell division is based upon the temporal and spatial considerations inherent in models of growth regulation. Stimulation of division of stationary phase cells[2] requires at least several hours from the time of serum addition until a significant increase is detected in the rate of DNA synthesis as measured by thymidine incorporation (10, 11). In addition, it seems likely that serum factors might act at the cell surface. Thus, a primary signal in the stimulation of DNA synthesis might be expected to 1) occur rapidly after the addition of serum, and 2) provide a means of transferring the signal from the cell surface to the cytoplasm and nucleus.

It was suggested as early as 1964 that changes in permeability of the cell membrane might be related to the control of cell division (12). Experimental confirmation of this hypothesis was shown for the uptake of P_i and uridine by 3T3 cells (13). We observed that uptake of these nutrients was roughly proportional to the rate of cell division. Further-

---

[1] In using the word "transport," we denote specifically and exclusively the rate of passage of a molecule across the plasma membrane. We will use the word "uptake" to denote experimental measurements of the initial rate of isotope accumulation into the acid-soluble fraction of the cell. Differences between transport and uptake will be further discussed in a later section.

[2] We will use the terms "stationary phase," "quiescent," and "arrested" interchangeably. In doing so, we refer to the reduction in rate of cell division and concurrent entrance into G_1 (or G_0) phase of the cell cycle observed in most nontransformed fibroblast-like cells.

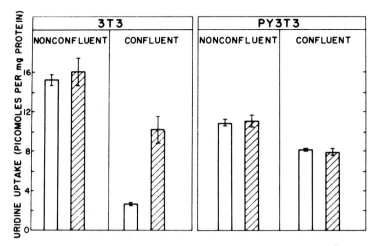

Fig. 1. Uridine uptake by 3T3 and polyoma virus-transformed 3T3 (Py3T3) cells. [³H] uridine up-take was measured during a 15-min incubation period. Uptake was linear during this time interval. Plain bars) control cultures. Lined bars) uridine uptake durint the 15-min interval immediately after addition of fresh serum to a final concentration of 10%. [Reprinted from Ref. 13, with permission]

more, addition of fresh serum to arrested cells resulted in an early (within 15 min) increase in uptake. These results for uridine are presented in Fig. 1.

Changes in uptake of these nutrients are selective since uptake of some amino acids does not vary with the rate of cell division in 3T3 cells (14), and since addition of fresh serum to quiescent 3T3 cells has no effect on adenosine uptake (13).

These results were extended by the observations that addition of serum or trypsin [which can act as a mitogen (15)] to chick embryo (CE) cells stimulated an early increase in the uptake of 2-deoxyglucose (16). Furthermore, the growth rate of these cells correlated with both uridine (17, 18) and 2-deoxyglucose (16, 18) uptake. An additional observation suggesting a close relationship between chick cell division and 2-deoxyglucose uptake concerned the effect of pH on these parameters (19). In arrested cultures, raising the pH for 24 h resulted in large and parallel increases in the rate of DNA synthesis and 2-deoxyglucose uptake. However, in rapidly dividing, low density cultures, changes in pH had little effect on either parameter (19). Another finding which suggests a close relation-ship between initiation of cell division and early increases in nutrient uptake is the fact that both responses can be elicited by the addition of highly purified mitogenic agents (20, 21).

These and similar observations suggest a general mechanism for growth regulation. It was hypothesized that a decline in nutrient uptake as rapidly growing cells began to approach quiescence would decrease the availability of nutrients inside the cell. This, in turn, would lead to specific arrest of cell growth in G_1 (or G_0). Addition of serum or other mitogens to these stationary phase cells would cause an immediate increase in nutrient uptake, followed by increased nutrient availability inside the cell, eventually leading to initiation of DNA synthesis and cell division (22).

The idea that intracellular nutrient availability might act as a direct signal to initiate or arrest cell proliferation was lent further credence by the observation that Chinese hamster ovary cells could be arrested in G_1 phase by lowering the concentration of isoleucine or

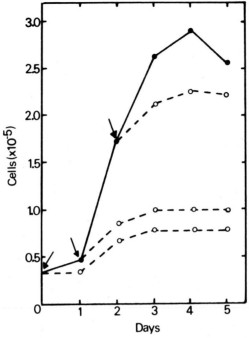

Fig. 2. Effect of reduction of P_i on 3T3 cell number. 3T3 cells were plated at a density of 5.5×10^3 per cm^2 in medium containing 10% serum. Shortly after the cells had attached, the medium was aspirated and replaced by medium containing 10% dialyzed serum and either 1.0 mM P_i or 0.005 mM P_i. On the following 2 days, some of the cultures in 1.0 mM P_i were switched to conditioned medium containing 0.005 M P_i (taken from parallel cultures) at the times indicated by the arrows. •—•) change to 1.0 mM P_i; o - - - o) changed to 0.005 mM P_i. [Reprinted from Ref. 26, with permission.]

glutamine in the medium (23). Addition of isoleucine or glutamine could then synchronously reinitiate cell division (24). Similar observations were reported for P_i and glucose in the control of 3T3 cell proliferation (25). Results from our laboratory (26) confirmed these reports and further demonstrated that arrest of 3T3 cells by lowering P_i in the medium was quantitatively dependent on initial cell density (Fig. 2).

At this point, then, we are left with the following observations: 1) Uptake of P_i, 2-deoxyglucose, and uridine is roughly proportional to the rate of cell division in a variety of cell types. Early stimulation of uptake by serum or highly purified mitogenic agents in arrested cells may fulfill a temporal expectation for a signal involved in the initiation of DNA synthesis. 2) Varying the extracellular concentrations of a variety of nutrients can specifically arrest and subsequently initiate cell division. This may be indicative of a spatial link between changes in transport at the cell surface, and direct control of DNA synthesis from within the cell.

INCREASES IN P_i AND GLUCOSE UPTAKE CAUSED BY ADDITION OF SERUM OR OTHER MITOGENS TO STATIONARY PHASE CULTURES

There are 2 phases of cell proliferation where changes in nutrient uptake could act as a regulatory signal. The first is the decline in proliferation as cells approach a quiescent monolayer; the second is the stimulation of quiescent cells by serum or other mitogens. The stimulation of quiescent cells is often the easier system to study since cell division can be synchronously initiated in almost the entire population.

Do Uptake Changes Reflect Changes in Membrane Transport?

One of the first questions that must be considered is whether or not changes in uptake of P_i or glucose represent actual changes in membrane transport. Serum addition causes a variety of metabolic "posttransport" effects. This raises the possibility that observed changes in P_i or glucose uptake may actually reflect changes in metabolism of these nutrients. For example, in 3T3 cells, incorporation of P_i into phospholipids (27) and total organic phosphate (28) rapidly increases two- to threefold after serum addition. In the case of glucose, the analog 2-deoxyglucose is frequently used as an indicator of transport activity since it is not significantly metabolized beyond the initial phosphorylation step (29). However, the initial phosphorylation is increased in both CE (30) and 3T3 (31) cells following addition of fresh serum. This observation is consistent with the idea that observed changes in glucose uptake may actually be a result of changes in posttransport metabolism (31, 32). This idea is supported by the observation that phosphofructokinase activity in CE cells is stimulated more than 10-fold by the addition of fresh serum (33).

One way to demonstrate that a change in uptake reflects an actual change in membrane transport is to use a nonmetabolizable analog. In the case of glucose, 3-O-methyl glucose, which is not phosphorylated at all (29, 34), has been reported to be transported by the same carrier as glucose and 2-deoxyglucose (29, 35).[3] Using this analog, several investigators have found that serum does stimulate uptake of 3-O-methyl glucose in CE (30) and 3T3 (36) cells.

Another way to demonstrate changes in membrane transport rather than "posttransport" metabolism of nutrients is to measure uptake with membrane vesicles (37). A difference in nutrient uptake by membrane vesicles prepared from cells in different proliferative states is clearly indicative of a transport change. However, the absence of an observable change does not in itself show that changes in transport might not occur in the whole cell since the vesicle isolation procedure could alter the transport properties of the membrane. In the case of P_i, vesicles derived from transformed 3T3 cells have higher transport activity than those derived from normal, quiescent 3T3 cells (38). With glucose, however, no changes in transport were observed between vesicles derived from growing, quiescent, or transformed 3T3 cells (39).

Serum Stimulation of P_i and Glucose Uptake: Quantitative Aspects

The early stimulation of both P_i and glucose uptake by serum addition to quiescent cells raises the question that these changes might be brought about by the same process. This has been termed the pleiotypic response (40, 41). One approach to answering this question has involved a quantitative characterization of uptake activation by serum. In the

---

[3]One disadvantage of using 3-O-methyl glucose as an indicator of glucose transport is that this analog is not "trapped" inside the cell and, therefore, is easily washed out during the rinsing procedure that follows incubation. In contrast, after 2-deoxyglucose is phosphorylated, efflux is very low. This results in an extended "initial rate" period for 2-deoxyglucose, where gross cellular accumulation of radioactivity remains linear for as long as 10–15 min. In the case of 3-O-methyl glucose, the initial rate period is usually less than 5 min. This requires extremely short incubation times to obtain an accurate indication of transport.

case of P_i, early uptake stimulation in 3T3 cells appears to be a result of a change in the observed V_{max} (42, 43). Following addition of 25% dialyzed serum to quiescent 3T3 cells, an increase in the rate of P_i uptake can be detected within 5 min (42). The uptake rate continues to increase up to about 10 min after serum addition at which time it remains constant at a level two- to fourfold over the control level (13, 42, 43). This increase appears to be energy dependent since it does not occur in glucose-free medium with oligomycin (28), or in the presence of iodoacetate and cyanide (13). The early increase does not require protein synthesis, however, since it occurs in the presence of 10 μg/ml cycloheximide (42). The early serum stimulation of glucose uptake in 3T3 cells follows a similar pattern to that of P_i, with the exception that glucose uptake appears to be mediated by facilitated diffusion in a variety of cell types and is not energy-dependent (29, 34, 44). Thus, in 3T3 cells, the rate of glucose uptake increases for 30 min to 1 h following serum addition (45, 46), and is not prevented by cycloheximide (46). In CE cells, one laboratory reported a partial inhibition by cycloheximide of early serum-stimulated glucose uptake (16), while another laboratory reported no effect of cycloheximide (30). The reasons for this discrepancy are unclear. In any case, it appears that serum does "induce" very rapid increases in P_i and glucose uptake both of which can be attributed to a change in the observed V_{max}. In some cells, this increase does not require synthesis of new protein, suggesting that the change in V_{max} is not simply a result of new carrier synthesis (30, 46).

The data cited above are consistent with the possibility that the early increase in serum activation of glucose and P_i uptake are brought about by the same process and thus are involved in the postulated pleiotypic response. This concept must be modified since hypotheses based on the pleiotypic response usually invoke one or more cyclic nucleotides as a universal regulator of transport, metabolism, and cell growth (40, 41). However, it has been demonstrated by several methods that early serum-stimulated changes in P_i (42, 43) and glucose (46, 47) uptake are not mediated by changes in cAMP.

Any consideration of the possibility that serum activation of nutrient uptake leads to increased intracellular pool sizes and DNA synthesis must take into account levels of uptake not only immediately after serum addition, but up until the time of commitment to DNA synthesis. Several laboratories have examined these parameters, and although there are some minor discrepancies in their results, it is clear that in 3T3 cells, both P_i and glucose uptake remain elevated for at least 4 and 8 h, respectively (45, 46). In CE cells there is no stimulation of P_i uptake by serum (48), but glucose uptake remains elevated for at least 8 h (16, 30). Some laboratories have reported secondary increases in rates of uptake for P_i in 3T3 cells (46), and for glucose in 3T3 (45, 46), and CE (30) cells. These secondary increases in uptake occur 1−2 h after the first increases in uptake and without exception are dependent on protein synthesis. Thus, it appears that 1) serum activates immediate increases in uptake which are not transient, 2) the total activation may be biphasic in nature with the second phase dependent on protein synthesis, and 3) the similarity of the changes in P_i and glucose uptake is consistent with the idea that these processes may be brought about by a similar mechanism.

Other Mitogenic Agents That Stimulate Uptake

In addition to serum there are other mitogenic agents that stimulate P_i and glucose uptake in several types of cells. This may be fundamentally different from stimulation by serum alone. Trypsin (16, 49, 50) and neuraminidase (51) both stimulate 2-deoxyglucose uptake in CE cells. These enzymes are similar to serum in their effect on the time course of 2-deoxyglucose uptake stimulation. However, levels of uptake stimulated by optimal enzyme concentrations are less than half that of uptake levels stimulated by optimal serum concentrations (50). In addition, mixing experiments suggest that neuraminidase and trypsin stimulate uptake by a common mechanism. However, low levels of trypsin and serum have a synergistic effect on 2-deoxyglucose uptake, suggesting different mechanisms of stimulation by these 2 agents (50).

In 3T3 cells, prostaglandin $F2\alpha$ and insulin both stimulate P_i uptake (52). Prostaglandin $F2\alpha$ has an effect similar to that of serum in terms of both temporal and protein synthesis dependent aspects of uptake stimulation. Insulin, however, appears to be quite different from serum in both of these aspects (52). There are other agents that stimulate uptake in different types of cells. For example, multiplication stimulating activity (MSA) (53) and fibroblast derived growth factor (FDGF) (21) stimulate 2-deoxyglucose uptake in CE and 3T3 cells, respectively. Insulin (51), Zn^{2+}, Mn^{2+}, Cd^{2+}, and the carcinogenic hydrocarbon DMBA (54), all stimulate 2-deoxyglucose uptake in CE cells. At this time, however, it is still unclear how, if at all, uptake stimulation by these agents differs from that by serum alone.

Are Increases in P_i or Glucose Uptake Causal Events in the Initiation of Cell Proliferation?

The primary question that we wish to consider is whether or not these increases in uptake caused by serum or other mitogenic agents are causal events in the initiation of DNA synthesis and cell division. There are several approaches to this problem, each of which answers a subtly different question.

One approach involves a careful analysis of the dose of serum on the response in both DNA synthesis and uptake (36). Data for 2-deoxyglucose uptake in 3T3 cells is shown in Fig. 3. Here, we have shown that 0.5% serum, which maximally stimulates 2-deoxyglucose uptake (circles), has no effect whatsoever on thymidine incorporation (triangles) or cell number (squares). Thus, a maximal increase in 2-deoxyglucose uptake is not sufficient to initiate DNA synthesis or cell division under these conditions.

A second approach to analyze the putative causality between increased uptake and initiation of DNA synthesis involves the use of agents other than serum, that when used under the appropriate conditions, stimulate DNA synthesis but not uptake (36). This directly shows that under the experimental conditions used, increased uptake is not necessary for initiation of DNA synthesis. An example of this approach is shown in Fig. 4 for the stimulation by cortisol of DNA synthesis in 3T3 cells. As shown, levels of cortisol causing a fivefold increase in DNA synthesis actually caused a decrease in 2-deoxyglucose uptake. It should be pointed out that this level of stimulation in DNA synthesis is the same as that caused by the addition of 5% fresh serum and represents a 25% increase in cell number (36). Using a similar technique with trypsin and insulin to stimulate CE cell division, we have shown that an increase in P_i uptake is not necessary for initiation of DNA synthesis (48).

A third approach to determine if a transport increase is necessary for initiation of cell division is very direct and applicable to any nutrient or mitogen. This involves controlling nutrient uptake by adjusting its concentration in the growth medium (26, 48, 55). Specifically, it is possible to lower the nutrient concentration such that its uptake, even after stimulation by serum or another mitogenic agent, remains below uptake in control quiescent cells. If, under these conditions, dialyzed serum initiates cell division to the same extent as in medium containing the "normal" nutrient concentration, we can con-

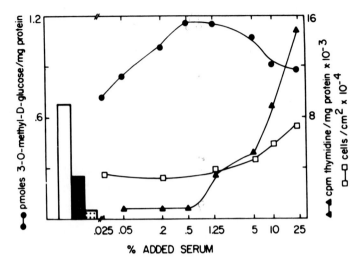

Fig. 3. Effect of concentration of added fresh serum on 3-O-methyl glucose uptake (•——•), DNA synthesis (▲——▲), and 3T3 cell number (□——□). Hexose uptake was measured during a 1-min incubation period 4 h after adding fresh serum to quiescent 3T3 cells. DNA synthesis was measured at 24 h and cell number was monitored at 72 h. Vertical bars show control values for quiescent cultures: plain bar) hexose uptake; solid bar) cell number; stippled bar) DNA synthesis. [Reprinted from Ref. 36, with permission.]

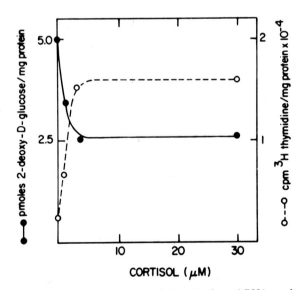

Fig. 4. Effect of cortisol concentration on 2-deoxyglucose uptake and DNA synthesis by quiescent 3T3 cells. Uptake of 2-deoxyglucose (•——•) was measured 4 h after cortisol addition; DNA synthesis (o- - -o) was measured 24 h after cortisol addition. [Reprinted from Ref. 36, with permission.]

clude that increased uptake is not necessary for initiation of cell division. The data in Fig. 5 show the use of this approach to analyze the serum stimulated increase in P_i uptake (48). Quiescent 3T3 cells were stimulated by switching either to medium containing fresh serum and the "normal" amount of P_i (circles), or to medium containing fresh serum and 5% of the "normal" amount of P_i (triangles). The upper panel of Fig. 5 shows that this low level of P_i kept uptake below the level of unstimulated control cultures. When we examined the subsequent effects on cell proliferation (middle panel), we found that this low concentration of P_i did not inhibit the initiation of cell division. Thus, the increase in P_i uptake that occurs under normal culture conditions is not necessary for initiation of cell division. Similar studies on CE and mouse embryo cells have led to the same conclusion (48). We have also used this approach to demonstrate that under normal culture conditions, an increase in glucose uptake is not necessary for initiation of division in 3T3, CE, or human foreskin (HF) cells (55).

Although increased uptake of P_i or glucose was not required for initiation of DNA synthesis, the possibility remained that intracellular nutrient availability (as measured by intracellular pools of P_i or glucose) might still increase under our experimental conditions even though extracellular nutrient concentration and nutrient uptake were below the control or unstimulated level. Increased intracellular availability of P_i or glucose could result from alterations of certain metabolic events. It could also result from decreased rates of P_i or glucose efflux, since serum stimulates influx of P_i more than efflux (13). Thus, it was possible that even though increased uptake of P_i or glucose was not necessary for initiation of proliferation, increased nutrient availability might still be required. When we made direct chemical measurements of intracellular inorganic P_i, however (Fig. 5, lower panel), we found that P_i pool size in the cells stimulated by fresh serum in low P_i media (triangles) was not significantly different from unstimulated control cultures (squares) (48). Moreover, intracellular P_i pool size in the cells stimulated by fresh serum in normal media remained virtually constant. We have also shown that intracellular pool size of glucose did not significantly change after quiescent CE cells in medium containing the usual amount of glucose are stimulated with fresh serum (55). Taken together, these results demonstrate that under normal culture conditions, neither an increase in the uptake of P_i or glucose, nor an increase in the intracellular pools of these nutrients is required for the initiation of cell proliferation.

CHANGES IN P_i AND GLUCOSE UPTAKE CORRELATED WITH GROWTH TO QUIESCENCE

Although changes in nutrient uptake and intracellular pool size were not causal events in the initiation of DNA synthesis in nonproliferating cultures, the possibility remained that the observed decrease in P_i or glucose uptake as proliferating cells began to approach a quiescent monolayer might limit cell division. As mentioned earlier, there is less information available on this aspect of growth regulation since an unsynchronized population of low density cells is difficult to study. Most reports dealing with the involvement of uptake in the control of division in proliferating cells have only dealt with differences in uptake between exponentially growing and stationary phase cultures. Generally, most investigators have found that exponentially growing cells take up P_i and/or glucose at rates three- to fourfold higher than their stationary phase counterparts (13, 16, 35, 56–58). In addition, the change in uptake, as with the serum stimulation of

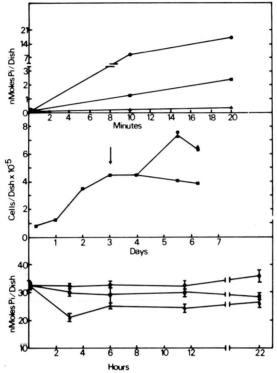

Fig. 5. Effect of reduction of P_i concentration on P_i uptake, initiation of cell division, and pool size of P_i in 3T3 cells. 3T3 cells were seeded at a density of 1.0×10^4 cells/cm^2 in medium containing 10% calf serum. After 3 days, at the time indicated by the arrow (middle panel), the medium was changed to one containing 25% dialyzed calf serum and 1.1 mM P_i (●), or 25% dialyzed calf serum and 0.05 mM P_i (▲). The medium was not changed on control cultures (■). After the medium change, P_i uptake (upper panel), cell number (middle panel), and intracellular pool sizes of P_i (lower panel) were measured at the indicated times. Error bars in pool size measurements designate one standard deviation from the mean. The absence of an error bar means that it was smaller than the symbol. [Reprinted from Ref. 48, with permission.]

quiescent cells, usually involves a change in the observed V_{max} of the transport system (26, 35). Representative data for P_i uptake by 3T3 cells is shown in Fig. 6 (26). A Lineweaver-Burk plot of this data (lower panel) shows a twofold change in the observed V_{max} with no apparent change in the observed K_m. What is particularly interesting, however, is the observation that diffusion of P_i, as indicated in the upper panel by the plain solid line (quiescent cells) and plain dashed line (proliferating cells), two- to threefold greater in the proliferating cells than in the quiescent cells. This probably reflects a greater exposed surface area per μg of protein in the proliferating cells.

In any case, what is of primary concern is whether or not this decrease in diffusion as well as carrier-mediated uptake as proliferating cells begin to approach quiescence is causally involved in the inhibition of DNA synthesis. A first requirement for a primary causal signal is a temporal one. That is, if we ask what is the earliest time that proliferating cells are committed to entering stationary phase, any putative signal controlling DNA synthesis must precede that time. Under normal culture conditions, a decline in thymidine

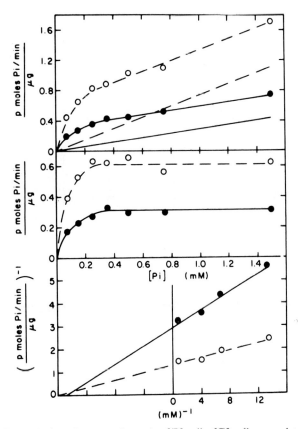

Fig. 6. Kinetics of P_i uptake in quiescent and growing 3T3 cells. 3T3 cells were plated in medium containing 10% serum at a density of either 1.1×10^3 or 1.1×10^4 per cm^2. Three days later, the cultures set up at the higher density were quiescent ($\bullet$——$\bullet$). To remove residual P_i, the cultures were preincubated for 2 h at 37°C in dialyzed depleted medium. After the preincubation, the cells were incubated for 15 min with $^{32}P_i$ in dialyzed depleted medium containing the indicated concentrations of P_i. Upper panel: total uptake with diffusion estimated by drawing a line through the original that was parallel to total uptake at regions where uptake appeared to be linear (0.4–1.5 mM). Middle panel: carrier-mediated uptake calculated by subtracting diffusion from total uptake. Lower panel: Lineweaver-Burk plot of carrier-mediated uptake data. [Reprinted from Ref. 26, with permission.]

incorporation in proliferating cells occurs some time before these cells actually show a decrease in their rate of division. This probably indicates the first sign of decreased DNA synthesis and must, therefore, follow any putative regulatory signal. Experiments that compared the decrease in nutrient uptake with the decline in thymidine incorporation have suggested that a decrease in P_i (26) and glucose (16, 55, 59) uptake usually slightly precedes the decline in thymidine incorporation. This is consistent with the possibility that a decreased rate of uptake signals a decline in DNA synthesis.

To directly determine if this decrease in uptake is a causal signal in the entrance of cells into stationary phase, we used a similar technique to the one used in Fig. 5. After plating the cells at a low density, we changed them to media containing a lowered nutrient concentration and subsequently monitored cell number, DNA synthesis, and nutrient uptake (55). Representative data for glucose uptake in 3T3, HF, and CE cells are shown

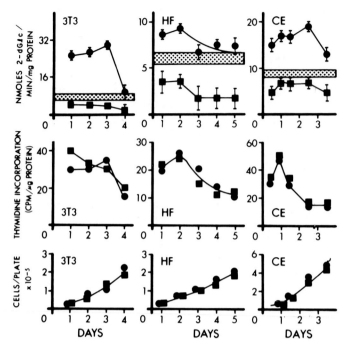

Fig. 7. Effect of reduced concentration of glucose on glucose uptake, DNA synthesis, and cell number. 3T3, HF, and secondary CE cells were plated at densities of 4.0×10^3, 5.0×10^3, and 1.0×10^4 cells, respectively per cm^2, The medium contained 10% calf serum (3T3 cells), 5.0% calf serum (HF cells), or 2.0% calf serum (CE cells). Shortly after the cells attached, the medium was changed to fresh medium containing the same concentrations of dialyzed fresh calf serum and either 22 mM glucose (•) or 2.8 mM glucose (■). Time zero indicates the time at which the cells were plated. Glucose uptake, thymidine incorporation, cell number, and cell protein were measured at the indicated times. Each glucose uptake data point represents an initial velocity of uptake determined from a time course of uptake using 4 cultures and incubation times between 10 sec and 4 min. The shaded areas in the panels for glucose uptake designate the uptake rates characteristic of quiescent cells plus and minus one standard deviation from the mean. Error bars for the measurements of glucose uptake show one standard deviation from the mean. The absence of an error bar means that it was smaller than the symbol. [Reprinted from Ref. 55, with permission.]

in Fig. 7. After changing to medium containing 10% of the normal amount of glucose (squares), we found no differences from control cultures (circles) in either thymidine incorporation (middle panel) or cell number (lower panel). Glucose uptake (upper panel) in the low glucose cultures, however, was kept below the level characteristic of stationary phase cultures (as indicated by the shaded bars). We found similar results for P_i uptake in these cells (26). Since levels of uptake below those characteristic of quiescent cells had no effect on DNA synthesis or cell division over a several day period, we concluded that the decrease in P_i or glucose uptake that very slightly precedes the decline in DNA synthesis was not regulating cell division during growth to quiescence (26, 55). As with the case of serum stimulation of quiescent cells, however, the possibility remained that intracellular pool size might be regulating proliferation under our experimental conditions. When we measured intracellular pool sizes of both P_i (26) and glucose (55) during growth to quiescence, however, we found either no changes or actual increases (in the case of P_i in 3T3 cells) during growth to quiescence. This directly demonstrated that changes in intracellular pool size were 1) not exclusively controlled by changes in uptake, and 2) not controlling DNA synthesis under normal culture conditions.

It should be pointed out that one group of investigators has reported changes in intracellular P_i pool sizes in 3T3 cells during growth to quiescence and after serum stimulation that are at variance with our results (60). However, these investigators measured intracellular P_i pools using an isotope equilibration technique with a 2-h incubation. If a radioisotope equilibration method is used, it is exceedingly important to make sure that a long enough incubation time is employed to allow the isotope to become completely equilibrated. In the case of P_i, uptake appears to be linear for at least 3 h (48, 56), indicating that a 2-h incubation probably reflects measurements of uptake rather than intracellular pool sizes. The results presented here are from experiments which utilized a direct chemical method to measure pool sizes (26, 48). We have also used a radioisotope equilibration method with an incubation period of 2 days and obtained essentially identical results (27).

In conclusion, it appears that under normal culture conditions neither uptake nor intracellular pool size of P_i or glucose serves as a causal signal in the regulation of proliferation.

CHANGES IN UPTAKE OF OTHER MOLECULES

A number of investigators have described changes in the uptake of other molecules that appear to be related to the control of cell proliferation. At the outset, we can conclude that for some molecules changes in uptake are not critical for the regulation of cell division since these molecules are not required for growth and are not present in serum or the usual cell culture media. On the other hand, some of the changes in uptake involve molecules that are required for cell growth. There is not enough information at the present time to decisively determine whether these latter changes participate as causal events in the regulation of DNA synthesis or cell division.

Molecules Not Required for Cell Proliferation and Not Present in Serum or Most Cell Culture Media

Although changes in uptake for this class of molecules are not causally involved in the control of cell proliferation, it is noteworthy that some of the changes are closely correlated with certain aspects of growth control. As noted in the background section, uptake of uridine decreases four- to fivefold during growth of 3T3 or CE cells to quiescence (13, 17). This same difference in uridine uptake has been observed with membrane vesicles prepared from growing and nongrowing 3T3 cells, indicative of a transport change (37). Addition of fresh serum to quiescent 3T3 or CE cells brings about a severalfold increase in uridine uptake that can be detected within several minutes (13, 42, 61). Although this increase slightly follows the increase for P_i and glucose uptake (42), it is one of the earliest events that has been detected following initiation of division in arrested cells. As mentioned earlier, serum addition to quiescent 3T3 cells has no effect on adenosine uptake (13). In view of these findings about uridine and adenosine uptake, it is curious that in 3T3 and BHK cells (62–64) added adenosine potentiates serum stimulated DNA synthesis while uridine does not (62).

Uptake of putrescine by cultured human fibroblasts is also closely related to the proliferative state of these cells. Uptake decreases as the cells grow to quiescence and markedly increases within 30 min after initiating proliferation with fresh serum (65). Although these cells do not require added putrescine in the culture medium for growth, it is significant that they produce putrescine, and that putrescine can function as a growth factor under certain conditions (66).

Uptake of choline by cultured CE cells also increases soon after initiating proliferation by some but not all methods (61). Serum, elevated pH, excess Zn^{2+}, DMBA (a carcinogenic hydrocarbon), and insulin accelerate the progress of growth-inhibited CE cells into the S phase of the cell cycle. All of these treatments except Zn^{2+} addition increase choline uptake (61).

In addition to these early increases in uptake after initiation of proliferation, increased uptake of pyrimidine deoxynucleosides occurs at a time which coincides closely with initiation of DNA synthesis (67, 68). This increase is selective for uptake of thymidine and deoxycytidine; uptake of other DNA precursors such as deoxyadenosine, deoxyguanosine, and P_i does not increase at this time (68). The increase in thymidine uptake is not a result of higher rates of DNA synthesis since inhibition of DNA synthesis does not prevent the increase in uptake. In addition, it is not a result of increased phosphorylation of thymidine, since it takes place in cells which lack thymidine kinase. Thus, it is probably a direct change in membrane transport of thymidine (68). These increases in thymidine and deoxycytidine uptake at the time of DNA synthesis suggest that increased pool sizes of these compounds or their derivatives might influence the initiation of DNA synthesis. However, we could detect no significant change in either the timing or extent of DNA synthesis after adding varying levels of deoxynucleosides to quiescent cells or to quiescent cells initiated with low levels of fresh serum (69).

Nutrients Required for Growth

It is particularly important to examine the changes in uptake that occur for this class of compounds since changes that correlate with growth rate might be involved in the regulation of cell proliferation. In this section, we will examine changes in amino acid and potassium ion uptake that correlate with some aspects of cell division.

Two amino acid analogs are commonly used to measure amino acid uptake. α-Aminoisobutyric acid (AIB) is taken up by the Na^+-dependent A system, and cycloleucine is transported by the Na^+-independent L system. Transport of amino acids by these 2 systems has been extensively studied, although many important questions about the fundamental aspects of amino acid transport remain to be answered (70).

Early studies on growing and nongrowing 3T3 cells revealed that AIB and cycloleucine uptake decreased about 30–50% after growth to a quiescent state, while uptake of arginine, glutamic acid, and glutamine did not change (14). Similar density-dependent changes in AIB uptake have recently been described for membrane vesicles prepared from cells at different densities (39, 71). Studies on synchronized cell populations have revealed that uptake of AIB and cycloleucine are generally low during early G_1 and then increase about two- to threefold late in G_1 or during S, although the exact patterns are somewhat different for the 2 amino acid analogs (72, 73). Addition of fresh serum to quiescent fibroblasts brings about a rapid decrease in the uptake of leucine and lysine (74) and an increase at about 3 h in the uptake of cycloleucine (75). It is noteworthy that stimulation of human lymphocytes with phytohemagglutinin also brings about an increase in AIB uptake (76). Furthermore, injection of rats with growth hormone brings about a twofold increase in the maximum velocities of the AIB and cycloleucine transport systems in the perfused liver under conditions where growth (protein synthesis) is stimulated (77). Taken together, these studies show that rates of uptake of some amino acids are related to the growth state of the cell and that there are cell cycle-dependent changes in the uptake of certain amino acids.

Similarly, there are now several lines of evidence showing that changes in potassium ion uptake are associated with alterations in cell growth. Early experiments showed that relatively high intracellular K^+ levels are necessary for protein synthesis and growth in cultured mammalian cells (78) and that intracellular K^+ levels vary during the cell cycle of cultured mouse lymphoblasts (79). Studies on human lymphocytes have revealed that the stimulation of DNA synthesis by phytohemagglutinin is quite sensitive to ouabain, an inhibitor of membrane Na^+, K^+-ATPase. Importantly, excess extracellular K^+ reverses the inhibitory effect of ouabain on DNA synthesis (80, 81). It has also been shown that treatment of human lymphocytes with phytohemagglutinin leads to an early stimulation of K^+ uptake (82). These studies have led to the conclusion that increased uptake of K^+ seems to be a necessary condition for the biosynthetic and morphological events of lymphocyte stimulation (82). More recent studies have shown that Na^+, K^+-ATPase fluctuates during the cell cycle of cultured hamster fibroblasts (83) and that there are cell cycle-dependent changes in K^+ transport in cultured Ehrlich ascites tumor cells (84). Also, active K^+ uptake decreases in 3T3 and SV40-transformed 3T3 cells when the growth rate of both cell types decreases (85). Stimulation of DNA synthesis in 3T3 cells brings about rapid increases in K^+ uptake (86) and Na^+, K^+-ATPase (52). All of these findings point to the importance of further investigations to determine if the changes in K^+ uptake and intracellular levels of K^+ are causal events in the regulation of DNA synthesis and cell division in these cases.

ACKNOWLEDGMENTS

This work was supported by Grant CA-12306 from the National Cancer Institute. D.D.C. is a recipient of Research Career Development Award CA-00171 from the National Cancer Institute.

REFERENCES

1. Hatanaka M: Biochim Biophys Acta 355:77, 1974.
2. Pardee AB: Biochim Biophys Acta 417:153, 1975.
3. Pardee AB, Rozengurt E: In Fox CF (ed): "Biochemistry of Cell Walls and Membranes." Baltimore: University Park Press, 1975, p 155.
4. Roblin R, Chou I, Black P: Adv Cancer Res 22:203, 1975.
5. Nicolson GL: Biochim Biophys Acta 457:1, 1976.
6. Nicolson GL: Biochim Biophys Acta 458:1, 1976.
7. Mora PT: "Fogarty Symposium on Cell Surfaces and Malignancy." Washington, DC: US Government Printing Office, 1976.
8. Forgarty Symposium on Cellular Regulation of Transport and Uptake of Nutrients: J Cell Physiol 89:493–863, 1977.
9. Perdue JF: In Nicolau C (ed): "Verus-Transformed Cell Membranes." London: Academic Press, (In press).
10. Todaro GJ, Lazar GK, Green H: J Cell Comp Physiol 66:325, 1965.
11. Rubin H: In Wolstenhome GEW, Knight J (eds): "Growth Control in Cell Cultures." London: Churchill Livingstone, 1971, p 127.
12. Pardee AB: Natl Cancer Inst Monogr 14:7, 1964.

13. Cunningham DD, Pardee AB: Proc Natl Acad Sci USA 64:1049, 1969.
14. Foster DO, Pardee AB: J Biol Chem 244:2675, 1969.
15. Sefton BM, Rubin H: Nature (London) 227:843, 1970.
16. Sefton BM, Rubin H: Proc Natl Acad Sci USA 68:3154, 1971.
17. Weber MJ, Rubin H: J Cell Physiol 77:157, 1971.
18. Rubin H: Proc Natl Acad Sci USA 72:1676, 1975.
19. Rubin H: J Cell Biol 51:686, 1971.
20. Rudland PS, Seifert W, Gospodarowicz D: Proc Natl Acad Sci USA 71:2600, 1974.
21. Bourne HR, Rozengurt E: Proc Natl Acad Sci USA 73:4555, 1976.
22. Holley RW: Proc Natl Acad Sci USA 69:2840, 1972.
23. Tobey RA, Ley KD: J Cell Biol 46:151, 1970.
24. Ley KD, Tobey RA: J Cell Biol 47:453, 1970.
25. Holley RW, Kiernan JA: Proc Natl Acad Sci USA 71:2942, 1974.
26. Barsh GS, Greenberg DB, Cunningham DD: J Cell Physiol 92:115, 1977.
27. Cunningham DD: J Biol Chem 247:2464, 1972.
28. Renner ED, Plagemann PGW, Bernlohr RW: J Biol Chem 247:5765, 1972.
30. Kletzien RF, Perdue JF: J Biol Chem 249:3383, 1974.
31. Hassel JA, Colby C, Romano AH: J Cell Physiol 86:37, 1975.
32. Colby C, Romano AH: J Cell Physiol 85:15, 1974.
33. Fodge DW, Rubin H: Nature (London) 246:181, 1973.
34. Weber MJ: J Biol Chem 248:2978, 1973.
35. Kletzien RF, Perdue JF: J Biol Chem 249:3366, 1974.
36. Thrash CR, Cunningham DD: Nature (London) 252:45, 1974.
37. Quinlan DC, Hochstadt J: Proc Natl Acad Sci USA 71:5000, 1974.
38. Nilsen-Hamilton M, Hamilton RT: J Cell Physiol 89:795, 1976.
39. Lever J: J Cell Physiol 89:779, 1976.
40. Kram R, Mamont P, Tomkins GM: Proc Natl Acad Sci USA 70:1432, 1973.
41. Kram R, Tomkins GM: Proc Natl Acad Sci USA 70:1659, 1973.
42. Jiménez de Asua L, Rozengurt E, Dulbecco R: Proc Natl Acad Sci USA 71:96, 1974.
43. Hilborn DA: J Cell Physiol 87:111, 1975.
44. Venuta S, Rubin H: Proc Natl Acad Sci USA 70:653, 1973.
45. Bradley WEC, Culp LA: Exp Cell Res 84:335, 1974.
46. Jimenez de Asua L, Rozengurt E: Nature (London) 251:624, 1974.
47. Sheppard JR, Plagemann PGW: J Cell Physiol 85:163, 1974.
48. Greenberg DB, Barsh GS, Ho T-S, Cunningham DD: J Cell Physiol 90:193, 1977.
49. Blumberg PM, Robbins PW: Cell 6:137, 1975.
50. Hale AH, Weber MJ: Cell 5:245, 1975.
51. Vaheri A, Ruoslahti E, Nordling S: Nature (London) 238:211, 1972.
52. Lever JE, Clingan D, Jimenez de Asua L: Biochem Biophys Res Commun 71:136, 1976.
53. Smith GL, Temin HM: J Cell Physiol 84:181, 1974.
54. Rubin H, Koide T: J Cell Physiol 81:387, 1972.
55. Naiditch WP, Cunningham DD: J Cell Physiol 92:319, 1977.
56. Weber MJ, Edlin G: J Biol Chem 246:1828, 1971.
57. Schultz AR, Culp CA: Exp Cell Res 81:95, 1973.
58. Harel L, Jullien M, Blat C: Exp Cell Res 90:201, 1975.
59. Bose SK, Zlotnick BJ: Proc Natl Acad Sci USA 70:2374, 1973.
60. Gray PN, Cullum ME, Griffin MJ: J Cell Physiol 89:225, 1976.
61. Rubin H, Koide T: J Cell Physiol 86:47, 1975.
62. Schor S, Rozengurt E: J Cell Physiol 81:339, 1973.
63. Clarke GD, Smith C: J Cell Physiol 81:125, 1973.
64. Brooks RF: J Cell Physiol 86:369, 1975.
65. Pohjanpelto P: J Cell Biol 68:512, 1976.
66. Pohjanpelto P: Nature (London) 235:247, 1972.
67. Nordenskjöld BA, Skoog L, Brown NC, Reichard P: J Biol Chem 245:5360, 1970.
68. Cunningham DD, Remo RA: J Biol Chem 248:6282, 1973.
69. Matzinger P, Cunningham DD: Unpublished observations.
70. Christensen HN, de Cespedes C, Handogten ME, Ronquist G: Biochim Biophys Acta 300:487, 1973.

71. Parnes JR, Garvey TO, Isselbacher KJ: J Cell Physiol 89:789, 1976.
72. Sander G, Pardee AB: J Cell Physiol 80:267, 1972.
73. Tupper JT, Mills B, Zorgniotti F: J Cell Physiol 88:77, 1976.
74. Wiebel F, Baserga R: J Cell Physiol 74:191, 1969.
75. Costlow M, Baserga R: J Cell Physiol 82:411, 1973.
76. Mendelsohn J, Skinner SA, Kornfeld S: J Clin Invest 50:818, 1971.
77. Jefferson LS, Schworer CM, Tolman EL: J Biol Chem 250:197, 1975.
78. Lubin M: Nature (London) 213:451, 1967.
79. Jung C, Rothstein A: J Gen Physiol 50:917, 1967.
80. Quastel MR, Kaplan JG: Nature (London) 219:198, 1968.
81. Quastel MR, Kaplan JG: Exp Cell Res 62:407, 1970.
82. Quastel MR, Kaplan JG: Exp Cell Res 63:230, 1970.
83. Graham JM, Sumner MCB, Curtis, DH, Pasternak CA: Nature (London) 246:291, 1973.
84. Mills B, Tupper JT: J Cell Physiol 89:123, 1976.
85. Kimelberg HK, Mayhew E: J Biol Chem 250:100, 1975.
86. Rozengurt E, Heppel LA: Proc Natl Acad Sci USA 72:4492, 1975.

Journal of Supramolecular Structure 7:277–285 (1977)
Molecular Aspects of Membrane Transport 443–451

A Role for Anion Transport in the Regulation of Release From Chromaffin Granules and Exocytosis From Cells

Harvey B. Pollard*, Christopher J. Pazoles*, Carl E. Creutz*, Avner Ramu*, Charles A. Strott*, Probhati Ray*, Edward M. Brown†, G. D. Aurbach†, Karen M. Tack-Goldman**, and N. Raphael Shulman**

*Reproduction Research Branch, National Institute of Child Health and Human Development, †Metabolic Diseases Branch, **Clinical Hematology Branch, National Institute of Arthritis, Metabolism, and Digestive Diseases, National Institutes of Health, Bethesda, Maryland 20014

Release of epinephrine from isolated adrenergic secretory veiscles from the adrenal medulla (chromaffin granules) was found to be inhibited by a number of anion transport blocking agents, including SITS, probenecid, pyridoxal phosphate, and Na-isethionate. High concentrations of permeant anion, such as chloride, are required for granule release and the drugs were found to be competitive inhibitors with respect to chloride. The anion transport blockers were also found to suppress exocytosis of serotonin from human platelets and parathyroid hormone from dissociated bovine parathyroid cells. By contrast, they had no effect on ACTH-activated corticosterone secretion from dissociated rat adrenocortical cells, a process which occurs by diffusion rather than exocytosis. The important anion in the medium for human platelets was hydroxyl ion, rather than chloride, and the most effective drug on platelets was suramin. Isethionate was inactive. In the case of PTH secretion, both chloride and hydroxyl ions were important anions and were both competitively inhibited by anion blocking drugs including Na-isethionate. We conclude from these studies that the chemistry of exocytosis appears to be quite similar to the chemistry of release from isolated secretory vesicles. We suggest that when vesicles are fused to plasma membranes prior to exocytosis they are exposed to higher chloride and hydroxyl ion concentrations of the medium, and that inward anion flux into the vesicle promotes release, possibly by local osmotic lysis. Blockade of exocytosis by anion transport blocking drugs would occur by inhibition of inward anion flux into the fused vesicle, by analogy with previous results from studies on isolated chromaffin granules.

Key words: anion transport, chromaffin granules, exocytosis, platelets, parathyroid hormone

Many neurotransmitters, hormones, and enzymes are stored in intracellular secretory vesicles and, in response to appropriate stimuli, are released into the extracellular compartment by exocytosis (1). The process seems well defined ultrastructurally but the chemical and energetic basis for exocytosis remains obscure.

Received June 17, 1977; accepted June 24, 1977.

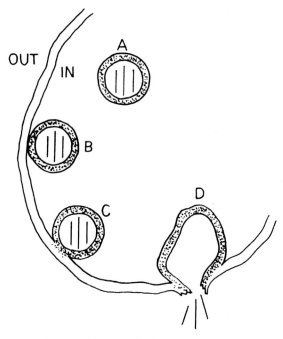

Fig. 1. Schematic representation of possible steps in the process of exocytosis. Isolated secretory granule in the cytoplasm (A) is recruited into juxtaposition (or "fusion") with the plasma membrane (B). The fusion state becomes more intimate, and a trilaminar membrane becomes the only structure separating the intragranular space and the extracellular medium (C). Finally granular contents are released when the fusion complex undergoes breakage or "fission." The nomenclature is based on that suggested by Palade (3).

In secretory systems such as the exocrine pancreas (2, 3), frog neuromuscular junction (4), and mast cells (5, 6), electron microscopy has been intensively applied to visualize exocytosis. As indicated in Fig. 1, the process has appeared to proceed by discrete steps in which isolated secretory vesicles first contact the plasma membrane forming a "pentalaminar" contact or "fusion" complex by an as yet poorly understood process possibly involving calcium. The fusion becomes more intimate, and in some cases it has been shown that the submembrane particles in the plasma membrane portion of the fusion complex move aside. A single bilayer structure then forms dividing vesicle contents and extracellular space. "Fission" of the bilayer finally occurs, resulting in secretion. The mechanism for the latter step is also not known.

As a departure point for our own studies, we assumed that the chemistry of release as defined by isolated secretory vesicles might also apply to the fission step in exocytosis (see Fig. 1D). We therefore devoted some of our subsequent efforts to the study of release of epinephrine from adrenal medullary secretory vesicles, or chromaffin granules. We have recently found that anion transport blocking drugs such as SITS (4-acetamido-4'-isothiocyanostilbene-2,2'-disulfonic acid), probenecid, pyridoxal phosphate, and others are able to block release of epinephrine from isolated chromaffin granules. We have also found that the blocking drugs inhibit secretion from several types of cells that release by exocytosis, but not by one cell type that secretes by diffusion. In this paper we suggest that permeant anions in the external medium, such as Cl^- and OH^-, may play an important role in regulating exocytosis.

MATERIALS AND METHODS

Chromaffin Granule Release

Chromaffin granules were prepared by differential centrifugation as previously described (7, 8) from 10–12 pairs of fresh bovine adrenal glands. For analysis of release, granules were incubated in an isotonic medium (335 mosmol) containing 500 μg granule protein, 1 mM $MgSO_4$, 1 mM ATP, 50 mM Hepes-NaOH buffer, pH 6.0, and 90 mM KCl in a total volume of 1.75 ml. After incubation for 10 min at 37°C, samples were mixed with 1 ml cold 0.33 M sucrose and centrifuged at 20,000 $\times$ g for 20 min. Aliquots of supernatant solution were then assayed for released endogenous epinephrine using the trihydroxy indole reaction (9). In experiments with isethionate, chloride was kept constant and 0.3 M sucrose was varied reciprocally with 0.15 M isethionate to generate different isethionate concentrations. SITS was obtained from British Drug House, and pyridoxal phosphate and probenecid were obtained from Sigma Chemical Company (St. Louis, Missouri). Suramin was obtained from Imperial Chemical Industries (London, England).

Human Platelets

Platelet-rich plasma in 0.1% EDTA was isolated from fresh human blood by differential centrifugation and platelets were labeled with [^{3}H] serotonin (final concentration = 0.2 μM) by incubation at 18°C for 45 min (10). The platelets were then chilled, centrifuged, washed twice with cold Rossi's medium (11), and resuspended in ice cold 0.3 M sucrose containing 1 mg/ml human serum albumin. Platelets in a 50-μl volume were then mixed with 450 μl of a prewarmed medium containing 130 mM NaCl, 25 mM Mes-NaOH buffer, pH 7.32, and thrombin (0.16 units/ml). After 1 min at 37°C, 10 μl of 10% glutaraldehyde were added to terminate the reaction, and the platelets were sedimented at 1,030 $\times$ g for 10 min. Samples of supernatant were subjected to liquid scintillation counting to determine released [^{3}H] 5HT, and the percent release was determined from the relation, % R = 100 (released cpm − blank)/(maximum released cpm − blank). The maximum release was determined by incubation of platelets in 100-fold excess thrombin or A23187 for 1 min, and the blank was determined by incubating platelets in the absence of thrombin. [^{3}H] Serotonin was obtained from New England Nuclear Corporation (Boston, Massachusetts). Suramin was obtained from Imperial Chemical Industries (United Kingdom).

Isolated Parathyroid Cells

Dispersed parathyroid cells were prepared from fresh bovine parathyroid glands by mincing and digestion with 0.2% collagenase and 50 μg/ml DNase (12). Washed cells (150,000–200,000/0.5 cm^3) were incubated for 30–60 min at 37°C with 0.5 mM $CaCl_2$, 0.5 mM $MgSO_4$, and 0.2% bovine serum albumin in either a) Eagle's medium number 2 (bicarbonate deleted) with 20 mM Hepes, pH 7.5, or b) 150 mM NaCl, 1 g/liter dextrose, 0.25 g/liter K_2HPO_4, 0.3 g/liter KCl, and 20 mM Hepes, pH 7.5. The drugs indicated above were added at the beginning of the incubation and the reaction was terminated by sedimenting the cells for 30 sec in a microfuge (Beckman). Parathyroid hormone (PTH) in supernatant samples was determined by radioimmunoassay (12). Results are expressed as percent of release occurring at 0.5 mM calcium without added drugs.

Isolated Rat Adrenocortical Cells

Adrenals were obtained from male Holtzman rats (260–300 g). The medulla-reticularis-fasciculata tissue was expressed and the remaining tissue minced with a razor

TABLE I. Influence of Isethionate on Kinetics of Release of Epinephrine From Isolated Chromaffin Granules

Isethionate, mM	Cl^- $K_{1/2}$ mM	ISETH K_i mM	V_{max} [a]
0	69.2	–	7.1
10	81.2	57.3	7.1
20	105.9	37.7	7.1
30	131.8	33.1	7.1

[a]% total catecholamine released/min.

blade. Adrenocortical cells were mechanically dispersed following enzymatic digestion with collagenase at room temperature. Isolated cells ($1-3 \times 10^5$/ml) were incubated in Ham's F-10 medium containing 2.5 mM calcium and 10% fetal calf serum at $37°C$ for 2 h. Following the incubations the cells were removed by centrifugation and the corticosterone in the medium was assayed directly by radioimmunoassay. The steroid assay was accurate (90–110%) and precise (intraassay cultivariant = 7%; interassay cultivariant = 10%). A medium blank determined in each assay was invariably zero.

RESULTS

Chemistry of Release From Isolated Chromaffin Granules

Chromaffin granules are 2,000 Å diameter secretory vesicles from adrenal medulla which contain large amounts of epinephrine, ATP, and specific proteins (13). The isolated granules are stable in isotonic sucrose, but release their total contents when exposed to Mg-ATP and high concentrations of permeant anion such as chloride at pH 6, and $37°C$ (see Ref. 7 for details and complete references). The mechanism of release in this process is osmotic lysis since granule release is suppressed by increased extracellular osmotic strength (8, 14, 15). The anion requirement atracted our attention since most cells have anion gradients across their plasma membranes (120 mM external Cl^- versus 5–30 mM internal Cl^-; $pH_{out} > pH_{in}$), and secretory vesicles in the "fusion" state (Fig. 1B, C) would naturally be exposed to these gradients.

In the case of release from isolated chromaffin granules, we found that release was a saturable function of $[Cl^-]$. By contrast, impermeant anions would not support release but rather would competitively inhibit release with respect to $[Cl^-]$. Isethionate ($HO-C_2H_4-SO_3^-$), a typical impermeant anion, was a competitive inhibitor of release with respect to Cl, in that it changed the K_m but not the V_{max} of Cl^--induced release (Table I).

We had previously found that Cl^- (as ^{36}Cl) actually entered the granule in the presence of Mg-ATP (15), and that the role of Mg-ATP was to provide a positive electric potential within the granule to attract the anion (8). This suggested that permeant anion entry was competitively blocked by impermeant anions, and we therefore concluded that a specific anion transport site might exist on the granule membrane.

Anion transport across red cell membranes occurs by exchange and is sensitive to specific, impermeant drugs (16–18). Examples are shown in Fig. 2, and we decided to test these compounds on Cl^--dependent ATP-mediated release from granules. As shown in Table II, these drugs blocked granule release in a dose-dependent fashion and kinetic analysis revealed that they also competed with Cl^-.

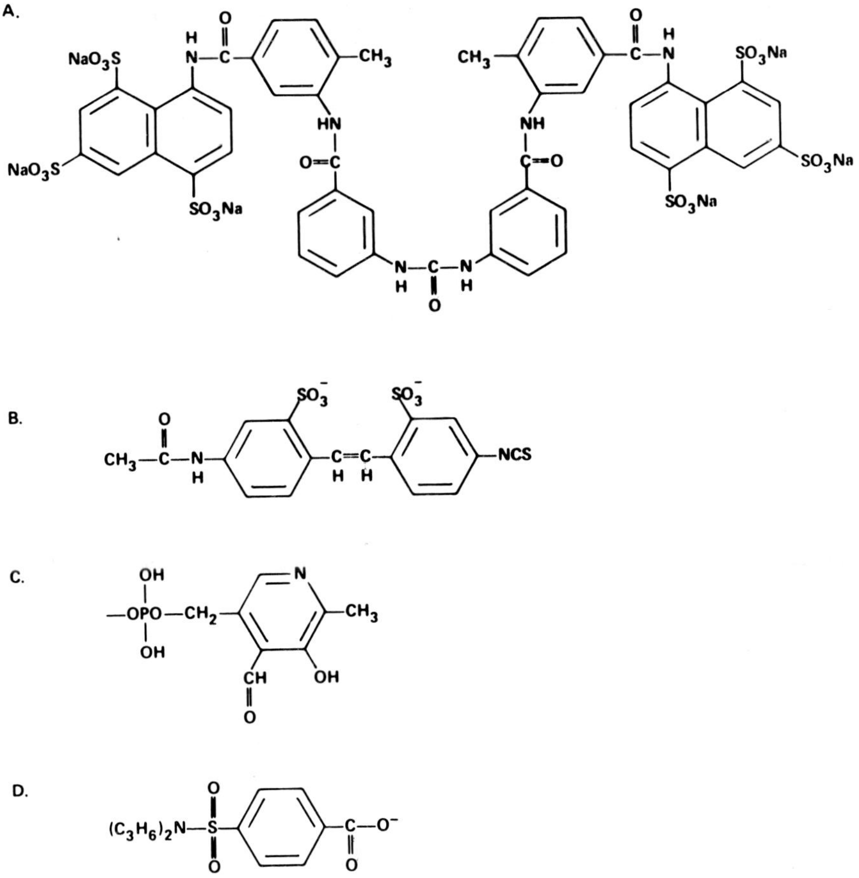

A.

B.

C.

D.

Fig. 2. Chemical structures of anion transport blocking drugs. A) Suramin, B) SITS, C) pyridoxal phosphate, D) probenecid.

TABLE II. Influence of Anion Transport Blockers on ATP, Cl-Induced Release of Epinephrine From Isolated Chromaffin Granules

Additions	Concentration	% Release in 10 min	% Inhibition
None	—	67 ± 2	—
SITS	50 μM	48 ± 1	28
	100 μM	31 ± 2	54
	500 μM	10 ± 4	85
Probenecid	50 μM	58 ± 2	13
	100 μM	48 ± 2	28
	500 μM	18 ± 3	73
Pyridoxal Phosphate	2 mM	53 ± 2	21
	5 mM	35 ± 2	48

[a]The reaction mixture was as described in Methods, except that the chloride concentration in all cases was 90 mM.

TABLE III. Influence of Anion Transport Blocking Drugs on Serotonin Secretion From Human Platelets

Additions	Concentration	% Release in 1 min	% Inhibition
None	–	58	–
Suramin	1 μM	46 ± 2	20
	5 μM	19 ± 2	66
Pyridoxal Phosphate	10 μM	48 ± 3	18
	100 μM	8 ± 2	86
SITS	10 μM	52 ± 3	10
	100 μM	16 ± 2	72
Probenecid	1 mM	41 ± 3	30
	5 mM	6 ± 2	90

TABLE IV. Influence of Anion Transport Blocking Drugs on PTH Secretion From Isolated Bovine Parathyroid Cells

Additions	Concentration	% Release[a]	% Inhibition
None	–	100	–
SITS	100 μM	87 ± 4	13
	1 mM	52 ± 2	48
	10 mM	10 ± 4	90
Probenecid	100 μM	88 ± 4	12
	1 mM	66 ± 4	34
	10 mM	15 ± 3	85
Na Isethionate (as an	100 mM	55 ± 5	45
isotonic replacement for NaCl)	150 mM	30 ± 4	70
Sucrose (as an isotonic	300 mM	28 ± 2	72
replacement for NaCl)			

[a]Reactions were carried out at pH 7.5.

Influence of Anion Transport Blocking Drugs on Secretion From Platelets

Human platelets secrete serotonin from storage sites in secretory vesicles by exocytosis (19–21) and this process was also found to be blocked by the anion transport blocking drugs. As shown in Table III, the effect was dose-dependent. Suramin (bis(m-amino-m-aminobenzoyl-p-methylbenzyl)-1-naphthylamine-7,6,8-trisulfonate carbamide) was found to be the most potent blocker of platelet release, and also suppressed chromaffin granule release though at somewhat higher concentrations. The data in Table III were collected using thrombin as the stimulus, and similar drug effects were obtained when A23187, a calcium ionophore, was used instead.

Platelets proved to be insensitive to removal of NaCl from the medium or substitution of Na-isethionate for NaCl, though they were quite sensitive to small reductions in the pH. Detailed kinetic analysis revealed that the anion transport blocking drugs were in fact competitive inhibitors with respect to OH^- ions.

Influence of Anion Transport Blocking Drugs on Secretion of Parathyroid Hormone

Parathyroid hormone (PTH) is also stored in secretory vesicles and is released from dispersed bovine parathyroid cells when stimulated either with low calcium (0.5 mM) or

TABLE V. Influence of SITS on ACTH-Activated Corticosterone Release From Isolated Rat
Adrenal Cortical Cells

| Condition | Corticosterone released in 2 h (ng/ml) | |
	Experiment 1	Experiment 2
Control	104 ± 4	195 ± 47
ACTH (1 mU/ml)	830 ± 125	$2,240 \pm 464$
SITS (0.1 mM)	200 ± 43	282 ± 67
ACTH + SITS	$1,050 \pm 173$	$1,831 \pm 243$

β-adrenergic agonists (12). We decided to test the anion transport blocking drugs on
exocytosis in this second cell type, and found that they also inhibited PTH secretion (see
Table IV). Inhibition was found regardless of whether stimulation was evoked with low
calcium or isoproterenol. Probenecid and SITS were not as potent in the case of the PTH
cells as in the chromaffin granules or platelet systems, but they did prove to be competitive inhibitors with respect to $[Cl^-]$. As expected from the similar chloride sensitivity of
PTH cells and chromaffin granules, isethionate also proved to be an effective inhibitor of
PTH secretion.

As indicated in Table IV, replacement of NaCl by sucrose, or replacement of
chloride by isethionate, decreased PTH release at pH 7.4 to only 30% of the original
release level. However, nearly complete inhibition was obtained by, in addition, lowering
the pH down to 6.5. This result suggested that hydroxyl ions might also be important in
PTH release, and probenecid also proved to be a competitive inhibitor of PTH release with
respect to $[OH^-]$, just as it did with serotonin release from the platelets.

Influence of Anion Transport Blocking Drugs on Corticosterone Secretion

Steroids are presumed to be released by diffusion rather than by exocytosis. We
therefore studied the influence of SITS on corticosterone release from isolated rat adrenal
cortical cells. As shown in Table V, SITS did not inhibit ACTH-activated corticosterone
release. In fact, a small increment in cortisol secretion was noted in one experiment.

DISCUSSION

These data have led us to consider anion transport as a key regulatory event in
secretion by exocytosis. Both hydroxyl ions and chloride ions appear to be implicated as
anionic substrates for the transport sites. The fact that isolated chromaffin granules are
also sensitive to these agents suggests that the site of action of the impermeant blocking
drugs in secreting cells may be at regions of the external cell surface that contain secretory
vesicle anion transport sites. These regions might be the "fusion" complexes described in
Fig. 1 (B and C).

By analogy with the results from studies on isolated chromaffin granules, anion
transport in secreting cells may ultimately be directed into the fused vesicle interior (14).
Chromaffin granules have relatively low internal concentrations of chloride [approxi-
mately 30 mM (7)] and have relatively acidic interior compartments (pH 5.5–6.25)
(8, 22, 23), but little is known about the interior environment of other secretory vesicles.
It is perhaps not surprising that the concentrations of OH^- and Cl^- are higher outside

cells than inside. In this regard it is possible that other secretory vesicle interiors are similar to the chromaffin granule interior and that they would also support local [OH$^-$] or [Cl$^-$] gradients when fused to the plasma membrane of the secreting cell. These anion chemical potentials may be related to electric potentials, and anion transport may be accompanied by a counterion, but we have no definitive information on this point. Based on chromaffin granule studies, it is likely that anion transport in exocytosis represents flux rather than simply exchange.

It is evident that the anion transport blocking drugs inhibit release with different potencies in different secreting systems. This may be due either to structural heterogeneity of anion transport sites in different cell types, or to variation in the accessibility of the drugs to transport sites in different cells. That anion transport is specifically involved in exocytotic secretion, but not in other mechanisms of secretion, is indicated by the failure of SITS to inhibit steroid secretion. The latter is believed to occur by immediate diffusion of newly synthesized hormone.

One of the most important conclusions from these studies is that the chemistry of exocytosis appears to be quite similar to the chemistry of release from isolated secretory vesicles. Both processes are blocked by the same ion transport blocking drugs and depend on specific permeant anions. In the paradigm for the exocytotic process represented in Fig. 1, the "fission" step (D) involves breakage of the trilaminar membrane separating the granule interior from the extracellular medium. This may be the anion-dependent step equivalent to release in the chromaffin granule system. The actual motive force for release from chromaffin granules is osmotic lysis (8, 14, 15, 24), and a similar force may operate in secretory cells as well.

REFERENCES

1. Poste G, Allison AC: Biochim Biophys Acta 300:421, 1973.
2. Jamieson J, Palade G: J Cell Biol 50:135, 1974.
3. Palade G: Science 189:347, 1975.
4. Heuser JE, Reese TS: J Cell Biol 57:315, 1973.
5. Chi EY, Lasunoff D, Koehler JK: Proc Natl Acad Sci USA 73:2823, 1976.
6. Lawson D, Raff MC, Gomperts B, Fewtrell C, Gilula NB: J Cell Biol 72:242, 1977.
7. Hoffman PG, Zinder O, Bonner WM, Pollard HB: Arch Biochem Biophys 176:375, 1976.
8. Pollard HB, Zinder O, Hoffman PG, Nikodejevik O: J Biol Chem 251:4544, 1976.
9. Anton AH, Sayers DF: J Pharmacol Exp Ther 138:360, 1962.
10. Hirschman RJ, Shulman NR: Br J Haem 24:793, 1973.
11. Rossi EC: J Lab Clin Chem 70:240, 1972.
12. Brown EM, Hurwitz H, Aurbach GD: Endocrinology 99:1582, 1976.
13. Smith AD: In Campell PN (ed): "The Interaction of Drugs and Subcellular Components in Animal Cells." London: Churchill, 1967, p 239.
14. Casey RD, Njus D, Radda GK, Sehr PA: Biochem J 158:383, 1976.
15. Pollard HB, Pazoles CP, Zinder O, Hoffman PG, Nikodejevik O: In "Cellular Neurobiology: Progress in Clinical and Biological Research." New York: Alan R. Liss, 1977, vol 15, pp. 269.
16. Cabantchik ZI, Rothstein A: J Membr Biol 15:207, 1974.
17. Cabantchik ZI, Balshin M, Breuen W, Rothstein A: J Biol Chem 250:5130, 1975.
18. Motais R, Cousin JL: Biochem Biophys Acta 419:309, 1976.
19. Detweiler TC, Martin BM, Feinman RD: In "Biochemistry and Pharmacology of Platelets." Ciba Foundation Symposium 35 (new series). Amsterdam: Elsevier-Excerpta Medica-North Holland, 1975, p 77.

20. Costa JL, Detweiler TC, Feinman RD, Murphy DL, Patlak CS, Pettigrew KD: J Physiol (London) 264:297, 1975.
21. Costa JL, Murphy DL, Kafka M: Biochem Pharmacol 26:517, 1977.
22. Casey RP, Njus D, Radda GK, Sehr PA: Biochemistry 16(5):972, 1977.
23. Johnson RG, Scarpa A: J Gen Physiol 68:601, 1976.
24. Ferris RM, Viviros OH, Kirshner N: Biochem Pharmacol 19:505, 1970.

Journal of Supramolecular Structure 7:287–300 (1977)
Molecular Aspects of Membrane Transport 453–466

Reconstitution of D-Glucose Transport in Vesicles Composed of Lipids and Intrinsic Protein (Zone 4.5) of the Human Erythrocyte Membrane

Arthur Kahlenberg and Cedric A. Zala

Laboratory of Membrane Biochemistry, Lady Davis Institute for Medical Research, Jewish General Hospital, Montreal, Quebec, Canada H3T 1E2

Elucidation of the mechanism of facilitated D-glucose transport in human erythrocytes is dependent on the identification and isolation of the membrane protein(s) mediating this process. Based on the fact that stereospecific D-glucose transport is reconstituted in liposomes prepared by sonication of a lipid suspension with ghosts or fractions derived from ghosts, a quantitative assay for the stereospecific D-glucose transport activity of these fractions was developed (Zala CA, Kahlenberg A: Biochem Biophys Res Commun 72:866, 1976). This assay was used to monitor the purification of ghosts. The solubilized membrane protein fraction was chromatographed on a column of diethylaminoethyl cellulose which was eluted stepwise with NaCl-phosphate buffers of increasing ionic strength. A fraction, eluted at an ionic strength of 0.1, displayed a 13- and 27-fold increase in reconstituted transport activity relative to ghosts and to the unfractionated Triton X-100 extract, respectively. This fraction, when analyzed by sodium dodecyl sulfate-polyacrylamide gel electrophoresis, consisted predominantly of the ghost proteins with an apparent molecular weight of 55,000, commonly designated as zone 4.5; periodic acid-Schiff-sensitive membrane glycoproteins 1–4 were absent. Transport reconstituted by this preparation of zone 4.5 membrane proteins was almost completely abolished by 1-fluoro-2,4-dinitrobenzene, mercuric chloride, and p-chloromercuribenzene sulfonate, but was unaffected by sodium iodoacetate. Extra- and intraliposomal phloretin and cytochalasin B, respectively, exhibited partial inhibition. The stereospecificity and inhibition characteristics of the reconstituted transport imply that all the components of the erythrocyte D-glucose transport system are contained in the zone 4.5 membrane protein preparation.

Key words: erythrocytes, glucose transport, glucose transport protein, liposomes, reconstitution

Abbreviations: SDS-PAGE – sodium dodecyl sulfate-polyacrylamide gel electrophoresis; DEAE – diethylaminoethyl; PAS – periodic acid-Schiff; FDNB – 1-fluoro-2,4-dinitrobenzene; NEM – N-ethylmaleimide; PMBS – p-chloromercuribenzene sulfonate.

Received April 6, 1977; accepted July 26, 1977.

INTRODUCTION

The human erythrocyte monosaccharide transport system is a typical example of facilitated diffusion (1, 2), whereby a stereospecific mechanism equilibrates the concentration of the permeant across the cell membrane. This transport system is inhibited by protein-alkylating reagents and exhibits a high degree of substrate structural specificity, a property which is not mimicked by erythrocyte membrane phospholipids (for reviews see Refs. 3, 4). Consequently, all models of the erythrocyte monosaccharide transport system invoke a membrane protein(s), hereafter referred to as the D-glucose transport protein, possessing one or more sugar binding sites (3, 4).

Although 3 distinct amino acid residues of the D-glucose transport protein essential for its activity have been identified (5), the results of recent attempts at the identification of this membrane component have not been in agreement. In inhibitor binding studies, band 3 protein (6), bands 3 and 4 (7), membrane proteins of zone 4.5 (8), or protein approximately 180,000 in molecular weight (9) were implicated in the erythrocyte sugar transport system [nomenclature of membrane proteins separated in SDS-PAGE and stained with Coomassie blue is according to Steck (10)]. Other attempts to identify the D-glucose transport protein involved the use of selective extraction procedures coupled to the measurement of stereospecific D-glucose uptake by the residual membrane preparations (11) or to the reconstitution of D-glucose transport catalyzed by Triton X-100 extracts of these preparations when incorporated in sonicated liposomes (12, 13). Based on several considerations (reviewed in Ref. 11), band 3 protein, which represented the major polypeptide component of the membrane preparations or Triton X-100 extracts thereof, was suggested to contain the erythrocyte D-glucose transport protein (11–13).

In this report however, we show that following fractionation of a Triton X-100 extract of erythrocyte membrane protein on DEAE-cellulose, reconstitution of D-glucose transport is associated with a column eluate which is devoid of band 3 protein and PAS-sensitive glycoproteins, PAS 1–4; this fraction consists predominantly of the proteins of the broad, complex region designated zone 4.5 on SDS-PAGE.

MATERIALS AND METHODS

All ^{14}C- and ^{3}H-labeled sugars were obtained from New England Nuclear Corporation (Boston, Massachusetts); SDS from Pierce; cytochalasin B from Aldrich; polyacrylamide gel electrophoresis reagents and Bio-Gel P4 from Bio-Rad; inorganic salts and organic solvents from Fisher or Baker; and all other reagents from Sigma Chemical Company (St. Louis, Missouri). Diaflo PM 10 ultrafiltration membranes and Minicon B15 macrosolute concentrators were obtained from Amicon, and DE 52 DEAE-cellulose from Whatman.

Hypotonic phosphate buffer contained 6.20 mM Na_2HPO_4 and 0.70 mM NaH_2PO_4 and was adjusted to pH 7.5 at 5°C. Solution 1 ($\mu = 0.02$) contained 2 mM NaN_3, 0.56 mM NaH_2PO_4, 4.44 mM $NaHPO_4$, 4.18 mM NaCl, and 0.125% Triton X-100 and was adjusted to pH 8.0 at 5°C. Solution 2 ($\mu = 0.02$) was identical to solution 1 except that the concentration of Triton X-100 was 0.100%. Solutions 3 ($\mu = 0.1$) and 4 ($\mu = 0.5$) were identical to solution 2 except that the concentrations of NaCl were 84 mM and 484 mM, respectively. Liposome-forming buffer contained 20 mM $MgCl_2$, 0.03 mM $CaCl_2$, 5 mM Na_2HPO_4, 3 mM NaN_3, and 115.0 mM NaCl and was adjusted to pH 7.4. A stock solution of 5 mM D-[2-^{3}H] glucose (specific activity 160 μCi/mmole) and 5 mM L-[1-^{14}C] glucose (specific activity 100 μCi/mmole) was prepared in liposome-forming buffer and used for all experiments except the reverse isotope experiments, where a similar solution of 5 mM D-[U-^{14}C]-

glucose (specific activity 100 μCi/mmole) and 5 mM L-[1-^{3}H] glucose (specific activity 100 μCi/mmole) in liposome-forming buffer was used.

Hemoglobin-free erythrocyte ghosts were prepared from recently outdated transfusion blood by hypotonic hemolysis (14) as previously described (15). Extraction of ghosts was performed at 5°C as follows: Twenty-four milliliters of ghosts at a protein concentration of 3–4 mg/ml were added to 96 ml of solution 1; after mixing and incubating for 15 min, the supernatant and pellet fractions were separated by centrifugation at $10^5 \times$ g for 1 h. The pellet was washed once with a 10-fold volume of hypotonic phosphate buffer and made up to 10 ml in this buffer; aliquots of the pellet and supernatant were removed for protein determination (16) and analysis by SDS-PAGE (11, 17, 18). The Triton X-100 extract (105 ml) was adjusted to pH 8.0 at 5°C with 0.5 N NaOH and applied at a flow rate of 50 ml/h to a DE 52 DEAE-cellulose column (2.3 × 7 cm), equilibrated with solution 2 at 5°C. After loading, the column was washed with 10 ml of solution 2, 50 ml of solution 3 ($\mu = 0.1$), followed, in initial experiments, by 50 ml of solution 4 ($\mu = 0.5$). Aliquots of 0.25 ml were removed from each column fraction (4 ml) for protein analysis. The 4 fractions (15 ml) under each peak containing the highest protein values (Fig. 2) were separately pooled, concentrated to 3.5–4.0 ml using an Amicon PM 10 membrane and assayed for protein and reconstitution of D-glucose transport. In the case of the 0.5-μ fraction, polypeptide composition was determined by SDS-PAGE. Prior to similar analysis of the polypeptide composition of the 0.1-μ fraction, SDS was added to a concentration of 1% and the fraction was further concentrated 10-fold in a Minicon B15 macrosolute concentrator. The addition of SDS prevented the appearance on Coomassie blue-stained gels of a second broad band with an apparent molecular weight of 120,000, presumably resulting from an aggregation of approximately one-third of the zone 4.5 proteins during concentration.

Reconstitution of D-glucose transport, based on the stereospecific efflux of D- relative to L-glucose from sonicated liposomes, was measured by a slight modification of the method previously described (12). Briefly, the membrane protein fraction is incorporated into liposome bilayers by sonication with a suspension of erythrocyte lipids containing both D-[2-^{3}H]- and L-[1-^{14}C]glucose. The sonicated suspension is then passed through a Bio-Gel P4 column which retards extraliposomal glucose, resulting in a concentration gradient between the inside and the outside of the liposomes. Thus, liposomes reconstituted in the presence of D-glucose transport protein lose their D-glucose while retaining the L-isomer; this D-glucose is in turn separated from the liposomes during further gel filtration. The liposomes are collected in the void volume and analyzed for ^{3}H and ^{14}C and protein content. The nmoles of D-glucose stereospecifically lost from the liposomes per mg protein is then calculated.

The present procedure differed from that previously described (12) in that a solution containing 25 mg of lipid was evaporated to dryness on the bottom of each Quickfit tube; 1.67 ml of a stock solution of 5 mM D- and L-glucose was then added followed by 0.40 ml of various membrane protein fractions. Appropriate volumes of liposome buffer and inhibitor solution, where indicated, were added to give a final volume of 2.5 ml. Protein-free control samples contained the stock sugar solution, 0.8 ml of liposome buffer, and 0.02 ml of 10% Triton X-100. Tubes were sonicated at 30°C for 12 min, and 1-ml duplicates removed for gel filtration. The concentration of D- and L-glucose before gel filtration was therefore 3.33 mM so that the specific activity of D-glucose transport is equal to

$$\left[\frac{\text{dpm }^{14}\text{C after}}{\text{dpm }^{14}\text{C before}} - \frac{\text{dpm }^3\text{H after}}{\text{dpm }^3\text{H before}} \right] \times \frac{3,330 \text{ nmoles D-glucose/ml}}{[\text{protein}] \text{ (mg/ml)}}$$

The inhibition experiments were performed as follows. For the covalent inhibitors, solutions of 150 mM FDNB or NEM in ethanol or 150 mM sodium iodoacetate (pH 7.5) in distilled water were prepared; 0.133 ml of these solutions was added to the sonicated liposome suspension. The samples were then incubated for 1 h at 25°C and duplicate aliquots were assayed for reconstituted D-glucose transport activity. For $HgCl_2$ or PMBS, 0.05 ml of 5 mM aqueous solutions of these inhibitors were added to the liposome suspension prior to sonication. This latter procedure was used also for the noncovalent inhibitors, 10 mM phloretin or 0.5 mM cytochalasin B, which were dissolved in ethanol. In all cases, control samples contained the equivalent volume of the corresponding solvent. In experiments where phloretin was present outside the liposomes during gel filtration on Bio-Gel P4, the columns were previously equilibrated with 15 ml of liposome buffer containing 0.2 mM phloretin and 2% ethanol; control samples were chromatographed on columns equilibrated with 2% ethanol in liposome buffer.

RESULTS

Fractionation of Erythrocyte Membrane Proteins and Reconstitution of D-Glucose Transport

Table I summarizes the data on the fractionation of ghost proteins and reconstitution of D-glucose transport catalyzed by each fraction, the polypeptide composition of which is shown in Fig. 1. Extraction of ghosts with 0.1% Triton X-100 ($\mu = 0.02$) resulted in the solubilization of 22% of the membrane protein and 13% of the D-glucose transport activity. The membrane proteins solubilized were portions of bands 3, 4.2, 5, 6, 7, zone 4.5, and all of the PAS-sensitive glycoproteins. The remainder of the reconstituted D-glucose transport activity of ghosts was recovered in the membrane pellet after Triton X-100 extraction, which contained each of the major Coomassie blue-staining polypeptide bands of ghosts but none of the PAS-sensitive glycoproteins, PAS 1–4.

TABLE I. Reconstituted D-Glucose Transport Activity and Protein Content of Various Erythrocyte Membrane Fractions*

Membrane fraction	Protein (mg)	Protein (%)	D-Glucose transport specific activity (nmoles/mg)[a]	Total D-glucose transport activity (nmoles)	Total D-glucose transport activity (%)
Whole ghosts	87.3	100	93.2 ± 8.2 (23)	8,140	100
0.1% Triton X-100 pellet	56.3	65	131.0 ± 32.7 (6)	7,370	90
0.1% Triton X-100 supernatant	19.9	22	52.5 ± 8.4 (6)	1,038	13
0.1-μ fraction	1.1	1	311.6 ± 30.2 (23)	377	4
0.5-μ fraction	6.2	7	13.7 ± 1.7 (6)	85	1

*Membrane fractions were prepared and incorporated into liposomes which were assayed for D-glucose transport activity as described in Methods. The polypeptide composition of each fraction is shown in Fig. 1. Transport specific activity refers to the nmoles of D-glucose stereospecifically lost from the liposomes per mg of membrane protein associated with the liposomes following chromatography on Bio-Gel P4. Total transport activity represents the product of the specific activity of D-glucose transport and the protein content of the membrane fraction.
[a]In this and subsequent tables, values for D-glucose transport activity are the mean ± standard error of the results from the number of experiments shown in parentheses.

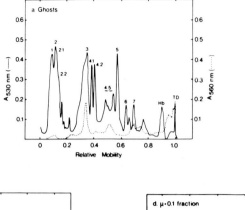

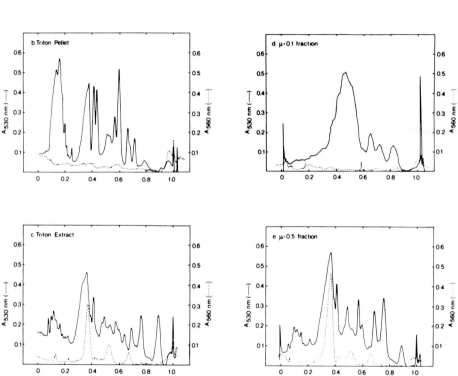

Fig. 1. Protein composition of erythrocyte membrane fractions incorporated into liposomes. Ghosts were extracted with 0.1% Triton X-100 and the resultant solubilized membrane proteins were separated into the 0.1- and 0.5-μ fractions by chromatography on DEAE-cellulose (see Methods for details). Samples from the designated membrane fractions were analyzed electrophoretically on 5% polyacrylamide gels in 0.2% SDS. Solid lines) scans of Coomassie blue-stained gels at 530 nm; dashed lines) scans of periodic acid-Schiff-stained gels at 560 nm. The major Coomassie blue-stained bands were enumerated according to increasing electrophoretic mobility as described by Steck (10). TD) inked needle stab recording the position of the tracking dye; Hb) hemoglobin.

Figure 2 shows the results of chromatography of the Triton X-100 extract of ghosts on a column of DEAE-cellulose. A small amount of protein (12% of that applied) was eluted by solution 3 ($\mu = 0.1$) while 76% was eluted by solution 4 ($\mu = 0.5$). The 0.1-μ fraction consisted primarily of zone 4.5 proteins, with no detectable PAS-sensitive glycoproteins, PAS 1–4. In contrast, the polypeptide and PAS-sensitive glycoprotein composi-

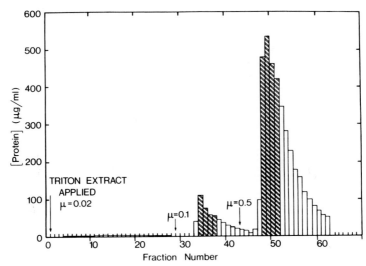

Fig. 2. DEAE-cellulose chromatography of Triton X-100-solubilized erythrocyte membrane proteins. The Triton X-100 extract (105 ml) was applied to a column of DEAE-cellulose and eluted stepwise with NaCl-phosphate buffers of increasing ionic strength (μ = 0.1 and 0.5) as described in Methods. The 4 fractions under each peak containing the highest protein values (hatched lines) were separately pooled and concentrated to yield the 0.1- and 0.5-μ fractions.

tion of the 0.5-μ fraction was similar to that of the original Triton X-100 extract, with some zone 4.5 protein still present (Fig. 1).

In previous reconstitution studies (12), both the membrane residue and Triton X-100 extract of 2,3-dimethylmaleic anhydride-treated ghosts were capable of catalyzing stereospecific D-glucose efflux. The proteins common to these 2 fractions were those of bands 3 and 7 and zone 4.5. The present fractionation of the Triton X-100 solubilized membrane proteins by column chromatography on DEAE-cellulose represents a substantial purification of the D-glucose transport protein. The 0.1-μ fraction displayed a 3.3- and 6-fold increase in D-glucose transport specific activity relative to ghosts and the unfractionated Triton X-100 extract, respectively (Table I). In contrast, the residual Triton X-100 solubilized membrane proteins eluted in the 0.5-μ fraction had a transport specific activity amounting to only 15% of that of the original ghost preparation. Further analysis of the extent of the present purification of the erythrocyte D-glucose transport protein is described below.

Properties of the Reconstituted D-Glucose Transport System

The nmoles of D-glucose lost from the reconstituted liposomes was measured as a function of the amount of membrane protein associated with the liposomes collected in the void volume of the Bio-Gel P4 columns. The lipid suspensions were supplemented with 0.02–0.8 ml of ghosts or the 0.1-μ fraction and sufficient 10% Triton X-100 was added to the latter samples to give a final concentration of 0.16%. With both types of reconstituted liposome preparations, a small amount of membrane protein produced a large efflux of D-glucose (Fig. 3). However, as the protein content of the liposomes was increased, stereospecific D-glucose efflux displayed an asymptotic approach to a limiting value of approximately 30 and 15 nmoles for ghosts and the 0.1-μ fraction, respectively. Thus, the effect on D-glucose transport of incorporation of functional D-glucose transport protein into liposomes formed in the presence of a fixed amount (25 mg) of lipid is a saturable process.

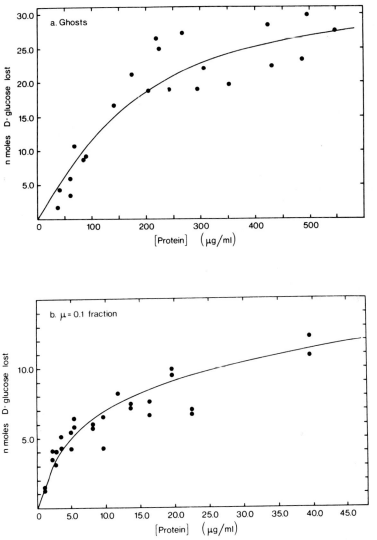

Fig. 3. Reconstituted D-glucose transport activity as a function of membrane protein content of liposomes. Liposomes were reconstituted in the presence of increasing amounts of ghosts (a) or the 0.1-μ fraction (b). The nmoles of D-glucose lost from the reconstituted liposomes was measured as a function of the amount of membrane protein associated with the liposomes collected in the void volume of the Bio-Gel P4 columns (see Methods).

The specific activity values recorded in Table I for the reconstituted transport of D-glucose catalyzed by ghosts or the 0.1-μ fraction were obtained from experiments utilizing a liposomal protein concentration (after gel filtration) of 150–300 and 10–20 μg per ml, respectively. As indicated in Fig. 3, these protein values were in excess of those required for maximum specific activities of reconstituted transport. Since a comparison of the transport specific activity of liposomes reconstituted in the presence of these 2 membrane preparations should be made under conditions in which the liposomal protein concentration is the limiting factor in the amount of stereospecific D-glucose efflux, the values of the initial slopes of the curves in Fig. 3 were used to calculate specific activity.

The slope of the initial linear portion of each curve (8 data points), determined by regression analysis, gave D-glucose transport specific activity values for liposomes reconstituted with ghosts and the 0.1-μ fraction of 111 and 1,420 nmoles per mg membrane protein, respectively. Thus relative to ghosts, Triton X-100 extraction and DEAE-cellulose fractionation results in a 13-fold purification of the erythrocyte D-glucose transport protein, with the recovery of 16% of the initial total transport activity of ghosts in the 0.1-μ fraction based on the protein recoveries listed in Table I.

The concentrated 0.1-μ fraction was unstable and had to be rapidly reconstituted to prevent losses in activity; incubation for 2 h at 20°C decreased activity to 48% of that observed following immediate reconstitution. In addition, incubation for 10 min at 40°C or higher temperatures resulted in complete loss of activity. However, once the fractions had been incorporated into liposomes and stored at 5°C prior to gel filtration, activity was slowly lost, with a half-time of about 1 day (data not shown).

The use of D-[U-^{14}C] glucose and L-[1-^{3}H] glucose as tracers instead of D-[2-^{3}H]-glucose and L-[1-^{14}C] glucose did not affect the D-glucose transport activity values. In 6 experiments, the difference in D-glucose efflux measured by the reverse isotope group (6.76 ± 1.15 nmoles) and the usual mixture of radioactive monosaccharides (8.32 ± 1.89 nmoles) was not significant (P > 0.1). This result indicates that the reconstituted D-glucose transport activity measured in the present experiments cannot be due to an isotope effect.

Effect of Inhibitors on the Reconstituted Transport System

Reconstituted transport activity catalyzed by the 0.1-μ fraction was tested for other properties of the erythrocyte D-glucose transport system by exposing the liposomes to various known inhibitors and a noninhibitor of transport. Incubation of the sonicated liposomes with 10 mM FDNB for 1 h at 25°C essentially abolished transport, while 10 mM NEM gave partial inhibition; 10 mM iodoacetate failed to inhibit (Table II). These results are in agreement with those obtained with intact red cells (19).

The entrapment of mercuric chloride and PMBS (0.1 mM) within liposomes during sonication (extraliposomal HgCl$_2$ and PMBS are retarded during gel filtration) strongly inhibited D-glucose transport. Phloretin, however, must be present on the outside face of the liposomes during gel filtration to inhibit D-glucose efflux (Table II). This agrees with the observation of Benes et al (20) that phloretin inhibits transport in ghosts only if present at the face to which diffusion is proceeding. The failure to completely inhibit transport by the high concentration of phloretin (0.2 mM; 80–400 × K$_i$) (21, 22) may be due to adsorption of the inhibitor to the very high surface area of the liposomes (23), with the concomitant drastic lowering of its effective concentration. Cytochalasin B, a competitive inhibitor of erythrocyte D-glucose transport (24, 25), when present intraliposomally during gel filtration at a concentration of 100 × K$_i$ (24, 25) partially inhibited transport (Table II).

Effect of Increasing Concentrations of Substrate on Reconstituted D-Glucose Transport

The assay for measuring the reconstitution of D-glucose transport activity is based on the fact that the time required for the filtration of liposomes on Bio-Gel P4 is slow enough to allow complete loss of D-glucose from reconstituted liposomes when present at an initial concentration of 3.33 mM. Using liposomes loaded with higher concentrations of D-and L-glucose, one might expect that above a certain substrate concentration the time (2–4 min) of gel filtration might not be sufficient to allow for the complete loss of D-glucose from reconstituted liposomes. Figure 4 shows the results obtained when the

TABLE II. Effect of Known Inhibitors of D-Glucose Transport on the Reconstituted Transport System*

Inhibitor	Transport activity relative to control (%)
1-Fluoro-2,4-dinitrobenzene (10 mM)	3.2 ± 1.7 (4)
N-Ethylmaleimide (10 mM)	56.7 ± 12.6 (4)
Iodoacetate (10 mM)	96.3 ± 6.9 (4)
HgCl$_2$, inside[a] (0.1 mM)	4.2 ± 1.5 (4)
p-Chloromercuribenzene sulfonate, inside (0.1 mM)	7.9 ± 3.5 (4)
Phloretin (0.2 mM)	
inside	110.8 ± 7.6 (6)
inside and outside	44.9 ± 9.6 (4)
outside	61.8 ± 1.4 (4)
Cytochalasin B, inside (0.01 mM)	56.8 ± 8.7 (4)

*The 0.1-μ fraction, recovered from the DEAE-cellulose column (Fig. 2), was used for reconstitution. Mercuric chloride, PMBS, phloretin, and cytochalasin B were added to the lipid suspension before sonication followed immediately by gel filtration; FDNB, NEM, and iodoacetate were added to the liposomes after sonication and the treated samples were incubated for 1 h at 25°C prior to gel filtration. Reconstituted D-glucose transport activity was measured as described in Methods.
[a]Inside and outside refers to the presence during gel filtration of the inhibitor in the intra- and extraliposomal space, respectively.

0.1-μ fraction was sonicated with lipid suspensions containing increasing, but equal, concentrations of D- and L-glucose. Above a concentration of 25 mM, the D-glucose transport activity (solid line) fell below that predicted by extrapolation of the relation defined by the 4 points between 3.33 and 25 mM (dashed line), thus indicating a trend toward substrate saturation of the transport system. At 100 mM, the highest sugar concentration tested, the observed transport activity was 67% of the extrapolated vlaue. It is significant that the substrate concentration used for all other experiments (3.33 mM) is well below that at which nonlinearity in transport activity is apparent.

Substrate Specificity of the Reconstituted Transport System

For the same reasons that an apparent saturation of reconstituted D-glucose transport activity is evident with increasing concentrations of substrate, a varying degree of inhibition of D-glucose transport by sugars with high affinities for the erythrocyte monosaccharide transport system should be observed. The degree of inhibition is a function of the affinity of the sugar analogue, relative to that of D-glucose, for the transport protein. The addition of 100 mM concentrations of sugar analogues to the lipid suspensions ([D- and L-glucose] each equal 3.33 mM) prior to sonication in the presence of the 0.1-μ fraction on the inhibition of D-glucose transport activity is shown in Table III. The most effective inhibition of transport is displayed by the nontransported sugar, maltose (22). Of the transported sugar analogues tested, only 2-deoxy-D-glucose and D-galactose, which have the highest affinities for the D-glucose transport protein (26), inhibited D-glucose efflux.

For the transported sugar analogues, there are 2 effects to consider which contribute to the resultant inhibition of reconstituted D-glucose transport and make quantitative interpretation of the data difficult. 2-Deoxy-D-glucose, by virtue of its high affinity

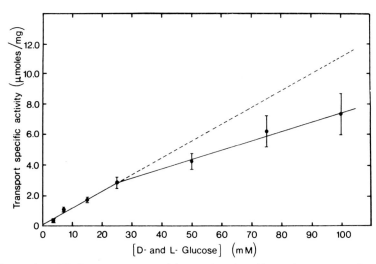

Fig. 4. Reconstituted D-glucose transport activity as a function of increasing concentrations of D-glucose. Liposomes reconstituted in the presence of the 0.1-μ fraction of membrane protein and increasing but equal concentrations of D- and L-glucose were prepared and assayed for D-glucose transport activity (solid line) as described in Methods. The dashed line is a linear extrapolation of the transport activity measured between 3.33 and 25 mM D-glucose. Transport specific activity values are the mean ± standard error of 6 experiments.

together with its high concentration relative to D-glucose, will initially be bound to most of the binding sites on the transport protein and strongly inhibit D-glucose efflux. As efflux progresses, however, the intraliposomal 2-deoxy-D-glucose concentration decreases and with it, the proportion of binding sites occupied by 2-deoxy-D-glucose. With time there will be an increasing efflux of D-glucose. Ultimately, the D-glucose lost would be equal to that lost in the absence of 2-deoxy-D-glucose. However, the time of gel filtration is too short to allow these processes to reach equilibrium, and inhibition of D-glucose efflux by 2-deoxy-D-glucose is observed. A similar argument applies to D-galactose. Sugars of low affinity, despite their high concentration, do not compete with D-glucose for sufficient binding sites to inhibit transport activity under the conditions of the assay. However, since maltose is a nontransported sugar, its intraliposomal concentration remains constant during gel filtration and thus it competes with D-glucose. Therefore, maltose, despite its lower affinity, inhibited D-glucose transport to the same extent as 2-deoxy-D-glucose. As would be expected from its apparent affinity, the degree of inhibition by 2-deoxy-D-glucose is about the same as that observed for D-glucose (Fig. 4) at a concentration of 100 mM.

DISCUSSION

The D-glucose transport system of the human erythrocyte membrane has been the subject of intensive kinetic analysis and many functional properties of the system have been defined (1–4). However, the identity and structure of the D-glucose transport protein(s), its interaction with substrates and membrane lipids, and the molecular processes resulting in the transport event all remain to be elucidated. A first goal in realizing these objectives is the identification of the membrane protein(s) involved in D-glucose transport

TABLE III. Effect of Sugar Analogues on the Activity of the Reconstituted D-Glucose Transport System*

Sugar added	K_m	Transport activity relative to control (%)
Maltose	14	59.4 ± 26.4
2-Deoxy-D-glucose	4	67.1 ± 13.1
D-Galactose	35	73.2 ± 24.8
L-Arabinose	130	112.0 ± 23.4
D-Fucose	240	111.2 ± 22.0
L-Fucose	2,500	112.0 ± 15.4

*Reconstituted D-glucose transport activity, catalyzed by the 0.1-μ fraction, was measured in the presence of 3.33 mM D- and L-glucose as described in Methods. When present, the sugar analogues were added to the lipid suspensions at a concentration of 100 mM prior to sonication. Maltose is a nontransported, competitive inhibitor of the erythrocyte monosaccharide transport system (22). The other transported sugars are listed in a descending order of their apparent affinities (K_m) for the D-glucose transport protein (26).

and the reconstitution of a minimal system capable of catalyzing D-glucose transport. The present study represents a step in this direction.

Upon chromatography of a Triton X-100 extract of erythrocyte ghosts on DEAE-cellulose, a fraction eluted at an ionic strength of 0.1 and consisting predominantly of the polypeptides of zone 4.5, was shown to catalyze D-glucose transport when reconstituted into liposomes. The transport activity of various membrane preparations correlated with their relative amounts of zone 4.5 protein: by SDS-PAGE analysis, zone 4.5 protein accounted for 11, 80, and 12% of the major Coomassie blue-stained membrane polypeptides of ghosts, the 0.1-μ, and the 0.5-μ fractions, respectively. Comparison of the transport specific activities of liposomes reconstituted in the presence of ghosts and the 0.1-μ fraction, under conditions where the amount of membrane protein is limiting, suggests that a 13-fold purification of the D-glucose transport protein had been achieved. However, it should be noted that this degree of purification is an apparent value, dependent on an evaluation of the precise relation between the number of transport proteins in various membrane preparations and their efficiency of reconstitution and consequent ability to support D-glucose transport.

In addition to exhibiting stereospecificity, D-glucose transport reconstituted by the 0.1-μ fraction was shown to have other properties characteristic of the erythrocyte monosaccharide transport system. a) Phloretin, cytochalasin B, and sulfhydryl reagents which inhibit D-glucose transport in erythrocytes (1–4, 26), inhibited the reconstituted transport system. b) There was a tendency to exhibit substrate saturation at high D-glucose concentrations. c) An inhibition of D-glucose transport activity by high-affinity sugars was observed. These data taken together constitute strong evidence that a partially purified, functionally intact D-glucose transport system has been reconstituted in liposomes.

At a Triton X-100 concentration of 0.1%, only 13% of the total D-glucose transport activity of ghosts was extracted. Although we attempted to improve the extraction of this activity by using higher concentrations of Triton X-100, activity was found to decrease, possibly because of detergent-induced denaturation; the optimal concentration for extraction was found to be 0.1% (unpublished data). Furthermore, use of higher concentrations of Triton X-100 results in a decrease in the selectivity of extraction of intrinsic membrane proteins (27), so that subsequent purification of the transport protein might prove more difficult. Since extraction and recovery of D-glucose transport activity were incomplete, the possibility that a distinct subpopulation of D-glucose transport proteins may have been selectively purified cannot be excluded.

The present assay for reconstituted D-glucose transport does not provide a measure of the rate of efflux but rather monitors the amount of D-glucose completely lost from functionally reconstituted liposomes during gel filtration. The assay is therefore not suitable for measurement of kinetic parameters and other rate-dependent properties such as competitive inhibition. However, the method is well suited for estimation of the relative amounts of functional transport protein and is especially useful in the measurement of irreversible inhibition.

Sonicated lipid suspensions are known to contain 2 general classes of liposomes: a population of small, fairly uniform, single-walled spherules 25–30 nm in diameter and a more heterogeneous population of larger size (28). The predominant presence of the former in our sonicated ghost-lipid suspensions was confirmed by electron microscopy (unpublished data). The aqueous volume contained by a small liposome of 11 nm internal radius is 5.6×10^{-18} ml. At a D-glucose concentration of 3.33 mM, the average small liposome would therefore contain about 11 molecules of sugar. This type of calculation helps explain several of our observations if it is assumed that the bulk of transport observed is due to loss of D-glucose from these small liposomes. First, a single transport protein need only catalyze the net efflux of 11 molecules of D-glucose during the time of filtration on Bio-Gel P4 (2–4 min). Even if the reconstituted transport protein had an unusually low turnover number, such a small number of molecules could be lost extremely rapidly. Second, the presence of 2 or more functional transport proteins per liposome would have no further effect on the loss of D-glucose, which would approach a maximum value as an increasing number of transport proteins are incorporated into a constant number of liposomes (Fig. 3). Third, at a concentration of 100 mM, approximately 340 molecules of a high-affinity sugar are present inside the small liposomes; most, if not all, would be lost during gel filtration, and inhibition of D-glucose transport, if seen at all, would be weak (Table III).

Similar considerations apply to the inhibition by chemical reagents (Table II). If $HgCl_2$ and PMBS, at an overall concentration of 0.1 mM, were uniformly distributed throughout the suspension, mercuric and PMBS ions would be present in one out of every 3.4 small liposomes; similarly, at 0.01 mM, cytochalasin B molecules would be included in one out of every 34 liposomes. The observed complete inhibition by mercurials may be explained on the basis of their extremely high reactivity with sulfhydryl groups, so that they would remain complexed with these groups on the D-glucose transport protein during gel filtration. On the other hand, binding of cytochalasin B is rapidly reversible (24), so that essentially all cytochalasin B would be removed from the extraliposomal space during gel filtration. Only intraliposomal cytochalasin B would therefore cause inhibition. If cytochalasin B had bound to a substantial number of sites on the intraliposomal exposed portion of the D-glucose transport protein, this intraliposomal cytochalasin B would cause substantial inhibition of D-glucose efflux (Table II). In this situation, the value of inhibition would be between the small value predicted by liposomal entrapment of cytochalasin B on the basis of its overall concentration and the value of about 100% expected if the binding equilibrium were very slowly reversible.

Collectively, the polypeptides of zone 4.5 account for approximately 11% of the Coomassie blue-staining material of ghosts; with an average apparent molecular weight of 55,000 (10), they would be present in a total of 7.3×10^5 copies per cell. This is in two- to fourfold excess of previous estimates of the number of D-glucose transport proteins per erythrocyte (6, 11). As evident from the SDS-PAGE of ghosts, the broad zone 4.5 region has a complex structure, comprising at least 3 polypeptides, each of which is present in sufficient quantity to account for $2-3 \times 10^5$ D-glucose transport proteins per cell (6, 11). The resolution of these components on SDS-PAGE, however, is impaired by the purification procedure so that a single broad band results. Consistent with the exposure of zone 4.5 proteins on the cytoplasmic surface of the erythrocyte membrane (29), preliminary studies indicate that these proteins contain 4–8% carbohydrate by weight (data not shown).

Although zone 4.5 protein accounts for 80% of the Coomassie blue-stained protein of the 0.1-μ fraction, band 7 and lower-molecular-weight proteins are present as minor contaminants. As indicated previously (6, 11), since a protein exposed on the extracellular surface of the erythrocyte membrane is part of the transport mechanism (8, 30), it is unlikely that band 7 protein which is exposed only on the cytoplasmic membrane surface (10) contains the D-glucose transport protein. However, the possibility that a minor protein contaminant or some non-Coomassie blue-staining component of the 0.1-μ fraction is involved in D-glucose transport has not been definitely eliminated in the present study. In this connection, Batt et al. (8), employing a two-step differential labeling procedure, have recently identified an exofacial component of the erythrocyte hexose transport system which migrated with the polypeptides of zone 4.5 on SDS-PAGE. Both D-glucose and cytochalasin B protected this exofacial transport site from alkylation by impermeant maleimides which irreversibly inhibit sugar transport. In addition, Hinkle and Kasahara (31) have recently demonstrated that the incorporation of zone 4.5 protein in sonicated liposomes results in the reconstitution of D-glucose transport. The present results support and extend these findings. The stereospecificity and inhibition characteristics of the reconstituted D-glucose transport catalyzed by the 0.1-μ fraction imply that all the components of the transport system are contained in this membrane preparation consisting predominantly of zone 4.5 polypeptides. Thus, this soluble membrane fraction should prove invaluable for further pruification of the human erythrocyte monosaccharide transport system in a functional form, suitable for study at the molecular level.

ACKNOWLEDGMENTS

This work was supported by a grant (MT-3120) from the Medical Research Council of Canada. C.A. Zala was the recipient of an MRC Postdoctoral Fellowship. The excellent technical assistance of Randi Greenberg is gratefully acknowledged. We wish to thank Doctors Norman Kalant and James Perdue for their critical review of this manuscript, David Saxe and Karen Cadham for the preparation of the illustrations, and Hoda Karam for the typing of this manuscript.

REFERENCES

1. Miller DM: In Jamieson GA, Gleenwalt TJ (eds): "Red Cell Membrane Structure and Function." Philadelphia: JB Lippincott Company, 1969, p 240.

2. LeFevre PG: In Hokin LE (ed): "Metabolic Pathways." New York: Academic Press, 1972, vol, XI, p 438.
3. LeFevre PG: In Bronner F, Kleinzeller A (eds): "Current Topics in Membranes and Transport." New York: Academic Press, 1975, vol 7, p 109.
4. Jung CY: In MacN Surgenor D (ed): "The Red Blood Cell." Academic Press, 1975, vol 2, p 705.
5. Bloch R: J Biol Chem 249:1814, 1974.
6. Lin S, Spudich JA: Biochem Biophys Res Commun 61:1471, 1974.
7. Taverna RD, Langdon RG: Biochem Biophys Res Commun 54:593, 1973.
8. Batt ER, Abbott RE, Schachter D: J Biol Chem 251:7184, 1976.
9. Jung CY, Carlson LM: J Biol Chem 250:3217, 1975.
10. Steck TL: J Cell Biol 62:1, 1974.
11. Kahlenberg A: J Biol Chem 251:1582, 1976.
12. Zala CA, Kahlenberg A: Biochem Biophys Res Commun 72:866, 1976.
13. Kasahara M, Hinkle PC: Proc Natl Acad Sci USA 73:396, 1976.
14. Dodge JT, Mitchell C, Hanahan DJ: Arch Biochem Biophys 100:119, 1963.
15. Kahlenberg A, Urman B, Dolansky D: Biochemistry 10:3154, 1971.
16. Wang CS, Smith RL: Anal Biochem 63:414, 1975.
17. Fairbanks G, Steck TL, Wallach DFH: Biochemistry 10:2606, 1971.
18. Steck TL, Yu J: J Supramol Struc 1:220, 1973.
19. Stein WD: In Stein WD (ed): "Movement of Molecules Across Cell Membranes." New York: Academic Press, 1967, p 290.
20. Benes I, Kolinska J, Kotyk A: J Membr Biol 8:303, 1972.
21. LeFevre PG, Marshall JK: J Biol Chem 234:3022, 1959.
22. Krupka RM: Biochemistry 10:1143, 1971.
23. Jennings ML, Solomon AK: J Gen Physiol 67:381, 1976.
24. Lin S, Spudich JA: J Biol Chem 249:5778, 1974.
25. Taylor NF, Cagneja FL: Can J Biochem 53:1078, 1975.
26. LeFevre PG: Pharm Rev 13:39, 1961.
27. Yu J, Fischman DA, Steck TL: J Supramol Struct 1:233, 1973.
28. Huang C: Biochemistry 8:344, 1969.
29. Abbott RE, Schachter D: J Biol Chem 251:7176, 1976.
30. Vanstevenink J, Weed RI, Rothstein A: J Gen Physiol 48:617, 1965.
31. Hinkle PC, Kasahara M: J Supramol Struct (Suppl) 1:614, 1977.

Journal of Supramolecular Structure 7: 301–306 (1977)
Molecular Aspects of Membrane Transport 467–472

Interaction Between Cytoplasmic (Ca^{2+}—Mg^{2+}) ATPase Activator and the Erythrocyte Membrane

Frank F. Vincenzi and Martha L. Farrance

Department of Pharmacology, School of Medicine, University of Washington, Seattle, Washington 98195

Human red blood cells (RBC) contain a cytoplasmic, nonhemoglobin protein which activates the (Ca^{2+}-Mg^{2+})ATPase of isolated RBC membranes. Results presented in this paper confirm that activation of (Ca^{2+}-Mg^{2+})ATPase is associated with binding of the cytoplasmic activator to the membrane. Binding of the cytoplasmic activator is reversible and dependent on ionic strength and Ca^{2+}. Cytoplasmic activator is sensitive to trypsin but is not degraded when intact RBC are exposed to trypsin. Cytoplasmic activator does not modify the (Ca^{2+}-Mg^{2+})-ATPase of membranes from RBC exposed to activator prior to hemolysis. Thus, the activator is located in the cell and appears to act by binding to the inner membrane surface.

Key words: cytoplasmic activator, red blood cells, membrane ATPase, Ca^{2+} transport, (Ca^{2+}-Mg^{2+})ATPase

Several years ago, Bond and Clough (1) reported on a curious kind of activity which was found in the membrane-free hemolysate of the human red blood cell (RBC). The activity, which was due to a nonhemoglobin protein, increased the (Ca^{2+}-Mg^{2+})ATPase but not the Mg^{2+}-ATPase or (Na^{+}-K^{+}-Mg^{2+})ATPase activity of isolated RBC membranes. The material will be referred to as "cytoplasmic activator" or "activator" in this communication.

(Ca^{2+}-Mg^{2+})ATPase of RBC membranes has been associated with active transport of Ca^{2+} (2–4). On the other hand, some workers have also suggested that RBC (Ca^{2+}-Mg^{2+})-ATPase may be associated with a system of actomyosin-like fibers which maintain the flexibility and normal biconcave shape of the RBC (5–7). Thus, Bond and Clough (1) sug-

Martha L. Farrance is presently with the Department of Physiology and Biophysics, University of Washington, Seattle, Washington 98195.

Received April 11, 1977; accepted August 11, 1977

This work is taken in part from a dissertation submitted by Martha L. Farrance in partial fulfillment of the requirements of the Doctor of Philosophy Degree.

gested that the cytoplasmic activator could be related to either Ca^{2+} transport, or cell flexibility and shape, or both. Luthra and co-workers partially purified the cytoplasmic activator (8) and demonstrated its existence in the RBC of newborn and adult animals of various species (9). They found it "provocative to consider that this activator may be related to Ca^{2+} transport in erythrocytes which is associated with $(Ca^{2+}\text{-}Mg^{2+})$ATPase or may be a part of a system of actomyosin-like fibers thought to maintain the flexibility and characteristic biconcave shape of the cell" (8).

We were led to investigate the cytoplasmic activator because of widely different $(Ca^{2+}\text{-}Mg^{2+})$ATPase activity of membranes from RBC hemolyzed in various buffers. The explanation is that the cytoplasmic activator can bind to the membrane in high ionic strength. Once binding occurs it is not reversed, even with repeated washings in low ionic strength buffer (10). We demonstrated that the cytoplasmic activator is present in excess in the RBC, that it binds physically to the membrane, and that binding is reversible and dependent on Ca^{2+} (10, 11). Evidence in this paper provides further data compatible with the idea that the cytoplasmic activator is a regulatory protein.

METHODS

Outdated packed human RBC were obtained from the local blood bank. All preparation steps were performed at 0–5°C. The packed cells were washed 3 times with 155 mM NaCl, removing any remaining buffy coat at each step. RBC were hemolyzed in 1 of 2 buffers (pH 7.4): 20 ideal milliosmolar (imosM) imidazole (I20), or 310 imosM imidazole (I310). Calculation of osmolarity was done using the Henderson-Hasselbalch equation assuming a pK_a value for imidazole of 7.05 (an Advanced osmometer gave a measured osmolality of 290 mOs/kg for I310). Hemolysis buffer was added rapidly and vigorously to a measured volume of packed, washed RBC with a buffer:cells ratio of 14:1. The resulting hemolysate was mixed thoroughly and centrifuged at 48,000 × g for 20 min. The pellet of packed ghosts was washed 4 times with I20 (pH 7.4) whether hemolyzed in I20 or I310. After each wash and centrifugation, the supernatant fluid was aspirated, and the tube was rotated to allow the loosely packed membranes to slide off the small pellet of unlysed cells and/or debris; the pellet was then aspirated. Membranes were washed once with 40 mM histidine-40 mM imidazole buffer, pH 7.1 (HI-40). An equal volume of HI-40 buffer was then added to the membranes, and the resulting suspension was stored in the refrigerator on ice. Protein content of the membrane suspension was determined by the method of Lowry et al. (12) using bovine serum albumin as the standard.

The ATPase incubation medium contained (in a final volume of 1 or 2 ml) $\sim$ 100 μg membrane protein, 3 mM ATP (Na_2 ATP, neutralized to pH 7.1), 18 mM histidine-18 mM imidazole buffer (pH 7.1), 3 mM $MgCl_2$, 80 mM NaCl, 15 mM KCl, and 0.1 mM ouabain. $CaCl_2$ (0.1 mM) was added to appropriate tubes for determination of $(Ca^{2+}\text{-}Mg^{2+})$ATPase, which was taken as "extra" ATP splitting induced by Ca^{2+} addition in the presence of Mg^{2+} and ouabain. Tubes without Ca^{2+} gave Mg^{2+}-ATPase activity and also served as a blank to correct for nonenzymatic breakdown of ATP and for the presence of inorganic phosphate in the membrane preparation. All assay tubes (in duplicate) were incubated at 37°C in a shaking water bath for 1 h. The reaction was started by addition of substrate and stopped by the addition of one-half the incubation medium volume of ice-cold 1.5 N perchloric acid. After thorough mixing and centrifugation, 0.5 or 1 ml of supernatant was analyzed for inorganic phosphate using the method of Fiske and SubbaRow (13). Results are expressed as μmoles of inorganic phosphate (P_i) released·mg membrane protein$^{-1} \cdot h^{-1}$.

TABLE I. RBC Membrane $(Ca^{2+}\text{-}Mg^{2+})$ATPase Activity: Influence of Hemolysis Buffer*

Hemolysis buffer	n	$(Ca^{2+}\text{-}Mg^{2+})$ATPase activity
I310	87	3.25 ± 0.07[a]
I20	46	0.76 ± 0.04
TP310	3	0.03 ± 0.02
U310	3	0.37 ± 0.01

*RBC were hemolyzed in various buffers and membranes were prepared and assayed according to Methods. Activities are expressed as μmole $PO_4{}^{3-} \cdot$mg protein$^{-1} \cdot h^{-1}$, mean $\pm$ SEM.
[a]Significantly greater than I20 membrane value, $P < 0.01$.

Endogenous $(Ca^{2+}\text{-}Mg^{2+})$ATPase activator was partially purified by the method of Luthra et al. (8). The sole modification of that method was in the use of imidazole buffers for hemolysis. In experiments in which activator was added to the incubation medium (0.2 ml activator fraction/1ml incubation medium), an equal amount of 20 mM Tris-maleate buffer, pH 6.8 at 25°C, was added to the control.

In all assays, conditions were such that less than 15% of total substrate was utilized during any incubation period. Enzymatic activity was linear with time and proportional to the amount of membrane protein added. Membranes were usually assayed on the day after preparation. Data were analyzed for statistical significance by Student's paired or unpaired t test as appropriate.

RESULTS

We recently found that the cytoplasmic activator of RBC membrane $(Ca^{2+}\text{-}Mg^{2+})$-ATPase (or simply cytoplasmic activator) can bind to the RBC membrane and that binding correlates with high $(Ca^{2+}\text{-}Mg^{2+})$ATPase activity (10, 11). Binding can occur during hemolysis under certain conditions. Thus hemolysis of RBC in isoosmotic (310 imosM) imidazole buffer (I310) results in membranes with very high $(Ca^{2+}\text{-}Mg^{2+})$ATPase activity (Table I). "I310 membranes" have bound cytoplasmic activator whereas "I20 membranes" (prepared from RBC hemolyzed in 20 imosM imidazole) have no associated cytoplasmic activator and have low $(Ca^{2+}\text{-}Mg^{2+})$ATPase activity (Table I). A question which arose as to whether an isoosmotic environment (such as that provided by I310) is sufficient to promote binding of cytoplasmic activator to the RBC membrane during hemolysis. RBC were hemolyzed in isoosmotic urea (U310) or 2,4,6-trimethylpyridine (TP310) solutions, pH 7.4. Membranes derived from these hemolysates did not have an enhanced $(Ca^{2+}\text{-}Mg^{2+})$ATPase activity compared to I20 membranes (Table I). Thus, an isoosmotic environment at the time of hemolysis is not sufficient to promote binding of cytoplasmic activator to the RBC membrane. The results in Table I might be interpreted to mean that imidazole promotes cytoplasmic activator binding in a somewhat specific fashion. However, ionic strength per se appears more important for binding of activator than does the presence of a high concentration of imidazole. This can be demonstrated by a somewhat different approach to hemolysis. RBC in 0.9% NaCl were freeze-thawed 3 times using dry ice/$CHCl_3$−37°C H_2O bath. The broken RBC were then washed with I20 or I310 and membranes were prepared. As shown in Table II, there was no difference in activity between these treatments and each of these "freeze-thaw" membrane preparations had much higher $(Ca^{2+}\text{-}Mg^{2+})$ATPase activity than regular I20 membranes prepared from the same cells. Thus, the presence of 0.9% NaCl during lysis caused by freeze-thawing was apparently sufficient to promote significant activator binding.

TABLE II. RBC Membrane $(Ca^{2+}-Mg^{2+})$ATPase Activity: Influence of Freeze-Thawing*

Type of membrane	Pretreatment of intact cells	$(Ca^{2+}-Mg^{2+})$ATPase activity
I310	None	3.08
I20	None	0.87
I20	1 ml packed RBC in 14 ml 0.9% NaCl, freeze-thaw 3x in dry ice/CHCl$_3$ and 37°C H$_2$O	2.55
I310	1 ml RBC in 14 ml 0.9% NaCl, freeze-thaw 3x in dry ice/CHCl$_3$ and 37°C H$_2$O	2.21

*RBC were freeze-thawed, or not, and membranes were prepared and assayed according to Methods. Activities expressed as μmole PO$_4^{3-}\cdot$mg protein $^{-1}\cdot$h^{-1}.

TABLE III. RBC Membrane $(Ca^{2+}-Mg^{2+})$ATPase Activity: Influence of Cytoplasmic Activator Pretreatment of RBC*

Type of membrane	$(Ca^{2+}-Mg^{2+})$ATPase activity
I20	0.65
I20, from RBC treated with activator in 0.62% NaCl	0.68
I20 (plus added activator in incubation)	2.39
I20, from RBC treated with activator in 0.62% NaCl (plus activator added in incubation)	2.78

*Cytoplasmic activator was partially purified according to Luthra et al. (8). Intact RBC were treated, or not, with cytoplasmic activator in 0.62% NaCl. I20 membranes were then prepared and assayed according to Methods. ATPase activity is expressed as μmole PO$_4^{3-}\cdot$mg protein $^{-1}\cdot$h^{-1}.

Binding of cytoplasmic activator to membranes was shown to be reversible and, under appropriate conditions, to be dependent on Ca^{2+}. The presence of small amounts of Ca^{2+} promotes cytoplasmic activator binding. Chelation of Ca^{2+} with EGTA causes removal of cytoplasmic activator from membranes (11). We considered that reversible Ca^{2+}-dependent binding of cytoplasmic activator might represent a means by which intracellular Ca^{2+} could regulate the Ca^{2+} pump. Of course, in considering that the cytoplasmic activator might be a regulator of the plasma membrane Ca^{2+} pump, it seemed reasonable to predict that it acts on the inner membrane surface. However, neither our results nor those of other workers ruled out that the activator might bind to the outer membrane surface of broken RBC membranes to cause increased $(Ca^{2+}-Mg^{2+})$ATPase activity. The following experiment suggests that activator does not act at the outer membrane surface: Exposure of RBC to partially purified activator in 0.62% NaCl before preparation of I20 membranes neither increases $(Ca^{2+}-Mg^{2+})$ATPase activity nor does it prevent the $(Ca^{2+}-Mg^{2+})$ATPase activity of the resultant I20 membranes from being increased by activator during incubation (1) (Table III). The concentration of NaCl (0.62%) employed is sufficient both to allow the ionic-strength dependent interaction between activator and membrane and to prevent hemolysis. Since there appears to be no significant change in $(Ca^{2+}-Mg^{2+})$ATPase activity from this pretreatment, these data suggest that the activator does not interact with the outside of the membrane and that access to the cytoplasmic surface must be achieved.

TABLE IV. RBC $(Ca^{2+}-Mg^{2+})$ATPase Activity: Influence of Pretreating RBC With Trypsin or Trypsin Plus Trypsin Inhibitor*

Type of membrane	n	$(Ca^{2+}-Mg^{2+})$ATPase activity
I310	2	2.17 ± 0.03
I20	2	1.03 ± 0.06
I20, RBC pretreated with trypsin	2	0.23 ± 0.07^a
I20, RBC pretreated with trypsin plus trypsin inhibitor	2	1.30 ± 0.12

*RBC were pretreated with trypsin or trypsin plus trypsin inhibitor as noted in the text. Membranes were then prepared and assayed as in Methods. ATPase activity is expressed as μmole $PO_4^{3-} \cdot$mg protein^{-1}, mean ±SEM.
[a]Significantly less than I20 membrane value.

The question arose as to whether cytoplasmic activator is really cytoplasmic. Since hemolysis is necessary to detect cytoplasmic activator binding, it is possible that the activator is an outer surface protein and is released from the RBC surface upon hemolysis. It could then gain access to the inner surface (and to the hemolysate). To rule this out, we carried out experiments on partially purified activator and on intact RBC using trypsin and trypsin inhibitor. Activity of partially purified activator was completely lost upon incubation with trypsin (0.5 mg/ml) for 2 h at 25°C. RBC were incubated at 25°C in trypsin, 0.5 mg/ml, in 0.9% NaCl for 2 h, after which trypsin inhibitor, 1 mg/ml, was added. No hemolysis occurred during this procedure or in the control preparation of RBC incubated with trypsin plus trypsin inhibitor; 0.5 and 1.0 mg/ml, respectively. Both groups of RBC were washed twice in 0.9% NaCl and then I20 membranes were prepared. The respective hemolysates were collected and the activator was isolated by the method of Luthra et al (8). Results demonstrate that treatment of intact RBC with trypsin did not decrease activator effectiveness of the subsequent hemolysate (no difference from trypsin plus trypsin inhibitor control). Determination of the $(Ca^{2+}-Mg^{2+})$ATPase activity of the respective I20 membranes derived from these RBC showed that trypsin pretreatment significantly decreased $(Ca^{2+}-Mg^{2+})$ATPase activity of I20 membranes (Table IV). RBC pretreatment with trypsin plus trypsin inhibitor caused a small, nonsignificant ($P > 0.05$) increase in $(Ca^{2+}-Mg^{2+})$ATPase activity. These data suggest that the $(Ca^{2+}-Mg^{2+})$ATPase, but not the activator, is accessible to trypsin in intact RBC.

DISCUSSION

Results of this and related studies demonstrate the existence in human RBC of a cytoplasmic activator of $(Ca^{2+}-Mg^{2+})$ATPase. Effectiveness of the activator is associated with its binding to the RBC membrane, apparently at the inner membrane surface. Treatment of intact RBC with activator results in no effect, and trypsin treatment in intact RBC results in no loss of activator. Thus, the activator appears to be located in the cell and is almost certainly a cytoplasmic protein since it is so readily soluble (1, 8, 10).

Because at least some of the $(Ca^{2+}-Mg^{2+})$ATPase of the RBC membrane is associated with Ca^{2+} transport (2–4), it is suggested that the cytoplasmic activator is a regulator of the Ca^{2+} pump. Support for this suggestion can be found in the recent work of Macintyre and Green (14) who added membrane-free RBC hemolysate to inside-out RBC membrane vesicles. The uptake of $^{45}Ca^{2+}$ into the inside-out vesicles was increased by 50% under their conditions. It must be admitted that there may be more than one $(Ca^{2+}-Mg^{2+})$ATP-ase in the RBC membrane (15–19). Thus, it is possible that some additional, or other, function of the RBC membrane is influenced by cytoplasmic activator. We prefer to think of the cytoplasmic activator as an "activator" of the Ca^{2+} pump but recognize that substantiation of this view will require further work.

ACKNOWLEDGMENTS

The work was supported by USPHS grants AM 16436, GM 00109, and GM 07270.

REFERENCES

1. Bond GH, Clough DL: Biochim Biophys Acta 323:592, 1973.
2. Schatzmann HJ, Vincenzi FF: J Physiol (London) 201:369, 1969.
3. Lee KS, Shin BC: J Gen Physiol 54:713, 1969.
4. Olson EJ, Cazort RJ: J Gen Physiol 53:311, 1969.
5. Weed RI, LaCelle PL, Merrill EW: J Clin Invest 48:795, 1969.
6. LaCelle PL, Kirkpatrick FH, Udkow MP, Arkin B: In Bessis M, Weed RI, LeBlond PF (eds): "Red Blood Cell Shape." New York: Springer-Verlag, 1973.
7. Weed RI, Chailley B: Nouv Rev Fr Hematol 12:775, 1972.
8. Luthra MG, Hildenbrandt GR, Hanahan DJ: Biochim Biophys Acta 419:164, 1976.
9. Luthra MG, Hildenbrandt GR, Kim HD, Hanahan DJ: Biochim Biophys Acta 419:180, 1976.
10. Farrance ML, Vincenzi FF: Biochim Biophys Acta (In press)
11. Farrance ML, Vincenzi FF: Biochim Biophys Acta (In press).
12. Lowry OH, Rosebrough NJ, Farr AL, Randall RJ: J Biol Chem 193:265, 1951.
13. Fiske CH, SubbaRow Y: J Biol Chem 66:375, 1925.
14. Macintyre JD, Green JW: Fed Proc Fed Am Soc Exp Biol 36:27, 1977.
15. Horton CR, Cole WQ, Bader H: Biochem Biophys Res Commun 40:505, 1970.
16. Schatzmann HJ, Rossi GL: Biochim Biophys Acta 241:379, 1971.
17. Bond GH, Green JW: Biochim Biophys Acta 241:393, 1971.
18. Quist EE, Roufogalis BD: Arch Biochem Biophys 168:240, 1975.
19. Quist EE, Roufogalis BD: Biochem Biophys Res Commun 72:673, 1976.

Journal of Supramolecular Structure 7:443–461 (1977)
Molecular Aspects of Membrane Transport 473–491

Energetics and Molecular Biology of Active Transport in Bacterial Membrane Vesicles

H. R. Kaback, S. Ramos, D. E. Robertson, P. Stroobant, and H. Tokuda

Laboratory of Membrane Biochemistry, Roche Institute of Molecular Biology, Nutley, New Jersey 07110

Bacterial membrane vesicles retain the same sidedness as the membrane in the intact cell and catalyze active transport of many solutes by a respiration-dependent mechanism that does not involve the generation of utilization of ATP or other high-energy phosphate compounds. In E. coli vesicles, most of these transport systems are coupled to an electrochemical gradient of protons ($\Delta\bar{\mu}_H +$, interior negative and alkaline) generated primarily by the oxidation of D-lactate or reduced phenazine methosulfate via a membrane-bound respiratory chain. Oxygen or, under appropriate conditions, fumarate or nitrate can function as terminal electron acceptors, and the site at which $\Delta\bar{\mu}_H +$ is generated is located before cytochrome b_1 in the respiratory chain.

Certain (N-dansyl)aminoalkyl-β-D-galactopyranosides (Dns-gal) and N(2-nitro-4-azidophenyl)aminoalkyl 1-thio-β-D-galactopyranosides (APG) are competitive inhibitors of lactose transport but are not transported themselves. Various fluorescence techniques, direct binding assays, and photoinactivation studies demonstrate that the great bulk of the *lac* carrier protein (ca. 95%) does not bind ligand in the absence of energy-coupling. Upon generation of a $\Delta\bar{\mu}_H +$ (interior negative and alkaline), binding of Dns-gal and APG-dependent photoinactivation are observed. The data indicate that energy is coupled to the initial step in the transport process, and suggest that the *lac* carrier protein may be negatively charged.

Key words: bioenergetics, membrane sidedness, electrochemical proton gradient, D-lactate dehydrogenase, dansylgalactosides, azidophenylgalactosides

Over the past decade or so, it has become increasingly apparent that membrane vesicles isolated from bacteria as well as eucaryotes (1) provide a unique and useful system for the study of certain aspects of active transport. These vesicles are devoid of the cytoplasmic constituents of the intact cell and their metabolic activities are restricted to those provided by the enzymes of the membrane itself, constituting a considerable advantage over intact cells. Since transport by membrane vesicles per se is practically nil, the driving force for transport of a particular substrate can be determined by studying which compounds or experimental manipulations stimulate its accumulation. In addition, metabolic conversion of the transport substrate and the energy source is minimal, allowing clear definition of the reactions involved. Finally, removal or disruption of the cell wall allows

Received for publication March 14, 1977; accepted June 22, 1977 the use of certain probes, inhibitors, and ionophores which are normally inaccessible to the plasma membrane.

Since many aspects of the bacterial membrane vesicle system have been reviewed (2–6), this contribution is intended as more of a progress report than a review. Thus, various aspects of the work will be summarized, but emphasis is placed upon more recent observations dealing with respiration-linked active transport. In a general sense, however, it should be stressed that studies with this system are relevant not only to active transport, but to the general problem of energy transduction and possibly other membrane-related phenomena (7–12). As opposed to mitochondria and chloroplasts where respiratory energy and light are converted to another form of chemical energy (i.e., ATP), in this experimental system respiratory energy is converted primarily into work in the form of solute concentration against an electrochemical or osmotic gradient.

GENERAL ASPECTS OF VESICULAR TRANSPORT

Transport assays with membrane vesicles are performed in various ways (3, 4). The most widely used method is a filtration assay in which vesicles are incubated with a radioactive transport substrate in the presence of an appropriate energy source, and at a given time, the reaction mixtures are diluted to terminate the uptake reaction. The vesicles are then immediately separated from the medium by means of rapid membrane filtration. Although this method is rapid and convenient, it suffers from the disadvantage that the reaction mixtures must be diluted prior to filtration, and in many instances, this operation results in significant losses of certain solutes (13). The vesicles can also be separated from the reaction mixtures by centrifugation, but in this case, samples cannot be assayed rapidly, corrections must be made for radioactive solute trapped in the pellet, and the reaction mixtures become anaerobic even during rapid centrifugation (13). Very recently, the flow dialysis technique devised by Colowick and Womack (14) has been adapted to measure transport (13, 15, 16), and it is presently clear that this is the method of choice in many ways. By this means, changes in the external concentration of solute are determined continuously, accurately, and highly reproducibly under conditions which require no manipulation of the experimental system.

Evidence was presented as early as 1960 (17) which indicated that bacterial membrane vesicles would provide a useful model system for the study of active transport. Subsequent studies demonstrated that the vesicles catalyze the transport of a plethora of metabolities in the presence of appropriate energy sources (5), and that initial rates of transport of many of these metabolites are comparable to those of the intact cell (18, 19). Moreover, the vesicles accumulate many solutes to concentrations markedly in excess of those in the external medium. Initial progress with the system was slow primarily because of preconceived ideas regarding the physical nature of the vesicles and the energetics of active transport. It was generally thought that transport would be driven by ATP or other nucleoside triphosphates, and considerable effort was expended in an effort to provide evidence for this notion. However, as it turns out, the energy source for transport in isolated membrane vesicles varies with the organism, with the substance transported, and ATP or other nucleoside triphosphates do not drive transport in the vesicle system (see Ref. 5 for a review of the evidence).

In general, the transport systems elucidated in the vesicle system fall into 3 categories: i) group translocation mechanisms in which a covalent change is exerted upon the transported molecule so that the reaction itself results in the passage of the molecule through the diffusion barrier (2,5), ii) active transport in which solute is accumulated against an electrochemical or osmotic gradient, and iii) passive diffusion of certain weak acids and lipophilic ions, followed by equilibration with the pH gradient and the electrical potential, respectively, across the membrane.

Transport of many sugars, amino acids, organic acids, and ions by E. coli and S. typhimurium membrane vesicles occurs by active transport. These transport systems are coupled primarily to the oxidation of D-lactate to pyruvate catalyzed by a flavin adenine dinucleotide-linked, membrane-bound D-lactate dehydrogenase (D-LDH) which has been purified to homogeneity (20, 27). Electrons derived from D-lactate are passed to oxygen via a membrane-bound respiratory chain, and during this process, respiratory energy is converted into work in the form of active transport (18, 22–25). Although other oxidizable substrates stimulate transport to some extent, they are not nearly as effective as D-lactate unless ubiquinone-1 (CoQ_1) is added to the vesicles (25). It should be emphasized, however, that generation of NADH from inside the vesicles stimulates transport in the absence of exogenous CoQ_1 (26) and that D-lactate is not an effective electron donor for active transport in all bacterial membrane systems (5). Active transport in the vesicles is also driven very dramatically by the nonphysiologic electron donors reduced phenazine methosulfate [PMS] (27) or pyocyanine (25), both of which donate electrons to the respiratory chain at a site prior to the cytochromes. The use of these nonphysiologic electron donors has allowed the generalization of the vesicle system to many bacteria (5, 27).

E. coli membrane vesicles also catalyze active transport in the absence of oxygen when the appropriate anaerobic electron transfer systems are present (28–30). Lactose and amino acid transport under anaerobic conditions can be coupled to the oxidation of α-glycerol-P with fumarate as an acceptor or to the oxidation of formate utilizing nitrate as electron acceptor. Both of these anaerobic electron transfer systems are induced by growth of the organism under appropriate conditions, and components of both systems are loosely bound to the membrane.

Finally, active transport can be driven by artificially induced potassium gradients of appropriate polarity (31, 32). When membrane vesicles prepared in potassium-containing buffers are diluted into solutions lacking this cation, and the ionophore valinomycin is added, the diffusion of potassium out of the vesicles creates an electrical potential across the vesicles membrane (interior negative), and uptake of certain substrates is observed. Although the extent of transport observed under these conditions is considerably less than that observed with D-lactate and reduced PMS, the finding that solute accumulation occurs under these conditions has great implications with respect to the energetics of respiration-linked active transport (see below).

SIDEDNESS OF MEMBRANE VESICLES AND SPECIFICITY OF D-LACTATE AS AN ELECTRON DONOR FOR ACTIVE TRANSPORT

One of the most striking and therefore controversial aspects of the vesicle system is the degree of specificity of the physiologic electron donors which drive active transport. In E. coli vesicles, of a large number of potential energy sources tested, very few replace D-lactate to any extent whatsoever, and none is as effective although many are oxidized at least as rapidly (18, 22, 23, 33, 34). It should also be emphasized in passing that incubation of the vesicles with radioactive D-lactate, L-lactate, succinate, or α-glycerol-P results in stoichiometric conversion of these compounds to pyruvate, fumarate, or dihydroxyacetone-P, respectively (33–35). Thus, in each case, the ability of these compounds to drive transport is related to a clearly defined enzymatic reaction.

Since each electron donor reduces the same membrane-bound cytochromes, both qualitatively and quantitatively (22, 35), it was suggested that the energy-coupling site for active transport in E. coli and Staphylococcus aureus vesicles is located in a relatively specific segment of the respiratory chain between D-LDH and cytochrome b_1, the first cytochrome in the common portion of the respiratory chain. Some critics have argued, however, that a significant number of vesicles become inverted during preparation, and that these inverted vesicles oxidize NADH and other electron donors but do not catalyze active transport (36–38). This is an extremely important consideration, but there is now a large body of evidence demonstrating that it is extremely unlikely. That is, the membrane of each vesicle retains the same orientation as the membrane in the intact cell. Another explanation that has been proposed for the inability of certain electron donors to drive transport is dislocation of dehydrogenases from the inner to the outer surface of the vesicle membrane during preparation (39–41). In the one instance in which this possibility has been studied extensively (i.e., D-LDH, see below), it is clearly not the case (42–44). Some of the evidence supporting these contentions are as follows:

i) Initial rates of transport in the vesicles are, in many cases, similar to those observed in whole cells (18, 19). Moreover, in most instances, the steady-state level of accumulation of transport substrates is comparable to that observed in the intact cell, and the electrochemical proton gradient generated by E. coli vesicles (see below) is at least as great as that of the intact cells (compare the data presented in Refs. 13, 15, 16 with that in Ref. 45).

ii) Freeze fracture studies of membrane vesicles in at least 3 different laboratories (3, 5, 39, 46) demonstrate that the "texture" of the convex surface of the vesicles is distinctly different than that of the concave surface and that the vesicles are homogeneous in this respect. Moreover, the texture observed on the respective surfaces is exactly the same as that observed in the intact cell.

iii) As mentioned previously, all of the electron donors which are oxidized by the vesicles reduce the same cytochromes both qualitatively and quantitatively (22, 35). If a percentage of the vesicles were inverted, and only these inverted vesicles oxidized NADH, it is difficult to understand how NADH could reduce all of the cytochrome in the preparations.

iv) Although NADH is generally a poor electron donor for transport in E. coli vesicles, it is the best physiologic electron donor for transport in B. subtilis vesicles which are prepared in a similar manner (47). Moreover, recent experiments carried out in this laboratory (25) demonstrate that addition of ubiquinone (CoQ_1) to E. coli ML 308–225 vesicles in the presence of NADH results in rates and extents of lactose and amino acid transport which are comparable to those observed with D-lactate. Since this effect of CoQ_1 is not observed in the presence of NADPH nor in vesicles lacking NADH dehydrogenase activity, it seems apparent that CoQ_1 is able to shunt electrons from NADH dehydrogenase to an energy-coupling site which is not located in that portion of the respiratory chain between NADH dehydrogenase and the cytochromes. As such, these observations provide direct evidence for specific localization of the energy-coupling site.

v) Studies by Reeves et al. (48) demonstrate that fluorescence of 1-anilino-8-naphthalene sulfonate (ANS) is dramatically quenched upon addition of D-lactate to E. coli ML 308-225 membrane vesicles, an observation similar to that observed in energized mitochondria and ethylenediaminotetraacetic acid-treated intact E. coli. In chloroplasts and submitochondrial particles, in which the polarity of the membrane is opposite to that of intact mitochondria, ANS fluorescence is enhanced upon energization. It follows that any inverted membrane vesicles in the preparations would exhibit enhanced ANS fluorescence in the presence of D-lactate. Thus, if 50% of the vesicles were inverted, no net change in ANS fluorescence should have been observed by Reeves et al. because half of the vesicles would exhibit quenching and half would exhibit enhancement.

Similarly, Rosen and McClees (49) have demonstrated that inverted membrane preparations catalyze calcium accumulation but do not catalyze D-lactate dependent proline transport. In contrast, vesicles prepared by osmotic lysis (3) do not exhibit calcium transport but accumulate proline effectively in the presence of D-lactate.

vi) Although D-LDH mutants exhibit normal transport and vesicles prepared from these mutants do not exhibit D-lactate-dependent transport, addition of succinate to these vesicles drives transport to the same extent as D-lactate in wild-type vesicles (50). Since succinate oxidation by both wild-type and mutant vesicles is similar, it seems apparent that the coupling between succinate dehydrogenase and transport is increased in the mutant vesicles. In vesicles prepared from double mutants defective in both D-LDH and succinate dehydrogenase, the coupling between L-lactate dehydrogenase and transport is increased, and L-lactate is the best physiologic electron donor for transport (F. Grau, J.-S. Hong, and H. R. Kaback, unpublished information). Moreover, in vesicles prepared from a triple mutant defective in D-LDH, succinate dehydrogenase, and L-lactate dehydrogenase, the coupling between NADH dehydrogenase and transport is markedly increased, and NADH drives transport as well as D-lactate in wild-type vesicles (F. Grau, J.-S. Hong, H. R. Kaback, unpublished information). In addition, it is noteworthy that vesicles prepared from galactose-grown E. coli exhibit high rates and extents of lactose transport in the presence of NADH. These observations indicate that the coupling between a particular dehydrogenase and the energy-coupling site for transport is subject to regulation, and that it may be difficult, if not impossible, to demonstrate specificity of energy-coupling in the intact cell. In some bacteria, however, evidence in favor of this hypothesis has been presented with intact cells. In Arthrobacter pyridinolis (51), hexose transport in both intact cells and membrane vesicles is coupled to malate dehydrogenase; and in a marine pseudomonas (52), it has been shown that amino acid transport in whole cells and membrane vesicles is coupled to alcohol dehydrogenase.

vii) As will be discussed below (see section on Reconstitution), studies with antibodies against D-LDH demonstrate that this membrane-bound enzyme is present exclusively on the inner surface of the vesicle membrane (42–44). Moreover, D-lactate oxidation drives transport normally in dld^- membrane vesicles reconstituted with D-LDH, and in this system, the enzyme is located exclusively on the outer surface of the vesicle membrane. Thus, none of the wild-type vesicles can be inverted or sufficiently damaged to allow access of antibody to the interior surface of the membrane, and D-LDH can drive transport normally even when it is present on the wrong side of the membrane.

viii) 2-Hydroxy-3-butenoic acid (vinylglycolate) is an analogue of lactate which is actively transported by the lactate transport system and oxidized by D- and L-LDHs. As opposed to normal substrates, however, oxidation of this compound yields a reactive electrophile (2-keto-3-butenoate) which is attacked by many sulfhydryl-containing proteins on the membrane. Although there is considerable evidence supporting these conclusions (53–56) only 2 points are critical for this discussion: a) Vinylglycolate transport is the limiting step for labeling the membrane proteins, and b) almost all of the vinylglycolate taken up is covalently bound to the vesicles. In experimental terms, the rate of covalent binding of vinylglycolate is stimulated at least tenfold by ascorbate-phenazine methosulfate (ascorbate-PMS); and stimulation is completely abolished by uncoupling agents or phospholipase treatment, neither of which affect vinylglycolate oxidation.

Using extremely high specific activity [³H] vinylglycolate, vesicles have been labeled for an appropriate time in the presence of ascorbate-PMS and examined by radioautography in the electron microscope (53). Each vesicle that takes up vinylglycolate is overlaid with exposed silver grains, and examination of the preparations reveals that 85–90% of the vesicles are labeled. It should be emphasized that this is a minimal estimation. Virtually all of the large vesicles are labeled, while the size of the smaller vesicles is such that their proximity to individual silver grains in the emulsion may be limiting. Moreover, essentially identical radioautographic results are obtained with [³H] acetic anhydride, a reagent which reacts nonspecifically with the vesicles. Thus most, if not all, of the vesicles in the preparations catalyze active transport.

ENERGETICS OF ACTIVE TRANSPORT

An initial model proposed by Kaback and Barnes (23) depicted the carriers as electron transfer intermediates in which a change from the oxidized to the reduced state results in translocation of the carrier-substrate complex to the inner surface of the membrane and a concomittant decrease in the affinity of the carrier for substrate. The model was posed as a tentative working hypothesis that could provide a role for sulfhydryl groups in translocation and at the same time, account for the observation that only certain electron transfer inhibitors cause efflux of accumulated solutes. A very different hypothesis, one that emphasizes the positioning of respiratory chain components within the matrix of the membrane giving rise to an electrochemical gradient of protons as the immediate driving force for active transport, was proposed by Peter Mitchell (36, 37, 57–62), and over the past few years, it has become eminently clear that Mitchell's so-called "chemiosomotic hypothesis" provides the best explanation for active transport to date.

As visualized by Mitchell, oxidation of electron donors via the membrane-bound respiratory chain or hydrolysis of ATP catalyzed by the membranous Ca^{2+}, Mg^{2+}-stimulated ATPase complex is accompanied by the expulsion of protons into the external medium, leading to an electrochemical gradient of protons ($\Delta\bar{\mu}_H+$) which is composed of an electric and a chemical parameter according to the following relationship:

$$\Delta\bar{\mu}_H+ = \Delta\Psi - \frac{2.3RT}{F}\Delta pH \qquad (1)$$

where $\Delta\Psi$ represents the electric potential across the membrane, and ΔpH is the chemical difference in proton concentrations across the membrane (2.3RT/F is equal to 58.8 mV at room temperature). According to this hypothesis, it is the electrochemical gradient of protons or one of its components which is the immediate driving force for the inward movement of transport substrates. Transport of organic acids is postulated to be dependent upon the pH gradient (ΔpH) (i.e., the undissociated acid is transported through the membrane and is presumed to accumulate in the ionized form due to the relative alkalinity of the internal milieu), while the transport of positively charged compounds such as lysine or potassium is purportedly coupled to the electric component ($\Delta\Psi$), and the uptake of neutral substrates such as lactose or proline is thought to be coupled to $\Delta\bar{\mu}_H+$ and to occur via symport with protons.

Using lipophilic cations and rubidium (in the presence of valinomycin), it has been demonstrated that E. coli membrane vesicles generate a $\Delta\Psi$ (interior negative) of approximately -75 mV in the presence of reduced PMS or D-lactate (31, 63, 64). Furthermore, the potential causes the appearance of high affinity binding sites for dansylgalactosides, azidophenylgalactosides, and p-nitrophenyl-α-D-galactopyranoside on the surface of the vesicles membrane (65, 66) and is partially dissipated as a result of lactose accumulation (64). Although these findings lend strong support to the chemiosmotic hypothesis, it was apparent that $\Delta\Psi$ in itself is insufficient to account for the magnitude of solute accumulation by the vesicles if it is assumed that the stoichiometry between protons and solute is 1:1 (36). This deficiency, in addition to the apparent absence of a transmembrane pH gradient, left reasonable doubt as to the quantitative relationship between $\Delta\bar{\mu}_H+$ and solute accumulation (64). Recent experiments from this laboratory (13, 15, 16) have resolved this problem to a large extent.

Utilizing flow dialysis, a technique uniquely suited to the measurement of ΔpH across isolated membrane vesicles, it has been shown that membrane vesicles isolated from E. coli grown under various conditions generate a transmembrane pH gradient of about 2 units (interior alkaline) under appropriate conditions. Using the distribution of weak acids (i.e., acetate, propionate, butyrate, and 5,5-dimethyloxazolidine-2, 4-dione) to measure ΔpH and the distribution of the lipophilic cation triphenylmethylphosphonium to measure $\Delta\Psi$, the vesicles are demonstrated to develop a $\Delta\bar{\mu}_H+$ of almost -200 mV (interior negative and alkaline) at pH 5.5 in the presence of reduced PMS or D-lactate, the major component of which is a ΔpH of about -120 mV. As external pH is increased, ΔpH decreases, reaching 0 at about pH 7.5 and above, while $\Delta\Psi$ remains at about -75 mv and internal pH remains at pH 7.5—7.8. To some extent, these variations in ΔpH are probably caused by changes in the oxidation of reduced PMS or D-lactate, both of which vary with external pH in a manner similar to that described for ΔpH. However, it should also be mentioned that recent experiments (66a; S. Ramos, H. Rottenberg, and H. R. Kaback, unpublished information) suggest the operation of a mechanism which catalyzes the exchange of external protons for intravesicular sodium or potassium at relatively alkaline pH. Finally, and importantly, ΔpH and $\Delta\Psi$ can be varied reciprocally in the presence of valinomycin and nigericin with little change in $\Delta\bar{\mu}_H+$ and no apparent change in respiratory activity. In addition to providing direct support for some of the general predictions of the chemiosmotic hypothesis, these results provide a powerful experimental framework within which to test the relationship between $\Delta\bar{\mu}_H+$, ΔpH, and $\Delta\Psi$ and the accumulation of specific transport substrates.

Addition of lactose or glucose-6-P to membrane vesicles containing the appropriate transport systems results in partial collapse of ΔpH, demonstrating that respiratory energy can drive active transport via the pH gradient across the membrane. Titration studies with valinomycin and nigericin lead to the conclusion that at pH 5.5, there are 2 general classes of transport systems: Those that are coupled primarily to $\Delta\bar{\mu}_H+$ (lactose, proline, serine, glycine, tyrosine, glutamate, leucine, lysine, cysteine, and succinate) and those that are coupled primarily to ΔpH (glucose-6-P, lactate glucuronate, and gluconate). Strikingly, however, it is eminently clear that at pH 7.5, all of the transport systems are driven by

$\Delta\Psi$ which comprises the only component of $\Delta\bar{\mu}_H+$ at this external pH. In addition, when the effect of external pH on the steady-state level of accumulation of various transport substrates is examined, none of the pH profiles corresponds to those observed for $\Delta\bar{\mu}_H+$, $\Delta\Psi$, or ΔpH, and at external pH values exceeding 6.0–6.5, $\Delta\bar{\mu}_H+$ is insufficient to account for the concentration gradients observed for most of the substrates. This finding and the observation that the accumulation of organic acids is coupled to $\Delta\Psi$ at relatively high external pH values indicate that the stoichiometry between protons and transport substrates may vary as a function of external pH, exhibiting a value of 1 at relatively low external pH and increasing to 2 or more as external pH is increased. Experimental evidence which provides direct support for this conclusion has been presented (66b).

One attractive conceptual aspect of the chemiosmotic hypothesis for bacterial active transport is its analogy to the mechanism suggested for sugar and amino acid transport in many eucaryotic cells (67). In these systems, an electrochemical gradient of sodium rather than protons in generated through the action of the membranous sodium, potassium-dependent ATPase, and accumulation of sugars and amino acids occurs via coupled movements with sodium (this process is referred to traditionally as cotransport rather than symport).

Although it is almost certain that many bacterial transport systems catalyze proton/substrate symport, several instances have been reported in which the transport of a specific solute is dependent upon the presence of sodium or lithium ion (see Ref. 68 for a review). Moreover, some of these studies, in particular those of Stock and Roseman (69) and Lanyi et al. (70), indicate that symport or cotransport mechanisms may be operative. Since the basic energy-yielding process in bacteria is thought to be proton extrusion and bacteria apparently do not possess a sodium, potassium-dependent ATPase or a primary sodium pump, the existence of such transport systems presents certain obvious problems, among which are: i) the relationship between the proton electrochemical gradient and these transport systems, ii) the mechanism by which the internal sodium concentration is maintained at a low level.

Tokuda and Kaback (68) have recently shown that membrane vesicles isolated from Salmonella typhimurium G-30 grown in the presence of melibiose catalyze methyl-1-thio-β-D-galactopyranoside (TMG) transport in the presence of sodium or lithium as shown initially with intact cells by Stock and Roseman (69). TMG-Dependent sodium uptake is also observed, but only when a potassium diffusion potential (interior negative) is induced across the vesicle membrane. Cation-dependent TMG accumulation varies with the electrochemical gradient of protons generated as a result of D-lactate oxidation, and the vesicles catalyze D-lactate-dependent sodium efflux in a manner which is consistent with the operation of a proton/sodium exchange mechanism. Although the stoichiometry between sodium and TMG appears to be 1:1 when transport is induced by a potassium diffusion potential, evidence is presented which indicates that the relationship may exceed unity under certain conditions. The results are consistent with a model in which TMG/sodium (lithium) symport is driven by an electrochemical gradient of protons which functions to maintain a low intravesicular sodium or lithium concentration through proton/sodium (lithium) antiport. A similar mechanism has been suggested for light-dependent glutamate transport in vesicles from Halobacterium halobium (70).

RECONSTITUTION OF D-LACTATE DEHYDROGENASE-DEPENDENT FUNCTIONS IN D-LACTATE DEHYDROGENASE MUTANTS

The membrane-bound D-LDH of E. coli has been solubilized and purified to homogeneity (20, 21). The enzyme has a molecular weight of 75,000 ± 7%, contains approximately 1 mole of flavin adenine dinucleotide per mole of enzyme, and exhibits low activity towards L-lactate. Oxidized diphosphopyridine nucleotide (NAD) has no effect on the catalytic conversion of D-lactate to pyruvate. Finally, recent work carried out in collaboration with Drs. John Salerno and Tomoko Ohnishi of the Johnson Foundation of The University of Pennsylvania indicates that the enzyme probably does not contain a nonheme iron center (P. Stroobant, J. Salerno, H. R. Kaback, and T. Ohnishi, unpublished information).

While much of this work was in progress, Reeves et al. (71) demonstrated that guanidine·HCl extracts from wild-type membrane vesicles containing D-LDH activity are able to reconstitute D-lactate-dependent oxygen consumption and active transport in membrane vesicles from E. coli and S. typhimurium mutants defective in D-LDH (dld^-). These studies have been confirmed and extended by Short et al. (72) using the homogeneous preparation of D-LDH described above, and Futai (73) has independently confirmed many of the observations.

Reconsituted dld^- vesicles carry out D-lactate oxidation and catalyze the transport of a number of substrates when supplied with D-lactate. D-Lactate is not oxidized, and will not support transport of any of these substrates in unreconstituted dld^- membranes. Binding of enzyme to wild-type membranes produces an increase in D-lactate oxidation but has little or no effect on the ability of the membranes to catalyze active transport. Reconstitution of dld^- membranes with increasing amounts of D-LDH produces a corresponding increase in D-lactate oxidation, and transport approaches an upper limit which is similar to the specific transport activity of wild-type membrane vesicles. However, the quantity of enzyme required to achieve maximum initial rates of transport varies somewhat with different transport systems.

Binding of 2-(N-dansyl)aminoethyl-β-D-thiogalactoside (DG_2) to membrane vesicles containing the lac transport system is dependent upon D-lactate oxidation, and this fluorescent probe can be utilized to quantitate the number of lac carrier proteins in the membrane vesicles (see subsequent discussion). When dld-3 membrane vesicles are reconstituted with increasing amounts of D-LDH, there is a corresponding increase in the binding of DG_2. Assuming that each lac carrier protein molecule binds one molecule of DG_2, it can be estimated that there is at least a seven- to eightfold excess of lac carrier protein relative to functional D-LDH in reconstituted dld^- vesicles. A similar determination can be made for wild-type vesicles. These vesicles contain approximately 0.07 nmole of D-LDH per mg membrane protein (based on the specific activity of the homogeneous enzyme preparation), about 1.1 nmoles of lac carrier protein per mg membrane protein, yielding a ratio of about 15 for lac carrier protein relative to D-LDH.

Although the rate and extent of transport decreases dramatically with reconstitution, the rate and extent of labeling of dld^- vesicles with radioactive vinylglycolate remains constant. As discussed above, this compound is transported via the lactate transport system, and oxidized to a reactive product by D- and L-LDHs on the inner surface of the vesicle membrane. The observation that reconstituted *dld⁻* membranes do not exhibit enhanced labeling by vinylglycolate suggests that bound D-LDH is present on the outer surface of the vesicles. In this case, the reactive product released from D-LDH would be diluted into the external medium, whereas if the enzyme were on the inner surface of the vesicle membrane, the rate of labeling would be expected to increase with reconstitution, since the reactive product should accululate within the vesicles to higher effective concentrations.

The suggestion that D-LDH is localized on the outer surface of reconstituted *dld⁻* membrane vesicles, as opposed to the inner surface of native ML 308-225 vesicles has received strong support from recent experiments with antibody against D-LDH (42–44). Incubation of ML 308-225 membrane vesicles with anti-D-LDH does not inhibit D-LDH activity (assayed by tetrazolium dye reduction, oxygen uptake, and/or D-lactate-dependent transport) unless the vesicles are disrupted physically or spheroplasts are lysed in the presence of antibody. In contrast, treatment of reconstituted *dld⁻* vesicles with anti-D-LDH results in marked inhibition of D-LDH activity. The titration curves obtained with reconstituted *dld-3* membrane vesicles are almost identical quantitatively to that obtained with the homogeneous preparation of D-LDH. The conclusion that D-LDH is able to drive transport from the outer surface of the membrane is also consistent with recent experiments of Konings (74) and Short et al. (43) demonstrating that reduced 5-N-methylphenazonium-3-sulfonate, an impermeable electron carrier, drives transport as well as reduced PMS, its lipophilic analogue. In addition to providing information about the localization of D-LDH in native and reconstituted vesicles, the results with the native vesicles are consistent with other experiments which demonstrate that essentially all of the vesicles catalyze active transport (cf. above) and therefore cannot be inverted or sufficiently damaged to allow access of antibody to D-LDH. Given these conclusions and the suggested mechanism for generation of an electrochemical proton gradient (36, 37), it is amazing that oxidation of D-lactate by reconstituted *dld⁻* vesicles leads to an electrochemical proton gradient which is indistinguishable in polarity and magnitude from that observed in wild-type vesicles where the enzyme is on the inner surface of the vesicle membrane (S. Ramos, R. Schuldiner, and H. R. Kaback, unpublished information).

The flavin moiety of the holoenzyme appears to be critically involved in binding D-LDH to the membrane (72). Treatment with [1-¹⁴C] hydroxybutynoate leads to inactivation of D-LDH by modification of the flavin adenine dinucleotide coenzyme bound to the enzyme. Enzyme labeled in this manner does not bind to *dld⁻* membrane vesicles. The findings suggest that the flavin coenzyme itself may mediate binding or alternatively, that covalent inactivation of the flavin may result in a conformational change that does not favor binding. It is tempting to speculate on the relevance of this finding to the synthesis of membrane-bound dehydrogenases in the intact cell. Possibly, the apoprotein moiety of D-LDH is synthesized on cytoplasmic ribosomes, but is not inserted into the membrane until coenzyme is bound. If this is so, D-LDH mutants which are defective in the flavin binding site should exhibit soluble material which cross-reacts immunologically with native D-LDH.

MOLECULAR ASPECTS OF CARRIER FUNCTION

It seems apparent from the foregoing discussion that although many of the details are not yet completely clear, a unifying concept which explains the energetics of respiration-dependent active transport in bacterial systems is emerging. However, it should be emphasized that this is only the beginning, that ultimately, carrier molecules must be solubilized, purified, and reconstituted so that this concept can be studied on a molecular level. Initial promising developments in this respect have already been reported (75, 76). In any case, within the past few years, certain important insights into carrier function at a more refined level of resolution have been achieved. These studies (see Refs. 65, 77 for reviews) have involved a collaborative effort between the authors' laboratory and that of Dr. Rudolf Weil of Sandoz Forschungsinstitut in Vienna, Austria.

In addition to the wealth of information available with regard to the β-galactoside transport system, and the ability to manipulate it genetically, this transport system has another notable property which makes it particularly advantageous for study. As opposed to most other bacterial transport systems, the β-galactoside system is relatively nonspecific with respect to the nongalactosyl moiety of the disaccharide. Thus, reporter groups or chemically reactive species can be incorporated into this class of sugars without abolishing their affinity for the *lac* carrier protein.

Substituted Galactopyranosides

The structures of the compounds to be discussed are shown in Figs. 1 and 2. As shown, they fall into 2 general categories (N-dansyl)aminoalkyl-β-D-galactopyranosides [dansylgalactosides] (78, 83) and (2-nitro-4-azidophenyl)-β-D-galactopyranosides [azidophenylgalactosides] (84, 85). In both cases, a β-D-galactopyranoside (usually 1-thio-β-D-galactopyranoside) is linked directly or through an alkyl chain of varying length to the appropriate probe. In one instance, a fluorescent moiety, commonly referred to as a dansyl fluorophore is sensitive to solvent polarity, exhibiting an increase in quantum yield and a blue shift in the emission spectrum as the solvent becomes less polar. Thus, the dansyl moiety is useful as a reporter group, yielding information about the polarity of its environment.

FLUORESCENT β-GALACTOSIDES

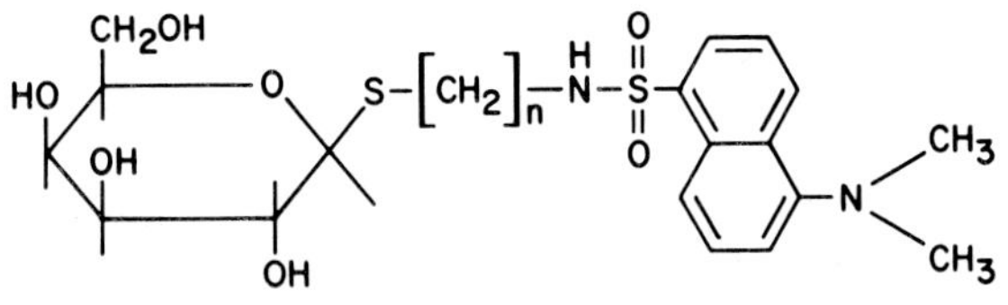

R=	
	DG_0
$S(CH_2)_2$	DG_2
$S(CH_2)_3$	DG_3
$S(CH_2)_4$	DG_4
$S(CH_2)_5$	DG_5
$S(CH_2)_6$	DG_6
$O-CH_2CH_2$	OXY DG_2

Fig. 1. Structural formulae of various dansylgalactosides.

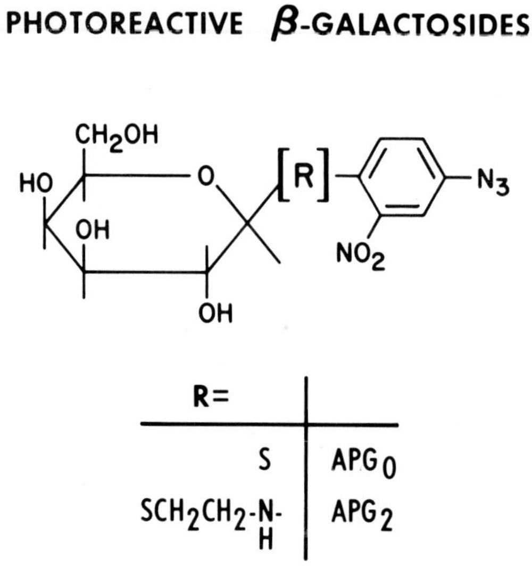

PHOTOREACTIVE β-GALACTOSIDES

R=	
S	APG_0
$SCH_2CH_2\text{-}N\text{-}H$	APG_2

Fig. 2. Structural formulae of axidophenylgalactosides.

The second group of compounds contains an arylazide moiety linked to the galatosyl portion of the molecule (Fig. 2). The rationale behind the use of this class of compounds is that irradiation with visible light causes photolysis of the azido group to form molecular nitrogen and a highly reactive nitrene which then reacts covalently with the macromolecule to which the nitrene-containing ligand is bound, resulting in irreversible inactivation.

Dansylgalactosides Are Not Transported.

Each of the dansylgalactosides shown in Fig. 1 is a competitive inhibitor of lactose uptake in membrane vesicles prepared from strains of E. coli which contain a functional *lac* carrier protein (K_i values for Dns[0,2,3,4,5, and 6]-Gal are approximately 550, 30, 12, 6, 3, and 5 μM, respectively). Strikingly, however, an exhaustive number of experiments indicates that the dansylgalactosides are not transported to a significant extent (78–80).

Binding Is Energy Dependent

The fluorescent properties of the dansylgalactosides in aqueous solution are similar to those of other dansyl derivatives, exhibiting emission maxima at approximately 540–550 nm and excitation maxima at approximately 340 nm. Significantly, no change in these parameters is observed in the presence of membrane vesicles. On addition of D-lactate, however, there is a marked increase in dansylgalactoside fluorescence which is absolutely dependent upon the presence of functional *lac* carrier protein in the vesicles. Moreover, no change in the fluorescence of 2′-(N-dansyl)aminoethyl-1-thio-β-D-glucopyrano-side (dansylglucoside) is observed, indicating that the effects observed are specific for the galactosyl configuration of the ligand. It is also highly significant that the effect of D-lactate can be mimicked by the imposition of an ionic diffusion potential (interior negative) or by a lactose diffusion gradient (inside ⟶ outside).

The fluorescence increase observed with each of the dansylgalactosides under the conditions described exhibits an emission maxima at 500 nm and excitation maxima at

345 and 292 nm. The blue shift in the emission maximum is equivalent to that observed when the dansylgalactosides are dissolved in 85% dioxane, indicating that the affected molecules find themselves in a relatively hydrophobic environment. The appearance of the new peak in the excitation spectrum at 292 nm indicates moreover that the bound dansylgalactoside molecules are excited by energy transfer from tryptophanyl residues in the membrane proteins.

The increase in dansylgalactoside fluorescence induced by D-lactate is blocked or rapidly reversed by addition of β-galactosides, sulfhydryl reagents, inhibitors of D-lactate oxidation, and proton conductors or other reagents which collapse the membrane potential. On the other hand, the fluorescence increase induced by imposition of ionic diffusion potentials is blocked by β-galactosides, sulfhydryl reagents, and proton conductors, but not by respiratory poisons. Moreover, the degree to which dansylgalactoside fluorescence is increased under these conditions is dependent upon the magnitude of the applied ionic diffusion potential. Since D-lactate oxidation leads to the generation of a membrane potential (interior negative), it seems clear that the increase in dansylgalactoside fluorescence observed with D-lactate or artificially imposed ion gradients occurs by a similar mechanism. Finally, the increase in dansylgalactoside fluorescence observed with a lactose diffusion gradient is inhibited by β-galactosides and sulfhydryl reagents, both of which block the binding site of the carrier, but not by respiratory poisons or reagents which collapse the membrane potential. Thus, it seems clear that the fluorescence changes induced under these conditions occur by a mechanism which is independent of the membrane potential.

Titration of vesicles containing the *lac* carrier protein with each dansylgalactoside demonstrates that the vesicles bind 1–2 nmol of each dansylgalactoside per mg of membrane protein, a value almost identical to that obtained from direct binding measurements with [^{3}H] DG$_6$ (82, 83). Assuming that one dansylgalactoside molecule is bound per molecule of *lac* carrier protein and that the molecular weight of the *lac* carrier protein is approximately 30,000, 1–2 nmol/mg membrane protein is equivalent to 3–6% of the membrane protein. This value is very similar to that reported by Jones and Kennedy (85) who used a completely independent method. In addition, these titration studies indicate that the affinity of the *lac* carrier for ligand is directly related to the length of the aklyl chain between the galactosyl and dansyl moieties of the dansylgalactosides. There is excellent agreement, moreover, between the affinity constants of the various dansylgalactosides as determined by fluorimetric titration or flow dialysis (78, 79, 82, 83) and their apparent K$_i$ values for lactose transport as given above.

There are at least 3 possible mechanisms by which energy might lead to dansylgalactoside binding to the *lac* carrier protein: i) The carrier is accessible to the external medium and binding occurs spontaneously. In this case, energy coupling results in partial translocation of the bound ligand, resulting in its exposure to the hydrophobic interior of the membrane, and thus to the fluorescence changes observed. This possibility seems unlikely because no changes in the emission or excitation spectra of the dansylgalactosides are observed in the absence of D-lactate and for reasons to be discussed below. ii) The carrier is accessible to the external medium in the absence of energy coupling, but its affinity is increased when energy is supplied. iii) The carrier is inaccessible to the external medium, and energy coupling casues a conformational change in the carrier such that high affinity binding sites appear on the external surface of the membrane. If it is postulated moreover that the *lac* carrier protein (or part of it) has a negative charge, the appearance of binding sites on the exterior surface of the membrane can be more easily conceptualized. Imposition of a membrane potential (interior negative) would cause "movement" of the negatively charged carrier to the external surface of the membrane and binding of dansylgalactosides. It should be emphasized that the last 2 possibilities are not mutually exclusive and that energy-coupling may well increase the accessibility of the carrier and its affinity for ligand simultaneously. In any case, the data suggest that energy is coupled to one of the initial steps in transport, and that facilitated diffusion therefore cannot represent the first step in the active transport of β-galactosides.

Further evidence for the proposition that the fluorescence changes observed upon "energization" of the membrane are due to binding of the dansylgalactosides per se rather than binding followed by translocation into the hydrophobic interior of the vesicle membrane has been provided by the use of 2 other independent techniques. Anisotropy of fluorescence can be used to assess binding specifically since changes in this parameter reflect alterations in the rotation of molecules in solution. Studies with Dns^2-Gal and Dns^6-Gal (81) demonstrate that there is a marked increase in fluorescence anisotropy in vesicles containing the *lac* carrier protein on addition of D-lactate. In the absence of this electron donor, anisotropy values are minimal and identical in vesicles with or without the *lac* carrier protein, and no increase in anisotropy is observed in vesicles devoid of the *lac* carrier protein when D-lactate is added. In addition to these changes in anisotropy, fluorescence lifetime studies with Dns^2-Gal and Dns^6-Gal are also consistent with the proposition that changes in dansylgalactoside fluorescence observed on "energization" of the appropriate vesicles reflect binding of the fluorescent probes to the *lac* carrier protein (81). Furthermore, it can be calculated from the anisotropy and lifetime values that the rotational relaxation time of Dns^2-Gal increases dramatically when the probe is bound to the *lac* carrier protein.

Finally, high specific activity $[^3H] Dns^6$-Gal has been synthesized, and its binding to membrane vesicles has been studied directly by flow dialysis (82, 83). With vesicles containing the *lac* carrier protein, little, if any, binding is detected in the absence of D-lactate or reduced PMS. In the presence of these electron donors, binding is observed and the binding constant and number of binding sites are approximately 5 μM and 1.5 nmol/mg membrane protein, respectively. Both values are in excellent agreement with those obtained by fluorescence titration. These results demonstrate almost unequivocally that the changes in dansylgalactoside fluorescence observed on energization of membrane vesicles containing the β-galactoside transport system reflect binding of the probe to the *lac* carrier protein.

ENERGY DEPENDENT PHOTOINACTIVATION

Studies with photoreactive azidophenylgalactosides (Fig. 2) have provided completely independent support for the conclusions derived from the dansylgalactoside experiments. 2-Nitro-4-azidophenyl-1-thio-β-D-galactopyranoside (APG_0) is a competitive inhibitor of lactose transport in ML 308-225 membrane vesicles, exhibiting an apparent K_i of 75 μM (84). The initial rate and steady-state level of $[^3H] APG_0$ accumulation are markedly stimulated by the addition of D-lactate to vesicles containing the β-galactoside transport system, and kinetic studies reveal an apparent K_m of 75 μM. Vesicles devoid of the β-galactoside transport system do not take up significant amounts of APG_0 in the presence or absence of D-lactate. When exposed to visible light in the presence of D-lactate APG_0 irreversibly inactivates the β-galactoside transport system. Strikingly, APG_0-dependent photoinactivation is not observed in the absence of D-lactate. Kinetic studies of the inactivation process yield a K_D of 77 μM, a value which is almost identical to the K_m and K_i values obtained with this compound. Moreover, lactose protects against APG_0 photoinactivation and significant inactivation of amino acid transport is not observed with APG_0. Thus, it is clear that these effects are specific for the *lac* carrier protein.

Analogous studies carried out with 2'-N(2-nitro-4-azidophenyl)aminoethyl-1-thio-β-D-galactopyranoside demonstrate that this compound behaves similarly with respect to photoinactivation of the β-galactoside transport system with 2 important exceptions (85). Like its analogue Dns^2-Gal1, APG_2 is not accumulated by the vesicles in the presence of D-lactate or ascorbate-PMS and it exhibits a higher affinity for the *lac* carrier protein than APG_0 (i.e., the K_1 for competitive inhibition of lactose transport and the K_D for photoinactivation in the presence of D-lactate are 35 μM). In addition, it has been demonstrated that an artificially imposed membrane potential (interior negative) also leads to APG_2-dependent photoinactivation of the *lac* carrier protein.

An Apparent Discrepancy

As discussed above, it seems clear from studies with dansyl- and azidophenyl-galactosides that binding of these ligands by the *lac* carrier protein is energy dependent. On the other hand, Kennedy et al. (87) have reported that binding of β-D-galactopyranosyl-1-thio-β-D-galactopyranoside (TDG) and p-nitrophenyl-α-D-galactopyranoside (NPG) to membrane particles prepared by ultrasonic disruption is independent of the presence of an energy source and is not inhibited by sodium azide. However, assuming that the purity of these membrane particles is comparable to that of isolated membrane vesicles,[1] that binding rather than trapping of ligand within an internal space was measured,[2] and that TMG permease II (88) was not present in the particles, the amount of binding observed by Kennedy et al. (87) is considerably less than the total amount of *lac* carrier protein (i.e., M protein) present in the membrane as determined by titration studies with the dansylgalactosides and as determined by Jones and Kennedy (85). In any case, a small but significant amount of nonspecific binding of dansylgalactosides is detected by fluorescence anisotropy and lifetime studies, allowing the possibility that the techniques described thus far are not sufficiently sensitive to detect less than 10% of the binding observed under energized conditions. For this reason, high specific activity [^{3}H] NPG was synthesized and its binding to membrane vesicles investigated by means of flow dialysis (66). These studies corroborate the findings of Kennedy et al. (87), but they also confirm the observations discussed above. There is a small amount of NPG binding by vesicles containing the *lac y* gene product (about 0.2 nmol per mg membrane protein at saturation) which is abolished and reversed by p-chloromercuribenzene-sulfonate (p-CMBS) but not by proton conductors, and this binding is not dependent upon the presence of D-lactate. On addition of D-lactate, however, approximately 2.3 nmol NPG are bound per mg membrane protein (i.e., an amount similar to that observed with the dansylgalactosides), and all of the bound ligand is displaced by p-CMBS.

---

[1] This assumption may not be justified as membrane particles prepared as described by Kennedy et al. (87) apparently contain as much as 30–40% of the total cellular protein (C. F. Fox, personal communication).

[2] Kennedy et al. (87) utilized the distribution of inorganic phosphate to correct for ligand which may have been trapped in the internal space of these particles. The use of phosphate in this respect is somewhat unorthodox, as this anion is generally felt to be impermeant.

Moreover, the K_D for binding under energized and nonenergized conditions is very similar (i.e., 6–9 μM). Thus, although there is a small amount of binding in the nonenergized state, it seems quite evident that the great bulk of the *lac* carrier protein is unable to bind ligand unless the membrane is energized.

Translocation and Accumulation of β-Galactosides

The observation that a small but significant amount of NPG binding to membrane vesicles containing the *lac* carrier protein occurs in the absence of an electrochemical gradient of protons or artificially applied ion diffusion potentials suggests that the *lac* carrier protein may exist in two forms which are in a state of dynamic equilibrium: i) a high affinity form which is accessible on the external surface of the membrane, and ii) a low affinity, cryptic form. In the absence of D-lactate or reduced PMS, 90% or more of the carrier is in the low affinity, cryptic form and only 10% or less is in the high affinity, accessible form. Upon generation of an electrochemical proton gradient across the vesicle membrane (interior negative and alkaline), one or more negatively charged groups in the low affinity, cryptic form of the protein might be influenced, resulting in a conformational change and a shift in the equilibrium. According to such a model, active transport would occur by binding of ligand to the high affinity form of the carrier on the external surface of the membrane, followed by conversion of the carrier to the low affinity, cryptic form and release of ligand from the inner surface of the membrane. However, since the carrier may be negatively charged, translocation of ligand would require neutralization of the high affinity form of the carrier on the external surface of the membrane. This might be accomplished if ligand binding increased the pK_a of a negatively charged functional group(s) in the carrier. The protein would then be uncharged and no longer under the influence of the electrochemical gradient. The protonated carrier-ligand complex would "relax" to the cryptic form and release proton(s) and ligand on the inner surface of the membrane, regenerating the charged form of the carrier, and the cycle could then be repeated. Clearly, this reaction sequence is consistent with many of the observations discussed above. It is also noteworthy that according to this model, the carrier would not translocate protons across the membrane in the absence of ligand, a stipulation that is necessary lest the carriers themselves dissipate the electrochemical gradient without performing work. It should also be emphasized that this formulation might account for low rates of facilitated diffusion without necessitating that the process represent the initial step in the active transport of β-galactosides.

Distance Measurements With the *lac* Carrier Protein

It should be clear from the foregoing discussion that the fluorescence changes observed with the dansylgalactosides are due specifically to binding of these ligands to a site in the *lac* carrier protein and not to a subsequent translocation event. Since the spectral properties of the dansyl group are sensitive to solvent polarity, varying the distance between the galactosyl and dansyl moieties of the dansylgalactosides might be informative with respect to the environment in the immediate vicinity of the binding site in the *lac* carrier protein.

Recently (83), each of the dansylgalactoside homologues shown in Fig. 1 was synthesized in radioactive form and binding to membrane vesicles containing the *lac* carrier protein measured directly by flow dialysis in the presence of D-lactate. The results were then compared with the D-lactate-induced fluorescence enhancement observed with each dansylgalactoside and with the ability of N-methylpicolinium perchlorate (89) to quench the fluorescence of the bound homologues. The following observations have been clearly documented (83): i) The binding affinity of the *lac* carrier protein is directly related to the length of the alkyl chain linking the galactosyl and dansyl ends of the molecules. ii) The maximum number of binding sites observed for each homologue is essentially identical. iii) The increase in fluorescence observed when the probes are bound to the *lac* carrier protein changes markedly when the distance between the galactosyl and dansyl moieties is varied. As the linkage is lengthened from 2 to 4 carbons, fluorescence decreases by a factor of 10 or more and then increases dramatically and progressively with Dns^5-Gal and Dns^6-Gal. These effects vary inversely with the ability of N-methylpicolinium perchlorate to quench the fluorescence of the bound probes.

In view of the specificity of dansylgalactoside binding and the fluorescence properties of the dansyl group, it seems apparent that the galactosyl end of these molecules is anchored at the binding site of the *lac* carrier protein, while the dansyl end proceeds from a hydrophobic environment to an aqueous environment to a hydrophobic environment as the alkyl linkage is lengthened. Possible interpretations of this behavior depend upon assumptions regarding the flexibility and hydrophobicity of the alkyl linkage in the molecules. If it is assumed simplistically that the linkage merely maintains a linear configuration, the variations in fluorescence might reflect differences in the polarity of the microenvironment within the membrane or on the surface in the vicinity of the *lac* carrier protein. An alternative interpretation is that the alkyl linkage is both flexible and hydrophobic, in which case, both parameters would vary directly with chain length. Under these circumstances, it seems reasonable to suggest that the dansyl moiety in Dns^2-Gal reflects a hydrophobic environment in the binding site of the *lac* carrier protein, and as the dansyl end of the molecule is removed 3 and then 4 carbons from the binding site, it becomes accessible to the aqueous solvent at the membrane interface. When the alkyl linkage is then elongated to 5 and subsequently 6 carbons, however, the molecules might become sufficiently flexible and hydrophobic such that the dansyl moiety and part of the alkyl chain adsorb to a hydrophobic site on the surface of the membrane which may or may not comprise part of the *lac* carrier protein. Although it is impossible to distinguish absolutely between these 2 alternatives at the present time, it should be apparent that the latter interpretation is favored by the results of the quenching studies with N-methylpicolinium perchlorate. Moreover, if the latter interpretation is correct, the binding site in the *lac* carrier protein is probably about 5–6 Å from the aqueous solvent at the surface of the membrane.

REFERENCES

1. Cellular Regulation of Transport and Uptake of Nutrients, Fogarty International Symposium, J Cell Physiol 89:495, 1976.
2. Kaback HR: Annu Rev Biochem 39:561, 1970.
3. Kaback HR: Methods Enzymol 22:99, 1971.
4. Kaback HR: Methods Enzymol 31: 698, 1974.
5. Kaback HR: Science 186:882, 1974.
6. Kaback HR: J Cell Physiol 89: 575, 1976.
7. Weissbach H, Thomas E, Kaback HR: Arch Biochem Biophys 147:249, 1971
8. Thomas E, Weissbach H, Kaback HR: Arch Biochem Biophys 150:797, 1972.
9. Thomas E, Weissbach H, Kaback HR: Arch Biochem Biophys 157:327, 1973.
10. Cox GS, Thomas E, Kaback HR, Weissbach H: Arch Biochem Biophys 158:667, 1973.
11. Cox GS, Kaback HR, Weissbach H: Arch Biochem Biophys 167:610, 1974.

12. Grollman E, Lee G, Ambesi-Impiombato FS, Meldolesi MF, Aloj SM, Coon HG, Kaback HR, Kohn LD: Proc Natl Acad Sci USA 74:2352, 1977.
13. Ramos S, Schuldiner S, Kaback HR: Proc Natl Acad Sci USA 73:1892, 1976.
14. Colowick SP, Womack FC: J Biol Chem 244:774, 1969.
15. Ramos S, Kaback HR: Biochemistry 16:848, 1977.
16. Ramos S, Kaback HE: Biochemistry 16:854, 1977.
17. Kaback HR: Fed Proc Fed Am Soc Exp Biol 19:130, 1960.
18. Lombardi FJ, Kaback HR: J Biol Chem 247:7844, 1972.
19. Short S, White D, Kaback HR: J Biol Chem 247:7452, 1972.
20. Kohn LD, Kaback HR: J Biol Chem 246:5518, 1971.
21. Futai M: Biochemistry 12:2468, 1973.
22. Barnes EM Jr, Kaback HR: J Biol Chem 246:5518, 1971.
23. Kaback HR, Barnes EM Jr: J Biol Chem 246:5523, 1971.
24. Kaback HR: Biochim Biophys Acta 265:367, 1972.
25. Stroobant P, Kaback HR: Proc Natl Acad Sci USA 72:3970, 1975.
26. Futai M: J Bacteriol 120:861, 1974.
27. Konings WN, Barnes EM Jr, Kaback HR: J Biol Chem 246:5857, 1971.
28. Konings WN, Kaback HR: Proc Natl Acad Sci USA 70:3376, 1973.
29. Boonstra J, Huttunen MT, Konings WN, Kaback HR: J Biol Chem 250:6792, 1975.
30. Konings WN, Boonstra J: Adv Microbiol Physiol (In press).
31. Hirata H, Altendorf K, Harold FM: Proc Natl Acad Sci USA 70:1804, 1973.
32. Hirata H, Altendorf K, Harold FM: J Biol Chem 249:2939, 1974.
33. Kaback HR, Milner LS: Proc Natl Acad Sci USA 66:1008, 1970.
34. Barnes EM Jr, Kaback HR: Proc Natl Acad Sci USA 66:1190, 1970.
35. Short SA, White D, Kaback HR: J Biol Chem 247:298, 1972.
36. Mitchell P: J Bioenerg 4:63, 1973.
37. Harold FM: Bacteriol Rev 36.172, 1972.
38. Hare JB, Olden K, Kennedy EP: Proc Natl Acad Sci USA 71:4843, 1974.
39. Altendorf K, Staehelin LA: J Bacteriol 117:888, 1974.
40. Weiner JH: J Membr Biol 15:1, 1974.
41. Futai M: J Membr Biol 15:15, 1974.
42. Short SA, Kaback HR, Hawkins T, Kohn LD: J Biol Chem 250:4285, 1975.
43. Short SA, Kaback HR, Kohn LD: J Biol Chem 250:4291, 1975.
44. Futai M: J Bacteriol 124:470, 1975.
45. Padan E, Zilberstein D, Rottenberg H: Eur J Biochem 63:533, 1976.
46. Konings WN, Bisschop A, Voenhuis M, Vermeulen CA: J Bacteriol 116:1456, 1973.
47. Konings WN, Freese E: J Biol Chem 247:2408, 1972.
48. Reeves JP, Lombardi FJ, Kaback HR: J Biol Chem 247:6204, 1972.
49. Rosen BP, McClees JS: Proc Natl Acad Sci USA 71:5042, 1974.
50. Hong J-s, Kaback HR: Proc Natl Acad Sci USA 69:3336, 1972.
51. Wolfson EB, Krulwich TA: Proc Natl Acad Sci USA 71:1739, 1974.
52. Thompson J, MacLoed RA: J Bacteriol 117:1055, 1974.
53. Short SA, Kaback HR, Kaczorowski G, Fisher J, Walsh CT, Silverstein S: Proc Natl Acad Sci USA 71:5032, 1974.
54. Walsh CT, Kaback HR: Ann NY Acad Sci 235:519, 1974.
55. Walsh CT, Kaback HR: J Biol Chem 248:5456, 1973.
56. Shaw L, Grau F, Kaback HR, Hong J-s, Walsh CT: J Bacteriol 127:1047, 1975.
57. Mitchell P: Biol Rev Cambridge Philos Soc 41:445, 1966.
58. Mitchell P: Adv Enzymol 29:33, 1967.
59. Mitchell P: "Chemiosmotic Coupling in Oxidative Phosphorylation and Photosynthetic Phosphorylation." Bodmin, England: Glynn Res. Ltd., 1968.
60. Mitchell P: In Bittar EE (ed): "Membranes and Ion Transport." New York: Wiley-Interscience, 1970, vol 1, p 192.
61. Mitchell P: In Charles HP, Knight BCJG (eds): "Organization and Control in Procaryotic and Eukaryotic Cells." Symposium of the Society of General Microbiology. London and New York: Cambridge University Press, 1970, vol 20, p 121.
62. Greville GD: Curr Top Bioenerg 3:1, 1969.

63. Altendorf K, Hirata H, Harold FM: J Biol Chem 249:4587, 1975.
64. Schuldiner S, Kaback HR: Biochemistry 14:5451, 1975.
65. Schuldiner S, Rudnick G, Weil R, Kaback HR: Trends Biochem Sci 1:41, 1976.
66. Rudnick G, Schuldiner S, Kaback HR: Biochemistry 15:5126, 1976.
66a.Eisenbach M, Cooper S, Garty H, Johnstone RM, Rottenberg H, Caplan SR: Biochim Biophys Acta (In press).
66b.Ramos S, Kaback HR: Biochemistry 16:4271, 1977.
67. Crane RK: In "Reviews on Physiology, Biochemistry, and Pharmacology," Heidelberg: Springer-Verlag (In press).
68. Tokuda H, Kaback HR: Biochemistry 16:2130, 1977.
69. Stock J, Roseman S: Biochem Biophys Res Commun 44:132, 1971.
70. Lanyi Y, Renthal R, MacDonald RI: Biochemistry 15:1603, 1976.
71. Reeves JP, Hong J-s, Kaback HR: Proc Natl Acad Sci USA 70:1917, 1973.
72. Short SA, Kaback HR, Kohn LD: Proc Natl Acad Sci USA 71:1461, 1974.
73. Futai M: Biochemistry 13:2327, 1974.
74. Konings WN: Arch Biochem Biophys 167:570, 1975.
75. Hirata H, Sone H, Yoshida M, Kazawa Y: Biochem Biophys Res Commun 69:655, 1976.
76. Altendorf K, Müller CR, Sandermann H: Eur J Biochem 73:545, 1977.
77. Schuldiner S, Kaback HR: Biochim Biophys Acta (In press).
78. Reeves JP, Shechter E, Weil R, Kaback HR: Proc Natl Acad Sci USA 70:2722, 1973.
79. Schuldiner S, Kerwar G, Weil R, Kaback HR: J Biol Chem 250:1361, 1975.
80. Schuldiner S, Kung H, Weil R, Kaback HR: J Biol Chem 250:3679, 1975.
81. Schuldiner S, Spencer RD, Weber G, Weil R, Kaback HR: J Biol Chem 250:8893, 1975.
82. Schuldiner S, Weill R, Kaback HR: Proc Natl Acad Sci USA 73:109, 1976.
83. Schuldiner S, Weil R, Robertson D, Kaback HR: Proc Natl Acad Sci USA 74:1851, 1977.
84. Rudnick G, Weil R, Kaback HR: J Biol Chem 250:1371, 1975.
85. Rudnick G, Weil R, Kaback HR: J Biol Chem 250:6847, 1975.
86. Jones THD, Kennedy EP: J Biol Chem 244:5981, 1969.
87. Kennedy EP, Rumley MK, Armstrong JS: J Biol Chem 249:33, 1974.
88. Prestidge LA, Pardee AB: Biochim Biophys Acta 100:591, 1965.
89. Shinitzky M, Rivnay B: Biochemistry (In press).

Journal of Supramolecular Structure 7:463–480 (1977)
Molecular Aspects of Membrane Transport 493–510

The Molecular Mechanism of Dicarboxylic Acid Transport in Escherichia coli K 12

Theodore C. Y. Lo

Department of Biochemistry, University of Western Ontario, London, Ontario, Canada

It is the purpose of this communication to review the properties of the dicarboxylic acid transport system in Escherichia coli K12, in particular the role of various dicarboxylate transport proteins, and the disposition of these components in the cytoplasmic membrane. The dicarboxylate transport system is an active process and is responsible for the uptake of succinate, fumarate, and malate. Membrane vesicles prepared from the EDTA, lysozyme, and osmotic shock treatment take up the dicarboxylic acids in the presence of an electron donor. Genetic analysis of various transport mutants indicates that there is only one dicarboxylic acid transport system present in Escherichia coli K12, and that at least 3 genes, designated *cbt*, *dct A*, and *dct B*, are involved in this transport system. The products corresponding to the 3 genes are: a periplasmic binding protein (PBP) specified by *cbt*, and 2 membrane integral proteins, SBP 1 and SBP 2, specified by *dct B* and *dct A*, respectively. Components SBP 1 and SBP 2 appear to be exposed on both the inner and outer surfaces of the membrane, and lie in close proximity to each other. The substrate recognition sites of SBP 2 and SBP 1 are exposed on the outer and inner surfaces of the membrane respectively. The data presently available suggest that dicarboxylic acids may be translocated across the membrane via a transport channel. A tentative working model on the mechanism of translocation of dicarboxylic acids across the cell envelope by the periplasmic binding protein, and the 2 membrane carrier proteins is presented.

Key words: dicarboxylate transport, transport channel, membrane structure, membrane protein, periplasmic binding protein

In the past two decades considerable efforts have been devoted to the study of the relationship between membrane structure and function. In the area of membrane transport, most investigations have been limited to kinetic studies of uptake or the effect of environmental changes on the transport process. Only recently has significant progress been made in our understanding of the energy-coupling mechanisms for various transport processes. It has now been demonstrated by quite a number of workers that a membrane potential, and/or pH gradient are involved in a large number of active transport systems. Exactly

Abbreviations: EDTA – Ethylenediaminetetraacetic acid; DCCD – dicyclohexylcarbodiimide; CCCP – carbonyl cyanide-m-chlorophenylhydrazone; NEM – N-ethylmaleimide

Received June 1, 1977; accepted September 1, 1977

how this protonmotive force can affect the membrane structure, or more precisely how the membrane transport proteins respond to such changes, is still shrouded in darkness. Not until we have determined the number and spatial arrangement of membrane transport components involved in a transport process, can we begin investigating the mechanism(s) by which solute translocation is affected by the energized state of the membrane.

Generally speaking, in gram-negative bacteria, there are at least 2 different types of transport components. These are the periplasmic binding proteins and the intergral membrane transport proteins. Various periplasmic binding proteins have been isolated, purified, and characterized. Both biochemical and genetic evidence indicates that these proteins are indeed involved in the transport process. However, the disposition of these proteins in the periplasmic space or in the outer membrane is far from clear. It is not certain whether they are embedded in the outer membrane, or whether they exist in a free or bound form in the periplasmic space. Recently, Boos' laboratory indicated that some of these proteins may be exposed on the outer surface of the outer membrane of the cell envelope (1). Despite the enormous amount of work devoted to the characterization of the periplasmic binding proteins, the exact role of these proteins in the transport process is not known for certain. It has been suggested by Silhavy and Boos (2) that one of the functions of the periplasmic binding proteins may be the maintenance of a high concentration of substrate in the periplasmic space. On the other hand, G. Ames has demonstrated very elegantly by genetic means that specific physical interactions between the periplasmic binding protein, and specific membrane carrier protein(s) are required for the transport process (3). It was suggested by J. Singer (4) that the periplasmic binding proteins may actually be the loosely bound receptor proteins of the membrane transport proteins(s).

As far as the integral membrane transport components are concerned, even much less is known concerning their number, properties, and spatial arrangement in the membrane. Our ignorance in this area is mainly due to the inability to isolate "active membrane transport component(s)." The lactose transport component – the "M protein" – was ingeniously labeled with radioactive N-ethylmaleimide and isolated by Fox and Kennedy (5). However, since this protein is inactivated by the labeling procedure, not much information can be obtained on the mode of action of the M protein. Recently, Lo and Sanwal (6) have succeeded in isolating 2 active membrane-bound transport components which are involved in the translocation of dicarboxylic acids. More recent data (7) from this laboratory indicated that these 2 membrane transport components are transmembrane proteins and that they may form multimeric subunit aggregates traversing the entire thickness of the membrane. These findings suggest that dicarboxylic acids may be translocated across the membrane via a transport channel. In addition to these 2 membrane proteins, it was also found that a periplasmic binding protein is involved in dicarboxylate transport (8).

It is the purpose of this communication to review the properties of the dicarboxylic acid transport system, in particular the role of various dicarboxylate transport proteins, and the spatial arrangement of the membrane transport components. Finally, we will present a tentative working model for the translocation of dicarboxylic acids across the cell envelope by the periplasmic binding protein, and the 2 membrane carrier proteins.

METHODS

The methods used have been described in previous publications (6–12, 16) from this laboratory, and are referred to at the appropriate sections in this communication.

RESULTS

The Number of Dicarboxylate Transport Systems Present in Escherichia coli K12

In studying the transport of a given substrate, it is essential to establish the number of transport systems by which the organism can transport the substrate under a given experimental condition. This is very important in the interpretation of substrate specificity, transport kinetics, and in deciding the number of transport components which are involved in that particular transport system. The following biochemical and genetic studies on the transport system indicate that there is only one dicarboxylate transport system present. Using an Escherichia coli strain *(sdh, frd)* which cannot metabolize succinate, we found that there is only one K_m value (30 μM) for the transport of succinate, and no biphasic curve is observed when the data are plotted in the form of a double reciprocal plot. Transport studies with membrane vesicles also provide similar information (9, 10). These data indicate the presence of only one dicarboxylate transport system.

In the isolation of transport mutants, the frequency of spontaneous mutation is found to be around 2×10^{-6} (11). It is a well established fact that spontaneous mutations occur at frequencies around $10^{-5}-10^{-6}$. If there were 2 dicarboxylate transport systems present in the cell, transport defective mutants would be found only if mutations occur in both transport systems. The frequency of occurrence of such a double mutant would then be around 10^{-12}. The fact that we obtain a frequency of around 10^{-6} indicates that there is only one dicarboxylate transport system present.

Furthermore, as discussed later, at least 3 different genes are responsible for the transport process, and mutants defective in any one of these 3 genes are unable to transport the substrate (9). Again, this serves to indicate that only one transport system is present in the cell.

General Properties of the Dicarboxylic Acid Transport System

The dicarboxylic acid transport system is responsible for the uptake of succinate, fumarate, and malate. The uptake of succinate is competitively inhibited by fumarate and malate. Mutants defective in this transport system are unable to take up or to grow on succinate, fumarate, or malate (9). Membrane vesicles prepared from the EDTA, lysozyme, and osmotic shock treatment transport the 3 dicarboxylic acids in the presence of an electron donor, such as D-lactate. Transport studies carried out with these membrane vesicles show the same substrate specificity as the whole transport system (10).

Both whole cell and membrane vesicle studies indicate that dicarboxylic acids are transported against a concentration gradient. Using an *sdh, frd* mutant, we have demonstrated that at least 95% of the succinate taken up is not chemically modified (9). This would rule out the possibility of group translocation. Both uncouplers and inhibitors of the electron transport chain are found to inhibit succinate transport in membrane vesicles (12). Furthermore, membrane vesicles from mutants defective in lactate dehydrogenase are unable to utilize D-lactate as the electron donor for the transport process. These data suggest that the functioning of an electron transport chain is required for the generation of a proton gradient across the membrane before dicarboxylic acids can be translocated to the inside of the cell. Indeed, it has been demonstrated recently by Kaback that succinate transport is dependent on the protonmotive force (13) and by Rosenberg that approximately 2 protons enter the cell with each dicarboxylate molecule (14).

The dicarboxylate transport system appears to be more complicated than other active transport systems such as those for proline or lactose. Dicarboxylate uptake by membrane vesicles is also dependent on the presence of a functioning Ca^{2+},Mg^{2+}-ATPase. The ATPase inhibitors, such as DCCD or pyrophosphate, are found to inhibit dicarboxylate transport in energized membrane vesicles, although there is no effect on proline uptake (12). The involvement of the Ca^{2+},Mg^{2+}-ATPase is further substantiated by the construction of a Ca^{2+},Mg^{2+}-ATPase negative mutant (12). Whole cells and membrane vesicles from this mutant are unable to take up succinate, even though they can transport proline normally. These findings suggest that the Ca^{2+},Mg^{2+}-ATPase may play an indirect role in dicarboxylate transport. The mechanism by which the Ca^{2+},Mg^{2+}-ATPase exerts its effects is currently being investigated in our laboratory.

Transport Components Involved in Dicarboxylic Acid Transport

a. Genetic dissection of the transport system. Indications of the number of components involved in dicarboxylate transport may be obtained by studying the genotypic and phenotypic properties of various transport mutants. Phenotypically the transport mutants can be divided into 2 classes, *cbt* and *dct* (Table I). Both types are unable to grow on the transport substrates, succinate, fumarate, or malate. The *cbt* mutants differ further from the *dct* mutants in that they are unable to grow on D-lactate (Table I). It turns out that the *cbt* mutants are also defective in the D-lactate transport system. The *dct* mutants can be divided genotypically into the *dct A* and the *dct B* mutants. Genetic analysis of these mutants indicates that the *dct A* and *dct B* genes map at 78 min and 16 min of the E. ooli linkage map respectively (15, 16). The *cbt* gene is located at 16 min of the linkage map. The above findings suggest that at least 3 genes are involved in the dicarboxylate transport process (16).

b. Biochemical dissection of the transport system. At least 3 transport components are involved in dicarboxylate transport. They are comprised of a periplasmic binding protein (PBP), and 2 membrane transport proteins (SBP 1 and SBP 2). Active species of these molecules have been isolated through the use of aspartate-coupled Sepharose columns (6, 8).

i. The periplasmic binding protein (PBP). The involvement of PBP in the dicarboxylate transport process is based on the following findings: 1) The dicarboxylate transport system in whole cells is a shock-sensitive system. Transport activities are not observed after subjecting the cells to EDTA-osmotic shock treatment, and a periplasmic binding protein capable of binding with succinate can be isolated from the shock fluid (9). 2) PBP is found to have similar substrate specificity (i.e., same binding site for succinate, fumarate, and malate), and substrate affinity (K_d = 35 μM) as the whole cell transport system (8). 3) The whole cell transport system differs from the membrane vesicle transport system in that the former is inhibited by N-ethylmaleimide, whereas the latter is not. This suggests that an NEM-sensitive transport component may be present outside the cytoplasmic membrane. Indeed, it is found that the binding of succinate to PBP is inhibited by NEM (8–10). 4) Although the *cbt* mutant cannot transport dicarboxylic acids, membrane vesicles prepared from this mutant are able to do so (Table I). This means that the membrane components in this mutant are functioning normally, and it may be lacking an active periplasmic binding protein. 5) Biochemical analysis of the *cbt* mutant indicates that the PBP cannot be isolated from the osmotic shock fluid by the affinity column (Fig. 1) (8). This suggests that either the substrate recognition site of PBP is defective, or the protein is not synthesized. This serves as an important piece of evidence indicating that the *cbt* gene is responsible for the periplasmic binding protein. 6) It is found that D-lactate also binds to

TABLE I. Properties of Various Dicarboxylate Transport Mutants*

	Phenotype			Relative rates of succinate transport		
Strains	Acetate	D-Lactate	Succinate, malate, or fumarate	Whole cells	Membrane vesicles	Genetic loci
Wild type	+	+	+	+(710%)	+(52%)	
sdh, frd	–	+	(–)†	+(100%)	+(100%)	
cbt	+	–	–	–(8%)	+(58%)	16 min
dct A	+	+	–	–(9%)	–(2%)	78 min
dct B	+	+	–	–(12%)	–(5%)	16 min

*Under "PHENOTYPE", (+), (–) indicate growth or no growth respectively at 37°C for 48 h on minimal medium using the mentioned carboxylic acids as the sole carbon sources. Under "RELATIVE RATES OF SUCCINATE TRANSPORT",(+), (–) indicate the capabilities or incapabilities respectively of the whole cells or membrane vesicles to transport succinate. Transport studies with whole cells or membrane vesicles were carried out as described in Ref. 9 and 10. The initial rates of succinate uptake by the sdh, frd mutant were 0.9 nmoles/mg of cells (dry weight)/min for the whole cells, and 1.18 nmoles/mg protein/min for the membrane vesicles. These values were taken as 100% for the respective uptake systems. The mapping of various genes was described in Ref. 16. The genetic loci were revised according to the new map by Bachmann et al (15).

†The sdh, frd mutant is unable to grow on succinate or fumarate; however, it can grow on malate.

the substrate recognition site of PBP. If the cbt gene is responsible for PBP, then one would expect that the cbt mutant is unable to transport D-lactate. This indeed is the case. All the evidence presented above indicates that PBP is involved in the dicarboxylate transport process, and that the cbt gene is responsible for this protein. We will elaborate on the role of PBP in the "Discussion."

ii. The membrane transport components (SBP 1 and SBP 2). When membrane vesicles from an sdh, frd mutant or from a wild-type strain are treated with the nonionic detergent Lubrol 17A-10 (Imperial Chemical Industries Ltd., Blackley, Manchester, England), the succinate binding proteins can be solubilized. Fractionation of the solubilized membrane proteins on an aspartate-coupled Sepharose column using succinate elution yields 2 protein peaks (Fig. 2) (6). The following observations strongly suggest that these 2 components (SBP 1 and SBP 2) are involved in the translocation of the dicarboxylic acids across the membrane. 1) Treatment of the membrane vesicles with various detergents abolishes transport and releases succinate binding activity in the supernatant. 2) Both SBP 1 and SBP 2 are able to bind with succinate, fumarate, and malate. Malonate, which is a potent competitive inhibitor of succinate dehydrogenase, has no effect of the binding of succinate to these 2 proteins (6). 3) Like the whole cell transport system, SBP 1 has a K_d of 23 μM for succinate, and 47 μM for malate. However, it should be noted that SBP 2 has a K_d of 2 μM for succinate, and 7 μM for malate. No enzymatic activities can be detected in preparations of SBP 1 and SBP 2 (6). 4) Membrane vesicles from mutants defective in the dct A or dct B gene are unable to transport succinate in the

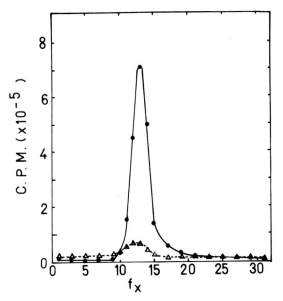

Fig. 1. Elution profile of PBP from the aspartate-coupled Sepharose. Cells were labeled with $^{35}SO_4^{2}$ for 12 h. After harvesting and washing, the cells were subjected to EDTA and osmotic shock treatment (8). The shock fluid was loaded onto the aspartate-coupled Sepharose column. After washing off the unbound proteins, 0.2 M succinate was then added to elute the bound proteins. (●) indicates elution profile from wild type cells (*cbt*[+]), (△) indicates elution profile from the *cbt* mutant.

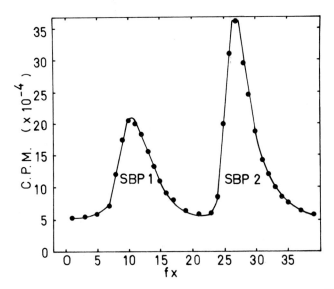

Fig. 2. Elution profile of membrane proteins from the aspartate-coupled Sepharose. $^{35}SO_4^{2}$-labeled membrane vesicles were prepared and solubilized according to Ref. 11. Fractionation of the solubilized proteins was carried out by affinity chromatography in the presence of 0.1% Lubrol-phosphate buffer, pH. 6.6.

presence of an electron donor (10). This points to the likelihood that the *dct* genes may be responsible for some membrane transport components. It will be shown later on in the membrane binding studies, that the *dct A* membrane vesicles do not possess the SBP 2 substrate recognition sites, and that the SBP 1 substrate recognition sites cannot be detected in the *dct B* membrane vesicles (Fig. 6) (11). 5) The properties of the *dct* membrane vesicles suggest that the SBP 1 and SBP 2 proteins may be altered or absent in the corresponding mutants. If so, one should be able to detect these changes using affinity chromatography. Figure 3 shows that the SBP 1 component cannot be detected when the solubilized membrane proteins from the *dct B* mutant are fractionated by affinity chromatography, and similarly SBP 2 protein cannot be detected in the *dct A* mutant (11). This is in agreement with the membrane binding studies with the mutant membranes. It should be noted that both *dct A* and *dct B* mutants contain a functioning succinate dehydrogenase. The fact that SBP 1 or SBP 2 is absent in the *dct* mutants again suggests that both of these proteins are different from succinate dehydrogenase. The above findings provide the essential evidence that both SBP 1 and SBP 2 are involved in dicarboxylate transport, and that the *dct A* and *dct B* genes are responsible for components SBP 2 and SBP 1 respectively.

The Orientation of the Substrate Recognition Sites of SBP 1 and SBP 2 in the Cytoplasmic Membrane

Having demonstrated that both SBP 1 and SBP 2 are involved in dicarboxylate transport, the next obvious question concerns the orientation of the substrate recognition sites on these 2 proteins. Formulation of reasonable molecular mechanisms for the membrane

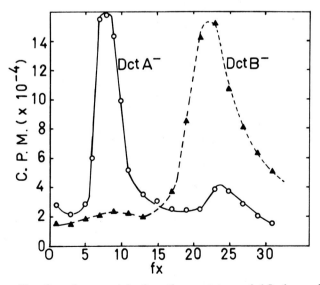

Fig. 3. Elution profiles of membrane proteins from the aspartate-coupled Sepharose. Membrane proteins were prepared according to procedure described in Fig. 2. o) The elution profile of membrane preparations from the *dct A* mutant, Dct A$^-$; ▲) the elution profile of membrane preparations from the *dct B* mutant, Dct B$^-$.

translocation process would demand information on whether the substrate recognition site(s) of a protein can be exposed to both surfaces of the membrane by oscillation of the protein across the membrane as predicted by the "mobile carrier model (2)" (Fig. 4), or if these sites are fixed and accessible on only one surface of the membrane, as required by models (1), (3), and (4). If the latter is the case, then one would have to determine whether the substrate recognition sites of both SBP 1 and SBP 2 are exposed to the same surface of the membrane. The above information would certainly be very useful in deciphering the molecular mechanisms of the transport process.

As mentioned earlier, the K_d values of SBP 1 and SBP 2 are 47 μM and 7 μM respectively for malate. This difference in the K_d values can be used as a means of distinguishing the substrate recognition sites of SBP 1 and SBP 2. One may thus determine the disposition of the SBP 1 and SBP 2 substrate recognition sites on the cytoplasmic membrane by measuring the binding affinities of spheroplasts, right-side-out (R.S.O), and inside-out (I.S.O.) vesicles. It has now been established that R.S.O. vesicles can be prepared by subjecting the cells to EDTA, lysozyme; and osmotic shock treatment, and that I.S.O. vesicles can be obtained by subjecting the cells to French-press treatment.

Several precautions were taken in carrying out the membrane binding studies. Firstly, one must assure that there is no substrate transport during the binding process. The presence of any residual transport activity would certainly make the binding data difficult to interpret. We have shown previously that dicarboxylic acids are transported across the membrane only when a proton gradient is established across the membrane by the addition of an electron donor (10, 12); furthermore, this process is possible only if both SBP 1 and SBP 2 are functioning normally (10). We have also demonstrated that hardly any transport activity can be detected at 4°C. Therefore, in order to ensure that there is no transport

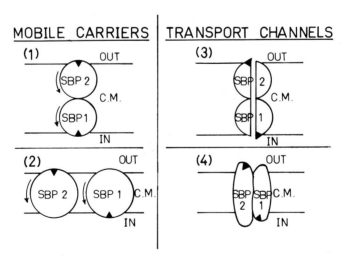

Fig. 4. Schematic diagram showing the various mechanisms by which dicarboxylic acids can be translocated across the cytoplasmic membrane. The disposition and mode of action of the 2 dicarboxylate membrane transport components (SBP 1 and SBP 2) in the cytoplasmic membrane are depicted in Models (1) to (4). Models (1) and (2) can be regarded as "mobile carrier" models; and Models (3) and (4) are referred to as "transport channel" models. In Models (2) and (4), both SBP 1 and SBP 2 are transmembrane proteins, whereas in Models (1) and (3), the membrane transport components are only exposed to one surface of the membrane. ▲) The substrate recognition sites(s) of the transport components; C.M.) cytoplasmic membrane.

activity, we carried out the binding studies in the absence of any electron donor and in the presence of an uncoupler, CCCP (which was shown to collapse to proton gradient). In the case of R.S.O. and I.S.O. vesicles, binding studies were carried out at 4°C; however, binding studies with spheroplasts were carried out at 23°C, so as to prevent lysis of spheroplasts at low temperatures (17). Results presented in Fig. 5 suggest that there is no detectable transport activity during the binding process. Our binding studies and previous report (10) from this laboratory indicate that the K_d value for binding to R.S.O. membrane preparations is 4 μM for malate, and the K_m value for transport is around 45 μM. If residual transport activity were occurring during the binding process, the resulting curve in a double reciprocal plot should show a biphasic behaviour. It is, however, clearly evident from Fig. 5 that the binding plot is linear and yields only one K_d value. This indicates that our procedure is capable of measuring exclusively the binding of the substrate to the membrane surface, and not the uptake of the substrate. This conclusion is strengthened by the results of binding studies with membrane transport mutant (*dct A* and *dct B*) vesicles (Fig. 6). These mutant membrane vesicles are unable to transport the substrate even after energization. Figure 6 indicates that essentially the same K_d values as those with the wild-type membrane vesicles are obtained for the normal components in these mutant membranes. Thus, one can eliminate the possibility of facilitated diffusion, or residual active transport of the substrate across the membrane during the binding process.

A second problem that arises in the interpretation of the binding data concerns the specificity of the binding to membrane preparations. It is well established that in Escherichia coli succinate dehydrogenase is a membrane-bound enzyme, whereas malate dehydrogenase

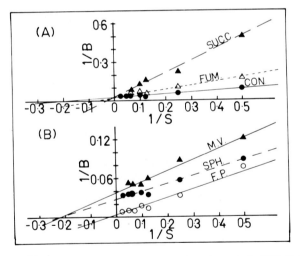

Fig. 5. The binding of [^{14}C] malate to membrane preparations from strain CBT 43 (*sdh,frd*). A) The binding of [^{14}C] malate to spheroplasts in the control (CON.) or in the presence of 0.1 mM succinate (SUCC.) or 0.1 mM fumarate (FUM.). Spheroplasts were kept in 20% sucrose, 0.05 M phosphate buffer, pH 6.6. They were preincubated with 10 μM CCCP, and binding studies were carried out as described in Ref. 11. Incubation was carried out at 23°C to prevent lysis of spheroplasts at low temperature. B) The binding of [^{14}C] malate to spheroplasts, right-side-out membrane vesicles (M.V.) and inside-out membrane vesicles (F.P.). Binding studies with R.S.O. and I.S.O. vesicles were carried out as in the case of spheroplasts (SPH.) (11) except that the binding studies were carried out in 0.05 M phosphate buffer, pH 6.6, at 4°C; the membrane preparations were preincubated with 10 μM CCCP as in the case of spheroplasts.

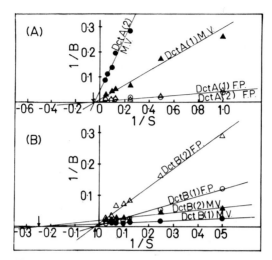

Fig. 6. The binding of [^{14}C] malate to membrane preparations from various *dct* mutants. Binding studies were carried out essentially as indicated in Fig. 5B. A) Malate binding by right-side-out membranes, (M.V.) and inside-out membrane vesicles (F.P.) with 2 independently isolated *dct A* mutants, Dct A (1) and Dct A (2). B) Similar binding studies with 2 independently isolated *dct B* mutants, Dct B (1) and Dct B (2).

is not (20). Although succinate dehydrogenase has quite a different optimum pH value and substrate specificity than the dicarboxylate transport components, it is still possible that succinate dehydrogenase may bind with succinate at pH 6.6; this would certainly complicate the interpretation of the binding data. Figure 5A shows that membrane vesicles have the same binding site for succinate, malate, and fumarate. Therefore in order to avoid the complications incurred with succinate binding, we carried out the binding studies using malate as the substrate. It will be indicated later on in the studies with various *dct* mutants, that the binding of malate to various membrane preparations is not due to the presence of malate dehydrogenase, or succinate dehydrogenase.

Binding studies with membrane preparations from "wild-type" cells indicate that both succinate and fumarate competitively inhibit the binding of malate to the spheroplasts (Fig. 5A) and membrane vesicles (results not presented). It is also found that both R.S.O. vesicles and spheroplasts have K_d values of around 4 μM for malate, and the I.S.O. vesicles have a value of around 30 μM (Fig. 5B). We have indicated earlier that SBP 1 and SBP 2 have K_d values of 47 μM, and 7 μM, respectively, for malate. This similarity between the binding affinities of the R.S.O. vesicles (or spheroplasts) and the SBP 2 component seems to indicate that the substrate recognition sites of SBP 2 are exposed to the outer surface of the membrane. Similarly, binding studies with the I.S.O. vesicles suggest that the substrate recognition sites of SBP 1 are exposed to the inner surface of the membrane. If this were the case, then one should not be able to detect any binding with the I.S.O. membrane vesicles from the *dct B* mutant (defective in SBP 1), and with the R.S.O. vesicles or spheroplasts from the *dct A* mutant (defective in SBP 2). Indeed this is what we found (Fig. 6). Although the R.S.O. membrane vesicles from the *dct B* mutants demonstrate the normal affinity of binding (i.e. 4 μM), the I.S.O. membrane vesicles can hardly bind with the substrate. This substantiates our findings that the substrate recognition sites of SBP 1 are exposed to the inner surface of the membrane. Binding studies with the *dct A* membrane vesicles show that the R.S.O. membrane vesicles have a much lower affinity for the substrate as compared with the wild type, and I.S.O. membrane vesicles show the normal affinity of binding (Fig. 6A). Again, this confirms our findings that the substrate recognition sites of SBP 2 are exposed to the outer surface of the membrane, and this is altered in the *dct A* mutants. It should be pointed out here that spheroplasts prepared from the wild type and various *dct* mutants also provide the same results as the R.S.O. vesicles. This may serve as another indication that membrane vesicles prepared by the EDTA, lysozyme, and osmotic shock treatment are mainly R.S.O. vesicles and the amount of I.S.O. vesicles in the R.S.O. membrane preparation is negligible. Since membrane vesicles from the *dct* mutants are unable to transport malate even in the presence of an electron donor, and since binding studies with wild-type membrane vesicles agree with those from the transport mutant vesicles, we are quite confident that we are measuring the binding of the substrate to the surface of the membrane, and not some residual transport activities in the presence of CCCP; in fact, similar results are obtained even in the absence of CCCP, or in the presence of azide (11).

The above findings have several important implications for models of the molecular mechanisms of transport. If the membrane vesicles are not sealed, or if the substrate recognition sites of both SBP 1 and SBP 2 are present on the same surface of the membrane, then one would expect to obtain a biphasic curve on a double reciprocal plot giving binding constants representing those of SBP 1 and SBP 2. The fact that such curves are not observed with wild-type spheroplasts, R.S.O., and I.S.O. vesicles suggests that this is not likely to be the case. Using the appropriate mutant membrane vesicles, SBP 1 and SBP 2 substrate recognition sites cannot be detected on the R.S.O. and I.S.O. membrane vesicles respectively. This observation corroborates the above findings in that both R.S.O. and I.S.O. vesicles are sealed vesicles, and that both SBP 1 and SBP 2 substrate recognition sites cannot be exposed to the same surface of the membrane; and more important this suggests that the transport components are not likely to oscillate from one surface of the membrane to another – as suggested by Model (2) (Fig. 4). Furthermore, one can conclude that the substrate recognition sites of SBP 1 and SBP 2 are only exposed to the inner and outer surfaces of the membrane respectively. It should be noted that these findings cannot be used to distinguish between Models (1), (3), or (4). Finally, binding studies with various membrane preparations indicate that the inner surface of the membrane has a much lower substrate binding affinity as compared with the outer surface.

It may be evident from Fig. 4 that the major difference between Models (1), (3), and (4) is that in Models (1) and (3), the transport proteins are not transmembrane proteins, whereas in Model (4) both proteins are exposed on both surfaces of the membrane. One should be able to distinguish between these 2 possibilities through the use of nonpenetrating covalent labeling reagents. Using R.S.O. or I.S.O. membrane vesicles, one should be able to label only one transport component in the case of Models (1), and (3). However, if Model (4) were applicable, then one should be able to label both transport components. By labeling spheroplasts or I.S.O. membrane vesicles with the lactoperoxidase system (Fig. 7A), or with the pyridoxal phosphate-sodium [³H] borohydride system (Fig. 7B), we found that both components can be labeled on the inner or outer surface of the membrane (7). Similar findings are observed when R.S.O. membrane vesicles are used. Binding studies with various membrane preparations indicate that these are sealed vesicles and that they have the proper orientations. Since both lactoperoxidase and pyridoxal phosphate cannot penetrate the membrane, one can then conclude that both SBP 1 and SBP 2 are exposed on the 2 surfaces of the membrane. This would rule out the arrangements of the transport components as indicated in Models (1) and (3). As we have seen earlier, binding studies

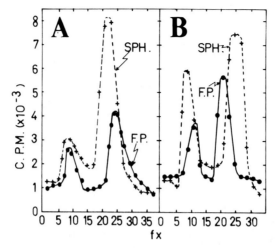

Fig. 7. Labeling the membrane transport components with nonpenetrating covalent labeling reagents. Spheroplasts (SPH.) and inside-out-membrane vesicles (F.P.) were prepared and labeled as described in Ref. 7. The labeled membrane preparations were then solubilized with 4% Lubrol 17A-10, and the solubilized proteins were fractionated by aspartate-coupled Sepharose as described in Fig. 2. A) The elution profiles of spheroplasts (+ – – – +) and I.S.O. vesicles (●—●) labeled with lactoperioxidase and ^{125}I; B) the elution profiles of spheroplasts (+ – – – +) and I.S.O. vesicles (●—●) labeled with pyridoxal phosphate and NaB^3H_4.

with membrane vesicles have eliminated Model (2), so it seems Model (4) may be the likely mechanism by which dicarboxylic acids are translocated across the membrane.

SBP 1 and SBP 2 Are Lying in Close Proximity to One Another

So far, our findings suggest that Model (4) seems to be the most feasible transport model. Let us explore the properties of this model one step further. This model predicts that the transport components may form multimeric subunit aggregates transversing the entire thickness of the membrane, thereby creating an aqueous transport channel. If this were the case, then one would expect that SBP 1 and SBP 2 should be in close proximity to each other. Hence one should be able to cross-link these 2 proteins. However, if the proteins are sitting one on top of the other, as depicted by Models (1) and (3), then one should not be able to cross-link these 2 proteins using nonpenetrating cross-linking reagents. We used a cleavable, nonpenetrating cross-linking reagent, tartaryl diazide, for such an experiment. This is a polar reagent, so it should not penetrate the membrane, and it has been demonstrated by Lutter (18) that the azide activated carbonyl groups react readily with amino groups to produce amide linkages. This tartaryl diazide is capable of cross-linking 2 protein amino groups which are 6 Å apart. Another advantage of this reagent is that the cross-linked proteins can be cleaved by mild treatment with periodic acid. Figure 8 shows that both SBP 1 and SBP 2 can be cross-linked with tartaryl diazide (7). However, when the cross-linked complex is cleaved with periodic acid, both SBP 1 and SBP 2 can be recovered. In the covalent labeling experiments, the validity of the results depends on the fact that the membranes vesicles are sealed structures, and we have demonstrated through binding studies that this is the case. However, in the cross-linking experiments, the reaction does not depend on the intactness of the membrane vesicles, it depends only on the

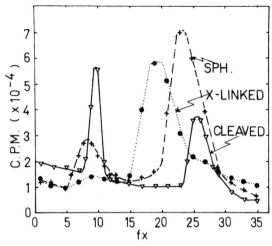

Fig. 8. Elution profile of cross-linked, cleaved, and un-cross-linked membrane transport components with tartaryl-diazide. Cross-linking experiments are carried out as indicated in the text. ^{35}S-labeled spheroplasts were cross-linked with freshly prepared 0.05 mmoles of tartaryl diazide. The cross-linked membranes were washed and solubilized with Lubrol 17A-10. Half of the solubilized proteins were fractionated by aspartate-Sepharose. The other half of the proteins were treated with periodic acid which was then removed by running through a Sephadex G-25 column. The protein peaks were then loaded onto an aspartate-sepharose column. The bound proteins were eluted with 0.2 M succinate. +) Elution profile of un-cross-linked spheroplasts; •) elution profile of cross-linked spheroplasts, and ▽) elution profile of the cleaved cross-linked complex.

fact that the transport components have to be in close proximity to each other on the membrane surface. Since we can cross-link these proteins with a nonpenetrating reagent, it indicates that both SBP 1 and SBP 2 must be exposed on the same surface of the membrane and that they are in close proximity to each other. Again this agrees with our previous postulation that Model (4) seems to be the most feasible transport mechanism.

Transport Studies With Cross-Linked Membrane Vesicles

According to the "mobile carrier model (2)," once the carrier protein has been cross-linked with other membrane surface components, transport should not be possible, as oscillation of the transport components across the membrane would no longer be possible. However, according to Model (4), the cross-linked protein may still be able to carry out the transport process — depending on the nature and the extent of the conformation changes (7). In order to test this hypothesis, we carried out transport studies with membrane vesicles that had been cross-linked with tartaryl-diazide under identical conditions as indicated earlier. Figure 9 indicates that membrane vesicles that have been cross-linked with tartaryl diazide can transport dicarboxylic acids to the same extent as the unmodified vesicles. This suggests that large conformational changes are not required for the transport process to occur. Consequently, it is unlikely that the substrate is translocated across the membrane through oscillation of the transport components across the membrane. Again, in agreement with the binding studies, one can eliminate the transport mechanisms depicted by Model (2).

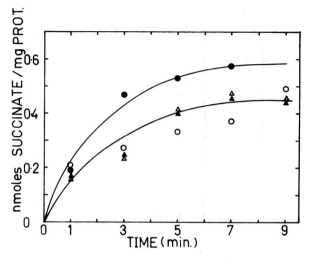

Fig. 9. Succinate transport by membrane vesicles cross-linked with different amounts of tartaryl diazide. Membrane vesicles from an *sdh, frd* strain (CBT 43) were prepared by the EDTA-lysozyme-osmotic shock method (10). Membrane vesicles were cross-linked under identical conditions as described in Fig. 8. Transport studies were carried out by the standard procedure using phenazine methosulfate and ascorbate as the electron donors (10). ●) Succinate uptake by the normal membrane vesicles; ○) transport by membrane vesicles treated with the cross-linking buffer A (0.05 M triethanolamine-HCl pH 8.5, 0.005 M $MgCl_2$ and 0.1 M KCl); ▲ and △) transport by membrane vesicles treated with 0.025 mmoles and 0.05 mmoles of tartaryl diazide respectively in the cross-linking buffer A.

Role of Specific Phospholipids in the Transport Process

So far, we have been concentrating on the properties and the spatial arrangement of the membrane transport components; not much attention has been directed to the role of phospholipids in the transport process. Conventionally, most membrane functions have been attributed to the presence of certain specific proteins, and the role of phospholipids has often been ignored or underrated. Phospholipids play at least 2 conceivable roles in biological membranes. Firstly, they maintain the uniqueness and the integrity of the membrane, and they also serve as the matrix in which membrane proteins are embedded. Secondly, studies with various membrane-associated enzymes indicate that phospholipids are required for the normal functioning of these proteins. They generally play the role of physical cofactors, activating the enzyme system but not themselves participating in the reaction. They may activate the enzyme by inducing a conformational change in the protein. A well documented example of this kind is the very specific requirement of phosphatidylglycerol for both phosphorylation and transport of α-methylglucoside mediated by the phosphotransferase system of gram-negative bacteria. However, not all transport systems require phosphatidylglycerol for activities. It is therefore very important for our understanding of the molecular mechanism of the dicarboxylate transport process, and for our eventual reconstitution experiments, to determine whether any specific phospholipids are required for dicarboxylic acid transport.

A cursory examination of the phospholipid requirement for dicarboxylate transport was carried out by studying the effect of phospholipase on the transport process. Milner and Kaback (19) demonstrated quite clearly that phospholipase D (cabbage) acts specifically on phosphatidylglycerol of E. coli membranes releasing phosphatidic acid and glycerol.

Phosphatidylethanolamine, phosphatidylserine, cardiolipin, and lysophosphatidylethanolamine are not affected at all. The effect of phospholipase D (cabbage) on dicarboxylate transport is presented in this section.

Before using this commercially available enzyme, we ascertained that phospholipase D (cabbage) had no proteolytic activities by testing its effect on pyruvate kinase (rabbit muscle), and lactate dehydrogenase, and demonstrating that even after prolonged incubation with the phospholipase D preparation, these enzymes retained all of their activities. Figure 10 shows the effect of phospholipase D on the rates of uptake and efflux of succinate in membrane vesicles. Like the α-methylglucoside transport system, the dicarboxylate transport system is inhibited by phospholipase D (cabbage). Figure 10 also shows the rates of proline uptake and efflux. These results are similar to those reported by Milner and Kaback (19) who showed that proline uptake was only slightly affected by these concentrations of phospholipase D (cabbage) (7). Thus, these findings suggest that phosphatidylglycerol may be required for the transport of dicarboxylic acids. Further work is being carried out to determine the specificity of this requirement.

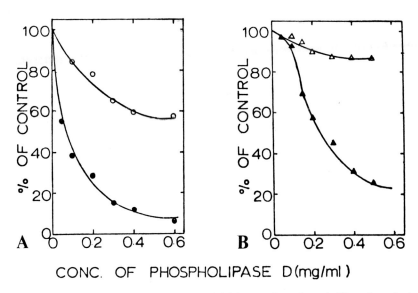

CONC. OF PHOSPHOLIPASE D(mg/ml)

Fig. 10. Effect of phospholipase D (cabbage) on the initial rates of uptake and efflux of succinate and proline. Membrane vesicles prepared from strain CBT 43 at a concentration of 1.5 mg/ml were used. Phospholipase D (cabbage) (31 units/mg) was added to the membrane at zero time. The reaction was carried out at 23°C.

A) The effect of phospholipase D on the initial rates of uptake of succinate and proline. D-Lactate (20 mM) was used as the electron donor. Samples were taken at 20, 40, and 60 sec. ●) The initial rate of uptake of succinate (2×10^{-5}M); ○) the initial rate of uptake of proline (2×10^{-5}M). In the absence of phospholipase D, the initial rates of succinate and proline uptake were 1.14 and 0.24 nmoles/mg protein/min, respectively. These values were taken as 100% for the respective uptake systems. B) The effect of phospholipase D on the initial rates of efflux of succinate and proline. The same concentrations of succinate (▲) and proline (△) were used as in Fig. 10A. Membrane vesicles were first preloaded with the respective radioactive ligands for 15 min using D-lactate as the electron donor, then phospholipase D was added at zero time. At zero time, the amount of succinate and proline accumulated in the vesicles were 2.03 nmoles/mg protein, and 0.37 nmoles/mg protein, respectively. These values were taken as 100% for the respective efflux systems. Percent of control indicates the percentage of radioactive ligands retained in the membrane vesicles after 20 sec.

DISCUSSION

At least 3 different transport components are found to be involved in dicarboxylate transport system of E. coli K12 — one periplasmic binding protein, and 2 membrane transport components. Transport of dicarboxylic acids across the cell envelope can best be described by the following tentative working model (Fig. 11). The substrate is first captured by PBP, which may be exposed on the outer surface of the outer membrane, or which may be located in the periplasmic space. After binding with PBP, the substrate is transferred to the substrate recognition site of SBP 2. It is quite possible that this process is carried out by direct specific physical interactions between PBP and the membrane carrier proteins, as found to be the case in the histidine transport system (3).

One may question the necessity of PBP in dicarboxylate transport, especially when it has been demonstrated that membrane vesicles are able to take up the substrate in the absence of PBP. Our findings indicate that in intact cells, PBP is required for delivering the substrate to the membrane transport components. As mentioned earlier, the whole cell transport system, but not the membrane vesicle transport system, is inhibited by N-ethyl-maleimide. It is also found that the binding of succinate to PBP is inhibited by N-ethyl-maleimide. This observation suggests that PBP is essential for the whole cell uptake system but not for the membrane system. This finding is corroborated by the properties of the cbt mutants. The cbt mutants are defective in PBP. As indicated earlier, although intact cells of the cbt mutant are unable to transport, cbt membrane vesicles take up the substrate normally. One may explain these observations by postulating that in the case of membrane vesicles, in which most of the cell wall materials are removed, the substrate is readily accessible to the membrane carrier proteins, and so PBP is not essential for the process.

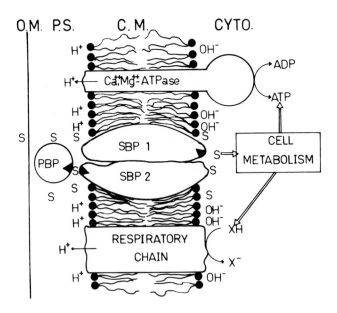

Fig. 11. Schematic diagram of the tentative working model for dicarboxylate transport system in Escherichia coli K12. S) Transport substrate; ▲) substrate recognition sites of the transport components; O.M.) outer membrane; P.S.) periplasmic space; C.M.) cytoplasmic membrane; CYTO.) cytoplasm.

However, in the case of intact cells, PBP is required to deliver the substrate across the cell wall to the membrane transport components. So far there is no indication that metabolic energy is required for this process.

The present evidence indicates that the substrate is translocated across the membrane via a transport channel formed by both SBP 1 and SBP 2. Both SBP 1 and SBP 2 are found to have the same substrate specificity as the transport system, i.e., they have the same binding site for succinate, fumarate, and malate. SBP 1 had a K_d of 23 μM for succinate and 47 μM for malate which are similar to the K_m values for transport. SBP 2 differs from SBP 1 in that it has a K_d of 2 μM for succinate and 7 μM for malate. Binding studies with spheroplasts and R.S.O. and I.S.O. vesicles indicate that the substrate recognition sites of only one membrane transport component are exposed on one surface of the membrane. The substrate recognition sites of SBP 2 and SBP 1 are exposed on the outer and inner surfaces of the membrane respectively. Of the 3 types of membrane preparations used, spheroplasts were subjected to only very mild treatment; therefore the binding capacity of spheroplasts for malate would likely be the best reflection of the number of binding sites in whole cells. Figure 5B indicates that spheroplasts have a binding capacity of 38 pmoles of malate per mg of cellular protein. In our laboratory, we found that 1 mg of cellular protein is equivalent to 7.83×10^9 cells, which is comparable to that reported by Jones and Kennedy (21). Assuming one molecule of substrate binds with one substrate recognition site on the transport component, we found that each bacterium contains around 3,000 binding sites on the outer surface of the membrane.

Both cross-linking and covalent-labeling experiments with nonpenetrating reagents point to the likelihood that both SBP 1 and SBP 2 are transmembrane proteins and that they lie in close proximity to each other. Transport studies with cross-linked membrane vesicles suggest that large conformational changes are not required for the translocation of the substrate across the membrane. This would tend to rule out the involvement of a mobile carrier mechanism. Therefore, the present available data points to the involvement of multimeric transport channels in the translocation of dicarboxylic acids across the membrane. This is depicted in Fig. 11. It is speculated that the transmembrane motion of the substrate is brought about by local conformational changes on the membrane transport components.

It should be noted that the transport channel model proposed is slightly different from that postulated by Singer for the shock-sensitive transport systems (4). In his model, the periplasmic binding protein is postulated as the loosely bound substrate recognition site of the transport components, and the integral proteins do not possess any substrate recognition site per se. However, in the present model, all 3 transport components have been demonstrated unequivocally to possess their own substrate recognition sites. Presumably the SBP 2 substrate recognition site is responsible for influx, and that of SBP 1 is responsible for both influx as well as efflux.

Cursory examination of the effects of phospholipases indicate that both the influx and efflux of dicarboxylic acids are inhibited after incubation of membrane vesicles with phospholipase D (cabbage). Thus, like the PEP-phosphotransferase system, phosphatidylglycerol seems to be required for the normal functioning of SBP 1 and SBP 2. It is quite possible that phosphatidylglycerol may play a role as a "physical cofactor," maintaining the transport components in the proper conformation.

It has been well established in our laboratory and by other workers, that a proton gradient is required for the translocation of dicarboxylic acids across the membrane. A proton gradient can be generated by the functioning of either the electron transport chain or the Ca^{2+},Mg^{2+}-ATPase. How, at a molecular level, this proton gradient affects the functioning of the transport components is far from clear. Transport studies carried out with membrane vesicles in the presence of Ca^{2+},Mg^{2+}-ATPase inhibitors or with membrane vesicles prepared from a Ca^{2+},Mg^{2+}-ATPase mutant indicate that the functioning of Ca^{2+},Mg^{2+}-ATPase is required for the uptake of dicarboxylic acids by membrane vesicles. This is different from the proline transport system in which Ca^{2+},Mg^{2+}-ATPase is not required. It is possible that the Ca^{2+},Mg^{2+}-ATPase has a role in the transport of dicarboxylic acids other than maintaining a proton gradient in whole cells.

ACKNOWLEDGMENTS

The author wishes to thank Drs. B. D. Sanwal and P. Galsworthy for critically reviewing the manuscript. This investigation was supported by an operational grant from the Medical Research Council of Canada. T.C.Y. Lo is a recipient of a Scholarship from the Medical Research Council of Canada.

REFERENCES

1. Argast M, Schumacher G, Boos W: J Supramol Struct 6:135, 1977.
2. Silhavy TJ, Szmelcman S, Boos W, Schwartz M: Proc Natl Acad Sci USA 72:2120, 1975.
3. Ames GF, Spudich EN: Proc Natl Acad Sci USA 73:1877, 1976.
4. Singer SJ: Annu Rev Biochem 43:805, 1974.
5. Kennedy EP: The lactose Permease system of Escherichia coli. In "The Lactose Operon" (J.R. Beckwith and D. Zipser, Eds). Cold Spring Harbor Laboratory p. 49.
6. Lo TCY, Sanwal BD: Biochem Biophys Res Commun 63:278, 1975.
7. Lo TCY, Bewick MA: Manuscript submitted.
8. Lo TCY, Sanwal BD: J Biol Chem 250:1600, 1975.
9. Lo TCY, Rayman MK, Sanwal BD: J Biol Chem 247:6328, 1972.
10. Rayman MK, Lo TCY, Sanwal BD: J Biol Chem 247:6332, 1972.
11. Lo TCY, Bewick MA: Manuscript submitted.
12. Lo TCY, Rayman MK, Sanwal BD: Can J Biochem 52:854, 1974.
13. Ramos S, Kaback HR: Biochemistry 16:854, 1977.
14. Gutowski SJ, Rosenberg H: Biochem J 152:647, 1975.
15. Bachmann BJ, Low KB, Taylor AL: Bacteriol Rev 40:116, 1976.
16. Lo TCY, Sanwal BD: Mol Gen Genet 140:303, 1975.
17. Leive L, Kollin V: Biochem Biophys Res Commun 28: 229, 1967.
18. Lutter LC, Ortanderl F, Fasold H: FEBS Lett 48:288, 1974.
19. Milner LS, Kaback HR: Proc Natl Acad Sci USA 65:683, 1970.
20. Sanwal BD: J Biol Chem 244:1831, 1969.
21. Jones THD, Kennedy EP: J Biol Chem 244:5981, 1969

Journal of Supramolecular Structure 7:481–487 (1977)
Molecular Aspects of Membrane Transport 511–517

Reconstitution of Neutral Amino Acid Transport From Partially Purified Membrane Components From Ehrlich Ascites Tumor Cells

Gary Cecchini, Gregory S. Payne, and Dale L. Oxender

Department of Biological Chemistry, Medical Science I, University of Michigan, Ann Arbor, Michigan 48109

Solubilized protein fractions have been obtained from plasma membranes of Ehrlich ascites cells either by extraction with 0.5% Triton X-100 or by extraction with 2% cholate. Partial purification of the solubilized protein fraction has been obtained by utilizing a combination of ammonium sulfate precipitation and column chromatography. Leucine-binding activity has been detected in the Triton X-100 solubilized membrane fraction. The leucine-binding activity was measured by equilibrium dialysis and was saturable with high levels of leucine or phenylalanine and is not strongly effected by alanine. These properties are similar to those previously identified as System L. In addition, the cholate extracted protein fraction was partially purified and reconstituted into liposomes. Sodium dependent uptake of alanine and leucine could be demonstrated in the reconstituted vesicles. Concentrative uptake was dependent upon a sodium gradient. A membrane potential produced by valinomycin mediated potassium diffusion in the presence of sodium also stimulated amino acid transport in reconstituted liposomes.

Key words: reconstitution, transport, Ehrlich cell, amino acid, liposome

The transport of neutral amino acids in animal tissues is carried out by several distinct transport systems. Early studies on the detailed interactions for transport in Ehrlich ascites cells among a large group of natural and synthetic amino acids showed that the affinities of the neutral amino acids clustered into 2 groups which were transported by the A and the L transport systems (1). We recently demonstrated that the A and L transport systems are present in mammalian cells grown in tissue culture as well as Ehrlich ascite cells (2). System A (alanine-preferring) serves mainly for such amino acids as alanine, glycine, and serine while System L (leucine-preferring) shows a preference for the branched chain and aromatic amino acids.

Several groups of investigators have successfully examined transport in vesicles derived from plasma membranes of animal cells (3–6). There have also been several reports of the successful reconstitution of transport systems from several mammalian cell types (7–10) as well as a thermophilic bacterium (11).

Received June 30, 1977; accepted August 2, 1977

In this report, we have used the 2 detergents Triton X-100 and cholate to solubilize plasma membrane fractions from Ehrlich ascites cells. We found that the Triton extract retains leucine-binding activity with a specificity similar to that of the L system but this preparation was not as useful for reconstitution studies. We were able to use the cholate solubilized membrane preparations to reconstitute transport activity for alanine and leucine by incorporating partially purified extracts into soybean phospholipid liposomes. The cholate extract could be partially purified by ammonium sulfate precipitation and column chromatography.

METHODS

L-[^{3}H] leucine and L-[^{3}H] alanine were purchased from New England Nuclear Corporation (Boston, Massachusetts), and [^{14}C]-α-methylglucopyranoside from Amersham-Searle. Triton X-100 was obtained from Research Products International, and sodium cholate from Calbiochem (La Jolla, California). Soybean phospholipid (L-α-phosphatidyl-choline, Type II-S) and Concanavalin A-Sepharose were obtained from Sigma Chemical Company (St. Louis, Missouri). Membrane filters were from Millipore Company and Biogel P-60 was from Biorad Corporation (Richmond, California).

Preparation of Plasma Membrane Fraction

A plasma membrane fraction enriched approximately 25-fold based upon Na$^+$, K$^+$-ATPase activity was obtained from the Ehrlich ascites cells by a slight modification of procedures described by Im et al. (12).

Triton-solubilized Leucine-binding Component of Plasma Membrane Vesicles

Plasma membrane vesicles (25 mg of protein) in 15 mM sodium phosphate buffer, pH 7.4, were incubated with 0.5% Triton X-100 for 30 min at 4°C. After centrifugation at 35,000 × g for 50 min the supernatant fluid was brought to 10% saturation by addition of solid ammonium sulfate at 4°C. This suspension was then centrifuged at 30,000 × g for 20 min. Following centrifugation a pellicle containing the majority of the protein was found in the centrifuge tube and this was carefully removed and resuspended in 2 ml of 15 mM sodium phosphate buffer, pH 7.4. The protein solution was then placed on a Biogel P-60 column (1 × 16 cm) previously equilibrated with the sodium phosphate buffer and 0.1% Trition X-100.

The majority of the protein applied to the column eluted in 1 broad peak near the void volume of the column. This material was pooled and placed on a Concanavalin A-Sepharose 4B affinity column (0.5 × 7 cm) equilibrated with 15 mM sodium phosphate, pH 7.4, 1 mM CaCl$_2$ and MgCl$_2$, and 0.1% Trition X-100. About 5 mg of protein was applied to the column. About 3 mg of this protein was rapidly eluted from the column. The remaining protein was removed by addition of the same buffer with 0.1 M α-methyl-mannopyranoside (Pfansthiel). The leucine-binding activity was contained in the fraction that did not bind to the Concanavalin A-Sepharose 4B column. No binding activity for leucine was detected in the material eluted with α-methylmannopyranoside.

Equilibrium Dialysis for Measurement of Leucine-binding Activity

Binding activity was measured by equilibrium dialysis at 4°C in 15 mM sodium phosphate, pH 7.4, with 0.1% Triton X-100. Dialysis was carried out in 50-ml flasks containing the above buffer with the addition of the appropriate labeled amino acid. Fractions containing from 1 to 5 mg of protein in a final volume of 1 ml of the same buffer were placed inside the dialysis bags. Dialysis proceeded with constant stirring of the solution until equilibrium had been reached (12–24 h). Triplicate samples were then removed from inside and outside the dialysis bag and the radioactivity was determined in a Packard scintillation counter. Sodium azide (0.03%) was included in all dialysis solutions to prevent bacterial contamination.

Reconstitution of Transport Activity From a Cholate Extract of Plasma Membrane Vesicle

Extraction and purification. Ehrlich ascites plasma membrane vesicles (75 mg in 8 ml of 0.4 M potassium phosphate buffer, pH 7.4) were incubated with 2% sodium cholate, 1 mM dithiothreitol, and 0.5 mM EDTA for 30 min at 5°C with gentle mixing. The extract was then centrifuged at 100,000 × g for 60 min. The clear supernatant fluid contained from 34 to 45 mg of protein. The supernatant fraction was then brought to 20% saturation with solid ammonium sulfate (114 g/liter) at 4°C and slowly mixed for 30 min before centrifugation at 35,000 × g in a Sorvall RC2-B centrifuge. The supernatant fraction containing 27–37 mg of protein was dialyzed against 0.4 M potassium phosphate (pH 7.4) containing 1% cholate, 1 mM dithiothreitol for 12 h. This material was then placed on a Biogel P-60 column (1 × 18 cm) and eluted with the same buffer as used for dialysis. The majority of the protein (21–27 mg) eluted as a large aggregate in the void volume of the column. This protein peak was then placed on a DEAE-cellulose column (1 × 22 cm) and eluted with a linear gradient of 0.05 M to 0.4 M potassium phosphate buffer (pH 7.4) containing 1% cholate and 1 mM dithiothreitol. The initial fractions containing approximately 10 mg of protein were pooled and used for reconstitution studies. The variation in different preparations resulted in a recovery of 15–30% of the protein initially extracted from the plasma membrane vesicles.

Reconstitution. The reconstitution of the protein fraction into artificial lipid vesicles was accomplished by a minor a modification of the cholate-dilution procedure of Racker (13). Crude soybean lecithin was partially purified according to the method described by Kagawa and Racker (14). The phospholipids were dispersed by sonication under N_2 at 25–30°C in 2% cholate, 0.4 M potassium phosphate (pH 7.4), and 1 mM dithiothreitol until a clear suspension was obtained. The solubilized protein fraction described above was then added to the liposomes in a ratio of 20:1 (phospholipid to protein) and the mixture was dialyzed for 36 h at 4°C against 0.4 M potassium phosphate buffer (pH 7.4) containing 1 mM dithiothreitol. To determine if sealed liposomes were formed by this procedure an aliquot of the above was treated identically except [^{14}C]-α-methylglucopyranoside was included. At the end of the dialysis the reconstituted liposomes with the [^{14}C]-α-methylglucopyranoside were passed over a Biogel P-60 column (1 × 18 cm) equilibrated with 0.4 M potassium phosphate (pH 7.4). We found that 90% of the protein and a significant proportion of the labeled α-methylglucopyranoside eluted in the void volume of the column indicating that the labeled sugar had been trapped by the reconstituted vesicles.

To measure uptake of radioactive amino acid the reconstituted vesicles (0.8–1.2 mg protein in 1 ml) containing high levels of potassium (0.4 M) were resuspended in 9 ml of 0.4 M sodium phosphate (pH 7.4) containing 0.1 mM labeled amino acid (2μCi/ml). The liposomes were incubated at 37°C and in some cases 20 μg of valinomycin in 10 μl of 95% ethanol was added. At intervals 1-ml samples were removed and filtered through membrane filters (0.45 μm) and washed with 2 5-ml portions of 0.4 M potassium phosphate (pH

TABLE I. Leucine Binding Activity Extracted With Triton X-100

Addition	Ratio cpm inside/cpm outside
–	1.65
100 μM Leucine	1.01
50 μM Phenylalanine	1.03
1 μM Alanine	1.35

Binding activity was measured by equilibrium dialysis as discussed in Methods. L-[^{3}H] leucine at 10 nM was included in all above dialysis experiments and cold amino acid at the concentration indicated in the table was added to each dialysis chamber.

7.4), 1 mM $CaCl_2$ and $MgCl_2$ at 4°C. The filters were then rapidly removed, dried, and radioactivity was assayed in a Packard liquid scintillation counter. It required about 20–25 sec for filtering and washing the samples.

Protein in all cases was determined by the method of Lowry et al. (15).

RESULTS AND DISCUSSION

Leucine-binding Activity Extracted From Plasma Membranes of Ehrlich Cells With Triton X-100

Extraction of plasma membrane vesicles of Ehrlich cells with Triton X-100 as described in the Methods section usually solubilized 50–70% of the membrane proteins. Preliminary studies had already established that binding activity for leucine could be detected in the Triton X-100 extracts from the Ehrlich cell plasma membrane fraction (16). A further purification of the leucine-binding activity was accomplished by ammonium sulfate precipitation and Concanavalin A-Sepharose-affinity chromatography as described in the Methods section. These purification procedures yielded a protein fraction which upon analysis with sodium dodecyl sulfate-polyacrylamide gel electrophoresis showed from 4 to 7 Coomassie blue staining bands (data not shown). Utilizing various protein markers the bands have an apparent molecular weight range of 20,000–80,000. The amount of Triton X-100 remaining bound to the protein in the preparation is unknown; therefore, the molecular weight values obtained should be viewed with appropriate caution. The binding activity does not involve a high affinity Concanavalin A receptor from the plasma membrane of Ehrlich cells because the material which bound tightly to the Concanavalin A-Sepharose column and had to be removed by elution with methyl α-$\underline{D}$-mannoside had no detectable leucine-binding activity. The binding activity for leucine was eluted from the Concanavalin A-Sepharose column in the early fractions which had little affinity for the lectin and were only slightly retarded.

The properties of the partially purified leucine-binding activity obtained from the Ehrlich cell plasma membrane are shown in Table I. A binding ratio of 1.65 was obtained during equilibrium dialysis with 10 μM leucine. The addition of a 10,000-fold excess of unlabeled leucine decreases the ratio to 1.0 indicating that the binding activity was saturable. Phenylalanine is transported in intact Ehrlich cells by System L to about the same extent as leucine (1); therefore, phenylalanine would also be expected to inhibit leucine-binding activity. Table I shows that this was the case. Alanine also partially inhibits leucine-binding. Oxender and Christensen (1) reported that alanine, although mostly transported by System A, will compete to some extent with leucine for entry into Ehrlich cells. The binding activity for leucine described here has properties similar to those observed for leucine transport into intact Ehrlich cells.

Attempts to demonstrate binding activity for System A amino acids such as L-alanine and glycine were unsuccessful using the Triton X-100 solubilized plasma membrane fractions. Thus the binding activity seems to be present only for System L amino acids and not System A. We were also not able to reconstitute leucine transport activity into lipid vesicles using this preparation. This failure may be related to the difficulty of removing the Triton X-100. Treatment of the preparation with Bio Beads SM-2 to remove the Triton X-100 resulted in a complete loss of detectable binding activity for leucine.

Reconstitution of Neutral Amino Acid Transport Using Cholate Extracted Plasma Membrane Vesicles of Ehrlich Cells

Cholate extraction of Ehrlich cell plasma membrane vesicles as described in the Methods section solubilized 40–60% of the membrane proteins. The extract was used to reconstitute phospholipid vesicles from either soybean phospholipid or a mixture of purified phospholipids composed of a 4:1 ratio of phosphatidylcholine to phosphatidylethanolamine. Analysis of the plasma membrane showed 50–60 Coomassie blue staining bands on sodium dodecyl sulfate slab gel electrophoresis. The reconstituted liposomes were used to demonstrate uptake of alanine or leucine which was dependent upon a gradient of sodium ions. Since Johnston and Bardin (7) had already shown that cholate extracts of Ehrlich cell plasma membrane vesicles could be used to form reconstituted liposomes that would concentrate System A amino acids such as α-aminoisobutyric acid, it was of primary interest in this investigation to attempt to purify the components responsible for transport activity.

The Methods section describes the procedures utilized to obtain a partial purification of the components required for the transport of alanine and leucine. The protein fraction obtained by these procedures contains 10–15 Coomassie blue bands when subject to gel electrophoresis (data not shown). The molecular weight ranges of the proteins present in this fraction appear to be between 30,000 and 130,000 using standard proteins on the sodium dodecyl sulfate gels.

Table II shows the time course of the uptake of alanine and leucine by soybean phospholipid reconstituted with the partially purified extract. The uptake of alanine and leucine requires the addition of sodium to the medium. Maximum uptake occurs 3 min after the initiation of uptake and then appears to diminish. This loss of amino acid probably represents a decay in the sodium gradient. If sodium is not added to the reconstituted liposomes but is replaced by potassium, amino acid uptake does not take place (data not shown). Thus, sodium is required for the transport of both leucine and alanine by the reconstituted liposomes.

Generation of Membrane Potential

Lever has previously shown that membrane potential can be used to drive α-aminoisobutyric acid uptake in membrane vesicles of 3T3 cells (6). We were able to demonstrate increased uptake for alanine in the reconstituted K^+-loaded vesicles when valinomycin was added to the medium (Table III). The uptake of alanine was more rapid when valinomycin was included than that observed for sodium alone. The "overshoot" is also much more

TABLE II. Uptake of Alanine and Leucine by Reconstituted Soybean Phospholipid Vesicles

Time (min)	Uptake Alanine	Leucine
	nmoles/mg protein	
1	0.05	0.07
2	0.25	0.11
3	0.42	0.23
4	0.40	0.20
5	0.21	0.18
7	0.23	0.15

Uptake of 0.1 mM L-alanine and 0.05 mM L-leucine by reconstituted soybean phospholipid vesicles. Each time point represents 100 μg of reconstituted protein. Uptake was initiated by the addition of 0.4 M sodium phosphate, pH 7.4 to phospholipid vesicles prepared in potassium phosphate. Other assay conditions are described in the Methods section.

TABLE III. Alanine Uptake in Reconstituted Soybean Phospholipid Vesicles

Time (min)	37°C A.	B.	2°C C.
		nmoles/mg protein	
1	0.11	0.20	0.01
2	0.33	0.43	0.05
3	0.30	0.42	0.03
4	0.28	0.23	0.09
5	0.29	–	0.04
7	0.26	0.23	–

Uptake of 0.1 mM L-alanine by reconstituted soybean phospholipid vesicles preloaded with 0.4 M potassium phosphate, pH 7.4.

A. 0.4 M sodium phosphate, pH 7.4 added with L-alanine to initiate uptake at 37°C.

B. Same as A with the addition of 2 μg/ml of valinomycin added at time zero.

C. Same as B except uptake was measured at 2°C.

pronounced and this may be due to a more rapid dissipation of the gradient of membrane potential when coupled to valinomycin-mediated potassium diffusion. Table III shows that transport at 37°C is severalfold higher than at 2°C.

Accurate measurements of initial rates are rather difficult in the reconstituted system since considerable variation was obtained for the low transport rates observed. The sodium-dependent alanine uptake was saturable with alanine concentrations greater than 1 mM (data not shown). Leucine uptake in this reconstituted system appears to be sodium dependent; however, it is difficult to show sodium-dependent uptake of leucine in whole cells (1) presumably because of the dominance of the sodium-independent System L. It may be that leucine uptake observed in this study is the result of entry by System A. These results are consistent with our failure to observe System L leucine-binding activity in the cholate extracted material in contrast to the earlier studies with the Triton X-100 extracted material. In addition, when transport was measured in liposomes that were reconstituted without protein added, active uptake was not observed.

The present study confirms and extends the studies of Johnstone and Bardin (7) on the reconstitution of neutral amino acid transport from Ehrlich cells in liposomes. The purification of the transport components has been greatly facilitated by the ability to reconstitute transport into artificial lipid vesicles. We are currently attempting to further purify the membrane transport activity.

ACKNOWLEDGMENTS

This research was supported by the National Institutes of Health, grant GM 20737 for D. L. O.

REFERENCES

1. Oxender DL, Christensen HN: J Biol Chem 238:3686, 1963.
2. Oxender DL, Lee M, Moore PA, Cecchini G: J Biol Chem 252:2675, 1977.
3. Colombini M, Johnstone RM: J Membr Biol 15:261, 1974.
4. Hamilton RT, Nilsen-Hamilton M: Proc Natl Acad Sci USA 73:1907, 1976.
5. Quinlan DC, Hochstadt J: J Biol Chem 251:344, 1976.
6. Lever JE: J Cell Physiol 89:779, 1976.
7. Johnstone RM, Bardin C: J Cell Physiol 89:801, 1976.
8. Kasahara M, Hinkle PC: Proc Natl Acad Sci USA 73:396, 1976.
9. Shertzer HG, Racker E: J Biol Chem 251:2446, 1976.
10. Zala CA, Kahlenberg A: Biochem Biophys Res Commun 72:866, 1976.
11. Hirata H, Sone N, Yoshida M, Kagawa Y: Biochem Biophys Res Commun 69:665, 1976.
12. Im WB, Christensen HN, Sportés B: Biochim Biophys Acta 436:424, 1976.
13. Racker E: J Biol Chem 247:8198, 1972.
14. Kagawa Y, Racker E: J Biol Chem 246:5477, 1971.
15. Lowry OH, Rosebrough NJ, Farr AL, Randall RJ: J Biol Chem 193:265, 1951.
16. Cecchini G, Lee M, Oxender DL: J Supramol Struct 4:441, 1976.

Journal of Supramolecular Structure 7:489—497 (1977)
Molecular Aspects of Membrane Transport 519—527

The Structure of Intrinsic Membrane Proteins

Guido Guidotti

The Biological Laboratories, Harvard University, Cambridge, Massachusetts 02138

Intrinsic membrane proteins are embedded in the lipid bilayer so that the polypeptides come in contact with the non-polar region of the bilayer. There are two major types of intrinsic proteins: those with most of their mass outside the cytoplasm (Type I) and those with most of their mass inside the cytoplasm (Type II). In the latter group are the membrane transport systems. The anion exchange system of the human erythrocyte is a dimer of band 3 polypeptides. These polypeptides span the bilayer, have most of their mass in the cytoplasm, and are glycosylated. About 20—25% of the polypeptide, however, is in the bilayer. Arguments are presented to support the view that the intramembrane segments of the protein are α-helical and that the major protein-protein inter-actions between the subunits are in the cytoplasmic portion of the protein.

Key words: membrane proteins, anion exchange, band 3 polypeptide, red cell membrane, transport

Intrinsic membrane proteins are by definition those proteins which cannot be re-moved from the membrane without the use of detergents. This property suggests that these proteins are embedded in or interact strongly with the nonpolar region of the bilayer.

This hypothesis is supported by the evidence obtained on a small number of intrinsic membrane proteins. This evidence strongly indicates that all intrinsic proteins examined so far span the bilayer. In these proteins, there clearly must be parts of the polypeptide which are in contact and thus interact with the nonpolar part of the bilayer. Thus, we may start with the assumption that all intrinsic proteins are transmembrane proteins.

There seem to be 2 general types of intrinsic membrane proteins which differ in their properties and arrangement in the membrane. They are called Type I and Type II proteins (Fig. 1).

Type I intrinsic proteins have most of their mass and all of their functional proper-ties in the aqueous environment outside the cytoplasm. The intramembrane portion serves only to anchor these proteins to a particular membrane and it consists of a short amino acid sequence sufficient to span the bilayer as an α helix. Examples of Type I proteins are the major sialoglycoprotein of the red blood cell membrane, several proteins of enveloped viruses like Semliki Forest virus and Sindbis virus, and the histocompatability antigens of human cells (HL-A). The best studies of these is the major sialoglycoprotein of the human erythrocyte.

Received for publication August 25, 1977; accepted September 1, 1977

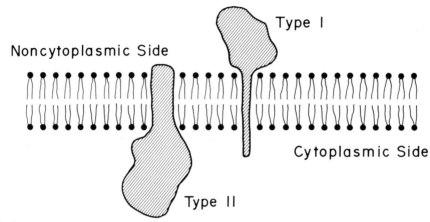

Fig. 1. Arrangement in the membrane of the 2 classes of intrinsic membrane proteins: Type I and Type II.

This protein, whose amino acid sequence and arrangement in the membrane have been described (1), is composed of 131 amino acids of which 72 at the NH$_2$-terminal end are located outside the cell, 23 form the intramembrane and transmembrane portion, and the 36 amino acid residues at the COOH-terminal end of the protein are located inside the cell in the cytoplasm. In addition, the protein is a glycoprotein and the carbohydrate comprises two-thirds of the mass of the protein. All the sugar residues, approximately 100 of them, are attached to the amino acid residues on the outside of the membrane. Thus, all the protein-bound carbohydrate is extracellular (1).

The intramembrane portion is 23 amino acids long. Importantly, almost all of these residues are nonpolar and none of them is an acidic or basic amino acid. One can visualize that this stretch of residues could be accommodated in the lipid bilayer. As will be discussed later, the only likely structure for this portion of the protein is an α helix.

This protein illustrates clearly the salient features of the intrinsic proteins of Type I: The proteins are basically aqueous proteins with an anchor which attaches them to a membrane. The anchor comprises a small fraction of the polypeptide chain. All the functional properties of the protein are in the aqueous compartments on the exterior of the membrane.

Type II intrinsic proteins have 2 major characteristics: The major part of their polypeptide mass is in the cytoplasm and a very small fraction of their amino acid residues is exposed to the noncytoplasmic side of the bilayer. This means that over 90% of the amino acid residues are either in the bilayer or in the cytoplasm. Clearly these proteins are drastically different in their membrane arrangement from the Type I proteins. It is also apparent that the mechanism by which the Type I proteins are inserted in the bilayer must be different from that used by Type II proteins.[1]

[1] Type I intrinsic proteins resemble secreted proteins in all aspects except that a small COOH-terminal segment remains inserted in the bilayer and in the cytoplasm. It is likely that the proteins are synthesized by the same mechanism involved in the synthesis and secretion of extracellular proteins, which has recently been described by Blobel (Blobel G, Dobberstein B: J Cell Biol 67:835–851, 1975). On the other hand, Type II proteins are probably made on normal, soluble ribosomes, and fold to produce a structure similar to that shown in Fig. 2. Since the nonpolar area on the outside of the polypeptide is localized in one discrete area, while most of the protein surface is polar, the protein will insert in a distinct way in the bilayer. In this view, Type II proteins will become attached to the membrane after synthesis and there will be no necessary localization of the NH$_2$- and COOH-terminal residues.

TABLE I. Properties of Membrane Transport Proteins*

	Na^+,K^+- ATPase	Ca^{2+}- ATPase	Anion- exchange protein	Acetyl- choline receptor	Rhodopsin
Molecular weights of component polypeptides	α-90,000 β-40,000	α-100,000	α-90,000	α- 40,000 β- 48,000 γ- 58,000 δ- 64,000 ε-105,000	α-38,000
Glycoproteins	β		α	$α(β - ε)$?	α
Probable structure and molecular weight of the protein part of the enzyme	$α_2β_2$ 260,000	$(α_2$?) (200,000?)	$α_2$ 180,000	$ε_2$ or $α_2β_2$ 240,000	($α_2$ to $α_4$?) (76,000– 152,000?)
Transmembrane arrangement	α		α		α
Detergent binding (mg/mg of protein)	0.28	0.20	0.77	0.7	1.10
Relative hydrophobic surface area of subunit	0.20–0.24	0.2–0.25	0.5–0.65	0.5–0.6	0.54

*Reprinted from Guidotti (9), with permission of the publisher.

There appear to be 2 categories of Type II proteins. One group is represented by cytochrome b_5, cytochrome b_5 reductase, and stearyl-CoA desaturase (2). These proteins are similar to the Type I proteins in that their functional domains are entirely in the aqueous compartment, in this case in the cytoplasm. Cytochrome b_5 is attached to the membrane through a hydrophobic region of the molecule which is COOH-terminal and contains 44 amino acids. It has not been established whether or not the intramembrane portion of this polypeptide traverses the bilayer entirely or is localized to the inner leaflet. The other proteins in this category resemble cytochrome b_5 in their arrangement on the membrane. One concludes that the attachment to the membrane of these proteins has the principal function of concentrating the enzymes on the endoplasmic reticulum.

The second category of Type II proteins is represented by proteins involved in trans-membrane processes, for example, transport. There are only a few well-characterized proteins in this group: the Na^+,K^+-ATPase (3), the Ca^{2+}-ATPase (4), band 3 of the human erythrocyte membrane which is involved in anion exchange (5,6), vertebrate rhodopsin (7), and the acetylcholine receptor (8). All of these intrinsic proteins have explicit or putative transport functions and thus are involved in the transfer of material across the bilayer. It is likely that all transmembrane transport of both material and information (i.e., hormone reception) is catalyzed by proteins that have properties similar to those of the proteins described above. These proteins are all transmembrane oligomers which are glyco-proteins if attached to the plasma membrane, as is shown in Table I and discussed in a previous publication (9).[2]

Let us focus now on a particular transport system — the anion exchange system of human erythrocytes. The diagram shown in Fig. 2 indicates the relevant features of this protein with regard to its arrangement in the membrane. The transport system is composed of dimers of Type II polypeptides: There is very little protein outside the cytoplasm; there is a substantial amount in the bilayer; the major fraction of the protein is inside the cytoplasm.

[2]It is likely that transport across bacterial membranes is catalyzed by oligomeric transmembrane pro-teins which resemble eukaryotic systems. This suggestion is supported by recent evidence that bacterial rhodopsin, which is a light-activated proton pump, probably is a transmembrane protein which may be oligomeric. [Henderson R, Unwin PNT: Nature 257:28–32 (1975)].

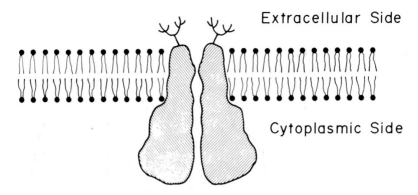

Anion Exchange System (Band 3)

Fig. 2. Oligomeric structure and arrangement in the membrane of the anion exchange systems of human erythrocyte membranes (band 3 polypeptide).

The evidence for this arrangement is fairly straightforward. In the first place, proteolysis of the exterior surface of intact red blood cells causes no degradation of the protein (trypsin) or at best cleavage of a single peptide bond (6). This experiment shows that while a portion of the polypeptide must span the bilayer, very few amino acid residues are present on the outer surface.

The fraction of the polypeptide present in the bilayer is between 0.2 and 0.4. Two experiments bear on this point. Clarke (10) has shown that in some cases the extent of Triton X-100 binding to a protein gives an estimate of the surface area of the protein which is nonpolar and thus presumably in contact with the nonpolar part of the bilayer. The estimate is likely to be correct if the protein binds detergent below the critical micellar concentration (cmc) of the detergent, and thus can be shown not to insert into a micelle. The band 3 polypeptide behaves in this way. Thus the data shown in Table II can be interpreted as giving an estimate of the hydrophobic surface of the protein: It is approximately 40% of the total.

The second experiment involves the removal by extensive proteolysis of all segments of the protein which are not in the bilayer, and then isolation and characterization of the intramembrane segments. The data are shown in Table III, and they indicate that approximately 20% of the polypeptide remains in the bilayer. The fraction of nonpolar residues in these segments is very high and it substantiates the view that they are located in the bilayer.

Finally, the localization of the major part of the polypeptide inside the cytoplasm, which can be deduced by exclusion, is compatible with the experiments of many investigators which show that there is extensive destruction of the polypeptide by proteolytic enzymes which can attack the inner surface of the membrane (11).

Let us consider now the structure of the oligomers. There are 3 areas of interest: the symmetry of the oligomer, the protein-protein interactions which stabilize the oligomers, and the structure of the intramembrane part.

Symmetry

The evidence shown in Table II, along with the cross-linking data of Yu and Steck, indicate that the band 3 polypeptide is a dimer (12).

The oligomeric arrangement of the subunits together with the membrane asymmetry of polypeptide arrangement means that the axis of symmetry of the oligomer must be

TABLE II. Size of the Triton X-100 Complex of Band 3 Protein*

$s_{20,w}$	6.9S	f/f_0	1.7
$\bar{v}$	0.81 cm^3/g	g Triton/g protein	0.77
$D_{20,w}$	2.7×10^{-7} cm^2/sec	g carbohydrate/g protein	0.08
M_r complex	320,000	M_r protein portion	175,000
Moles Triton/Moles protein	208	M_r protein in SDS	90,000
Fraction of surface area covered by detergent[a]	0.41	M_r (Triton)/M_2 (SDS)	1.95

*Data taken from Clarke (10).

[a]This calculation was done in the following way. The total surface area of the protein was determined as that of a sphere with a mass of the polypeptide chain and a $\bar{v}$ of 0.73 cm^3/g. The surface area occupied by the measured number of bound detergent molecules was calculated from the data in this Table assuming that each detergent molecule occupies an area of 0.5 nm^2 (data for an air to water interface: Technical bulletin, Rohn and Haas).

TABLE III. Membrane-Bound Proteolytic Fragments of Band 3*

Material	% of total	% nonpolar amino acids
Intact band 3	100	52
Tryptic fragments (mol. wt. 7,000)	20	63
Papain fragments (mol. wt. 8,000)	18	61

*Proteolysis of membrane-bound band 3 was done at 23°C for 24 h. The membranes were separated from the solution, washed, and the fragments isolated by SDS gel electrophoresis.

perpendicular to the plane of the membrane. If the oligomer is a dimer, homologous bonding between the subunits can satisfy this requirement. It should be emphasized that homologous bonding is the only arrangement which can close at a dimer stage. On the other hand, if the oligomer is larger than a dimer, say a tetramer or hexamer (for example, the Ca^{2+}-ATPase of sarcoplasmic reticulum has been envisaged as a tetramer (13)], then the requirements stated above can only be satisfied by heterologous bonding between the subunits. This means that an n-mer will have an n-fold axis of symmetry perpendicular to the bilayer. This also means, since heterologous bonding is less frequent than is homologous bonding, that the most likely structure of membrane proteins is dimeric.

PROTEIN-PROTEIN INTERACTIONS

The possible interactions between subunits in oligomeric membrane proteins can involve the 3 parts of the polypeptide — that part which is exposed outside the membrane to the extracellular fluid, the intramembrane part, and the major cytoplasmic component. Since evidence suggests that very little of the polypeptide is exposed to the outer environment, extracellular interactions should be negligible. On the other hand the intramembrane part and the cytoplasmic portion are likely to have a major role.

The intramembrane piece of the polypeptide, which, as we shall discuss, is likely to be α helical, will interact with the lipid by nonpolar interactions, but will interact with the segments of the polypeptide and with another polypeptide by hydrogen bonds: These clearly are more stable in a nonpolar environment and will strongly stabilize protein-protein interactions. This means that any contact surface between 2 polypeptides will be stabilized by polar interaction; i.e., there is likely to be a polar surface at the contact between membrane polypeptides. This is the likely surface which will interact with polar solutes. Thus, I suggest that the active site of membrane transport systems is at the interface between 2 subunits. Necessarily this means that the number of active sites is less than the number of polypeptides, indeed the requirement for half-of-the-sites activity is a necessity for membrane transport. If the intramembrane segments of the polypeptide are the main ones involved in subunit-subunit interactions and these are necessarily polar interactions, they should vanish if these areas of the helices are exposed to water. Thus, under conditions which eliminate the lipid bilayer, for example with detergents like Triton X-100 or deoxycholate, transmembrane oligomers stabilized by the intramembrane segments should dissociate into protomers. This is not the case with the band 3 oligomer (10). Therefore one might conclude that the major part of the stabilizing interactions between the monomers does not involve the intramembrane portion. However, the cytoplasmic portion of the protein, which is large for the band 3 polypeptide, can have a major role in the intersubunit interactions. These can be both polar and nonpolar interactions, and can contribute the main stabilization for the oligomeric structure. Since the cytoplasmic portion of the protein is normally present in the aqueous environment, removal of the protein from the bilayer with detergents should have no effect on the oligomeric structure of the protein.

In fact, the band 3 polypeptide exists at least as a dimer in the membrane, and as a dimer in solutions containing certain detergents (Triton X-100) (10). This fact suggests that the main stabilizing energy is not affected by weak detergents any more than are the usual water-soluble proteins.

INTRAMEMBRANE PART

Approximately 25% of the mass of the band 3 polypeptide is in the bilayer (10). Results obtained from this laboratory by K. Drickamer (6) show the location in the linear sequence of the polypeptide of at least one region which must span the bilayer (Fig. 3): It is located between the NH_2-terminal 10,000 dalton fragment which is on the cytoplasmic side of the membrane and the 7,000 dalton fragment which is 30,000 daltons from the COOH-terminal end and on the outside of the membrane. Since Steck's group (11) has suggested that there is a 17,000 dalton fragment which is embedded in the membrane, this piece must be located in the region described above and illustrated in Fig. 3. Evidence from another laboratory (14, 15) suggests that the polypeptide of band 3 has at least 2 different regions which span the bilayer. One concludes that a substantial fraction of this protein is in the bilayer.

It is interesting and important that careful studies of the structure of the proteins of erythrocyte membrane by optical rotatory dispersion (ORD) and circular dichroism measurements indicate fairly conclusively that the proteins exist as random coils and α helices and that there is very little if any β structure (16). Accordingly, the intramembrane sections of band 3 are most likely in an α-helical conformation.

There is a rationale for this observation on the intramembrane portions of the band

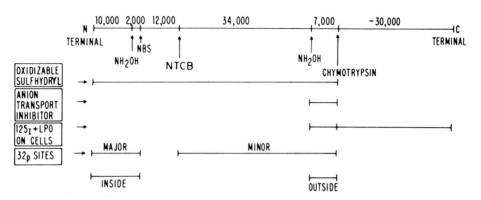

Fig. 3. The sites of chemical and enzymatic cleavage of the 95,000 dalton polypeptide from the human erythrocyte membrane (Band 3 polypeptide). The regions of the polypeptide which are located on the outer surface of the bilayer and on the inner surface of the bilayer are indicated, as are the regions to which specific labels have been attached. NBS) N-Bromosuccinimide; LPO) lactoperoxidase; NTCB) 2-nitro 5-thiocyanobenzoic acid. (Reprented from Drickamer [6], with permission.)

$$\Delta G^0 \quad (Kcal/mole)$$

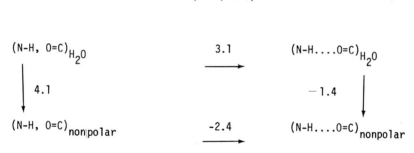

Fig. 4. Free energy changes for the formation of a hydrogen bond between the atoms of an amide group (peptide bond) and for the transfer of the group from water into a nonpolar solvent.

3 polypeptide. A polypeptide chain is composed of 2 parts: the peptide units which make up the backbone and the R groups attached to the α carbons of the backbone. While several R groups are hydrophobic and can interact with the nonpolar part of the bilayer, the peptide group is extremely polar and essentially insoluble in the bilayer. The polarity of the peptide group can be decreased if it is hydrogen-bonded. This is demonstrated by the results obtained by Klotz on the association of substituted acetamides in CCl_4 (17), The results are depicted in Fig. 4, which shows the standard free energy changes for the transfer of a hydrogen-bonded and a non-hydrogen-bonded peptide group from water to a nonpolar solvent. Thus, even a nonpolar amino acid residue will not be stable in a nonpolar environment if its peptide group is not hydrogen-bonded. This can be accomplished in 2 ways: The polypeptide can have an α-helical structure or a β structure. In the former case, the hydrogen bonds are formed between residues in the same segment of the polypeptide chain, while in the β structure the hydrogen bonds are between different strands of the polypeptide. As mentioned above, however, there is no appreciable β structure in the pro-

teins of the erythrocyte membrane. I conclude, therefore, that the transmembrane regions of the band 3 polypeptide are in the α-helical conformation.

The α-helical transmembrane segments are a simple and economical way to anchor a protein in the bilayer. Furthermore, specific aggregates of helices are well suited to form ordered structures, as happens, for example, in the protein tropomyosin, that have the unique feature of forming a hydrophilic active site which is not in contact with the nonpolar part of the lipid bilayer. This view derives from the following arguments. One would expect that the helix-helix interactions in the bilayer would be mainly polar ones since these interactions would be strong ones in the nonpolar environment of the bilayer. Thus the complex of helices would be stabilized by 2 types of interactions: Polar interactions for the helix-helix contacts, and nonpolar ones to stabilize the helix complex in the nonpolar region of the bilayer. Obviously, such an arrangement would require a very specific amino acid sequence for the regions of the polypeptide that span the bilayer. For example for a 2 helix complex in the bilayer the sequence should be nonpolar-nonpolar-polar-nonpolar-nonpolar-nonpolar-polar. A simple extension of this view suggests that a polar channel which spans the bilayer can be formed in the membrane by the association of 2 or more polypeptides which are in contact at the transmembrane helical segments. The stabilizing interactions between the transmembrane segments of the polypeptides would certainly be polar ones, and would thus create a well-defined polar region which spans the bilayer, i.e., a transmembrane polar active site for the catalysis of transport.

The view that α helices are the main, if not the only, ways in which a polypeptide traverses the bilayer is supported by a consideration of the free energy of interaction of these structures with the nonpolar region of the bilayer. If indeed all peptide bonds and polar residues are neutralized in the bilayer by the formation of hydrogen bonds, then the only parts of the polypeptide exposed to the solvent (the lipid) are the nonpolar ones. One can calculate the free energy of interaction of a helix or a group of helices with the nonpolar region of the bilayer by estimating the surface area of the structure in the bilayer and using the value of 25 calories mole^{-1} A^{-2} for the stabilizing energy which has been obtained by Clothia (18). The stabilizing free energy is -30 Kcal/mole for one helix and -300 Kcal/mole for a group of 7 helices. Clearly, the stabilizing energy is sufficient to prevent the polypeptide from dissociation from the membrane.

ACKNOWLEDGMENTS

This work was supported by NIH grant HL 08893 and National Science Foundation grants BMS 73-06752 and BMS 75-09919. I thank Michael Ho for some of this work.

REFERENCES

1. Marchesi VT, Furthmayr H, Tomita M: Annu Rev Biochem 45:667–698, 1976.
2. Enock HG, Catala A, Strittmatter P: J Biol Chem 251:5095–5103, 1976.
3. Dahl JL, Hokin LE: Annu Rev Biochem 43:327–356, 1974.
4. McLennan DH, Holland DC: Annu Rev Biophys Bioeng 4:377–404, 1975.
5. Ho MK, Guidotti G: J Biol Chem 250:675–683, 1975.

6. Drickamer LK: J Biol Chem 251:5115–5123, 1976.
7. Cone RA: In Schmitt FO, Schneider DM, Crothers DM (eds): "Functional Linkage in Biomolecular Systems." New York: Raven Press, 1975, pp 234–246.
8. Karlin A: Life Sci 14:1385–1415, 1974.
9. Guidotti G: Trend Biochem Sci 1:11–13, 1976.
10. Clarke S: J Biol Chem 250:5459–5464, 1975.
11. Steck TL, Ramos B, Strapozon E: Biochemistry 15:1154–1161, 1976.
12. Yu J, Steck TL: J Biol Chem 250:9176–9184, 1975.
13. Murphy AJ: Biochem Biophys Res Commun 70: 160–166, 1976.
14. Jenkins RE, Tanner MJA: Biochem J 147:393–399, 1975.
15. Jenkins RE, Tanner MJA: Biochem J 161:134–147, 1976.
16. Schneider AS, Schneider MJT, Rosenheck K: Proc Natl Acad Sci USA 66:793–798, 1970.
17. Kresheck GC, Klotz IM: Biochemistry 8:8–12, 1969.
18. Clothia C: J Mol Biol 105:1–14, 1976.

Journal of Supramolecular Structure 7:499–513 (1977)
Molecular Aspects of Membrane Transport 529–543

Regulation of Glucose Carriers in Chick Fibroblasts

Harold Amos, Thomas A. Musliner, and Hovan Asdourian

Department of Microbiology and Molecular Genetics, Harvard Medical School, Boston, Massachusetts 02115

The derepression of glucose transport initiated by removing glucose from the incubation medium requires both protein and RNA synthesis. The synthesis and accumulation of putative mRNA for the carrier protein(s) can be demonstrated by inhibiting protein synthesis with cycloheximide (2 μg/ml). Release from inhibition with simultaneous addition of actinomycin D (1–5 μg/ml) results in a burst of carrier synthesis that achieves virtually maximal derepression in 4–6 h. An external energy source provided by a "nonrepressive" sugar (D-fructose, D-xylose) or by pyruvate is required to accomplish carrier synthesis. Previous failure to demonstrate mRNA accumulation was due to the depletion of energy in the starved cells. Glucose acts as a repressor at a posttranscriptional step, probably at the level of turnover of formed carrier.

The protection of formed carrier in the absence of glucose and by inhibitors of protein synthesis even in the presence of glucose has encouraged conjecture that a protease is activated by a metabolic product of glucose that is analogous to a corepressor. The glucose metabolite either activates the protease by direct interaction with it or alters the conformation of the carrier to expose a critical region to protease attack. Indeed the regulation of carrier density in the membrane of chick fibroblasts may be achieved entirely by carrier inactivation, the rate of which is a function of glucose concentration in the culture medium.

Key words: glucose, carrier, regulation, transport

The derepression of glucose transport that follows when cultivated mammalian and avian cells are deprived of a carbon source, has provided a means of examining the mechanisms controlling hexose carrier synthesis and turnover (1–7). The most detailed analysis of elements involved in determining carrier concentration has been contributed by Christopher and Kalckar (8, 9). They have proposed that regulation involves the turnover of components of hexose uptake systems on the one hand and hexose carrier synthesis which may or may not be alternatively accelerated and restrained on the other.

Thomas A. Musliner is presently with the National Heart, Lung, and Blood Institute, National Institutes of Health, Bethesda, Maryland 20014.
Hovan Asdourian is presently with the Department of Microbiology, Harvard School of Public Health, Boston, Massachusetts 02115.

Received April 12, 1977; accepted October 4, 1977.

Since the phenomenon of "deprivation derepression" first described in glucose transport by chick fibroblasts (1) has now been extended to amino acid transport in mammalian cells (10) and to the leucine-binding protein of E. coli (11), some aspects of the mechanisms involved may have a broader interest than that of hexose transport in vertebrate cells.

In chick fibroblasts (1, 6, 7, 12, 13) and in Nil (hamster) fibroblasts (8, 9) the - increase in carrier molecules during starvation is blocked at virtually any stage by inhibitors of protein synthesis (cycloheximide and puromycin) as well as by inhibitors of RNA synthesis (actinomycin D and cordycepin). The inhibition of protein synthesis is readily explained as blocking the formation of new carriers while carrier destruction proceeds at some finite rate. Why inhibitors of RNA synthesis are effective after derepression is initiated has been difficult to reconcile with the commonly observed features of the synthesis of protein. Generally when mRNA molecules are permitted to accumulate (14, 15), translation proceeds and is unaffected by the later addition of inhibitors of RNA synthesis. Indeed the so-called "paradoxical effect" of actinomycin D is often observed (16).

Several observations with respect to repression and derepression of glucose transport await a more general hypothesis to be explained. Among these are the derepression that is characteristic of cells cultivated on D-galactose, D-fructose, and D-xylose as sole carbon source. Certain glucose analogues such as 2-deoxy-D-glucose and D-glucosamine simulate glucose as "repressors" of transport in repressed cells but have no "repressor" effect when added to derepressed cells. Glucose, on the other hand, effects a sharp repression of derepressed cells (8, 9, 13)

In the course of studies of ATP conservation by normal cell lines and early passage cell isolates, we have rediscovered the dramatic lowering of the ATP level in a variety of mammalian and avian cells by 2-deoxy-D-glucose and D-glucosamine (30).

This led to a series of experiments that provided presumptive evidence for the synthesis and accumulation of putative mRNA for hexose carrier protein. In addition a posttranscriptional negative control by glucose emerged. This last provides additional evidence for the Christopher-Kalckar model of hexose carrier turnover (8, 9).

MATERIALS AND METHODS

Cell Cultures

Both primary and secondary chick embryo fibroblast (CEF) cultures were prepared by trypsinization as previously described (12). Cells were grown in small (30–50 cm^2) glass bottles, T-flasks, roller bottles, or multiwell plates. The medium employed in all experiments was Eagle's basal medium (BME) (19) without $NaHCO_3$ and supplemented with 3% or 4% calf serum. The cultures were incubated in sealed vessels in incubators not gassed with CO_2. This lack of $NaHCO_3$ and CO_2 is critical to the growth and assay of glucose uptake in chick cells (20) in that the depression observed in starved cells is far greater when cells are maintained in the absence of $NAHCO_3$ and CO_2.

Both primary and secondary chick cells were inoculated into multiwell plates at approximately $2-3 \times 10^4$ cells per well. At 3–4 days as the cells were approaching confluence, they were used for the experiments described.

Measurement of D-glucose Entry into Cells

a. **Bottles and T-flasks.** The following procedures were conducted at $37°C$ with reagents equilibrated at $37°C$. Monolayers were washed twice with 20 ml of balanced salt solution (BSS) without glucose or with normal saline. D-[^{3}H] Glucose (2 μC/ml, 4 $\times$ 10^{-6} M) or D-[U-^{14}C] glucose (0.5 μC/ml, 4 $\times$ 10^{-6} M) was added in 1.0 ml BSS Glc$^-$ (glucose-free) and the monolayers were incubated on a rotatory shaker (80 oscillations per minute) for 5 min. The reaction was stopped by washing twice with 40 ml of normal saline. After draining, the cells were covered with 1.5 ml cold 5% trichloroacetic acid (TCA) and allowed to stand 15–25 h at 4°C. The TCA solution was transferred quantitatively into scintillation vials and neutralized with NaOH. Samples were counted in a liquid scintillation counter after addition of 15 ml of Bray's scintillation fluid. The protein content of each cell population (which remained fixed to the culture vessel after exposure to 5% TCA) was measured by the method of Lowry et al. (21). The acid-insoluble fraction, as reported previously (1), contained less than 1% of the total cell-associated counts.

    b. Multiwell plates. Sugar Uptake Assays. At the end of the "preconditioning" period, the monolayers were washed by rinsing them twice with BSS Glc$^-$ and chilled on ice, whereupon 0.3 ml of ice cold radiolabeled sugar solution was added per well. The plates were allowed to float on a 37°C water bath for the pulse period (usually 5 min) desired.

    The uptake was stopped in one motion by immersing the plate in a beaker containing cold PBS. After draining, 0.5 ml of 5% TCA was added to each well and the plates were kept at 4°C overnight. The TCA-soluble phase was obtained, mixed with 10 ml scintillation fluid, and counted on a Beckman LS-230 scintillation counter.

    Materials. [U-^{14}C] Glucose 196 mCi/mmole; D-[2-^{3}H] glucose 540 mCi/mmole; D-[3-^{3}H] glucose 5 Ci/mmole; [^{3}H] 3-O-methyl glucose (140 mCi/ mmole); D-glucose-6-[^{3}H] amine 10 Ci/mmole; and D-[1-^{14}C] mannose 50 mCi/mmole were purchased from New England Nuclear Corporation (Boston, Massachusetts); D-glucose, D-galactose, and D-fructose were obtained from Sigma Chemical Company (St. Louis, Missouri); 2-deoxy-D-glucose and 3-O-Methyl glucose from Calbiochem (La Jolla, California); and 6-deoxy-D-glucose from Schwarz-Mann (Orangeburg, New York). Bovine insulin was purchased from Schwarz-Mann, Calbiochem, and Sigma Chemical Company, dissolved in dilute acid (0.01 N HCl) at a concentration of 10 units per milliliter, sterilized by filtration through Millipore filters, and neutralized upon dilution before use. Cycloheximide, actinomycin D, and cordycepin were purchased from Sigma Chemical Company. Purified yeast hexokinase was obtained from Sigma Chemical Company and from Schwarz-Mann.

RESULTS AND DISCUSSION

Deprivation Derepression of Glucose Transport

    When CEF cells growing on D-glucose are washed and resuspended in complete BME with or without dialyzed serum and without glucose their capacity to transport glucose, 2-deoxyglucose and 3-O-methyl glucose (1, 17, 12) is progressively increased. Within 24 h of incubation at 37°C such cells can transport glucose and certain of its analogues 15 to 25 times more rapidly than at the start of the starvation. If glucose is included in the medium at the usual concentration (5.5 mM), the derepression of glucose uptake is not observed and at the end of 24 h the rate of transport is little changed from that at the start (1). At equimolar concentrations, 2-deoxy-D-glucose (2DOG), D-mannose, and D-glucosamine are about equally effective in maintaining the repressed level of glucose transport (1, 13).

Growth or maintenance on certain other naturally occurring sugars results in de-repression equal to or surpassing that induced by starvation. Among the sugars of the latter category are D-fructose and D-xylose (1, 8, 9, 22). D-Galactose at times resembles D-glucose at concentrations of 5.5 mM or higher and at others permits maximal derepression. At concentrations lower than 5.5 mM the repressive effect of several sugars on glucose transport is reduced or is not exerted at all. Such is the case for 2DOG, D-glucosamine, D-galactose, and D-mannose. This is true of cells grown either in medium containing serum or in serum-free medium. Glucose on the other hand remains an effective repressor and even concentrations as low as 0.055 mM have considerable repressive action over a 24-h period (13) (Fig. 1).

Asymmetry of Repression of Transport of Different Hexoses by the Same Treatment

Somewhat surprisingly (13) the hexoses previously thought to use the same carrier as glucose are not equally derepressed by the same pretreatment of the cells. It should be pointed out that the differential contribution of phosphorylation by hexokinase, gluco-kinase, galactokinase, and fructokinase to what is deemed "uptake" is far from clear. Nonetheless while starvation derepresses glucose uptake 30-fold, 2DOG will be derepressed somewhat less (20-fold), followed in descending order by mannose (15-fold), galactose (5–7-fold), and glucosamine (2-fold). Glucose represses the uptake of all 5 sugars effectively at a concentration of 5.5 mM (Fig. 2). A 10-fold reduction in glucose concentration in the con-ditioning medium remains repressive for glucose uptake, but mannose uptake escapes repression. Glucosamine is a good repressor of glucose transport, but mannose transport is derepressed at the same glucosamine concentration. Thus while it is not clear what

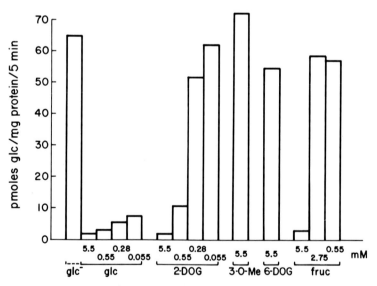

Fig. 1. Pattern of derepression of glucose uptake as a function of the concentration of carbon source. Chick embryo fibroblasts (CEF) were seeded at an initial inoculum density of 10^5 cells per culture vessel. After 4 days of growth, the monolayers were washed and provided with fresh medium free of serum and containing the sugars indicated at the concentrations given. Twenty-four hours later the rate of uptake of radioactive glucose ([U-^{14}C]glucose) was determined on triplicate culture vessels for each. Uptake is expressed as picomoles per milligram of cell protein accumulated during a 5-min assay.

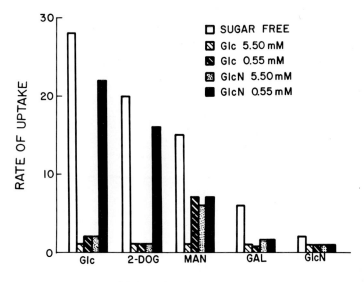

Fig. 2. Asymmetry of derepression of transport of related hexoses. Cells prepared as described for Fig. 1. Twenty-four hours of conditioning was accomplished without glucose or with glucose or glucosamine at concentrations indicated on the bars. The rate of uptake of each of the 5 radioactive hexoses was determined on triplicate cultures. Uptake is expressed as picomoles of hexose accumulated per milligram of cell protein during a 5-min assay. The values were normalized with the rate of uptake for each hexose by cells fed glucose at 5.5 mM concentration arbitrarily set as 1. [Reprinted from Musliner et al (13), with permission.]

elements are responsible for the transport of the several sugars, the complex of transport and phosphorylation is subject to subtle differences suggesting that common and unique elements are involved.

High- and Low-Affinity Carriers

Cells derepressed for glucose transport by starvation proved to possess a second carrier for glucose. The K_m of the derepressible carrier (12) proved to be about 20-fold lower (K_m = 40–50 μM) than that of the constitutive carrier (K_m = 1 mM) (Table I). The high-affinity carrier is inhibited by N-ethylmaleimide (12) while the constitutive carrier is unaffected by that reagent. Moreover, the high-affinity carrier has a much lower relative affinity for 2DOG and 3-O-methyl glucose than the constitutive carrier that favors 2DOG. In addition to derepression of a high-affinity carrier, there is also evidence for derepression of the low-affinity carrier in the change in V_{max} demonstrated for that carrier (Table I) (12).

Derepression Blocked by Cycloheximide and Actinomycin D

Both protein and RNA synthesis appear to be required for derepression initiated by starvation (1, 7, 12, 13). Cycloheximide (Fig. 3) added at any point to cells deprived of glucose blocks the further increase in transport rate. Puromycin, actinomycin D, and cordycepin (13) have a similar effect. As reported in previous publications (13, 17, 20) insulin further increases the transport rate when it is used to treat cells deprived of sugar.

TABLE I. Properties of Glucose Carriers in Chick Fibroblasts*

Conditioning (18–24 h)	Sugar transported	Carrier	K_m (mM)	V_{max} (nmol/min/mg)
Glc (5.5 mM)	D-Glucose	I	1.0	2–3
	2-deoxy-Glc		1–2	6–7
	3-O-methyl Glc		2–3	3–4
	D-Glucose	II	0.04–0.05	0.4
	2-deoxy-Glc		—	—
	3-O-methyl Glc		—	—
Sugar free	D-Glucose	I	1.0	10
	2-deoxy-Glc		1–2	15–20
	3-O-methyl Glc		3–4	20–25
	D-Glucose	II	0.04–0.05	4–5
	2-deoxy-Glc		—	—
	3-O-methyl glc		—	—

*CEF cultures were grown on BME for 3–4 days and, after washing, reincubated in medium containing dialyzed serum and either no added sugar or glucose (5.5 mM) for 18–24 h. The transport assays were initial rates from 10 second assay periods (12). The concentrations of sugars ranged between 5 μM and 1.25 mM for each sugar. The computations of K_m and V_{max} were as explained in the earlier report of Christopher, Kohlbacher, and Amos (12).

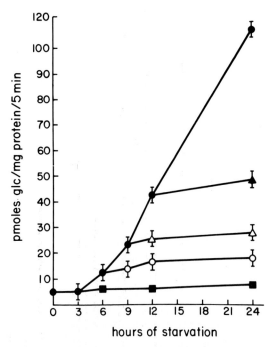

Fig. 3. Cycloheximide inhibition of derepression., Conditions as in Fig. 1. •) No cycloheximide; ■) cycloheximide (10 μg/ml) at 0 time; ○) cycloheximide at 6 h; △) cycloheximide at 9 h; ▲) cycloheximide at 12 h.

The stimulation attributed to insulin, like that induced by starvation, is blocked by inhibitors either of RNA or protein synthesis (Fig. 4).

Evidence for Putative Carrier mRNA

If indeed the removal of glucose from the culture medium sets in motion the synthesis of a new carrier or increases the rate of synthesis of an existing carrier, the accumulation of mRNA for one or both carriers should be demonstrable. Many efforts to achieve continued increase in transport after addition of actinomycin to starved cultures failed. Since the interruption of putative mRNA synthesis should not theoretically abruptly terminate translation, we sought other explanations for the blockage by actinomycin D. Indeed we found that if derepression is effected by the replacement of glucose with D-xylose (6.6 mM) or D-fructose (2.75 mM), actinomycin D is not inhibitory except in the starved cells (Fig. 5). D-Xylose and D-fructose were chosen because they have been shown to be "nonrepressive" sugars (1, 18) for chick cell glucose transport at the concentrations employed here (Fig. 1). This we took as presumptive evidence for the involvement of energy depletion in the actinomycin D inhibition. Interference with glucose metabolism has been reported as a side effect of actinomycin D treatment of some mammalian cells (25).

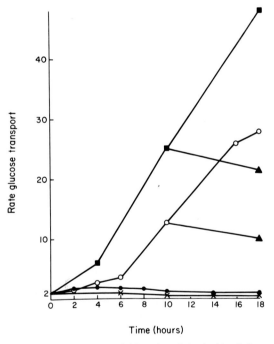

Time (hours)

Fig. 4. Inhibition of derepression by actinomycin D and cycloheximide. Cells were grown for 4 days as described for Fig. 1. The monolayers were washed and provided with fresh medium with the sugars indicated or sugar-free. Some monolayers received insulin (10^{-1} U/ml) without sugar. At 10 h actinomycin D (2 μg/ml) or cycloheximide (10 μg/ml) was added to some insulin-stimulated as well as to some starved cultures. As in Fig. 2 the rate of glucose uptake in cells provided with glucose has been arbitrarily set as 1. ○) Starved cultures; ■) insulin (10^{-1} U/ml; ✕) cycloheximide (10 μg/ml) alone; ●) glucose (5.5 mM).

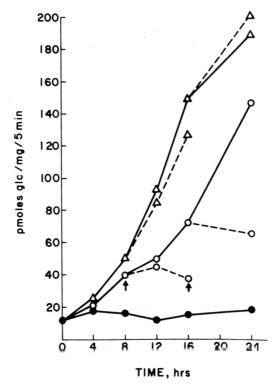

Fig. 5. Refractoriness of derepression to inhibition by actinomycin D. Chick embryo fibroblast (CEF) cells were grown for 3 days on BME, glucose (5.5 mM), and 4% calf serum. At that time all cultures were washed then provided with fresh serum-free BME containing: △—△) D-xylose (6.6 mM); ●—●) D-glucose (5.5 mM); ○—○) no sugar. Glucose uptake was measured at intervals (5-min assay) with triplicate cultures per time point. Actinomycin D (5 μg/ml) was introduced at the 8th and 16th hours (indicated by arrows): △- - -△) D-xylose BME; ○- - -○) sugar-free BME.

Evidence for increased potential to derepression was obtained by employing cyclo-heximide to inhibit carrier synthesis while permitting mRNA to accumulate as has been shown to occur in enzyme induction in chick cells by hydrocortisone (14, 15). Thus cells deprived of glucose and simultaneously inhibited with cycloheximide (10 μg/ml) for 20 h exhibited a rapid derepression of glucose uptake during the 4–6 h following removal of cycloheximide (Fig. 6). The increase in glucose uptake was observed only when pyruvate (data not shown), xylose, or fructose was provided in the incubation medium, ostensibly as an energy source. The failure to provide an exogenous sugar blocked derepression as did the addition of glucose as the energy source.

Actinomycin D at concentrations ranging from 1 to 5 μg/ml did not inhibit the de-repression observed after cycloheximide removal (Fig. 7).

Post-Cycloheximide Derepression of Cells Maintained on Glucose

Whereas cells starved of glucose while inhibited by cycloheximide for 20 h responded to the removal of cycloheximide by rapidly increasing their rate of glucose uptake (Figs. 5–7), cells maintained on glucose without inhibition of protein synthesis showed little response to removal of glucose over a 6-h period whether or not a nonrepressive energy source was provided (Fig. 8).

If on the other hand glucose (5.5 mM) and cycloheximide were both supplied for 20 h, the removal of both resulted in derepression with a delay of about 2 h (Fig. 8). The extent of derepression at 6 h was considerably greater than that achieved by cells treated with cycloheximide alone (Figs. 6, 7) because the combination of cycloheximide and glucose reduces uptake to a very low level after 20 h (8, 9). The absolute rate of entry at 6 h was not significantly different in the 2 sets. It is also to be noted that the addition of actinomycin D did not alter the derepression of glucose-cycloheximide treated cells, evidence for mRNA accumulation during the 20-h period of inhibition (Fig. 8). Apparently inhibition of protein synthesis by cycloheximide or some other metabolic consequence of cycloheximide treatment results in the availability of mRNA for carrier synthesis whether or not glucose is provided during the 20-h period. Once again the addition of glucose to the post-cycloheximide medium blocked the increase in rate of entry (Fig. 8). In contrast to cells starved during cycloheximide treatment, substantial derepression was achieved without the addition of an energy source to the post-cycloheximide incubation medium (data not shown). Evidently sufficient energy reserves were available in cells provided with

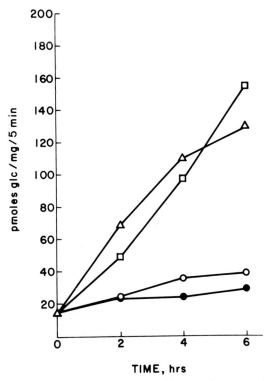

Fig. 6. Derepression after 20-h inhibition by cycloheximide. For all sets except one, cells were conditioned for 20 h with cycloheximide (10 μg/ml) without glucose. At 20 h all cultures were washed 3 times to remove the cycloheximide and were provided with fresh medium with sugars as indicated below: △) D-xylose (6.6 mM); □) D-fructose (5.5 mM); ○) no sugar; ●) D-glucose (5.5 mM).

glucose during the preincubation to permit synthesis of carriers over a 6-h period. The 2-h lag was reproduced in each of the experiments conducted and may well reflect residual "corepressor" formed in the metabolism of glucose.

Evidence for Repression of Glucose Carrier Formation

Let us assume that the blocking of derepression by reagents that inhibit protein synthesis (cycloheximide, puromycin) is in fact due to the inhibition of the formation of new glucose carrier molecules, themselves proteins. When protein synthesis is inhibited the capacity of the cells to transport glucose falls with time such that it is lower at the end of 20 h than the level observed when glucose alone is provided. Christopher and Kalckar have attributed this reduction to the failure to synthesize new carrier molecules coupled with the normal turnover of existing carrier molecules (8, 9). Since, as they have pointed out, the rate of carrier loss is much greater when glucose is present during inhibition by cycloheximide, it is not unreasonable to propose that glucose or a metabolic product of glucose has a role in the rate of carrier turnover. This inference gains support from the failure of glucose to allow the mRNA that accumulates after 20 h of inhibition of protein synthesis to be expressed as new carrier protein. The glucose or a product of glucose alone or in conjunction with a regulatory protein acts: 1) to destroy the carrier mRNA, 2) to block its translation, 3) to inactivate the carrier molecules as they are formed on the polyribosomes, or 4) to destroy formed carrier in transit or after its insertion into the plasma membrane.

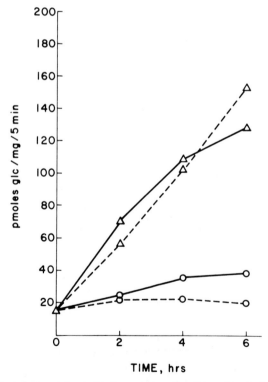

Fig. 7. Actinomycin and post-cycloheximide derepression. Conditions identical to those described for Fig. 6. △—△) D-Xylose (6.6 mM); △ - - - △) D-xylose + Act D (5 μg/ml); ○—○) no sugar; ○ - - - ○) Act D (5 μg/ml).

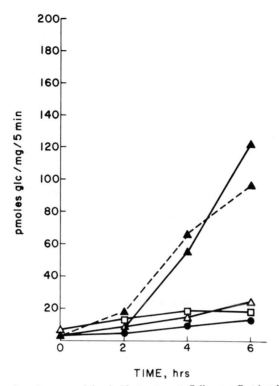

Fig. 8. Derepression after glucose-cycloheximide treatment. Cells were first incubated either with glucose (5.5 mM) alone or with glucose and cycloheximide (10 μg/ml) for 20 h. The post-cyclohexi-mide incubation medium contained sugars as indicated below: △—△) glucose alone followed by D-xylose (6.6 mM) after washing; □—□) glucose alone followed by D-fructose (5.5 mM); ▲—▲) glucose + cycloheximide (10 μg/ml) followed by D-xylose (6.6 mM); ▲- - -▲) glucose + cyclo-heximide followed by D-xylose (6.6 mM) + actinomycin D (5 μg/ml); ●) glucose + cycloheximide followed by glucose (5.5 mM).

A Model to Account for a Two-Component "Inactivator" of Carrier Protein

The evidence presented above can be reconciled by a model assuming that carrier synthesis is blocked at a posttranscriptional stage and/or that formed carrier is inactivated by a protein (Protein A, Fig. 9) in concert with a cofactor product of glucose metabolism. The protein in question is impotent except when modified by the cofactor to become the "inactivator." The protein can be considered to have a finite half-life, perhaps a short half-life, so that it is reduced significantly in concentration when protein synthesis is arrested. Thus after an extended period of inhibition of protein synthesis little Protein A exists, though its mRNA may well have accumulated.

Glucose-Induced Decay of Transport

Evidence lending support to the "Protein A-Cofactor-Inactivator" model has been presented by Christopher and Kalckar (8, 9) in experiments in which they demonstrate the reversal of derepression in Nil cells by addition of glucose to highly derepressed cells. Carrier function is sharply reduced with a half-decay time estimated at approximately 2.5 h. Cycloheximide protects the carriers from loss if it is added before or simultaneous-ly with glucose.

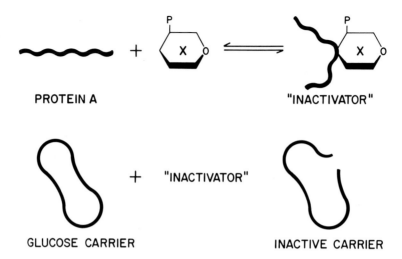

PROTEIN A   "INACTIVATOR"

GLUCOSE CARRIER   INACTIVE CARRIER

Fig. 9. Two-component "Inactivator" of glucose carrier.

We have been able to confirm the essential findings of Christopher and Kalckar using chick cells (13) (Fig. 9). The rate of decay of transport function is dependent upon the concentration of glucose, as is the protective effect of cycloheximide. Glucose at a concentration of 5.5 mM reduces uptake with a half-decay time of less than 15 min. A 10-fold reduction in the concentration of glucose extends the half-decay time to approximately 3 h. The decay of transport effected by glucose at either concentration is not reversed by removal of the sugar or its replacement by D-xylose or D-fructose (data not shown). Cycloheximide added before glucose or simultaneously with it reduces the rate of loss of carrier function (Figs. 10 and 11). Nearly full protection is achieved when the glucose concentration is lowered to 0.55 mM. Actinomycin D, on the other hand, is unable to block the loss of carrier function even with the lowered glucose concentration (Fig. 11).

We interpret the loss of carrier function to represent the inactivation of carrier already inserted into the plasma membrane. During starvation much less Protein A is formed, more in chick fibroblasts than in Nil cells. The addition of glucose permits an early burst of Protein A synthesis from accumulated mRNA. Cycloheximide blocks the formation of Protein A and the system for carrier inactivation is not operative despite a high level of the glucose cofactor. Actinomycin D does not have a protective effect because mRNA for Protein A synthesis is available and is translated as soon as glucose is supplied. 2-Deoxy-D-glucose (8) does not mimic glucose in initiating carrier destruction in Nil cells. With chick fibroblasts the cells are further derepressed when 2DOG is supplied in lieu of glucose. There is as yet no clear explanation of the 2DOG effect. Among the most interesting aspects of the putative carrier inactivating system is the conjecture that it is a model for assigning specificity to protease-mediated protein degradation in mammalian cells. Inhibitors of protease action may prove helpful in further study of glucose carrier decay.

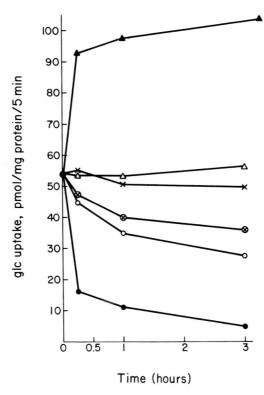

Fig. 10. Transport decay in derepressed cells. Cells were starved for 24 h, washed, and provided with fresh medium with the components indicated. At intervals after initiation of the experiment triplicate monolayers were assayed for glucose uptake with [U-^{14}C]glucose (5-min assay). •) glucose (5.5 mM); ○) glucose (0.55 mM); ▲) DOG (0.55 mM); △) cycloheximide (10 μg/ml); ⊗) glucose (5.5 mM) + cycloheximide (10 μg/ml); ×) glucose (0.55 mM) + cycloheximide (10 μg/ml).

DISCUSSION

The results presented here are considered presumptive evidence for a regulatory mechanism operating to control the density of glucose carriers in the chick fibroblast membrane. It is now well established that the uptake measured in chick fibroblasts is indeed a measure of transport and not of rate-limiting phosphorylation (7, 8, 9, 12, 13). In fact the derepression resulting from removal of glucose from the cell culture medium or its replacement by "nonrepressive" sugars such as xylose and fructose (1, 13, 18) has also been shown by the use of 3-O-methyl glucose to represent an increase in transport rather than of coupled metabolism of glucose or 2-deoxy-D-glucose (7, 12, 13). Transport appears to be limiting both in the repressed and derepressed chick fibroblast.

Christopher et al. (12) concluded that starvation for glucose resulted in the derepression of 2 carrier systems for glucose in the chick cell. One, a constitutive carrier (K_m = 1 mM) and a second higher affinity carrier (K_m = 0.04–0.05 mM) were derepressed, the second severalfold in the absence of glucose. Banjo and Perdue (26) identified 2 membrane

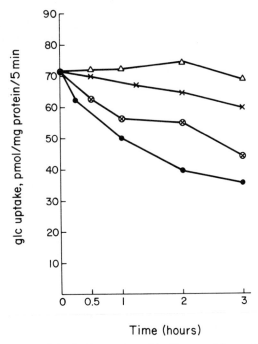

Fig. 11. Actinomycin versus cycloheximide as antagonist of transport decay, Conditions same as for Fig. 10. Test additions as indicated below: ●) glucose (0.55 mM); ×) glucose (0.55 mM) + cycloheximide (10 μg/ml); ⊗) glucose (0.55 mM) + Act D (5 μg/ml); △) cycloheximide (10 μg/ml).

proteins (approximate molecular weights 95,000 and 75,000) whose rate of synthesis was accelerated when chick cells were starved of glucose. Pastan and collaborators have confirmed and extended (27) the observations of Banjo and Perdue. The implication of the specific repression of both proteins that results from maintaining high glucose concentrations in the culture medium is that both proteins may be involved in glucose transport.

The previously reported inhibition of derepression by actinomycin D (7, 13) is herein shown to be attributable in all probability to an energy deficit. Supplying a "nonrepressive" source of energy such as xylose, fructose, or pyruvate makes derepression of glucose transport refractory to actinomycin D while retaining the sensitivity to inhibitors of protein synthesis.

By inhibiting glucose carrier formation with cycloheximide for an extended period, rapid derepression has been shown to occur upon removal of cycloheximide. The introduction of actinomycin D to block further RNA synthesis does not inhibit the rapid increase in transport during a 6-h period. The derepression achieved in 6 h virtually equals that observed only after 24 h of starvation. Again a "nonrepressive" source of energy is required for full expression. Whether or not glucose is present during the long period of interruption, the release from cycloheximide results in a rapid increase in transport function.

Significantly, glucose supplied as energy source with or without actinomycin blocks derepression on removal of cycloheximide. The tentative interpretation is that glucose exerts a posttranscriptional effect: 1) by destruction of preformed mRNA, 2) by inhibiting translation of mRNA, or 3) by a mechanism that results in inactivation of the carrier as it is being formed or after its insertion into the membrane. Acceleration of the inactivation of carrier already resident in the membrane is observed when glucose is added to derepressed cells (8, 9, 13). The rate of transport reduction is a function of glucose concentration. 2-Deoxyglucose is not active in accelerating the inactivation of carrier. It in fact promotes a paradoxical enhancement of transport.

Thus the role of glucose or a close metabolic derivative of glucose in the regulation of glucose carrier density has 2 potential facets: 1) that of posttranscriptional corepressor and 2) that of signal for carrier inactivation.

Derepression of glucose transport in Neurospora crassa (28, 29) parallels that observed with chick and mammalian cells. In Neurospora a second carrier of lower K_m and capable of active transport is derepressed when cells are starved of glucose or provided with fructose in lieu of glucose for several hours. The inactivation of the carriers is likewise triggered by the addition of glucose to the incubation medium and is concentration dependent. Evidence has been presented implicating protease action in the inactivation of the transport system with kinetics that resemble those observed with chick cells.

REFERENCES

1. Martineau R, Kohlbacher M, Shaw S, Amos H: Proc Natl Acad Sci USA 69:3407, 1972.
2. Hatanaka M: Proc Natl Acad Sci USA 70:1364, 1973.
3. Hatanaka M: Biochim Biophys Acta 355:77, 1974.
4. Ullrey D, Gammon MT, Kalckar HM: Arch Biochem Biophys 167:410, 1975.
5. Kalckar HM, Ullrey D: Proc Natl Acad Sci USA 70:2502, 1973.
6. Kletzien RF, Perdue JF: J Biol Chem 249:3366, 1974.
7. Kletzien RF, Perdue JF: J Biol Chem 250:593, 1975.
8. Christopher CW, Ullrey D, Colby W, Kalckar HM: Proc Natl Acad Sci USA 73:2429, 1976.
9. Christopher CW, Colby W, Ullrey D: J Cell Physiol 89:683, 1976.
10. Peck WA, Rockwell LH, Lichtman MA: J Cell Physiol 89:417, 1976.
11. Oxender DL, Quay SC: J Cell Physiol 89:517, 1976.
12. Christopher CW, Kohlbacher M, Amos H: Biochem J 158:439, 1976.
13. Musliner TA, Chrousos GP, Amos H: J Cell Physiol 91:155, 1977.
14. Reif-Lehrer L, Amos H: Biochem J 106:425, 1968.
15. Granner DK, Hayashi S, Thompson EB, Tomkins GM: J Mol Biol 35:291, 1968.
16. Tomkins GM, Thompson EB, Hayashi S, Gelehrter T, Granner D, Peterkofsky B: Cold Spring Harbor Symp Quant Biol 31:349, 1966.
17. Shaw SN, Amos H: Biochem Biophys Res Commun 53:357, 1973.
18. Rossow P, Radeos M, Amos H: Arch Biochem Biophys 168:520, 1975.
19. Eagle H: J Biol Chem 214:839, 1955.
20. Amos H, Christopher CW, Musliner TA: J Cell Physiol 89:669, 1976.
21. Lowry OH, Rosebrough NJ, Farr AL, Randall RJ: J Biol Chem 193:265, 1951.
22. Rossow PW: Effect of Sugar Source on Nucleotide Sugar Metabolism in Cultured Mammalian Cells. PhD Thesis, Harvard University, Cambridge, Massachusetts, 1975.
23. Young M, Oger J, Blanchard MH, Asdourian H, Amos H, Arnason BGW: Science 187:361, 1975.
24. Hanks JH, Wallace RE: Proc Soc Exp Biol Med 71:196, 1949.
25. Soiero R, Amos H: Biochim Biophys Acta 129:406, 1966.
26. Banjo B, Perdue JF: J Cell Biol 70:270a, 1976.
27. Adams SL, Sobel ME, Howard BH, Olden K, Yamada KM, DeCrombrugghe B, Pastan I: Proc Natl Acad Sci USA 74:3399, 1977.
28. Scarborough GA: J Biol Chem 245:3985, 1970.
29. Neville MM, Suskind SR, Roseman S: J Biol Chem 246:1294, 1971.
30. Brown J: Metabolism 11(10):1098, 1962.

Molecular Aspects of Membrane Transport 545–552 (1978)

Some Effects of Trypsin on the Subunits of the Membrane ATPase From Escherichia coli

Jeffrey B. Smith, Paul C. Sternweis, and Robert J. Larson

Section of Biochemistry, Molecular and Cell Biology, Cornell University, Ithaca, New York 14853

Five-subunit ($\alpha, \beta, \gamma, \delta, \epsilon$) F_1-ATPase from E. coli (ECF_1) was reconstituted by combining 2 mixtures of inactive subunits (an $\alpha\beta$ fraction and a $\gamma\epsilon$-rich one) and the purified δ subunit. The combination of $\alpha\beta$ and $\gamma\epsilon$-rich fractions was sufficient for reconstituting ATPase activity, which was achieved by dialyzing the subunits at 23°C in the presence of Mg-ATP. Addition of the purified δ subunit to the reconstituted ATPase restored coupling factor activity to the enzyme. The reconstituted enzyme was as active as the native enzyme in restoring ATP-driven transhydrogenase activity to F_1-depleted vesicles. Incubation of the $\gamma\epsilon$-rich fraction with trypsin decreased markedly the reconstitution of ATPase activity, whereas treatment of the $\alpha\beta$ fraction with trypsin under the same conditions had no significant effect. ECF_1 containing only the α and β subunits, which was prepared by trypsin digestion, was highly active hydrolytically but remained inactive as a coupling factor even after the addition of the $\gamma\epsilon$-rich fraction and δ. The inhibition of ECF_1 by the purified ϵ subunit was reversed by trypsin. Thus, while trypsin inactivates the γ and ϵ subunits of ECF_1, the α and/or β subunits are altered in a way which only inactivates the coupling factor activity of the enzyme without affecting its ATPase activity or the reconstitution of ATPase activity after cold inactivation. Since exposure of the α and β subunits to trypsin does not appear to alter their interaction with added γ and ϵ, it may be that the loss of coupling factor activity is due to a modification in α or β which disrupts their interaction with the membrane attachment subunit (δ).

Key words: E. coli membrane ATPase, ATPase subunits, energy coupling, oxidative phosphorylation, proton pump, trypsin

The proton-pump ATPase located in the inner membrane of E. coli consists of 2 morphologically and functionally distinct components. Both are oligomeric proteins and referred to as F_0 and F_1 following the terminology used for the corresponding portions of the homologous ATPase present in mitochondria and chloroplasts (1, 2). Some properties of the enzyme complex and its assembly are depicted in Fig. 1. The lipophilic F_0 portion is intrinsic to the membrane and exhibits ionophore activity for protons. F_0 also contains a proteolipid that reacts with a covalent inhibitor of the ATPase, dicyclohexylcarbodiimide (3, 4). Although the proton ionophore activity of F_0 is not normally expressed, it becomes apparent once the F_1 subunits have been disconnected from F_0 by genetic or biochemical

Received April 18, 1977; accepted June 6, 1977.

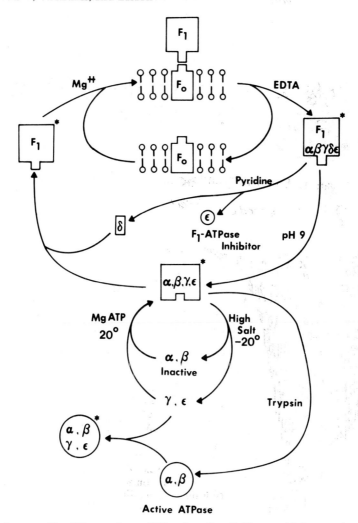

Fig. 1. In vitro assembly of the membrane ATPase from E. coli. The asterisk denotes that the ATPase is sensitive to the inhibitory action of the ϵ subunit.

manipulations (5, 6). F_0 also fulfills a structural role in the complex by anchoring the relatively hydrophilic F_1 subunits to the membrane.

By contrast to F_0, F_1 is extrinsic to the membrane and does not appear to interact with membrane lipids. F_1 contains the catalytic sites of the enzyme as well as high affinity sites for nucleotides (7, 8). Five distinct polypeptides are present in the F_1 purified from E. coli (9) and each is now known to be a subunit of the enzyme. Figure 2 shows an acrylamide gel of ECF_1 and each of the purified subunits after electrophoresis in the presence of sodium dodecyl sulfate (SDS). The subunits are called α, β, γ, δ, and ϵ and range in size from mol wt $\sim$ 60,000 for α (9) to mol wt 16,000 for ϵ (10). The exact number of copies of each of the polypeptides in a molecule of ECF_1 (mol wt 350,000) is still uncertain (11, 12), but the α, β, and γ chains compose the major portion of the molecule with the other 2 polypeptides (δ and ϵ) contributing only about 5% each to the total ECF_1 protein (11, 13).

The 2 smaller subunits apparently fulfill separate and distinct functions in ECF_1. This has been shown directly with active δ and ϵ which were purified to apparent homogeneity (10). Epsilon is apparently a regulatory subunit since it is a powerful inhibitor of the ATPase activity of ECF_1 (10, 14), although the importance of ϵ in the control of the enzyme in vivo has not yet been assessed. The δ subunit (mol wt 18,500) is required for the attachment of F_1 to F_0 (10, 14, 15). ECF_1 missing δ does not reattach to F_0 unless δ is added and it appears that only one δ is required to bind one ECF_1 molecule to F_0 in the membrane (13). In Fig. 1, δ is depicted as a component of the stalk between F_1 and F_0 as seen in electron micrographs of negatively stained enzymes (1, 16). Delta appears to be an elongated molecule with a relatively high content of α-helical structure (13). Purified δ readily binds to δ-deficient ECF_1 in the absence of the F_0 portion of the enzyme (13).

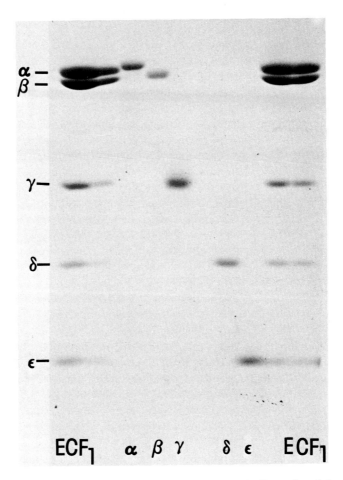

Fig. 2. Acrylamide gel electrophoresis of the purified subunits of the F_1 portion of the membrane ATPase from E. coli in the presence of sodium dodecyl sulfate. The 2 smaller (δ and ϵ) subunits were purified in an active state (10) but the 3 larger ones (α, β, and γ) were denatured before purification (see Methods). The proteins were incubated with 1% SDS and 2% β-mercaptoethanol for 3 min in a boiling water bath and then electrophoresis was carried out for 18 h at 70 V in a 7.5–25% acrylamide gel (30).

The δ and ϵ subunits of F_1 from E. coli and chloroplasts (17–19) appear to be functionally similar.

The α and β subunits of ECF_1 appear to be sufficient for catalyzing ATP hydrolysis
(20). After incubating ECF_1 with trypsin, highly active ATPase was obtained which contained only α and β chains when examined on SDS gels (20). This 2-subunit ECF_1 was not
sensitive to inhibition by the ϵ subunit (10, 14) but became sensitive to ϵ after addition of
an excess γ fraction (21) indicating that γ is required for inhibition by ϵ, as depicted in
Fig. 1. Subunit γ is also required for the assembly of the α and β chains of ECF_1 into the
catalytic unit (21). Here we describe the reconstitution of ECF_1 from subunits and some
effects of trypsin on the subunits.

MATERIALS AND METHODS

Five-subunit ECF_1 was purified from the ML308-225 strain of E. coli as previously
described (10), as were the δ and ϵ subunits of ECF_1 (10). The 3 larger α, β, and γ polypeptides were prepared in a denatured form by slicing the individual protein bands from
cylindrical acrylamide gels (1.6×18.0 cm). The γ subunit was sliced from 11% SDS gels
(22). Since the resolution of the α and β chains is rather poor on SDS gels, (Fig. 1) these
subunits were sliced from 6% acrylamide gels after electrophoresis in 6 M urea (23). The
proteins were eluted electrophoretically from the gel slices. Minor impurities in the α subunit from the urea gels were removed by electrophoresis in an SDS gel.

The ECF_1 which was purified from the K12(λ) strain usually lacked the δ subunit
(13). ECF_1 containing chiefly the 2 larger α and β subunits was prepared by treating the
purified enzyme from the K12(λ) strain with TPCK-trypsin (Worthington Biochemical
Corporation) according to the method of Nelson et al. (20).

Four-subunit enzyme from the K12(λ) strain was used for all of the cold inactivation experiments. To cold inactivate ECF_1 ~ 10 mg of enzyme was precipitated 4 times
with 3 volumes of saturated ammonium sulfate, pH 7, to free the enzyme of glycerol and
ATP which are used to stabilize the enzyme for storage at $-80°C$. After each precipitation
the enzyme was dissolved with 1 ml of 100 mM Tris-SO_4, pH 7.5, containing 4 mM
(ethylenedintrilo)tetraacetic acid (EDTA). The final precipitate was dissolved with 1 ml
of 50 mM Tris-succinate, pH 6.0, containing 1 M KCl, 0.1 M KNO_3, 1 mM EDTA, and
0.1 mM dithiothreitol (DTT). The enzyme was frozen and stored overnight at $-20°C$
which destroyed over 95% of ATPase activity.

The inactive subunits of ECF_1 were fractionated on an hydroxylapatite column. The
enzyme was diluted with 9.0 ml of SED (50 mM succinate-Tris, pH 6.0, containing 0.1 mM
EDTA and 0.1 mM DTT) and applied to a column containing 1.0 g (dry weight) of Bio-
Gel HTP hydroxylapatite (Bio-Rad Laboratories) which was equilibrated with SED. The
protein was eluted in a stepwise fashion with SED containing 0, 50, 100, 150, 200, and
250 mM potassium phosphate. Ten milliliters were used for each step and 2-ml fractions
were collected.

ATPase activity was reconstituted by dialyzing the subunit containing fractions from
the hydroxylapatite column at $23°C$ against reconstitution buffer: 50 mM Tris-Cl, pH 7.3
with 10% glycerol, 5 mM Mg-ATP, 1 mM DTT, and 0.05 mM EDTA.

ATPase activity (24) and its inhibition by the ϵ subunit (15) and ATP-driven transhydrogenase (14) in inverted membrane vesicles (24) were measured by the published
procedures. Protein was measured by the method of Lowry et al. (25), using bovine serum
albumin as a standard.

RESULTS AND DISCUSSION

The F_1 portion of the ATPase from E. coli (ECF_1) is cold labile and may be split into subunits by freezing in high salt (12, 26). The α and β chains are dissociated to monomers by this treatment (21, 26). After cold inactivating 4-subunit ECF_1 (no δ) the dissociated subunits were separated on an hydroxylapatite column (see Methods). The subunits eluted from the column in 2 major peaks: The first, containing approximately equal amounts of α and β, is called the $\alpha\beta$ fraction. The other peak which eluted at a higher KP_i concentration also contained α and β chains, again in equal amounts, as well as about a threefold excess of subunits γ and ϵ, relative to the native enzyme. The composition of the $\alpha\beta$ and $\gamma\epsilon$-rich fractions was assessed by SDS gel electrophoresis (21). Dialysis at $23°C$ against Mg-ATP restored ATPase to $\gamma\epsilon$-rich fraction, but the $\alpha\beta$ fraction remained inactive (Table I). Combining the 2 fractions produced a synergistic restoration of ATPase activity. With an excess of the $\alpha\beta$ fraction, about threefold more activity was reconstituted than was obtained with the $\gamma\epsilon$-rich fraction alone (Table I). Purified δ and ϵ subunits, either separately or together, were ineffective in restoring activity to the $\alpha\beta$ fraction (21). These results indicate that subunit γ is essential for the reconstitution of the ATPase from subunits. Since ECF_1 without an intact γ is highly active as an ATPase, γ is not part of the ATPase catalytic center (20). Furthermore, once the catalytic unit has been assembled, it is not disrupted when γ is removed (20). The γ chain of the F_1 from a thermophilic bacterium was recently shown to be required for renaturing hydrolytic activity using subunits isolated in urea and SDS (27).

Incubating the $\gamma\epsilon$-rich fraction with trypsin strongly reduced its capacity to restore ATPase activity to the mixture of α and β subunits, whereas a similar treatment of the $\alpha\beta$ fraction with trypsin had no effect on the reconstitution (Table I). Thus the γ subunit is apparently more sensitive to trypsin than the α and β subunits, which is consistent with the earlier finding that trypsin digests the γ and ϵ subunits off of 4-subunit ECF_1 without destroying its ATPase activity (20).

The purified ϵ subunit strongly inhibits the ATPase activity of ECF_1 (10, 14). Figure 3 shows that trypsin reversed the inhibition of ECF_1 by ϵ. Although ϵ is destroyed by trypsin (14), it is unclear whether trypsin reverses ϵ inhibition by modifying γ, or ϵ, or

TABLE I. Effect of Trypsin on the Reconstitution of ATPase Activity

Sample	Total ATPase activity (units)	Net ATPase activity (units)
1. $\alpha\beta$ fraction	< 0.05	0
2. $\gamma\epsilon$-rich fraction	0.75	0
3. $\alpha\beta$ fraction + $\gamma\epsilon$-rich fraction	2.75	2.00
4. Trypsin-treated $\alpha\beta$ fraction + $\gamma\epsilon$-rich fraction	2.95	2.20
5. $\alpha\beta$ fraction + trypsin-treated $\gamma\epsilon$-rich fraction	1.55	0.78

The $\alpha\beta$ and $\gamma\epsilon$-rich fractions (0.3 mg/ml) were each treated separately with TPCK-trypsin (0.5 μg/ml) for 12 min at room temperature before terminating the digestion with soybean trypsin inhibitor (2 μg/ml). The fractions were then combined (34 μg of $\alpha\beta$ and 10 μg of $\gamma\epsilon$-rich) as indicated and dialyzed at room temperature against reconstitution buffer for about 10 h. ATPase activity was assayed. The net activity due to the $\alpha\beta$ fraction alone was calculated by subtracting the amount of activity contributed by the $\gamma\epsilon$-rich fraction.

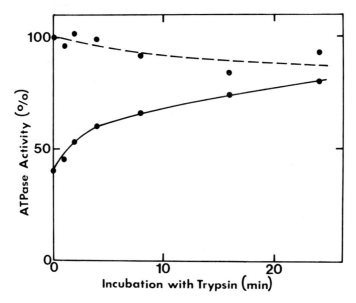

Fig. 3. Reversal by trypsin of the inhibition of ECF$_1$ by the ϵ subunit. ECF$_1$ (0.8 μg) in the presence (solid curve) or absence (dotted curve) of the ϵ subunit was incubated with 2.5 μg of TPCK-trypsin (Worthington Biochemical Corporation) at 23°C. Samples were withdrawn at the indicated times, mixed with 2.5 μg of trypsin inhibitor (Worthington Biochemical Corporation) and assayed for ATPase activity.

both, as subunit γ is required for ϵ inhibition (21). The 2-subunit $\alpha\beta$-ATPase, prepared by trypsin digestion, is insensitive to ϵ (10, 14), but became sensitive after the addition of $\gamma\epsilon$-rich fraction (21). This result suggests that the interaction of the γ and ϵ subunits with α and β was not altered by the treatment of ECF$_1$ with trypsin. Moreover the treatment of the $\alpha\beta$ fraction with trypsin also did not affect the reconstitution of these subunits by the $\gamma\epsilon$-rich fraction (Table I). Therefore, it seems unlikely that the reversal of inhibition is due to an alteration in the α or β subunits.

The activity of the membrane-bound ATPase from E. coli is stimulated about two-fold when the membranes are incubated with trypsin (28). This stimulation of the ATPase may be due to the inactivation of the ϵ subunit. Although the protease detaches ECF$_1$ from the membrane, the activation seems to occur after the ECF$_1$ becomes soluble (29).

The activity of the reconstituted ATPase as a coupling factor in energy transduction is shown in Fig. 4. The enzyme reconstituted from the $\gamma\epsilon$-rich fraction and the $\alpha\beta$ fraction was as active in restoring ATP-driven transhydrogenase to ECF$_1$-depleted membrane vesicles as the purified enzyme which had not been dissociated to subunits (Fig. 4). This restoration of coupling factor activity required the addition of subunit δ since it is essential for at-taching ECF$_1$ to the F$_0$ portion of the enzyme (10, 14, 15). The essentially complete restoration of coupling factor activity from inactive subunits plus δ is somewhat better than that reported by Vogel and Steinhart (12). The dilution method of reconstituting the ATPase which was used by these authors is not as effective as our dialysis method which yields enzyme with nearly the same specific activity as that observed before cold inactivation (21, 26). Also there may have been some loss of δ during the purification of ECF$_1$ which frequently happens with the K12 strain of E. coli (13, 20). We circumvented

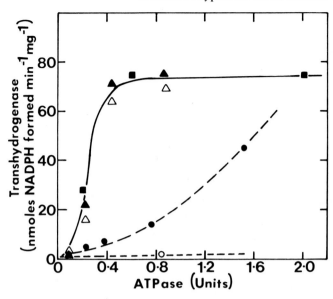

Fig. 4. Reconstitution of ATP-driven transhydrogenase in ECF_1-depleted membrane vesicles. Depleted membrane vesicles were incubated at $37°C$ as previously described (10) with the indicated amount of 5-subunit ECF_1 from the ML308-225 strain (■), ECF_1 reconstituted from a mixture $\alpha\beta$ and $\gamma\epsilon$-rich fractions in a 3:1 ratio of protein (▲), ECF_1 reconstituted from the $\gamma\epsilon$-rich fraction by itself (△), $\alpha\beta$ enzyme prepared by treating ECF_1 with trypsin (○), and $\alpha\beta$-enzyme mixed with the $\gamma\epsilon$-rich fraction (●). Purified δ subunit (0.4 μg) was added to attach ECF_1 to F_0. The $\alpha\beta$ and $\gamma\epsilon$-rich fractions were mixed together or with the $\alpha\beta$ enzyme and dialyzed at $23°C$ overnight against 5 mM ATP to reconstitute ATPase activity.

this problem by using δ-deficient ECF_1 for the cold inactivation experiments and adding purified δ when reconstituting coupling factor activity.

The 2-subunit $\alpha\beta$ enzyme, obtained by trypsin digestion, remained inactive as a coupling factor even after adding the $\gamma\epsilon$-rich fraction and δ (Fig. 4). The coupling factor activity observed after mixing the $\alpha\beta$ enzyme with the $\gamma\epsilon$-rich fraction is due to the $\gamma\epsilon$-rich fraction (Fig. 4). Since the $\gamma\epsilon$-rich fraction restored the sensitivity of the $\alpha\beta$ enzyme to the inhibitory subunit, ϵ (21), the interaction of the $\alpha\beta$ enzyme with subunits γ and ϵ would appear to be intact. Therefore, the defect in the $\alpha\beta$ enzyme responsible for its lack of coupling factor activity may be in its interaction with the membrane attachment subunit (δ). Recently Abrams et al. (31) reported that chymotrypsin removed a short segment from the α chain of the F_1 from Streptococcus faecalis and destroyed the binding of enzyme to depleted membranes.

ACKNOWLEDGMENTS

We thank Sharon L. Johnston for steadfast technical assistance and Professor Leon A. Heppel who provided enthusiasm and encouragement and also helped with the experiments. Christofer Wilkowski purified the α, β, and γ chains.

This work was supported by National Science Foundation grant BMS 75-20287 and grant AM-11789 from the National Institutes of Health to L.A.H., and a National Research Service Award (5 F32 GM02419-02) to J.B.S.

REFERENCES

1. Racker E: "A New Look at Mechanisms in Bioenergetics." New York: Academic Press, 1976.
2. Harold FM: Curr Top Bioenerg 6:83, 1977.
3. Fillingame RH: J Bacteriol 124:870, 1975; Fillingame RH: J Biol Chem 251:6630, 1976.
4. Altendorf K, Zitzmann W: FEBS Lett 59:268, 1975; Altendorf K: FEBS Lett 73:271, 1977.
5. Rosen BP: Biochem Biophys Res Commun 53:1289, 1973; Altendorf K, Harold FM, Simoni RD: J Biol Chem 249:4587, 1974.
6. Tsuchiya T, Rosen BP: J Biol Chem 250:8409, 1975.
7. Smith JB, Abrams A: (Abstract) Fed Proc Fed Am Soc Exp Biol 33:599, 1973.
8. Maeda M, Kobeyashi H, Futai M, Anraku Y: Biochem Biophys Res Commun 70:228, 1976.
9. Bragg PD, Hou C: FEBS Lett 28: 309, 1972.
10. Smith JB, Sternweis PC: Biochemistry 16:306, 1977.
11. Bragg PD, Hou C: Arch Biochem Biophys 167:311, 1975.
12. Vogel G, Steinhart R: Biochemistry 15:208, 1976.
13. Sternweis PC, Smith JB: Manuscript submitted.
14. Smith JB, Sternweis PC, Heppel LA: J Supramol Struc 3:248, 1975.
15. Smith JB, Sternweis PC: Biochem Biophys Res Commun 62:764, 1975.
16. Hertzberg EL, Telford JN, Hinkle PC: Unpublished results.
17. Nelson N, Nelson H, Racker E: J Biol Chem 247:7657, 1972.
18. Nelson N, Karny O: FEBS Lett 70:249, 1976.
19. Younis HM, Winget GD, Racker E: J Biol Chem 252:1814, 1977.
20. Nelson N, Kanner BI, Gutnick DL: Proc Natl Acad Sci USA 71:2720, 1974.
21. Larson RJ, Smith JB: Manuscript submitted.
22. Laemmli UK: Nature (London) 227:680, 1970.
23. Andreu JM, Carreira J, Munoz E: FEBS Lett 65:198, 1976.
24. Futai M, Sternweis PC, Heppel LA: Proc Natl Acad Sci USA 71:2725, 1974.
25. Lowry OH, Rosebrough NJ, Farr AL, Randall RJ: J Biol Chem 193:265, 1951.
26. Smith JB, Sternweis PC, Larson RJ, Heppel LA: J Cell Physiol 89:567, 1976.
27. Kagawa YJ: Cell Physiol 89:569, 1976.
28. Neiuwenhuis FJRM, vd Drift JAM, Vogt AB, Van Dam K: Biochim Biophys Acta 368:461, 1974.
29. Sternweis PC: Unpublished results.
30. Studier W: J Mol Biol 79:237, 1973.
31. Abrams A, Morris D, Jensen C: Biochemistry 15:5560, 1976.

Molecular Aspects of Membrane Transport 553–565 (1978)

Bacteriorhodopsin: Lipid Environment and Conformational Changes

Evert P. Bakker, Michael Eisenbach, Haim Garty, Carmela Pasternak, and
S. Roy Caplan

Department of Membrane Research, The Weizmann Institute of Science, Rehovot, Israel.

The polar lipids of the purple membrane were exchanged for different phosphatidylcholine species. The resulting complexes had the same protein to lipid-phosphorus ratio as the natural membrane, but only about 0.5–1.0 mole of original lipid was still present per mole of bacteriorhodopsin. In such complexes the bacteriorhodopsin photocycle is slowed down 10–20 times, but the strong protein-protein interaction is not abolished. Due to the slow rate of the photocycle we were able to measure in the light the ratio between net proton release and net accumulation of the last intermediate of the photocycle, the unprotonated M_{412}. This ratio was not constant and equal to 1.0, as expected for a single deprotonation reaction, but varied with pH from 1.5 to 0.4. The variable ratio suggests that light-induced conformational changes occur in the nonchromophore part of the protein, which shift the pK_a values of unidentified groups so as to cause binding or release of additional protons. A similar conclusion was drawn from experiments on the kinetics of proton transfer by bacteriorhodopsin in subbacterial particles of Halobacterium halobium and in reconstituted bacteriorhodopsin proteoliposomes. However, in this case light-induced association and dissociation of additional protons occurs simultaneously on different sides of the membrane.

Key words: bacteriorhodopsin, lipid environment, conformational changes, purple membrane, lipid substitution

Bacteriorhodopsin is synthesized by cells of Halobacterium halobium and related species under conditions of strong illumination and low oxygen tension (1–3). It is located in the purple patches of the cell membrane, where it is the only protein present (2, 4). Bacteriorhodopsin functions as a light-driven proton pump (5) and forms a pathway for

Abbreviations: bR_{570}, K_{590}, L_{550}, M_{412}, N_{520}, and O_{660} – respectively the light-adapted form of bacteriorhodopsin, which absorbs maximally at 570 nm, and the intermediates of the bacteriorhodopsin photocycle absorbing maximally at the indicated wavelengths; DLPC – dilauroylphosphatidylcholine; DMPC – dimyristoylphosphatidylcholine; DPPC – dipalmitoylphosphatidylcholine; EPC – egg phosphatidylcholine; EPE – egg phosphatidylethanolamine; $TPMP^+$ – the cation of the salt triphenylmethylphosphonium bromide; preparation $(x)^n$-purple membrane refers to one in which the lipids were exchanged n times for lipid x; ΔpH – pH gradient across the membrane; $\Delta\bar{\mu}_{H^+}$ – electrochemical potential difference of protons; $\Delta\psi$ – electrical potential difference.

Received April 19, 1977; accepted August 16, 1977.

Evert P. Bakker is presently at National Jewish Hospital and Research Center, Division of Molecular and Cellular Biology, 3800 East Colfax Avenue, Denver, CO 80206.

proton transfer through the membrane parallel to that of the respiratory chain (6), which is located in the nonpurple part of the membrane. Isolated patches, the so-called purple membrane, have the form of open sheets (1). In these a strong protein-protein interaction has been detected both by electron diffraction (7) and circular dichroism (8—10).

Bacteriorhodopsin has a molecular weight of about 25,000 (11). Its chromophore is retinal, which is covalently linked to a lysine residue (2, 11). A good deal of attention has been paid to the bacteriorhodopsin photocycle. The light-adapted form of bacterio-rhodopsin [bR_{570}, with retinal in the all-trans form (12)], undergoes a reversible light reaction to form the K_{590} intermediate, which decays successively to L_{550} and M_{412} before the cycle is completed by regeneration of bR_{570} (Refs. 13—16, nomenclature after Ref. 13). We believe that the other intermediates reported, i.e., N_{520} and O_{660}, may not be located on the main path of the cycle, since these intermediates are formed at a slower rate than M_{412}, O_{660} falls back at a similar rate as M_{412} to bR_{570} (13, 16), and N_{520} and O_{660} occur in relatively low amounts (13, 16). The half-time of completion of the whole photocycle is 5 msec at $20°C$ for isolated purple membranes (16), but it is longer in intact cells (6, 18, 19).

Not much is known about the molecular mechanism of proton transfer by bacterio-rhodopsin. The nitrogen atom of the lysine residue, which is bound to retinal via the Schiff-base linkage, is protonated in bR_{570}, becomes deprotonated upon formation of M_{412}, and binds a proton upon completion of the cycle (20). It seems very likely that this cyclic deprotonation process has a role in the transfer of protons by bacteriorhodopsin across the membrane, since the steps of the photocycle involved show a large deuteron-isotope effect (21), and bleaching of bacteriorhodopsin to M_{412} is accompanied by the appearance in solution of about one proton per mole of M_{412} formed (13, 22). Bacterio-rhodopsin spans the purple membrane (7, 23, 24). It seems extremely unlikely, however, that the chromophore functions as a mobile carrier during the photocycle, transferring protons from the one side of the membrane to the other, since the chromophore orienta-tion with respect to the membrane is completely fixed (16, 25).

In this communication we will describe some experiments on the mechanism of action of bacteriorhodopsin. Firstly, we will focus on the question of whether there exists a special lipid requirement for bacteriorhodopsin, and whether its properties are affected by an altered lipid environment. Secondly, we will briefly discuss some data on biphasic proton transfer by bacteriorhodopsin in subbacterial particles of H. halobium and in reconstituted bacteriorhodopsin proteoliposomes (26). Since from both types of experi-ments some evidence was obtained that light-induced conformational changes occur in bacteriorhodopsin, which might have a function in the process of proton transfer, we shall present a schematic model for proton transfer by bacteriorhodopsin, which accounts for the observations reported here.

MATERIALS AND METHODS

Preparations

Strains R_1 and $R_1 M_1$ of H. halobium were grown purple as described previously (27, 28). The purple membrane was isolated from R_1 cells according to the method of Oesterhelt and Stoeckenius (2). The method of substitution of the purple membrane lipids was adapted from a similar method for lipid substitution of the sarcoplasmic reticulum ATPase (29). In essence the procedure was as follows: 5—10 mg of purple membrane were incubated for 1 h at room temperature with 25 mg of a particular phospholipid and 30–60 mg sodium cholate at pH 6.5. After this period the purple membrane was separated from excess lipid on a sucrose gradient and dialyzed extensively in order to remove most of the cholate still present. We shall write $(x)^n$-purple membrane for a preparation in which the lipids had been exchanged n times for lipid x in this way. Subbacterial particles were prepared from strain R_1M_1 as described by MacDonald and Lanyi (30) and Eisenbach et al. (31). Reconstituted bacteriorhodopsin proteoliposomes were prepared according to the method of Racker (32) from purple membranes and a lipid mixture of 99% egg phosphatidylcholine and 1% dicetylphosphate (by weight) (26).

Measurements

Light-induced pH changes were measured in the various preparations as described in Ref. 26. Steady state concentrations of M_{412} in the light were measured in a Cary 1605 spectrophotometer equipped with a device for side illumination of the sample cuvette. For illumination a 24 V, 150 W slide projector was used. Its light was passed through a Corning 3-69 cutoff filter. The photomultiplier was protected from this light by a combination of Corning 7-59 and 4-96 filters. For calculations of the amount of M_{412} a differential molar extinction coefficient at 412 nm of 27,000 M^{-1} cm^{-1} was assumed (13, 16).

RESULTS AND DISCUSSION

Substitution of the Purple Membrane Lipids

To study the role of lipids in the function of bacteriorhodopsin, we replaced the natural lipids of the purple membrane [mainly diphytanylether analogs of phospho- and glycolipids (3, 33, 34)] by different phosphatidylcholine species. The complexes obtained after 2 substitution procedures had about the same lipid phosphorus to protein ratio as the natural membrane (Table I). (From the known lipid composition of the purple membrane (3, 34) and its measured lipid-phosphorus content (Table I) it was calculated that the natural membrane contains about 7.4 moles of polar lipids per mole of bacteriorhodopsin.) Treatment of the purple membrane with cholate leads to partial delipidation of it (Table I). In this case, or in that when the purple membrane lipids were exchanged, no preferential removal could be detected of either the analogs of phospholipids or of those of glycolipids.

TABLE I. Substitution of Different Phosphatidylcholine Species for the Polar Lipids of the Purple Membrane

Preparation	Lipid phosphorus content (moles/mole of bacteriorhodopsin)	% exchange of original lipids
purple membrane	10	—
$(DLPC)^2$-purple membrane	9	70
$(DMPC)^2$-purple membrane	11	80
$(DPPC)^2$-purple membrane	9–11	90
$(EPC)^2$-purple membrane	11–12	95
$(cholate)^2$-purple membrane	6	—

The extent of substitution increased with the chain length of the fatty acids of the phosphatidylcholine species used. This may indicate that the chains of DLPC and DMPC, which substitute poorly, are too small to provide the membrane span of 45 Å required by bacteriorhodopsin (7). DPPC and EPC replaced 90 and 95% of the polar lipids of the native membrane, respectively; i.e., in these preparations less than 1.0 and 0.5 moles of original lipid were present per mole of bacteriorhodopsin.

Bacteriorhodopsin in Phospholipid Substituted Purple Membranes

In the complexes listed in Table I the major absorption band of bacteriorhodopsin is shifted from 570 nm to a slightly shorter wavelength. A normal cis-trans isomerization of retinal occurs when these complexes are preilluminated, as was inferred from small shifts of the main absorption band towards longer wavelengths (12). The half-time of decay in the dark to the equilibrium mixture was about 1 h at room temperature, except for $(DPPC)^2$-purple membrane where it was twice as long. No traces could be detected in these complexes of the form of bacteriorhodopsin having an absorption band centered at 470 nm (12). This form is an artefact induced by treating purple membrane either with dimethyl sulfoxide (12) or with cholate plus fluid phospholipids at alkaline pH (Ref. 23, and Bakker EP, unpublished observations).

The bacteriorhodopsin photocycle of the complexes was similar to that of the native membrane, except that its rate was slower, and the O_{660} intermediate was not observed (Sherman WV, Bakker EP, unpublished observations). The slowest rates were observed for $(DPPC)^2$-purple membrane, where the half-time of decay of M_{412} to bR_{570} was about 80 msec at room temperature, compared to about 30 msec for the other complexes of Table I and about 5 msec for the native membrane. We measured polarization spectra of the phototransients by modulation-excitation spectrophotometry (17) in the native and lipid-exchanged membranes (Pasternak C, Bakker EP, unpublished observations). It was found that at room temperature the polarization of M_{412} in the native membrane was high (about 0.45) and approached the theoretical upper limit of 0.5. For $(cholate)^2$-purple membrane and $(DPPC)^2$-purple membrane the polarization was also high, but the other complexes of Table I approached complete depolarization. For the $(DPPC)^2$-purple membrane a Perrin depolarization plot was drawn for temperatures up to 40°C; from about 35°C the polarization dropped gradually to a low value. These data indicate that in the latter cases the rotational mobility of the protein-chromophore complex in the membrane increases. However, calculations of the size of the moving unit for $(DPPC)^2$-purple membrane showed that it is still much larger than that of a single bacteriorhodopsin molecule (Pasternak C, unpublished observations). This may indicate that the 2-dimensional lattice of bacteriorhodopsin is destroyed in part during the lipid substitution procedure but that clusters of it are still present in the complexes in which bacteriorhodopsin exerts its strong protein-protein interaction. This conclusion seems to be confirmed by the following observations: 1) circular dichroism spectra of the complexes show a negative band at approximately 590 nm indicating that exciton interaction between neighboring bacteriorhodopsin molecules (8–10) still exists (Brith-Lindner M, Bakker EP, unpublished observations), and 2) the sheet structure as present in the native membrane and in $(cholate)^2$-purple membrane is only partially distorted to more bent but not closed membrane structures in the other complexes of Table I (Bakker EP, unpublished observation).

From these observations it is clear that there exists some influence of lipid environment on the properties of bacteriorhodopsin: $(DPPC)^2$-purple membrane behaves at room temperature differently from the other complexes studied, probably indicating that at this

temperature the fatty acids of DPPC are in a solid state. From experiments on the temperature dependence of net light-induced proton release by $(DPPC)^2$-purple membrane
a melting temperature of about $30°C$ was inferred (Bakker EP, unpublished observations).
On the other hand, there seems not to be much influence of lipid environment on the
spectral properties of bacteriorhodopsin and neither is there a dramatic effect on the net
light-induced release of protons by the protein due to bleaching to M_{412} (see below, Table
II). These observations confirm the conclusion that bacteriorhodopsin is a rather rigid
membrane protein (7) which is able to function inside a solid lipid environment (35).

The Stoichiometry Between Proton Release and M_{412} Formation

The complexes of Table I all release protons in the light. The kinetics of this process
are too fast to measure with a conventional pH measuring system ($t_{1/2} < 0.8$ sec). The
dependence on light intensity of net proton release and net M_{412} formation in $(DPPC)^2$-
purple membrane is shown in Fig. 1. The constant stoichiometry of the 2 processes at
different light intensities suggests a close relation. However, this stoichiometry is equal to
only about 0.6 at pH 6.5 (Fig. 1), not 1.0 as expected for the deprotonation reaction:

$$bR_{570} \rightarrow M_{412} + H^+. \tag{1}$$

For the other complexes the stoichiometries were also lower than 1.0 at pH 6.5 (Table II).
Proton conductors did not influence the values found, in agreement with the observation
that these complexes have the form of bent open sheets (Bakker EP, unpublished observations). Figure 2 shows that the stoichiometry of proton release is strongly pH dependent.

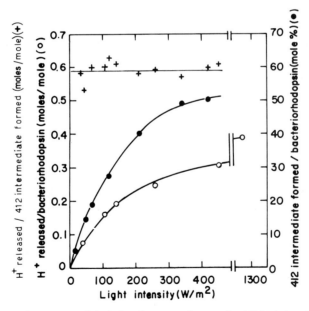

Fig. 1. Stoichiometry between net light-induced proton release and net light-induced M_{412} formation
by $(DPPC)^2$-purple membrane as a function of light intensity. 31 nmoles of bacteriorhodopsin were
present in 3 ml of a medium containing 150 mM KCl at pH 6.5. ○——○) Net proton release;
●——●) net M_{412} formation measured as a peak at $_{412}$ nm in the difference spectrum of illuminated
versus nonilluminated purple membrane. A value of 27,000 $M^{-1} \cdot cm^{-1}$ was assumed for the differential molar extinction coefficient at 412 nm (16); + — +) stoichiometry between the 2 processes.

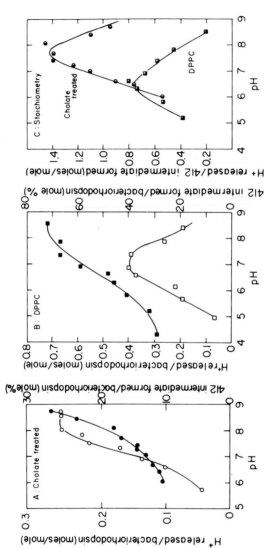

Fig. 2. pH Dependence of the stoichiometry between net proton release and M_{412} formation. (Cholate)^2-purple membrane (A) or (DPPC)^2-purple membrane (B) was suspended in 3.0 ml medium containing 150 mM KCl. The pH of the suspension was varied by addition of small amounts of HCl or KOH. The suspension was illuminated with light of intensity 500 W/m^2. Open symbols) net proton release; filled symbols) net M_{412} formation. C) Stoichiometries of the 2 processes: ●, (cholate)^2-purple membrane; ■, (DPPC)^2-purple membrane.

TABLE II. The Stoichiometry Between Net Light-Induced Proton Release and M_{412} Formation by Different Phospholipid-Substituted Purple Membranes Under Conditions of Continuous Illumination

Preparation	Proton release (H^+/bacterio-rhodopsin in moles/mole)	M_{412} (as a fraction of total bacterio-rhodopsin)	Stoichiometry (H^+ released/M_{412} formed in moles/mole)
(DPPC)^2-purple membrane	0.40	0.60	0.67
+ SF 6847 (10^{-5} M)	0.38	0.60	0.63
+ gramicidin (10 μg/ml)	0.37	0.60	0.61
(DLPC)^2-purple membrane	0.16	0.28	0.57
(DMPC)^2-purple membrane	0.20	0.35	0.57
(EPC)^2-purple membrane	0.14	0.30	0.47
+ SF 6847 (10^{-5} M)	0.13	0.30	0.43
(cholate)^2-purple membrane	0.11	0.13	0.84

20–50 nmoles of bacteriorhodopsin were present in a medium of 3 ml containing 150 mM KCl, final pH 6.5–6.7. Temperature 23–25°C. Light intensity 500 W/m^2.

For (cholate)^2-purple membrane it reaches a maximal value of 1.5 at pH 8 and decreases to 0.5 at pH 6.0 and to 0.9 at pH 9.0. This observation indicates that, depending on pH, association or dissociation of protons occurs in addition to the deprotonation process of reaction (1). These processes of additional binding or release of protons may be caused by a light-induced conformational change in the nonchromophore part of the protein, which could lead to shifts in pK_a values of unidentified groups. However, at present it cannot be excluded that reaction (1) is itself pH dependent, having a pK in the pH range studied, and that this could contribute to the pH-dependent stoichiometry observed in Fig. 2 [no data are available on the pH dependence of reaction (1) (Ref. 20)].

Bacteriorhodopsin Proteoliposomes

Bacteriorhodopsin has been reconstituted by cosonication of purple membrane fragments and excess phospholipids (32). The resulting proteoliposomes show light-induced proton uptake: i.e., the direction of proton transfer is inverted in comparison to that in intact cells (5, 6, 28) or in subbacterial particles (26, 30) due to an inverted orientation of bacteriorhodopsin in the membrane of the proteoliposomes (22). It would be of interest to determine whether a strong protein-protein interaction of bacteriorhodopsin is still present in the proteoliposomes. However, proteoliposomes obtained by the sonication method have a light scattering due to the relatively high lipid content too large to enable one to do spectroscopic studies with such preparations. When egg phosphatidylethanolamine was substituted for the polar lipids of the purple membrane, a preparation was obtained of which the lipid content was increased to 25 moles of phospholipids per mole of bacteriorhodopsin (compare with Table I). This ratio is still much lower than that of proteoliposomes normally used, where it equals about 1,000 (26, 32, 35). The EPE purple membrane preparation consists of closed vesicles with a diameter of 300–500 Å, and, most important, it behaves completely like proteoliposomes as to light-induced proton uptake (Bakker EP, unpublished observations). In this preparation M_{412} is completely depolarized (Pasternak C, unpublished observations). However, a negative band was observed in the circular dichroism spectrum at around 590 nm (Brith-Lindner, M, unpublished observations) indicating still existing exciton interaction in the proteoliposomes. Whether these strongly interacting bacteriorhodopsin molecules have a function in or are even required for the process of proton pumping cannot be decided, since the molar ellipticity of the circular dichroism spectra was only about one half of that of the native membrane, and the nondetectable bacteriorhodopsin molecules, which interact less strongly, may have a function in proton transport as well.

Biphasic Proton Transfer in Subbacterial Particles and Proteoliposomes

As described above, the variable ratio between net light-induced proton release and net M_{412} formation (Fig. 2) might indicate that a light-induced conformational change occurs in bacteriorhodopsin, leading to additional binding or release of protons. A similar conclusion was drawn from experiments where the kinetics were studied of light-induced proton release by subbacterial particles of H. halobium and of light-induced proton uptake by bacteriorhodopsin proteoliposomes. It was observed that in these 2 systems the light-induced proton transfer and the following dark-induced return flux can be described by a combination of 2 first-order processes (26). This is illustrated for proteoliposomes in Fig. 3. The data points of the fast phase of proton uptake were obtained by subtracting the extrapolated data points of the slow phase of proton uptake from those of total proton uptake (Fig. 3, lower part, see Ref. 26). The 2 phases of proton transfer both show the action spectrym of bR_{570} and represent 2 essentially distinct processes, since the extent of each of the 2 phases can be influenced separately by varying the experimental conditions (see below, Figs. 5 and 6). In both systems studied the slow phase of proton transfer

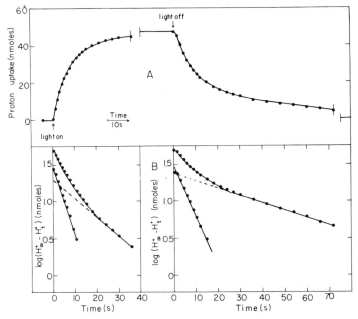

Fig. 3. Light-induced proton uptake by bacteriorhodopsin proteoliposomes. Proteoliposomes were prepared in a medium of 5 ml containing 5 mmoles NaCl, purple membrane (25 nmoles of bacteriorhodopsin), and 25 mg of a mixture of 99% EPC and 1% dicetylphosphate; final pH 6.25. 3.0 ml of the suspension were tested for light-induced proton uptake at 30°C. The sample was preilluminated and, after stabilization of pH in the dark, illuminated with light of intensity of about 1,000 W/m². A) Time course of light-induced proton uptake; B) separation of the slow phase of proton uptake from the fast one (see Ref. 26).

turned out to be much more sensitive to variations in ion composition of the medium than was the fast phase. In subbacterial particles conditions of low external pH or the presence of permeant cations like TPMP$^+$ increase the extent of the slow phase of proton release. This effect is compatible with the formation at such conditions of an increased pH gradient (ΔpH) across the membrane formed at the cost of the electrical potential gradient ($\Delta\Psi$) across it (36). A high concentration of chloride ions at low external pH has a similar effect on the extent of the slow phase of proton uptake by proteoliposomes (Fig. 4). This indicates that at such conditions Cl$^-$ may function as a permeant anion.

The dependence on pH of the parameters describing proton uptake by proteoliposomes at low chloride concentration is shown in Fig. 5. At low external pH the extent of the slow phase of proton uptake was much larger than that of the fast phase; it progressively decreased with increasing pH and was completely abolished at pH = 7.5 (Fig. 5). This may be correlated to the fact that the buffer capacity inside the proteoliposomes is centered at low pH and that there is almost none present at weakly acidic or neutral pH; thus, at high external pH a large pH difference across the membrane is generated which progressively decreases at lower external pH. [for the experiment of Fig. 5 at an external pH of 7.5, we estimated a ΔpH of 4.3 pH units with the 9-aminoacridine method (37, 38).] Assuming that the electrical potential across the membrane is pH independent, $\Delta\bar{\mu}_H$+ will increase with external pH and eventually cause the back leakage of protons to become a limiting factor at higher external pH. Because of kinetic considerations, the slow phase of proton uptake will be more affected by this limitation than the fast phase.

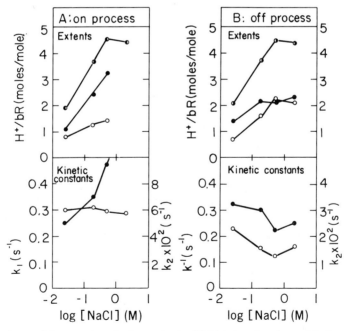

Fig. 4. Dependence of the extents and rate constants of light-induced proton uptake by proteoliposomes on NaCl concentration. Proteoliposomes were prepared and tested at pH 5.2 in NaCl media of different concentrations. Further conditions are as described in the legend to Fig. 3. Upper part extents of proton transfer; lower part, kinetic constants. ○—○) Parameters of the fast process; ●—●parameters of the slow process; ◖—◗) extents of total proteon transfer.

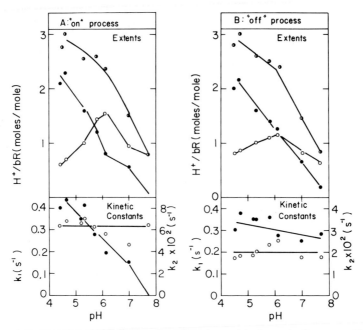

Fig. 5. Dependence on external pH of parameters of proton uptake by proteoliposomes. Different preparations were prepared in media of 150 mM NaCl, final pH as indicated in the figure. For mode of presentation and symbols, see legend to Fig. 4.

The experiments of Figs. 4 and 5 and those mentioned in the text all indicate that the slow phase of proton transfer represents the proton-pump function of bacteriorhodopsin. Although the fast phase of proton transfer can also produce a light-induced pH gradient across the membrane, it seems not to represent a pumping process in which bacteriorhodopsin has a catalytic role. If this were the case, one would expect that the extent of this phase would reach a maximal value at low external pH too, where the extent of total proton uptake is maximal, due to formation of a relatively small ΔpH (Fig. 5). In order to obtain more information on the nature of the fast process, the dependence was studied of light intensity on the parameters describing proton uptake by proteoliposomes (Fig. 6). The fast phase required a much higher light intensity than did the slow phase, and vanished below a light intensity of 100 W/m^2 (Fig. 6). In addition, we observed that the extent of the fast phase of proton uptake decreased with temperature, in contrast to that of the slow phase, which increased with temperature (not shown). The last 2 experiments might indicate a relationship between the occurrence of the fast phase of proton transfer on the one hand and of that of one of the intermediates of the photocycle, possibly M_{412} on the other: steady-state formation of M_{412} was also dependent upon a relatively high light intensity (Fig. 1) and favored by a low temperature.

A Schematic Model for Proton Transfer by Bacteriorhodopsin

Since the fast phases of proton transfer do not reflect the process of pumping protons across the membrane (see above), we assume that they reflect association or dissociation of protons at the surface of the membrane. Fast proton transfer is observed both in subbacterial particles and in proteoliposomes, which have opposite orientations of bacteriorhodopsin in the membrane; therefore it seems likely that this association and dissociation

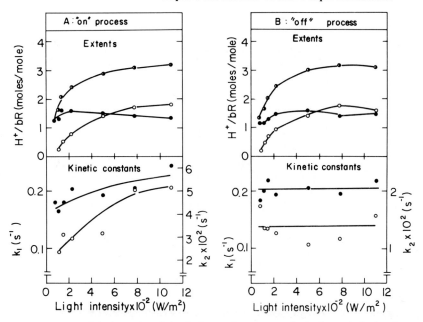

Fig. 6. Dependence on light intensity of the 2 phases of proton uptake by bacteriorhodopsin proteolip-osomes. Preparation and conditions as in Fig. 3. Symbols: see legend to Fig. 4.

occurs simultaneously on opposite sides of the membrane. Such a simultaneously occurring process of proton association-dissociation will lead to the formation of a ΔpH across the membrane, but it does not represent a pumping process in which bacteriorhodopsin has a catalytic role (see above).

We propose that this fast proton association-dissociation process is either triggered by the photocycling activity of bacteriorhodopsin or by net M_{412} formation which leads to conformational changes in the protein with resulting pK_a shifts of unidentified groups and additional binding of m protons on one side and release of n protons on the other side of the membrane (Fig. 7, upper part). The proton-pump function of bacteriorhodopsin is given in the lower part of Fig. 7. The photocycling chromophore induces asymmetric proton transfer from a relatively acidic to a relatively basic part of the protein. These 2 parts are supposed to be separated from each other by the chromophore and are either in direct or indirect contact with the aqueous solution at the 2 sides of the membrane.

The model of Fig. 7 accounts for the observation that the stoichiometry of proton release by phospholipid-substituted or cholate-treated purple membrane is a strong function of pH (Fig. 2). In these (uncoupled) preparations net proton release will amount to $[n - m + (M_{412}/bR)]$ protons, where the ratio M_{412}/bR equals that fraction of total bacteriorhodopsin which is bleached to M_{412} [cf. Fig. 7 and reaction (1)].

Support for our assumption that light-induced conformational changes do occur in bacteriorhodopsin comes from several reports in the literature. In the light the access to free NH_2 groups and to tryptophane residues in the protein is altered (39), deuterons reach the chromophore more easily (observations of A. Lewis, cited in Ref. 40), and tryptophane fluorescence is altered in bleached (i.e., M_{412}) purple membrane (41). In the model of Fig. 7 we analyze a possible relationship between such conformational changes and the mechanism of proton transfer by bacteriorhodopsin.

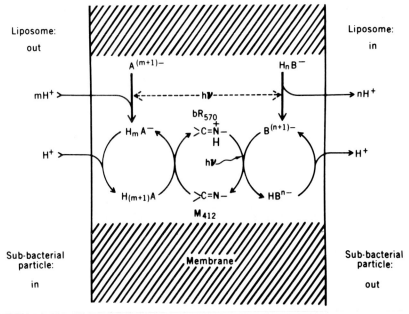

Fig. 7. A model for biphasic proton transfer by bacteriorhodopsin. For details, see text.

NOTE ADDED IN PROOF

A full account of the work on lipid substitution of the purple membrane will be published elsewhere (42).

ACKNOWLEDGMENTS

The work on lipid substitution of the purple membrane was started during E.P.B.'s short term EMBO Fellowship at the B. C. P. Jansen Institute, University of Amsterdam, The Netherlands. This work was supported in part by grants from the United States-Israel Binational Science Foundation (BSF), Jerusalem, Israel, and from the National Council for Research and Development, Israel, and the KFA Jülich, W. Germany.

REFERENCES

1. Stoeckenius W, Rowen R: J Cell Biol 34:365, 1967.
2. Oesterhelt D, Stoeckenius W: Nature (London) New Biol 233:149, 1971.
3. Kushwaha SC, Kates M, Martin WG: Can J Biochem 53:284, 1975.
4. Blaurock A, Stoeckenius W: Nature (London) New Biol 233:152, 1971.
5. Oesterhelt D, Stoeckenius W: Proc Natl Acad Sci USA 70:2853, 1973.
6. Bogomolni RA, Baker RA, Lozier RH, Stoeckenius W: Biochim Biophys Acta 440:68, 1976.
7. Henderson R, Unwin PNT: Nature (London) 257:28, 1975.
8. Heyn MP, Bauer PJ, Denscher NA: Biochem Biophys Res Commun 67:897, 1975.
9. Bauer PJ, Denscher NA, Heyn MP: Biophys Struct Mech 2:79, 1976.
10. Becher B, Ebrey TG: Biochem Biophys Res Commun 69:1, 1976.
11. Bridgen J, Walker ID: Biochemistry 15:792, 1976.
12. Oesterhelt D, Meentzen M, Schumann L: Eur J Biochem 40:453, 1975.

13. Lozier RH, Bogomolni RA, Stoeckenius W: Biophys J 15:955, 1975.
14. Kung M Chu, DeVault D, Hess B, Oesterhelt D: Biophys J 15:907, 1975.
15. Denscher NA, Wilms M: Biophys Struct Mech 1:259, 1975.
16. Sherman WV, Slifkin MA, Caplan SR: Biochim Biophys Acta 423:238, 1976.
17. Slifkin MA, Caplan SR: Nature (London) 253:56, 1975.
18. Sherman WV, Caplan SR: Nature (London) 258:766, 1975.
19. Wagner G, Hope AB: Aust J Plant Physiol 3:665, 1976.
20. Lewis A, Spoonhower J, Bogomolni RA, Lozier RH, Stoeckenius W: Proc Natl Acad Sci USA 71:4462, 1974.
21. Sherman WV, Korenstein R, Caplan SR: Biochim Biophys Acta 430:454, 1976.
22. Lozier RH, Niederberger W, Bogomolni RA, Hwang S-B, Stoeckenius W: Biochim Biophys Acta 440:545, 1976.
23. Henderson R: J Mol Biol 93:123, 1975.
24. Blaurock AE: J Mol Biol 93:139, 1975.
25. Razi Naqvi K, Gonzales-Rodrigues J, Cherry RJ, Chapman D: Nature (London) 245:249, 1973.
26. Eisenbach M, Bakker EP, Korenstein R, Caplan SR: FEBS Lett 71:228, 1976.
27. Danon A, Stoeckenius W: Proc Natl Acad Sci USA 71:1234, 1974.
28. Bakker EP, Rottenberg H, Caplan SR: Biochim Biophys Acta 440:557, 1976.
29. Warren GB, Toon PA, Birdsall NJM, Lee AG, Metcalfe JC: Proc Natl Acad Sci USA 71:622, 1974.
30. MacDonald RE, Lanyi JE: Biochemistry 14:2882, 1975.
31. Eisenbach M, Cooper S, Garty H, Johnstone RM, Rottenberg H, Caplan SR: Biochim Biophys Acta 465:599, 1977.
32. Racker E: Biochem Biophys Res Commun 55:224, 1973.
33. Kates M: In Snyder F (ed): "Ether Lipids, Chemistry and Biology." New York and London: Academic Press, 1972, pp 351–397.
34. Kushwaha SC, Kates M, Stoeckenius W: Biochim Biophys Acta 426:703, 1976.
35. Racker E, Hinkle PC: J Membr Biol 17:181, 1974.
36. Renthal R, Lanyi JK: Biochemistry 15:2136, 1976.
37. Schuldiner S, Rottenberg H, Avron M: Eur J Biochem 25:64, 1972.
38. Deamer DW, Prince RC, Crofts AR: Biochim Biophys Acta 274:323, 1972.
39. Konishi T, Packer L: Biochem Biophys Res Commun 72:1437, 1976.
40. Stoeckenius W, Bogomolni RA, Lozier RH: In Kaback HR, Neurath H, Radda GK, Schwyzer R, Wiley WR (eds): "Molecular Aspects of Membrane Phenomena." Berlin and New York: Springer-Verlag, 1975, pp 306–315.
41. Oesterhelt D, Hess B: Eur J Biochem 37:316, 1973.
42. Bakker EP, Caplan SR: Biochim Biophys Acta, in press.

Molecular Aspects of Membrane Transport 567–578 (1978)

The Kinetics of Oxygen-Induced Proton Efflux and Membrane Energization in Escherichia coli

J. Michael Gould

Department of Biological Sciences, Purdue University, West Lafayette, Indiana 47907

The kinetics of respiration-dependent proton efflux and membrane energization have been studied in intact cells of logarithmic phase Escherichia coli. Proton efflux following a small O_2 pulse is slow ($t_{1/2} \simeq 10$ sec) and inefficient ($H^+/O \simeq 0.5$), taking 5–10 times longer than expected from the time required for the cells to reduce the O_2 added in the pulse. A much closer agreement is found in cells treated to enhance counter ion fluxes and eliminate the transmembrane electric potential ($\Delta\Psi$). In cells treated with SCN^-, or with colicin E1 (which enhances K^+ permeability), the rates of proton efflux are much faster ($t_{1/2} \leqslant 1$ sec) than in untreated cells. The kinetics of formation and dissipation of $\Delta\Psi$ were estimated from changes in the fluorescence properties of the cell envelope bound probe N-phenyl-1-naphthylamine. In untreated cells, a small O_2 pulse induces a rapid ($t_{1/2} \leqslant 0.5$ sec) decrease in fluorescence intensity followed by a slower ($t_{1/2} \simeq 40$ sec) return of the fluorescence to the original level. The extent of the initial fluorescence decrease is proportional to the amount of O_2 added, although the half-time for the relaxation is independent of the amount of O_2 added. Colicin E1 (plus K^+) and the uncoupler FCCP greatly decrease the half-time of the relaxation, while only slightly affecting the extent of the initial decrease, indicating that the initial fluorescence decrease is reporting the energization of the membrane while its relaxation is reporting the subsequent deenergization of the membrane resulting from counterion redistributions. The fact that the efflux of H^+ into the medium after an O_2 pulse is small and much slower ($t_{1/2} \simeq 10$ sec) than the actual energization of the membrane ($t_{1/2} \leqslant 0.5$ sec) suggests that the current of respiratory H^+ involved in membrane energization is confined within the bacterial cell envelope.

Key words: proton (H^+) translocation, membrane energy transduction, membrane potential, oxygen pulse, H^+/O ratio, ion fluxes

There is now a large body of evidence which indicates that proton fluxes associated with energy transducing membranes play an important and obligatory role in the mechanism of energy conservation. According to the chemiosmotic hypothesis (1–3), as applied to the bacterial cell, the electrically uncompensated transfer of a proton from the cytoplasm of the cell into the external medium, catalyzed by the respiratory chain or an H^+-pumping ATPase, results in the formation of a large membrane potential ($\Delta\Psi$) which serves as an energy source to drive endergonic reactions such as active transport and ATP formation. These endergonic reactions are often coupled to the influx of a proton or protons (see Ref. 4), so that an effective current of protons is established across the cell membrane.

J. Michael Gould is now at the Department of Chemistry, Program in Biochemistry and Biophysics, University of Notre Dame, Notre Dame, IN 46556.
Received April 22, 1977; accepted August 15, 1977.

568   Gould

The basic predictions of the chemiosmotic hypothesis are consistent with a great deal of recent data (reviewed in Ref. 5) and the existence of a transmembrane electric potential of 100–140 mV has been measured in E. coli (6, 7, 27).

Alternatively, Williams has proposed that the actual charge separation and proton movement occurs within the membrane, and that the functional current of protons does not necessarily include the osmotic or aqueous phase (8–10). This model is consistent with recent studies on energy transduction in chloroplasts (11–14a), and has led to the reexamination of the involvement of osmotic phase proton activities in bacterial energy transduction presented here.

Scholes and Mitchell (15) were the first to note that an anaerobic suspension of bacteria will extrude protons into the medium when given a small pulse of oxygen. These authors noted that, in the absence of permeant charged ions (e.g., SCN^-), the rate of H^+ efflux was sluggish, and the apparent efficiency (H^+/O ratio) was lower than in the presence of the mobile counterions.

In this paper, and elsewhere (16), we have confirmed and extended these observations on the nature of the oxygen pulse induced proton efflux from bacteria. In addition, we have made use of a recently described structural change in the E. coli cell envelope, which is apparently modulated by the energy level of the cytoplasmic membrane (17–21), to monitor the kinetics of the energization of the bacterial membrane. This structural change can be conveniently monitored as changes in the binding and fluorescence properties of the lipophilic probe N-phenyl-1-naphthylamine (NPN).

The data presented here indicates that the reduction of the oxygen added in an oxygen pulse and the concomitant energization of the membrane occur very much more rapidly than does the appearance of H^+ ions in the external aqueous phase, suggesting that the functional proton current involved in bacterial energy transduction may in fact be confined within the cell envelope.

EXPERIMENTAL METHODS

The experiments reported in this paper were performed using cells of the Escherichia coli strain B/1,5, which was a generous gift of Dr. S. Silver. Complete details of the medium used to grow the cells and of the cell harvesting and washing procedure have been published elsewhere (16). The carbon source used for growth was 1% succinate. Changes in pH of anaerobic cell suspensions were monitored as described earlier (16, 22), as were the rates of oxygen reduction. Changes in N-phenyl-1-naphthylamine (NPN) fluorescence intensity were measured essentially as described by Helgerson and Cramer (20). All experiments were performed in thermostated vessels maintained at $33°C$ by a constant temperature circulating water bath.

Oxygen pulses were added as small aliquots of $23°C$ air-saturated glass-distilled water (generally ≤ 10 μl containing ≤ 5.5 ng atoms O) rapidly injected into the anaerobic sample with a long-needled microliter syringe. The time required to inject a 10 μl pulse was $\simeq 0.1$ sec. Anaerobiosis of the samples was achieved and maintained by passing a stream of water-saturated N_2 or Ar gas across the surface of the cell suspension as described elsewhere (16, 22).

RESULTS

Oxygen Pulse-Dependent Proton Efflux

The addition of a small pulse of oxygen to an anaerobic suspension of succinate-grown, logarithmic-phase E. coli induces an efflux of protons from the cells into the suspending medium. The resulting pH change decays very slowly with a half-time > 10 min. Under normal conditions, where counterion fluxes are minimal, the number of protons extruded per oxygen atom added is small ($H^+/O \simeq 0.5$), and the rate of appearance of protons in the medium is relatively slow ($t_{1/2} \simeq 10$ sec) (Fig. 1a,b). If the cells are incubated in the presence of the permeant anion SCN^- or in the presence of colicin E1 [which increases the cell permeability to K^+ but not to H^+ (22–24)] a very different pattern of proton efflux is observed. Under these conditions, where the movement of electrically compensating counterions is enhanced, the rate of H^+ efflux is very much faster ($t_{1/2} < 1$ sec), and the H^+/O ratio is increased to values of 2–2.5 (Fig. 1d, e). The initial rapid acidification of the medium is followed by a somewhat slower alkalization which can be attributed to the reentry of the extruded protons into the cells (15). The addition of FCCP in either the presence or absence of mobile counterions abolishes the net efflux of protons into the medium (Fig. 1c).

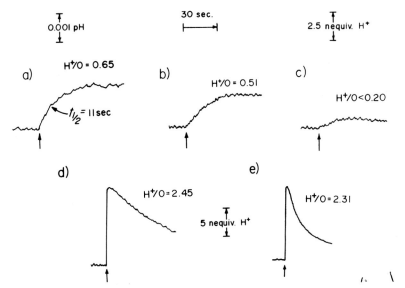

Fig. 1. Oxygen-pulse induced proton extrusion by an anaerobic E. coli suspension and the effects of SCN^-, colicin E1, and FCCP. Cells of E. coli B/1,5 were grown to logarithmic phase, washed twice, and resuspended in 2 ml 150 mM KCl, 0.5 mM MOPS-KOH (pH 7.0) at a concentration of 6×10^9 cells/ml (a) or 3×10^9 cells/ml (b–e). The oxygen pulses (upward arrows) contained 5.5 ng atoms O. An upward deflection of the pH trace represents an acidification of the medium. a,b) Untreated cells; c) cells plus 5 μM FCCP; d) cells washed and suspended in 100 mM KCl, 50 mM KSCN, 0.5 mM MOPS-KOH (pH 7); e) cells pretreated with 1 μg/ml colicin E1.

Previously the effects of permeant counterions on the kinetics and extent of proton efflux following an oxygen pulse have been explained, within the context of the chemiosmotic hypothesis, as resulting from respiratory control attributable to a large membrane potential ($\Delta\Psi$) generated across the membrane by the uncompensated, unidirectional proton translocation (15). That is, the initial translocation of protons is electrogenic, and the resulting $\Delta\Psi$ acts to slow further respiration so that some protons reenter the cell before the reduction of the oxygen added in the pulse is complete, resulting in a lower apparent H^+/O ratio. However, this explanation is not consistent with a number of experimental findings. For example, the rate of oxygen consumption needed to explain the slow proton efflux rate is approximately one fifth to one tenth the observed steady-state respiration rate obtained in the absence of substrate (Table I). In the presence of 10 mM lactate as substrate, the half-time for proton efflux remains unchanged, although the respiration rate is increased three- to fourfold. Furthermore, it is not possible to demonstrate significant levels of respiratory control, in either the presence or absence of added substrate, by conditions which enhance proton or counterion permeability (Table I).

Even though measurements of steady-state respiration give no indication of significant control of electron transport by $\Delta\Psi$ in these cells, such control might exist under the conditions of the oxygen-pulse experiments since the amount of the oxygen added in the pulse is, in most experiments, roughly equivalent to the size of the ubiquinone pool (~ 5 nmoles ubiquinone/mg protein, Ref. 25), which would be mostly reduced under anaerobic conditions. Thus, most of the reducing equivalents transferred to the oxygen in the pulse may be derived from ubiquinone oxidation. If a rate-limiting step for overall electron transport were present in the dehydrogenase region of the transport chain, or in the substrate transport system, any control by $\Delta\Psi$ of respiration coupled to substrate oxidation could be masked. The transient electron flow from the quinone region of the chain to oxygen, such as might occur after a small oxygen pulse, could still be subject to respiratory control, and the explanation (15) for the effects of permeant ions on H^+ efflux might then apply.

TABLE I. Absence of Respiratory Control in Logarithmic Phase E. coli*

Experiment	D-lactate (10 mM)	Additions	O_2 consumption rate (nmoles $O_2 \cdot min^{-1} \cdot$ mg protein^{-1})	$t_{1/2}(O_2)$[a] (sec)	$t_{1/2}(\Delta H^+)$[b] (sec)
I	−	none	36	1.5	11
	−	colicin E1	33		
	+	none	110	0.5	9
	+	colicin E1	82		
II	−	none	17	3.0	10
	−	SCN^-	17		
	+	none	93	0.6	11
	+	SCN^-	74		

*E. coli strain B/1,5 was grown on 1% succinate to mid-logarithmic phase before harvesting as described in (16). The final cell concentration was 3×10^9 cells/ml. The sample chamber containing 3 ml of the cell suspension was maintained at 33°C. When colicin was added the cells were pretreated for 4 min before the start of the measurement. The concentration of O_2 in the 150 mM KCl, 0.5 mM MOPS-KOH (pH 7.0) buffer was assumed to be 235 μM. The conditions were otherwise as described in the legend to Fig. 1. When added, colicin E1 was 1 μg/ml, and SCN^- was 50 mM.
[a]Calculated half-time for the reduction of a 5.5 ng atom O pulse assuming the steady state rate of oxygen consumption.
[b]Measured half-time for proton efflux induced by a 5.5 ng atom O pulse.

By decreasing the size of the oxygen pulse and increasing the number of cells in the suspension it should be possible to arrive at an O_2/cell ratio where the $\Delta\Psi$ generated will be below the level required to inhibit respiration, and the apparent H^+/O ratio will be increased. However, even at very high cell densities and with very small oxygen pulses, the rate of proton extrusion remains slow, and the number of protons extruded remains a linear function of the number of oxygen atoms added with an apparent $H^+/O \simeq 0.7$ (Fig. 2). From the data presented in Fig. 2 it is possible to calculate the maximum membrane potential ($\Delta\Psi_{max}$) which could be generated assuming that all of the H^+ efflux observed is electrically uncompensated. The value for $\Delta\Psi_{max}$ (in volts) becomes

$$\Delta\Psi_{max} = \frac{en}{C}$$

where e is the charge on the proton (1.6×10^{-19} coulombs); n is the number of protons translocated per cell, and C is the capacitance of the cell [3×10^{-14} farads assuming a membrane specific capacitance of 1 μfarad/cm^2 (26) and a membrane area $\simeq 3 \times 10^{-8}$ cm^2/cell] . The calculated values for $\Delta\Psi_{max}$, shown on the right hand axis in Fig. 2, are well below the steady-state values for $\Delta\Psi$ of 100–140 mV which have been measured

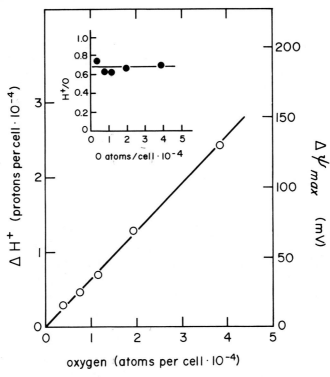

Fig. 2. Relationship between the number of protons extruded per cell (ΔH^+), the amount of oxygen added per cell, and the calculated maximum value for the generated transmembrane potential ($\Delta\Psi_{max}$). Cells of E. coli strain B/1,5 were grown on 1% succinate to mid-logarithmic phase before harvesting. The final cell concentration in the reaction vessel (2 ml) was 4.3×10^{10} cells/ml. The amount of oxygen added in the oxygen pulses was varied from 0.55 ng atoms to 5.5 ng atoms O. Values for $\Delta\Psi_{max}$ were calculated as described in the text.

(6, 7, 27) and show clearly that even the smallest H^+ efflux measured (2.9×10^3 H^+/cell), which is sufficient to generate a maximum membrane potential of only 16 mV, still exhibits an apparent H^+/O ratio of only 0.7. It should be emphasized that the values for $\Delta\Psi_{max}$ represent the maximum energy available, and any ion redistributions or counterion movements will decrease the actual $\Delta\Psi$ generated. The implication of all of these data is that the proton efflux observed under the experimental conditions employed here is either independent of the $\Delta\Psi$ generated, or the level of $\Delta\Psi$ which is sufficient to limit respiration is exceedingly small.

These data show that the efflux of protons into the external aqueous phase following an oxygen pulse can be much slower than the reduction of the oxygen added in the pulse, and suggest that the protons translocated by the redox reactions of the respiratory chain do not necessarily need to enter the external aqueous phase in order to participate in membrane energy transduction. Such a model has been previously proposed by Williams (8–10), and is further supported by the data presented below.

Change in Membrane Energy Level Monitored by N-phenyl-1-naphthylamine Fluorescence

Recently several authors have described a structural change in the E. coli envelope which is apparently regulated by the energy-level of the cytoplasmic membrane. This structural change can be conveniently monitored as changes in the fluorescence intensity of the probe N-phenyl-1-naphthylamine (NPN) (17–20). Nieva-Gomez et al. (18) and Helgerson and Cramer (20) have shown that when the cytoplasmic membrane is de-energized by the addition of FCCP, KCN, azide, colicin E1, K, or Ia, or by the lack of oxygen, there is a large increase in NPN fluorescence resulting in part from increased probe binding to the cells. This increase in probe binding (as well as the increase in fluorescence intensity) can be effectively reversed when the cytoplasmic membrane is reenergized (18, 19, 21). Helgerson and Cramer (20) have also shown that there is a change in the nature of the probe-binding site with the net rotational freedom of the probe molecules becoming restricted upon membrane deenergization. Furthermore, Cramer et al. (19) have shown that the intensity of NPN fluorescence is most sensitive to small energy levels, so that even a small reenergization of the cytoplasmic membrane can cause the reversal of a major portion of the fluorescence increase originally observed upon deenergization.

When a suspension of succinate-grown E. coli harvested in mid-logarithmic phase is allowed to become anaerobic, there is a large (approximately threefold) increase in NPN fluorescence intensity (Fig. 3). The length of the lag period preceding the fluorescence increase is partially dependent upon the endogenous substrate level in the cells, and is significantly shorter with suspensions of starved cells (not shown). A small oxygen pulse (5.5 ng atoms O) added to the anaerobic suspension causes a rapid, partial reversal of the fluorescence increase, followed by a slower relaxation of the fluorescence intensity to the original high level. If the atmosphere in the cuvette is changed from argon to air, the fluorescence intensity decreases to the original, preanoxic level.

The maximum extent of the rapid decrease in NPN fluorescence following the addition of an oxygen pulse ($\Delta F\downarrow$) is dependent upon the amount of oxygen present in the pulse, although the half-time for the subsequent relaxation of the fluorescence decrease ($\Delta F\uparrow$) is independent of the amount of oxygen added (Fig. 4). This suggests that the magnitude of $\Delta F\downarrow$ is correlated with the degree of energization of the membrane, while $\Delta F\uparrow$ monitors properties intrinsic to the membrane which are involved in deenergization. This concept is further supported by the observation that in cells treated with colicin E1 (to increase K^+ permeability) or FCCP the extent of $\Delta F\downarrow$ is inhibited by only about 50%

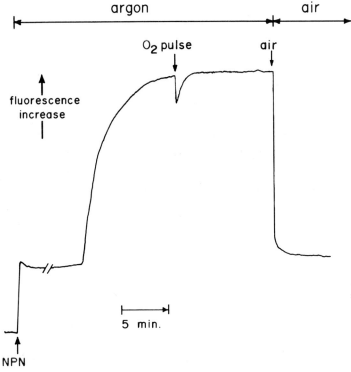

Fig. 3. Changes in the fluorescence intensity of N-phenyl-1-naphthylamine (NPN) associated with changes in oxygen tension of an E. coli suspension. Cells of E. coli strain B/1,5 were grown on 1% succinate to mid-logarithmic phase before harvesting. The cell concentration in the cuvette (2.5 ml) was 3×10^9 cells/ml. The suspending medium contained 150 mM KCl, 0.5 mM MOPS-KOH (pH 7.0). The cuvette was sealed and gassed with H_2O-saturated argon as indicated. NPN (4 μM) was added immediately after sealing the cuvette. The time between the hash marks represents about 12 min in this experiment but was appreciably less in suspensions of mildly starved cells (not shown). Note the abrupt increase in fluorescence intensity associated with anaerobiosis, the transient, partial reversal of this fluorescence increase by a 5.5 ng atom oxygen pulse, and the complete reversal of the fluorescence increase when the stream of argon gas passing through the cuvette is abruptly changed to air.

whereas the relaxation of the fluorescence decrease ($\Delta F\uparrow$) is now very much accelerated (Fig. 5). In the presence of these inhibitors, changing the cuvette atmosphere from argon to air causes only a small ($\sim$ 10%) reversal of the fluorescence intensity (not shown), and active transport is inhibited > 90% (19). All of these results suggest that the decrease in NPN fluorescence reflects the process of energization of the cytoplasmic membrane, while the relaxation of the fluorescence decrease is correlated with the dissipation of the membrane energy.

When one compares the rate of membrane energization, monitored as the decrease in NPN fluorescence (Fig. 6), and the rate of appearance of protons in the medium following an oxygen pulse (Fig. 1a) under similar conditions, it becomes clear that the fluorescence decrease is completed in approximately one tenth the time required for proton efflux. Furthermore, there does not appear to be any kinetic correlation between the energy-dependent changes in NPN fluorescence and the appearance of protons in the external aqueous phase.

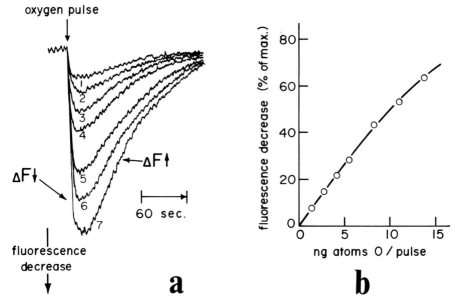

Fig. 4. Dependence of the oxygen-pulse induced decrease in NPN fluorescence of an anaerobic E. coli suspension upon the amount of oxygen present in the pulse. E. coli B/1,5 cells were grown on 1% succinate and harvested in mid-logarithmic phase growth. The final cell concentration in the cuvette (2.5 ml) was 4.2×10^9 cells/ml. The concentration of NPN was 4 μM. The fluorimeter amplifier half-response time was 0.7 sec in this experiment. a) Superimposed traces for the fluorescence decreases (ΔF↓) and subsequent increases (ΔF↑) induced by oxygen pulses containing 1) 1.38, 2) 2.75, 3) 4.13, 4) 5.5, 5) 8.3, 6) 11, and 7) 13.75 ng atoms O. b) Plot of the maximum extent of the fluorescence decrease (ΔF↓) versus the amount of oxygen present in the pulse.

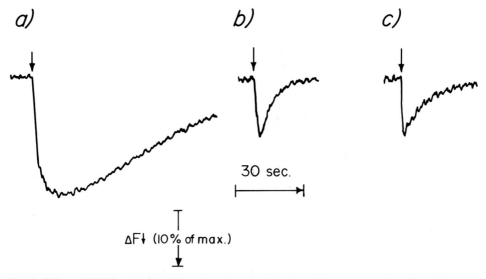

Fig. 5. Effects of FCCP and colicin E1 on the relaxation kinetics of the oxygen-pulse induced decrease in NPN fluorescence in an anaerobic E. coli suspension. The reaction conditions were essentially as described in Fig. 4 except the cell concentration was 3×10^9 cells/ml. The oxygen pulses (downward arrows) contained 5.5 ng atoms O. a) Untreated cells; b) cells plus 4.2 μM FCCP; c) cells treated with 1 μg/ml colicin E1.

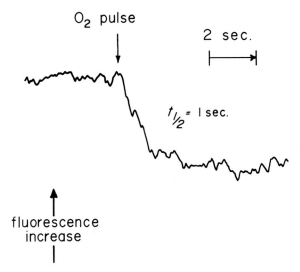

Fig. 6. Kinetics of the decrease in NPN fluorescence intensity following the addition of an oxygen pulse to an anaerobic E. coli suspension. Cells of E. coli strain B/1,5 were grown on 1% succinate to mid-logarithmic phase before harvesting. The final cell concentration in the cuvette (2.5 ml) was 3 $\times$ 10^9 cells/ml. The fluorimeter amplifier half-response time was 0.1 sec in this experiment. The concentration of NPN in the sample was 4 μM. An oxygen pulse containing 5.5 ng atoms O in 10 μl air saturated distilled water was rapidly injected (injection time $\simeq$ 0.1 sec) at the downward arrow.

DISCUSSION

The energy-dependent changes in NPN fluorescence studied here are used in an empirical manner to indicate the direction and relative degree of membrane energization and are not meant to imply that NPN can directly indicate the presence and magnitude of a transmembrane electric field as has been reported for some other probes (28). It should be emphasized that the fluorescence changes observed are largely the result of NPN binding and unbinding to the cell envelope (18). Helgerson and Cramer (20) have shown that the permeability of the E. coli outer membrane to large, lipophilic molecules such as NPN is modulated by the energy-level of the cytoplasmic membrane by an as yet unknown mechanism. Furthermore, Cramer et al. (19) have shown that the changes in NPN fluorescence can act as an amplifier of the actual level of membrane energy, since large decreases in fluorescence intensity arise from small levels of reenergization. This would suggest that the structural change in the E. coli envelope which is monitored by NPN fluorescence is in fact also very sensitive to the cytoplasmic membrane energy level and may be reversed with only the presence of a very small amount of energy (20). These things considered, it is clear that the kinetics of the changes in NPN fluorescence can only be taken as a lower limit for the kinetics of membrane reenergization. That is, the actual energization of the membrane following an oxygen pulse could be significantly faster than the observed kinetics for $\Delta F\downarrow$. However, the fact that the changes in $\Delta F\downarrow$ occur with a half-time very close to the mixing half-time (not shown) suggests that in these experiments the rate-limiting process for membrane energization may be the mixing of the oxygen added in the pulse into the anaerobic suspension.

The very obvious discrepancy between the rate of membrane energization (monitored as $\Delta F\downarrow$) and the rate of proton efflux does not support a model for energy transduction in which the transfer of protons from one osmotic phase to another represents the primary energization event (1–3). Furthermore, the data presented here and elsewhere indicate that the FCCP-sensitive efflux of protons continues unabated for a long period of time after the reduction of the oxygen added in the pulse is complete (16). In the past, the phenomenon of respiratory control has been invoked to explain the observed slow H^+ efflux kinetics, low H^+/O ratios, and the effects of permeant counterions. However, we have shown that the log-phase cells used in this study do not exhibit respiratory control and that the calculated maximum levels of $\Delta\Psi$ attained after an oxygen pulse can be far below the observed steady-state values for $\Delta\Psi$ of 100–140 mV (6, 7, 27), with no significant changes in the properties of the H^+ efflux. That the actual $\Delta\Psi$ generated by a 5.5 ng atom O_2 pulse is well below the calculated $\Delta\Psi_{max}$ value is also shown by the observation that such an oxygen pulse is sufficient to cause only 10–20% of the NPN fluorescence decrease ($\Delta F\downarrow$) induced by air (Figs. 3, 4).

All of the data presented here and elsewhere (16) are consistent with a model for bacterial membrane energization in which the primary energizing event is the transfer of a proton from the cytoplasm to some location closely associated with the membrane and not accessible to the pH electrode (Fig. 7). The slow equilibration of the protons tightly associated with the membrane with protons in the aqueous medium would account for the observed small H^+ efflux. The low H^+/O ratio implies that the membrane does in fact have the capacity to retain ~ 75% of the protons translocated after a small O_2 pulse so that they may be utilized in endergonic energy transduction reactions (e.g., active transport) or leak back into the cell without ever entering the external osmotic phase. Furthermore, it is apparent that the small $\Delta\Psi$ generated after an oxygen pulse sufficient to induce

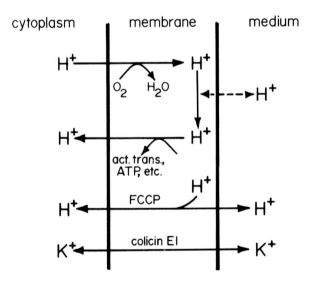

Fig. 7. A simplified scheme for energy-linked ion fluxes in E. coli. According the chemiosmotic hypothesis (1–3), the transfer of a proton from one osmotic phase (cytoplasm) to the other (medium) represents the energizing event associated with the membrane. The model presented above, based on the data presented here and elsewhere (16), and upon discussions already in the literature (8–10, 11–14a), posulates that the translocated proton remains in close association with the membrane and does not appear in the external medium except through a relatively slow equilibration or "leak" (dashed line). See the text for further discussion.

even a small H^+ translocation is sufficient to retain the majority of the translocated H^+ ions in close association with the membrane. It is only when the H^+ translocation generated $\Delta\Psi$ is rapidly collapsed by the movement of permeant counterions that all of the protons actually transported appear in the external medium.

The model presented above is very similar to the model for membrane energy transduction proposed earlier by R.J.P. Williams (8–10), which postulates that the primary current of protons involved in energy transduction occurs within the membrane rather than across it and between the osmotic phases separated by it as proposed in the chemiosmotic hypothesis by Mitchell (1–3). The data presented here are also quite consistent with recent findings concerning the role of osmotic phase pH in energy transduction in chloroplast membranes (11–15) and with the demonstration of localized pH changes within the mitochondrial membrane (29).

It should be mentioned that we do not yet know where the protons translocated but not detected by the pH electrode are actually going. One possibility is that the periplasmic space between the inner and outer membranes is serving as a reservoir for these protons and that the outer membrane of the intact cell constitutes a barrier to the extruded H^+ ions (33). This appears unlikely, however, since a) partial removal of lipopolysaccharide from the outer membrane by EDTA treatment does not alter the time course or amplitude of H^+ efflux (not shown); b) the number of buffering groups that would be required in the outer membrane is large ($\geqslant 5 \times 10^7$/cell) since the number of H^+ extruded $\simeq 5 \times 10^6$/cell for the largest pulses; and C) the effects of SCN^- on the kinetics and amplitude of the H^+ extrusion could not be explained without assuming that SCN^- somehow decreases the buffering capacity of the outer membrane.

One can, however, speculate on other possible reservoirs for these "missing" protons. For example, the E. coli envelope consists largely of phosphatidylethanolamine ($\sim 70\%$), and is known to have a net negative surface charge (30, 31). It is therefore possible that many of the translocated protons remain closely associated with the negative charges in the polar head group region of the phospholipid bilayer. Alternatively, the protons may accumulate in an unstirred layer of structured water at a membrane/water interface. This possibility is intriguing since the lateral mobility of protons within the structured water layer would be expected to be quite high (32). Finally, it should be pointed out that at present there is no evidence to rule out the possibility that the "missing" protons translocated across the inner (cytoplasmic) membrane are not actually associated in some way with the outer membrane, or with an as yet unidentified chemical intermediate of energy transduction.

In any case, it is clear from the data presented in this paper and elsewhere (16) that there exists in bacteria a functional barrier between the majority of the H^+ ions translocated by the respiratory chain after a small oxygen pulse and the external aqueous phase. While the nature of this barrier is not known, recent experiments have shown that very low concentrations of FCCP (5×10^{-8} M), which have no effect on the energy level of the inner membrane, will effectively eliminate the barrier allowing all of the protons translocated to appear in the external medium.

ACKNOWLEDGMENTS

This work was supported by National Institutes of Health grant GM-18457 and National Science Foundation grant BM75-1693X to W. A. Cramer and by a National Research Service Award (1-F32GM00913) from the National Institute of General Medical

Sciences, NIH, to the author. I also wish to thank Suzanne Middendorf for valuable technical assistance and Mona Imler for typing the manuscript.

REFERENCES

1. Mitchell P: Nature (London) 191:144, 1961.
2. Mitchell P: Biol Rev 41:445, 1966.
3. Mitchell P: In Quagliariello E et al. (eds): "Electron Transfer Chains and Oxidative Phosphorylation." Amsterdam: Elsevier, 1975, pp 305–316.
4. West IC, Mitchell P: Biochem J 132: 6:83, 1977.
5. Harold FM: Curr Top Bioenerg 6:83, 1977.
6. Hirata H, Altendorf K, Harold FM: Proc Natl Acad Sci USA 70:1804, 1973.
7. Griniuviene B, Chieliauskaite V, Grinius L: Biochem Biophys Res Commun 56:206, 1974.
8. Williams RJP: J Theor Biol 1:1, 1961.
9. Williams RJP: Curr Top Bioenerg 3:79, 1969.
10. Williams RJP: In Quagliariello E et al. (eds): "Electron Transfer Chains and Oxidative Phosphorylation." Amsterdam: Elsevier, 1975, pp 417–422.
11. Ort DR, Dilley RA: Biochim Biophys Acta 449:95, 1976.
12. Ort DR, Dilley RA, Good NE: Biochim Biophys Acta 449:108, 1976.
13. Gould JM: FEBS Lett 66:312, 1976.
14. Ort DR FEBS Lett 69:81, 1976.
14a. Izawa S, Ort DR, Gould JM, Good NE: In Avron M (ed): "Proceedings of the Third International Congress on Photosynthesis." Amsterdam: Elsevier, 1974, pp 449–461.
15. Scholes P, Mitchell P: J Bioenerg 1:309, 1970.
16. Gould JM, Cramer WA: J Biol Chem 252:5875, 1977.
17. Phillips SK, Cramer WA: Biochemistry 12:1170, 1973.
18. Nieva-Gomez D, Konisky J, Gennis RB: Biochemistry 15:2747, 1976.
19. Cramer WA, Postma PW, Helgerson SL: Biochim Biophys Acta 449:461, 1976.
20. Helgerson SL, Cramer WA: J Supramol Struct 5:291, 1976.
21. Nieva-Gomez D, Gennis RB: Proc Natl Acad Sci USA (In press).
22. Gould JM, Cramer WA, van Thienen G: Biochem Biophys Res Commun 72:1519, 1976.
23. Gould JM, Cramer WA: J Biol Chem 252:5491, 1977.
24. Feingold D: J Membr Biol 3:372, 1970.
25. Cox GB, Newton NA, Gibson F, Snoswell DM, Hamilton JA: Biochem J 117:551, 1970.
26. Cole NS, Cole RH: J Gen Physiol 19:609, 1936.
27. Ramos S, Schuldiner S, Kaback HR: Proc Natl Acad Sci USA 73:1892, 1976.
28. Sims PJ, Waggoner AS, Wang C-H, Hoffman JF: Biochemistry 13:3315, 1974.
29. Huang C-S, Kopacz SJ, Lee C-P: Biochim Biophys Acta 459:241, 1977.
30. Dyer MT, Ordal EJ: J Bacteriol 51:149, 1946.
31. McQuillen K: Biochim Biophys Acta 5:463, 1950.
32. Klotz IM: In Kasha M, Pullman B (eds): "Horizons in Biochemistry." New York: Academic Press, 1962, pp 523–550.
33. Stock JB, Rauch B, Roseman S: J Biol Chem 252:7850, 1977.

Molecular Aspects of Membrane Transport 579–589 (1978)

Photoreactive Probes for High Resolution Mapping of Membrane Proteins

Kenneth K. Iwata, Carol A. Manweiler, John Bramhall, and
Bernadine J. Wisnieski

*Department of Bacteriology and The Molecular Biology Institute, University of California,
Los Angeles, California 90024*

The preparation and characterization of a novel series of radioactively labeled membrane probes is described. These probes are carbohydrate derivatives of fatty acids which contain a photosensitive azide moiety at a specified distance along the alkyl chain. The function of the carbohydrate group is to restrict the azide function to the outer surface monolayer of sealed membrane systems. These azide probes have been used in several well-characterized membrane systems including erythrocyte ghosts, membrane-enveloped viruses, and artificial vesicles. Upon activation, the probes attach to integral proteins to form a stable, covalent complex which may be extracted and identified. The activation protocol is outlined and some of the preliminary results are discussed.

Key words: photoreactive membrane probes, membranes, membrane proteins

Recently considerable attention has been focused on the concept of membrane asymmetry. There is a wealth of evidence to demonstrate that membrane components are distributed asymmetrically between the 2 bilayers of the membrane, and it is clear that in the case of proteins and carbohydrates this asymmetry is absolute in that no protein or glycoprotein has been found to be distributed symmetrically across the membrane, or to be totally embedded within the membrane, unexposed on either surface (1). It appears, too, from the work of Bretscher (2) and others(3) that lipid asymmetry exists in biological membranes. Although the mammalian red blood cell has been used to provide much of the information on this topic, it is probable that compositional asymmetry of this sort is a property of most biological membranes. Studies with viral particles (such as influenza) which bud from their host cell to become enveloped in a bilayer of host membrane lipid support this thesis (4, 5).

Vertical asymmetry of protein disposition has been investigated using a variety of different techniques such as accessibility to iodination (6) or proteolytic digestion (7), but a more detailed study of the extent of penetration of integral proteins into the lipid bilayer

Abbreviations: ESR – electron spin resonance; SDS – sodium dodecyl sulfate; PBS – phosphate buffered saline (5 mM NaPO$_4$, 150 mM NaCl); BSA – bovine serum albumin; NDV – Newcastle disease virus; SDS-PAGE – sodium dodecyl sulfate-polyacrylamide gel electrophoresis; 12-APSGA – 12-(4-azido-2-nitrophenoxy) stearoylglucosamine; 6-ASGA – 6-azidostearoylglucosamine; VP – viral protein; RBC – red blood cell.

Received April 22, 1977; accepted August 15, 1977.

has been delayed because of a lack of suitable technology to allow the interior of the membrane to be probed without undue perturbation to the system under investigation. Recent evidence from our laboratory (8) using glycosylated ESR probes which are the direct counterparts of the azide probes described here, has extended the concept of compositional asymmetry to include a physical asymmetry in the behavior of the 2 membrane monolayers.

The technology developed during the synthesis of the ESR probes has been adapted to provide a means of studying compositional asymmetry in greater detail. This approach is one which we feel will facilitate high resolution mapping of membrane proteins in the vertical plane of the bilayer. The approach utilizes a series of azide derivatives of fatty acids which can be inserted into a specified monolayer of a membrane with the photoreactive azide function immersed at different depths. Following activation, these probes attach to membrane components located in the plane of the functional group. If these labeled components are membrane integral proteins, then information regarding their vertical positioning may be obtained.

METHODS

1. Synthesis of Probes

General methods of synthesis are described below for representative probes. All probes were radioactively labeled to permit easy identification of tagged portions. Typically, glycosylated probes were labeled in the carbohydrate moiety. The free fatty acid azides were synthesized from radioactively labeled fatty acid precursors.

a). Fatty Acid Azides

9, 10-Epoxystearic acid. A modification of the method described by Farmer et al. (9) was adopted. m-Chloroperbenzoic acid (2.0 g) was dissolved in dichloromethane (100 ml), and the resulting solution was added dropwise to a flask containing a solution of [9,10-^{3}H]oleic acid (2.8 g) in dichloromethane (100 ml). The mixture was stirred vigorously during this addition, and the temperature was maintained at 10°C. After the addition, the reaction was allowed to equilibrate to room temperature, neutralized with a 10% (wt/vol) solution of sodium sulfite, then washed successively with 5% (wt/vol) sodium bicarbonate (3 × 1 vol), distilled water (1 vol) and saturated sodium chloride solution (2 × 1 vol). The organic phase was dried over sodium sulfate, filtered, then evaporated to dryness under reduced pressure. The residue was recrystallized from n-hexane to yield a white crystalline mass. (Yield 2.3 g; mp 58°C.)

9/10-Hydroxystearic acid. 9,10-Epoxystearic acid (2.0 g) was dissolved in glacial acetic acid (10 ml). Palladium-carbon catalyst (250 mg) was added to this solution according to the method of Mack and Bickford (10). The mixture was stirred vigorously for 8 h at room temperature under hydrogen at 3 atmospheres pressure. The reaction mixture was subsequently filtered and the filtrate diluted with cold water (10 vol) before being extracted with diethyl ether (3 × 2 vol). The pooled ethereal phases were washed with water (2 × 1 vol) then dried over sodium sulfate before being evaporated to dryness under reduced pressure. The residue was recrystallized from ethanol to yield coarse white crystals. (Yield 1.2 g; mp 74°C.)

9/10-Azidostearic acid. 9/10-Hydroxystearic acid (200 mg) was dissolved in a 14% (wt/vol) solution of boron trifluoride in methanol as described by Mitchell et al. (11). The mixture was boiled for 2 min then diluted with petroleum ether (30 ml) and water (20 ml).

This mixture was agitated then allowed to separate into 2 phases. The organic phase was dried over sodium sulfate and evaporated to dryness (in vacuo). The residue was redissolved in pyridine (5 ml), together with methane sulfonyl chloride (80 mg). This mixture was stirred for 3 h at room temperature. Subsequently, the solvent was removed under vacuum, and the residue was extracted with diethyl ether (10 ml). The ethereal solution was washed with cold water (3 × 1 vol) and the washed organic phase was dried over sodium sulfate and evaporated to dryness under vacuum. The residue was redissolved in dimethylformamide (10 ml) and to this stirred solution was added a solution of sodium azide (160 mg) in water (1 ml). The mixture was stirred for 50 h at room temperature as described by Chakrabarti and Khorana (12). After this incubation the mixture was evaporated to dryness in vacuo and then hydrolyzed in 1 N methanolic NaOH (3 ml) for 24 h at room temperature. The reaction mixture was diluted with water, acidified with 1 N HCl, and extracted with ether. This ethereal solution was washed with cold water (2 × 1 vol), dried over sodium sulfate, and evaporated to dryness in vacuo. The product was purified by thin layer chromatography and identified by IR and mass spectroscopy. (Yield 73%.)

b). Azido Fatty Acid Glycosides.

12-(4-azido-2-nitrophenoxy)oleic acid. Ricinoleic acid (250 mg) was dissolved in dry diethyl ether (5 ml). This mixture was agitated during the addition of potassium tert-butoxide (250 mg). Stirring was continued for 4 h at room temperature before the addition (under dark conditions) of a solution of 4-fluoro-3-nitrophenyl azide (200 mg) in dry diethyl ether (2 ml). This mixture was stirred, in the dark, at room temperature for 48 h; subsequently, the solution was acidified with HCl and diluted with water (100 ml). This aqueous solution was extracted exhaustively with diethyl ether. The organic phases were pooled, dried over sodium sulfate, and evaporated to dryness in vacuo. The residue was redissolved in chloroform/methanol and purified as described previously (12). The product, a light red oil, was obtained in 28% yield.

N-hydroxysuccinimide ester of 12-(4-azido-2-nitrophenoxy)oleic acid. 12-(4-Azido-2-nitrophenoxy)oleic acid (50 mg) was added to a solution of N-hydroxysuccinimide (28 mg) in dry ethyl acetate (20 ml). To this mixture was added a solution of dicyclohexylcarbodiimide (49 mg) in dry ethyl acetate (10 ml). This mixture was stirred in the dark at room temperature for 16 h. The mixture was filtered to remove dicyclohexylurea and evaporated to dryness under vacuum. The residue was recrystallized from ethanol to yield pale yellow crystals in 92% yield. The methodology involved was adapted from Lapidot et al. (13).

12-(4-azido-2-nitrophenoxy)oleoylglucosamine. $[1\text{-}^{14}C]$ Glucosamine-HCl (1.0 mg) was dissolved in 1% (wt/vol) sodium bicarbonate (36 µl). To this solution was added the N-hydroxysuccinimide ester of 12-(4-azido-2-nitrophenoxy)oleic acid (3.0 mg) in dimethyl formamide (120 µl) and the reaction mixture was stirred at room temperature, in the dark, for 24 h. Following this incubation, the reaction mixture was acidified with 1 N HCl, then extracted with chloroform/methanol according to the method of Bligh and Dyer (18). The organic extract was evaporated to dryness yielding a fine golden crystalline product which was purified by thin layer chromatography and identified by IR and mass spectroscopy as well as the colorimetric assay of Reissig et al. (14). The product was stored as an ethanolic solution at 4°C in the dark. (Yield 54%.)

2. Photoactivation Procedures

Optimal activation protocol was established with the use of a radiation source composed of a 250-W xenon arc lamp collimated by a multiple slit assembly and monochromated by a diffraction grating blazed at 300 nm. This excitation source provided a flexible, controlled illumination system. Having established optimal photoactivation conditions, it was apparent that the line emission spectra of low and medium pressure mercury vapor lamps offered a convenient source of energy for use with the photoreactive azide probes.

Alkyl azides were activated using the 254-nm line emission from a low pressure mercury vapor source, yielding illumination of 80 $\mu W \cdot cm^{-2}$ at a distance of 45 cm from the discharge tube. Aromatic azide probes were activated using the 366-nm line emission from a medium pressure mercury lamp of similar power rating.

The suspension to be irradiated was contained in an optical cuvette. The contents were agitated with an overhead-drive stirrer and maintained at an appropriate temperature with a circulating water temperature block. Quartz cuvettes were used when irradiating at 254 nm; glass cuvettes, with a cutoff adsorption starting at 340 nm, were used in conjunction with the 366-nm source.

3. Protein Labeling

a). General

Erythrocyte ghosts were prepared from fresh human blood according to the method of Fairbanks et al. (15). Newcastle disease virus was prepared according to the method of Samson and Fox (16) with slight modification. Protein determinations were made according to the method of Lowry et al. (17). Protein-lipid separations were performed by a modification of the method of Bligh and Dyer (18) or by solvent extraction of lyophilized samples.

Electrophoretic protein separations were performed on 10% acrylamide slab gels, using a Tris-glycine reservoir buffer containing 1% SDS (19). Developed gels were stained with Coomassie blue when visualization was necessary. Radioactive sections of unstained duplicate gels were identified either by autoradiography or by direct counting of gel slices in Aquasol fluor, using a Packard Tri-Carb liquid scintillation counting spectrometer. Ultraviolet and visible spectra were obtained using a Beckman LS 25 spectrophotometer. Infrared spectra were obtained using a Perkin-Elmer 137 spectrophotometer.

b). Red Blood Cells

Erythrocyte ghosts were suspended in phosphate buffer (0.5 mM, pH 8.0) and diluted to a protein concentration of 1 mg·ml^{-1}. To 2 ml of this cell suspension was added azide probe (200 μg) dissolved in ethanol (20 μl). This mixture was incubated for 30 min at room temperature in the dark. After incubation, the mixture was divided into 2 (1 ml) aliquots. One aliquot was irradiated in a quartz cuvette at 22°C; the other was stored in the dark at 22°C to serve as a control. After irradiation the samples were transferred to centrifuge tubes. The ghosts were harvested and washed twice with buffer before being subjected to lipid extraction followed by liquid scintillation counting.

c). Newcastle Disease Virus (Strain HP-16)

Preparations of NDV were made essentially as described by Samson and Fox (16). Fertile hens' eggs were inoculated with seed virus and incubated at 41°C for 11 days. Following incubation, the amniotic fluid was collected using a Cornwall syringe and the virus particles harvested from this fluid by centrifugation at 10^5 X g for 90 min. The viral preparation was purified by density gradient centrifugation, using successive gradients of sucrose (64–16% wt/wt), Renograffin (64–16% wt/vol) and Renograffin (54–14% wt/vol). The pure viral preparation was stored in Tris/saline/EDTA at −70°C. [Protein determinations were made according to the method of Lowry et al. (17), and the purity of the viral preparation was determined by gel electrophoresis of the sample protein.] For protein labeling experiments viral particles prepared as above were suspended in phosphate-buffered saline (pH 7.4) and diluted to a protein concentration of 100 μg·ml^{-1}. To 2 ml of the virus suspension was added azide probe (100 μg) dissolved in ethanol (10 μl). This mixture was incubated for 30 min at room temperature in the dark prior to being divided into 2 (1 ml) aliquots. One aliquot was irradiated; the other was stored in the dark to serve as a control. Both aliquots were maintained at 22°C. After irradiation the samples were transferred to nitrocellulose centrifuge tubes and spun at 183,000 × g for 90 min in order to pellet the viral particles. The supernatents were removed and the pellets resuspended in buffer (20 μl), then incubated with SDS prior to being loaded onto acrylamide gels. Gel radioactivity profiles were determined either by autoradiography or by direct liquid scintillation counting of gel slices.

d). M13 Vesicles

Vesicles containing M13 coat protein were prepared according to the method of Wickner (20). These vesicles are comprised of a sealed lipid bilayer system into which has been inserted viral coat protein monomers derived from the M13 coliphage which infects E. coli. Vesicles were suspended in 0.1 M phosphate buffer (pH 7.0) to a final protein concentration of 500 μg·ml^{-1} and treated with azide probe according to the protocol established for NDV (above). Vesicles were harvested by centrifugation, lyophilized, then subjected to Bligh-Dyer solvent extraction (18) followed by acetone washes. Solvent fractions were assayed for radioactivity by liquid scintillation counting.

RESULTS

Representative photoreactive probes are illustrated in Fig. 1. Structures were identified by IR, UV, and mass spectroscopy. Figure 2 demonstrates that photolysis of the aryl

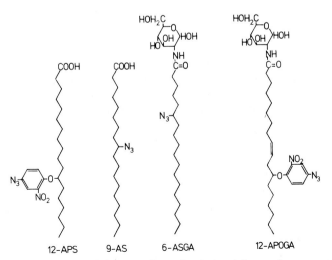

Fig. 1. Representative structures of photoreactive probes designed for membrane mapping. The probes illustrated are: 12-(4-azido-2-nitrophenoxy)stearic acid (12-APS), 9-azidostearic acid (9-AS), 6-azido-stearoylglucosamine (6-ASGA) and 12-(4-azido-2-nitrophenoxy)oleoylglucosamine (12-APOGA).

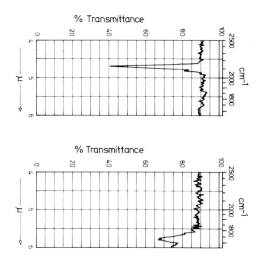

Fig. 2. A chloroform solution of 4-fluoro-3-nitrophenyl azide (5.5 × 10⁻⁴M) was irradiated for 15 min with radiation of wavelength 366 nm (mercury vapor source). Total energy flux intercepted by sample was 10 mW·min⁻¹. This irradiation was sufficient to degrade all azide present in the sample as evidenced by infrared spectroscopy (A: before irradiation; B: after irradiation). The ring resonances associated with aryl azides are minimized by the presence of the m-nitro function, and all the phenyl azide probes display a very intense and clearly differentiated asymmetric stretching absorption at 2160–2090 cm⁻¹. This peak is highly characteristic for the azido group. (The symmetric stretch at 1340–1180 cm⁻¹ and bend at 680 cm⁻¹ frequencies are less intensely absorbed, and are often more confused.)

azide, 4-fluoro-3-nitrophenyl azide, is complete within 15 min during irradiation at 366 nm. Similar infrared analysis was performed on the various alkyl and aryl fatty acid azides used in our studies to confirm their susceptibility to the irradiation conditions employed.

Two model protein systems, bovine serum albumin (essentially fatty acid free; Sigma Chemical Company, St. Louis, Missouri) in water, and erythrocyte ghosts in PBS buffer, were analyzed to determine whether proteins could be linked to radioactive analogues of the probes upon irradiation. Table I shows the results from a series of experiments. The following 2 systems used were: a) BSA in which a solution of BSA (10^{-2} g·ml⁻¹) was incubated with [1-¹⁴C]12-APSGA and b) RBC in which erythrocyte ghosts were incubated with a variety of radioactively labeled probes (as described in the preceding section). Following irradiation, samples were solvent-extracted to remove lipid, and protein fractions were assayed for radioactivity. The results are expressed as cpm per unit weight of protein for both control and irradiated samples. The differences between control and irradiated protein-specific radioactivities reflects the extent of protein labeling during photoactivation. (The differences between experimental sets reflect the varied specific radioactivities of the probes used). [1-¹⁴C]Oleoylglucosamine was included as a light insensitive control. The specific radioactivity of this control probe was the same as that of the [1-¹⁴C]12-APSGA. Table I shows that protein samples irradiated in the presence of glycosylated probes showed a 10-fold increase in protein-associated radioactivity when compared with nonirradiated controls. (The control samples contained azide probe but were maintained in the dark.)

TABLE I. Results From a Series of Experiments Measuring Protein Labeling During Photoactivation

Systems	Probes	(cpm) Irradiated	(cpm) Control	Irradiated (% increase over control)
BSA	[1'-^{14}C] 12-APSGA	1,200	105	1,043
RBC	[1'-^{14}C] 12-APSGA	2,290	229	1,000
RBC	[1'-^{14}C] 12-APSGA	1,960	140	1,300
RBC	[1'-^{14}C] 12-APOGA	451	41	1,000
RBC	[1'-^{14}C] oleoylglucosamine	127	119	7
RBC	[9,10-^{3}H] 9-APS	90	31	190

In comparison with the glycosylated probes, the free fatty acid derivatives were less reactive with protein, possibly because of the formation of micelles or aggregates of probe, which may have been more prevalent with these hydrophobic fatty acids. ESR spin label studies with the nitroxide counterparts of these probes under similar sample conditions indicate that this is a very reasonable explanation. (Micelle formation leads to a distinct spin-broadening of the signal indicating spin-spin coupling with hydrophobic ESR probes.) Radioactive oleoylglucosamine was used as a control probe to show that irradiation of a nonazide fatty acid is not sufficient to induce an association between radioactive fatty acid and protein.

Another test system employed was Newcastle disease virus (NDV). Incubation and irradiation of NDV (200 μg protein) with 12-APSGA (200 μg) had no effect on the protein profile (SDS-PAGE) of the virus as compared with control (untreated, unirradiated) virus. This was demonstrated from densitometer scans of Coomassie blue stained gels (not shown). Both treated and untreated viral samples were washed twice before solubilization for SDS-PAGE. This experiment indicates that the amphiphilic probes exert no detergent effect on the membrane samples. Similar results were obtained using erythrocyte ghost preparations.

Since our immediate goal is to characterize the ability of glycosylated azide probes to distinguish between inner and outer monolayer membrane proteins, we have initiated investigations to compare the SDS-PAGE protein profiles of NDV to determine which specific viral proteins are most susceptible to attack by the azide probes. Preliminary results, shown in Fig. 3, indicate that we do indeed effect a linkage between viral proteins and azide probe upon irradiation and this linkage appears to be more pronounced for the surface glycoproteins VP-1 and VP-2.

Ultimately, we hope to employ photoreactive probes for high resolution mapping of membrane proteins. To this end we have designed a model experimental system to pinpoint the precise segment, and subsequently the precise amino acids, "hit" by the azide moiety as it is placed deeper into an artificial membrane bilayer which contains the small membrane protein (5,000 daltons) called M13 coat protein. Not only is the amino acid sequence of this coliphage protein known, but it has recently been shown to orient in synthetic membrane vesicles (20). Table II shows the results from an experiment in which a suspension of M13 vesicles was irradiated in the presence of [1-^{14}C] 9,10-ASGA according to the procedures described in the preceding section. The results from duplicate experiments show that considerably more radioactivity (cpm) was associated with the protein extracted from the irradiated vesicles than with the protein extracted from the controls.

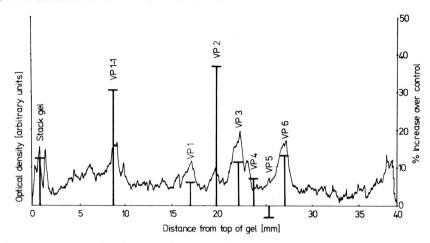

Fig. 3. SDS-polyacrylamide gel pattern of Newcastle disease virus proteins after irradiation in the presence of 12-APSGA (NDV, 200 μg as protein plus 200 μg of 12-APSGA). This densitometer tracing of the Coomassie blue stained gel was comparable to that obtained when untreated, control virus was similarly examined. Vertical columns represent the percentage increase in radioactivity associated with gel slices from an identically treated, irradiated sample compared with gel slices from an adjacent treated, but unirradiated, sample. The columns marked VP refer to the locations of viral proteins 1 to 6, and VP1-1 refers to the dimeric VP1. The results show that significantly more radioactivity was associated with proteins VP1 and VP2 after irradiation in the presence of [1-^{14}C] 12-APSGA, a surface-restricted probe. The radioactivity near the top of the gel is attributed to aggregated material which did not enter the gel proper.

The radioactivity associated with the solvent fractions represents [1-^{14}C] 9,10-ASGA which had not attached to M13 protein during the irradiation (presumably because of prior attachment to vesicle lipids). The vesicle protein concentration was determined using radioactively labeled M13 coat protein. Data presented in Table II show that we obtain a 50% increase in counts associated with M13 protein after photolysis of the photoreactive probe. This information is encouraging and we are actively pursuing these studies at present.

DISCUSSION

Photoactivation of the alkyl azide probes can be effected at 216 nm or 287 nm. Both wavelengths are far from ideal, the first because of its inaccessibility, the second because of its closeness to the wavelength of maximal absorption of proteins. In addition, the transitions associated with both these energy levels are highly symmetry forbidden, hence displaying low extinction and restricting the otherwise general applicability of the alkyl azide probes.

The aryl azides are very much more useful in this context. On irradiation, energy is absorbed in the entire aromatic system. Decomposition of the azide is preceded by excitation transfer from the hydrocarbon to the azido group. The chromophore absorption of the aryl azide probes described here (m-nitrophenyl azide) is the direct superposition of nitrobenzene and phenyl azide, and as such displays 2 principal wavelengths of maximal absorption at 248 nm ($\epsilon = 1.7 \times 10^4$) and 360 nm ($\epsilon = 3 \times 10^3$). The first, and more intensely absorbed wavelength is also of an energy which is likely to be absorbed by elements of membranes other than the probe itself; however, 360 nm is well removed from the absorption profiles of most membrane components, is easily accessible, and is a favored transi-

TABLE II. Results of Irradiation of a Suspension of M13 Particles in the Presence of [^{14}C] 9,10-ASGA

Fraction	Experiment I		Experiment II	
	Control (cpm)	Irradiated (cpm)	Control (cpm)	Irradiated (cpm)
chloroform/methanol acetone wash	3,523	3,364	3,525	3,387
2nd acetone wash	333	215	328	221
3rd acetone wash	86	82	87	88
protein	785	1,266	774	1,229
total	4,727	4,927	4,714	4,925
protein (% total)	17%	26%	16%	25%

tion for the azide. As a consequence, the 366-nm mercury vapor emission energy can be exploited to provide a convenient and highly specific stimulus for generation of the reactive species of the aryl azide probe.

It has been proposed (21) that light absorption promotes the conversion of ground-state azide to a singlet excited state, which results in the rapid generation of a nitrene function. This may well prove to be the dominant process in which case the reactive species can be considered to be largely indiscriminate in its choice of target, selectivity being solely a function of diffusion or collision processes within the membrane. Intersystem crossing is known to occur in the excited state to yield a triplet state of rather less reactivity and slightly greater functional specificity but this is unlikely to be a significant element at room temperature. Far more interesting is the observation that with aryl azides the photo-generated nitrene may react intramolecularly to form a transient azadiene which would then become the true active species. If this were the case, then the active probe would be directed largely towards nucleophilic centers, and would thus display a significant element of functional specificity. Such specificity would have the effect of decreasing probe-lipid interaction, and would emphasize probe-protein complex formation without generating specificity towards particular types of protein. This situation would be compatible with the purposes envisaged for the probes described here.

The results we have described in this paper show that the photoactivated probes do indeed react with proteins both in aqueous medium and in the hydrophobic environment of the membrane. Typically we observed that the radioactivity associated with membrane proteins after irradiation in the presence of probe was an order of magnitude higher than that in control preparations which had been maintained in the dark. The radioactivity associated with the protein from the controls is most likely to result from the copartitioning of unreacted azide probe with the polar fraction of the extraction mixture. This effect is enhanced in the case of the glycosylated probes by the polar head group of the molecule.

The general extraction experiments with soluble proteins and with erythrocyte ghosts served to show that the photoactivation was indeed leading to the formation of covalent linkages between probe and protein; however, for mapping of membrane protein distribution some form of separation of the protein components is essential in order to resolve which proteins have been labeled. For this purpose we have used electrophoretic separation on slab gels in the presence of detergent (sodium dodecyl sulfate). Once separated it is easy to distinguish protein bands on the gel which bear radioactive label from those which do not. From this profile detailed information concerning the position of the proteins in the membrane can be derived.

We have previously developed ESR (paramagnetic) spin labels (8) which are the counterparts of the azide probes described here. Use of these spin labels in membrane systems indicated that glycosylation of the spin-label probes restricts the distribution of the probe to a single (outer) lipid monolayer in sealed membrane systems; furthermore, the equilibrium partitioning of the probe between hydrophobic (membrane) and polar (medium) compartments is relatively slow. This means that once the glycosylated probe molecule has entered the membrane environment it is difficult to remove. The free acid forms of the probes, in contrast, are not restricted to a single monolayer, are easily removed by washing, and seem to partition relatively evenly between the 2 monolayers of a membrane system.

In the red cell membrane it is known that the vast majority, if not all, of the externally exposed proteins span the entire thickness of the lipid bilayer (22) and are also exposed at the inner surface of the membrane. It has been suggested (23) that as a consequence of their mode of insertion into the membrane all externally exposed integral proteins may extend across the entire thickness of the bilayer. This thesis is difficult to test with any of the currently available surface labeling techniques but could be readily investigated with the surface-restricted azide probes described in this paper.

Clearly, in a sealed vesicle or ghost system an integral protein which is embedded in the inner monolayer of the lipid matrix, and which does not penetrate the outer monolayer, will not be available for reaction with a glycosylated azide probe which is restricted in its distribution to the outer monolayer. As a consequence, there will be no radioactive label associated with the protein even after photoactivation of the membrane probe. In contrast, with the free acid forms of the probes, which distribute throughout both monolayers, the protein would become labeled. By conjoining the use of these 2 species of probe it will be possible to determine not only whether the protein is accessible for labeling but also under what conditions and in which regions of the membrane. The first point is an important consideration in studies of this sort.

We have extended the sensitivity of our approach by incorporating the photosensitive azide moiety at different positions along the length of the alkyl chain. This ensures that when the probe is equilibrated with a membrane system the reactive function will be localized in a specific and predetermined plane of the lipid monolayer. Thus, the azide function of 6-ASGA will be inserted into the membrane monolayer to a depth plane centered on 7.8 Å, while that of 12-APSGA will be inserted at a depth of 15.6 Å. Naturally, these depths are not absolute and will have a 3-dimensional deviation associated with them because of the free movement afforded to the alkyl chains of the lipid which will be communicated to the azide function. This zone-broadening will be most evident with probes whose azide substituent is located towards the methyl terminal of the alkyl chain, and will be affected by changes in the phase state of the lipid bilayer. This zone broadening is likely to be the limiting factor to the sensitivity of our analytical approach and may govern the resolution which can be obtained using these probes, but we believe that this is merely a reflection of the dynamic aspects of biological membranes which are an important and integral characteristic of the system under investigation. We have developed an approach for calibrating the position of the azide function of the probe under a variety of different lipid conditions. This procedure utilizes M13 vesicle membranes, an artificial viral assembly which has been investigated by Wickner (20) and which has been well characterized. It represents an ideal calibration system for the membrane probes described here. Our preliminary findings demonstrate that the M13 coat protein can be covalently linked to the azide probes in this vesicle system. Having established our ability to attach fatty acid probes to proteins via a photoactivated azide function, we are currently engaged in improving the resolution so that we can begin to pinpoint the vertical disposition of proteins in membranes.

ACKNOWLEDGMENTS

This work was supported by USPHS grant GM22240 (BJW) and University of California Academic Senate grant 3183. K. K. Iwata is supported by USPHS Training Grant CA09056 and C.A. Manweiler by a UCLA Regents Fellowship. J.S. Bramhall is a Fulbright-Hayes Scholar supported by USPHS grant GM22240. B.J. Wisnieski is the recipient of a USPHS Research Career Development Award GM00228.

We thank Dr. William Wickner for generously supplying the M13 viral coat protein used in these studies and Dr. Christopher Foote for helpful discussion.

REFERENCES

1. Bretscher MS, Raff MC: Nature (London) 258:43, 1975.
2. Bretscher MS: Nature (London) New Biol 236:11, 1072.
3. Rothman JE, Lenard J: Science 195:743, 1977.
4. Tsai K, Lenard J: Nature (London) 253:554, 1975.
5. Lenard J, Rothman JE: Proc Natl Acad Sci USA 73:391, 1976.
6. Morrison M: In Fleischer S, Packer L (eds): "Methods in Enzymology." New York: Academic Press, 1974, vol 32, p 103.
7. Bender WW, Garan H, Berg HC: J Mol Biol 58:783, 1971.
8. Wisnieski BJ, Iwata KK: Biochemistry 16:1321, 1977.
9. Farmer JH, Friedlander BT, Hammond LM: J Org Chem 38:3145, 1973.
10. Mack CH, Bickford WG: J Org Chem 18:686, 1953.
11. Mitchell J, Smith DM, Bryant WMD: J Am Chem Soc 62:4, 1940.
12. Chakrabarti P, Khorana HG: Biochemistry 14:5021, 1975.
13. Lapidot Y, Rappaport S, Wolman Y: J Lipid Res 8:142, 1967.
14. Reissig JL, Strominger JL, Leloir LF: J Biol Chem 217:959, 1955.
15. Fairbanks G, Steck TL, Wallach DFH: Biochemistry 10:2606, 1971.
16. Samson ACR, Fox CF: J Virol 12:579, 1973.
17. Lowry OH, Rosebrough NJ, Farr AL, Randall RJ: J Biol Chem 193:265, 1951.
18. Bligh EG, Dyer WJ: Can J Biochem Physiol 37:911, 1959.
19. O'Farrell PH: J Biol Chem 250:4007, 1975.
20. Wickner W: Proc Natl Acad Sci USA 73:1159, 1976.
21. Fleet GWJ, Knowles JR, Porter RR: Biochem J 128:499, 1972.
22. Steck TL: J Cell Biol 62:1, 1974.
23. Guidotti G: J Supramol Struct (In press).

Author Index

Subject Index